REVIEWS in MINERALOGY
AND GEOCHEMISTRY

Volume 85   2019

Reactive Transport in Natural and Engineered Systems

EDITORS

Jennifer Druhan
University of Illinois Urbana Champaign, *USA*

Christophe Tournassat
BRGM, Orléans, France
Lawrence Berkeley National Laboratory, Berkeley, USA
ISTO, Orléans, France

*Series Editor: **Ian Swainson***

MINERALOGICAL SOCIETY of AMERICA
GEOCHEMICAL SOCIETY

Reviews in Mineralogy and Geochemistry, Volume 85
Reactive Transport in Natural and Engineered Systems

ISSN 1529-6466 (print)
ISSN 1943-2666 (online)
ISBN 978-1-946850-01-0

COPYRIGHT 2019

THE MINERALOGICAL SOCIETY OF AMERICA
3635 CONCORDE PARKWAY, SUITE 500
CHANTILLY, VIRGINIA, 20151-1125, U.S.A.
WWW.MINSOCAM.ORG

Reactive Transport in Natural and Engineered Systems

85   *Reviews in Mineralogy and Geochemistry*   **85**

PREFACE

Open system behavior is predicated on a fundamental relationship between the timescale over which mass is transported and the timescale over which it is chemically transformed. This relationship describes the basis for the multidisciplinary field of reactive transport. In the 20 years since publication of RiMG volume 34: *Reactive Transport in Porous Media*, reactive transport principles have expanded beyond early applications largely based in contaminant hydrology to become broadly utilized throughout the Earth Sciences. Reactive transport is now employed to address a wide variety of natural and engineered systems across diverse spatial and temporal scales, in tandem with advances in computational capability, quantitative imaging and reactive interface characterization techniques. This breadth of application demonstrates an increasing recognition among the Earth Science community of (1) the balance between chemical reactivity and transport of (multiple) phases and (2) the coupled nature of these processes in the terrestrial environment.

The extraordinary expansion in application and diversity of reactive transport models developed over the past 20 years motivates this timely update to the 1996 RiMG volume. In the last two decades, the community has organized a series of benchmarking efforts, resulting in multiple demonstrations of confidence and reliability between independent simulation frameworks. Perhaps the most broadly utilized set of test cases, the MoMaS benchmark initiated in 2002, allows code developers and users to step through a set of sequential generalized simulations featuring increasing dimensions, reactive pathways and mobile/immobile species, culminating in a simulation that is both highly non-linear and heterogeneous. Since 2011, the Subsurface Environmental Simulation Benchmarking (SeSBench) initiative has built upon this basis to propose multiple code benchmark exercises for which the model descriptions are contextualized in natural or engineered systems of interest. In tandem with these capability-driven code comparisons, multiple scientific community workshops have been cultivated in the interest of expanding reactive transport principles within a given discipline. For example, a 2014 workshop held in the United States, *"Expanding the Role of Reacting Transport Modeling within the Biogeochemical Sciences"*, invited a variety of modeling specialists and critical zone scientists to discuss reactive transport principles, opportunities, limitations and needs. Subjects such as chemical weathering across climate gradients, the storage, cycling and stability of carbon in soils, the functioning of the rhizosphere in the development of soil, the signatures of isotope partitioning in chemically complex systems and the leveraging of omics data within biogeochemically dynamic systems were identified as critical areas standing to benefit from quantitative reactive transport frameworks. The publication resulting from this gathering has served as a guide towards future reactive transport software development as these principles are applied to an ever-expanding diversity of Earth systems. In 2017, a workshop held in France emphasized *"Reactive transport in the Earth and Environmental Sciences for the 21st Century"*. This gathering reviewed state of the art capabilities and future directions of reactive transport in application to Earth systems and served as the starting point for the present RiMG volume.

1529-6466/18/0085-0000$00.00 (print)
1943-2666/18/0085-0000$00.00 (online)

http://dx.doi.org/10.2138/rmg.2019.85.0

The progress made in such workshops has been complemented by an increasingly frequent and broad diversity of conference sessions dedicated to reactive transport modeling which now routinely reach large international audiences. Only a few months ago, *Elements* magazine dedicated their April 2019 issue to the subject of reactive transport modeling. These short-format articles intended for a general geoscience audience proved an ideal and timely complement to the present RiMG volume, where interested readers may alternate between accessible overviews and the extensive detail provided herein. As noted in this *Elements* issue, the interface between scientists and software is evolving and the so-called 'black box' era of model operation segregated from all but the most advanced user is coming to an end. Educational material, short courses and training programs are increasingly commonplace and broadly available. Computer literacy is closer to a native language in the next generation of Earth scientists. Most importantly the ability to leverage process in a predictive framework is now more vital than ever as we face a rapidly changing Earth system challenged with the need to sustain an increasing global population. In total, the necessity of forward computational frameworks as a platform for the mathematical representation of such complex, multi-component, coupled and coevolving systems as soil, crops, contaminants and water is paramount to the next century of resource management and environmental sustainability.

In this review, we depart from the structure of the prior RiMG reactive transport volume, which was primarily sectioned based on modeling capabilities, in favor of scientific topics describing recent applications, techniques and current requirements necessary to foster future advances. At the widest view these chapters fall into three categories. The first set emphasize the state of reactive transport simulation capability to address challenging new frontiers such as multi-scale hybrid approaches, fractured and nanoporous media, multiphase systems, and evolving physical structures. The second set then focus on a variety of novel or expanded reactive transport applications in Earth Science such as stable isotope partitioning, microbial catalysis, vadose zone systems, soils, and watersheds. Finally, the third set emphasizes industrial applications of reactive transport modeling in the fields of waste repositories, carbon sequestration, and sub-surface resource recovery. In total this volume serves as a significant update to its predecessor, describing the diversity of applications that reactive principles are now used to quantify, and highlighting the key areas of reactive transport software development necessary to continue advancing these fields.

We wish to extend our sincere thanks to all chapter authors who contributed to this volume, as well as Series Editor Ian Swainson and Managing Editor Rachel Russell. Special thanks are also owed to our chapter reviewers, including Bernhard Ahres, Adam Atchley, Pascal Audigane, Felipe de Barros, Shaun Brown, Jacques Diederik, Thomas Gimmi, Noah Jemison, Georg Kosakowski, Praveen Kumar, Marina Lebedeva, Marco de Lucia, Linda Luquot, Sergi Molins, Joel Moore, Doug La Rowe, Delphine Roubinet, Dave Savage, Josep Soler, Dongkook Woo, and Xiaofan Yang.

Jennifer L. Druhan, University of Illinois Urbana Champaign

Christophe Tournassat, BRGM, ISTO & LBNL

Reactive Transport in Natural and Engineered Systems

85   *Reviews in Mineralogy and Geochemistry*   **85**

TABLE OF CONTENTS

1   **Reactive Transport at the Crossroads**

Carl I. Steefel

2   Multiscale Approaches in Reactive Transport Modeling

Sergi Molins, Peter Knabner

3   Modeling Reactive Transport Processes in Fractures

Hang Deng, Nicolas Spycher

4

Reactive Transport Modeling of Coupled Processes in Nanoporous Media

Christophe Tournassat, Carl I. Steefel

5

Mixing and Reactive Fronts in the Subsurface

Massimo Rolle, Tanguy Le Borgne

6   Multiphase Multicomponent Reactive Transport and Flow Modeling

Irina Sin, Jérôme Corvisier

7   Reactive Transport in Evolving Porous Media

Nicolas Seigneur, K. Ulrich Mayer, Carl I. Steefel

8   Stable Isotope Fractionation by Transport and Transformation

Jennifer L. Druhan, Matthew J. Winnick, Martin Thullner

9   Microbial Controls on the Biogeochemical Dynamics in the Subsurface

Martin Thullner, Pierre Regnier

10   Understanding and Predicting Vadose Zone Processes

Bhavna Arora, Dipankar Dwivedi, Boris Faybishenko,
Raghavendra B. Jana, Haruko M. Wainwright

11 Abiotic and Biotic Controls on Soil Organo–Mineral Interactions: Developing Model Structures to Analyze Why Soil Organic Matter Persists

Dipankar Dwivedi, Jinyun Tang, Nicholas Bouskill,
Katerina Georgiou, Stephany S. Chacon, William J. Riley

12 Reactive Transport Processes that Drive Chemical Weathering: From Making Space for Water to Dismantling Continents

Kate Maher, Alexis Navarre-Sitchler

13 Watershed Reactive Transport

Li Li

14   RTM for Waste Repositories

Olivier Bildstein, Francis Claret, Pierre Frugier

15   Acid Water–Rock–Cement Interaction and Multicomponent Reactive Transport Modeling

Jordi Cama, Josep M. Soler, Carles Ayora

16   Industrial Deployment of Reactive Transport Simulation: An Application to Uranium *In situ* Recovery

Vincent Lagneau, Olivier Regnault, Michaël Descostes

RiMG Series

Reviews in Mineralogy & Geochemistry
Vol. 85 pp. 1-26, 2019
Copyright © Mineralogical Society of America

1

Reactive Transport at the Crossroads

Carl I. Steefel

Energy Geosciences Division
Lawrence Berkeley National Laboratory
1 Cyclotron Road
Berkeley, CA 94720
USA

CISteefel@lbl.gov

INTRODUCTION

Reactive transport in the Earth and Environmental Sciences is at a crossroads today. The discipline has reached a level of maturity well beyond what could be demonstrated even 15 years ago. This is shown now by the successes with which complex and in many cases coupled behavior have been described in a number of natural Earth environments, ranging from corroding storage tanks leaking radioactive Cs into the vadose zone (Zachara et al. 2002; Steefel et al. 2003; Lichtner et al. 2004), to field scale sorption behavior of uranium (Davis et al. 2004; Li et al. 2011; Yabusaki et al. 2017) to the successful prediction of mineral and pore solution profiles in a 226 ka chemical weathering profile (Maher et al. 2009), to the prediction of ion transport in compacted bentonite and clay rocks (Tournassat and Steefel 2015; Soler et al. 2019; Tournassat and Steefel 2019, this volume). Yet for those thinking deeply about Earth and Environmental Science problems impacted by reactive transport processes, it is clear that many challenges remain.

A common theme here is that improved scientific understanding of complex Earth and environmental processes become possible as new conceptual and numerical models for reactive transport are developed. In some examples, this success has hinged on demonstrating that reactive transport models were computationally feasible. In others, it was also critical to employ the flexible framework of these multi-component, multi-dimensional models to show that it was possible to parameterize them for application to complex field systems. Certainly, the computational burden increased with the use of these models, as well as the considerable effort and knowledge required of researchers to develop and apply them. The benefits that have resulted include a much improved and generalized set of model capabilities, as well as a deeper level of understanding of the underlying coupled physical and biogeochemical processes of Earth systems. The maturity of the discipline has also been demonstrated by recent benchmarking activities that have shown that as many as ten different software packages can simulate complex natural reactive transport problems and achieve essentially the same results (Steefel et al. 2015).

So, is the "reactive transport problem" solved? Where do we go from here? This is the crossroads we are at now as we decide what are the challenges that need to be faced so as to continue advancing the field. Arguably the most significant challenges we now face are associated with the huge range of length scales that need to be addressed, since these extend from the molecular to nanoscale to pore scale all the way up to the watershed and continental scale (Molins and Knabner 2019, this volume). Across this extreme range of scales, the constitutive equations and parameters that are used to describe reactive transport processes often change as well, thus requiring mathematical and numerical models to become "scale aware". Charged porous media offers special challenges, since ion mobility can be strongly affected by electrostatic interactions, and this can lead to effects such as anion exclusion that are not captured by Fick's Law

1529-6466/19/0085-0001$05.00 (print)
1943-2666/19/0085-0001$05.00 (online)

http://dx.doi.org/10.2138/rmg.2019.85.1

(Appelo and Wersin 2007; Appelo et al. 2010; Tournassat and Steefel 2015). Where the charged porous media involves nanoscale porosity, off-diagonal coupling effects on transport between such master variables as fluid pressure, electrical current and chemical composition may become important (Tournassat and Steefel 2019, this volume). At the watershed and continental scales, reactive transport is further complicated by the coupling with diverse Earth surface processes, including subsurface and surface water, vegetation, and the atmosphere, all played out typically in highly heterogeneous and transient settings (Li 2019, this volume). This can lead to "hot spots" and "hot moments" that may have an outsized effect on system function even though they represent a limited percentage of the total land surface area (Dwivedi et al. 2018a).

HISTORICAL DEVELOPMENT

Geochemical modeling of Earth and environmental systems began with equilibrium descriptions of the thermodynamic state of a particular multi-component solution, and such approaches are still in wide use today for the purposes of interpreting the chemistry of natural waters. An important next step was the development of reaction path models that captured the sequence of chemical/mineralogical states resulting from such processes as chemical weathering and hydrothermal alteration. It could be argued, in fact, that these reaction path models represent the first "reactive transport" models insofar as they address irreversible geochemical/mineralogical processes for the first time. Early versions of the reaction path models, pioneered by Helgeson and co-workers, did not include an explicit treatment of real-time kinetics, but rather quantified the system evolution instead as a function of reaction progress (Helgeson 1968; Helgeson et al. 1969). These models provided a way of interpreting quantitatively the sequence of minerals observed in nature as the natural consequence of the dissolution of some primary phase (e.g., feldspar), which is itself out of equilibrium due to either the initial state of the system, or more commonly, due to the flux of reactive constituents.

An explicit treatment of transport processes is not factored into these reaction path approaches. The method can be used to describe chemical processes in a batch or closed system (e.g., a laboratory beaker), or for exceedingly simplified transport. However, such conditions are of limited interest in the geosciences where the driving force for most reactions is often transport. Lichtner (1988) clarified the application of the reaction path models to water–rock interaction by demonstrating that they could be used to describe pure advective transport through porous media. By adopting a reference frame which followed the fluid packet as it moved through the medium, the reaction progress variable could be thought of as travel time instead.

Multi-component reactive transport models that could treat any combination of transport and biogeochemical processes date back to the mid-1980s. Lichtner (1985) outlined much of the basic theory of a continuum model for reactive transport. Yeh and Tripathi (1989) also presented the theoretical and numerical basis for the treatment of reactive contaminant transport, demonstrating for the first time the inapplicability of classical linear distribution, or K_d, approaches in predicting contaminant dynamics. Steefel and Lasaga (1994) presented a reactive flow and transport model for non-isothermal, kinetically controlled water–rock interaction and fracture sealing in hydrothermal systems based on simultaneous numerical solution of both reaction and transport. This study was the first to consider multicomponent reactive transport in the context of non-isothermal flow fields, an important subject for geothermal and hydrothermal systems. It was apparently also the first to consider how reaction-induced permeability change (clogging) could alter the behavior of a hydrothermal or other flow system. Along with earlier studies of mineral dissolution related feedbacks to the flow system through permeability (Ortoleva et al. 1987; Steefel and Lasaga 1990), this work introduced the important theme of coupled processes in reactive transport analysis (Seigneur et al. 2019, this volume).

In what follows, we briefly review the state of the science in terms of conceptual and mathematical models for reactive transport, considering how the equations change as we proceed down scale from continuum to pore to molecular models. This is followed with an overview of the numerical methods that are used to solve these potentially complex problems. We then proceed to discussions of individual topics where reactive transport has had a significant impact since the publication of *Reviews in Mineralogy* volume 34 in 1996 (Lichtner et al. 1996) including an update to the persistent issue of contaminants.

MATHEMATICAL FORMULATION

Presentations of the basic reactive transport equations can take different directions depending on the interests of the author, but for this Rev Mineral Geochem volume, it seems appropriate to emphasize the geochemical and mineralogical aspects. The primary variables in the reactive transport problem are the concentrations of component i (Ca, Na, …) in phase γ (aqueous, gas, and mineral). Typically, reactive transport equations are developed for each of the phases, or they are combined with an overall mass balance that covers all of the phases present, with differing transport properties for each of the phases. The concentrations, along with temperature and pressure, are the primary variables. Various derived or secondary variables follow from the primary variables, including the reaction rates (for minerals, dependent on the aqueous and/or surface concentrations and the mineral surface area), the mineral volume fractions (essentially linear combinations of the component concentrations in the mineral phase), and the mineral surface areas (related to the volume fractions through the specific surface area). Whether the mineral volume fractions are formally included within a single nonlinear solve depends on the software considered. Often the time scale separation resulting from slow kinetics of common mineral reactions in many cases justifies the update of these at the end of the time step. In the cases where all of the components are in equilibrium in the multiple phases present, we regain the Gibbs Phase Rule, with degrees of freedom equal to the number of components plus 2 (for T and P).

The mineral volumes are further related to the porosity of the medium through the relation:

$$\phi = 1 - \sum_{m=1}^{N_m} \phi_m \tag{1}$$

The permeability of the medium, which determines the flow rate for a given pressure or hydraulic head gradient (thus completing the feedback circle), is determined by a porosity–permeability relationship applicable to the medium in question. Typically, the Kozeny–Carman equation is used for porous media (Bear 1972):

$$k = \frac{\phi^3}{(1-\phi)^2} \frac{1}{5M^2} \tag{2}$$

where M is the fluid–solid interfacial area. The applicability of this relation has been questioned in many cases, particularly for reaction-induced porosity–permeability change. For fractured media, the cubic law is used:

$$k = \frac{b^3}{12d} \tag{3}$$

where b is the fracture aperture and d is the separation of the fractures perpendicular to the fracture plane (Witherspoon et al. 1980; Steefel and Lasaga 1994; Deng et al. 2016). The effective diffusivity in porous media, which includes the effect of tortuosity, also depends on the porosity, often through Archie's Law (Seigneur et al. 2019, this volume).

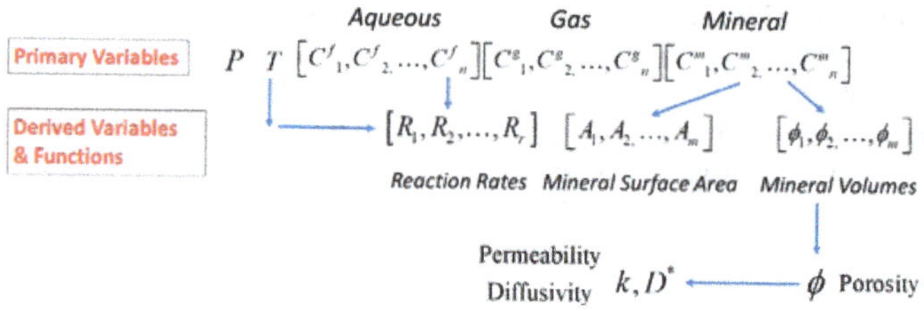

Figure 1. Primary and derived variables and functions for geochemical and reactive transport modeling. The primary variables are those that are typically solved for within a single nonlinear iteration, although in some cases the mineral concentrations (volume fractions) may be updated only at the end of a time step.

The geochemical relations in Figure 1 form the basis for the reactive transport equations. For a fully kinetic formulation in which each aqueous species is treated separately (i.e., they are not assumed to be in equilibrium), the equation is given by (Steefel et al. 2015):

$$\underbrace{\frac{\partial\left(\phi S_L C_i\right)}{\partial t}}_{\text{Accumulation Term}} = \underbrace{\nabla\cdot\left(\phi S_L \mathbf{D}_i^*\nabla C_i\right)}_{\text{Dispersion}} - \underbrace{\nabla\cdot\left(\mathbf{q}C_i\right)}_{\text{Advection}} - \underbrace{\sum_{r=1}^{N_r}v_{ir}R_r}_{\substack{\text{Aqueous}\\\text{Reactions}}} - \underbrace{\sum_{m=1}^{N_m}v_{im}R_m}_{\substack{\text{Mineral}\\\text{Reactions}}} - \underbrace{\sum_{g=1}^{N_g}v_{ig}R_g}_{\substack{\text{Gas}\\\text{Reactions}}} \qquad (4)$$

In Equation (4), S_L is the liquid saturation in the porous media, $\mathbf{D}^*$ is the dispersion tensor, $\mathbf{q}$ is the Darcy velocity, and the v_{ir}, v_{im} and v_{ig} refer to the stoichiometric coefficients for component i in aqueous reaction r, mineral reaction m, or gas reaction g respectively (Steefel et al. 2015).

For the partial equilibrium case, where some number of aqueous and/or surface complexes are assumed to be in equilibrium and thus described by algebraic mass action equations, we have:

$$\frac{\partial\left(\phi S_L \Psi_i\right)}{\partial t} = \nabla\cdot\left(\phi S_L \mathbf{D}_i^*\nabla\Psi_i\right) - \nabla\cdot\left(\phi S_L v\Psi_i\right) - \sum_{r=1}^{N_r}v_{ir}R_r - \sum_{m=1}^{N_m}v_{im}R_m - \sum_{g=1}^{N_g}v_{ig}R_g \qquad (5)$$

where the total concentration, Ψ_i is defined as a linear combination of the concentrations of the primary and secondary species:

$$\underbrace{\Psi_i}_{\substack{\text{Total}\\\text{Concentration}}} = \underbrace{C_i}_{\substack{\text{Primary}\\\text{Species}}} + \underbrace{\sum_{l=1}^{N_s}v_{i,l}C_l}_{\substack{\text{Secondary}\\\text{Species}}} = C_i + \sum_{l=1}^{N_s}v_{i,l}\left[\prod\gamma_p C_p^{v_{p,l}}K_l^{-1}\right] \qquad (6)$$

where γ_p is the activity coefficient for the primary species, and K_l is the equilibrium constant for the secondary species reaction. Surface complexes can be added similarly. The right most term in Equation (6) is the total concentration rewritten entirely in terms of primary species—the so-called "Direct Substitution Approach". The Direct Substitution Approach (of DSA) involves making use of the laws of mass action for secondary species assumed to be in equilibrium with the primary species.

Most of the software in use today for continuum reactive transport makes use of some form of Equation (6), with or without the Direct Substitution Approach, although alternative formulations have been proposed (Yeh et al. 2014). The total concentration approach (or "canonical formulation", Lichtner 1985) lends itself to treatment of reactions with a mixed equilibrium and kinetic approach (see Steefel and MacQuarrie 1996 for a more detailed discussion). Other approaches, particular involving the use of Gibbs free energy minimimization (Steefel and MacQuarrie 1996; Leal et al. 2014) are now in common use, although they require the use of separate kinetic routines if such chemical processes are to be included.

It is also possible to treat fully equilibrium reaction networks with the canonical formulation, as described in Steefel et al. (2015). In this case, all of the reaction rates are eliminated, replaced typically with the mass action equations in logarithmic form:

$$\frac{1}{\Delta t}\left[C_i + \sum_{l=1}^{N_s} v_{il} C_l + \sum_{m=1}^{N_m} v_{im} C_m\right]^{n+1} - \left[C_i + \sum_{l=1}^{N_s} v_{il} C_l + \sum_{m=1}^{N_m} v_{im} C_m\right]^{n} = 0 \quad i = 1,\cdots N_c \quad (7)$$

$$\log C_l = \sum_{l=1}^{N_s} v_{il} \log(\gamma_i C_i^{n+1}) - \log \gamma_l^{n+1} - \log K_l \quad 1 = 1,\cdots N_s$$

$$\log C_m^{n+1} = \sum_{m=1}^{N_m} v_{im} \log(\gamma_i^{n+1} C_i^{n+1}) - \log K_m \quad m = 1,\cdots N_m$$

(8)

where n and $n+1$ refer to the present and future time level respectively, and the γ_i's and γ_j's are the activity coefficients for the secondary and primary species respectively, and K_l and K_m are the equilibrium constants for the secondary species and minerals, respectively. A similar approach can be taken to include equilibrium surface complexes and gases.

Details of the formulations for the transport terms (Darcy's Law, Fick's Law) are provided in other chapters in this volume, as well as Steefel et al. (2015).

NUMERICAL FORMULATION

The numerical treatment of reactive transport was described in some detail in Steefel and MacQuarrie (1996) and in Steefel et al. (2015). In what follows, we first consider the treatment of the typically nonlinear reaction term first without transport (i.e., a set of nonlinear ordinary differential equations). Then we proceed to a brief discussion of how reactions and transport can be coupled over the spatial domain (a set of nonlinear partial differential equations).

Reaction terms

Steefel and MacQuarrie (1996) reviewed the range of fully kinetic formulations and equilibrium or mixed equilibrium–kinetic formulations for the reaction terms. Steefel et al. (1996) presented the mixed equilibrium–kinetic approach based on the canonical formulation (Lichtner 1985), essentially the numerical approach commonly employed by many modern reactive transport software packages such as CrunchFlow, MIN3P, and PFLOTRAN. Other mathematical and numerical treatments, however, have been considered. For example, Yeh and co-workers (Yeh and Tsai 2014; Yeh et al. 2014) have proposed and made use of an alternative to the canonical formulation.

The accumulation and reaction terms are discretized at the present and future time step, giving rise to a set of nonlinear ordinary differential equations that must be solved numerically in most cases. Various methods can be used to treat the time derivatives, such as backwards Euler (the simplest) and Runge–Kutta (the most common). The nonlinear ordinary differential equations can be solved with Newton's method given by:

$$\sum_{k=1}^{N_c} \frac{\partial f_i}{\partial C_k} \delta C_k = -f_i \quad (9)$$

where f_i are the function residuals (mass balance equations written typically in terms of the total concentrations in which the sum of the terms equals zero), and $\partial f_i / \partial C_k$ are elements of the Jacobian matrix (the derivatives of the function residuals, or mass balance equations, with respect to the primary k unknowns). For the case where only the accumulation and reaction terms are considered for the sake of simplicity (no transport terms), we have the ordinary differential equations given by:

$$\phi S_L \frac{\left[\psi_{i,jx}^{n+1} - \psi_{i,jx}^{n}\right]}{\Delta t} + \sum_{m=1}^{N_m} \left[R_{i,m}^{n+1}\right] = 0 \quad (10)$$

Using the Direct Substitution Approach (rewriting the secondary equilibrium species appearing in the total concentration in terms of primary species), Equation (10) can be written for a single Newton iteration as:

$$\underbrace{\frac{\phi}{\Delta t}\left[\delta_{i,k}C_i + \sum_{l=1}^{N_x}\frac{v_{kl}}{C_k}v_{il}\left[\prod_{p=1}^{N_c}C_p^{v_{pl}}K_{\text{eq}}^{-1}\right] + A_m k_m \frac{v_{km}}{C_k}\prod_{p=1}^{N_c}a_i^{v_{pm}}K_m^{-1}\right]^{n+1}}_{\text{Jacobian Matrix}} =$$

$$\underbrace{-\frac{\phi}{\Delta t}\left(\left[C_i + \sum_{l=1}^{N_x}v_{il}\left[\prod_{p=1}^{N_c}C_p^{v_{pl}}K_{\text{eq}}^{-1}\right]^{n+1}\right]_i - \psi_i^n\right) - A_m k_m\left(\left[1 - \prod_{p=1}^{N_c}a_i^{v_{pm}}K_m^{-1}\right]\right)}_{\text{Function Residuals}}$$

(11)

Note that even software employing an operator splitting approach (discussed below) to reaction and transport typically carries out a nonlinear solve of the accumulation and reactions terms. As discussed, the Gibbs free energy routines are based on minimization of free energy rather than root finding and thus do not follow this method.

Coupling of reaction and transport terms

To handle the coupling of reaction and transport processes over the spatial domain (now partial differential equations), the normal choices are between operator splitting approaches (either with or without iteration) and the global implicit or one-step approach.

The operator splitting approach, whether in continuum or pore scale models, is the most widely used and in non-iterative form begins with a solve of the conservative transport equations followed by the reaction terms using the transported concentrations in the accumulation term. To simplify the presentation, we introduce the differential operator

$$L(\Psi_j) = \left[\nabla \cdot \left(\mathbf{q} - \mathbf{D}^*\nabla\right)\right]\Psi_j,$$

(12)

and write the operator splitting approach as

$$\phi S_L \frac{(\psi_i^{\text{transp}} - \psi_i^n)}{\Delta t} = L(\psi_i)^{n+1}, \quad (i = 1,...,N_c),$$

(13)

followed by a reaction step using the transported total concentrations:

$$\phi S_L \frac{(\psi_i^{n+1} - \psi_i^{\text{transp}})}{\Delta t} = \sum_{i=1}^{N_m}R_i^{n+1} \quad (i = 1,...,N_c)$$

(14)

where ψ_i^n and ψ_i^{n+1} are the concentrations at the current (n) and future ($n + 1$) time levels, respectively, and ψ_i^{transp} is the transported total concentration calculated with Equation (13). The right hand side of Equation (13) could be written in terms of concentrations at the current time step (n), in which case the transport is said to be treated with an explicit in time approach. Alternatively, one can solve Equation (13) implicitly in time, in which case a speciation of the total concentration in the reaction step in Equation (14) is required, either using the Direct Substitution Approach as in Equation (6), or by including the mass action equations for the secondary species in logarithmic form, as in Equation (8).

In the global implicit or one–step approach, the transport and reaction terms are solved simultaneously. In the Direct Substitution Approach, we solve simultaneously for the complexation (which are assumed to be at equilibrium), the heterogeneous reactions, and transport terms. This means that the primary species (C_j's) rather than the total concentrations are the unknowns in the transport equations:

$$\frac{\phi S_L}{\Delta t}\left[\left(C_i+\sum_{l=1}^{N_x}v_{il}\left[\prod_{p=1}^{N_c}C_p^{v_{pl}}K_{eq}^{-1}\right]_i-\psi_i^n\right)\right]^{n+1}=$$

$$L\left[C_j+\sum_{i=1}^{N_x}v_{ij}\gamma_i^{-1}K_i^{-1}\prod_{j=1}^{N_c}(\gamma_jC_j)^{v_{ij}}\right]+-A_mk_m\left(\left[1-\prod_{p=1}^{N_c}a_i^{v_{pm}}K_m^{-1}\right]\right) \tag{15}$$

In the case of a global implicit treatment, the size of the Jacobian matrix which must be constructed and solved becomes larger, since each function will include contributions from the concentrations both in the grid cell itself and from neighboring grid cells that are used in the discretization. For example, in the case of one-dimensional transport and N_c unknown concentrations at each nodal point, the form of the Newton equations to be solved is:

$$\sum_{k=1}^{N_c}\frac{\partial f_{i,jx}}{\partial \ln C_{k,jx}}\delta\ln C_{k,j}+\sum_{k=1}^{N_c}\frac{\partial f_{i,jx}}{\partial \ln C_{k,jx+1}}\delta\ln C_{k,j+1}+\sum_{k=1}^{N_c}\frac{\partial f_{i,jx}}{\partial \ln C_{k,jx-1}}\delta\ln C_{k,jx-1}=-f_{i,jx} \tag{16}$$

where i refers to the component number, k is the unknown component species number, and jx, $jx+1$, and $jx-1$ are the nodal points, and the logarithms of the concentrations are solved for because of the improved numerical stability this provides. The Jacobian matrix in the case of one-dimensional transport takes a block tridiagonal form

$$\begin{bmatrix} A_{1,1} & A_{1,2} & 0 & \cdots & & & \\ A_{2,1} & A_{2,2} & A_{2,3} & \cdots & & & \\ & & \cdots & & & & \\ & & \cdots & A_{N-1,N-2} & A_{N-1,N-1} & A_{N-1,N} \\ & & \cdots & 0 & A_{N,N-1} & A_{N,N} \end{bmatrix}\begin{bmatrix} \delta\ln C_1 \\ \delta\ln C_2 \\ \cdots \\ \delta\ln C_{N-1} \\ \delta\ln C_N \end{bmatrix}=-\begin{bmatrix} f_1 \\ f_2 \\ \cdots \\ f_{N-1} \\ f_N \end{bmatrix} \tag{17}$$

where the entries in the Jacobian matrix (the A) are submatrices of dimension N_c by N_c in the case where there are N_c unknowns per nodal point. The entries $\delta\ln C_i$ refer here to the entire vector of unknown concentration corrections in logarithmic form at any particular nodal point, while the functions f_i include the entire vector of equations for the unknown concentrations at each nodal point.

The chief advantage of the operator splitting approach is the ability to use modular reaction routines without implementing a potentially complicated global implicit solve. There are no Jacobian entries for the transport terms in the operator splitting treatment of reactive transport, so less computation and less code development is involved. The approach works reasonably well where there is scale separation of processes, thus less coupling between reaction and transport. The disadvantage of the approach is that operator splitting error occurs when the time step is greater than the Courant number (Valocchi and Malmstead 1992). Applying the Courant condition would require that mass not be transported more than a single grid cell in any single time step—with a Courant number greater than 1, the transport skips over entire grid cells without reaction. This can be an issue in heterogeneous domains where locally very fast transport rates occur, driving the time step based on the local Courant condition to a very small value.

The chief advantage of the global implicit approach is that, unlike the operator splitting method for the reactive transport problem, one is not restricted to time steps less than the Courant condition. The other advantage of the global implicit is in systems where reaction and transport are strongly coupled. In such systems, an operator splitting approach may fail to converge to a fully coupled solution quickly. The quadratic convergence of the Newton method thus may provide substantially improved numerical stability and even faster execution time. The diffusion-dominant electrostatic calculations discussed later in the chapter and in Tournassat and Steefel (2019, this volume) are prime examples where the global implicit method dramatically outperforms the operator splitting approaches.

REACTIVE TRANSPORT ADVANCES

In the following sections, the author's view as to the advances in the development and use of reactive transport models since the publication of *Reviews in Mineralogy* Volume 34 in 1996 (Lichtner et al. 1996) is given. The list is far from exhaustive, however, and a more comprehensive discussion of advances in the application of reactive transport modeling are given in the other chapters in this volume.

Pore scale and continuum models

Historically, most of the developments and applications in reactive transport have centered on continuum scale models. Continuum models are those that treat the medium as having continuous properties, such as reactive surface area, mineral volume fractions, porosity permeability, and diffusivity (Fig. 2). This approach forms the basis of most of the reactive transport codes in use today (Steefel et al. 2015). However, an important recent development has been that of pore scale reactive transport models (Molins et al. 2012, 2014; Emmanuel et al. 2015). The pore scale models are distinguished from continuum models by the fact that interfaces between fluid, minerals, and gases are explicitly resolved.

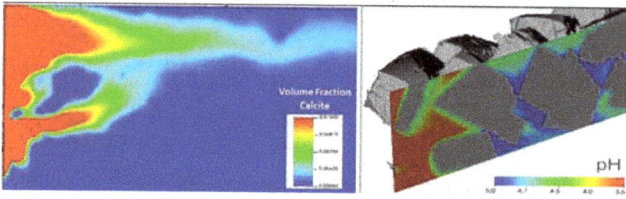

Figure 2. Continuum **(left)** versus pore scale reactive transport modeling results **(right)**. The continuum model tracks mineral volume fractions, porosity, and permeability implicitly. The left figure shows the clogging effects of infiltration of a supersaturated solution into heterogeneous porous material. The pore scale models track mineral–fluid–gas interfaces and associated diffusion boundary layers explicitly. [Reprinted with permission from Molins et al. (2014) Pore-scale controls on calcite dissolution rates from flow-through laboratory and numerical experiments. Environmental Science & Technology 48:7453–7460. Copyright (2014) American Chemical Society.]

	Pore Scale	**Continuum Scale**
Flow	$\dfrac{\partial u}{\partial t} + (u \cdot \nabla)u + \nabla p = \nu \Delta u$ $\nabla \cdot u = 0$	$q = -\dfrac{k}{\mu} \nabla p$ $\nabla \cdot (k \nabla p) = 0$
Transport	$\dfrac{\partial c}{\partial t} = \nabla \cdot (D \nabla c) - \nabla \cdot (uc)$	$\theta \dfrac{\partial c}{\partial t} = \nabla \cdot (D \nabla c) - \nabla \cdot (qc)$ where $\mathcal{D} = \theta \tau D + \alpha_L |u|$
Reaction	$-D \nabla c \cdot n = k\, f(c)$	$r = k\, A\, f(c)$

Figure 3. Pore versus continuum scale governing equations and associated parameters. For flow, the contrast is between the Navier–Stokes equation versus Darcy's Law, with permeability representing a continuum parameter. For reactions, the diffusive flux to discrete mineral interfaces is captured, while the continuum models typically use reactive surface area to represent heterogeneous reaction processes.

The constitutive equations for continuum and pore scale reactive transport are different. For pore scale, flow is described with the Navier–Stokes (or Stokes) equation, while Darcy's Law is used in the case of a continuum formulation. For transport, the pore scale will consider molecular diffusion in the pore fluid while the continuum model includes dispersion as an upscaled parameter (Fig. 3). Reactive surface area is used in continuum treatments, while diffusion to and from explicit reacting mineral surfaces are normally considered in the pore scale models.

Contaminant transport

Modern reactive transport methods have made significant contributions to the topic of contaminant transport in recent years. The contributions can be broadly divided into two important themes: 1) the more comprehensive and rigorous treatment of geochemistry in contaminant transport models, and 2) the role of aquifer heterogeneity on transport and mixing rates. The second of these topics is not considered further here, since it is primarily the purview of hydrology and transport disciplines—we focus here on the geochemical and mineralogical aspects of the problem.

The starting point for contaminant hydrogeology has always been the linear distribution coefficient (or K_d) models for sorption, which have the advantage that they can be easily incorporated into the transport equations without rendering them nonlinear. It has also been asserted that they have the advantage that the data required for the implementation is minimal, or at least less than that required by the multicomponent models discussed below. However, it should be pointed out that the K_d models, while potentially being as simple as the determination of the sorbed and aqueous concentration under a set of environmental conditions, typically require unique constraints for each and every site to which they are applied. In addition, as will be apparent from the discussion below, such K_d models might not even apply to a single site where conditions (temperature, salinity, competing ion concentrations, sorption site density) change over time. In other words, there is no real generality in the case of linear distribution coefficients, in contrast to the more rigorous ion exchange and surface complexation models (like that of metals on iron hydroxides) that can be applied quite widely.

The primary issue with the K_d models, and even somewhat more advanced sorption models like the Langmuir or Freundlich formulations, is that they do not consider competitive sorption. In contrast, surface complexation or multicomponent ion exchange models explicitly consider competitive effects. This is of course well known to the geochemical community, so perhaps the failure to embrace the more rigorous multicomponent sorption models is hard to understand. Multicomponent ion exchange and surface complexation models are briefly reviewed below, and then considered in the context of reactive transport.

Ion exchange and transport

Ion exchange reactions can be described via a mass action expression with an associated equilibrium constant. The exchange reaction can be written in generic form, assuming here a chloride solution, as:

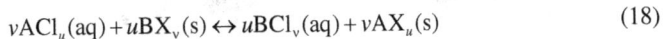

$$v\text{ACl}_u(\text{aq}) + u\text{BX}_v(\text{s}) \leftrightarrow u\text{BCl}_v(\text{aq}) + v\text{AX}_u(\text{s}) \tag{18}$$

where X refers to the exchange site occupied by the cations A^{u+} and B^{v+}. The equilibrium constant or selectivity coefficient, K_{eq}, for this reaction can be written as

$$K_{eq} = \frac{[\text{BCl}_v]^u [\text{AX}_u]^v}{[\text{ACl}_u]^v [\text{BX}_v]^u} \tag{19}$$

where the parentheses [] refer to the thermodynamic activities. Here it is clear that the sorption of any particular contaminant (for example, $^{137}\text{Cs}^+$ or $^{90}\text{Sr}^{2+}$) will be affected by the competing cation concentrations, for example, sodium via the reaction:

$$Na^+ + CsX \leftrightarrow NaX + Cs^+ \tag{20}$$

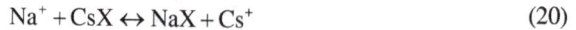

A strongly saline solution like that associated with the tank leaks at the Hanford site, arguably one of the most significant environmental problems facing the United States, will thus have an important role in determining the effective ^{137}Cs linear distribution coefficient. The real-world problem can be made more complicated (and potentially worse) when multiple exchange sites occur in the sediment or soil, as is the case at the Hanford 200 tank farm (Zachara et al. 2002). In this case, the effective K_d depends also on the Cs^+ concentration (Fig. 4), as noted by Zachara et al. (2002) and by Steefel et al. (2003).

An example 2D reactive transport simulation was presented in Steefel et al. (2005) to demonstrate the effect of the competing $NaNO_3$ (a major component of the leaking tank fluids) concentration. This simulation was carried out using ion exchange selectivities and site concentrations determined from Hanford bulk sediment (Steefel et al. 2003). Figure 5 shows the transport of the non-reactive nitrate, with the plume extending over the entire domain size considered. The fractional migration of ^{137}Cs at 1 M $NaNO_3$ is significantly less, in keeping with the expectation that some retardation of this contaminant will occur (although still more

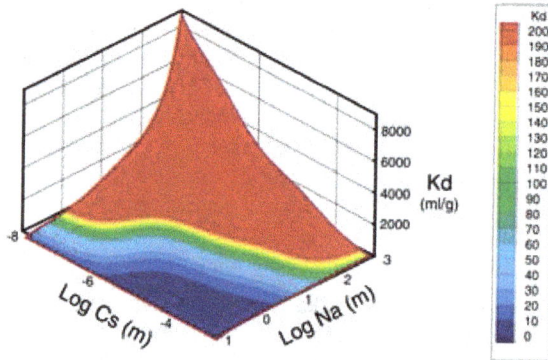

Figure 4. Effective linear distribution coefficient (K_d) for Cs^+ at the Hanford 200 tank farm (Zachara et al. 2002; Steefel et al. 2003). Note the dependence of the K_d on both the competing cation concentration (Na^+ associated with $NaNO_3$ in the tank wastes), and on Cs^+ itself because of the presence of at least two sites with different selectivities for Cs^+ versus Na^+. [Reprinted from Steefel et al. (2003) Cesium migration in Hanford sediment: a multisite cation exchange model based on laboratory. Journal of Contaminant Hydrology 67:219–246. Copyright (2003) with permission from Elsevier.]

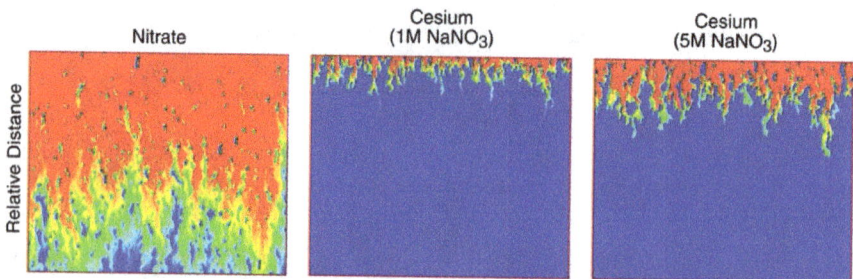

Figure 5. Relative migration of nitrate (non-reactive), and Cs^+ at 1 M $NaNO_3$ and 5 M $NaNO_3$. The high Na^+ concentrations in the tank leak explain the enhanced migration and thus weaker than expected retardation of ^{137}Cs at the Hanford 200 tanks. [Reprinted from Steefel et al. (2005) Reactive transport modeling: An essential tool and a new research approach for the earth sciences. Earth and Planetary Science Letters 240:539–558. Copyright (2005) with permission from Elsevier.]

than would be the case for ^{137}Cs in a typically dilute soil or vadose zone water). At 5 M NaNO$_3$, the migration of the ^{137}Cs is significantly farther, even if still retarded relative to the nitrate. The enhanced ^{137}Cs observed below many of the Hanford 200 tanks can thus be explained largely on the basis of a classical ion exchange model that accounts for the competing NaNO$_3$ concentrations in the plume, and on the elevated concentrations of Cs$^+$ that exceed the number of high affinity sites that are available for strong sorption. The model also suggests that as leaking of highly concentrated NaNO$_3$ tank fluids ceases, further migration of the ^{137}Cs is unlikely.

Surface complexation and transport

Perhaps an even more convincing demonstration of the inadequacy of the classical K_d models is provided by reactive transport analyses of metal and radionuclide migration influenced by surface complexation on iron hydroxides. In fact, an entire generation of geochemists have investigated the surface complexation behavior of metals on ferric hydroxides over many years (Dzombak and Morel 1990; Davis et al. 1998, 2004), but only more recently has the use of such models been demonstrated convincingly in reactive transport frameworks at the field scale.

Perhaps the first successful demonstration of the use of surface complexation models to describe field-scale reactive transport was presented by Davis and co-workers for the Naturita, Colorado uranium-contaminated site (Curtis et al. 2004, 2006; Davis et al. 2004). Davis et al. (2004) used both electrostatic and non-electrostatic surface complexation models to describe uranium sorption on the Naturita sediment, but finally settled on the non-electrostatic models because of their flexibility in treating natural and complex multi-mineralic sediments. In order to match the pH dependence of sorption in particular, it was necessary to use a three site SCM consisting of weak, strong, and very strong sites. Figure 6 shows the match with the experimental data at a variety of CO$_2$ partial pressures, an important variable because of the strong competition between uranium carbonate complexes in solution and the surface complexes developed on the Naturita sediment surfaces (Hsi and Langmuir 1985; Prikryl et al. 2001; Davis et al. 2004). Calcium uranium carbonate complexes in solution can further reduce the sorption of uranium, especially Ca$_2$UO$_2$(CO$_3$)$_3$(aq) (Bernhard et al. 2001; Brooks et al. 2003; Fox et al. 2006).

Curtis et al. (2006) used the non-electrostatic surface complexation model presented in Davis et al. (2004) to investigate field-scale transport at the Naturita, Colorado site (Fig. 7). They demonstrated that the surface complexation model, when combined with realistic flow and transport parameters for the aquifer, accurately reproduced the available data. They also demonstrated (again) the inadequacy of a constant K_d model (Fig. 7).

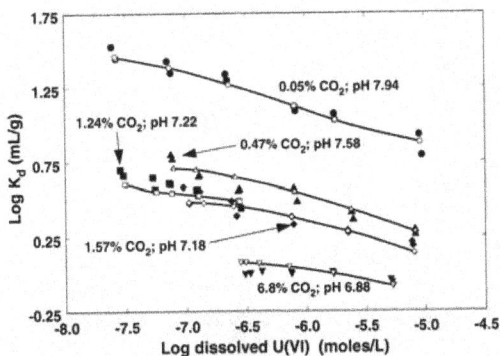

Figure 6. Fraction U(VI) adsorption on the <3 mm NABS composite sample as a function of the partial pressure of carbon dioxide, pH, and solid/liquid ratio. [Reprinted from Davis et al. (2004) Approaches to surface complexation modeling of Uranium(VI) adsorption on aquifer sediments. Geochimica et Cosmochimica Acta 68:3621–3641. Copyright (2004) with permission from Elsevier.]

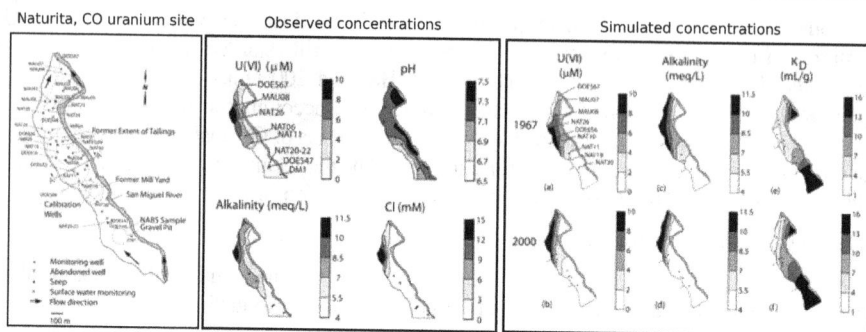

Figure 7. Left: Map of the Naturita uranium-contaminated site, southwestern Colorado. **Center:** Observed U(VI) and Cl concentrations with pH and alkalinity. **Center:** Measured U(VI) concentrations and associated chemical constituents interpolated across the site. **Right:** Simulated U(VI) concentrations and alkalinities, with calculated K_d values based on surface complexation model. [Reprinted with permission of John Wiley & Sons from Curtis et al. (2006) Simulation of reactive transport of uranium(VI) in groundwater with variable chemical conditions. Water Resources Research 42:W04404. Copyright (2006).]

A number of other studies have been carried out demonstrating the ability of the surface complexation models to describe field scale behavior, including those at the Rifle CO uranium contaminated site (Yabusaki et al. 2007, 2017; Zachara et al. 2013), and at the contaminated Savannah River site (Bea et al. 2013; Arora et al. 2018).

The laboratory–field rate discrepancy

The laboratory–field rate discrepancy has been a longstanding topic of discussion in the geochemical literature (White and Brantley 2003). The suggestion has been that field rates are three to five orders of magnitude slower than rates constrained in laboratory settings for what is argued to be the same reactive pathway, but is this really the case? What (if anything) does reactive transport analysis have to contribute to this debate?

Although it is not straightforward to quantify, it is a reasonable conclusion that with the use of a modern reactive transport model that considers multiple reactive minerals and perhaps most importantly the approach to equilibrium with respect to various phases, one can reduce this apparent discrepancy by roughly one order of magnitude. Beyond this reconciliation, it appears that careful application of a rigorous reactive transport analysis that considers detailed reaction mechanisms and physical/chemical heterogeneity can eliminate most or all of the remaining disparity. The sources of the discrepancy (which in fact may or may not be real) can be attributed to at least three possible factors:

1. Geochemical effects associated with inhibiting constituents in solution, or with incongruent reactions and a nonlinear dependence on the Gibbs free energy of the pore fluid in contact with the reacting phases (Zhu et al. 2004; Maher et al. 2006).

2. Effects of mineralogical heterogeneity on reactivity (Li et al. 2006) and physical heterogeneity operating at the pore scale (Molins et al. 2012; Beckingham et al. 2017).

3. Effects of flow field heterogeneity leading to non-uniform fluid travel times through structurally complex systems, thus leading to an apparent reaction rate that is the flux-weighted average of these distinct travel times (Steefel and Maher 2009; Maher 2010, 2011).

Maher et al. (2009) carried out a systematic reactive transport analysis of Terrace 5 (226 ka) in the Santa Cruz weathering chronosequence and demonstrated that there was essentially no lab-field discrepancy when several factors were accounted for in their 1D model (Fig. 8).

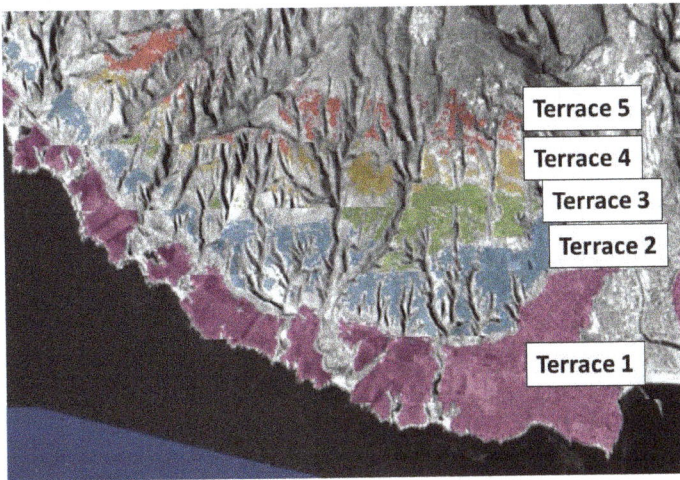

Figure 8. Santa Cruz, California chronosequences with Terraces 1–5 shown.

An advantage of the Santa Cruz field site was that the mostly granitic unconsolidated sediment allowed for truly 1D downward flow of pore fluids, so that the common issue of lateral diversion of flow in low-permeability crystalline rocks does not arise and confound the interpretation of the effective fluid–rock ratio along a flow path. To simulate the weathering in Terrace 5 (the oldest at 226 ka), Maher et al. (2009) set up a 1D reactive transport model that included K-feldspar and albite and allowed for kaolinite precipitation. Smectite was present as a primary phase, but did not have a large impact on the simulation results except for the role the mineral played in cation exchange. Much of the weathering took place in the variably saturated zone where it was important to include the diffusion of reactive species (mainly CO_2) in the gas phase.

The key to resolving the "discrepancy", however, was found in the suggestions from Zhu and co-workers (Zhu et al. 2004) that slow clay precipitation might explain why the system did not remain far from equilibrium where the rates of dissolution for the feldspars are at their maximum. Slow clay precipitation in the model was implemented so as to match observed supersaturation with respect to kaolinite, and then combined with the nonlinear dependence of feldspar dissolution rates on the reaction affinity (or ΔG) observed in experiments by Hellmann and Tisserand (2006). With these two coupled effects, the rates of dissolution of the feldspar were much slower than what would be expected with either fast (equilibrium) clay precipitation or a linear TST mineral dissolution model. Thus, it was possible to closely match the observed mineral profiles after 226 ka (Fig. 9, upper row). In addition, the model also matched present day mineral saturation states calculated based on pore water chemistry (Fig. 9, lower row).

Incorporating isotope fractionation

The incorporation of isotope fractionation into modern reactive transport models holds the potential for a large impact on many Earth systems applications, and such isotope-enabled simulations are beginning to gain recognition (Druhan and Winnick 2019). The reasons for this potential are illustrated by the significance of isotopes as an analytical component of modern geochemistry, namely their sensitivity to tracking reaction and transport processes far beyond the precision with which such processes can be monitored using other methods (e.g., conventional chemical analysis). This is particularly true for isochemical (or nearly so) systems like incongruent mineral dissolution and precipitation, where isotopes may be the only method available to unravel the underlying mechanisms of fluid–mineral interaction (DePaolo 2011).

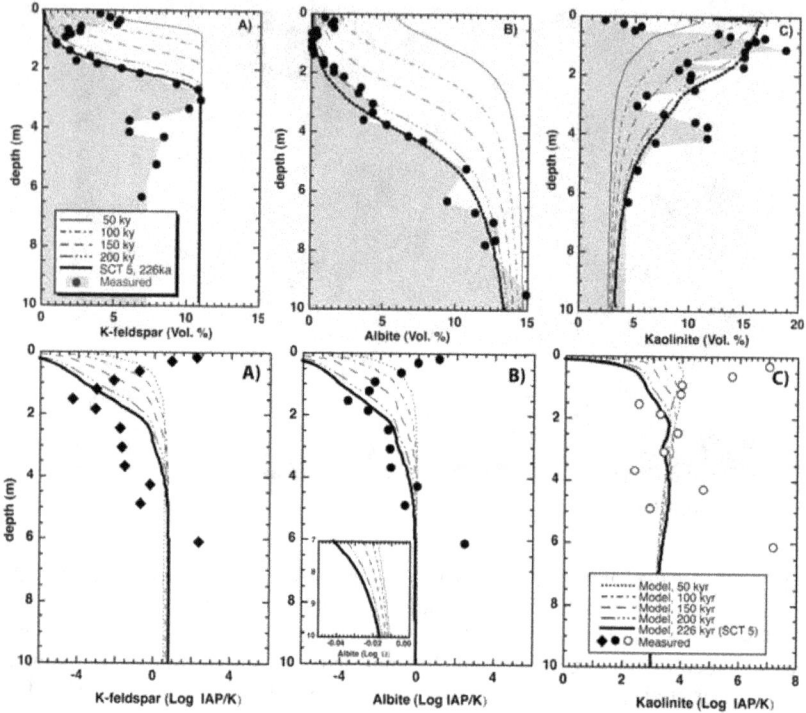

Figure 9. Upper: Simulated (solid lines) and observed (points) mineral volume percentages in 226 ka Terrace 5 weathering profile. Lower: Simulated and observed mineral saturation states. [Reprinted from Maher et al. (2009) The role of reaction affinity and secondary minerals in regulating chemical weathering rates at the Santa Cruz Soil Chronosequence, California. Geochimica et Cosmochimica Acta 73, 2804–2831. Copyright (2009) with permission from Elsevier.]

Reactive transport modeling in turn can also be used to evaluate some of the time-honored models of isotope geochemistry. For example, a distillation or Rayleigh model has been used to describe the change in isotopic ratio as a function of reaction progress in both open and closed systems (Druhan and Maher 2017) according to:

$$r = r_0 \left[f^{(\alpha-1)} \right] \qquad (21)$$

where r is the isotopic ratio, r_0 is the original ratio, f is the fraction of the original reactant remaining, and α is the fractionation factor. As pointed out by van Breukelen and Prommer (2008), the Rayleigh model underestimates the fractionation in the case of heterogeneous flow systems. This conclusion was reinforced by Druhan and Maher (2017) in their systematic study. The Rayleigh model also overestimates the fractionation for systems involving sorption (van Breukelen and Prommer 2008). In fact, in any system where the fractionation factor is not constant, the Rayleigh model fails. This could be expected to be the case where kinetic isotope fractionation occurs during mineral precipitation along a flow path. If the precipitating mineral approaches equilibrium (where the kinetic isotope fractionation should disappear), then one expects the fractionation factor α to change continuously.

At a first level, modern reactive transport software (e.g., PHREEQc, CrunchTope, FLOTRAN, TOUGHREACT, see Steefel et al. 2015) are fully capable of treating isotopes as if they were separate chemical components (Druhan et al. this volume). For example, the rate laws for the reduction of total aqueous $^{53}\text{CrO}_4^{2-}$ and $^{52}\text{CrO}_4^{2-}$ can be written respectively as:

$$^{53}r_{\text{total}} = k_{53} \left(^{53}\text{CrO}_4^{2-} \right)_{\text{total}} \qquad (22)$$

$$^{52}r_{\text{total}} = k_{52} \left(^{52}\text{CrO}_4^{2-} \right)_{\text{total}} \qquad (23)$$

with the kinetic fractionation factor given by:

$$\alpha_k = \frac{k_{53}}{k_{52}} \qquad (24)$$

The formulation in Equations (22) and (23) can be adapted to microbially mediated reactions, as for the reduction of the isotopologues of $^{34}\text{SO}_4^{2-}$ and $^{32}\text{SO}_4^{2-}$ to $^{34}\text{HS}^-$ and $^{32}\text{HS}^-$ respectively (Druhan et al. 2012, 2014), but this requires a coupling of the rate expressions:

$$r_{34} = B_{\text{SRB}}\mu_{34}X_{34} \left[\frac{C_{\text{SO}_4^{-2}}}{C_{\text{SO}_4^{-2}} + K_{\text{SO}_4^{2-}}} \right] F_K^{\text{acetate}} F_T \qquad (25)$$

$$r_{32} = B_{\text{SRB}}\mu_{32}X_{32} \left[\frac{C_{\text{SO}_4^{2-}}}{C_{\text{SO}_4^{2-}} + K_{\text{SO}_4^{2-}}} \right] F_K^{\text{acetate}} F_T \qquad (26)$$

where B_{SRB} is the biomass of sulfate-reducing bacteria, μ_{34} and μ_{32} are the maximum specific rates, and X_{34} and X_{32} are the mole fractions of $^{34}\text{SO}_4^{2-}$ and $^{32}\text{SO}_4^{2-}$, respectively. F_K^{acetate} is a Monod term for acetate and F_T is a thermodynamic function (Jin and Bethke 2005; Dale et al. 2006). This formulation is simplified from the full expression given by Druhan et al. (2012) for the case where the half saturation constant, $K_{\text{SO}_4^{2-}}$, for the two isotopologues is the same. The kinetic fractionation is captured by the use of slightly different maximum specific rate constants for the two isotopologues.

Using the formulations given in Equations (25) and (26), it is possible to track the isotopic evolution of the fluid and mineral phases as a function of space and time in sediment undergoing sulfate reduction due to electron donor injection (Williams et al. 2011). Druhan et al. (2014) simulated the time evolution of the effluent concentrations of iron and sulfate as well as the isotopologues of ^{34}S and ^{32}S as a result of sulfate reduction in a one meter column subject to a constant flow rate with high concentrations of the electron donor acetate (Fig. 10). The high Fe^{2+} concentrations are produced by the abiotic reaction of H_2S with ferric hydroxide, which then acts also to titrate out the sulfide as precipitated FeS.

As discussed above, the treatment of kinetic fractionation of aqueous species is relatively straightforward. Handling precipitation of a mineral phase, as in the case of FeS precipitated in the Rifle example above, is more complex. Where mineral precipitation occurs, it is necessary to treat the resulting mineral phase as a solid solution, which has the effect of coupling the isotopologues through their activities in the solid. As an example, we consider the case of calcite precipitation with the isotopologues ^{44}Ca and ^{40}Ca (Druhan et al. 2013). The rates are given by:

$$^{44}R = A_{cc} \, ^{44}k_b \, ^{44}X \left(\frac{[^{44}\text{Ca}][\text{CO}_3^{2-}]}{K_{eq} \, ^{44}X} - 1 \right) \qquad (27)$$

$$^{40}R = A_{cc} \, ^{40}k_b \, ^{40}X \left(\frac{[^{40}\text{Ca}][\text{CO}_3^{2-}]}{K_{eq} \, ^{40}X} - 1 \right) \qquad (28)$$

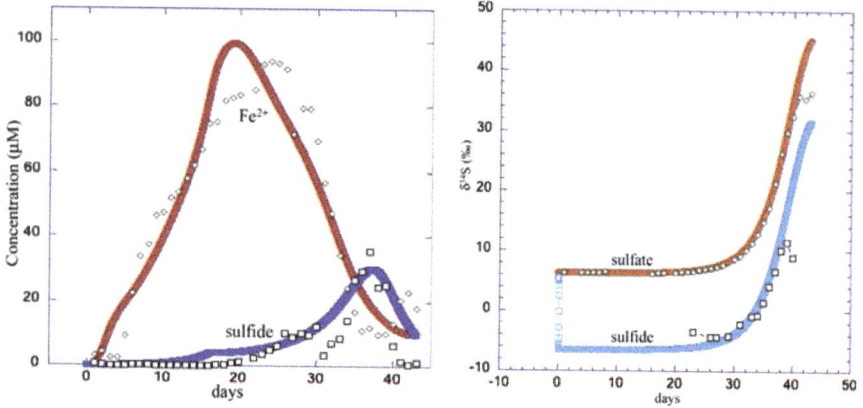

Figure 10. Left: Evolution of ferrous ion and sulfide in the effluent from a one meter column containing Rifle aquifer sediment injected with a continuous supply of acetate. **Right:** Co-evolution of the isotopic ratios of ^{34}S and ^{32}S in aqueous sulfate and sulfide (lines are simulation results, symbols are data). [Reprinted from Druhan et al. (2014) A large column analog experiment of stable isotope variations during reactive transport: I. A comprehensive model of sulfur cycling and $\delta^{34}S$ fractionation. Geochimica et Cosmochimica Acta 124:366–393. Copyright (2014) with permission from Elsevier.]

A_{cc} is the reactive surface area for the mineral calcite, $^{44}k_b$ and $^{40}k_b$ are the backward rate constants for precipitation, and ^{44}X and ^{40}X are the mole fractions (used to describe activity) for the two isotopologue mineral end-members, respectively. K_{eq} is the equilibrium constant for the reaction, which if different in Equations (27) and (28), would result in equilibrium isotopic fractionation. A linear TST formulation has been assumed here for the sake of simplicity. Fractionation in more complex rate expressions is now being explored.

Note that the activities of the end-members effectively couple the mineral rate laws through the mole fraction:

$$^{44}X = \frac{[^{44}CaCO_3]}{[^{44}CaCO_3 + {}^{40}CaCO_3]}, \quad ^{40}X = \frac{[^{40}CaCO_3]}{[^{44}CaCO_3 + {}^{40}CaCO_3]} \tag{29}$$

which is true even in the case where there is no kinetic fractionation associated with precipitation, as in the Rifle example above (Druhan et al. 2014) and in the Cr isotope benchmark problem (Wanner et al. 2015). However, when different rate constants are used for precipitation of the isotopologues, then kinetic fractionation occurs. In addition to resulting in kinetic fractionation during precipitation, the rate law proposed by Depaolo (2011) and implemented by Druhan et al. (2013) and Steefel et al. (2014) allows for back reaction of the isotopes over similar time scales. As shown in Figure 11, isotopic re-equilibration occurs after the system has reached chemical equilibrium. This behavior follows from the formulation of the TST rate law in DePaolo (2011) based on the principle of detailed balancing (Lasaga 1984), but whether this is realistic in most natural systems is another question. Any straightforward implementation of the DePaolo (2011) TST rate law results in kinetic fractionation during mineral dissolution, so one possibility is to require that no fractionation occurs during dissolution (an option now in the code CrunchTope).

In some cases back-reaction (or re-equilibration) may occur, but much more slowly than is predicted by the TST rate law. Such behavior is suggested by modeling of diagenetic processes at Site 984 in the Mid-Atlantic, where matching of the isotope profiles over the millions of years associated with sediment burial requires re-equilibration rates about 4 orders of magnitude slower than the bulk precipitation rates of calcite (Fig. 11, right) used to match the total Ca profile (Fig. 12, left) (Steefel et al. 2014).

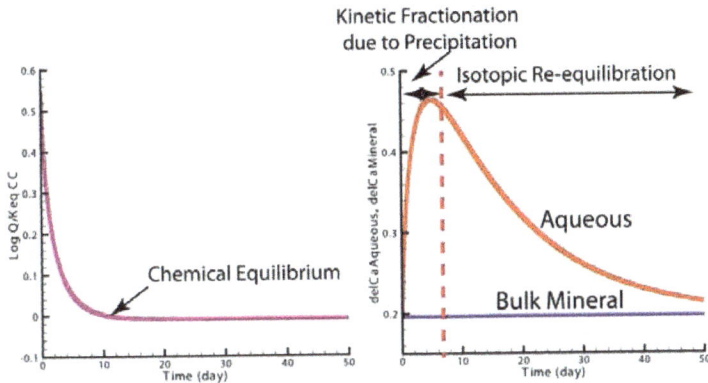

Figure 11. Left: Equilibration of a supersaturated solution in a batch system with calcite occurs after about 10 days if a conventional TST type precipitation rate law is used. **Right:** Kinetic fractionation of the isotopologues of calcium occurs during the same period, but this shift is followed by an extended period in which re-equilibration occurs if the linear TST model of DePaolo (2011) is used.[Modified from Steefel et al. (2014) Modeling coupled chemical and isotopic equilibration rates. Procedia Earth and Planetary Science 10:208–217. Copyright (2014) with permission from Elsevier.]

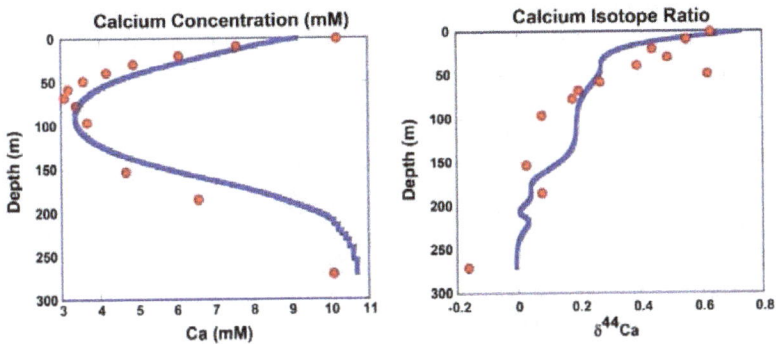

Figure 12. Left: Calcium concentration profile in the pore fluid (*symbols:* data, *solid lines:* model results) for ODP Site 984 in the Mid-Atlantic Ocean. The bulk Ca profile is modeled reasonably well with the conventional calcite precipitation rate laws assuming a linear TST formulation. **Right:** Calcium isotopic ratio simulated using an approximately 4 order of magnitude slower re-equilibration rate than is used for bulk calcite precipitation. Simulations from Steefel et al. 2014. Data from Turchyn and DePaolo (2011). [Reprinted from Steefel et al. 2014. Modeling coupled chemical and isotopic equilibration rates. Procedia Earth and Planetary Science 10:208–217.Copyright (2014) with permission from Elsevier.]

Electrostatic effects on reactive transport

There are two developments over the last 20 years concerning electrostatic effects on reactive transport that are worth noting. The first is provided by the use of the Nernst–Planck equation, which deals rigorously with transport effects associated with the development of diffusion potentials as ions diffuse at different rates (Steefel et al. 2015). The second effect is associated with transport within the diffuse or electrical double layer (EDL) bordering charged surfaces along pores (Tournassat and Steefel 2019, this volume).

Nernst–Planck equation

The Nernst–Planck equation accounts for electrochemical migration associated with the development of a diffusion potential as ions diffuse at different rates (Steefel et al. 2015;

Tournassat and Steefel 2015, 2019, this volume). The equation can be applied in both bulk (electrically neutral) solution, and as we shall see below and in Tournassat and Steefel (2019, this volume), within the electrical double layer. The Nernst–Planck equation can be derived from equation for the chemical potential μ_j (Steefel et al. 2015):

$$\mu_j = \mu_j^0 + RT \ln(\gamma_j C_j) + z_j F \psi \qquad (30)$$

where μ_j^0 is the chemical potential at standard state, R is the gas constant, T is the absolute temperature, γ_j is the activity coefficient for the jth species, z_j is its charge, F is Faraday's constant, and ψ is the electrical potential. Since the flux of an ion can be described in terms of the gradient of its chemical potential (Lasaga 1998):

$$J_j = -\frac{D_j C_j}{RT} \frac{\partial \mu_j}{\partial x} \qquad (31)$$

we can derive the Nernst–Planck equation by substituting Equation (30) into Equation (31) to give:

$$J_j = -D_j \nabla C_j - \frac{z_i F}{RT} D_i C_i \nabla \psi - D_j C_j \nabla \ln \gamma_j \qquad (32)$$

Note that the first term on the right hand side of Equation (32) is the Fickian term used in most hydrogeological transport models. The last term on the right hand side only applies where a gradient in ionic strength is present (Tournassat and Steefel 2019, this volume). The second term on the right hand side is usually referred to as the diffusion potential since it is generated by the diffusion of charged ions at unequal rates. It goes to zero when all of the ion diffusion coefficients are the same (in which case Fick's Law is recovered if no ionic strength gradients are present).

The Nernst–Planck equation has now been implemented in a number of reactive transport codes, including FLOTRAN, CrunchFlow, MIN3P, and PHREEQc (Steefel et al. 2015). It was used to model ion profiles in marine sediments with upwelling hydrothermal fluids (Giambalvo et al. 2002), a geological system where matching solute profiles would be impossible with the use of a simple Fickian diffusion model.

A transient test problem to show the effects of the diffusion potential term in the Nernst–Planck equation was proposed by Lichtner (1998). In this problem, Na^+ and Cl^- are the same concentration inside the domain and to the left of the Dirichlet boundary on the left side of the domain (the right hand boundary is treated as a no-flux boundary). So a simple application of Fick's Law would say there should be no flux of either Na^+ or Cl^- (since $\partial C / \partial x = 0$). But gradients in pH and NO_3^- develop because the concentrations outside (0.001 mM) and inside (0.1 mM) the domain are different for these ions. Through the effects of the Nernst–Planck equation (32), fluxes of both Na^+ and Cl^- are created through the diffusion potential term (Fig. 13). The more rapid loss of H^+ from the domain, which diffuses about 9 times faster than the NO_3^-, results in a net influx of Na^+ to compensate the charge. Cl^- is also removed from the internal domain to compensate for the higher out-flux of H^+ compared to NO_3^-.

Rolle et al. (2013) used the Nernst–Planck equation to model transverse diffusion/dispersion in a 2D flow field, and this problem became part of a benchmarking study of the Nernst–Planck equation described in Rasouli et al. (2015). Figure 14 shows the simulation results from CrunchFlow based on the problem described by Rolle et al. (2013), with injection of 0.29 mM KCl and $MgCl_2$ from the left side (flow left to right) over an interval of 1 cm (the remainder of the injection ports, designed to produce 1D flow, contained deionized water). Note the greater spreading of the K^+ plume resulting from its 2.8 times higher diffusivity than that of Mg^{2+}.

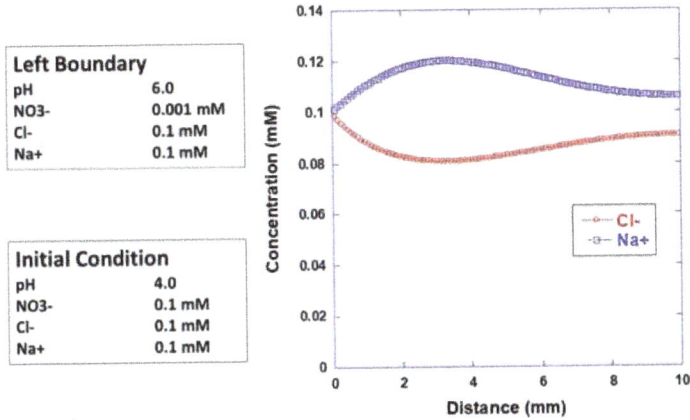

Figure 13. Demonstration of non-Fickian diffusion for Na^+ and Cl^-, which begin with no concentration gradient, driven by gradients in H^+ and NO_3^-. From a CrunchFlow simulation developed based on Lichtner (1998).

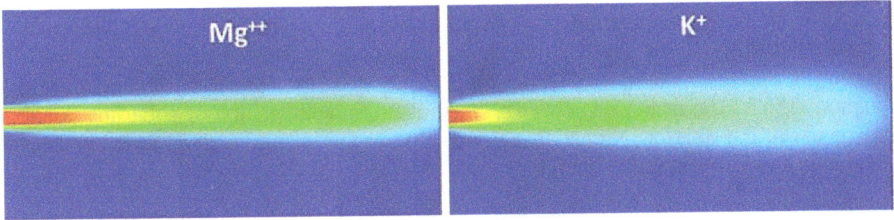

Figure 14. Transverse dispersion of K^+ versus Mg^{2+} in a flow field. Unidirectional flow is from left to right at a rate of 2.5625 cm/h for 16 h, with injection of 0.29 mM KCl and $MgCl_2$ over an interval of 1 cm on left side. The domain is 100 cm long and 12 cm wide. The 2.8 times higher diffusivity of the K^+ as compared to Mg^{2+} accounts for its greater spreading in the transverse direction. From a CrunchFlow simulation based on problem described by Rolle et al. (2013).

Transport in the EDL

Electrostatic effects on surface complexation have been considered for many years, even if the non-electrostatic models like those employed by Davis and co-workers are now more commonly applied at the field scale. A new class of models has been developed that considers double layer or electrical double layer (EDL) screening of the charge of mineral surfaces bordering the pores. The EDL is often conceptually subdivided into two regions: the Stern layer located within the first water monolayers of the interface in which ions adsorb as inner- and outer-sphere surface complexes, and a diffuse layer (DL) located beyond the Stern layer in which a diffuse cloud of ions screens the remaining uncompensated surface charge (Tournassat and Steefel 2019, this volume). At infinite distance from the charged surface, the solution is electrically neutral and is described as bulk (or free) water, contrasting with the diffuse layer where electrical neutrality does not prevail. In the diffuse layer, water is considered to have properties similar to bulk water and the distribution of ions predicted by molecular dynamic simulations is often in close agreement with predictions made with the Poisson–Boltzmann (PB) equation (Tinnacher et al. 2016).

The Poisson–Boltzmann equation is derived by combining the Poisson equation that relates the local charge imbalance at a position y in the direction perpendicular to the charged surface to the second derivative of the electrostatic potential (Ψ_{DL} in V) at the same position:

$$\frac{d^2\psi_{DL}(y)}{dy^2} = -\frac{1}{\varepsilon}\sum_i z_i FC_{i,DL} \tag{33}$$

where ε is the water dielectric constant ($78.3 \times 8.85419 \cdot 10^{-12}$ F·m^{-1} at 298 K), F is the Faraday constant (96 485 C·mol^{-1}), and z_i is the charge of ion i. The concentration of species i in the DL, $C_{i,DL}$, is given by the Boltzmann equation:

$$c_{i,DL} = c_{i,0}\exp\left(\frac{-z_i F\psi_{DL}(y)}{RT}\right) \tag{34}$$

where R is the gas constant (8.3145 J·mol^{-1}·K^{-1}), T is the temperature (K), and $c_{i,0}$ is the concentration of species i in the bulk water at equilibrium with the diffuse layer (Tournassat and Steefel 2019, this volume).

There is currently interest in solving the Poisson–Boltzmann equation numerically, but to do so is challenging because of the fine discretization required (typically on the order of nanometers). In addition, the well-known analytical solutions to the Poisson–Boltzmann are inaccurate in the case of asymmetric electrolytes, or where the surface potential is high (Schoch et al. 2008; Kirby 2010). Since the Poisson–Boltzmann equation applies to equilibrium or static systems, a superior treatment is provided by the Poisson–Nernst–Planck (PNP) equation (Schoch et al. 2008; Kirby 2010), although the same issues with discretization arise. The PNP equation is implemented by combining the Poisson equation (33) with the Nernst–Planck equation (32). To the author's knowledge, there is no implementation of either the Poisson–Boltzmann or the Poisson–Nernst–Planck equation in any of the modern reactive transport simulators for natural waters (Steefel et al. 2015).

An alternative to the PNP that has been now widely applied is the mean electrostatic potential (MEP) model, essentially a dual continuum model in which electrically neutral bulk water constitutes one continuum and the EDL the second (Revil and Leroy 2004; Appelo and Wersin 2007; Appelo et al. 2010; Revil et al. 2011; Tournassat and Steefel 2015; Tournassat et al. 2016). This model averages ion concentrations in the diffuse layer ($\overline{C_{i,DL}}$) by scaling them to a mean electrostatic potential (MEP, Ψ_M) that applies to the diffuse layer volume (VDL):

$$\overline{C_{i,DL}} = \frac{1}{V_{DL}}\iiint_{DL} z_i C_{i,0}\exp\left(\frac{-z_i F\psi_{DL}(x,y,z)}{RT}\right)dxdydz \approx C_{i,0}\exp\left(\frac{-z_i F\psi_M}{RT}\right) \tag{35}$$

The mean electrostatic potential value can be deduced from the charge balance in the diffuse layer:

$$\sum_i z_i F\overline{C_{i,DL}} = \sum_i z_i FC_{i,0}\exp\left(\frac{-z_i F\psi_M}{RT}\right) = -Q_{DL} \tag{36}$$

where Q_{DL} (in C·m^{-3}) is the charge that must be balanced in the diffuse layer (Tournassat and Steefel 2019, this volume).

The MEP model can be evaluated by comparing to an ingenious diffusion experiment conducted by Glaus et al. (2013). In this experiment, a gradient in concentration of NaClO$_4$ from 0.5 M to 0.1 M was applied across a diffusion cell containing Na-montmorillonite. After equilibrating for two days, both reservoirs were spiked with the same concentration of the isotope ^{22}Na. From Fick's Law alone, one would expect no change in the concentration of the ^{22}Na in either reservoir, since no gradient in concentration is initially present. In fact, the experiments showed that the concentration of the ^{22}Na increased continuously in the high concentration reservoir and decreased continuously in the low concentration reservoir. This was modeled with the MEP model implemented in CrunchClay (see Tournassat and Steefel 2019, this volume for figure showing the results).

Another recent demonstration of the MEP model is provided by a modeling study of the DR-A borehole diffusion experiment carried out in the Opalinus Clay in Switzerland (Soler et al. 2019). In this experiment, a cocktail of anions, cations, and non-reactive tracers were added to the borehole and their concentrations monitored over time as the cocktail solutes diffused out into the Opalinus Clay. Figure 15 shows the time evolution of two anions, iodide and bromide, as compared to the uncharged tritiated water (HTO). Note that over the same time period, the HTO drops to about 30% of its initial value, while the two anions decrease only to about 55% of their initial value. The MEP model captures the behavior of both the uncharged solute and the anions with a single model, which is not possible with a simple application of Fick's law. In addition, it captures accurately the phenomenon of anion exclusion (Soler et al. 2019).

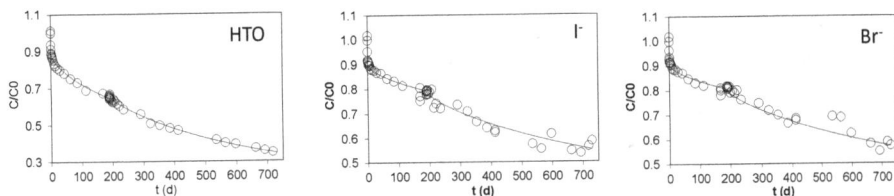

Figure 15. Time evolution of HTO (uncharged solute) and the anions iodide and bromide in the DR-A borehole diffusion test carried out in Opalinus Clay (Soler et al. 2019). Lines correspond to model results and symbols to measured data. [Reprinted from Soler et al. 2019. Modeling the ionic strength effect on diffusion in clay. The DR-A experiment at Mont Terri. ACS Earth and Space Chemistry 3, 442–451. Copyright (2019) with permission from American Chemical Society.]

Multiphase reactive systems

Arguably, one of the most significant achievements in reactive transport analysis in recent years has been the application to multiphase systems, including soils and the deep vadose zone where mostly water and air coexist (Dwivedi et al. 2019; Arora et al. 2019, both this volume), and the deep subsurface where other fluids and gases may be present (e.g., oil, gas, steam). It is well beyond the scope of this contribution to review these, but some discussion can be found in Sin and Corvisier (2019) and in Bildstein et al. (2019), both this volume. This is a challenging topic because of the presence of different phases, each of which has its own transport and chemical properties, and because of the complex, typically time-dependent interfaces between the phases.

Chemical–mechanical

Although some research has been initiated in recent years, particularly on the topic of pressure solution (Renard et al. 1997; Yasuhara et al. 2003, 2004, 2005; Yasuhara and Elsworth 2008; Taron and Elsworth 2010), and on coupled processes in nuclear waste respositories or geothermal systems (Taron et al. 2009; Gens et al. 2010; Taron and Elsworth 2010; Zheng et al. 2010; Rutqvist et al. 2014; Rutqvist 2015, 2017), much remains to be done to advance the topic together with the multicomponent chemistry we typically consider be a centerpiece of modern reactive transport analysis. Most of the analyses to date have been based on continuum models, but there is a new interest in incorporating chemistry into discontinuum mechanics model that can track the evolution of deforming and reacting interfaces explicitly (Hu et al. 2017).

Watersheds and beyond

Watershed reactive transport is another important frontier topic that is only now beginning to get the attention it deserves (Li 2019, this volume). Watersheds are situated in scale somewhere between the more local "Critical Zone" studies (e.g., Sullivan et al. 2019) and the larger regional and even global scale reactive transport models we can expect to see more of in the future (e.g., Goddéris et al. 2019). Progress is being made now in integrating reactive transport capabilities into the complex catchment scale hydrological models (Bao et al. 2017; Li et al. 2017; Zhi et al. 2019), which are complicated by the need to consider integrated surface and subsurface water flow, as well as the effects of both hydrological and biogeochemical hot spots (Dwivedi et al. 2018a,b).

CONCLUSIONS

The field of reactive transport has come a long way since the publication of Reviews in Mineralogy Vol. 34 in 1996. The modeling approach, which considers multicomponent chemical components and multiple reaction processes, has now been successfully applied in a large number of environments ranging from contaminated groundwater systems, hydrothermal and geothermal systems, nuclear waste repositories, chemical weathering, and now even watershed and regional scale river systems. Application of more sophisticated sorption models like those based on multicomponent ion exchange and surface complexation should have revolutionized the contaminant hydrogeology business, but so far the industry has resisted what appears to be the inevitable given the vastly superior performance of these models compared to the classical K_d model. Similarly, significant progress has been made in resolving the lab–field discrepancy using reactive transport modeling. The advent of pore scale models, which aim to capture explicitly the pore structure within which water–rock interaction occurs, is likely to provide further insights into the reactivity of natural soils, sediments, and rocks. The use of a new class of electrostatic models that consider the electrical double layer in charged media offers the promise of quantitatively explaining the migrations of ions in low permeability (especially clay) environments.

ACKNOWLEDGMENTS

This work was supported by the Director, Office of Science, Basic Energy Sciences, Chemical Sciences, Geosciences, and Biosciences Division, of the U.S. Department of Energy under Contract No. DE-AC02-05CH11231 to Lawrence Berkeley National Laboratory. This work was also supported as part of the Watershed Function Science Focus Area at Lawrence Berkeley National Laboratory funded by the U.S. Department of Energy, Office of Science, Biological and Environmental Research under the same contract. The author is grateful to the BRGM (Carnot project) and the University of Orleans (Labex Voltaire) in France for support for a research stay where at least some of the progress noted here was made, and where the view from the crossroads was first examined.

REFERENCES

Appelo CAJ, Wersin P (2007) Multicomponent diffusion modeling in clay systems with application to the diffusion of tritium, iodide, and sodium in opalinus clay. Environ Sci Technol 41:5002–5007

Appelo CAJ, Van Loon LR, Wersin P (2010) Multicomponent diffusion of a suite of tracers (HTO, Cl, Br, I, Na, Sr, Cs) in a single sample of Opalinus clay. Geochim Cosmochim Acta 74:1201–1219

Arora B, Davis JA, Spycher NF, Dong W, Wainwright HM (2018) Comparison of electrostatic and non-electrostatic models for U (VI) sorption on aquifer sediments. Groundwater 56:73–86

Arora B, Dwivedi D, Faybishenko B, Jana RB, Wainwright HM (2019) Understanding and predicting vadose zone processes. Rev Mineral Geochem 85:303–328

Bao C, Li L, Shi Y, Duffy C (2017) Understanding watershed hydrogeochemistry: 1. Development of RT–Flux–PIHM Water Resour Res 53:2328–2345

Bea SA, Wainwright H, Spycher N, Faybishenko B, Hubbard SS, Denham ME (2013) Identifying key controls on the behavior of an acidic-U (VI) plume in the Savannah River Site using reactive transport modeling. J Contam Hydrol 151:34–54

Bear J (1972) Dynamics of Fluids in Porous Media. Courier Dover Publications

Beckingham LE, Steefel CI, Swift AM, Voltolini M, Yang L, Anovitz LM, Sheets JM, Cole DR, Kneafsey TJ, Mitnick EH, Zhang S (2017) Evaluation of accessible mineral surface areas for improved prediction of mineral reaction rates in porous media. Geochim Cosmochim Acta 205:31–49

Bernhard G, Geipel G, Reich T, Brendler V, Amayri S, Nitsche H (2001) Uranyl (VI) carbonate complex formation: Validation of the $Ca_2UO_2(CO_3)_3$ (aq.) species. Radiochimica Acta 89:511–518

Bildstein O, Claret F, Frugier P (2019) RTM for waste repositories. Rev Mineral Geochem 85:419–457

Brooks SC, Fredrickson JK, Carroll SL, Kennedy DW, Zachara JM, Plymale AE, Kelly SD, Kemner KM, Fendorf S (2003) Inhibition of bacterial U (VI) reduction by calcium. Environ Sci Technol 37:1850–1858

Curtis GP, Fox P, Kohler M, Davis JA (2004) Comparison of in situ uranium K_D values with a laboratory determined surface complexation model. Appl Geochem 19:1643–1653

Curtis GP, Davis JA, Naftz DL (2006) Simulation of reactive transport of uranium (VI) in groundwater with variable chemical conditions. Water Resour Res 42:W04404

Dale A, Regnier P, Van Cappellen P (2006) Bioenergetic controls on anaerobic oxidation of methane (AOM) in coastal marine sediments: a theoretical analysis. Am J Sci 306:246–294

Davis JA, Coston JA, Kent DB, Fuller CC (1998) Application of the surface complexation concept to complex mineral assemblages. Environ Sci Technol 32:2820–2828

Davis JA, Meece DE, Kohler M, Curtis GP (2004) Approaches to surface complexation modeling of Uranium(VI) adsorption on aquifer sediments. Geochim Cosmochim Acta 68:3621–3641

Deng H, Molins S, Steefel C, DePaolo D, Voltolini M, Yang L, Ajo-Franklin J (2016) A 2.5 D reactive transport model for fracture alteration simulation. Environ Sci Technol 50:7564–7571

DePaolo DJ (2011) Surface kinetic model for isotopic and trace element fractionation during precipitation of calcite from aqueous solutions. Geochim Cosmochim Acta 75:1039–1056

Druhan JL, Maher K (2017) The influence of mixing on stable isotope ratios in porous media: A revised Rayleigh model. Water Resour Res 53:1101–1124

Druhan JL, Winnick MJ (2019) Reactive transport of stable isotopes. Elements 15:107–110

Druhan JL, Steefel CI, Molins S, Williams KH, Conrad ME, DePaolo DJ (2012) Timing the onset of sulfate reduction over multiple subsurface acetate amendments by measurement and modeling of sulfur isotope fractionation. Environ Sci Technol 46:8895–8902

Druhan JL, Steefel CI, Williams KH, DePaolo DJ (2013) Calcium isotope fractionation in groundwater: molecular scale processes influencing field scale behavior. Geochim Cosmochim Acta 119:93–116

Druhan JL, Steefel CI, Conrad ME, DePaolo DJ (2014) A large column analog experiment of stable isotope variations during reactive transport: I A comprehensive model of sulfur cycling and δ^{34}S fractionation. Geochim Cosmochim Acta 124:366–393

Dwivedi D, Arora B, Steefel CI, Dafflon B, Versteeg R (2018a) Hot spots and hot moments of nitrogen in a riparian corridor. Water Resour Res 54:205–222

Dwivedi D, Steefel CI, Arora B, Newcomer M, Moulton JD, Dafflon B, Faybishenko B, Fox P, Nico P, Spycher N, Carroll R (2018b) Geochemical exports to river from the intrameander hyporheic zone under transient hydrologic conditions: East River Mountainous Watershed, Colorado. Water Resour Res 54:8456–8477

Dwivedi D, Tang J, Bouskill N, Georgiou K, Chacon SS, Riley WJ (2019) Abiotic and biotic controls on soil organomineral interactions: Developing model structures to analyze why soil organic matter persists. Rev Mineral Geochem 85:329–348

Dzombak DA, Morel FMM (1990) Surface Complexation Modeling-Hydrous Ferric Oxide. John Wiley & Sons. New York

Emmanuel S, Anovitz LM, Day-Stirrat RJ (2015) Effects of coupled chemo-mechanical processes on the evolution of pore-size distributions in geological media. Rev Mineral Geochem 80:45–60

Fox PM, Davis JA, Zachara JM (2006) The effect of calcium on aqueous uranium (VI) speciation and adsorption to ferrihydrite and quartz. Geochim Cosmochim Acta 70:1379–1387

Gens A, Guimaräes L do N, Olivella S, Sánchez M (2010) Modelling thermo-hydro-mechano-chemical interactions for nuclear waste disposal. J Rock Mech Geotech Eng 2:97–102

Giambalvo ER, Steefel CI, Fisher AT, Rosenberg ND, Wheat CG (2002) Effect of fluid–sediment reaction on hydrothermal fluxes of major elements, eastern flank of the Juan de Fuca Ridge. Geochim Cosmochim Acta 66:1739–1757

Glaus MA, Birgersson M, Karnland O, Van Loon LR (2013) Seeming steady-state uphill diffusion of 22Na+ in compacted montmorillonite. Environ Sci Technol 47:11522–11527

Goddéris Y, Schott J, Brantley SL (2019) Reactive transport models of weathering. Elements 15:103–106

Helgeson HC (1968) Evaluation of irreversible reactions in geochemical processes involving minerals and aqueous solutions—I Thermodynamic relations. Geochim Cosmochim Acta 32:853–877

Helgeson HC, Garrels RM, MacKenzie FT (1969) Evaluation of irreversible reactions in geochemical processes involving minerals and aqueous solutions—II Applications. Geochim Cosmochim Acta 33:455–481

Hellmann R, Tisserand D (2006) Dissolution kinetics as a function of the Gibbs free energy of reaction: An experimental study based on albite feldspar. Geochim Cosmochim Acta 70:364–383

Hsi CD, Langmuir D (1985) Adsorption of uranyl onto ferric oxyhydroxides: application of the surface complexation site-binding model. Geochim Cosmochim Acta 49:1931–1941

Hu M, Rutqvist J, Wang Y (2017) A numerical manifold method model for analyzing fully coupled hydro-mechanical processes in porous rock masses with discrete fractures. Adv Water Resour 102:111–126

Jin Q, Bethke CM (2005) Predicting the rate of microbial respiration in geochemical environments. Geochim Cosmochim Acta 69:1133–1143

Kirby BJ (2010) Micro-and Nanoscale Fluid Mechanics: Transport in Microfluidic Devices. Cambridge University Press

Lasaga AC (1984) Chemical kinetics of water–rock interactions. J Geophys Res: Solid Earth 89:4009–4025

Lasaga AC (1998) Kinetic Theory in the Earth Sciences. Princeton University Press

Leal AM, Blunt MJ, LaForce TC (2014) Efficient chemical equilibrium calculations for geochemical speciation and reactive transport modelling. Geochim Cosmochim Acta 131:301–322

Li L (2019) Watershed reactive transport. Rev Mineral Geochem 85:381–418

Li L, Peters CA, Celia MA (2006) Upscaling geochemical reaction rates using pore-scale network modeling. Adv Water Resour 29:1351–1370

Li L, Gawande N, Kowalsky MB, Steefel CI, Hubbard SS (2011) Physicochemical heterogeneity controls on uranium bioreduction rates at the field scale. Environ Sci Technol 45:9959–9966

Li L, Bao C, Sullivan PL, Brantley S, Shi Y, Duffy C (2017) Understanding watershed hydrogeochemistry: 2. Synchronized hydrological and geochemical processes drive stream chemostatic behavior. Water Resour Res 53:2346–2367

Lichtner PC (1985) Continuum model for simultaneous chemical reactions and mass transport in hydrothermal systems. Geochim Cosmochim Acta 49:779–800

Lichtner PC (1988) The quasi-stationary state approximation to coupled mass transport and fluid–rock interaction in a porous medium. Geochim Cosmochim Acta 52:143–165

Lichtner P (1998) Modeling reactive flow and transport in natural systems. *In:* Proceedings of the Rome seminar on Environmental Geochemistry. Pacini Editorial Pisa, Italy

Lichtner PC, Steefel CI, Oelkers EH (Eds) (1996) Reviews in Mineralogy Volume 34: Reactive Transport in Porous Media. Min Soc Am, Washington DC

Lichtner PC, Yabusaki S, Pruess K, Steefel CI (2004) Role of competitive cation exchange on chromatographic displacement of cesium in the vadose zone beneath the Hanford S/SX tank farm. Vadose Zone J 3:203–219

Maher K (2010) The dependence of chemical weathering rates on fluid residence time. Earth Planet Sci Lett 294:101–110

Maher K (2011) The role of fluid residence time and topographic scales in determining chemical fluxes from landscapes. Earth Planet Sci Lett 312:48–58

Maher K, Steefel CI, DePaolo DJ, Viani BE (2006) The mineral dissolution rate conundrum: Insights from reactive transport modeling of U isotopes and pore fluid chemistry in marine sediments. Geochim Cosmochim Acta 70:337–363

Maher K, Steefel CI, White AF, Stonestrom DA (2009) The role of reaction affinity and secondary minerals in regulating chemical weathering rates at the Santa Cruz Soil Chronosequence, California. Geochim Cosmochim Acta 73:2804–2831

Molins S, Trebotich D, Steefel CI, Shen CP (2012) An investigation of the effect of pore scale flow on average geochemical reaction rates using direct numerical simulation. Water Resour Res 48:W03527

Molins S, Trebotich D, Yang L, Ajo-Franklin JB, Ligocki TJ, Shen C, Steefel CI (2014) Pore-scale controls on calcite dissolution rates from flow-through laboratory and numerical experiments. Environ Sci Technol 48:7453–7460

Molins S, Knabner P (2019) Multiscale approaches in reactive transport modeling. Rev Mineral Geochem 85:27–48

Ortoleva P, Chadam J, Merino E, Sen A (1987) Geochemical self-organization II; the reactive–infiltration instability. Am J Sci 287:1008–1040

Prikryl JD, Jain A, Turner DR, Pabalan RT (2001) UraniumVI sorption behavior on silicate mineral mixtures. J Contam Hydrol 47:241–253

Rasouli P, Steefel CI, Mayer KU, Rolle M (2015) Benchmarks for multicomponent diffusion and electrochemical migration. Comput Geosci 19:523–533

Renard F, Ortoleva P, Gratier JP (1997) Pressure solution in sandstones: influence of clays and dependence on temperature and stress. Tectonophysics 280:257–266

Revil A, Leroy P (2004) Constitutive equations for ionic transport in porous shales. J Geophys Res-Solid Earth 109

Revil A, Woodruff W, Lu N (2011) Constitutive equations for coupled flows in clay materials. Water Resour Res 47:1–21

Rolle M, Muniruzzaman M, Haberer CM, Grathwohl P (2013) Coulombic effects in advection-dominated transport of electrolytes in porous media: Multicomponent ionic dispersion. Geochim Cosmochim Acta 120:195–205

Rutqvist J (2015) Chapter 9—Coupled thermo-hydro-mechanical behavior of natural and engineered clay barriers. *In:* Tournassat C, Steefel CI, Bourg IC, Bergaya F (Eds.), Natural and Engineered Clay Barriers, Developments in Clay Science. Elsevier, pp. 329–357

Rutqvist J (2017) An overview of TOUGH-based geomechanics models. Comput Geosci 108:56–63

Rutqvist J, Zheng L, Chen F, Liu H-H, Birkholzer J (2014) Modeling of coupled thermo-hydro-mechanical processes with links to geochemistry associated with bentonite-backfilled repository tunnels in clay formations. Rock Mech Rock Eng 47:167–186

Schoch RB, Han J, Renaud P (2008) Transport phenomena in nanofluidics. Rev Modern Phys 80:839

Seigneur N, Mayer KU, Steefel CI (2019) Reactive transport in evolving porous media. Rev Mineral Geochem 85:197–238

Sin I, Corvisier J (2019) Multiphase multicomponent reactive transport and flow modeling. Rev Mineral Geochem 85:143–195

Soler JM, Steefel CI, Gimmi T, Leupin OX, Cloet V (2019) Modeling the ionic strength effect on diffusion in clay. The DR-A experiment at Mont Terri. ACS Earth Space Chem 3:442–451

Steefel CI, Lasaga AC (1990) Evolution of dissolution patterns—permeability change due to coupled flow and reaction. ACS Symposium Series 416:212–225

Steefel CI, Lasaga AC (1994) A coupled model for transport of multiple chemical-species and kinetic precipitation dissolution reactions with application to reactive flow in single-phase hydrothermal systems. Am J Sci 294:529–592

Steefel CI, MacQuarrie KTB (1996) Approaches to modeling of reactive transport in porous media. Rev Mineral Geochem 34:83–129

Steefel CI, Maher K (2009) Fluid–rock interaction: A reactive transport approach. Rev Mineral Geochem 70:485–532

Steefel CI, Carroll S, Zhao P, Roberts S (2003) Cesium migration in Hanford sediment: a multisite cation exchange model based on laboratory transport experiments. J Contam Hydrol 67:219–246

Steefel CI, DePaolo DJ, Lichtner PC (2005) Reactive transport modeling: An essential tool and a new research approach for the Earth sciences. Earth Planet Sci Lett 240:539–558

Steefel CI, Druhan JL, Maher K (2014) Modeling coupled chemical and isotopic equilibration rates. Procedia Earth Planet Sci 10:208–217

Steefel CI, Appelo CAJ, Arora B, Jacques D, Kalbacher T, Kolditz O, Lagneau V, Lichtner PC, Mayer KU, Meeussen JCL, Molins S, Moulton D, Shao H, Šimunek J, Spycher N, Yabusaki SB, Yeh GT (2015) Reactive transport codes for subsurface environmental simulation. Comput Geosci 19:445–478

Sullivan P, Goddéris Y, Shi Y, Gu X, Schott J, Hasenmueller E, Kaye J, Duffy C, Jin L, Brantley SL (2019) Exploring the effect of aspect to inform future earthcasts of climate-driven changes in weathering of shale. J Geophys Res: Earth Surf 124:974–993

Taron J, Elsworth D (2010) Constraints on compaction rate and equilibrium in the pressure solution creep of quartz aggregates and fractures: Controls of aqueous concentration. J Geophys Res: Solid Earth 115: B07211

Taron J, Elsworth D, Min K-B (2009) Numerical simulation of thermal-hydrologic-mechanical-chemical processes in deformable, fractured porous media. Int J Rock Mech Min Sci 46:842–854

Tinnacher RM, Holmboe M, Tournassat C, Bourg IC, Davis JA (2016) Ion adsorption and diffusion in smectite: molecular, pore, and continuum scale views. Geochim Cosmochim Acta 177:130–149

Tournassat C, Steefel CI (2015) Ionic transport in nano-porous clays with consideration of electrostatic effects. Rev Mineral Geochem 80:287–330

Tournassat C, Steefel CI (2019) Reactive transport modeling of coupled processes in nanoporous media. Rev Mineral Geochem 85:75–109

Tournassat C, Gaboreau S, Robinet J-C, Bourg IC, Steefel CI (2016) Impact of microstructure on anion exclusion in compacted clay media. CMS Workshop Lecture Series 21:137–149

Turchyn AV, DePaolo DJ (2011) Calcium isotope evidence for suppression of carbonate dissolution in carbonate-bearing organic-rich sediments. Geochim Cosmochim Acta 75:7081–7098

Valocchi AJ, Malmstead M (1992) Accuracy of operator splitting for advection-dispersion-reaction problems. Water Resour Res 28:1471–1476

van Breukelen BM, Prommer H (2008) Beyond the Rayleigh equation: reactive transport modeling of isotope fractionation effects to improve quantification of biodegradation. Environ Sci Technol 42:2457–2463

Wanner C, Druhan JL, Amos RT, Alt-Epping P, Steefel CI (2015) Benchmarking the simulation of Cr isotope fractionation. Comput Geosci 19:497–521

White AF, Brantley SL (2003) The effect of time on the weathering of silicate minerals: why do weathering rates differ in the laboratory and field? Chem Geol 202:479–506

Williams KH, Long PE, Davis JA, Wilkins MJ, N'Guessan AL, Steefel CI, Yang L, Newcomer D, Spane FA, Kerkhof LJ, McGuinness L, Dayvault R, Lovley DR (2011) Acetate availability and its influence on sustainable bioremediation of uranium-contaminated groundwater. Geomicrobiol J 28:519–539

Witherspoon PA, Wang JS, Iwai K, Gale JE (1980) Validity of cubic law for fluid flow in a deformable rock fracture. Water Resour Res 16:1016–1024

Yabusaki SB, Fang Y, Long PE, Resch CT, Peacock AD, Komlos J, Jaffe PR, Morrison SJ, Dayvault RD, White DC, Anderson RT (2007) Uranium removal from groundwater via in situ biostimulation: Field-scale modeling of transport and biological processes. J Contam Hydrol 93:216–235

Yabusaki SB, Wilkins MJ, Fang Y, Williams KH, Arora B, Bargar J, Beller HR, Bouskill NJ, Brodie EL, Christensen JN, Conrad ME (2017) Water table dynamics and biogeochemical cycling in a shallow, variably-saturated floodplain. Environ Sci Technol 51:3307–3317

Yasuhara H, Elsworth D (2008) Compaction of a rock fracture moderated by competing roles of stress corrosion and pressure solution. Pure Appl Geophys 165:1289–1306

Yasuhara H, Elsworth D, Polak A (2003) A mechanistic model for compaction of granular aggregates moderated by pressure solution. J Geophys Res: Solid Earth 108:2530

Yasuhara H, Elsworth D, Polak A (2004) Evolution of permeability in a natural fracture: Significant role of pressure solution. J Geophys Res: Solid Earth 109:B03204

Yasuhara H, Marone C, Elsworth D (2005) Fault zone restrengthening and frictional healing: The role of pressure solution. J Geophys Res: Solid Earth 110: B06310

Yeh GT, Tripathi VS (1989) A critical-evaluation of recent developments in hydrogeochemical transport models of reactive multichemical components. Water Resour Res 25:93–108

Yeh GT, Tsai CH (2014) User's manual for BIOGEOCHEM 1.5

Yeh GT, Tsai CH, Fang Y, Yabusaki S, Li MH (2014) BIOGEOCHEM 1.5: A numerical model to simulate BIOGEOCHEMical reactions under multiple phase system

Zachara JM, Smith SC, Liu C, McKinley JP, Serne RJ, Gassman PL (2002) Sorption of Cs^+ to micaceous subsurface sediments from the Hanford site, USA Geochim Cosmochim Acta 66:193–211

Zachara JM, Long PE, Bargar J, Davis JA, Fox P, Fredrickson JK, Freshley MD, Konopka AE, Liu C, McKinley JP, Rockhold ML (2013) Persistence of uranium groundwater plumes: contrasting mechanisms at two DOE sites in the groundwater–river interaction zone. J Contam Hydrol 147:45–72

Zheng L, Samper J, Montenegro L, Fernández AM (2010) A coupled THMC model of a heating and hydration laboratory experiment in unsaturated compacted FEBEX bentonite. J Hydrol 386:80–94

Zhi W, Li L, Dong W, Brown W, Kaye J, Steefel C, Williams KH (2019) Distinct source water chemistry shapes contrasting concentration–discharge patterns. Water Resour Res 55:4233–4251

Zhu C, Blum A, Veblen D (2004) Feldspar dissolution rates and clay precipitation in the Navajo aquifer at Black Mesa, Arizona, USA. *In:* Wanty R.B., Seal II R.R. (Eds.), Water Rock Interactions 11. Saragota Springs, New York, USA

Reviews in Mineralogy & Geochemistry
Vol. 85 pp. 27-48, 2019
Copyright © Mineralogical Society of America

2

Multiscale Approaches in Reactive Transport Modeling

Sergi Molins

Lawrence Berkeley National Laboratory
Energy Geosciences Division
Berkeley, California 94720
USA

smolins@lbl.gov

Peter Knabner

Friedrich-Alexander-Universität Erlangen-Nürnberg
Applied Mathematics
91058 Erlangen
Germany

knabner@math.fau.de

INTRODUCTION

The field of reactive transport lies at the intersection of several disciplines in the Earth and Environmental sciences, including hydrology, geochemistry, biology and geology. The processes in natural and engineered media that are the focus of study of these disciplines take place over a wide range of spatial and temporal scales. Specifically, geological media are characterized by their physical and mineralogical heterogeneity at spatial scales from nanometers to hundreds of meters and beyond. Flow and advection of solutes take place at the scale of individual pores but are commonly represented at the Darcy scale where the porous medium is treated as a continuum. A large contrast is often observed between fluid residence times in regions of enhanced permeability such as fractures or macropores and less permeable media where diffusion may be the dominating solute transport process. Understanding of reactive processes, including those mediated by microorganisms, is often developed at the molecular scale in the laboratory but their impact in the environment is observed at larger spatial scales.

In addition to considering the scales of the individual processes, reactive transport must also consider how these different scales interact with one another to give rise to the overall coupled behavior. In fact, in many instances considering the processes at the observation (or native) scales has limited applicability in subsurface environments. For example, reaction rates derived from laboratory studies show large discrepancies from those observed in natural environments, where transport processes and accessibility to reactive areas control effective rates. Reactive transport models, thus, even in a simple form, must make assumptions regarding the scales associated with each process and how they interact with each other. Implicit in any model is also the assumption that the models for each process are applicable at the same spatial scale as the other processes represented. For example, local geochemical equilibrium may only be assumed where reaction rates are faster than transport rates such that the solution reaches equilibrium over a characteristic spatial scale.

Reactive transport modeling, as a tool to integrate knowledge and develop mechanistic understanding, seeks to incorporate improved process model representations that reflect our advances in fundamental understanding (Druhan and Tournassat 2019 and references therein).

1529-6466/19/0085-0002$05.00 (print)
1943-2666/19/0085-0002$05.00 (online)

http://dx.doi.org/10.2138/rmg.2019.85.2

The multiscale nature of reactive transport is one of the most prominent aspects in models and, hence, modeling approaches that address this multiscale nature are an increasingly important component of the toolset needed by researchers (Scheibe et al. 2015a). In particular, multiscale approaches make it possible to incorporate process representations at the appropriate native scale in models intended to simulate the coupled problem at a different spatial or temporal scale. A variety of approaches have been brought to bear that range from conceptual to mathematical to numerical. Specific goals of multiscale models are also diverse and include using the appropriate coupling between processes, capturing the processes at the relevant scale in different regions, capturing the physical and mineralogical heterogeneity at multiple scales or incorporating fine-scale information in larger-scale applications. Ultimately, there is the need to identify what processes and at what scale are controlling overall system behavior, and hence the appropriate spatial and temporal scales to represent each process.

Multiscale modeling is a very broad topic with applications across many disciplines (Tomin and Lunati 2013, 2016; Scheibe et al. 2015a; Amanbek et al. 2019). In this chapter, we specifically review multiscale approaches for reactive transport modeling from the conceptual and mathematical perspectives. Many of the approaches have also been used in the individual disciplines reactive transport draws from. They are here discussed in a general manner here but also specifically in relation to reactive transport applications. We include approaches where the multiscale nature is reflected in a continuum-mechanics-based model and we discuss numerical aspects of these approaches where needed. Approaches that incorporate upscaling procedures in the numerical solution process such as multiscale finite element or finite volume methods, or based on numerical upscaling however are not included, e.g., Efendiev and Hou (2009).

We start by establishing the equations that describe the processes of interest at a single scale and discussing the multiscale aspects associated with process coupling at a single scale. We relate these equations to the two scales commonly identified in porous media—the pore scale and the Darcy scale—but are generally applicable at a range of spatial scales, from fluid in pores to streams and rivers. Derivation of effective models by upscaling pore-scale equations to the Darcy continuum scale is used specifically to motivate the need for multiscale approaches. First, we describe approaches that use different scale representations in different regions in the domain. We continue with approaches based on the existence of two or more porous continua in the same region of the domain. We give examples of the use of the multiscale approaches described in selected literature applications before making some concluding remarks.

SINGLE-SCALE DESCRIPTION OF REACTIVE TRANSPORT

Reactive transport models simulate flow, solute transport and geochemical reactions. In this chapter, for simplicity, we will consider only single-phase (aqueous) flow and transport. These processes are typically described by two sets of equations, one for the conservation of mass and momentum of the fluid and the other for the conservation of mass of the reactive components. The form of these equations depends on continuum of reference for which they are written: a fluid continuum, a solid continuum or a continuum that includes both fluid and solid phases. For example, flow in streams and rivers may be represented by considering the fluid phase as the continuum of reference. In porous and fractured media in the subsurface, when the fluid and the solid are both treated as separate continua, we refer to the scale of observation as the pore scale, while when the porous medium is the continuum, we refer to it as the Darcy scale. For convenience, in the derivation that follows we will focus on porous and fractured media, but these single-scale equations can be read more generally as applicable at a range of spatial scales. Further, we will not consider here the scales where the medium is not characterized as a continuum. For example, we do not discuss characterizations at the molecular or atomistic scale, or the organism level in the case of microbial processes.

Separate fluid and solid continua: Pore-scale equations

When the individual pores are represented explicitly, and the solid–fluid interfaces are the boundaries of the domain considered (Fig. 1), flow may be described with the Stokes equations

$$\nabla \cdot \boldsymbol{u} = 0 \tag{1}$$

$$\nabla^2 \boldsymbol{u} = \frac{1}{\mu} \nabla p \tag{2}$$

where $\boldsymbol{u}$ is the fluid velocity (with $|\boldsymbol{u}| = 0$ on the solid–fluid boundary), and μ, and p are the fluid viscosity, and pressure, respectively. The equations that describe the mass balance of chemical species subject to advective–diffusive transport and heterogeneous reactions at the fluid–solid surface may be written as

$$\frac{\partial c}{\partial t} = -\nabla \cdot (\boldsymbol{u}c) + \nabla \cdot (D\nabla c) \tag{3}$$

$$-D\nabla c \cdot \boldsymbol{n} = r \tag{4}$$

where c is the solute concentration, D is the diffusion coefficient of the solute in the solution, and r is the surface reaction rate. Equation (4) expresses the mass balance at the fluid–solid interface for the aqueous species involved in the heterogeneous reaction, where $\boldsymbol{n}$ denotes the unit normal pointing from solid to liquid.

Single porous continuum: Darcy-scale equations

When the porous medium is treated as a continuum, i.e., when within an elementary representative volume (or REV) properties that described the medium such as porosity (θ) or permeability ($\boldsymbol{k}$) may be assumed constant (Fig. 1), flow in porous media can be described by:

$$\frac{\partial \theta \rho}{\partial t} + \nabla q = 0 \tag{5}$$

$$q = -\frac{\boldsymbol{k}}{\mu}(\nabla p + \rho g z) \tag{6}$$

where q is the Darcy velocity vector, which is calculated with Darcy's law (Eqn. 6), θ is the porosity (water content in fully-saturated conditions), ρ is the fluid density, g is the gravitational constant, and z is the vertical coordinate. The equations that describe the mass balance of chemical species subject to advective-dispersive transport and heterogeneous reactions in porous media may be written as

$$\frac{\partial \theta C}{\partial t} = -\nabla \cdot (qC) + \nabla \cdot (\boldsymbol{D}^* \nabla C) + R \tag{7}$$

where C is the concentration at the Darcy-scale, $\boldsymbol{D}^*$ is the effective diffusion/dispersion tensor, and R is the bulk reaction rate. In the view presented in this section, the properties that characterize the porous medium, i.e., θ, $\boldsymbol{k}$, or $\boldsymbol{D}^*$, or the bulk rates are assumed known (for example, empirically) and applicable at this scale. In general, however, they encapsulate information of the processes that take place at the scale of individual pores. In the section devoted to upscaling, we discuss how one may formally derive these parameters from the pore-scale description.

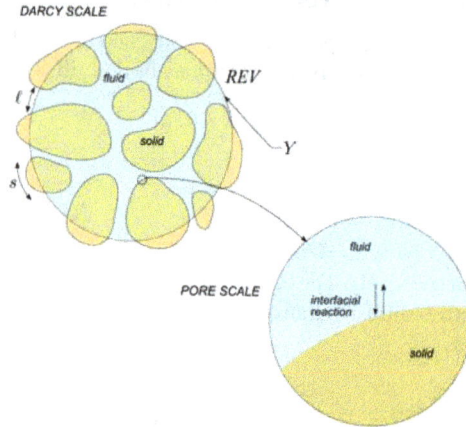

Figure 1. Conceptual representation of scales associated with interfacial (heterogeneous) reactions in porous media. The existence of an REV makes it possible to describe the medium as a continuum (left), while at the scale of individual pores interfacial reactions are explicitly described at the fluid–solid interface. Adapted from Wood et al. (2007).

Multiscale aspects of process coupling

The flow equations and the reactive transport equations are in general coupled via the composition of the fluid, as well as hydraulic and geometric properties of the porous medium. For example, changes in the fluid composition caused by geochemical reactions affect the fluid density, or dissolution–precipitation reactions change the pore space geometry which in turn affects the fluid velocity. The evolution of these properties is in many applications relatively slow compared to flow and transport and the coupling is assumed weak. In some other applications, the feedback processes are significant and are addressed specifically in the chapter devoted to porous media evolution in this volume (Seigneur et al. 2019, this volume).

Here we will briefly focus on the coupling of time scales associated with transport and reactions and how they affect the process representation at a specific spatial scale. For convenience, we re-write Equation (7) generically as

$$\frac{\partial \theta C}{\partial t} = \mathcal{L}(C) + \mathcal{R}(C) \qquad (8)$$

where $\mathcal{L}()$ is the transport operator and $\mathcal{R}()$ the reaction operator. Transport is important as it provides the driving force for reaction but also because it provides a characteristic time scale to which the time scale of the reaction is compared. In an open system, as implied by Equation (8), if the characteristic time of transport (for a given characteristic length scale), $\tau_{\mathcal{L}}$, is larger than that of the reaction, $\tau_{\mathcal{R}}$, the solution reaches equilibrium with itself or with a mineral phase. In Darcy-scale models, if this characteristic length scale is that over which the REV is defined, this makes it possible to assume local equilibrium. In the local equilibrium assumption (LEA) (Lichtner 1996), the rate of reaction is thus determined by the rate of transport of matter across the boundaries of the domain (Fig. 2a).

In pore-scale models (Eqns. 3–4 may also be written in a form similar to Eqn. 8), the concept of local equilibrium is different (Lichtner 1996). At the Darcy scale, geochemical equilibrium is attained at the REV scale which includes many pores. At the pore scale, the detailed pore space geometry plays a role. Although equilibrium may be attained at mineral surfaces, geochemical gradients may still be present in individual pores (Fig. 2b).

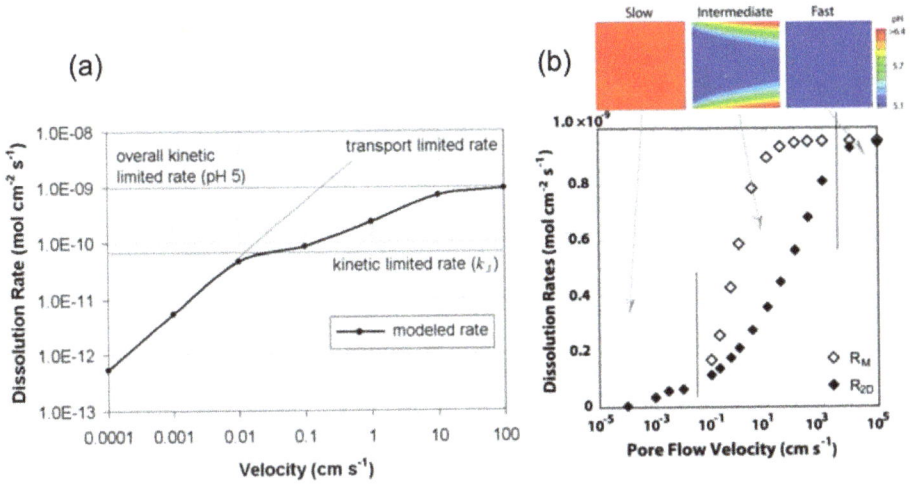

Figure 2. (a) Steady state calcite dissolution rates as a function of the flow velocity (using a double logarithmic scale) from a 1-D continuum model of transport and calcite dissolution (implemented with a kinetic rate expression). For velocities between 0.0001 and approximately 0.01 cm/s the dissolution rate is controlled by transport, hence it is proportional to velocity. For faster flow velocities, the rate is controlled by both transport and the kinetic reaction. [Reprinted from (a) From Molins S, Trebotich D, Steefel CI, Shen C (2012) An investigation of the effect of pore scale flow on average geochemical reaction rates using direct numerical simulation. Water Resources Research, 48(3):W03527, Figure 6, with permission] (b) Steady-state calcite dissolution rates in a single cylindrical pore calculated from a 2D pore-scale model (R2D) and from a Well-Mixed Reactor model (RM) as a function of the pore flow velocity for a pore of 100 μm in length and diameter. At low pore velocities, conditions in the pore are in equilibrium and R2D and RM produce the same results. Only under intermediate flow conditions where concentration gradients develop do the reaction rates depend on the spatial scale and the rate discrepancy between the two models reaches a maximum. [Reprinted from Li L, Steefel CI, Yang L (2008) Scale dependence of mineral dissolution rates within single pores and fractures. Geochimica et Cosmochimica Acta 72(2):360–377, Figure 5, with permission of Elsevier, Copyright 2008.]

UPSCALING AND EFFECTIVE MODELS

The macro- (Darcy) scale model in Equations (5–6) and (7) may be formally derived by scaling-up the micro-(pore) scale Equations (1–2) and (3–4). Upscaling has been the subject of intensive research for at least 50 years, with volume averaging being a commonly used approach for this purpose. In classical averaging theory, concentrations and fluxes are averaged over an REV composed of fluid and solid phases. The averaging of the pore-scale equations makes it possible to write conservation equations for these averaged quantities (see, e.g., Gray and Miller 2014). In these equations, however, new terms appear that still depend on pore-scale quantities and hence for which closure relations must be postulated. The mathematical theory of (periodic) homogenization is used to solve the closure problem (Hornung 1997).

In the most simple form, periodic homogenization relies on the spatial periodicity of the domain Ω, which is conceived to be composed of shifted and ε-scaled copies Y_ε of an REV Y (Fig. 3). The REV Y is made up of solid (Y_s) and liquid (Y_l), separated by an interface Γ. Accordingly, the porous media domain Ω_ε is composed of solid phase, with $Y_{s,\varepsilon}$ being the union of the shifted and ε-scaled Y_s and the liquid phase. The subdomain $Y_{l,\varepsilon}$ is defined similarly and assumed to be connected such that flow can take place. Let's assume a coefficient (φ) that oscillates with a periodicity of length ε such as the diffusion coefficient D, with $D>0$ in the fluid and $D=0$ in the solid. The aim of homogenization is to consider $\varepsilon \to 0$ (i.e., to zoom out) in order to see what description holds for the emerging homogeneous medium. Formally, this can be done assuming a two-scale asymptotic expansion for all unknown quantities φ in the form

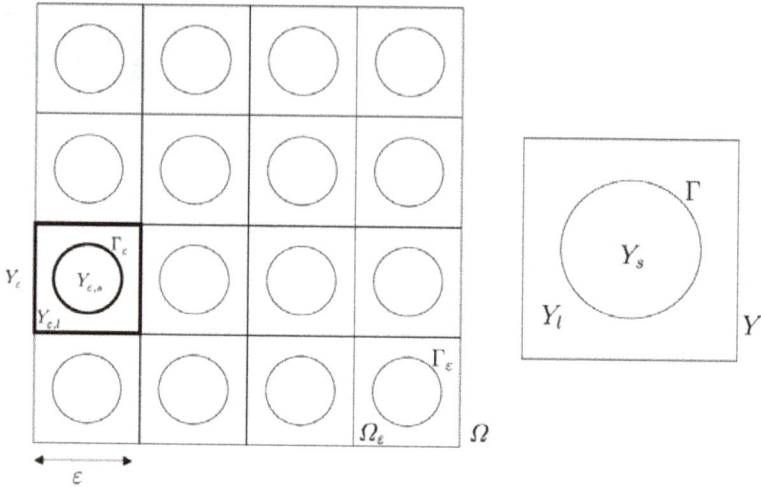

Figure 3. Two-dimensional domain Ω constructed by periodic repetition of a porous medium Ω_ε (left), represented by a unit cell Y composed of fluid and solid separated by a circular interface (right). Examples of other geometries used for unit cells are shown in Figure 4. [Modified from Ray N, van Noorden T, Frank F, Knabner P (2012) Multiscale modeling of colloid and fluid dynamics in porous media including an evolving microstructure. Transport in Porous Media 95(3):669–696, Figure 2, with permission of Springer Nature. Copyright 2012.]

$$\varphi_\varepsilon(x,t) = \varphi_0(x,t,y) + \varphi_1(x,t,y)\varepsilon + \varphi_2(x,t,y)\varepsilon^2 + \ldots \tag{9}$$

where $y := x/\varepsilon$ is a *fast* spatial variable in the sense that y covers Y, if x covers Y_ε. Depending on the problem, the interplay between the processes in ε and separation of scales, it is possible to derive equations in which only the Y-averaged 0-th order terms appear. This is done by inserting the expansion in the governing (pore-scale) equation such that each ε-power gives rise to a new equation for the coefficient that is being investigated.

For example, one can derive the upscaled, Darcy-scale form of the diffusion equation (i.e., Eqn. 7 with $q = 0$ and no reaction term) from the pore-scale diffusion counterpart (i.e., Eqn. 3) with $|u| = 0$). Solving the closure equations on Y numerically makes it possible then to obtain the effective diffusion tensor D^*, which encodes the information about the pore geometry, see, e.g., Ray et al. (2018). The closure equations are typically solved using idealized or randomly generated porous geometries (Fig. 4). From these calculations, it is possible to describe the dependence on pore geometry solely by a macroscopic parameter such as porosity (Fig. 5).

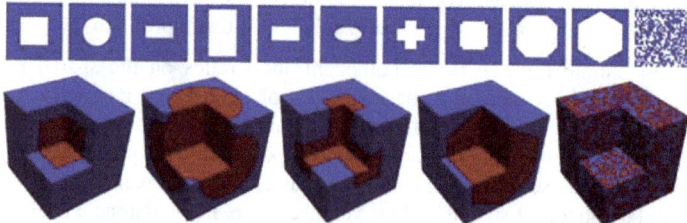

Figure 4. Representative elementary volumes in 2D: square, circle, rectangles, ellipse, crosses (type 1 and 2), octagon, hexagon, and random geometry (top), and representative elementary volumes in 3D: cube, sphere, 3D cross, hexagonal prism, and random (bottom) [Reprinted from Ray N, Rupp A, Schulz R, Knabner P (2018) Old and new approaches predicting the diffusion in porous media. Transport in Porous Media 124: 803–824, Figures 1 and 2, with permission of Springer Nature. Copyright 2018.]

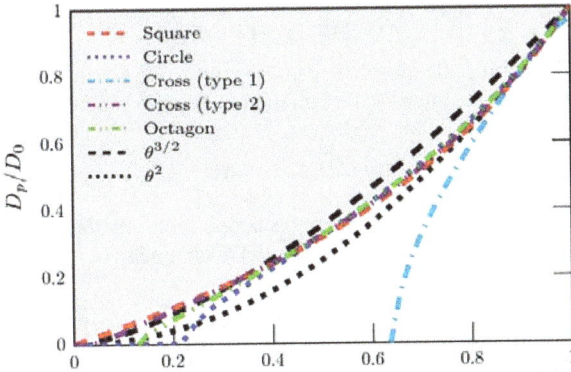

Figure 5. Scalar ratio between scalar representative effective diffusion (D_p) over intrinsic diffusion (D_0) as a function of porosity for isotropic geometries in 2D: square, circle, crosses (type 1 and 2, see Fig. 4), and octagon; Hashin–Shtrikman bound $\theta(2-\theta)$ (exclusion of gray area); and functional relations $\theta^{3/2}$ (Marshall 1959) and θ^2 (Buckingham 1904) [Reprinted from Ray N, Rupp A, Schulz R, Knabner P (2018) Old and new approaches predicting the diffusion in porous media. Transport in Porous Media 124: 803–824, Figures 1 and 2, with permission of Springer Nature. Copyright 2018]

The above procedure can be accomplished when there is scale separation (i.e., the micro- and macro-scales can be described independently from each other) but in general it also depends on the proper scaling of the parameters in the ε-problem. For example, to deduce Darcy's law from the Stokes equation at the pore scale (i.e., with a computable permeability), a scaling of the viscosity μ to $\varepsilon\mu$ is required. Further, when several different processes are in play (e.g., advection and diffusion), characteristic numbers such as the Péclet number (Pe) may be needed in the efficient coefficients, e.g., in the derivation of mechanical dispersion (Mikelić et al. 2006). For reactive transport problems, efforts in upscaling have focused on heterogeneous reactions. Here we present a short derivation of upscaled relationships for (first-order) sorption reactions and dissolution-precipitation reactions first with non-evolving geometries and then with evolving geometries.

Upscaling interfacial reactions

We consider a surface reaction such that Equations (3–4) may be written on Γ_ε as

$$\frac{\partial c_m}{\partial t} = -\nabla \cdot \left(\boldsymbol{u} c_m - \boldsymbol{D} \nabla c_m \right) \text{ in } Y_{l,\varepsilon} \tag{10}$$

$$\boldsymbol{D} \nabla c_m \cdot \boldsymbol{n} = \varepsilon \cdot r \text{ on } \Gamma_{l,\varepsilon} \tag{11}$$

where c_m is the concentration in the fluid, and the rate is expressed as an exchange between the fluid and the surface concentration, $r = \alpha(c_m - c_{im})$, with the immobile concentration (c_{im}) being a surface concentration such that

$$\frac{\partial c_{im}}{\partial t} = \alpha \left(c_m - c_{im} \right) \tag{12}$$

where α is the rate of mass transfer between the bulk fluid and the surface, with $\alpha > 0$. A no-flow boundary condition ($\boldsymbol{u} = 0$) is assumed on $\Gamma_{\ell,\varepsilon}$. The upscaled model then takes the following macroscale form

$$\frac{\partial \left(\theta C_m \right)}{\partial t} - \mathcal{L} \left(C_m \right) = -\sigma R \left(C_m, C_{im} \right) \tag{13}$$

where σ is the specific surface ($\sigma = \Gamma / Y$, which along with θ and $\boldsymbol{D}^*$ are computed from the pore geometry, and the macro-scale rate is described by

$$R\left(C_m, C_{im}\right) = \alpha\left(kC_m - C_{im}\right) \tag{14}$$

where micro- and macro-scale problems are separate. However, if surface diffusion is considered at Γ_ε such that the micro-scale Equation (12) is written as

$$\frac{\partial c_{im}}{\partial t} - \varepsilon^2 D_s \Delta_y c_{im} = \alpha\left(c_m - c_{im}\right) \tag{15}$$

where D_s, is the surface diffusion coefficient, it is not possible any longer to obtain the average of c_{im} and a coupled macro-scale/micro-scale model is obtained

$$\frac{\partial c_{im}\left(y, t; x\right)}{\partial t} - D_s \Delta_y c_{im} = \alpha\left(kC_m - c_{im}\right) \text{ for } y \in \Gamma, x \in \Omega \tag{16}$$

where the micro problem depends at every point x on the macroscopic concentration $C_m(x,t)$. (Note that the emerging equations for C_m and $C_{im} = C_{im}(y,t;x)$ read as Eqn. 35, with Γ in lieu of $\Omega_{\bar{x}}$).

Upscaling interfacial reactions with evolving geometries

Precipitation–dissolution reactions are of interest as they change the pore geometry and can have a positive feedback on flow and transport processes, leading in some case to clogging or wormholing. In the classical approach this is reflected in the change of porosity as described by Ray et al. (2015)

$$\frac{\partial \theta}{\partial t} = \frac{1}{\rho_s} R \tag{17}$$

where ρ_s is a surface density used as conversion factor between mass and volume. To close the model, the specific surface σ must be related to θ, which requires assumptions on the evolving micro geometry. In the case of known geometries (e.g., spheres, cubes, etc), this leads to the commonly used relation (Lichtner 1996)

$$\sigma = \beta C_{min}^{3/2} \tag{18}$$

where C_{min} is the mineral concentration and β is a factor to convert from units of mass to volume

The evolution of the interface as a result of the reaction can be described with a sharp interface approach (e.g., a level set) or approximated by a phase field model (Bringedal et al. 2019). The normal component of the velocity of the interface is denoted by $v_{n,\varepsilon}$

$$v_{n,\varepsilon} = \varepsilon \frac{1}{\rho_s} R\left(c_m\right) \tag{19}$$

This velocity is a function of the reaction rate R and the density of the precipitate ρ_s. Considering the reaction to be a single-component reaction for simplicity, Equation (11) can be replaced with

$$D\nabla c_m \cdot \boldsymbol{n} = v_{n,\varepsilon}\left(c_m - \rho_s\right) \tag{20}$$

where the precipitate can form the outer surface of the solid grain. By upscaling we can again recover Equation (13), but here the closure comes from a detailed description of the interface via the evolution of a level-set function L_0 in the REV Y, described by (Ray et al. 2015)

$$\frac{\partial}{\partial t} L_0 - \frac{1}{\rho} R\left(C_m\left(x,t\right)\right) \left|\nabla_y L_0\right| = 0 \text{ in } Y \tag{21}$$

yielding θ, σ, **D** needed in Eqn. 13 but dependent now on the macroscopic concentration $C_m(x, t)$. Thus, we obtain again a coupled micro-macro model. The solution of Equation (21) is not straight-forward and can be derived by a two-scale asymptotic procedure (Ray et al. 2015).

Extension of this approach to multicomponent problems that include homogeneous and heterogeneous reactions, in equilibrium or kinetic, is still in the early stages of development. While homogeneous or sorption reactions are tractable, development of passivation layers in systems with multiple mineral reactions, e.g., Daval et al. (2009), is considerably more difficult to handle in this framework.

COMBINING SCALE REPRESENTATIONS

The ability of pore-scale models to explicitly resolve individual pores make them suitable to simulate flow and reactive transport without using bulk parameters to characterize the medium. The computational cost of pore-scale simulations, however, is very high if one wants to cover volumes of porous media large enough for relevant applications. As noted in the previous section, the macro (Darcy)-scale problem can only be formulated separately from the micro (pore)-scale problem under a number of simplifying assumptions, see also Battiato and Tartakovsky (2011). Rather than formulating and solving closure equations for periodic unit cells of idealized geometries in these cases, one may want to solve the pore-scale problem directly. In applications where the pore-scale characterization is needed only in certain regions rather the entire domain, an attractive approach is then to combine a pore-scale description in these regions and revert to a Darcy-scale description elsewhere. Broadly, two approaches have been used for this purpose: hybrid models and the Darcy–Brinkman–Stokes approach.

Hybrid multiscale models

Hybrid multiscale models combine different scale representations in a single simulation (Battiato et al. 2011; Roubinet and Tartakovsky 2013; Yousefzadeh and Battiato 2017). In this approach, the domain is divided in two or more regions where different scale representations are used (Fig. 6a). Because often this implies that different process models, and thus potentially also the numerical solution method, are used in each of the regions, hybrid models are sometimes known as multi-algorithm or algorithm refinement models. An advantage of hybrid models from the multi-algorithm perspective is that it is possible to consider different spatial and temporal discretization in different portions of the domain, or the use of different dimensionality for each of the sub-domains (e.g., Fig. 6b).

A hybrid model of reactive transport on a domain Ω composed of a pore-scale domain Ω_p and a Darcy-scale domain Ω_D, such that $\Omega = \Omega_p \cup \Omega_D$, entails the solution of Equations (1–4) in Ω_p and Equations (5–7) in Ω_D. The pore-scale and Darcy-scale simulations are coupled by enforcing the continuity of mass (concentration) and mass flux (its normal component) along the interface Γ between Ω_p and Ω_D:

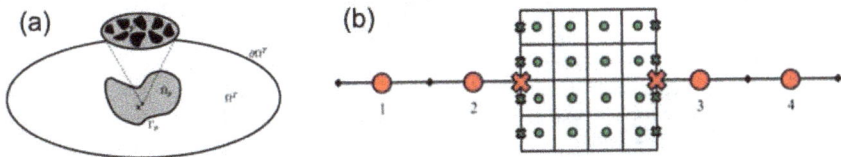

Figure 6. (a) Schematic representation of the pore-(Ω_p) and Darcy-scale ($\Omega_p = \Omega^\Gamma - \Omega_p$) domains [Reprinted from Battiato I, Tartakovsky DM, Tartakovsky AM, Scheibe TD (2011) Hybrid models of reactive transport in porous and fractured media. Advances in Water Resources, 34(9), 1140–1150, Figure 1, with permission of Elsevier. Copyright 2013.] (b) Hybrid finite-volume discretization. The (small green and big red) circles indicate nodes at which the (pore-scale and Darcy-scale) unknowns are computed with a regular finite-volume method. The (big red and small green) crosses denote extra nodes used to enforce the continuity conditions at the hybrid's interfaces. [Reprinted from Roubinet D, Tartakovsky DM (2013) Hybrid modeling of heterogeneous geochemical reactions in fractured porous media. Water Resources Research, 49(12), 7945–7956, Figure 1, with permission.]

$$C(x,t) = \frac{1}{|\Gamma|} \int_{\Gamma} c(X,t) ds \qquad (22)$$

$$Q^D(x,t) = \frac{1}{|\Gamma|} \int_{\Gamma} Q^p(X,t) ds \qquad (23)$$

where $Q^p(x, t)$ and $Q^D(x, t)$ are the normal components of the pore-scale and Darcy-scale mass fluxes , $Q^p(x, t) = uc - D\nabla c$ and $Q^D = qC - D^*\nabla C$, respectively.

From a numerical perspective, hybrid models need to consider the interface between the sub-domains explicitly and coupling the sub-problems at the interface. Typically, concentrations and fluxes are used as coupling unknowns for Equations (22) and (23). This adds some complexity in the implementation of hybrid models, especially when the dimensionality or discretization on both sides of the interfaces are different. An example of a numerical method designed to handle these exchanges in a general and flexible way is the mortar method (Balhoff et al. 2008; Mehmani et al. 2012). In this method, coupling between sub-domains is accomplished by using finite-element (FEM) spaces to determine interface conditions (Fig. 7). Using the flow problem as example (with pressure being the unknown of the problem), the pressure field in the mortar space (noted by p) is a linear combination of finite element (FE) basis functions (ϕ_i):

$$p = \Sigma \zeta_i \phi_i \qquad (24)$$

The basis functions may be constant, linear, quadratic, or higher order functions. The solution is obtained by determining the coefficients (ζ_i) that result in matching of fluxes at interface. The mortar solution then describes the pressure field only at the interface which is used as a boundary condition and projected onto the individual sub-domain. The sub-models are then solved using the appropriate algorithm.

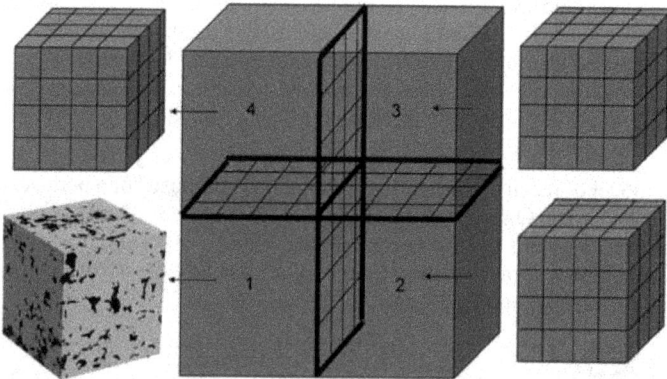

Figure 7. Schematic showing one pore-scale model and three Darcy-scale models arranged using 4×4 quadratic mortars. [Reprinted from Balhoff MT, Thomas SG, Wheeler MF (2008) Mortar coupling and upscaling of pore-scale models. Computational Geosciences 12(1):15–27, Figure 7, with permission of Springer Nature. Copyright 2008.]

Darcy–Brinkman–Stokes approach

An approach for combining pore- and Darcy-scale representations that has received increasing attention is the one conceptualized by the Darcy–Brinkman–Stokes equation (Golfier et al. 2002; Popov et al. 2009; Gulbransen et al. 2010; Yang et al. 2014; Soulaine and Tchelepi 2016; Soulaine et al. 2017). Darcy–Brinkman–Stokes describes flow in open pore space and in a porous continuum with a single equation (written here to recover the transient incompressible Navier–Stokes equations in the fluid domains in lieu of Eqn. 2):

$$\frac{1}{\varepsilon}\left(\frac{\partial \rho \boldsymbol{u}}{\partial t} + \nabla \cdot \left(\frac{\rho}{\varepsilon} \boldsymbol{u}\,\boldsymbol{u}\right)\right) = -\nabla p + \frac{\mu}{\varepsilon}\nabla^2 \boldsymbol{u} - \mu k^{-1}\boldsymbol{u} \tag{25}$$

where ε is the porosity of the medium, with $\varepsilon=1$ in the pore spaces, $0<\varepsilon<1$ in a porous continuum and $\varepsilon=0$ in the solid phase. The permeability k of the medium requires a constitutive relation linking it to ε, for example the Kozeny–Carman equation. As a result, the terms associated with porous-media flow become negligible in the pore scale, while the terms associated with pore-scale flow become negligible in the porous continuum or the solid phase (Fig. 8). The transport of the aqueous species is described by a locally averaged equation

$$\frac{\partial \varepsilon C}{\partial t} + \nabla \cdot \left(\boldsymbol{u} C\right) - \nabla \cdot \left(\varepsilon \boldsymbol{D}^* \nabla C\right) - R \tag{26}$$

where dissolution is here described as a source-sink term (R) as in Equation (7).

The use of a single equation simplifies the numerical implementation of this method. Further, it does not require explicit representation of an interface between pore-scale and Darcy-scale, which is especially convenient in problems with evolving media. In fact, the distinction between pore-scale and Darcy-scale is only conceptual. In the pore-scale limit, i.e., when ε and k are very small, Equations (25) and (26) recover the pore-scale description in Equations (1–3) and it may be used as a pore-scale method (Golfier et al. 2002; Soulaine et al. 2017).

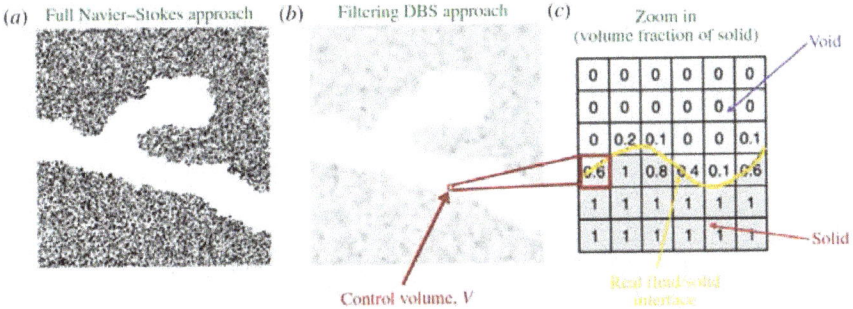

Figure 8. Conceptual representation of the void and solid in the pore-scale approach in contrast to the Darcy–Brinkman–Stokes (DBS) approach. **(a)** In the former, solid grains are explicitly described, the flow is governed by Navier–Stokes everywhere in the void. **(b)** In the latter, a cutoff length is introduced by means of the control volume V and the void is represented by the volume fraction ε, **(c)** discretized representation of the medium near the fluid–solid interface where the volume fraction of the solid ranges from 0 (pore space) to 1 (solid). [Reprinted from Soulaine C, Roman S, Kovscek A, Tchelepi HA (2017) Mineral dissolution and wormholing from a pore-scale perspective. Journal of Fluid Mechanics 827:457–483, Figure 2, with permission.]

MULTI-RATE AND MULTI-CONTINUA MODELS

Natural porous media are characterized by the existence of porosity at multiple spatial scales. The conceptual model presented in Figure 1 only considers inter-granular porosity, that is, porosity available between solid grains where in general fluid velocities are appreciable. We update here this model to include intra-granular porosity, that is, porosity that exists at small spatial scales in regions, which we will for convenience refer to as aggregates, where fluid flow is in general very slow and transport processes are dominated by diffusion (Fig. 9). In this media, the mass of each constituent is distributed between inter-porosity and intra-porosity, or mobile and immobile regions, between which mass is exchanged.

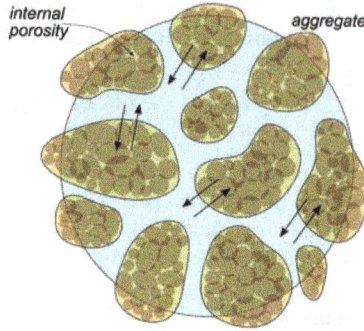

Figure 9. Conceptual representation of media with porosity at multiple spatial scales, showing mass exchange fluxes between internal porosity in aggregates and inter-granular porosity.

Multi-rate models

The concept of mass exchange between mobile and immobile regions is analogous to considering heterogeneous reactions between a mobile and immobile species, such as in sorption reactions, see Equations (10–12). We will use this idea here to present single- and multi-rate models, see, e.g., Haggerty and Gorelick (1995), Hollenbeck et al. (1999). We start by writing the mass balance of a species as (omitting the species index for simplicity)

$$\frac{\partial \theta_m C_m}{\partial t} + \frac{\partial \theta_{im} C_{im}}{\partial t} = \mathcal{L}(C_m) \qquad (27)$$

where C_m and C_{im} are the mobile and immobile concentrations of the species, where we assume the immobile concentration is not subject to transport $\mathcal{L}()$. The mobile (θ_m) and immobile (θ_{im}) porosities are the corresponding conversion factors to normalize the concentrations to the volume of an REV (if C_{im} is a surface concentration, then $\theta_{im} = \rho_b$, with ρ_b being the bulk density). Equation (27) needs a closure relation that links the two concentrations, which may be a quasi-stationary approximation

$$C_{im} = \kappa C_m \qquad (29)$$

describing a linear equilibrium sorption reaction ($\kappa \equiv K_d$), or a first-order kinetic exchange such as

$$\frac{\partial C_{im}}{\partial t} = \alpha \left(\kappa C_m - C_{im} \right) \qquad (29)$$

Integration of Equation (29) makes the memory effect of the linear mass transfer process evident

$$C_{im}(x,t) = e^{-\alpha t} C_{im}(0,t) + \int_0^t e^{-\alpha(t-s)} \alpha \kappa C_m(x,s) ds \qquad (30)$$

For more flexibility in the description of this memory (or tailing) effect, C_{im} can be further subdivided in different fractions (β_k) with different kinetic behavior in a multi-rate formulation. Equations (27) and (29) can be generalized to

$$\frac{\partial \theta_m C_m}{\partial t} + \sum_{k=1}^{N} \beta_k \frac{\partial \theta_{im} C_{im,k}}{\partial t} = \mathcal{L}(C_m) \qquad (31)$$

$$\frac{\partial C_{im,k}}{\partial t} = \alpha_k \left(\kappa_k C_m - C_{im} \right) \qquad k = 1,..,N \qquad (32)$$

with $\beta_k \geq 0, \sum_{k=1}^{N} \beta_k = 1$.

Applying (30) to each of the fraction and expanding the sum in (31), one can see that the exponential kernel in the memory term becomes multimodal. The model can be also used to describe physical non-equilibrium, where now the mass transfer is between the mobile region and each of the fractions of the immobile region (or classes of micro-porosity, each giving a different memory effect). All these fractions (or sorption sites in the sorption interpretation), exchange mass with the mobile region directly based on the same C_m.

A generalization of Equations (31) and (32) is given by extending them from a finite number of fractions (i.e., sorption sites/micro-porosity classes) to an infinite number (or a continuum) of them by re-writing them as

$$\frac{\partial \theta_m C_m}{\partial t} + \int_0^\infty \beta(k) \frac{\partial \theta_{im} C_{im,k}}{\partial t} dk = \mathcal{L}(C_m) \tag{33}$$

$$\frac{\partial C_{im}}{\partial t} = \alpha(k)(\kappa_k C_m - C_{im}) \qquad k \geq 0 \tag{34}$$

with $\int_0^\infty \beta(k) dk = 1$

Extension of multi-rate mass transfer models to multicomponent reactive transport is possible; see, e.g., Donado et al. (2009) where the equilibrium is assumed in each the mobile and immobile domains such that Equations (27) and (30) may be written in terms of total concentrations and speciation calculations performed once the equations are solved for these total concentrations.

Multi-continua models

Multi-rate models are derived for two continua: a mobile and immobile region. This can be generalized for any number of continua. One can transition from multi-rate models to multi-continua models (MC). We can start by assuming that in one aggregate class the diffusive transport within the aggregate is taken into account. For the ease of notation, we consider only one rate $(N=1)$. The averaged concentration $C_{im} = C_{im}(x,t)$ is now replaced by $c_{im} = C_{im}(y,t;\bar{x})$, $y \in \Omega_{\bar{x}}$ where $\Omega_{\bar{x}}$ is a representative aggregate and the total mass conservation from (31) reads

$$\frac{\partial \theta_m C_m}{\partial t} + \frac{\partial \left(\int_{\Omega_{\bar{x}}} \theta_{im} c_{im} dy \right)}{\partial t} = \mathcal{L}(C_m) \tag{35}$$

Equation (31) is substituted by a partial differential equation on $\Omega_{\bar{x}}$

$$\frac{\partial \theta_{im} c_{im}}{\partial t} - d_{im} \Delta_y c_{im} = 0 \tag{36}$$

$$d_{im} \nabla_y c_{im} \cdot n_y = \alpha(C_m(\bar{x},t) - c_{im}(y,t;\bar{x})) \quad \text{for } y \in \partial \Omega_{\bar{x}} \tag{37}$$

where $d_{im} > 0$ denotes the molecular diffusion coefficient in the aggregate and n_y is the unit normal at $\partial \Omega_{\bar{x}}$ pointing out of $\Omega_{\bar{x}}$.

If we have several points $\bar{x} = x_k$, where the detailed dynamics must be considered, we go from a two-continuum/region model to a multiple continuum/regions models with a considerable increase in numerical complexity. The extreme is to do this for every (discretization) point $\bar{x} = x \in \Omega$. In this case, we can again arrive at a micro-macro model as in Equations (13) and (16). To avoid enormous numerical complexity in simulating such a model, on can try solving Equations (36) and (37) analytically with appropriate initial conditions. This is possible in special cases, e.g., when $\Omega_{\bar{x}}$ is a sphere. This solution representation, which depends on $c_m(x,t)$, may be viewed as analogous to the multi-model extension of Equation (30), where now the integral in (30) has to be substituted by an integral of the type

$$\int_{\Omega_x} \theta_{im} c_{im}(y,t;x) dy = \int_0^t G(x,t-s) C_m(x,s) ds \qquad (38)$$

to close the relation in Equation (35), with $c_m(y,0,x) = 0$ for simplicity.

Multiple interacting continua

The multiple interacting continua (MINC) is a multi-continuum approach developed specifically as a discretization method for fractured media. In this approach, the matrix is represented by multiple continua, each one further away from the fracture continuum. The idea is that changes in fluid conditions propagate more slowly in less permeable matrix blocks compared to the smaller fracture volumes. This approach allows for accurate resolution of the gradients in pressures and concentrations into the matrix. It is distinct from the MRMT in that matrix continua are connected in series to account mass transfer between, while in the MRMT mass transfer is always between the mobile region and the immobile region according to a number of parallel rates. While the MINC method was developed for fractured media, it may be used as general multi-continuum discretization approach, and in some MINC implementations multi-rate mass transfer models can be obtained as a specific case.

MULTISCALE MODEL APPLICATIONS

In the previous sections, we have reviewed approaches to simulate reactive transport processes in media characterized by the multiplicity of scales. These approaches have been brought to bear on several applications in the field. Here we presented a selection of applications where one or more of these approaches have been used. In some of these applications, to evaluate the ability of multiscale methods to capture the processes of interest, they are compared to micro-scale simulations, macro-scale simulations, or simulations with other multiscale approaches. We organize this section by first distinguishing between applications in granular porous media and fractured media. We move on to applications in integrated hydrology that connect surface and subsurface compartments, which share similarities to the multiscale concepts discussed in this chapter.

Granular porous media

Dissolution of rocks often involves the development of altered rinds in the grains that make up the rock, e.g., Navarre-Sitchler et al. (2009). These layers may be characterized by changes in porosity or diffusivity that determine the rate at which reactant accesses the reactive mineral. Rates of dissolution observed at the large scale depend on diffusion-reaction processes that take place at the scale of micrometers. Although the method of multiple interacting continua was developed for fractured media it can be viewed as similar to the "shrinking core" model (Wunderly et al. 1996). In this view, the grains can be conceptualized as consisting of several continua connected in series. Aradóttir et al. (2013) used this approach to capture the microscopic dynamics of basaltic glass dissolution in a Darcy-scale model (Fig. 10). The MINC method involves dividing the system up to ambient fluid and grains, using a specific surface area to describe the interface between the two. The various grains and regions within grains can then be described by dividing them into continua separated by dividing surfaces. Millions of grains can thus be considered within the method without the need to explicitly discretizing them. Four continua were used for describing a dissolving basaltic glass grain; the first one describes the ambient fluid around the grain, while the second, third and fourth continuum refer to a diffusive leached layer, the dissolving part of the grain and the inert part of the grain, respectively.

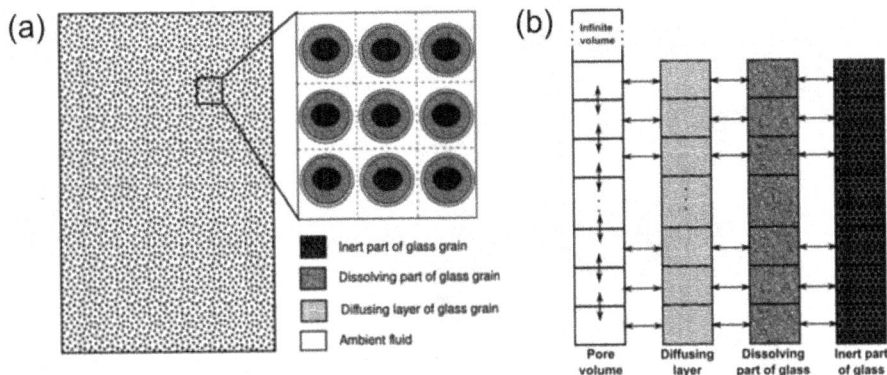

Figure 10. (a) Four-dimensional MINC interpretation of basaltic glass dissolution in the context of a column flow through experiment, (b) Schematic illustration of elements and connections in the four-dimensional MINC setup. Columns represent different continua, each of which has a number of elements (represented by boxes). Arrows show connections between elements and continua. [Reprinted from Aradóttir ESP, Sigfússon B, Sonnenthal EL, Björnsson G, Jónsson H (2013) Dynamics of basaltic glass dissolution–Capturing microscopic effects in continuum scale models. Geochimica et Cosmochimica Acta 121:311–327, Figures 2 and 5, with permission of Elsevier. Copyright 2013.]

The physical heterogeneity of natural porous media often leads to poorly mixed conditions such that different conditions exists within relatively small volumes of the media. From a reactive transport perspective, the existence of micro-environments implies that geochemical gradients can develop. For example, diffusion-dominated micro-environments conditions may be at equilibrium or close to equilibrium with certain minerals, while in advection-dominated pore spaces reactions are far from equilibrium. Mineralogical heterogeneity at different scales compounds to this effect where a correlation may exist between a certain mineralogy and enhanced reactivity due to availability of micro-porosity (Landrot et al. 2012). Poorly mixed conditions may also exist within individual pores, which under certain conditions lead to scale dependent rates (Li et al. 2008).

Although pore-scale simulations make it possible to reproduce geochemical gradients at the micrometer scale (Molins et al. 2012), multiscale models offer an alternative that is less computationally demanding. Liu et al. (2015) designed a micromodel experiment coated with hematite where macro- and micro-porosity domains where present (Fig. 11a). Three separate models were used to describe transport processes and reductive dissolution of hematite: a pore-scale model, a 1D single-continuum model, and a 1D triple-continua model. The predictions from the pore-scale reactive transport model predicted reasonably well the measured pore-scale rates of hematite reduction. Geochemical gradients within the domain (Fig. 11a) made it necessary to divide the domain in three continua: one that captured advection-dominated domain, one to capture the diffusive gradients within the macro-pore and a third on to capture diffusive limitations in the micro-pores. While the rate of hematite reduction in the advection-dominated and macro-pore domains was affected by the flow rate, the rate in the micropore domain was not, however, as reactant diffusion was rate-limiting. Results from the single domain model deviated significantly from the pore-scale results.

Pore-scale is not always available due to the large dimensions of the domain and the associated large computation costs. In these circumstances, multiscale approaches that retain a pore-scale description for part of the domain while using Darcy-scale for the rest help bridge the trade-off between process resolution and domain size. Yan et al. (2017) used a Darcy–Brinkman–Stokes-based approach to simulate biogeochemical reaction rates in heterogeneous sediments. An X-ray computed tomography image was used to construct a

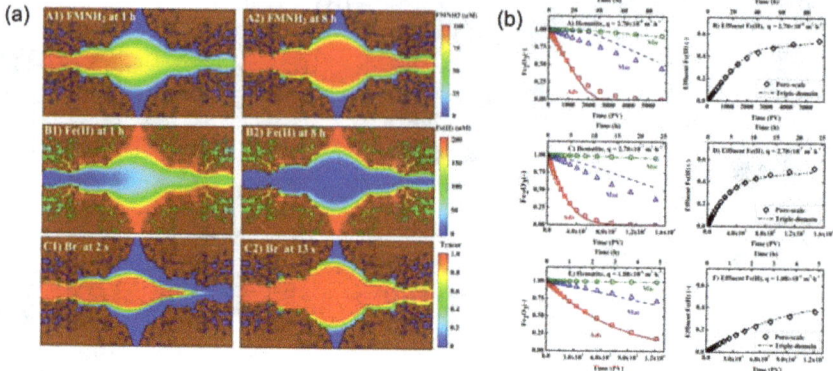

Figure 11 (a) Pore-scale simulation results for selected components and **(b)** Hematite concentration normalized to initial values in triple-continua domains (A, C and E), and accumulated Fe(II) mass in effluent (normalized to initial hematite-Fe in the micromodel) (B, D and F). Symbols denote calculated results from pore-scale simulations, and lines denote predicted results from the triple-domain model. [Reprinted from Liu Y, Liu C, Zhang C, Yang X, Zachara JM (2015) Pore and continuum scale study of the effect of subgrid transport heterogeneity on redox reaction rates. Geochimica et Cosmochimica Acta 163:140–155, Figures 4, 8, with permission of Elsevier. Copyright 2015.]

3D multiscale domain where ε was assigned a value from grayscale image (Fig. 12a), and in turn, a permeability value calculated from ε. A critical aspect of the model was, however, the assumption that the distribution of soil organic carbon (SOC) and biomass was correlated inversely with ε. That is, these variables were high near or on solid surfaces while low in large pore spaces. Additional simulations with single- and dual-domain models were performed to test this assumption. These simulations show that only when a large fraction of the soil organic carbon and biomass was placed in the immobile domain, dual-domain models were able to capture effluent concentrations of nitrate (Fig. 12b). In fact, single-domain models captured well effluent concentrations for non-reactive tracers. The multiscale aspect of the problem appeared only in the reactive transport component.

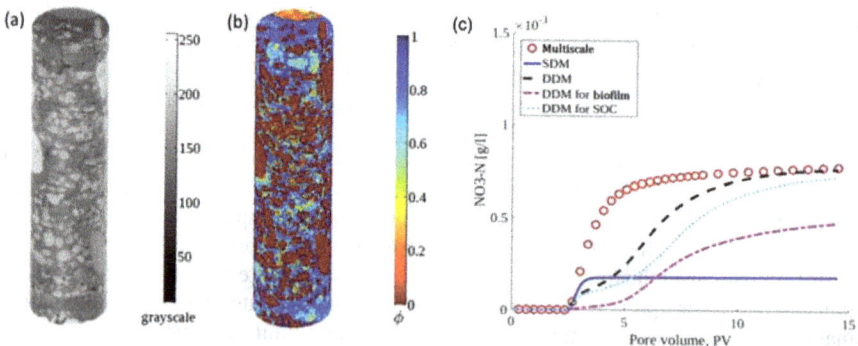

Figure 12 (a) An X-ray computed tomography image of a sediment column where a larger grayscale value indicates that the volume contains a higher content of solids; **(b)** porosity distribution converted from the left grayscale image where 0 denotes solid, 1 denotes pore, and other values between them denote the regions with mixed pores and solids, **(c)** Effects of biofilm and SOC heterogeneity on NO_3 reduction for the DBS-based model (Multiscale), the single-domain model (SDM), the dual-domain model (DDM) and two additional DDM, one where the positive correlation is only assumed for the biofilm (DDM for biofilm) and the other for the soil organic carbon (DDM for SOC). [Reprinted from Yan Z, Liu C, Liu Y, Bailey VL (2017) Multiscale investigation on biofilm distribution and its impact on macroscopic biogeochemical reaction rates. Water Resources Research 53(11):8698–8714, Figures 1, 9, with permission.]

These applications highlight the importance of mixing processes in reactive transport in heterogeneous porous media. In some instances, however, simulation of mixing processes in relatively homogeneous media may need of a multiscale approach when they are coupled to reactive processes. An example of this are mixing-controlled reactions. When two solutions mix such that a precipitate may form that has the potential to clog the pore space, it may be necessary to perform pore scale simulations. Scheibe et al. (2015b) presented a hybrid model that performed Darcy-scale simulations everywhere in the domain, and based on an incomplete mixing conditions, dynamically performed additional pore-scale simulations in a narrow region of the domain where precipitation occurred as a result of the mixing. This overlapping or hierarchical approach eliminated the need for matching boundary conditions between pore-scale and Darcy-scale domains. Hybrid simulations showed a sharper reaction front than equivalent Darcy-scale simulations, although some instabilities were observed in the hybrid approach (Fig. 13).

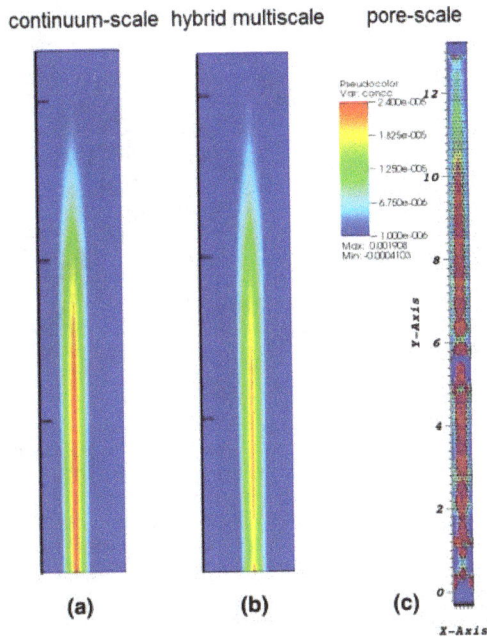

Figure 13. Concentration of product species C (mol/cm^3) in **(a)** a single-scale (Darcy-scale only) simulation, **(b)** hybrid multiscale simulation and (c) pore-scale simulation. [Reprinted from Scheibe, T. D., Schuchardt, K., Agarwal, K., Chase, J., Yang, X., Palmer, B. J., et al. (2015). Hybrid multiscale simulation of a mixing-controlled reaction. Advances in Water Resources 83:228–239, Figure 5, with permission of Elsevier. Copyright 2015.]

Fractured media

Flow and transport in fractured media occur primarily through a network of fractures, while flow in the matrix may be significantly slower with transport dominated by diffusive processes. While the fractures account for most of the flow and transport they typically make up a small portion of the overall volume of the medium. One could argue that to simulate fracture systems and incorporate this disparate scale, most fracture models has in one way or other multiscale aspects. Specific approaches to simulate fractures are reviewed in detail in a chapter of this volume (Deng and Spycher 2019, this volume). Here we describe the work of Molins et al. (2019) to develop a hybrid multiscale of fractured media as an example of the two separate scales. In this hybrid model, a pore-scale component captures Navier–Stokes flow, multi-component transport and aqueous equilibrium in the fracture, while a Darcy-scale component captures multi-component diffusive transport, aqueous equilibrium and mineral reactions in the porous matrix (Fig. 14). The interface between the sub-models, the fracture surface, is represented by an embedded-boundary. To simplify exchange of concentrations and fluxes at this interface, adaptive mesh refinement is used such that resolutions of the sub-models match at the interface while still using coarser resolution away from the interface when not needed in the Darcy-scale domain. The multiscale model is capable to capture flow channelization observed in an experimental fractured core and, at the same time, limitations in the dissolution of calcite by diffusive transport through an altered porous layer.

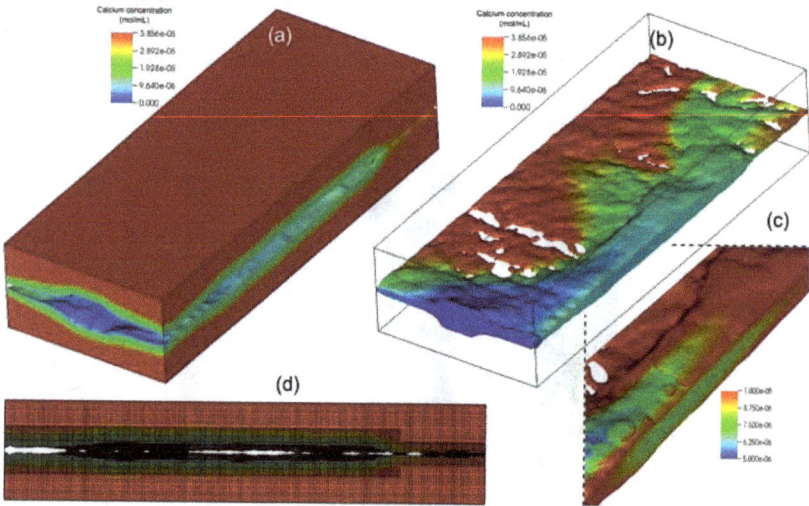

Figure 14. Steady-state calcium concentrations in the 3D simulations of the Duperow fracture experiment **(a)** in the Darcy-scale domain and **(b)** the pore-scale domain. **(c)** A close-up view of the pore-scale domain shows concentration gradients within the fracture opening, and **(d)** a side view of the Darcy-scale domains shows the embedded boundary and the mesh refinement around the fracture surface, where it interfaces with the pore-scale domain and steep concentration gradients develop. [Reprinted from Molins S, Trebotich D, Arora B, Steefel CI, Deng H (2019) Multi-scale model of reactive transport in fractured media: Diffusion limitations on rates, Transport in Porous Media 128:701–721, Figure 7, with permission.]

Surface–subsurface hydrologic coupling

Reactive transport of geochemical species in streams is result from an interplay between biogeochemical processes and mass exchange between the stream and the subsurface. The saturated sediment adjacent to the stream is therefore an important region for understanding the composition and evolution of water in the stream. For its role, the hyporheic zone has been the focus of study to understand flow and solute transport. Increasingly, there is interest to simulate reactive transport in the context of integrated surface–subsurface processes where both compartments are considered.

Although we have motivated the need for multiscale approaches in porous media from the pore- to Darcy-scale models in porous media, coupling of surface and subsurface processes requires the solution of similar equations in a coupled manner. The understanding of multiscale approaches in this sense is related to that of multi-physics, where the processes of interest are described by different equations. These processes may be characterized by different time scales, e.g., fast overland flow compared to long residence times for subsurface flow. Conceptually, these systems are similar to some of the subsurface systems considered in this chapter such as fractured media with fast flow in fractures compared to long residence times in the rock matrix. As a result, the approaches to coupling processes between the different compartments fall within those described in this review, including multi-rate approaches and hybrid approaches that require enforcement of continuity of mass and fluxes across interfaces. Examples of each of them are presented in what follows.

Painter (2018) use the residence time concept to develop a multiscale model for hyporheic exchange considering biogeochemical reactions. In this approach, the channel is a one-dimensional domain in which each cell of the discretization is connected to one dimensional sub-grid model for reactive transport (Fig. 15a), which are convolution representations of the exchange of solute with the hyporheic. This approach is mathematically equivalent to multi-rate mass transfer formulations such as Equations (33–34). In Painter (2018), the sub-grid model is generalized to include multicomponent reactive transport with general nonlinear reactions. Hyporheic zone denitrification is simulated with these non-linear models to demonstrate the approach (Fig. 15b).

Hybrid approaches are also being brought to bear on the surface–subsurface hydrologic exchange. Bao et al. (2018) developed a one-way coupled surface and subsurface water flow model to simulate a 7-km long reach along the Columbia River (Fig. 16). While the subsurface model is a Darcy-scale model, the surface component is solved with computational fluid dynamics software that solves a form of the Navier–Stokes equations with a free surface boundary. The model was employed to investigate surface water fluid dynamics and the impact of subsurface structures on the hydrologic exchange.

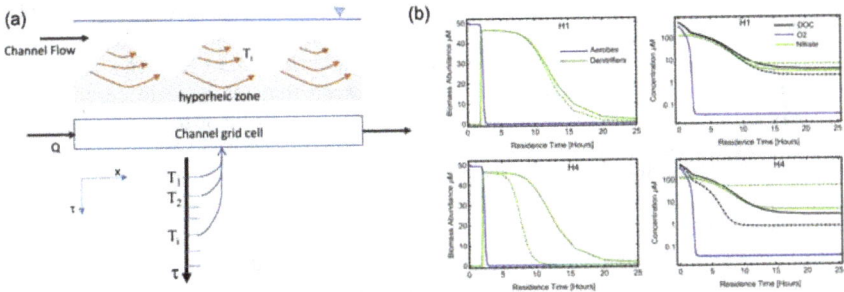

Figure 15. (a) Schematic of a multiscale representation of transport in a stream corridor. Hyporheic zone transport is represented by subgrid models in residence time formulation, which are coupled to the stream channel. The subgrid auxiliary equation associated with each channel grid cell is representative of the ensemble of pathways within the grid cell **(b)** Solutions of the subgrid reactive transport equations for biomass abundance (left column) and solute concentration (right column) versus residence time τ in the hyporheic zone at two channel locations and in steady state. Dissolved organic carbon (DOC) is consumed in the oxic zone at the downstream location leaving little DOC to fuel the denitrification reactions. [Reprinted from Painter SL (2018) Multiscale framework for modeling multicomponent reactive transport in stream corridors. Water Resources Research 54(10):7216–7230, Figures 1, 5, with permission.]

SUMMARY AND CONCLUSIONS

The issue of scale is central in reactive transport modeling and the different disciplines it draws from. It is often implicitly behind common assumptions used in most models, for example to couple different processes at the appropriate scale. The term *multiscale* is often used to specifically refer to applications that explicitly consider the different scales present in a given application. Here we have motivated the use of multiscale approaches by the need to scale-up processes that take place at the pore scale, where the medium is treated as composed of separate fluid and solid phases, to the Darcy-scale continuum, where the porous medium is characterized as a continuum. As shown, formally upscaling reactive transport only results in separate macro- and micro- problems under a limiting set of conditions. In general, however the macro and micro problem are coupled.

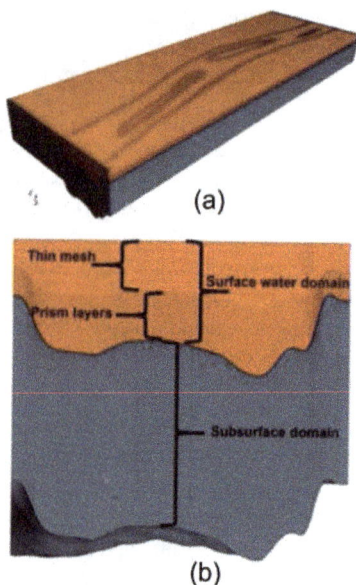

Figure 16. Meshes for the simulation domain of hydrologic exchange with a hybrid model. **(a)** Perspective view of the meshes. **(b)** Side view of the meshes for clarifying the mesh regions. [Reprint from Bao J, Zhou T, Huang M, Hou Z, Perkins W, Harding S, et al. (2018) Modulating factors of hydrologic exchanges in a large-scale river reach: Insights from three-dimensional computational fluid dynamics simulations. Hydrological Processes 32(23):3446–3463, Figure 2, with permission.]

Although insightful, mathematical methods for upscaling reactive transport rely on idealized representation of porous media and rarely used to investigate experimental or field systems. Simulation of these systems usually requires bringing to bear methods that use different scale descriptions within the same model such that micro-scale descriptions are used where it is needed; for example, to represent processes without making assumption of constitutive models for bulk parameters. Two approaches are prominent to simulate reactive transport in domains consisting of pore-scale and Darcy-scale sub-domains, one that uses a single equation to describe processes at both scales, and one that solves the problem in each domain separately and applies appropriate coupling conditions at the interface. These methods have also been used to find a compromise between the process and domain resolution and the size of domain that can be addressed with limited computational resources.

Multi-rate and multi-continua constitute another class of multiscale models that conceptualize the Darcy-scale porous medium as composed of two or more sub-regions that exists at each point in space and exchange mass according to a single-rate, multi-rate or continuum of rates. Mass exchanges between sub-regions or continua can represent micro-environments in which geochemical conditions may be different. Often these differences imply that a certain correlation exists between their hydrological accessibility and their mineralogical composition and reactivity.

Applications of these approaches in porous and fractured media show that often more than one approach can be used for a specific application. As application of reactive transport expands to consider surface processes, in addition to subsurface processes, available approaches are adapted and brought to bear on the surface–subsurface coupling. Generally multiscale approaches seek to represent processes at the appropriate scale, but different reasons are identified for this need. In some cases, the heterogeneous structure and composition of the porous medium necessitates micro-scale representation of the processes, while in other cases, the processes lead to the need for this micro-scale representation. For example, when solutions of different compositions mix, zones of active biogeochemical processes may require a micro-scale representation, especially when the media evolves as a result of it. While there is interest in capturing accurately these hot spots of biogeochemical activity, there is also interest in multiscale methods that can dynamically adjust process representation in time to capture periods of increased biogeochemical activity or hot moments.

ACKNOWLEDGMENTS

The contribution of Sergi Molins to this chapter was supported by the Interoperable Design of Extreme-scale Application Software Watersheds (IDEAS Watersheds) project funded by the U.S. Department of Energy, Office of Science, Office of Biological and Environmental Research under Award Number DE-AC02-05CH11231.

REFERENCES

Amanbek Y, Singh G, Wheeler MF, van Duijn H (2019) Adaptive numerical homogenization for upscaling single phase flow and transport. J Comput Phys 387:117–133

Aradóttir ESP, Sigfússon B, Sonnenthal EL, Björnsson G, Jónsson H (2013) Dynamics of basaltic glass dissolution – Capturing microscopic effects in continuum scale models. Geochim Cosmochim Acta 121:311–327

Balhoff MT, Thomas SG, Wheeler MF (2008) Mortar coupling and upscaling of pore-scale models. Comput Geosci 12:15–27

Bao J, Zhou T, Huang M, Hou Z, Perkins W, Harding S, Titzler S, Hammond G, Ren H, Thorne P, Suffield S, Murray C, Zachara J (2018) Modulating factors of hydrologic exchanges in a large-scale river reach: Insights from three-dimensional computational fluid dynamics simulations. Hydrol Process 32:3446–3463

Battiato I, Tartakovsky DM (2011) Applicability regimes for macroscopic models of reactive transport in porous media. J Contam Hydrol 120–121:18–26

Battiato, Tartakovsky DM, Tartakovsky AM, Scheibe TD (2011) Hybrid models of reactive transport in porous and fractured media. Adv Water Resour 34:1140–1150

Bringedal C, von Wolff L, Pop IS (2019) Phase field modeling of precipitation and dissolution processes in porous media: Upscaling and numerical experiments. UHasselt Computational Mathematics Preprint Nr. UP-19-01.

Buckingham E (1904) Contributions to Our Knowledge of the Aeration of Soils. U.S. Government Printing Office

Daval D, Martinez I, Corvisier J, Findling N, Goffé B, Guyot F (2009) Carbonation of Ca-bearing silicates, the case of wollastonite: Experimental investigations and kinetic modeling. Chem Geol 265:63–78

Deng H, Spycher N (2019) Modeling reactive transport processes in fractures. Rev Mineral Geochem 85:49–74

Donado LD, Sanchez-Vila X, Dentz M, Carrera J, Bolster D (2009) Multicomponent reactive transport in multicontinuum media. Water Resour Res 45:W11402

Druhan JL, Tournassat C (2019) Reactive Transport in Natural and Engineered Systems. Rev Mineral Geochem Vol 85

Efendiev Y, Hou T (2009) Multiscale Finite Element Methods: Theory and Applications. Springer-Verlag, New York

Golfier F, Zarcone C, Bazin B, Lenormand R, Lasseux D, Quintard M (2002) On the ability of a Darcy-scale model to capture wormhole formation during the dissolution of a porous medium. J Fluid Mech 457:213–254

Gray WG, Miller CT (2014) Introduction to the Thermodynamically Constrained Averaging Theory for Porous Medium Systems. Springer International Publishing, Cham

Gulbransen AF, Hauge VL, Lie K-A, ICT S (2010) A multiscale mixed finite-element method for vuggy and naturally fractured reservoirs. SPE J 15:395–403

Haggerty R, Gorelick SM (1995) Multiple-rate mass transfer for modeling diffusion and surface reactions in media with pore-scale heterogeneity. Water Resour Res 31:2383–2400

Hollenbeck KJ, Harvey CF, Haggerty R, Werth CJ (1999) A method for estimating distributions of mass transfer rate coefficients with application to purging and batch experiments. J Contam Hydrol 37:367–388

Hornung U (ed) (1997) Homogenization and Porous Media. Springer-Verlag, New York

Landrot G, Ajo-Franklin JB, Yang L, Cabrini S, Steefel CI (2012) Measurement of accessible reactive surface area in a sandstone, with application to CO_2 mineralization. Chem Geol 318–319:113–125

Li L, Steefel CI, Yang L (2008) Scale dependence of mineral dissolution rates within single pores and fractures. Geochim Cosmochim Acta 72:360–377

Lichtner PC (1996) Continuum formulation of multicomponent-multiphase reactive transport. Rev Mineral Geochem 34:1–81

Liu Y, Liu C, Zhang C, Yang X, Zachara JM (2015) Pore and continuum scale study of the effect of subgrid transport heterogeneity on redox reaction rates. Geochim Cosmochim Acta 163:140–155

Marshall TJ (1959) The diffusion of gases through porous media. J Soil Sci 10:79–82

Mehmani Y, Sun T, Balhoff MT, Eichhubl P, Bryant S (2012) Multiblock pore-scale modeling and upscaling of reactive transport: application to carbon sequestration. Transp Porous Media 95:305–326

Mikelić A, Devigne V, van Duijn C (2006) Rigorous upscaling of the reactive flow through a pore, under dominant peclet and damkohler numbers. SIAM J Math Anal 38:1262–1287

Molins S, Trebotich D, Steefel CI, Shen C (2012) An investigation of the effect of pore scale flow on average geochemical reaction rates using direct numerical simulation. Water Resour Res 48:W03527

Molins S, Trebotich D, Arora B, Steefel CI, Deng H (2019) Multiscale model of reactive transport in fractured media: Diffusion limitations on rates, Transp Porous Media 128:701–721

Navarre-Sitchler A, Steefel CI, Yang L, Tomutsa L, Brantley SL (2009) Evolution of porosity and diffusivity associated with chemical weathering of a basalt clast. J Geophys Res 114:F02016

Painter SL (2018) Multiscale framework for modeling multicomponent reactive transport in stream corridors. Water Resour Res 54:7216–7230

Popov P, Efendiev Y, Qin G (2009) Multiscale modeling and simulations of flows in naturally fractured karst reservoirs. Commun Comput Phys 162–184

Ray N, Elbinger T, Knabner P (2015) Upscaling the flow and transport in an evolving porous medium with general interaction potentials. SIAM J Appl Math 75:2170–2192

Ray N, Rupp A, Schulz R, Knabner P (2018) Old and new approaches predicting the diffusion in porous media. Transp Porous Media 124:803–824

Roubinet D, Tartakovsky DM (2013) Hybrid modeling of heterogeneous geochemical reactions in fractured porous media. Water Resour Res 49:7945–7956

Scheibe TD, Murphy EM, Chen X, Rice AK, Carroll KC, Palmer BJ, Tartakovsky AM, Battiato I, Wood BD (2015a) An analysis platform for multiscale hydrogeologic modeling with emphasis on hybrid multiscale methods. Groundwater 53:38–56

Scheibe TD, Schuchardt K, Agarwal K, Chase J, Yang X, Palmer BJ, Tartakovsky AM, Elsethagen T, Redden G (2015b) Hybrid multiscale simulation of a mixing-controlled reaction. Adv Water Resour 83:228–239

Seigneur N, Mayer KU, Steefel CI (2019) Reactive transport in evolving porous media. Rev Mineral Geochem 85:197–238

Soulaine C, Tchelepi HA (2016) Micro-continuum approach for pore-scale simulation of subsurface processes. Transp Porous Media 113:431–456

Soulaine C, Roman S, Kovscek A, Tchelepi HA (2017) Mineral dissolution and wormholing from a pore-scale perspective. J FLUID Mech 827:457–483

Tomin P, Lunati I (2013) Hybrid Multiscale Finite Volume method for two-phase flow in porous media. J Comput Phys 250:293–307

Tomin P, Lunati I (2016) Spatiotemporal adaptive multiphysics simulations of drainage-imbibition cycles. Comput Geosci 20:541–554

Wood BD, Radakovich K, Golfier F (2007) Effective reaction at a fluid–solid interface: Applications to biotransformation in porous media. Adv Water Resour 30:1630–1647

Wunderly MD, Blowes DW, Frind EO, Ptacek CJ (1996) Sulfide mineral oxidation and subsequent reactive transport of oxidation products in mine tailings impounds: A numerical model. Water Resour Res 32:3173–3187

Yan Z, Liu C, Liu Y, Bailey VL (2017) Multiscale investigation on biofilm distribution and its impact on macroscopic biogeochemical reaction rates. Water Resour Res 53:8698–8714

Yang X, Liu C, Shang J, Fang Y, Bailey VL (2014) A unified multiscale model for pore-scale flow simulations in soils. Soil Sci Soc Am J 78:108

Yousefzadeh M, Battiato I (2017) Physics-based hybrid method for multiscale transport in porous media. J Comput Phys 344:320–338

Reviews in Mineralogy & Geochemistry
Vol. 85 pp. 49-74, 2019
Copyright © Mineralogical Society of America

3

Modeling Reactive Transport Processes in Fractures

Hang Deng and Nicolas Spycher

Earth and Environmental Sciences Area
Lawrence Berkeley National Laboratory
Berkeley, CA, 94720
USA

hangdeng@lbl.gov; nspycher@lbl.gov

INTRODUCTION

Fractures are ubiquitous and important features in the Earth subsurface (Berkowitz 2002; Pyrak-Nolte et al. 2015). They are created as a result of rock failure when the critical stress (i.e., fracture toughness) is exceeded, or in the case of subcritical crack growth, when cracks propagate under stress conditions below fracture toughness, facilitated by chemical reactions (Atkinson 1984). The necessary conditions for fracture growth can be created by natural unloading from land erosion (Engelder 1987), tectonic events (Molnar et al. 2007), and crystal growth in presence of fluids supersaturated with respect to solid phases (Royne and Jamtveit 2015). Fractures can also be artificially created for enhanced energy recovery, through excess fluid pressure and change of thermal stress, as in the case of geothermal energy extraction and unconventional oil and gas production (McClure and Horne 2014; Lampe and Stolz 2015). Fractures can be observed by surveying rock outcrops, inferred by fluid flow and geochemical measurements, and detected using geophysical techniques (Berkowitz 2002; St Clair et al. 2015; Walton et al. 2015).When open, fractures act as preferential flow pathways because of their high permeability, and thus typically control fluid migration and solute transport in fractured rocks. For this reason, fractures are avoided when siting and designing geologic isolation systems, such as for nuclear waste and CO_2 storage, in order to prevent undesired fluid and chemical migration (Kovscek 2002; Lewicki et al. 2007; Birkholzer et al. 2012). In the Earth's critical zone, fractures control the availability of water for rock weathering and hence the development of the regolith layer (Brantley et al. 2017). It has also become accepted that weathering itself typically controls fracture permeability in hard rock aquifers (Lachassagne et al. 2011).

Simulations of flow and transport in fractured porous media are of prime importance in studies related to oil and geothermal exploration, waste disposal, and environmental stewardship of contaminated sites. With that aim, three broad categories of conceptual and numerical models have been developed, to capture the unique flow and transport properties of fractures: equivalent continuum models, dual or multiple continua models, and discrete fracture networks. The equivalent continuum model treats the fractures and rock matrix as a single continuum with a permeability corrected for the effect of fractures (Long et al. 1982). In the dual-continua (dual-porosity and dual-permeability) models (Warren and Root 1963), a separate fracture continuum is added, co-located with the rock matrix continuum, with its own hydraulic properties including porosity-permeability relationships that can differ from those applied to the matrix. The conceptualization of dual continuum models can vary, depending on the types of connections implemented between fracture and matrix grid blocks, with advantages and disadvantages, as discussed by Lichtner (2000). Multiple-continua models build upon the dual-continua concept by adding more co-located interacting media, thus allowing consideration of more hydrogeological components (Doughty 1999). For example,

1529-6466/19/0085-0003$05.00 (print)
1943-2666/19/0085-0003$05.00 (online)   http://dx.doi.org/10.2138/rmg.2019.85.3

multiple interacting continua (MINC) models can be used to capture transient processes resulting from mass transfer between fractures and rock matrices at different distances from the fracture surface (Pruess 1991). Similarly, a triple-continuum approach has been proposed to investigate the effect of small fractures on flow through fractured bedrock (Wu et al. 2004).

In contrast to the dual- and multiple-continua models, discrete fracture networks represent the individual fractures explicitly, although with reduced dimensionality, i.e., $(n-1)$-dimensional fractures in an n-dimensional domain (Hyman et al. 2015; Lei et al. 2017). Because the discrete fracture network is generated based on the geometrical properties of each individual fracture (e.g., aperture, length, orientation), and can account for internal aperture variability (Makedonska et al. 2016), the flow and transport results are arguably more accurate than with the dual/multiple-continua approach. However, the application of discrete fracture networks is typically not suitable for large-scale systems with numerous fractures, or when information on fracture geometrical properties is unavailable.

The morphology and thus hydrophysical properties of fractures evolve dynamically as a function of mineral dissolution and precipitation driven by fluid composition, temperature, and pressure variations. This is the case when advective flow in fractures introduces fluids that are out of chemical equilibrium with the wallrock, or displaces fluids to zones of different temperature and/or pressure. For example, solubility changes by cooling, and boiling through depressurization are key processes leading to fracture alteration in hydrothermal systems (e.g., Browne 1978). Mixing of working fluids in geothermal systems with shallower and cooler groundwater can also lead to significant mineral deposition (Griffiths et al. 2016). In the case of geologic carbon storage systems, the introduction of CO_2 into deep aquifers creates an acidic and carbonated fluid that reacts with minerals (Rochelle et al. 2004), especially carbonates, with the potential of impacting fracture permeability and caprock integrity (Fitts and Peters 2013). Acidic fluids are also introduced in hydraulic fracturing processes to dissolve minerals and thus create and maintain open fractures (Kalfayan 2008). In all these cases, a detailed understanding of reactive transport mechanisms leading to mineral precipitation and dissolution in fractures is crucial to the development of predictive tools that can be applied with confidence in studies such as engineered system design and optimization, energy recovery feasibility studies, and environmental impact assessments.

Much work has been done, for decades, towards the understanding and modeling of fluid flow in fractured systems (National Research 1996). Many studies have also been performed since the 1960's towards understanding the geochemical processes responsible for the mineralogical alteration of fractures (veins) observed in hydrothermal systems and ore deposits (e.g., Hemley and Jones 1964; Meyer 1967; Browne 1978; Barnes 1997; Parry 1998), including the concept of water/rock ratio to characterize mass transfer between fractures and rock matrix (Giggenbach 1984). Comparatively much fewer investigations have focused on understanding and modeling the actual mechanisms of coupled chemical reaction and transport within fractures and between fractures and rock matrix. Models of rock matrix–fracture interactions, including multicomponent geochemical effects, have been developed using both discrete fracture networks and dual/multi-continua approaches in the context of hydrothermal systems, geologic nuclear waste disposal and geologic carbon sequestration, both at the field scale (e.g., Steefel and Lichtner 1998; Sonnenthal et al. 2005; Xu et al. 2006; Gherardi et al. 2007; Liu et al. 2017) and smaller scales (e.g., Steefel and Lasaga 1994; Xu and Pruess 2001; Dobson et al. 2003). However these studies relied on simplified conceptualizations of fracture properties (Snow 1970; Witherspoon et al. 1980), reactive surface areas, and porosity-permeability relationships (Sonnenthal et al. 2005). Advances in computational power and microscopic imaging have since allowed the development of more sophisticated and mechanistic models that can more accurately capture the intricate interplay between fluid flow and reaction in fractures, including effects of local fluid dynamics, matrix diffusion, fluid chemistry variability and mixing, and the impacts of mineral precipitation/dissolution on reactive surface areas, as further discussed here.

This chapter focuses on the geochemical alteration of single fractures, which is critical for improving the conceptualization and prediction of flow and transport in fractured rocks at the reservoir scale. Here, we concentrate our discussions on single-phase saturated systems, whereas a detailed discussion of multiphase processes can be found in Sin and Corvisier (2019, this volume). First, experimental observations of fracture alteration caused by fluid–rock interactions are summarized. This is followed by a review of the modeling efforts dedicated to simulating reactive transport processes in individual fractures. We also discuss the major factors controlling fracture alteration as elucidated by the modeling efforts, and conclude the chapter with an outlook on directions where further research is needed.

FRACTURE ALTERATION DRIVEN BY GEOCHEMICAL REACTIONS

The detailed observation of fracture alteration driven by geochemical reactions is made possible by the advancement of non-invasive imaging techniques to characterize fracture surfaces and geometry, such as profilometry (Ameli et al. 2013), Nuclear Magnetic Resonance Imaging (Dijk et al. 1999), Positron Emission Tomography (Loggia et al. 2004; Tenchine and Gouze 2005), and Computed Tomography (Gouze et al. 2003). In particular, the combination of fracture-flow experimental apparatus with computed tomography imaging has enabled *in situ* visualization of fracture morphology changes due to reactive fluid flow (Deng et al. 2015; Ajo-Franklin 2017). Improvement of imaging detectors and resolutions, and development of new image processing algorithms, have also further improved our capability to quantify fracture geometries; this can be challenging because fracture aperture is orders of magnitude smaller than the dimensions of the fracture plane. For example, an inverse point-spread function was developed to accurately determine fracture apertures that are one-tenth of the voxel size (Ketcham et al. 2010), and filters that detect the unique characteristics of fracture morphology, i.e., the planar geometry, were used to distinguish fracture from large pores/vugs in the rocks (Deng et al. 2016a). Segmentation algorithms have also been developed and applied to differentiate the fracture void, the rock matrix immediately adjacent to the fracture that has gone through geochemical alteration, and the intact rock matrix (Noiriel et al. 2007; Deng et al. 2013). Furthermore, coupling of tomography imaging with microscopic techniques such as Scanning Electron Microscope (SEM), Energy Dispersive X-ray Spectroscopy (EDS) and X-Ray Fluorescence (XRF), and the application of techniques such as dual-energy microtomography provide spatially resolved geochemical data in addition to bulk information derived from fluid chemistry (Noiriel 2015). By correlating the grayscale values of the 3D tomography images with high resolution 2D SEM/EDS images, algorithms can be trained to extract mineralogical information from tomography images, and to map mineral composition on fracture surfaces (Ellis and Peters 2016). Such information is critical for investigating the controls of mineralogy on the changes in fracture flow properties.

Fracture opening due to mineral dissolution

The continuous replenishment of fresh reactive fluid can accelerate mineral reactions in fractures. When the fluid is under-saturated with respect to the minerals in contact, minerals dissolve, resulting in fracture aperture enlargement (Dijk et al. 2002; Detwiler et al. 2003). Fracture opening caused by the dissolution of carbonate minerals when exposed to acidic fluid is of particular interest because of its prevalence in the subsurface and its relevance to various natural processes (e.g., karst formation) and human activities (e.g., geologic carbon storage). Given the fast kinetics of carbonate minerals (Chou et al. 1989), especially calcite, considerable fracture opening within days or hours has been observed in experimental studies that use aggressive synthetic acidic fluids (Singurindy and Berkowitz 2004; Ellis et al. 2011; Elkhoury et al. 2013; Noiriel et al. 2013; Emmanuel and Levenson 2014; Deng et al. 2015; Garcia-Rios et al. 2015; Ajo-Franklin 2017). Figure 1a shows 3D reconstructions of a limestone fracture

(a)

(b)

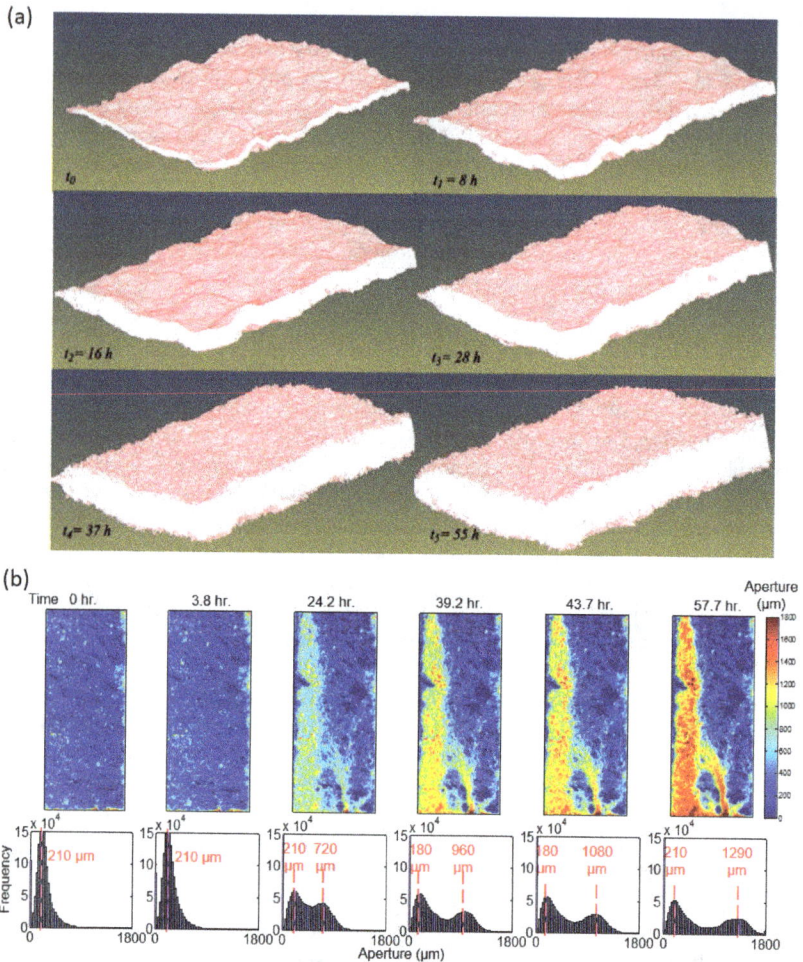

Figure 1. (a) 2 × 2 mm detail of the 3D fracture morphology at different stages of a fracture flow experiment exposing a slightly argillaceous limestone to a carbonated fluid (Noiriel et al. 2013). **(b)** Aperture maps and histograms of a limestone fracture after exposure to CO_2-rich fluid under confining stress. [Reprinted with permission from Deng H et al. (2015) Alterations of fractures in carbonate rocks by CO_2-acidified brines. Environ Sci Technol 49:10226–10234. Copyright (2015) American Chemical Society.]

at different time intervals of a flow-through experiment, demonstrating a uniform fracture opening over time caused by calcite dissolution (Noiriel et al. 2013). The average aperture increased from 48 µm to 346 µm after reacting with a CO_2-charged fluid (P_{CO_2}=0.1 MPa) for 55 hours. Figure 1b shows a series of fracture aperture maps (2D projections on the fracture plane of the 3D fracture volumes reconstructed from *in situ* tomography images) and histograms that highlight localized fracture enlargement (Deng et al. 2015). After reacting the limestone fracture with CO_2-rich brine (P_{CO_2}=77 bar) for 42 hours, the average aperture increased from 244 µm to 344 µm, while local apertures exceeded 1000 µm.

The intrinsic fracture permeability increases in response to fracture enlargement caused by mineral dissolution. For a parallel wall fracture, the volumetric flux flowing through the fracture (Q, [m³/s]) for a given pressure gradient (∇P, [Pa/m]) is proportional to the cube of the fracture aperture (b, [m]) (i.e., the cubic law, Eqn. 1), and thus fracture permeability (k_f, [m²]) is proportional to the square of fracture aperture (Eqn. 2) (Snow 1969):

$$Q = \frac{b^3 W \nabla P}{12\mu} \tag{1}$$

$$k_f = \frac{b^2}{12} \tag{2}$$

where W [m] is the width of the fracture, and μ [Pa s] is the dynamic viscosity of the fluid.

For non-parallel fractures, a hydraulic aperture (b_h, [m]) can be defined such that the cubic law still holds. The increase of the hydraulic aperture does not necessarily follow the geometric aperture. Fracture permeability increases more substantially than would be expected based on the cubic law if preferential flow channels develop (Deng et al. 2015; Garcia-Rios et al. 2017), while the increase in fracture permeability is mitigated when fracture enlargement is accompanied by an increase in fracture roughness and hydraulic tortuosity (Ellis et al. 2011; Deng et al. 2013; Noiriel et al. 2013). For example, unreacted mineral bands can make fracture surfaces rougher and serve as flow strictures (Fig. 2).

Fracture aperture enlargement caused by mineral dissolution may not be a linear function of time because of decreasing subsequent mineral dissolution. One mechanism that causes mineral dissolution to slow down is the development of preferential channels (Deng et al. 2015). As more reactive fluid is directed into these channels, the surface area in the nonchannelized regions becomes less accessible for reactions. Furthermore, as preferential channels grow, transverse diffusion (i.e., perpendicular to the fracture plane) in these channels becomes important and the reactions are increasingly limited by the transport process. Another mechanism that limits subsequent mineral dissolution is the retreat of the mineral front into the rock matrix (Noiriel et al. 2007; Abdoulghafour et al. 2013; Davila et al. 2016a; Ajo-Franklin 2017). As such, reactants and products of the dissolution reactions have to diffuse through the remaining rock matrix, creating a transport limitation on chemical reactions.

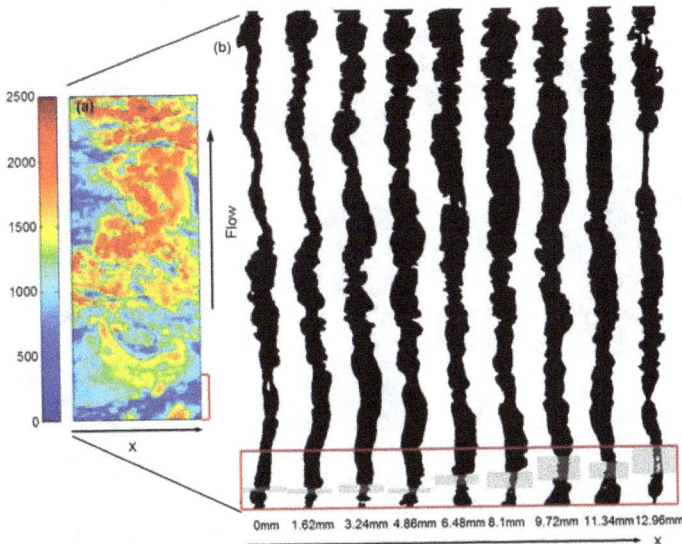

Figure 2. (a) Fracture aperture map (µm) of after-reaction geometry with degraded zones treated as fracture, and **(b)** sections of the fracture along the flow direction sampled every 1.62 mm. Gray boxes highlight the areas where aperture increase is minimal, corresponding to the blue band in (a) near the bottom of the fracture. [Reprinted with permission from Deng H et al. (2013) Modifications of carbonate fracture hydrodynamic properties by CO2-acidified brine flow. Energy & Fuels 27:4221–4231 Copyright (2013) American Chemical Society.]

The alteration of rock matrix bordering the fracture

When the rock matrix is composed of minerals of varying reactivity, preferential dissolution of the fast reacting minerals can result in the development of an altered porous layer on the fracture surface (Noiriel et al. 2007; Ellis et al. 2011; Elkhoury et al. 2015; Davila et al. 2016a; Ajo-Franklin 2017). For example, Figure 3 shows an altered layer composed of quartz and other silicates left behind after calcite dissolution (Davila et al. 2016a). Within the altered layer, removal of the fast reacting minerals also exposes more surface area of the remaining minerals (Garcia-Rios et al. 2015).

The impacts of the alteration in the near-fracture region on fracture permeability is complicated. The permeability of the altered layer is negligible compared to that of the open fracture (Noiriel et al. 2007; Chen et al. 2014). If the altered layer remains near the fracture surface, the initial flow path is mostly maintained and the fracture permeability change is limited (Davila et al. 2016a). Mineral grains in the altered layer, however, may reorganize and lead to local aperture decreases (Noiriel et al. 2007). The altered layer may also detach from the fracture surface, causing fracture aperture to increase locally (Noiriel et al. 2007; Andreani et al. 2008). Detachment of the altered layer may result from the decrease in cohesion relative to the shear stress imposed by the fluid flow (Noiriel et al. 2007), or due to the change of fluid pH at the grain boundaries (Pepe et al. 2010). The fracture permeability can increase following the detachment of the altered layer if the fracture flow flushes out the released particles (Deng et al. 2017); whereas the fracture permeability is decreased if the released particles re-deposit in the fractures and cause substantial fracture aperture reduction (Fig. 3d) (Noiriel et al. 2007; Ellis et al. 2013; Davila et al. 2016a).

Figure 3. ESEM (Environmental Scanning Electron Microscope) images of fractures after reaction with and acidic fluid in marl cores, which are composed of over 70% calcite, followed by quartz, illite, albite, gypsum, clinochlore, anhydrite and pyrite. In a first case (**d,e**) an S-free solution was initially at equilibrium with respect to calcite before reacting with CO_2 at 61 bar, and in a second case (**a,b,c,f**) an S-rich solution was initially at equilibrium with respect to calcite, dolomite and gypsum before reacting with CO_2 at 61 bar. (**a**) alteration at 8 mm from the inlet, (**b**) altered layer 12 mm from the inlet, showing gypsum precipitation highlighted by the white arrows, (**c**) gypsum precipitation 15 mm from the inlet, (**d**) fracture clogging by particles, (**e**) and (**f**) altered layer along the fracture walls from a high flow rate experiment using an S-free and S-rich solution, respectively. [Reprinted from Davila G et al. (2016) Interaction between a fractured marl caprock and CO_2-rich sulfate solution under supercritical CO_2 conditions. Int J Greenhouse Gas Control 48:105–119, copyright (2016) with permission from Elsevier.]

The alteration of the near-fracture region is also common in cements. Typically, three altered layers are observed (Fig. 4), including a Ca-depleted layer close to the unaltered cement, an amorphous Si-rich layer bordering the bulk fluid, and a calcite-enriched layer in between (Abdoulghafour et al. 2013; Luquot et al. 2013; Walsh et al. 2014a; Abdoulghafour et al. 2016). These altered layers are products of the reaction sequence of portlandite dissolution, the precipitation of amorphous silica, and calcite precipitation due to local enrichment of Ca^{2+} at the portlandite dissolution front, respectively. More details about acid–cement interactions is discussed in Cama et al. (2019, this volume.) While the net result of the reactions is the removal of materials, the degree of fracture permeability change is widely variable. This change depends on the relative thickness of the three layers, which is partly controlled by flow rate, and the dissolution pattern in the fracture, which is affected by the initial fracture geometry. Fracture permeability increases when channelization (Abdoulghafour et al. 2016) or net precipitation of a low porosity calcite layer (Luquot et al. 2013) dominates, and remains unchanged when the boundary of the secondary Si-rich layer tracks the initial geometry. In some cases, fracture permeability decreases. Because the amorphous silica layer has a large porosity, the thickness of this layer can be considerable even with a small amount of amorphous silica, leading to a decrease in fracture aperture and permeability (Abdoulghafour et al. 2013). The decrease in fracture permeability is also partly attributed to the weakened mechanical properties (Walsh et al. 2013, 2014a), i.e., deformation of the altered layer, or failure of the asperities under confining pressure.

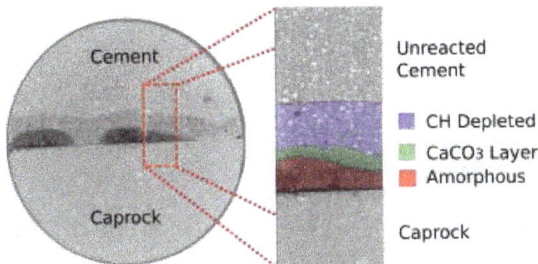

Figure 4. X-ray computed tomography image showing cement degradation by carbonated water. Three altered layers were identified and highlighted (Walsh et al. 2014a).

Fracture closing due to mineral precipitation

Fracture permeability can also decrease as a result of mineral precipitation when the injected fluids are supersaturated with respect to solid phases. The supersaturation state can be achieved by changing the temperature of the fluid, or by mixing two fluids of different compositions (Emmanuel and Berkowitz 2005). By injecting a well-mixed $CaCl_2$–$NaHCO_3$ solution supersaturated with respect to calcite into a transparent fracture analogue, Jones and Detwiler (2016) observed calcite precipitation in the artificial fracture, and showed that the precipitation patterns are sensitive to the initial mineral heterogeneity, i.e., seeding of calcite on the initial fracture surface.

Mineral precipitation is also possible when the necessary aqueous species are provided by simultaneous mineral dissolution, such as in the cases of calcite dissolution with gypsum precipitation (Garcia-Rios et al. 2015; Davila et al. 2016a) (Fig. 3b,c), and dolomite dissolution with calcite precipitation (Singurindy and Berkowitz 2004). Precipitation of carbonate minerals in basaltic fractures has also been observed, because the dissolution of minerals such as pyroxene and olivine releases a considerable amount of divalent cations in solution (Menefee et al. 2017).

In these cases, the direction and extent of fracture permeability change depends on the interplay between dissolution and precipitation. While fast dissolution supplies more reactants for the precipitation reaction, it also causes substantial fracture opening that can outcompete the impact of mineral precipitation and result in fracture permeability increase (Singurindy and Berkowitz 2005; Garcia-Rios et al. 2015; Davila et al. 2016a). It was also shown that transport limitation, such as experienced by dead ends of fractures and less connected fractures (Singurindy and Berkowitz 2005; Menefee et al. 2017), or due to slow flow velocity (Davila et al. 2016a), favors precipitation and fracture closing. Figure 5 summarizes experimental observations in a smooth fracture going through coupled calcite dissolution and gypsum precipitation, highlighting the important roles of fluid chemistry and flow velocity (Singurindy and Berkowitz 2005). In addition to the factors discussed above, the molar volume and the porosity of the precipitates can also impact fracture permeability change caused by precipitation. For the same amount of precipitates, a larger porosity in the precipitate corresponds to a sharper reduction in the fracture aperture and permeability, as in the case of the secondary Si-rich layer in cement fractures (Abdoulghafour et al. 2013).

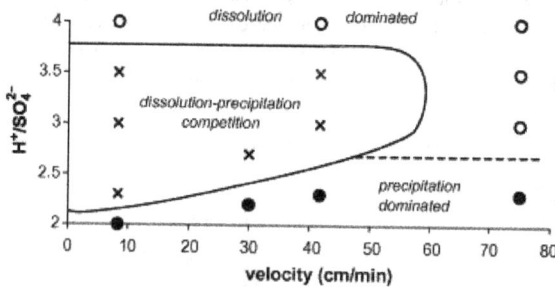

Figure 5. Summary of experimental observations of dissolution and precipitation in a smooth fracture over a range of fluid chemistry (injected H^+/SO_4^{2-} ratio) and velocity conditions. [Reprinted with permission from Singurindy and Berkowitz (2005) The role of fractures on coupled dissolution and precipitation patterns in carbonate rocks. Adv Water Resour 28:507–521 from Elsevier, copyright (2005).]

CONCEPTUAL AND NUMERICAL MODELS

As highlighted by experimental observations, the key processes involved in geochemical fracture alteration include: advective fluid flow, diffusion along the fracture plane, diffusion across the fracture walls and into/from the altered layer adjacent to the fracture, and accompanying mineral reactions (Fig. 6). In this section, we summarize reactive transport models of different complexities developed to examine one or more of these processes in single fractures. While theoretically, all these models can be applied to individual fractures of any length scale (typically between ~100 µm to ~100 m), in practice, the complexity of the model needs to be reduced as the length scale of the individual fracture being investigated increases. Another common practice is typically to apply pore-scale models to cases when the

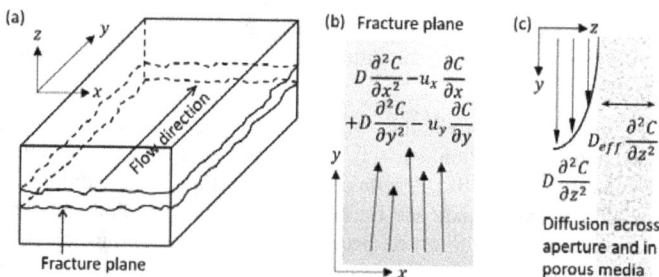

Figure 6. **(a)** 3D illustration of a fracture, **(b)** flow and diffusion in the fracture plane, and **(c)** flow and diffusion processes across the fracture surface and through the altered porous media.

processes across fracture aperture (with length scales on the order of μm–mm) are of interest, whereas continuum-scale models are used when processes along the flow direction or in the fracture plane are the focus of the investigation.

Generic formulation of the governing equations

Mesh-based direct numerical simulations of fracture alteration involves numerically solving a system of equations. The mathematical problem involves solving three governing equations (for continuity, momentum conservation, and mass conservation) and several constitutive relations (e.g., reaction rate laws, mass laws). The modeling approach can be categorized as either pore-scale or continuum-scale, depending on whether the fluid–rock interface (e.g., fracture surface) is explicitly resolved, or whether the solid phase and fluid phase co-exist in each model (numerical) grid cell, respectively. Here, we provide a summary of the key features of the pore-scale and continuum-scale modeling approach. More details can be found in Molins (2015) and Steefel et al. (2015), respectively.

Pore-scale models. With this type of modeling, the void space itself is discretized and the fluid–solid interface is tracked (Molins et al. 2012, 2014). Therefore, geometry evolution due to mineral precipitation and dissolution is explicitly modeled. The generic formulation of the governing equations of pore-scale models are given as follows:

$$\nabla \cdot \boldsymbol{u} = 0 \tag{3}$$

$$\rho \left(\frac{\partial \boldsymbol{u}}{\partial t} + \boldsymbol{u} \cdot \nabla \boldsymbol{u} \right) = -\nabla P + \mu \nabla^2 \boldsymbol{u} + \boldsymbol{F} \tag{4}$$

$$\frac{\partial C_i}{\partial t} = -(\boldsymbol{u} \cdot \nabla) C_i + \nabla \cdot D_{mi} \nabla C_i \tag{5}$$

where $\boldsymbol{u}$ is the velocity vector [m s^{-1}], ρ [kg m^{-3}] and μ [Pa s] are the density and dynamic viscosity of the fluid, respectively, P [Pa] is pressure, and $\boldsymbol{F}$ [N m^{-3}] is the body force such as gravity. C_i [mol m^{-3}] and D_{mi} [m^2 s^{-1}] are respectively the concentration and the molecular diffusion coefficient of aqueous species i. The mass flux of species i from reaction of mineral m (r_m [mol m^{-2} s^{-1}]) that is in contact with the fluid is treated as a boundary condition:

$$-D_{mi} \nabla C_i \cdot \boldsymbol{n} = \upsilon_{i,m} r_m \tag{6}$$

where $\upsilon_{i,m}$ is the stoichiometric coefficient of species i in the reaction of mineral m, and $\boldsymbol{n}$ is the unit normal vector.

Continuum-scale models. The continuum-scale governing equations are based on variables averaged over each grid cell where the void space and the solid phase co-exist, with its void space modeled using porosity. Assuming steady state flow, the continuity equation (Eqn. 7) and Darcy's law (Eqn. 8) are solved for the flow field:

$$\frac{\partial \phi \rho}{\partial t} + \nabla \cdot \boldsymbol{q} = 0 \tag{7}$$

$$\boldsymbol{q} = \phi \boldsymbol{v} = -\frac{k}{\mu} (\nabla P - \rho \boldsymbol{g}) \tag{8}$$

$$\frac{\partial \phi C_i}{\partial t} = -(\boldsymbol{q} \cdot \nabla) C_i + \nabla \cdot (\phi \boldsymbol{D} \nabla C_i) + R_i \tag{9}$$

where $\boldsymbol{v}$ [m s^{-1}] is the pore velocity, $\boldsymbol{q}$ [m^3 m^{-2}$_{porous\ medium}$ s^{-1}] is the Darcy flux, k [m^2] is the permeability, and $\boldsymbol{g}$ [m s^{-2}] is the gravitational acceleration. In the mass balance equation (Eqn. 9),

ϕ [$m^3_{\text{pore space}}$ $m^{-3}_{\text{porous medium}}$] is the porosity, D is the dispersion tensor, which can be used to account for dispersion in addition to molecular diffusion in porous media. Concentration changes due to mineral reactions are treated as source/sink terms (R_i, [mol $m^{-3}_{\text{porous medium}}$ s^{-1}]).

A common practice is to write the mass balance equation (Eqn. 9) with respect to the total concentration of component i (ψ_i), which is the summation of the concentration of the corresponding primary species p_i, and the concentrations of all secondary species (s_i) that are formulated using primary species p_i:

$$\Psi_i = C_{p_i} + \sum_{s_i} \upsilon_{s_i,p_i} C_{s_i} \qquad (10)$$

The concentrations of the secondary species (C_{s_i}) is related to the concentrations of primary species by the law of mass action (Eqn. 11), which assumes instantaneous equilibrium for aqueous reactions.

$$C_{s_i} = \gamma_{s_i}^{-1} K_{eq,s_i}^{-1} \prod_{p_i} \left(\gamma_{p_i} C_{p_i} \right)^{\upsilon_{s_i,p_i}} \qquad (11)$$

where K_{eq,s_i} and υ_{s_i,p_i} are the equilibrium constant and the stoichiometric coefficient of the aqueous reaction that relates the secondary species s_i to the primary species p_i, respectively, and γ is the activity coefficient.

Mineral reactions are typically treated kinetically. The reaction rate of a given mineral, r_m [mol $m^{-2}s^{-1}$], can be calculated as follows using the transition state theory rate laws (Lasaga 1984):

$$r_m = k_{\text{rxn},m} \exp\left(-\frac{E_a}{RT} \right) \prod a_i^{n_i} \left(1 - \left(\frac{IAP_m}{K_{eq,m}} \right)^{m_1} \right)^{m_2} \qquad (12)$$

where $k_{\text{rxn},m}$ [mol m^{-2} s^{-1}] is the kinetic coefficient (pre-exponential factor), E_a [kcal mol^{-1}] is the activation energy, R [kcal K^{-1} mol^{-1}] is the ideal gas constant, T [K] is the absolute temperature, a_i is the activity of any catalytic or inhibitory species, and n_i is the corresponding exponent. The last term is the chemical affinity that depends on the ionic activity product (IAP_m), the solubility constant ($K_{eq,m}$), and the empirical exponents m_1 and m_2. Note that theoretically, $1/m_1$ is the Temkin coefficient, which corresponds to the stoichiometric coefficient of the elementary reaction step (Lichtner 2016).

Unlike the pore-scale models, in which Equation (12) is used as a boundary condition (Eqn. 6) at the surface of each mineral, in the continuum-scale models, the reaction term (R_i, Eqn. 9) is treated as a source/sink term, accounting for the contribution from all minerals that are present in the grid cell and include species i. It relates to Equation (12) as:

$$R_i = \sum_m A_m \upsilon_{i,m} r_m \qquad (13)$$

where $\upsilon_{i,m}$ is the stoichiometric coefficient of species i in the reaction of mineral m, and A_m [m^2 $m^{-3}_{\text{porous medium}}$] is the reactive surface area, which is related to specific surface areas [m^2/g_{mineral} or m^2/m^3_{mineral}] through porosity and mineral molar volume.

In the continuum-scale models, porosity is updated based on the amount of mineral precipitation and dissolution taking place within each simulation time step:

$$\phi^{t+\Delta t} = \phi^t - \sum_m A_m V_m r_m \Delta t \qquad (14)$$

where V_m [m³/mol] is the molar volume of the mineral, and superscripts t and $t + \Delta t$ represent time before and after a time interval Δt, respectively. Permeability is updated from porosity based on user-specified constitutive relationships.

In continuum models, surface areas are updated based on simplified conceptualizations of geometric evolution, and typically follow a power law relationship with mineral volume fraction and/or porosity (Noiriel et al. 2009). In contrast, in pore-scale models, the fluid–solid interface is tracked as the reaction progresses and the reactive surface area is updated based on the evolution of the actual geometry. In addition, because the hydrodynamic processes are resolved at the pore scale, any transport limitation is accounted for explicitly instead of being lumped into an effective reactive surface area, as is common practice in continuum-scale models (Noiriel and Daval 2017).

1D modeling of near-fracture rock matrix alteration

One-dimensional continuum models have been typically used to investigate the alteration of rock matrix immediately adjacent to a fracture. Because this type of model focuses on reaction front propagation into an essentially impermeable rock matrix, a continuum approach is adopted. The only transport mechanism is diffusion, and fluid chemistry in the fracture defines the Dirichlet (fixed) boundary condition. The governing equation is therefore simplified to

$$\frac{\partial \phi C_i}{\partial t} = \frac{\partial}{\partial z}\left(\phi D_{eff}\frac{\partial C_i}{\partial z}\right) + R_i \tag{15}$$

where D_{eff} is the effective diffusion coefficient in the porous media, and is related to porosity (ϕ) by some correlation, such as Archie's Law:

$$D_{eff} = D_m a\phi^c \tag{16}$$

where a and c, which is typically known as the cementation exponent, are empirical coefficients depending on the textures of the rock.

This model has been used to investigate the alteration of rock matrix bordering a hyper-alkaline fluid-filled fracture under conditions relevant to nuclear waste repositories (Steefel and Lichtner 1994). The study predicted calcite precipitation in the rock matrix immediately adjacent to the fracture; the precipitated calcite was shown to serve as an armoring layer preventing the rock matrix from neutralizing the hyper-alkaline fluid in the fracture and from sorbing and retarding the transport of radionuclides in the long run. This type of 1D model has also been shown to qualitatively capture the three altered layers created by portlandite dissolution, calcite precipitation and calcite re-dissolution, and amorphous silica precipitation (Luquot et al. 2013).

An alternative approach for tracking the reaction front into the rock matrix neighboring a fracture treats the reaction front migration as a moving boundary problem (Ulm et al. 2003; Walsh et al. 2014a). In this approach, the transport between fronts is still governed by diffusion (Eqn. 17), and the front movement is described by mass conservation of the controlling species at the front (Eqn. 18). The concentrations of the aqueous species at each reaction front are controlled by local chemical equilibrium with mineral phases.

$$\frac{\partial \phi C_i}{\partial t} = \frac{\partial}{\partial z}\left(\phi D_{eff}\frac{\partial C_i}{\partial z}\right) \tag{17}$$

$$\left[\!\left[\rho_i\left(1-\phi\right)\right]\!\right]\frac{dZ_f}{dt} = -M_i\left[\!\left[D_{eff}\frac{\partial C_i}{\partial z}\right]\!\right] \tag{18}$$

In these equations $[\![\ ]\!]$ calculates the difference across the front, ρ_i [kg m⁻³] is the density of species i in the solid, Z_f is the location of the front, and M_i [kg mol⁻¹] is the molar mass of species i.

If the fluid chemistry boundary condition at the fracture surface is fixed, the 1D model implicitly assumes that the local fracture fluid chemistry does not vary with the mineral reactions and is independent of flow and transport in the fracture. This assumption is valid if the fluid velocity in the fracture is sufficiently fast or if the region of interest is close to the inlet/recharge zone (Steefel and Lichtner 1994). Alternatively, the 1D model can be coupled with modules that simulate the flow and transport within the fracture separately, and update the boundary conditions accordingly (Walsh et al. 2013, 2014a).

2D modeling of cross-aperture processes

Continuum-scale modeling of fracture flow and rock matrix alteration. In order to capture the changes of fluid chemistry along the flow direction, as a result of upstream reactions in addition to rock matrix alteration in the near-fracture region, a 2D domain that is perpendicular to the fracture plane needs to be simulated. Equation (9) is applicable to the entire domain if the fracture is treated as a continuum. However, properties of the fracture grid cells need to be defined in different ways: the porosity is calculated from the local fracture aperture (Eqn. 19), and the permeability is defined and updated following the cubic-law formulation (Eqn. 20),

$$\phi = \frac{b}{\Delta z} \tag{19}$$

$$k = k_0 \left(\frac{\phi}{\phi_0} \right)^3 \tag{20}$$

where Δz denotes the size of the fracture grid cells, other variables as defined previously, and the subscript 0 is used for the initial value (at time $t = 0$).

This type of 2D model has been used to successfully reproduce the fluid chemistry evolution observed in a series of flow-through experiments reacting fractured marl cores with carbonated water, as well as the thickness of the altered layer on the fracture wall due to calcite dissolution (Davila et al. 2016b). This type of 2D model would be the equivalent of a 1D dual-/multi-continua model (Dobson et al. 2003), in which the computational domain was discretized along the fracture and the interface area was configured to account for the mass transfer between the fracture and rock matrix continua.

Pore-scale modeling of fracture flow. In the continuum approach discussed above, because fracture aperture is not resolved explicitly, the underlying assumption is that fluid within the fracture is well-mixed and the concentrations of all aqueous species are homogeneous. This assumption does not necessarily hold because the fluid velocity is not uniform, and the velocity gradient can influence solute transport and create concentration gradients between the fracture walls. For example, in parallel fractures, fluid velocity follows a parabolic profile and approaches zero at the fracture surface (Poiseuille flow). Mineral reaction products tend to accumulate within this hydrodynamic boundary layer, where solute transport is dominated by diffusion.

To resolve the velocity gradient and the resulting concentration gradient, a pore-scale approach is needed. For parallel fractures, the analytical solution of Equation (4), i.e., the parabolic velocity profile, can be directly substituted into Equation (5):

$$\frac{\partial C_i}{\partial t} = -u_0 \left(1 - \left(\frac{2z}{b} \right)^2 \right) \frac{\partial C_i}{\partial x} + \frac{\partial}{\partial x} \left(D_{mi} \frac{\partial C_i}{\partial x} \right) + \frac{\partial}{\partial z} \left(D_{mi} \frac{\partial C_i}{\partial z} \right) \tag{21}$$

where u_0 is the average velocity. The reaction is given by the boundary condition, i.e., the diffusive flux at the fracture surface (at $z = \pm b/2$) is equal to the mass flux due to reactions.

For simple first-order surface reactions and equilibrium reactions, analytical solutions can be derived for the average concentration across the fracture aperture (Berkowitz and Zhou 1996). These analyses provide insights regarding the breakthrough of reactive solutes in the fractures and the effective dispersion coefficients that can be used for upscaling.

Numerical solutions of this model can be used to simulate systems with non-linear reaction rates. For example, Li et al. (2008) performed numerical simulations for several minerals under different kinetic dissolution constraints in parallel fractures, and investigated the flow and reaction conditions that would lead to the breakdown of the well-mixed assumption, i.e., the continuum treatment (Fig. 7a). The authors found two necessary conditions for the development of concentration gradient across the fracture aperture: (i) comparable advection and reaction rates, and (ii) reaction rates that are faster than diffusion across the aperture. The model can also be applied in rough fractures. Instead of assuming a parabolic velocity profile, the full Navier–Stokes equation is solved to capture the hydrodynamics that arise from surface roughness (Fig. 7b). As such, the impacts of surface roughness on local reaction at the fracture wall and overall reaction in the fracture can be investigated (Deng et al. 2018b). Fracture roughness has been observed to increase following mineral dissolution (Gouze et al. 2003; Ellis et al. 2011), therefore considering the compounded impacts and feedback of surface roughness on mineral reactions is of great importance.

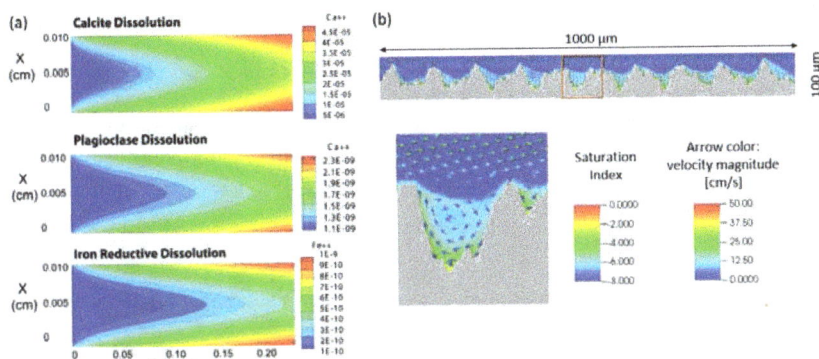

Figure 7. (a) Steady state concentration profile of the major species from calcite dissolution, plagioclase dissolution, and iron reductive dissolution in a fracture with an aperture of 100 μm and a length of 0.24 cm, at a flow velocity of 0.1 cm/s (Li et al. 2008). **(b)** Steady state flow field and saturation index for the case of calcite dissolution (as an indication of concentrations) in a rough fracture of 1000 μm and with an average aperture of 100 μm, at a flow velocity of 10 cm/s. [Reproduced with permission of Elsevier from Deng (2018) Pore-scale numerical investigation of the impacts of surface roughness: Upscaling of reaction rates in rough fractures. Geochim Cosmochim Acta 239:374–389. https://creativecommons.org/licenses/by-nc-nd/4.0/. https://doi.org/10.1016/j.gca.2018.08.005. 0016-7037/]

Hybrid and pore-scale modeling of fracture flow and rock matrix alteration. Multiscale models can be used to simultaneously capture pore-scale fluid flow in the fracture and diffusion-controlled reactive transport in the rock matrix (Molins et al. 2019). In this approach, the computational domain is divided into the fracture domain, in which the pore-scale governing equations are solved, and the porous media domain where the continuum-scale equations are used. At the boundary between the two domains, diffusive fluxes are balanced. The study of (Wen et al. 2016) represents a simplified implementation of the hybrid approach, in which fracture aperture is discretized and assigned velocity values according to a modified parabolic profile that corrects for roughness. Using this model, the authors investigated the evolution of the diffusivity of the rock matrix bordering a rough fracture as a result of preferential dissolution of calcite in the rock matrix, and its dependence on calcite abundance. More details of the multi-scale modeling approach can be found in Molins and Knaber (2019, this volume).

Pore-scale models can also be applied in both the fracture and rock matrix domains, which has the advantage of explicitly accounting for hydrodynamics and surface area change in the altered layer. This is, however, computationally expensive for finite-volume based reactive transport models. Alternative approaches, such as the Lattice Boltzmann method, have been explored (Chen et al. 2014; Fazeli et al. 2018). Chen et al. (2014) investigated fracture evolution in a binary mineral system using a pore-scale model, and showed the development of an altered layer, in which the flow velocity is negligible. This confirms that the dominant transport mechanism in the altered near-fracture region is diffusion, as has been assumed in the 1D continuum-scale modeling discussed above.

2D modeling of processes in the fracture plane

Flow variations in the fracture plane (i.e., along its surface) can arise from perturbations such as initial fracture aperture, and influence the diffusive transport of solutes in fractures (Nowamooz et al. 2013). Such flow instabilities and their positive feedback with geochemical reactions, as demonstrated by experimental observations, can result in self-organization phenomena, e.g., fracture channelization (Ortoleva et al. 1987; Deng et al. 2015). Models that discretize the two dimensions in the fracture plane have been used to explicitly capture these flow and transport processes (Hanna and Rajaram 1998; Detwiler and Rajaram 2007; Szymczak and Ladd 2012; Deng and Peters 2018). In this type of 2D model, the third dimension is considered by integrating over the fracture aperture (Eqn. 22):

$$\frac{\partial \left(b\overline{C}_i \right)}{\partial t} = -\nabla \cdot \left(\overline{q}\overline{C}_i \right) + \nabla \cdot \left(bD\nabla \overline{C}_i \right) + \overline{R}_i \qquad (22)$$

where $\overline{C}_i$ is the depth-averaged concentration, $\overline{R}_i$ is the effective reaction rate [mol m^{-2}s^{-1}], and $\overline{q}$ [m^{2}s^{-1}] is the local flux, solved using the Reynolds equation, i.e., depth-averaged Stokes equation:

$$\overline{q} = -\frac{b^3}{12\mu}\nabla P \qquad (23)$$

For single mineral systems, assuming the reactive surface area is the geometric fracture surface area (i.e., $2\Delta x\Delta y$), the aperture is updated using

$$b^{t+\Delta t} = b^t + 2\overline{R}_i V_m \Delta t / \upsilon_{i,m} \qquad (24)$$

It is noted that in reality the reactive surface area is affected by the topography of the fracture surfaces. Typically, it is corrected by multiplying the geometric fracture surface area by a surface roughness factor. The surface roughness factor can be determined independently if the surface geometry is mapped or BET surface area is measured. However, BET measurements may result in over-estimation because they capture, and are typically dominated by, pore space surface area that is not in direct contact with the fluid flow in the fracture. The surface roughness factor can also be inferred from the extent of reactions (Deng et al. 2016b). Overall, in the continuum approach, for which the geometry is not explicitly resolved, the determination of reactive surface area has been challenging and requires guidance based on knowledge of the system or pore-scale investigations as illustrated by Deng et al. (2018b).

Dividing Equation (22) by the thickness of the fracture domain considered, which is essentially the grid size in the z direction (Δz, [m]) in the 2D model and can be an arbitrary value larger than the local aperture, we get the mass balance equation of the continuum model (Eqn. 9) where the porosity follows the definition in Equation (19). The same pore velocity as calculated from Equation (23) can also be recovered from Darcy's law (Eqn. 8) by using a permeability that is the intrinsic fracture permeability (defined in Eqn. 2) weighted by the porosity (defined in Eqn. 19).

Modeling diffusion limitation due to the hydrodynamic boundary layer. Although the 2D fracture plane model does not explicitly resolve the non-uniform velocity profile across the fracture aperture, i.e., the hydrodynamic boundary layer close to the fracture wall, modifications can be made to account for the resulting impacts on mineral reactions. Because the major transport mechanism at the fracture surface is diffusion, for fast reacting minerals, the reaction rates are limited by solute diffusion to and from the fracture surface. The diffusive flux (i.e., diffusion controlled reaction rate R_{diff}, [mol m^{-2} s^{-1}]) can be expressed as

$$R_{\text{diff}} = \frac{D_{mi} Sh}{2b} \left(\bar{C}_i - C_{is} \right) \tag{25}$$

where Sh is the Sherwood number, which is a function of reaction rate and the velocity profile across the aperture and is bounded by two asymptotic limits 7.54 and 8.24 (Gupta and Balakotaiah 2001), and C_{is} is the concentration at the fracture surface. Assuming the concentration at the fracture surface is at equilibrium with respect to the mineral, R_{diff} can be evaluated independently from known variables. Hanna and Rajaram (1998) used the smaller of R_{diff} and the kinetically controlled reaction rate to account for the diffusion limitation caused by a strong hydrodynamic boundary layer.

Alternatively, an effective reaction rate coefficient ($k'_{\text{rxn,eff}}$ [m s^{-1}]) can be derived analytically for first order reactions with a stoichiometric coefficient of one, given that the diffusive flux has to be balanced by the mass flux from kinetic reaction (R_{rxn}):

$$R_{\text{rxn}} = k'_{\text{rxn}} \left(C_{is} - C_{ieq} \right) = k'_{\text{rxn,eff}} \left(\bar{C}_i - C_{ieq} \right) \tag{26}$$

$$R_{\text{rxn}} = R_{\text{diff}} \tag{27}$$

$$k'_{\text{rxn,eff}} = \frac{k'_{\text{rxn}}}{1 + 2kb / D_{mi} Sh} \tag{28}$$

where C_{ieq} is the equilibrium concentration, and k'_{rxn} is the kinetic rate coefficient for the first order reactions written in unit of [m s^{-1}], which can be estimated from k_{rxn} used in Equation (12). This formulation allows a smooth transition between the kinetically controlled and transport controlled reaction rate (Detwiler and Rajaram 2007; Szymczak and Ladd 2012).

2.5D modeling of the development and erosion of the altered layer. The so-called 2.5D model builds upon the 2D fracture plane model and accounts for the alteration of the near fracture region and the diffusion process through the altered layer (Deng et al. 2016b, 2017, 2018a). In the 2.5D model, each grid cell is a continuum where the fracture and rock matrix on the fracture wall co-exist with a total thickness of Δz. This model solves the same governing equations as the 2D fracture plane model, but with two important modifications.

The fracture aperture is defined for each mineral (m) and measures the distance between the reaction fronts of this mineral in the two fracture halves. It is calculated from $f_{i,j,m}$, the volume fraction of the mineral measured for the intact rock matrix and $V_{i,j,m}$, the volume fraction of the mineral in the grid (i, j):

$$b_{i,j,m} = \Delta z \left(1 - \frac{V_{i,j,m}}{f_{i,j,m}} \right) \tag{29}$$

Therefore, the reaction front for each mineral is tracked and the distance between the reaction front and the fracture surface (constrained by the mineral that retreats the slowest), d_{AL}, can be calculated.

The diffusion controlled reaction rate due to the presence of the altered layer (R_{ALdiff} [mol m^{-2} s^{-1}]) is calculated as:

$$R_{\text{ALdiff}} = -\frac{D_{\text{eff}}}{d_{\text{AL}}} \left(C_{i\text{eq}} - \overline{C}_i \right) \tag{30}$$

An effective reaction rate (R_{eff}) is then defined as the harmonic mean of the kinetically (R_{rxn}) and diffusion (R_{ALdiff}) controlled reaction rates, such that the lower one of the two dictates the reaction. This is similar to how the diffusion limitation caused by the hydrodynamic boundary layer is accounted for (Eqn. 28):

$$R_{\text{eff}} = \frac{1}{\dfrac{1}{R_{\text{ALdiff}}} + \dfrac{1}{R_{\text{rxn}}}} \tag{31}$$

The 2.5D model successfully reproduced channeling in the fracture plane, the development of the altered layer, and the resulting diffusion limitation and decreasing overall dissolution of calcite observed in a dolomite fracture reacting with carbonated water (Deng et al. 2016b). The conceptualization embodied in the 2.5D model can also be implemented using a 2D dual-contniua model, in which the mass transfer between the fracture continuum and the rock matrix continuum is explicitly defined by Equations (30) and (31).

By capturing the development of the altered layer, the 2.5D model also enables considering the erosion of the altered layer, which is proportional to the thickness of the altered layer (d_{AL}) as observed in previous experiments (Andreani et al. 2008). A phenomenological law (Eqn. 32) has been implemented in the 2.5D model to successfully simulate fracture opening due to detachment of the altered layer observed in a carbonate-rich shale (Deng et al. 2017):

$$E = \begin{cases} \eta \left(d_{\text{AL}} - d_{\text{ALc}} \right)^{\varepsilon} & d_{\text{AL}} > d_{\text{ALc}} \\ 0 & d_{\text{AL}} \leq d_{\text{ALc}} \end{cases} \tag{32}$$

where E [m s^{-1}] is the erosion rate, and d_{ALc} is the critical thickness, which has to be exceeded for the erosion to take place. Both η [m$^{1-\varepsilon}$ s^{-1}] and ε are empirical coefficients that are expected to vary with mineral composition of the rock matrix and can be well constrained by experimental observations.

3D pore-scale modeling of geochemical fracture alteration

Full 3D pore-scale modeling of the geochemical alteration of fractures involves numerically solving the pore-scale governing equations (Eqns. 3–5) in the actual fracture geometry. It captures the detailed fluid flow, diffusion and reaction dynamics, including flow instabilities in the fracture plane and the velocity variations across the fracture aperture. However, this modeling approach is computationally intensive because of the fine mesh resolution needed to resolve the fracture aperture and the relatively large domain as determined by the size of the fracture plane.

There are only a few studies performed in simple fracture geometries (Starchenko et al. 2016; Starchenko and Ladd 2018). Their model was implemented in OpenFOAM with customized libraries for boundary conditions and mesh updates. It was shown that the 3D model results deviate noticeably from the 2D fracture plane model results after channels are established, even with the implementation of the diffusion limited reaction rate in the 2D model. This was primarily attributed to the breakdown of Reynolds equation, which is used in the 2D model for fluid flow simulations. It has been shown that the use of the Reynolds equation may introduce large errors when the flow velocity is high, i.e., Reynolds number is large, or when the fracture has a large roughness (Brown et al. 1995; Zimmerman and Bodvarsson 1996; Oron and Berkowitz 1998).

There are even fewer 3D simulations on the evolution of fracture alteration with more complex, yet realistic geometries (e.g., Szymczak and Ladd 2009, using the Lattice Boltzmann method), let alone with mineralogical heterogeneity in the near fracture region.

CONTROLS ON DISSOLUTION-DRIVEN
REACTIVE TRANSPORT PROCESSES IN FRACTURES

Numerical models that capture the important reactive transport processes in fractures enable numerical experiments for systematic exploration of a broad parameter space. Flow rate, as it directly relates to residence time (Glassley et al. 2002), is the most widely studied controlling factor. The role of mineralogy, including mineral composition and spatial distribution, has received increasing attention. The following discussion focuses on dissolution-driven fracture alteration, based on the studies that are available.

The impact of flow and reaction rates

Discussions of the impacts of flow rate on fracture dissolution have been commonly framed using the dimensionless Peclet (Pe) and Damköhler (Da) numbers, which are defined for fractures as follows:

$$Pe = \bar{u}b / D_m \tag{33}$$

$$Da_I = k'_{rxn}L / \bar{u}b \tag{34a}$$

$$Da_{II} = Pe \cdot Da_I = k'_{rxn}b / D_m \tag{34b}$$

where $\bar{u}$ is the average velocity in fracture, L is the length of the fracture, and k'_{rxn} is the reaction rate coefficient in unit of $m\,s^{-1}$. Da_I captures the relative magnitude of reaction rate and advection rate, whereas Da_{II} measures the ratio between reaction rate and diffusive mass transfer rate.

These dimensionless numbers were originally derived from dimensional analysis of the governing equations (Berkowitz and Zhou 1996; Detwiler and Rajaram 2007), and involve simplifications of geometry and reaction kinetics. Therefore, it should be noted that the applicability of these dimensionless numbers is limited. As the impact of mineralogy will be discussed in the following subsection, we focus on a single mineral system for the discussion of flow rate, for which the traditional definitions of Pe and Da still provide a useful framework.

Consistent observations have been made on the impact of flow rate on fracture dissolution patterns (Detwiler and Rajaram 2007; Szymczak and Ladd 2009, 2011; Elkhoury et al. 2013; Deng et al. 2018a; Starchenko and Ladd 2018). Some of the data points from both numerical simulations and experimental observations are summarized in Figure 8 (Starchenko and Ladd 2018), where Da_{eff} is a variant of Da_I (Eqn. 34a). The choice of Da_I versus Da_{II} is somewhat arbitrary in constructing this type of figures (a similar alternative figure can be readily derived from Pe and Da_{II}), and the discussion of the dependence of dissolution patterns on either Da_I or Da_{II} can be convoluted when transport transitions from being advection-dominated to diffusion-dominated. At large Pe, when advection dominates, fracture dissolution patterns largely depend on Da_I, i.e., the relative magnitude of the reaction and advection rates. When the advection rate is large compared to the reaction rate (low Da_I), dissolution is uniform because the fast replenishment of reactive fluid homogenizes the concentration field and mineral reactions in the fracture. At large Da_I, the advection rate is slower than the reaction rate, and the reactants are consumed faster than they are transported. The resulting dissolution pattern is referred to as face dissolution (in flow-through experiments), as the dissolution is highly localized to the inlet, and does not result in much change in fracture permeability. At intermediate Da_I when the flow rate and the advection rate are comparable, the feedback between flow and reaction is most pronounced. Initial perturbations are magnified to create preferential channels in the fractures, leading to a dramatic fracture permeability increase. The morphology of the channels are also shown to be dependent on Da_I. The channels are more compact at slightly larger Da_I, and are more diffuse at smaller Da_I as fracture dissolution transitions into a uniform dissolution regime.

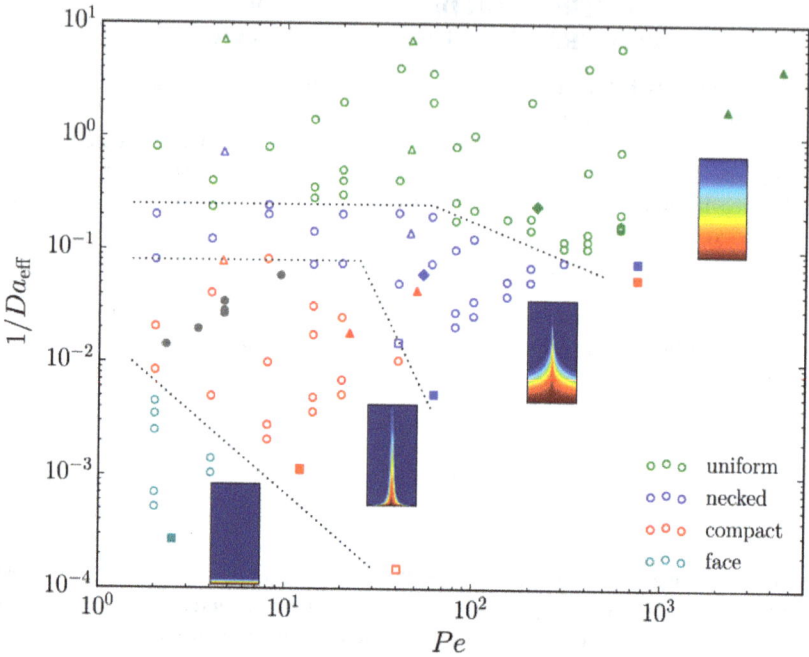

Figure 8. Diagram of fracture dissolution regimes in relation to Pe and $1/Da_{eff}$, $Da_{eff}=nk_{eff}L/q_0$, where n is the number of fracture surface (1 or 2), k_{eff} is the effective reaction rate coefficient as defined in Equation (28), L is the fracture length, and q_0 is average local flux (after Starchenko and Ladd 2018). The four color maps correspond to cases of face dissolution, compact wormhole, necked wormhole, and uniform dissolution, respectively, and illustrate the extent of aperture change, with the red (blue) color corresponding to significant (versus insignificant) aperture enlargement. Open symbols are data points from simulations. Circles: (Starchenko and Ladd 2018), open triangles: Elkhoury et al. (2013), squares: Starchenko et al. (2016). Solid symbols are data from experiments. Diamonds: Detwiler et al. (2003), triangles: Elkhoury et al. (2013), squares: Garcia-Rios et al. (2015), circles: Osselin et al. (2016). The solid grey circles are experiments that develop wormholes with evident instability in the dissolution front. Note: the experiments reported in Garcia-Rios et al. (2015) involved calcite dissolution and gypsum precipitation; however, dissolution was dominant and fracture permeability increased in all their cases, despite precipitation; therefore it was concluded that the dissolution patterns were not affected by precipitation in these experiments.

At small Pe and small Da_{II}, diffusion is the dominant transport process and tends to smooth out reaction instabilities and, therefore, results in a relatively uniform dissolution pattern. If the reaction rate is much faster than the diffusion rate (large Da_{II}), the reaction will be localized at the inlet, leading to face dissolution. Overall, this is consistent with the fact that fewer data points fall into the wormholing regime at low Pe on Figure 8.

The impact of mineralogy

Both mineral composition and the spatial distribution of various, especially reactive, minerals influence fracture alteration. In rocks where reactive minerals are well mixed with non-reactive or slow-reacting mineral phases, fracture alteration typically involves the development and erosion of the altered layer depending on the volume percentage of fast-reacting minerals. Deng et al. (2018a) performed numerical investigations of two multi-mineral scenarios. Based on these simulation results, conceptual models (Fig. 9) and multi-mineral Damköhler number (*mDa*) were developed for analyses of fracture alteration patterns in these systems. If fast-reacting minerals only account for a small volume fraction of the rock matrix, the altered layer in the near fracture region subsists or slowly dissolves. In this

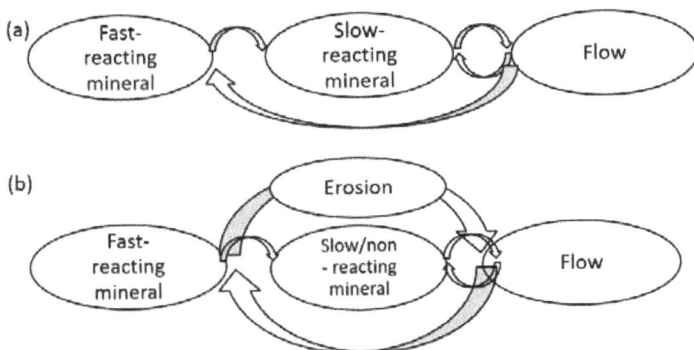

Figure 9. Conceptual models of the reaction-flow feedback loops for multi-mineral systems: **(a)** with development of an altered layer, and **(b)** with erosion of the altered layer. [Reproduced from Deng H et al. (2018) Fracture evolution in multimineral systems: the role of mineral composition, flow rate, and fracture aperture heterogeneity. ACS Earth Space Chem 2:112–124 with permission from American Chemical Society, https://pubs.acs.org/doi/abs/10.1021/acsearthspacechem.7b00130]

case, the fracture dissolution pattern and its dependence on the flow rate are controlled by the slow reacting minerals (Eqn. 35). In contrast, if the fast-reacting minerals account for a large volume fraction of the rock matrix, detachment of the porous altered layer is expected. The fracture geometry change is then controlled by the thickness of the altered layer, i.e., the fast reacting minerals (Eqn. 36).

$$mDa = \frac{V_{\text{frac}}\, 2k_{\text{rxn,slow}} V_{m\text{slow}}}{Q \bar{b} f_{r,\text{slow}}} \tag{35}$$

$$mDa = \frac{\left(\dfrac{V_{\text{frac}}}{Q}\right)}{\bar{b} f_{r,\text{fast}}\left(\dfrac{1}{2k_{\text{rxn, fast}} V_{m\text{fast}}} + \dfrac{d_{\text{ALc}}}{D_{\text{eff}} C_{i,\text{eq}} V_{m\text{fast}}}\right) + \dfrac{f_{r,\text{fast}} d_{\text{ALc}}}{2k_{\text{rxn,fast}} V_{m\text{fast}}}} \tag{36}$$

where V_{frac} is the volume of the fracture, Q is the volumetric flux, $\bar{b}$ is the average aperture, and f_r is the volume fraction of a given mineral.

The multi-mineral Da highlights the observation that the flow regimes corresponding to different dissolution patterns shift depending on the mineral composition.

If instead, the reactive minerals follow certain spatial patterns, the fracture alteration patterns are largely controlled by the spatial distribution of the reactive mineral phase (Deng and Peters 2018; Noiriel and Deng 2018). When the reactive minerals are discontinuous along the flow direction, fracture aperture enlargement is localized to where reactive minerals are present, but the change of hydraulic properties of the fracture is constrained by the non-reactive minerals. When the reactive minerals are continuous along the flow direction, reaction fronts tend to penetrate deeper into the fracture compared to the case when no mineral heterogeneity is present. The dissolution of the reactive minerals can thus result in considerable increase in fracture hydraulic properties (Fig. 10). Faster hydraulic opening is expected if the spatial distribution of the reactive minerals overlap with the preferential flow paths prescribed by the initial fracture geometry.

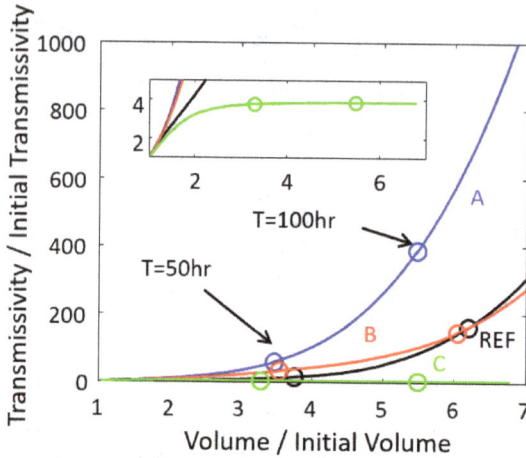

Figure 10. Transmissivity increase in relation to fracture volume increase for different mineral spatial patterns. The calcite content on the fracture surface is comparable in **A** and **B**, the spatial patterns are different but both continuous. The calcite content is lower and discontinuous in **C**. (Deng and Peters 2018, CC-BY).

CONCLUDING REMARKS

Overall, advancements in experimental techniques have provided valuable insights into important reactive transport processes in fractures. The experimental observations have also provided indispensable data sets for the development and testing of conceptual and numerical models, which have enabled systematic investigation of different controlling factors, and the development of more mechanistic approaches to improve our predictive capability. Enhanced understanding of the role of mineralogy on fracture alteration resulting from recent studies is of particular interest, as it is an important and necessary step in bridging fundamental understanding and studies of natural and engineered systems, where multiple minerals are typically present.

In this chapter, we summarized major reactive transport processes observed in single fractures. The morphology and hydraulic properties of fractures evolve as a result of the interplay between fluid flow in the fracture, diffusion in the fracture plane and across the fracture aperture, mineral dissolution/precipitation, and diffusion through altered zones adjacent to the fracture. Fracture permeability typically increases when the alteration is driven by mineral dissolution. The extent of the increase depends on the changes in fracture morphology, which is largely controlled by the relative magnitude of the time scale of transport and the time scale of reactions. While the increase of fracture permeability is limited when the dissolution is localized at the inlet, i.e., large Da, fracture permeability can increase sharply when the reaction rate and the transport rate are comparable and initial perturbations are amplified to produce self-organization features such as channels in the fractures. The altered layers bordering the fracture surface, which is created by the preferential dissolution of fast-reacting minerals in multi-mineral systems, can mitigate fracture permeability increase by increasing fracture surface roughness. The altered layers can also passivate subsequent mineral dissolution due to diffusion limitation. However, depending on the mechanical strength of the altered layers, these may detach from the fracture surfaces and result in fracture enlargement, while releasing fine particles that may eventually clog the fractures. Fracture permeability also decreases when precipitation dominates in the fractures.

In addition, we highlighted existing process-based models for the prediction of the alteration of an individual fracture driven by geochemical reactions in fully saturated rocks. This was not meant to be an exhaustive review of the literature but rather to provide some context for model selection. Model complexity, and, to a large extent, model accuracy, is inversely related to the computational requirement, and it is advised that model selection be based on the system and research question being investigated. Understanding the major reactive transport processes that each type of model addresses provides a basis for such consideration. Both the continuum- and pore-scale modeling approaches have been applied to simulate fracture alteration. 1D models along the flow direction or perpendicular to the fracture plane provide useful tools for investigating solute transport along a single direction through advection and diffusion, and their controls on mineral reactions. 2D models perpendicular to the fracture plane allow detailed examination of processes across the fracture aperture or into the bordering rock matrix, while considering flow and transport along the flow direction. When aperture and flow variations within the fracture plane cannot be ignored, 2D models within the fracture planes are necessary to capture the fracture morphological change and fracture permeability evolution from mineral reactions. The 2.5D models enables consideration of processes across the fracture aperture, such as diffusion limitation of the altered layers, while keeping the computational efficiency of a 2D model. This is especially useful in mineralogically complicated systems. 3D pore-scale models simulate explicitly all the physical and chemical processes that affect fracture alteration and thus are expected to produce more accurate results regarding fracture permeability change; but their computational costs are typically high. Therefore, their application in geochemically complex, yet realistic, systems where interactions between the fracture and the bordering rock matrix are important is unpractical. Hybrid models provide a promising alternative that balances the computational costs and the details of the processes involved.

By outlining the current landscape of this field, we also hope to help identify research gaps that remain. A few examples are discussed below.

The controlling factors of precipitation-driven fracture alteration are only partly understood. Precipitation is complicated by nature. It involves nucleation and crystal growth, both of which are highly sensitive to local heterogeneity and are dependent on a range of factors (Noiriel et al. 2016). The shape, size and spatial distribution of the neo-formed crystals are dependent on the material substrate (Noiriel et al. 2016), and the type of precipitate is affected by flow rate and local saturation state. For instance, aragonite was observed to precipitate in a fracture at faster flow rates, whereas calcite precipitated at slower flow rates (Singurindy and Berkowitz 2004). Therefore, while models described in this chapter may be applied directly to systems where mineral precipitation dominates, caution needs to be exercised. For example, it is expected that in some cases, the concentration gradient across the fracture aperture results in super-saturation at the fracture surface whereas the bulk fluid chemistry may indicate under-saturation.

Mineral reactions in fractures are typically accompanied by a redistribution of local stress and a change of the rock mechanical properties (Elkhoury et al. 2013, 2015). Therefore, a modeling capability that resolves coupled mechanical and geochemical processes in fractures is needed. Some groundbreaking efforts have been taken in this direction, and highlighted that geomechical processes can mitigate fracture opening caused by dissolution reactions (Yasuhara et al. 2004; Ameli et al. 2014; Walsh et al. 2014b; Kim et al. 2015; Spokas et al. 2018). A mechanistic understanding of mechanical weakening of near fracture regions (such as the detachment of the altered layer) and modeling of microscopic processes are still lacking.

Multiphase flow dynamics in factures are highly dependent on fracture geometry (Yang et al. 2016). This geometry dependence on the fracture surface mineralogy and the resulting contact between the reactive fluid phase and minerals are also important factors that affect fracture evolution in multiphase systems and need to be further investigated.

Most importantly, findings derived from experimental and modeling studies at an individual fracture scale need to be integrated into reservoir-scale studies. Implementation of conceptualizations developed from studies discussed in this chapter into the dual-/multi-continua models discussed earlier provides a potential upscaling framework for reservoir scale applications.

ACKNOWLEDGMENTS

The authors would like to thank the support by Laboratory Directed Research and Development (LDRD) funding from Berkeley Lab, provided by the Director, Office of Science, of the U.S. Department of Energy under Contract No. DE-AC02-05CH11231. We also would like to thank reviewers and the editors for their constructive comments.

REFERENCES

Abdoulghafour H, Luquot L, Gouze P (2013) Characterization of the mechanisms controlling the permeability changes of fractured cements flowed through by co_2-rich brine. Environ Sci Technol 47:10332–10338

Abdoulghafour H, Gouze P, Luquot L, Leprovost R (2016) Characterization and modeling of the alteration of fractured class-G Portland cement during flow of CO_2-rich brine. Int J Greenhouse Gas Control 48:155–170

Ajo-Franklin JV, Marco; Molins, Sergi; Yang, Li (2017) Coupled processes in a fractured reactive system: a dolomite dissolution study with relevance to GCS caprock integrity. *In:* Coupled Processes in a Fractured Reactive System: A Dolomite Dissolution Study with Relevance to GCS Caprock Integrity. Wiley Publishing

Ameli P, Elkhoury JE, Detwiler RL (2013) High-resolution fracture aperture mapping using optical profilometry. Water Resour Res 49:7126–7132

Ameli P, Elkhoury J, Morris J, Detwiler R (2014) Fracture Permeability Alteration due to Chemical and Mechanical Processes: A Coupled High-Resolution Model. Rock Mech Rock Eng 47:1563–1573

Andreani M, Gouze P, Luquot L, Jouanna P (2008) Changes in seal capacity of fractured claystone caprocks induced by dissolved and gaseous CO_2 seepage. Geophys Res Lett 35:L14404

Atkinson BK (1984) Subcritical crack-growth in geological-materials. J Geophys Res 89:4077–4114

Barnes HL (1997) Geochemistry of Hydrothermal Ore Deposits. Wiley, New York

Berkowitz B (2002) Characterizing flow and transport in fractured geological media: A review. Adv Water Resour 25:861–884

Berkowitz B, Zhou JY (1996) Reactive solute transport in a single fracture. Water Resour Res 32:901–913

Birkholzer J, Houseworth J, Tsang CF (2012) Geologic disposal of high-level radioactive waste: status, key issues, and trends. Ann Rev Environ Resour 37:79–106

Brantley SL, Lebedeva MI, Balashov VN, Singha K, Sullivan PL, Stinchcomb G (2017) Toward a conceptual model relating chemical reaction fronts to water flow paths in hills. Geomorphology 277:100–117

Brown SR, Stockman HW, Reeves SJ (1995) Applicability of the reynolds-equation for modeling fluid-flow between rough surfaces. Geophys Res Lett 22:2537–2540

Browne PRL (1978) Hydrothermal alteration in active geothermal fields. Ann Rev Earth Planet Sci 6:229–250

Cama J, Soler JM, Ayora C (2019) Acid water–rock–cement interaction and multicomponent reactive transport modeling. Rev Mineral Geochem 85:459–498

Chen L, Kang Q, Viswanathan HS, Tao W-Q (2014) Pore-scale study of dissolution-induced changes in hydrologic properties of rocks with binary minerals. Water Resour Res 50:9343–9365

Chou L, Garrels RM, Wollast R (1989) Comparative-study of the kinetics and mechanisms of dissolution of carbonate minerals. Chem Geol 78:269–282

Davila G, Luquot L, Soler JM, Cama J (2016a) Interaction between a fractured marl caprock and CO_2-rich sulfate solution under supercritical CO2 conditions. Int J Greenhouse Gas Control 48:105–119

Davila G, Luquot L, Soler JM, Cama J (2016b) 2D reactive transport modeling of the interaction between a marl and a CO_2-rich sulfate solution under supercritical CO_2 conditions. Int J Greenhouse Gas Control 54:145–159

Deng H, Peters CA (2019) Reactive transport simulation of fracture channelization and transmissivity evolution. Environ Eng Sci 36:90–101

Deng H, Ellis BR, Peters CA, Fitts JP, Crandall D, Bromhal GS (2013) Modifications of carbonate fracture hydrodynamic properties by CO_2-acidified brine flow. Energy & Fuels 27:4221–4231

Deng H, Fitts JP, Crandall D, McIntyre D, Peters CA (2015) Alterations of fractures in carbonate rocks by CO_2-acidified brines. Environ Sci Technol 49:10226–10234

Deng H, Fitts JP, Peters CA (2016a) Quantifying fracture geometry with X-ray tomography: Technique of Iterative Local Thresholding (TILT) for 3D image segmentation. Comput Geosci 20:231–244

Deng H, Molins S, Steefel C, DePaolo D, Voltolini M, Yang L, Ajo-Franklin J, Yang L (2016b) A 2.5D reactive transport model for fracture alteration simulation. Environ Sci Technol 50:7564–7571

Deng H, Voltolini M, Molins S, Steefel C, DePaolo D, Ajo-Franklin J, Yang L (2017) Alteration and erosion of rock matrix bordering a carbonate-rich shale fracture. Environ Sci Technol 51:8861–8868

Deng H, Steefel C, Molins S, DePaolo D (2018a) Fracture evolution in multimineral systems: the role of mineral composition, flow rate, and fracture aperture heterogeneity. ACS Earth Space Chem 2:112–124

Deng H, Molins S, Trebotich D, Steefel C, DePaolo D (2018b) Pore-scale numerical investigation of the impacts of surface roughness: Upscaling of reaction rates in rough fractures. Geochim Cosmochim Acta 239:374–389

Detwiler R, Rajaram H (2007) Predicting dissolution patterns in variable aperture fractures: Evaluation of an enhanced depth-averaged computational model. Water Resour Res 43:W04403

Detwiler R, Glass R, Bourcier W (2003) Experimental observations of fracture dissolution: The role of Peclet number on evolving aperture variability. Geophys Res Lett 30:1648–1651

Dijk P, Berkowitz B, Bendel P (1999) Investigation of flow in water-saturated rock fractures using nuclear magnetic resonance imaging (NMRI). Water Resour Res 35:347–360

Dijk PE, Berkowitz B, Yechieli Y (2002) Measurement and analysis of dissolution patterns in rock fractures. Water Resour Res 38:51–512

Dobson P, Kneafsey T, Sonnenthal E, Spycher N, Apps J (2003) Experimental and numerical simulation of dissolution and precipitation: implications for fracture sealing at Yucca Mountain, Nevada. J Contam Hydrol 62–63:459–476

Doughty C (1999) Investigation of conceptual and numerical approaches for evaluating moisture, gas, chemical, and heat transport in fractured unsaturated rock. J Contami Hydrol 38:69–106

Elkhoury JE, Ameli P, Detwiler RL (2013) Dissolution and deformation in fractured carbonates caused by flow of CO_2-rich brine under reservoir conditions. Int J Greenhouse Gas Control; The IEAGHG Weyburn-Midale CO_2 Monitoring and Storage Project 16, Suppl 1:S203–S215

Elkhoury JE, Detwiler RL, Ameli P (2015) Can a fractured caprock self-heal? Earth Planet Sci Lett 417:99–106

Ellis BR, Peters CA, Fitts J, Bromhal G, McIntyre D, Warzinski R, Rosenbaum E (2011) Deterioration of a fractured carbonate caprock exposed to CO_2-acidified brine flow. Greenhouse Gases Sci Technol 1:248–260

Ellis BR, Fitts JP, Bromhal GS, McIntyre DL, Tappero R, Peters CA (2013) Dissolution-driven permeability reduction of a fractured carbonate caprock. Environ Eng Sci 30:187–193

Ellis BR, Peters CA (2016) 3D Mapping of calcite and a demonstration of its relevance to permeability evolution in reactive fractures. Adv Water Resour 95:246–253

Emmanuel S, Berkowitz B (2005) Mixing-induced precipitation and porosity evolution in porous media. Adv Water Resour 28:337–344

Emmanuel S, Levenson Y (2014) Limestone weathering rates accelerated by micron-scale grain detachment. Geology 42:751–754

Engelder T (1987) 2 - Joints and shear fractures in rock. *In*: Fracture Mechanics of Rock. Atkinson BK, (ed) Academic Press, London, p 27–69

Fazeli H, Patel R, Hellevang H (2018) Effect of pore-scale mineral spatial heterogeneity on chemically induced alterations of fractured rock: a lattice Boltzmann study. Geofluids 28:6046182

Fitts JP, Peters CA (2013) Caprock fracture dissolution and CO_2 leakage. Rev Mineral Geochem 77:459–479

Garcia-Rios M, Luquot L, Soler JM, Cama J (2015) Influence of the flow rate on dissolution and precipitation features during percolation of CO_2-rich sulfate solutions through fractured limestone samples. Chem Geol 414:95–108

Garcia-Rios M, Luquot L, Soler JM, Cama J (2017) The role of mineral heterogeneity on the hydrogeochemical response of two fractured reservoir rocks in contact with dissolved CO_2. Appl Geochem 84:202–217

Gherardi F, Xu TF, Pruess K (2007) Numerical modeling of self-limiting and self-enhancing caprock alteration induced by CO_2 storage in a depleted gas reservoir. Chem Geol 244:103–129

Giggenbach WF (1984) Mass-transfer in hydrothermal alteration systems—A conceptual-approach. Geochim Cosmochim Acta 48:2693–2711

Glassley WE, Simmons AM, Kercher JR (2002) Mineralogical heterogeneity in fractured, porous media and its representation in reactive transport models. Appl Geochem 17:699–708

Gouze P, Noiriel C, Bruderer C, Loggia D, Leprovost R (2003) X-ray tomography characterization of fracture surfaces during dissolution. Geophys Res Lett 30:1267

Griffiths L, Heap MJ, Wang F, Daval D, Gilg HA, Baud P, Schmittbuhl J, Genter A (2016) Geothermal implications for fracture-filling hydrothermal precipitation. Geothermics 64:235–245

Gupta N, Balakotaiah V (2001) Heat and mass transfer coefficients in catalytic monoliths. Chem Eng Sci 56:4771–4786

Hanna RB, Rajaram H (1998) Influence of aperture variability on dissolutional growth of fissures in Karst Formations. Water Resour Res 34:2843–2853

Hemley JJ, Jones WR (1964) Chemical aspects of hydrothermal alteration with emphasis on hydrogen metasomatism. Econ Geol 59:538–569

Hyman JD, Karra S, Makedonska N, Gable CW, Painter SL, Viswanathan HS (2015) DFNWORKS: A discrete fracture network framework for modeling subsurface flow and transport. Computers Geosci 84:10–19

Jones TA, Detwiler RL (2016) Fracture sealing by mineral precipitation: The role of small-scale mineral heterogeneity. Geophys Res Lett 43:7564–7571

Kalfayan L (2008) Production Enhancement with Acid Stimulation. Pennwell Books

Ketcham RA, Slottke DT, Sharp JM (2010) Three-dimensional measurement of fractures in heterogeneous materials using high-resolution X-ray computed tomography. Geosphere 6:499–514

Kim J, Sonnenthal E, Rutqvist J (2015) A sequential implicit algorithm of chemo-thermo-poro-mechanics for fractured geothermal reservoirs. Computers Geosci 76:59–71

Kovscek AR (2002) Screening criteria for CO_2 storage in oil reservoirs. Pet Sci Technol 20:841–866

Lachassagne P, Wyns R, Dewandel B (2011) The fracture permeability of Hard Rock Aquifers is due neither to tectonics, nor to unloading, but to weathering processes. Terra Nova 23:145–161

Lampe DJ, Stolz JF (2015) Current perspectives on unconventional shale gas extraction in the Appalachian Basin. J Environ Sci Health Part A-Toxic/Hazard Subst Environ Eng 50:434–446

Lasaga AC (1984) Chemical-kinetics of water–rock interactions. J Geophys Res 89:4009–4025

Lei QH, Latham JP, Tsang CF (2017) The use of discrete fracture networks for modelling coupled geomechanical and hydrological behaviour of fractured rocks. Computers and Geotechnics 85:151–176

Lewicki JL, Birkholzer J, Tsang CF (2007) Natural and industrial analogues for leakage of CO_2 from storage reservoirs: identification of features, events, and processes and lessons learned. Environ Geol 52:457–467

Li L, Steefel CI, Yang L (2008) Scale dependence of mineral dissolution rates within single pores and fractures. Geochim Cosmochim Acta 72:360–377

Lichtner PC (2000) Critique of Dual Continuum Formulations of Multicomponent Reactive Transport in Fractured Porous Media. Dynamics of Fluids in Fractured Rock. Los Alamos National Laboratory LAUR-00-1097

Lichtner PC (2016) Kinetic rate laws invariant to scaling the mineral formula unit. Am J Sci 316:437–469

Liu PY, Yao J, Couples GD, Huang ZQ, Sun H, Ma JS (2017) Numerical modelling and analysis of reactive flow and wormhole formation in fractured carbonate rocks. Chem Eng Sci 172:143–157

Loggia D, Gouze P, Greswell R, Parker DJ (2004) Investigation of the geometrical dispersion regime in a single fracture using positron emission projection imaging. Transp Porous Media 55:1–20

Long JCS, Remer JS, Wilson CR, Witherspoon PA (1982) Porous-media equivalents for networks of discontinuous fractures. Water Resour Res 18:645–658

Luquot L, Abdoulghafour H, Gouze P (2013) Hydro-dynamically controlled alteration of fractured Portland cements flowed by CO2-rich brine. Int J Greenhouse Gas Control 16:167–179

Makedonska N, Hyman JD, Karra S, Painter SL, Gable CW, Viswanathan HS (2016) Evaluating the effect of internal aperture variability on transport in kilometer scale discrete fracture networks. Adv Water Resour 94:486–497

McClure MW, Horne RN (2014) An investigation of stimulation mechanisms in enhanced geothermal systems. Int J Rock Mech Min Sci 72:242–260

Menefee AH, Li PY, Giammar DE, Ellis BR (2017) Roles of transport limitations and mineral heterogeneity in carbonation of fractured basalts. Environ Sci Technol 51:9352–9362

Meyer C, Hemley JJ (1967) Wall rock alteration. In: Geochemistry of Hydrothermal Ore Deposits, pp 166–235

Molins S (2015) Reactive interfaces in direct numerical simulation of pore-scale processes. Pore-Scale Geochem Process 80:461–481

Molins S, Trebotich D, Steefel CI, Shen CP (2012) An investigation of the effect of pore scale flow on average geochemical reaction rates using direct numerical simulation. Water Resour Res 48:11

Molins S, Trebotich D, Yang L, Ajo-Franklin JB, Ligocki TJ, Shen C, Steefel CI (2014) Pore-scale controls on calcite dissolution rates from flow-through laboratory and numerical experiments. Environ Sci Technol 48:7453–7460

Molins S, Knabner P (2019) Multiscale approaches in reactive transport modeling. Rev Mineral Geochem 85:27–48

Molins S, Trebotich D, Arora B, Steefel CI, Deng H (2019) Multi-scale model of reactive transport in fractured media: diffusion limitations on rates. Transp Porous Media 128:701–721

Molnar P, Anderson RS, Anderson SP (2007) Tectonics, fracturing of rock, and erosion. J Geophys Res-Earth Surface 112

National Research C (1996) Rock Fractures and Fluid Flow: Contemporary Understanding and Applications. The National Academies Press, Washington, DC

Noiriel C (2015) Resolving time-dependent evolution of pore-scale structure, permeability and reactivity using X-ray microtomography. Pore-Scale Geochem Process 80:247–285

Noiriel C, Daval D (2017) Pore-scale geochemical reactivity associated with CO_2 storage: new frontiers at the fluid–solid interface. Acc Chem Res 50:759–768

Noiriel C, Deng H (2018) Evolution of planar fractures in limestone: The role of flow rate, mineral heterogeneity and local transport processes. Chem Geol 497:100–114

Noiriel C, Made B, Gouze P (2007) Impact of coating development on the hydraulic and transport properties in argillaceous limestone fracture. Water Resour Res 43:W09406

Noiriel C, Luquot L, Made B, Raimbault L, Gouze P, van der Lee J (2009) Changes in reactive surface area during limestone dissolution: An experimental and modelling study. Chem Geol 265:160–170

Noiriel C, Gouze P, Made B (2013) 3D analysis of geometry and flow changes in a limestone fracture during dissolution. J Hydrol 486:211–223

Noiriel C, Steefel CI, Yang L, Bernard D (2016) Effects of pore-scale precipitation on permeability and flow. Adv Water Resour 95:125–137

Nowamooz A, Radilla G, Fourar M, Berkowitz B (2013) Non-Fickian transport in transparent replicas of rough-walled rock fractures. Transp Porous Media 98:651–682

Oron AP, Berkowitz B (1998) Flow in rock fractures: The local cubic law assumption reexamined. Water Resour Res 34:2811–2825

Ortoleva P, Merino E, Moore C, Chadam J (1987) Geochemical self-organization.1. reaction-transport feedbacks and modeling approach. Am J Sci 287:979–1007

Osselin F, Kondratiuk P, Budek A, Cybulski O, Garstecki P, Szymczak P (2016) Microfluidic observation of the onset of reactive-infiltration instability in an analog fracture. Geophys Res Lett 43:6907–6915

Parry WT (1998) Fault-fluid compositions from fluid-inclusion observations and solubilities of fracture-sealing minerals. Tectonophysics 290:1–26

Pepe G, Dweik J, Jouanna P, Gouze P, Andreani M, Luquot L (2010) Atomic modelling of crystal/complex fluid/crystal contacts-Part II. Simulating AFM tests via the GenMol code for investigating the impact of CO_2 storage on kaolinite/brine/kaolinite adhesion. J Cryst Growth 312:3308–3315

Pruess K (1991) TOUGH2: A general-purpose numerical simulator for multiphase fluid and heat flow. Lawrence Berkeley Lab. LBL-29400 UC-251 CA,USA

Pyrak-Nolte LJ, DePaolo DJ, Pietraß T (2015) Controlling Subsurface Fractures and Fluid Flow: A Basic Research Agenda. ; USDOE Office of Science, United States

Rochelle CA, Czernichowski-Lauriol I, Milodowski AE (2004) The impact of chemical reactions on CO_2 storage in geological formations: a brief review. Geol Soc, London, Spec Publ 233:87

Royne A, Jamtveit B (2015) Pore-scale controls on reaction-driven fracturing. Pore-Scale Geochem Process 80:25–44

Sin I, Corvisier J (2019) Multiphase multicomponent reactive transport and flow modeling. Rev Mineral Geochem 85:143–195

Singurindy O, Berkowitz B (2004) Dedolomitization and flow in fractures. Geophys Res Lett 31:4

Singurindy O, Berkowitz B (2005) The role of fractures on coupled dissolution and precipitation patterns in carbonate rocks. Adv Water Resour 28:507–521

Snow DT (1969) Anisotropic permeability of fractured media. Water Resour Res 5:1273–1289

Snow DT (1970) The frequency and apertures of fractures in rock. Int J Rock Mech Min Sci Geomech Abstr 7:23–40

Sonnenthal E, Ito A, Spycher N, Yui M, Apps J, Sugita Y, Conrad M, Kawakami S (2005) Approaches to modeling coupled thermal, hydrological, and chemical processes in the Drift Scale Heater Test at Yucca Mountain. Int J Rock Mech Min Sci 42:698–719

Spokas K, Peters CA, Pyrak-Nolte L (2018) Influence of rock mineralogy on reactive fracture evolution in carbonate-rich caprocks. Environ Sci Technol 52:10144–10152

St Clair J, Moon S, Holbrook WS, Perron JT, Riebe CS, Martel SJ, Carr B, Harman C, Singha K, Richter DD (2015) Geophysical imaging reveals topographic stress control of bedrock weathering. Science 350:534–538

Starchenko V, Ladd AJC (2018) The development of wormholes in laboratory-scale fractures: perspectives from three-dimensional simulations. Water Resour Res 54:7946–7959

Starchenko V, Marra CJ, Ladd AJC (2016) Three-dimensional simulations of fracture dissolution. J Geophys Res-Solid Earth 121:6421–6444

Steefel CI, Lichtner PC (1994) Diffusion and reaction in rock matrix bordering a hyperalkaline fluid-filled fracture. Geochim Cosmochim Acta 58:3595–3612

Steefel CI, Lasaga AC (1994) A coupled model for transport of multiple chemical-species and kinetic precipitation dissolution reactions with application to reactive flow in single-phase hydrothermal systems. Am J Sci 294:529–592

Steefel CI, Lichtner PC (1998) Multicomponent reactive transport in discrete fractures: I. Controls on reaction front geometry. J Hydrol 209:186–199

Steefel CI, Appelo CA, Arora B, Jacques D, Kalbacher T, Kolditz O, Lagneau V, Lichtner PC, Mayer KU, Meeussen JC, Molins S (2015) Reactive transport codes for subsurface environmental simulation. Comput Geosci 19:445–478

Szymczak P, Ladd AJC (2009) Wormhole formation in dissolving fractures. J Geophys Res-Solid Earth 114:B06203

Szymczak P, Ladd AJC (2011) The initial stages of cave formation: Beyond the one-dimensional paradigm. Earth Planet Sci Lett 301:424–432

Szymczak P, Ladd AJC (2012) Reactive-infiltration instabilities in rocks. Fracture dissolution. J Fluid Mech 702:239–264

Tenchine S, Gouze P (2005) Density contrast effects on tracer dispersion in variable aperture fractures. Adv Water Resour 28:273–289

Ulm F-J, Lemarchand E, Heukamp FH (2003) Elements of chemomechanics of calcium leaching of cement-based materials at different scales. Eng Fracture Mech 70:871–889

Walsh SDC, Du Frane WL, Mason HE, Carroll SA (2013) Permeability of Wellbore-Cement Fractures Following Degradation by Carbonated Brine. Rock Mech Rock Eng 46:455–464

Walsh SDC, Mason HE, Du Frane WL, Carroll SA (2014a) Experimental calibration of a numerical model describing the alteration of cement/caprock interfaces by carbonated brine. Int J Greenhouse Gas Control 22:176–188

Walsh SDC, Mason HE, Du Frane WL, Carroll SA (2014b) Mechanical and hydraulic coupling in cement-caprock interfaces exposed to carbonated brine. Int J Greenhouse Gas Control 25:109–120

Walton G, Lato M, Anschutz H, Perras MA, Diederichs MS (2015) Non-invasive detection of fractures, fracture zones, and rock damage in a hard rock excavation —Experience from the Aspo Hard Rock Laboratory in Sweden. Eng Geol 196:210–221

Warren JE, Root PJ (1963) The behavior of naturally fractured reservoirs. Soc Pet Eng J 3:245–255

Wen H, Li L, Crandall D, Hakala A (2016) Where lower calcite abundance creates more alteration: enhanced rock matrix diffusivity by preferential dissolution. Energy & Fuels 30:4197–4208

Witherspoon PA, Wang JSY, Iwai K, Gale JE (1980) Validity of cubic law for fluid-flow in a deformable rock fracture. Water Resour Res 16:1016–1024

Wu YS, Liu HH, Bodvarsson GS (2004) A triple-continuum approach for modeling flow and transport processes in fractured rock. J Contam Hydrol 73:145–179

Xu TF, Pruess K (2001) On fluid flow and mineral alteration in fractured caprock of magmatic hydrothermal systems. J Geophys Res-Solid Earth 106:2121–2138

Xu TF, Sonnenthal E, Spycher N, Pruess K (2006) TOUGHREACT—A simulation program for non-isothermal multiphase reactive geochemical transport in variably saturated geologic media: Applications to geothermal injectivity and CO_2 geological sequestration. Computers Geosci 32:145–165

Yang Z, Neuweiler I, Méheust Y, Fagerlund F, Niemi A (2016) Fluid trapping during capillary displacement in fractures. Adv Water Resour 95:264–275

Yasuhara H, Elsworth D, Polak A (2004) Evolution of permeability in a natural fracture: Significant role of pressure solution. J Geophys Res: Solid Earth 109:B03204

Zimmerman RW, Bodvarsson GS (1996) Hydraulic conductivity of rock fractures. Transp Porous Media 23:1–30

Reviews in Mineralogy & Geochemistry
Vol. 85 pp. 75-109, 2019
Copyright © Mineralogical Society of America

4

Reactive Transport Modeling of Coupled Processes in Nanoporous Media

Christophe Tournassat

Lawrence Berkeley National Laboratory
1 Cyclotron Road
Berkeley, CA 94720
USA

and

BRGM
3 avenue Claude Guillemin
45060 Orléans
France

and

Université d'Orléans – CNRS/INSU – BRGM
UMR 7327 Institut des Sciences de la Terre d'Orléans
45071 Orléans
France

c.tournassat@brgm.fr; CTournassat@lbl.gov

Carl I. Steefel

Energy Geosciences Division
Lawrence Berkeley National Laboratory
Berkeley, CA 94720
USA

CISteefel@lbl.gov

INTRODUCTION

Nanoporous media consist of homogeneous or heterogeneous porous material in which a significant part of the pore size distribution lies in the nanometer range. Clayey rocks, sediments or soils are natural nanoporous media, and cementitious materials, the most widely used industrial materials in the world, are also nanoporous materials. Nanoporous materials also include compounds present at the interface between non-porous solids and a aqueous solution. For example, hydrous silica gel coatings are interfacial nanoporous media that control the weathering rates of silicate glasses and minerals (Bourg and Steefel 2012). The understanding of the reactive and transport properties of these interfacial phases is essential to our global understanding of the long-term evolution of natural systems, such as soil formation (Navarre-Sitchler et al. 2011), nutrient cycling in the oceans (Loucaides et al. 2010), and engineered applications, such as the prediction of radionuclides release in high-level radioactive waste disposals (Grambow 2006; Collin et al. 2018a,b; Frugier et al. 2018). In colloidal suspensions, the aggregation of nanoparticles can also lead to the formation of nanoporous aggregates in which the bulk properties of nanoparticles are strongly influenced by the surrounding nanopores. For example, the dynamics of contaminant retention in ferrihydrite aggregates can be slowed by diffusional processes in the into/out of aggregates (Beinum et al. 2005).

1529-6466/19/0085-0004$05.00 (print)
1943-2666/19/0085-0004$05.00 (online) http://dx.doi.org/10.2138/rmg.2019.85.4

The bulk fluid transport properties and the *in situ* chemical reactivity properties of nanoporous media are notoriously difficult to characterize. Because of their large specific surface area, most of the fluid volume in nanoporous media is influenced by the close proximity of mineral[1] surfaces, which explains the very low transmissivity of these materials. As a consequence, the experimental characterization of their permeability requires special techniques (Neuzil and Person 2017). Also, the large specific surface area of nanoporous material provides them with very high adsorption capacity. The strong adsorption and resulting retardation of many contaminants by nanoporous material make them ideal for use in natural or engineered barrier systems or in filtration technologies. A good understanding of their chemical reactivity coupled to their transport properties is necessary to predict the long-term evolution of these properties of interest as a function of a range of physical and chemical conditions and processes. In this regard, reactive transport modeling can help bridging the gap between current process knowledge and predictions of the long term evolution of natural and engineered nanoporous materials in geological and industrial settings. However, nanoporous media exhibit a remarkable array of macro-scale properties with marked departures from those observed in "conventional" porous media such as permeable aquifers, for the study of which reactive transport models and codes have been historically developed. These properties arise from the interactions of charged mineral surfaces with water and solutes present in the nanopores, which leads to coupling between flux terms. These couplings manifest themselves in macroscopic observations that have intrigued geologists for more than one century, such as geologic ultrafiltration, i.e., the accumulation of solutes on the inflow side of clay-rich lithologies (Lynde 1912; Neuzil and Person 2017).

The vast majority of published reactive transport studies dealing with clay and cement materials are related to the evaluation of the long-term stability of surface and underground radioactive waste storage systems (Claret et al. 2018 and references therein; Bildstein et al. 2019, this volume, and references therein). In these types of simulation, the modeling effort has been focused primarily on the reactivity of the nanoporous materials rather than on their transport properties. Traditionally, Fickian diffusion has been typically considered, i.e., without taking into account advection, and without taking into account the anomalous transport properties of nanoporous media. In the last decade, special capabilities have been developed in a limited number of reactive transport codes that make them able to model part of these unconventional properties, with consideration of coupled processes that go beyond the traditional coupling between advective flow, dispersion, diffusion and reactions. The present chapter reviews these recent developments and explores the need to develop additional code capabilities to encompass relevant properties of nanoporous media in a holistic model from the micro-continuum to macro-continuum scales. The micro-continuum scale is restricted here to the definition given by Steefel et al. (2015b), i.e., a scale with resolution intermediate between true pore scale models and macro-continuum models, and in which parameters and properties such as permeability or reactive surface area need to be averaged or upscaled in some fashion. This corresponds to the matrix domains in the hybrid micro-continuum scale description defined by Soulaine et al. (2016, 2018), in which flow in true pore scale domains are described with the Navier–Stokes equation, whereas flow in matrix is described with Darcy's Law. Clayey materials are, by far, the most studied nanoporous media using reactive transport modeling. Consequently, most of the examples described in this chapter deal with clays. Also, this review is limited to coupled processes in water-saturated nanoporous media.

This chapter begins with a description of nanoporous media and the semi-permeable behavior that arises from the combination of their microstructure and their surface properties. Then, non-coupled and coupled transport processes in porous media are defined. In the next section,

---

[1] In the following, we will use the term "mineral" as a generic term designing true minerals (e.g., clay minerals), but also other types of solid phases (e.g., cementitious phases).

a quantitative description of the ion concentration distribution in the porosity of nanoporous materials is given, together with the approximations used in reactive transport codes to calculate it effectively. The reactive transport treatment of coupled transport processes is then described in the two last sections, starting from a description of transport in the diffusive regime, and ending with the challenges associated with the consideration of advective flow in nanoporous media.

NANOPOROUS MEDIA: SMALL PORES, LARGE SURFACES AND MEMBRANE PROPERTIES

Surface and microstructural properties of nanoporous materials

Most nanoporous media found in geological and engineered settings are comprised of an assembly of layered minerals. Bentonite is used in geosynthetic clay liners for isolation of landfills, and is planned as a barrier material in radioactive waste repository sites (Gates et al. 2009; Guyonnet et al. 2009). Bentonite is composed primarily of clay minerals of the smectite group, particularly montmorillonite. Clayey rocks (claystone, mudstone, argillite, shales) are being investigated as cap-rocks for underground gas sequestration technologies, geological barriers for radioactive waste disposal and source rocks for unconventional hydrocarbons (Neuzil 2013; Bourg 2015). These rocks usually contain large amounts of illite and illite/smectite mixed-layer minerals (Gaucher et al. 2004). The fundamental structural unit of smectite and illite consists of layers made of a sheet of edge-sharing octahedra fused to two sheets (2:1 or TOT layers) of corner-sharing tetrahedra with a thickness of about 9.5 Å (Fig. 1). The cations in the octahedral sheet of illite and montmorillonite consist predominantly of Al^{3+}. In montmorillonite, isomorphic substitutions take place mostly in the octahedral sheets where Al^{3+} is replaced by Fe^{3+} or a cation of lower charge (Mg^{2+}, Fe^{2+}), whereas in illite isomorphic substitutions take place in the octahedral sheet as in montmorillonite, and in the tetrahedral sheet where Si^{4+} is partially replaced by Al^{3+}. Isomorphic substitutions by cations of lower charge results in a permanent negative layer charge (Brigatti et al. 2013). The permanent negative layer charge of clay minerals is compensated for the most part by the accumulation of compensating cations between the layers. These interlayer cations can be solvated, as in montmorillonite interlayer space, or not, as in the case of K^+ in illitic interlayer space (Fig. 1). In the former case, interlayer cations can be readily exchanged with other cations present in the surrounding pore water, reflecting thus the composition of the pore water. Illite and montmorillonite particles are formed by a stacking of a small number of TOT layers. Illite particles typically consist of 5 to 20 stacked TOT layers, while for smectite the number of layers per particle increases with decreasing water chemical potential and with the valence of the charge-compensating cation (Banin and Lahav 1968; Shainberg and Otoh 1968; Schramm and Kwak 1982; Saiyouri et al. 2000). The nanometric dimension of a small number of stacked layers must be compared with the basal plane dimensions that range from 50 to 100 nm for illite (Poinssot et al. 1999; Sayed Hassan et al. 2006), and from 50 to 1000 nm for montmorillonite (Zachara et al. 1993; Tournassat et al. 2003; Yokoyama et al. 2005; Le Forestier et al. 2010; Marty et al. 2011). Clay mineral particles are thus highly anisotropic, and their spatial organization in the matrix of a clayey materials can yield a wide distribution of pore sizes, the dimension of which range from less than one nanometer to several micrometers (Fig. 2). Inter-particle and inter-aggregate pores are usually elongated in the bedding direction or perpendicular to the compaction direction in remolded samples (Pusch 2001; Robinet et al. 2012; Gu et al. 2015; Gaboreau et al. 2016). These pores are usually represented by slit-shaped or cylindrical-shaped pores for modeling purposes. The smallest pores, i.e., the interlayer pores in swelling clay minerals, are logically usually described as slit-shaped pores (Birgersson and Karnland 2009; Appelo et al. 2010; Tournassat and Appelo 2011; Tournassat et al. 2016b; Appelo 2017; Wigger and Van Loon 2017).

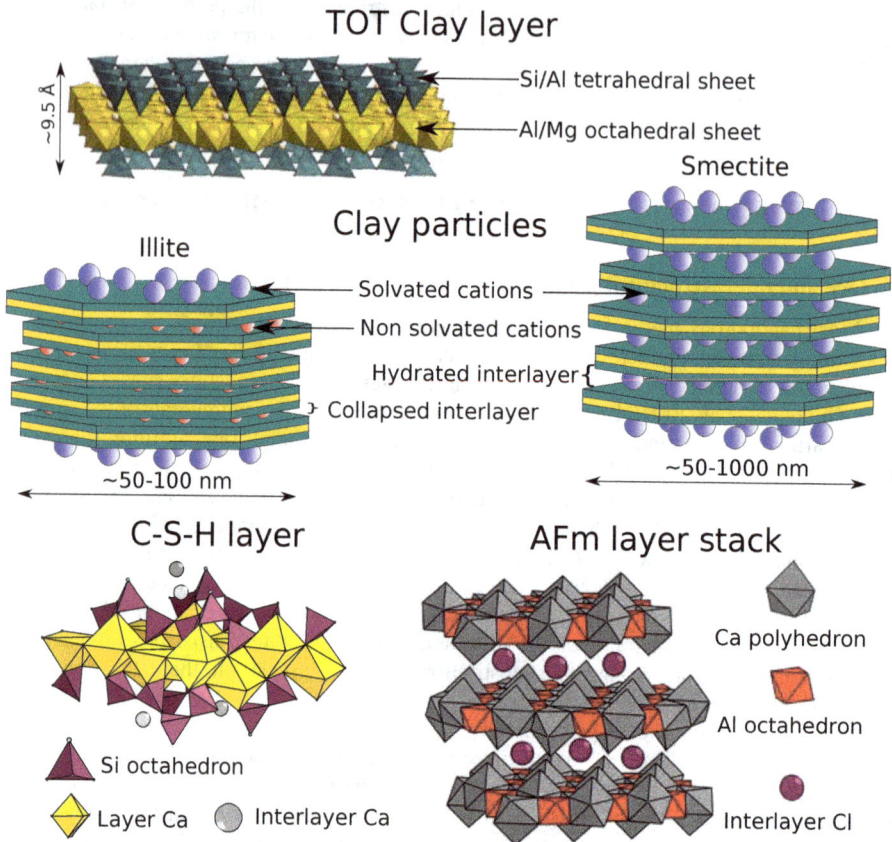

Figure 1. Examples of layered phases in clayey and cementitious materials. [Clay mineral structures reprinted from Tournassat et al. (2015a) Chapter 1—Surface properties of clay minerals. *In:* Tournassat C, Steefel CI, Bourg IC, Bergaya F (Eds.), Natural and Engineered Clay Barriers, Developments in Clay Science. Elsevier, p 5–31, Copyright 2015, with permission from Elsevier. C–S–H layer structure according to Grangeon et al (2016). AFm structure according to Marty et al. (2018).]

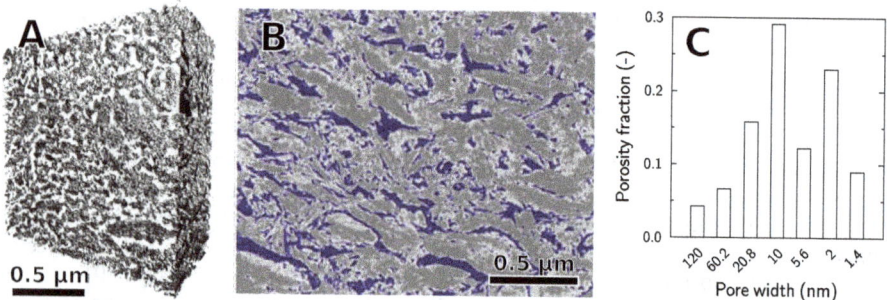

Figure 2. A: 3-D visualization of the microstructure of an illite sample compacted at 1.7 kg·dm^{-3} (back-scattered electrons focused ion beam nanotomography stack images $350 \times 350 \times 180$ voxels with voxel resolution size of $5 \times 5 \times 5$ nm^3); **B:** Segmented image of a TEM micrograph of the same illite sample. The blue color represents pores and light and dark gray are pore/solid mixtures and solids, respectively. **C:** Pore-width distribution obtained from a combination of bulk and microscopic characterization techniques. [Figure modified from Tournassat et al. (2016b).]

Cementitious materials are complex assemblages of a range of solid phases. Among them, nanocrystalline calcium silicate hydrate (C–S–H) is often the main hydration product (Richardson 2008), and constitutes the "glue material" of the cement (Pellenq and Van Damme 2004; Masoero et al. 2012). C–S–H have a complex chemistry, with calcium to silicon (Ca/Si) atomic ratio that vary as a function of pore water chemical conditions, and especially pH. C–S–H have layered structures that can be described as nanocrystalline and defective tobermorite, in which the Si chains depolymerize as the Ca/Si atomic ratio increases (Grangeon et al. 2013, 2016, 2017; Richardson 2014). The depolymerization reaction creates silanol groups at the surface having amphoteric (protonation/deprotonation) properties and surface complexation capabilities (Fig. 1). The resulting surface charge of C–S–H can take positive or negative values as a function of pore water chemistry following layer structure changes and surface site reactivity (Haas and Nonat 2015). A C–S–H particle is composed of a stack of a small number of layers, which are separated by a hydrated interlayer space that may also contain cations (Grangeon et al. 2016; Roosz et al. 2016). The spatial distribution of C–S–H particles in the cement matrix creates a heterogeneous distribution of inter-particulate pore sizes (Ioannidou et al. 2016). Other major cement phases include the hydrated calcium aluminate phase (AFm), which is a member of the layered double hydroxides (LDHs) group. Its structure consists of stacked positively charged layers of Al and Ca polyhedra separated from each other by hydrated interlayer spaces, which contain charge compensating hydrated anions (Fig. 1). Similarly to interlayer cations in swelling clay minerals, interlayer anions in AFm are exchangeable with the surrounding pore water, making them an important contributor to the retention of anions that enter into contact with cement-based materials (Ma et al. 2017, 2018; Marty et al. 2018).

The layered nature of the major components of clayey and cementitious materials confer to these materials very high specific surface area, and, consequently, a large fraction of their porosity lies within the nanometer range. C–S–H, AFm, illite and smectite layers exhibit surface charges that are compensated in the adjacent porosity. Between the layers, these charges are mostly compensated by counter-ions (ions with charge sign opposite to that of the surface) in the interlayer porosity. The layer charge present at the outer surfaces of the particles (hereafter named basal surfaces), which border the interparticle pores, must also be compensated. This is also true for non-permanent charges present on the border surfaces of other solid phase (e.g., silica) that arise from the cleavage of the grains. The charge compensation mechanisms and their consequences for the water and ion transport in nanoporous media are explored in the next section.

Ion distribution in the vicinity of charged surfaces

The charge of the surfaces bordering the pores is responsible for the presence of a double layer or electrical double layer (EDL), i.e., the layers of interfacial water and electrolyte ions that screen the surface charge. The EDL is often conceptually subdivided into two regions, in agreement with spectroscopic results (Lee et al. 2010, 2012b), direct force measurements (Zhao et al. 2008; Siretanu et al. 2014) and molecular dynamics (MD) calculations (Marry et al. 2008; Tournassat et al. 2009a; Rotenberg et al. 2010; Bourg and Sposito 2011): the Stern layer located within the first water monolayers of the interface in which ions adsorb as inner- and outer-sphere surface complexes (ISSC, OSSC) and a diffuse layer (DL) located beyond the Stern layer in which a diffuse cloud or swarm of ions screens the remaining uncompensated surface charge (Fig. 3). The concentrations of ions in this region depend on the distance from the surface considered. At infinite distance from the charged surface, the solution is neutral and is commonly described as bulk or free solution (or water). The exact ionic charge distribution in the EDL is related to the potentials of mean force for the various ions, and those potentials are for the most part related to the local magnitude of the electrostatic potential (Delville 2000). It is important to note that the ion distribution in the EDL cannot be probed directly by measurements because EDL features are inherently disturbed by direct measurement techniques (Bourg et al. 2017). A range of experimental techniques have been used to

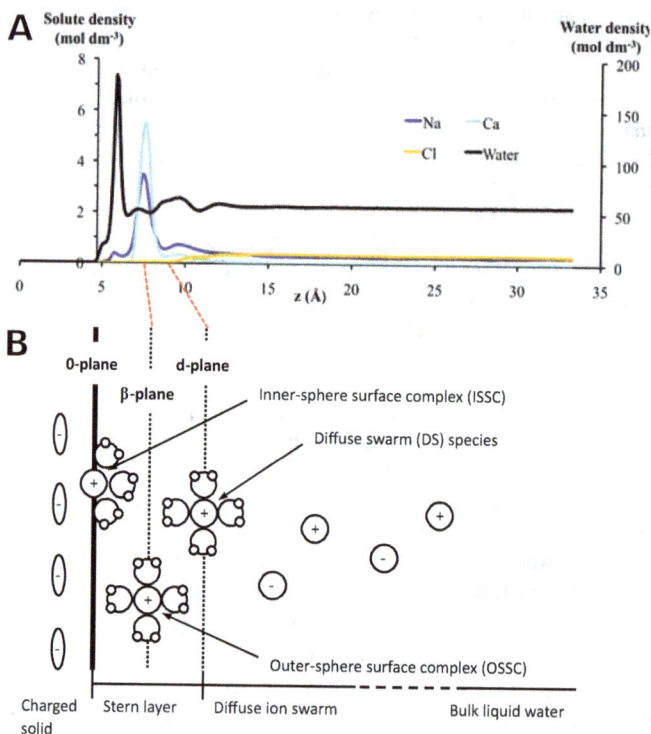

Figure 3. **A:** Water and ion distribution at the vicinity of a montmorillonite surface according to MD simulations. Average density (mol·dm⁻³) of water (black), Na (dark blue), Ca (light blue) and Cl (orange) are plotted as a function of distance from the mid-plane of the montmorillonite layer at a NaCl/CaCl₂ brine concentration of 0.34 mol·dm⁻³. **B:** Conceptual model of EDL structure. [Reprinted from Bourg and Sposito (2011) Molecular dynamics simulations of the electrical double layer on smectite surfaces contacting concentrated mixed electrolyte (NaCl–CaCl₂) solutions. *Journal of Colloid and Interface Science* 360:701–715 Copyright 2011, with the permission of Elsevier.]

characterize indirectly the EDL properties, but the interpretation of their results is always sensitive to an *a priori* choice of an EDL model. Fortunately, in the last two decades, atomistic level models, such as molecular dynamics, have provided a growing set of information on the properties of the EDL. It is now commonly acknowledged that the diffuse layer part of the EDL is well understood, while the characterization of the Stern layer remains a challenging issue (Bourg et al. 2017). In the following, we shall concentrate our attention on the diffuse layer, which has a large influence on transport properties in nanoporous media.

In the diffuse layer, water is often considered to have properties similar to bulk-liquid water and the distribution of ions predicted by molecular dynamic simulations is often in close agreement with predictions made with the Poisson–Boltzmann (PB) equation (Tinnacher et al. 2016). The Poisson equation relates the local charge imbalance at a position y in the direction perpendicular to the charged surface to the second derivative of the electrostatic potential (x in V) at the same position:

$$\frac{d^2\psi_{DL}(y)}{dy^2} = -\frac{1}{\varepsilon}\sum_i z_i F c_{i,DL} \qquad (1)$$

where ε is the water dielectric constant ($78.3 \times 8.85419 \cdot 10^{-12}$ F·m⁻¹ at 298.15 K), F is the Faraday constant (96 485 C·mol⁻¹), and z_i is the charge of ion i. The concentration of species i in the DL, $c_{i,DL}$, is given by the Boltzmann equation:

$$c_{i,DL} = c_{i,0} \exp\left(\frac{-z_i F \psi_{DL}(y)}{RT}\right) \qquad (2)$$

where R is the gas constant (8.3145 J·mol^{-1}·K^{-1}), T is the temperature (K), and $c_{i,0}$ is the concentration of species i in the bulk water at equilibrium with the diffuse layer. At infinite distance from the surface, the electrostatic potential vanishes and the concentration in the diffuse layer is equal to that in the bulk water, where electro-neutrality prevails. However, if pores are not large enough, there will be no regions within the pore that meet this electro-neutrality condition because of the overlap of the diffuse layers. The size of the diffuse layer depends primarily on the concentration of ions, and on the valence of the counter-ions (Carnie and Torrie 1984; Sposito 2004). As the ionic strength (I, here in mol·L^{-1}) increases, the diffuse layer thickness decreases, and, as the valence of the counter-ions increases, the diffuse layer thickness decreases as well. So too does the overlap of the diffuse layers at two neighboring surfaces (Fig. 4, top and middle).

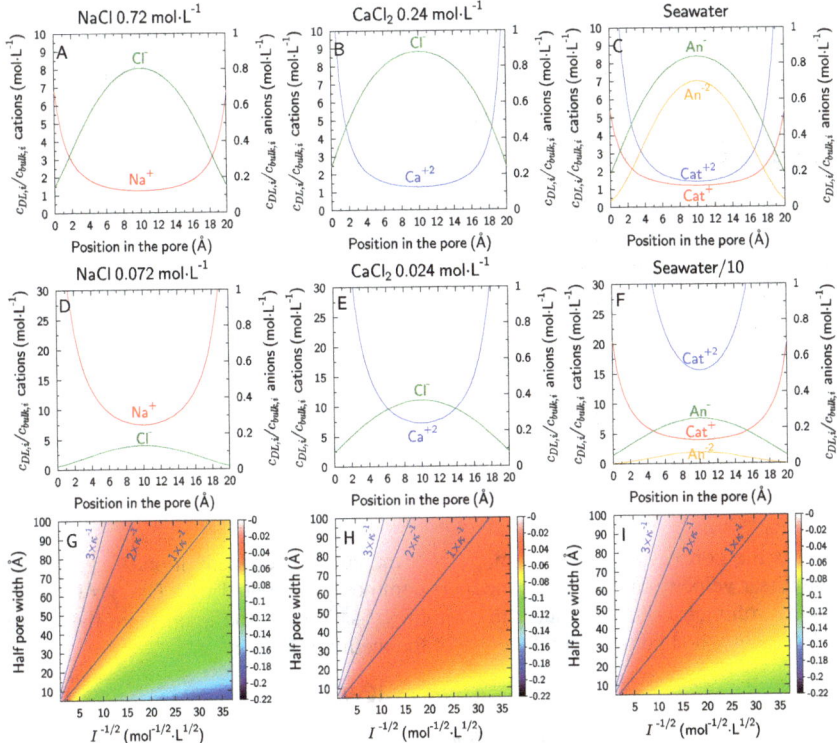

Figure 4. *Top and middle figures*: Ion enrichment/depletion according to the Poisson–Boltzmann equation in a diffuse layer in contact with negatively charged surfaces ($\sigma_0 = 0.11$ C·m^{-2}) bordering a slit-shaped pore with a distance of 2 nm between the surfaces. **A** and **D**: NaCl electrolyte; **B** and **E**: CaCl$_2$ electrolyte; **C** and **F**: diluted seawater. **A**, **B** and **C** correspond to an ionic strength of 0.72 mol L^{-1}, while **D**, **E** and **F** correspond to a ionic strength of 0.072 mol L^{-1}. The seawater composition was taken from Holland (1978) and was simplified into Na$^+$: 481 mmol·kg^{-1}; Mg^{2+}: 55 mmol·kg^{-1}; Ca^{2+}: 11 mmol·kg^{-1}; K$^+$: 11 mmol·kg^{-1}; Cl$^-$ 564 mmol·kg^{-1}; SO$_4^{2-}$: 29 mmol·kg^{-1}; HCO$_3^-$: 2 mmol·kg^{-1}. No aqueous complexation in bulk solution was considered for this calculation. In figures **C** and **F**, the enrichment/depletion are indicated for monovalent cations (Cat$^+$), divalent cations (Cat^{+2}), monovalent anions (An$^-$) and divalent anions (An^{-2}). Calculations were done at 298 K. *Bottom figures* (**G**, **H** and **I**): Poisson–Boltzmann prediction of the electrostatic potential (in V) at the middle of a slit-shaped pore as a function of the inverse of the square root of ionic strength (*x*-axis) and half distance between the surfaces (*y*-axis) ($\sigma_0 = -0.11$ C·m^{-2}). The blue lines indicate 1, 2 and 3 times the Debye length (κ^{-1}, Eqn. 21), which is commonly assumed to be proportional to the thickness of the diffuse layer.

Semi-permeable properties

The presence of overlapping diffuse layers in charged nanoporous media is responsible for a partial or total repulsion of co-ions from the porosity. In the presence of a gradient of bulk electrolyte concentration, co-ion migration through the pores is hindered, as well as the migration of their counter-ion counterparts because of the electro-neutrality constraint. This explains the salt-exclusionary properties of these materials. These properties confer these media with a semi-permeable membrane behavior: neutral aqueous species and water are freely admitted through the membrane while ions are not, giving rise to coupled transport processes. Semi-permeable membrane properties of nanoporous materials have been extensively studied for decades. An overwhelming majority of these studies have dealt with clay materials, in the form of compacted powder samples of purified clay, or natural materials, such as soils or clayey rocks, with a strong emphasis on chemo-osmotic membrane properties (Lynde 1912; McKelvey and Milne 1960; Kemper 1960, 1961a,b; Olsen 1962; Kemper and Evans 1963; Young and Low 1965; Kemper and Rollins 1966; Bresler 1973; Hanshaw and Coplen 1973; Neuzil 2000; Cey et al. 2001; Garavito et al. 2007; Horseman et al. 2007; Rousseau-Gueutin et al. 2009; Gonçalvès et al. 2015; Sun et al. 2016).

NON-COUPLED AND COUPLED TRANSPORT PROCESSES

Transport equations in traditional reactive transport modeling

Single phase numerical reactive transport models aim at describing the fluxes $\mathbf{J}_i$ ($mol \cdot m_{medium}^{-2} \cdot s^{-1}$) of aqueous chemical species i in space and time, as well as the mineralogy and sorbed species evolution. The constitutive equations describing these fluxes at the continuum scale are thermodynamic and kinetic equations for chemical reactions coupled to advection (*adv*), mechanical dispersion (*disp*) and diffusion equations (*diff*) (Steefel and Maher 2009; Steefel et al. 2015a):

$$\mathbf{J}_i = \mathbf{J}_{adv,i} + \mathbf{J}_{disp,i} + \mathbf{J}_{diff,i} \qquad (3)$$

Advection, dispersion and diffusion are non-steady, irreversible processes expressed in the form of partial differential field equations (Bear 1972). The advective flux is related to the average linear velocity, $\mathbf{v}$ ($m \cdot s^{-1}$), in the media:

$$\mathbf{J}_{adv,i} = \phi C_i \mathbf{v} \qquad (4)$$

where ϕ is the porosity (–), C_i ($mol \cdot m_{water}^{-3}$) is the aqueous concentration of species i. The average linear velocity in the porous media is commonly evaluated with Darcy's Law, which relates the volumetric flux of water ($m_{fluid}^3 \cdot m_{medium}^{-2} \cdot s^{-1}$), $\mathbf{q}$, to the gradient in the hydraulic head, h (m), and the hydraulic conductivity of the medium, $\mathbf{K}$ ($m \cdot s^{-1}$):

$$\mathbf{q} = \phi \mathbf{v} = -\mathbf{K} \nabla h \qquad (5)$$

Molecular diffusion in porous media is usually described in terms of Fick's First Law:

$$\mathbf{J}_{diff,i} = -\mathbf{D}_{e,i} \nabla C_i \qquad (6)$$

where $\mathbf{D}_{e,i}$ is the effective coefficient of species i. The effective diffusion coefficient differs from the diffusion coefficient in water, because diffusion takes place only in the aqueous solution present in the medium. Consequently, the diffusion coefficient must be corrected using the porosity value and from a geometrical factor, τ_i (–), that accounts for the difference of paths lengths that the aqueous species would follow in water alone compared to the tortuous paths lengths it would follow in the porous medium (tortuosity), and for the constrictivity factor of the porous medium (Bear 1972):

$$\mathbf{D}_{e,i} = \phi \tau_i D_{0,i} \qquad (7)$$

where $D_{0,i}$ is the diffusion of species i in the solution alone. In the above formulation, the diffusive fluxes depend on the nature of the chemical species. This dependency would result in charge imbalance in conventional numerical reactive transport models if only a Fickian description of diffusion is used, so in practice, Equation (7) is usually simplified into:

$$\mathbf{J}_{\text{Diff,i}} = -\phi\tau D_0 \nabla C_i \qquad (8)$$

where τ and D_0 are mean representative values.

The coefficient of hydrodynamic dispersion is defined as the sum of molecular diffusion and mechanical dispersion. Mechanical dispersion is a scale-dependent process, with larger dispersivities observed for larger observation scales (Steefel 2008). It is usually computed in reactive transport codes as the product of the fluid velocity and dispersivity with longitudinal and transverse components (Steefel et al. 2015a; Rolle and Le Borgne 2019, this volume and references therein). The low permeability of nanoporous media limits mass transport by advection, with the result that hydrodynamic dispersion is dominated in this case by molecular diffusion (Patriarche et al. 2004a,b; Mazurek et al. 2011). Consequently, the contribution of mechanical dispersion to the total flux is usually neglected for the modeling of nanoporous media (Gonçalvès et al. 2015; Neuzil and Person 2017).

In traditional numerical reactive transport models applied to nanoporous media, transport of solutes is thus described with the above two constitutive equations, Darcy's and Fick's laws. While Fick's and Darcy's Laws constitute a modeling framework that is adequate to deal with a wide range of reactive transport problems for various geological settings and processes such as those described in Maher and Navarre-Stichler (2019, this volume), Cama et al. (2019, this volume), and Lagneau et al. (2019, this volume) these laws are phenomenological equations rooted in the results of physical experiments carried out on systems involving salts in water and packed sand columns respectively (Fick 1855; Darcy 1856). These systems are very different from the nanoporous media we are addressing in the present contribution. Because of the charge imbalance in the solution present in the diffuse layer, the rigorous modeling of transport processes in nanoporous media is not possible by considering the advective flux equation based on the classical Darcy equation and the diffusive flux based on Fick's equations.

Coupled transport processes and thermodynamics of irreversible processes

The thermodynamics of irreversible processes (or non-equilibrium thermodynamics), founded by Onsager (1931a,b), provides the theoretical background for the existence of coupled flows (Kjelstrup and Bedeaux 2008).

Many observations of fluid and solute transport in geological media cannot be related to the Darcian and Fickian modeling framework because of the existence of coupling of forces of one type and flows of another type (Bear 1972). For example, chemical osmosis is a flow of water driven by a gradient of water chemical potential, which is usually created by a gradient of electrolyte concentration across a semi-permeable membrane. Water flow can also result from the presence of gradients in the electric potential (electro-osmosis) or temperature (thermo-osmosis). Osmotic processes have been extensively characterized on natural materials ranging from soils to biological membranes (Medved and Cern 2013).

Non-equilibrium thermodynamics is based on the second law of thermodynamics and the principle of microreversibility. In any system, the rate of entropy production per unit volume, σ, can be written:

$$\sigma = \frac{1}{V}\frac{dS}{dt} = \sum_i J_i X_i \qquad (9)$$

where S is the entropy of the system, V its volume, t the time, and J_i is a flux of type i associated with a driving force X_i for that particular flux (Lasaga 1998). The second law of thermodynamics predicts that:

$$\sum_i J_i X_i \geq 0 \qquad (10)$$

With the assumption that all driving forces X_i needed to describe fully the evolution system are taken into account in Equation (10), each flux J_i is a function f_i of the forces X_i, and takes a value of zero if all X_i are zero (reference point). Accordingly, the functions f_i can be expanded in Taylor series of the form (Lasaga 1998):

$$J_i = f_i(X_1, X_2, ..., X_n) = 0 + \sum_j \frac{\partial f_i(X_1, X_2, ..., X_n)}{\partial X_{j,X=0}} X_j + ... \qquad (11)$$

where $X=0$ indicates that the derivative of the function is evaluated at the reference point. Keeping only the linear terms of the Taylor series expansion, i.e., making the assumption that higher order terms are negligible, Equation (11) yields:

$$J_i = \sum_j \frac{\partial f_i(X_1, X_2, ..., X_n)}{\partial X_{j,X=0}} X_j = \sum_j L_{i,j} X_j \qquad (12)$$

where $L_{i,j}$ are termed phenomenological coefficients. Equation (12) indicates that a driving force X_j not only influences the conjugate flux J_j, but also the other non-conjugate fluxes $J_{i \neq j}$ as in the case of the various osmosis processes. The use of irreversible thermodynamics theory is thus dependent on the identification of a set of extensive independent variable X_j that defines completely the system of interest (de Groot and Mazur 1984; Kjelstrup and Bedeaux 2008). Under these conditions, Onsager (1931b) demonstrated that the following reciprocal relations apply:

$$L_{i,j} = L_{j,i} \qquad (13)$$

Transport problems investigated in Earth Sciences are focused mostly on the fluxes of fluids, chemical components, and heat. We shall see in the following section that in the case of aqueous solutions present in porous media in general, and in nanoporous media in particular, charge (or electrical) fluxes must also be considered. Ideally, all types of conjugate and coupled flux listed in Table 1 should be considered. However, Equations (12) and (13) do not give any information about the way fluxes and conjugate forces must be chosen, and the set of equations is not unique for a given system. Consequently, the problem one wants to describe dictates the form that is most convenient (Kjelstrup and Bedeaux 2008). Numerical methods for modeling macroscopic transport properties of nanoporous media, and especially clayey materials, with the explicit consideration of the presence of a diffuse ion swarm have received growing interest in diverse research communities in the past years (Revil 1999; Revil and Leroy 2004; Leroy et al. 2006; Gonçalvès et al. 2007, 2015; Jougnot et al. 2009; Revil et al. 2011). If we assign the value X_1, X_2, and X_3 to the fluid pressure, temperature and electrical potential gradients respectively, and the value X_4 to X_{n+3} to each of the driving forces associated to the presence of n chemical species, then a complete description of coupled flux processes is given by:

$$\begin{aligned}
J_1 &= L_{1,1}X_1 + L_{1,2}X_2 + L_{1,3}X_3 + L_{1,4}X_4 + \cdots + L_{1,n+3}X_{n+3} \\
J_2 &= L_{2,1}X_1 + L_{2,2}X_2 + L_{2,3}X_3 + L_{2,4}X_4 + \cdots + L_{2,n+3}X_{n+3} \\
J_3 &= L_{3,1}X_1 + L_{3,2}X_2 + L_{3,3}X_3 + L_{3,4}X_4 + \cdots + L_{3,n+3}X_{n+3} \\
J_4 &= L_{4,1}X_1 + L_{4,2}X_2 + L_{4,3}X_3 + L_{4,4}X_4 + \cdots + L_{4,n+3}X_{n+3} \\
&\vdots = \vdots \\
J_{n+3} &= L_{n+3,1}X_1 + L_{n+3,2}X_2 + L_{n+3,3}X_3 + L_{n+3,4}X_4 + \cdots + L_{n+3,n+3}X_{n+3}
\end{aligned} \qquad (14)$$

Table 1. Terminology of coupled fluxes (after de Marsily 1986)

Flux	Gradients			
	Fluid pressure ∇p	**Temperature** $\nabla T/T$	**Electrical potential** $\nabla \psi$	**Chemical potential** $\nabla \mu_i$
Fluid	Darcy	Thermo-osmosis	Electro-osmosis	Chemical osmosis
Heat	Heat filtration or mechano-caloric effect	Fourier	Peltier effect	Dufour effect
Electrical	Streaming potential (Rouss effect)	Seebeck effect	Ohm	Diffusion current
Aqueous species i (relative to the fluid)	Ultrafiltration	Thermal diffusion (Soret effect)	Electrophoresis	Fick

It is clear that the evaluation of the phenomenological coefficients becomes increasingly more cumbersome as the numbers of fluxes and driving forces increase (Malusis et al. 2012), explaining why most of the effort has been put into predictive models restricted to simple electrolytes conditions (e.g., NaCl). If the electrolyte is not simple, then it is frequently assumed that the mobility of the different ions does not differ significantly from each other (Revil and Linde 2006). This approximation simplifies greatly the set of equations, but is a significant departure from reality, especially in systems at low or high pH, because the diffusion coefficient of OH^- and H_3O^+ are five to ten times greater than the diffusion coefficients of other ions (Li and Gregory 1974). The models and the formulation of the phenomenological coefficients are thus very dependent on the chemical conditions of the systems studied. Reactive transport modeling approaches, with their unique ability to deal with transient stages and chemically complex systems, may be able to overcome some of these limitations if it is possible to define equations linking the values of the phenomenological coefficients to the basic properties of the media of interest and to the chemical composition of the pore water. In the following, we shall explore in detail the sets of general equations that have been or could be implemented in reactive transport codes to deal with coupled transport processes in the presence of a diffuse layer. We have also highlighted that, in certain cases, the hindered mobility of co-ions controls the transport properties of all ions because of the requirement of charge balance at the macro-scale. A quantitatively correct description of ion concentration distribution in the diffuse layer volume is thus required to quantify ionic fluxes in nanoporous media. Consequently, the following section focuses on the methods used to compute the diffuse layer composition.

POISSON–BOLTZMANN EQUATION AND ION CONCENTRATION DISTRIBUTION CALCULATIONS IN REACTIVE TRANSPORT CODES

The mean electrostatic potential model

The Poisson–Boltzmann equation can be solved analytically, without approximations, in a very limited number of cases (Andrietti et al. 1976; Borkovec and Westall 1983; Chen and Singh 2002; Sposito 2004), and its solution for systems having complex electrolyte composition and/or geometries is possible only with numerical approaches (Leroy and Maineult 2018). In addition, nanoporous materials such as clayey and cementitious materials typically exhibit a range of pore size and shapes. For example, clay mineral particles are often segregated into aggregates delimited by inter-aggregate spaces, the size of which is usually larger than the inter-particle spaces inside the aggregates. In clayey rocks, the presence of non-clay minerals (e.g., quartz, carbonates, pyrite) also influences the structure of the pore network and the pore size distribution. Pores as large as few micrometers are frequently observed (Keller et al. 2011, 2013; Robinet et al. 2012; Philipp et al. 2017) and these co-exist with pores having a width as narrow as the clay interlayer spacing, i.e., one nanometer.

In practice, the information about ion concentrations in the diffuse layer must be upscaled so that calculations can be carried out at the continuum scale. A common upscaling approach relies on the use of a mean electrostatic potential model (Revil and Leroy 2004; Appelo and Wersin 2007; Appelo et al. 2010; Revil et al. 2011; Tournassat and Steefel 2015; Tournassat et al. 2016b). This model averages ion concentrations in the diffuse layer ($\overline{c_{i,DL}}$) by scaling them to a mean electrostatic potential (MEP, ψ_M) that applies to the diffuse layer volume (V_{DL}):

$$\overline{c_{i,DL}} = \frac{1}{V_{DL}} \iiint_{DL} z_i c_{i,0} \exp\left(\frac{-z_i F \psi_{DL}(x,y,z)}{RT}\right) dx\,dy\,dz \approx c_{i,0} \exp\left(\frac{-z_i F \psi_M}{RT}\right) \tag{15}$$

The mean electrostatic potential value can be deduced from the charge balance in the diffuse layer:

$$\sum_i z_i F \overline{c_{i,DL}} = \sum_i z_i F c_{i,0} \exp\left(\frac{-z_i F \psi_M}{RT}\right) = -Q_{DL} \tag{16}$$

where Q_{DL} (in $C \cdot m^{-3}$) is the volumetric charge that must be balanced in the diffuse layer. The accuracy of this method compared to a full resolution of the Poisson–Boltzmann equation can be appreciated by calculating the difference of average concentrations in the diffuse layer volume computed by the two methods in a simple geometry such as a slit-shaped pore (Fig. 5). When applied to the entire pore volume, the mean electrostatic potential model underestimates the average concentration of co-ions, and the accuracy of its predictions degrades as the size of the pore increases and as the ionic strength of the solution decreases (Fig. 5 A, B, C, D). The prediction of the average concentration of counter-ions is good for the simple NaCl and CaCl$_2$ electrolytes. It must be noted, however, that the average concentration value is governed primarily by the requirement of charge balance with the surface in the diffuse layer. Because the surface charge is mostly balanced by the increase of concentration of counter-ions in the diffuse layer, it is not surprising to find a good agreement between the two models. In the presence of a more complicated electrolyte composition (seawater dilution), the mean electrostatic potential model overestimates the contribution of the monovalent counter-ions in the diffuse layer, and underestimates the contribution of the divalent counter-ions. The mean electrostatic potential model is often referred as a Donnan model in the literature (Revil and Leroy 2004; Appelo and Wersin 2007; Birgersson and Karnland 2009; Appelo et al. 2010; Tournassat and Appelo 2011; Alt-Epping et al. 2015, 2018; Gimmi and Alt-Epping 2018). The Donnan equilibrium model, however, relies on two fundamental assumptions (Babcock 1960): (1) that chemical equilibrium exists between two distinct aqueous phases, a hydrated clay phase and a bulk aqueous solution phase, and (2) that Henry's Law applies at infinite dilution to all of the ions in both phases. The Donnan model gives strictly equivalent results to the mean electrostatic potential model only if two important conditions are met: first, the mean electrostatic potential model must be applied to the entire pore volume, and second, in the Donnan equilibrium model, the activity coefficients of chemical species present in the Donnan volume must be set at the same value as those in bulk water (Tournassat et al. 2016a). The consideration of a Donnan equilibrium model in clay media faces two challenges. First, the activity of chemical species in the Donnan volume cannot be measured, and thus the choice to set the activity coefficients of species in the Donnan volume to the same values as in bulk water

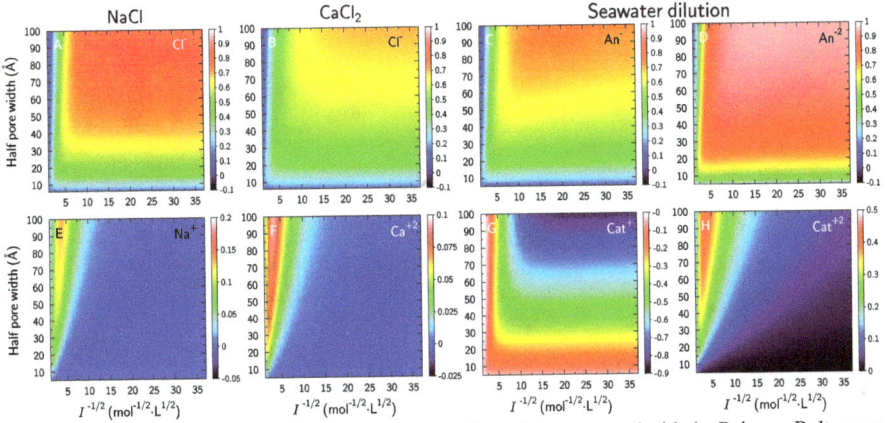

Figure 5. Difference of average diffuse layer concentration values computed with the Poisson–Boltzmann (c_{PB}) equation and the mean electrostatic potential (c_{MEP}) model (Eqns. 15 and 16). The difference, given by the color scale, is computed with the equation $(c_{PB}-c_{MEP})/c_{PB}$ as a function of the inverse of the square root of ionic strength ($I^{-1/2}$ on x-axis) and as a function of the half pore width (y-axis). The equivalence between the two models corresponds to $(c_{PB}-c_{MEP})/c_{PB}=0$. The geometry consists in a diffuse layer in contact with negatively charged surfaces (σ_0 =0.11 C·m^{-2}) bordering a slit-shaped pore. In figure C, D, G and H, the seawater composition was taken from Holland (1978) and was simplified into Na$^+$: 481 mmol·kg^{-1}; Mg^{2+}: 55 mmol·kg^{-1}; Ca^{2+}: 11 mmol·kg^{-1}; K$^+$: 11 mmol·kg^{-1}; Cl$^-$ 564 mmol·kg^{-1}; SO$_4^{2-}$: 29 mmol·kg^{-1}; HCO$_3^-$: 2 mmol·kg^{-1}. This composition corresponds to the lowest value if $I^{-1/2}$ on the figures. Other values of $I^{-1/2}$ correspond to the dilution of this seawater composition with pure water. No aqueous complexation in bulk solution was considered for this calculation. The calculation results are shown for monovalent anions (An$^-$) in C, divalent anions (An^{-2}) in D, monovalent cations (Cat$^+$) in G and divalent cations (Cat^{+2}) in H. Calculations were done at 298 K.

must be considered as arbitrary. This choice can be relaxed by choosing other relationships for the calculation of the activity coefficients, but experimental data are lacking to calibrate such relationships (Birgersson 2017). Second and more importantly, a strict application of Henry's Law (and, hence, of the Donnan equilibrium model) would require that the counterions be dissociated fully from the mineral surface, such that mean ionic concentrations in the hydrated clay phase be unambiguously determined (Babcock 1963). This second hypothesis is not in agreement with the observation of cations condensation at clay mineral surfaces, using diffraction techniques or molecular dynamics simulations (Marry et al. 2002; Schlegel et al. 2006; Tournassat et al. 2009a; Lee et al. 2010, 2012b; Holmboe and Bourg 2014; Bourg et al. 2017). The Poisson–Boltzmann model does not rely on the validity of Henry's Law. Consequently, the mean electrostatic model predictions remain valid within the limits of the underlying molecular-scale hypotheses and approximations of the model, while the Donnan model does not (Tournassat et al. 2016a).

Dual continuum representation of the pore space

We have highlighted in a previous paragraph that the semi-permeable membrane properties of nanoporous media originate from the exclusion of co-ions in the diffuse layer. It is thus desirable to build a model that reproduces as well as possible the mean co-ion concentration in the diffuse layer. The mean electrostatic potential model can be adapted to do so, following the basic principle of a subdivision of the pore space into two compartments, one being electroneutral, and the other being influenced by a non-zero mean electrostatic potential value (Fig. 6). This type of representation will be referred to as "Dual Continuum" in the following.

Using f_{DL} as the volumetric fraction of the porosity to which the mean electrostatic potential is applied, Equations (15) and (16) transform into:

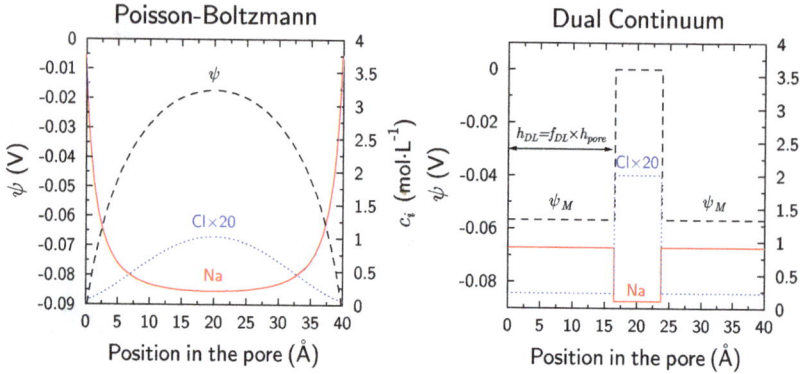

Figure 6. Comparison of the Poisson–Boltzmann (left figure) and MEP (right figure) models in a slit-shaped pore. The specific surface charge is -0.11 C·m^{-2} and the bulk concentration of NaCl electrolyte is 0.1 mol·L^{-1}. The half pore width (h_{pore}) is 20 Å. The width of the diffuse layer in the dual continuum model is set to $h_{DL}=f_{DL}\times h_{pore}$, and the value of f_{DL} is adjusted to obtain the same average pore concentrations of Cl$^-$ and Na$^+$ with the two models. Dashed curves = electrostatic potential, ψ; plain lines = Na$^+$ concentration; dotted line = Cl$^-$ concentration.

$$\overline{c_{i,DL}} = \frac{1}{V_{DL}}\iiint_{DL} z_i c_{i,0}\exp\left(\frac{-z_i F\psi_{DL}(x,y,z)}{RT}\right)dx\,dy\,dz \approx c_{i,0}\exp\left(\frac{-z_i F\psi_M}{RT}\right) \quad (17)$$

$$\sum_i z_i F\overline{c_{i,DL}} = \sum_i z_i F c_{i,0}\exp\left(\frac{-z_i F\psi_M}{RT}\right) = -Q_{DL} \quad (18)$$

where $\overline{c_{Cl,pore}}$ is the average concentration of species i in the pore. Note that $(1-f_{DL})F\sum_i z_i c_{i,0}$ is equal to zero because bulk water is electroneutral. The value of f_{DL} can be adjusted as a function of the conditions (pore volume, surface charge, electrolyte composition) to achieve consistency with the theoretical PB predictions or with experimental constraints. It should be also noted here that this model refinement does not imply necessarily that an electroneutral bulk water is present at the center of the pore in reality. This can be appreciated in Figure 6, which shows that the Poisson–Boltzmann predicts an overlap of the diffuse layers bordering the two neighboring surfaces, while the dual continuum model divides the same system into a bulk and a diffuse layer water volume in order to obtain an average concentration in the pore that is consistent with the Poisson–Boltzmann model prediction. Consequently, the pore space subdivision into free and DL water must be seen as a convenient representation that makes it possible to calculate accurately the average concentrations of ions, but it must not be taken as evidence of the effective presence of bulk water in a nanoporous medium. This modified MEP model was introduced in PHREEQC to model diffusion properties of Opalinus Clay at the Mont Terri underground research laboratory (URL) and of Callovian-Oxfordian argillite at the Laboratoire Souterrain de Meuse Haute Marne (LSMHM) URL (Appelo and Wersin 2007; Appelo et al. 2008). The method to calculate f_{DL} on the basis of anion (Cl$^-$) exclusion experimental data in clay samples was later developed in Appelo et al. (2010). Average Cl$^-$ concentrations in pore water ($c_{Cl,pore}$) can be obtained from Cl$^-$ concentration measured in leaching experiments ($c_{Cl,leach}$), in which a mass m_{sample} of clay sample is dispersed in a volume V_{leach} of milli-Q water. As $c_{Cl,leach}$ is proportional to the ratio m_{sample}/V_{leach} in clay samples, it follows that (Tournassat et al. 2015b):

$$\overline{c_{Cl,pore}} = c_{Cl,leach}\frac{V_{leach}}{m_{sample}\times\omega_w} \quad (19)$$

where ω_W is the water content of the sample (in $kg \cdot kg^{-1}$). The value of $\overline{c_{Cl,pore}}$ is further compared to the Cl$^-$ concentration value obtained from core squeezing experiments or seepage water composition from in situ equipped borehole, which is assumed to be representative of the bulk concentration $c_{Cl,0}$ (Vinsot et al. 2008a,b; Pearson et al. 2011; Tournassat et al. 2011; Fernández et al. 2014; Mazurek et al. 2015). The value of $\overline{c_{Cl,pore}}$ is always lower than that of $c_{Cl,0}$, demonstrating the depletion of Cl$^-$ in the diffuse layer. From simple mass balance consideration, it follows that:

$$\frac{\overline{c_{Cl,pore}}}{c_{Cl,0}} = \left(1 - f_{DL}\right) + f_{DL}\exp\left(\frac{z_{Cl}F\psi_M}{RT}\right) \qquad (20)$$

in which the unknown parameters are f_{DL} and ψ_M. If the bulk concentrations of the other species present in solution are also known, e.g., from analysis of squeezing or seepage waters or from (bulk) pore water composition modeling (Gaucher et al. 2009; Pearson et al. 2011; Beaucaire et al. 2012; Gailhanou et al. 2017), Equation (18) provides the necessary constraints to calculate ψ_M as a function of the volumetric charge that is compensated in the diffuse layer, Q_{DL}. The value of Q_{DL} can be itself estimated on the basis of macroscopic measurements (e.g., cation exchange capacities of clays) or mineralogical considerations. However, not all of the surface charge is usually compensated in the diffuse layer, and it may be necessary to consider the contribution of the Stern layer to the charge compensation (Appelo et al. 2010). Geochemical calculation codes capabilities make the calculation possible with the use of surface complexation models (Steefel et al. 2015a), although then this surface complexation model must be coupled (loosely or tightly) to the calculation of the diffuse layer charge. In absence of additional constraints, the experimental value of $\overline{c_{Cl,pore}}/c_{Cl,0}$ can thus be fitted by adjusting the values of f_{DL}, Q_{DL}, or both (Leroy and Revil 2004; Leroy et al. 2007; Appelo et al. 2010; Appelo 2017). The contribution of the Stern layer to the total charge compensation can be estimated with molecular dynamic simulations, which provides a growing set of independent information that can be used to calibrate models of ion distribution and mobility in nanopores (Rotenberg et al. 2007, 2014; Jardat et al. 2009; Tournassat et al. 2009b; Bourg and Sposito 2011; Obliger et al. 2014; Tinnacher et al. 2016; Bourg et al. 2017). The dual continuum model was first developed in PHREEQC, but is also now available in CrunchClay (Tournassat and Steefel 2015, 2019; Soler et al. 2019).

Influence of ionic strength on the dual continuum model predictions

Changes in ionic strength influence the ion concentration distribution in the diffuse layers (Fig. 4), and thus f_{DL} changes also as a function of ionic strength. In the absence of overlap, the effective thickness of the diffuse layer is proportional to the Debye length (κ^{-1} in m) (Fig. 4, bottom) (Sposito 2004; Hunter 2013):

$$\kappa^{-1} = \sqrt{\frac{\varepsilon RT}{2F^2 1000 I}} \qquad (21)$$

In PHREEQC and CrunchClay, the volume of the diffuse layer (V_{DL} in m^3), and hence the f_{DL} value, can be defined as a multiple of the Debye length in order to capture this effect of ionic strength on f_{DL}:

$$V_{DL} = \alpha_{DL}\kappa^{-1}S$$
$$f_{DL} = \frac{V_{DL}}{V_{pore}} \qquad (22)$$

where V_{pore} is the total volume of the pore (m^3), S is the surface area that borders the pore (m^2), and α_{DL} is an empirical multiplying factor. This relationship has, however, severe limitations. The value of f_{DL} is linearly related to the Debye length for a limited range of conditions only. For example, if $\alpha_{DL}=2$, a strict equality between the dual continuum model and the Poisson–Boltzmann equation predictions is observed in slit-shaped pores that are filled with NaCl electrolyte, the width of which is large compared to $4\kappa^{-1}$ (Sposito 2004; Tournassat and Appelo 2011; Tournassat et al. 2015a; Appelo 2017). As the width of these pores decreases, the value of f_{DL} approaches the limiting value of one, as shown by the good agreement of the original MEP model with the PB model under these conditions (Fig. 5). Also, it is obvious that f_{DL} cannot exceed 1. Equation (22) must then be seen as an approximation, the validity of which may be limited to small variations of ionic strength compared to the conditions at which f_{DL} is determined experimentally. This can be appreciated by looking at the results obtained with a simple model where:

$$\alpha_{DL} = 2 \text{ if } 4\kappa^{-1} \le \frac{V_{pore}}{S} \text{ and,} \tag{23}$$

$$f_{DL} = 1 \text{ otherwise.} \tag{24}$$

In a homogeneous nanoporous medium represented by a single slit-shaped pore, these model parameters make it possible to improve estimations of ion concentration distribution compared to the MEP model: concentrations of counter-ions are calculated very accurately (less than 2 % error, except for monovalent cations in the "seawater dilution" case on Fig. 7) while the prediction of co-ion concentrations is notably improved compared to the MEP model applied to the entire pore volume (Fig. 7). Still, some conditions cannot be modeled satisfactorily, especially at low ionic strength (large values of $I^{-1/2}$ on Fig. 7).

It is frequently argued that the Poisson–Boltzmann equation, and hence the mean electrostatic potential model or the dual continuum model, are not always accurate to model the diffuse layer composition, particularly in the presence of complex electrolytes or high surface charges (Torrie and Valleau 1982; Carnie and Torrie 1984; Luo et al. 2006; Wang and Chen 2007; Lima et al. 2008; Lee et al. 2012a). In PHREEQC, corrections can be applied by assigning an enrichment factor in the diffuse layer to each of the chemical species (Appelo et al. 2010). This correction makes it possible to fit any target ion concentration distribution in the diffuse layer, but it must be in turn calibrated with additional experimental or theoretical inputs.

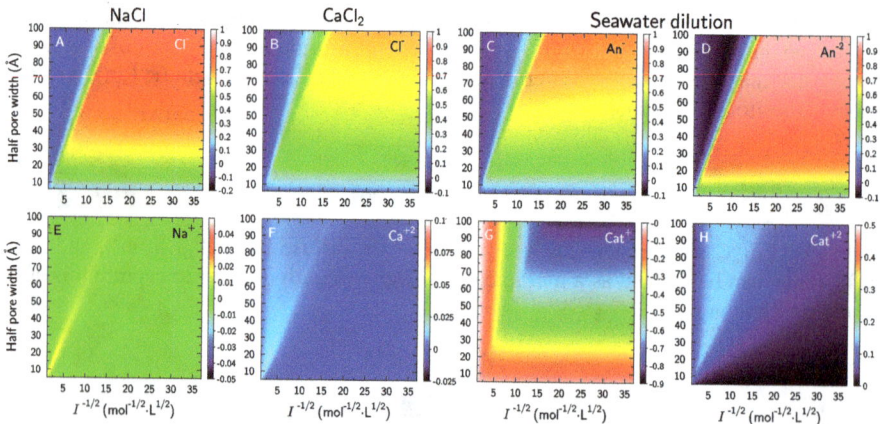

Figure 7. Difference of average diffuse layer concentration values computed with the Poisson–Boltzmann (c_{PB}) equation and the Dual continuum model (c_{Dual}) model described in Figure 6 and Equations (23) and (24). The difference, given by the color scale, is computed with the equation ($c_{PB}-c_{Dual})/c_{PB}$ as a function of the inverse of the square root of ionic strength ($I^{-1/2}$ on the x-axis) and as a function of the half pore width (y-axis). The equivalence between the two models corresponds to ($c_{PB}-c_{Dual})/c_{PB}$=0. Same caption as in Fig. 5 otherwise.

The dual continuum model introduced in the present section is currently the only model available to calculate the diffuse layer composition in PHREEQC and CrunchClay, i.e., the two reactive transport codes that consider explicitly the presence of a diffuse layer in aqueous species transport calculation. Recently, the MEP model has been added to an in-house version of Flotran by Gimmi and Alt-Epping (2018) with the consideration of fixed charges in solution, as opposed to surface charge in PHREEQC and CrunchClay.

In the following section, we focus on the transport equations in the presence of a diffuse layer. Additional work is certainly needed to improve the accuracy of the prediction of average concentration in the diffuse layer. However, the type of method used for this calculation does not influence the concepts and equations used in transport equations, and in the following, the concentration c_i of the species i in bulk or DL water is defined as a function of its concentration in bulk water and of an accumulation (or depletion) factor A_i, which takes the value of 1 in bulk water:

$$c_i = c_{i,0} A_i \qquad (25)$$

and with the following charge balance constraints, which is equivalent to Equation (18):

$$\sum_i z_i F \overline{c_{i,pore}} = F\left(1 - f_{DL}\right)\sum_i z_i c_{i,0} + F f_{DL} \sum_i z_i c_{i,0} A_{i,DL} = F f_{DL} \sum_i z_i c_{i,0} A_{i,DL} = -Q_{DL} \qquad (26)$$

This notation makes it possible to treat the transport equations without *a priori* choice of a diffuse layer model.

COUPLED TRANSPORT PROCESSES IN REACTIVE TRANSPORT CODES— THE ISOTHERMAL NO-FLOW CONDITION

Nernst–Planck equation in the isothermal, no-flow, and no external electric field condition

Tournassat and Steefel (2015) reviewed the basics of the application of the Nernst–Planck equation to the diffuse layer bordering charged surfaces in reactive transport codes in the isothermal, no-flow and no current condition. The basic hypothesis of this application is that, at each time step, the diffuse layer equilibrates infinitely rapidly with a bulk water composition in each of the grid cells that describe the system. As explained in the previous section, the bulk water present in the grid cell can be real or fictitious, depending on the microstructural meaning of the dual continuum representation (Fig. 6). The size of a grid cell, which is at least a few micrometers for the reactive transport calculations with the highest spatial resolution, is far larger than the size of a diffuse layer, which is usually less than a few nanometers. This difference of length scale justifies the assumption of rapid equilibrium between the bulk water and the diffuse layer. The equilibrium condition is written as:

$$\mu_{i,0} = \mu_{i,DL} \qquad (27)$$

$$\mu_{i,0} = \mu_i^\circ + RT \ln \gamma_i c_{i,0} + z_i F \psi_e$$

where $\mu_{i,0}$ is the electro-chemical potential of species i in bulk water, $\mu_{i,DL}$ is the electro-chemical potential of species i in the diffuse layer, μ_i° is the standard electro-chemical potential of species i (in J·mol^{-1}), γ_i is its activity coefficient (here in L·mol^{-1}), and ψ_e is the electrical potential. Note that ψ_e does not correspond to the potential in the diffuse layer, but to a diffusion potential or an external electric potential. According to the Nernst–Planck equation, in the absence of fluid flow, the flux of species i is (in one dimension):

$$J_i = -u_i c_i \frac{\partial \mu_i}{\partial x} = -u_i c_i RT \frac{\partial \ln\left(\gamma_i c_{i,0}\right)}{\partial x} - u_i z_i F c_i \frac{\partial \psi_e}{\partial x} \qquad (28)$$

where u_i is the mobility of species i. Equation (28) applies to the bulk water and to the diffuse layer, and introducing A_i, the accumulation factor defined in Equation (25), it follows:

$$J_i = -u_i c_{i,0} A_i RT \frac{\partial \ln(\gamma_i c_{i,0})}{\partial x} - u_i z_i F c_{i,0} A_i \frac{\partial \psi_e}{\partial x} \tag{29}$$

In the absence of an external electric field, there is no electrical current and so:

$$\sum_i z_i J_i = 0 = -\sum_i \left(u_i z_i c_{i,0} A_i RT \frac{\partial \ln(\gamma_i c_{i,0})}{\partial x} + u_i z_i^2 F c_{i,0} A_i \frac{\partial \psi_e}{\partial x} \right) \tag{30}$$

It follows:

$$\frac{\partial \psi_e}{\partial x} = \frac{-\sum_i u_i z_i c_{i,0} A_i RT \frac{\partial \ln(\gamma_i c_{i,0})}{\partial x}}{\sum_i u_i z_i^2 F c_{i,0} A_i} \tag{31}$$

The mobility of ion i in the porosity is:

$$u_i = \frac{\phi \cdot \tau_i \cdot D_{0,i}}{RT} \tag{32}$$

and Equation (29) becomes:

$$J_i = -\phi \tau_i D_{0,i} c_{i,0} A_i \frac{\partial \ln(\gamma_i c_{i,0})}{\partial x} + \phi \tau_i D_{0,i} z_i c_{i,0} A_i \frac{\sum_j \tau_j D_{0,j} z_j c_{j,0} A_j \frac{\partial \ln(\gamma_j c_{j,0})}{\partial x}}{\sum_k \tau_k D_{0,k} z_k^2 c_{k,0} A_k} \tag{33}$$

Equation (33) can be further rearranged to yield:

$$J_i = -\phi \tau_i D_{0,i} c_{i,0} A_i \left(1 - \frac{\tau_i D_{0,i} z_i c_{i,0} z_i^2}{\sum_k \tau_k D_{0,k} z_k^2 c_{k,0} A_k} \right) \frac{\partial \ln(\gamma_i c_{i,0})}{\partial x}$$

$$+ \sum_{j \neq i} \frac{\phi \tau_i \tau_j D_{0,i} D_{0,j} z_i z_j c_{i,0} c_{j,0} A_i A_j \frac{\partial \ln(\gamma_j c_{j,0})}{\partial x}}{\sum_k \tau_k D_{0,k} z_k^2 c_{k,0} A_k} \tag{34}$$

With isothermal, no-flow, and no external electric field conditions assumed here, the only driving forces are chemical potential gradients, and the phenomenological coefficients can be obtained directly from Equation (34):

$$L_{i,i} = -\phi \cdot \tau_i \cdot D_{0,i} c_{i,0} A_i \left(1 - \frac{\tau_i \cdot D_{0,i} z_i c_{i,0} z_i^2}{\sum_k \tau_k \cdot D_{0,k} z_k^2 c_{k,0} A_k} \right)$$

$$L_{i,j \neq i} = \frac{\phi \cdot \tau_i \tau_j \cdot D_{0,i} D_{0,j} \cdot z_i z_j \cdot c_{i,0} c_{j,0} \cdot A_i A_j}{\sum_k \tau_k \cdot D_{0,k} z_k^2 c_{k,0} A_k} \tag{35}$$

The Onsager relationships $L_{i,j}=L_{j,i}$ are verified for these phenomenological coefficients. These coefficients apply to the bulk water and to the diffuse layer water defined in the dual continuum model. These coefficients can also be used to characterize the entire pore volume by substituting an average accumulation factor $\overline{A_i}$ to the bulk and diffuse layer accumulation factors A_i:

$$\overline{A_i} = 1 - f_{DL} + f_{DL}A_i \tag{36}$$

This substitution requires however that average geometrical factors $\overline{\tau_i}$ can also be defined in some fashion.

In classical isotopic tracer diffusion experiments (Molera and Eriksen 2002; Van Loon et al. 2003, 2005, 2007; Descostes et al. 2008; Wersin et al. 2008; Tachi and Yotsuji 2014; Glaus et al. 2015; Dagnelie et al. 2018), the concentration of the diffusing species i is very low compared to the constant electrolyte background concentration. Under these conditions, Equation (35) reduces to:

$$L_{i,i} \approx -\phi \cdot \tau_i \cdot D_{0,i}c_{i,0}A_i$$
$$L_{i,j\neq i} \approx 0 \tag{37}$$

In addition, the activity coefficients are also constant, and thus Equation (33) reduces to Fick's law, augmented by the accumulation term A_i that accelerate the diffusive flux of counter-ions and decelerate that of co-ions. In the presence of an electrolyte background concentration gradient, i.e., a condition that is encountered in salt diffusion experiments (Shackelford 1991; Shackelford and Daniel 1991; Malusis et al. 2015), the coupling terms between the ionic species cannot be neglected, explaining thus the need to distinguish inter-diffusion (or counter-diffusion, or mutual-diffusion) coefficients from salt diffusion coefficients. The experimental observation that effective salt-diffusion coefficients in clay membrane are dependent on both the salt concentration gradient and the effective stress applied to the clay (Malusis et al. 2015) is directly traceable from Equation (35). The dependence on salt concentration gradient is related to the non-zero values of $L_{i,j}$ terms, while the dependence on effective stress is related to the variations of A_i values as a function of changes in pore size distribution following the compaction of the material. For diffusion processes, it is thus necessary to take into account the electrophoretic processes corresponding to the coupling between the charge imbalance in the diffuse layer, the differences in concentration gradients for each species having different charges and thus also different accumulation factors in the diffuse layer, together with the diffusion coefficient of each individual species (Eqn. 31).

Reactive transport codes such as PHREEQC and CrunchClay are able to solve Equation (33) (without the consideration of the activity coefficient gradient in CrunchClay) under transient and stationary conditions with a diffuse layer dual continuum model, making it thus possible to calculate diffusion membrane properties from species dependent self-diffusion coefficients and from the intrinsic properties of the material, i.e., diffusion pathway geometry and ion concentration distribution. In theory, these last two parameters could be calculated from the exact knowledge of microstructure and mineralogical composition. In practice, the geometrical factor values of diffuse layer and bulk water, as well as the concentration distribution between the bulk porosity and the diffuse layer in the dual continuum model, are fitted on experimental data. Reactive transport calculations have been essential to decipher and quantify the mechanisms explaining the contrasting diffusional behavior of cations, anions and neutral species in clay materials (Appelo and Wersin 2007; Appelo et al. 2010; Glaus et al. 2013, 2015; Tournassat and Steefel 2015; Tinnacher et al. 2016; Soler et al. 2019). The fact that the accumulation of cations in the diffuse layer enhances their effective diffusion coefficient according to $D_{e,i} = \phi\tau_i D_{0,i}A_i$, when present at trace concentrations, is an important result for the prediction of the fate of radionuclides such as radioactive Cs^+ or Sr^{2+} migrating from an underground radioactive waste disposal (Altmann et al. 2012). It is therefore surprising that reactive transport codes have not yet been used extensively to interpret experimental data obtained with geosynthetic clay liners used for the isolation of landfills.

Counter-intuitive results obtained with cationic tracer experiments in the presence of a gradient of salinity can also be explained and quantified using the dual-continuum model. In our opinion, the best example of such counter-intuitive results was published by Glaus et al. (2013). The authors carried out a $^{22}Na^+$ diffusion experiment through a compacted Na-montmorillonite in the presence of a $NaClO_4$ (stable isotopes) background electrolyte concentration gradient from 0.5 mol·L^{-1} to 0.1 mol·L^{-1}. At time t_0, after pre-equilibration of the clay with the two background electrolytes at each end of the clay sample, the two reservoirs were spiked with the same $^{22}Na^+$ concentration. A simple application of Fick's Law would predict that $^{22}Na^+$ diffuses into the clay and then $^{22}Na^+$ concentration stabilizes rapidly at the same value in both reservoirs. The actual results evidenced a continuous increase of the $^{22}Na^+$ concentration in the $NaClO_4$ 0.5 mol·L^{-1} reservoir and a decrease in the $NaClO_4$ 0.1 mol·L^{-1} reservoir. This result can be explained by the difference of accumulation factor $A_{^{22}Na^+}$ in the porosity at both ends of the sample. The value of $A_{^{22}Na^+}$ was anti-correlated with the concentration of stable Na^+:

$$A_{^{22}Na^+} = A_{Na^+} \frac{c_{^{22}Na^+}}{c_{Na^+}} \tag{38}$$

and the gradient in accumulation factor $A_{^{22}Na^+}$ drove the seemingly "up-hill" diffusion of $A_{^{22}Na^+}$ through the clay sample (Glaus et al. 2013) (Fig. 8).

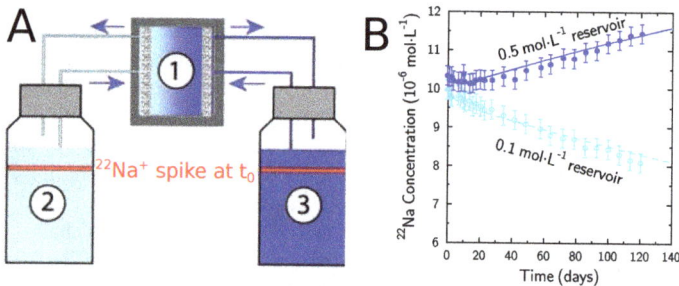

Figure 8. $^{22}Na^+$ "Up-hill" diffusion experiment of Glaus et al. (2013). **A:** experimental setup. The clay sample (**1**) is in contact with a low salinity reservoir (**2**) and high salinity reservoir (**3**) by liquid lines. Blue colors represent the concentration of the background electrolyte; the red bars indicate that the experiment started with equal initial concentrations of the $^{22}Na^+$ tracer in the two reservoirs. **B:** Experimental (symbols) and modeling (lines) results obtained with CrunchClay for the experiment carried out with with a ~5 mm thick clay plug and reservoir volumes of ~250 mL [Modified after Tournassat and Steefel (2015) Rev Mineral Geochem 80:287–330 CCBY and with permission from Glaus et al. (2013) Seeming steady-state uphill diffusion of $^{22}Na^+$ in compacted montmorillonite Environmental Science & Technology 47:11522–11527, copyright 2013 American Chemical Society.]

Nernst–Planck equation in the isothermal and no-flow condition in the presence of an external electric field

Electro-migration experiments consist in applying an external electric potential with two electrodes at both ends of a diffusion cell in order to accelerate the migration of ionic species in very resistive porous media such as cementitious or clayey materials (Goto and Roy 1981; Maes et al. 1999; Truc et al. 2000a,b; Samson et al. 2003; Friedmann et al. 2004, 2008a; Beauwens et al. 2005; Narsilio et al. 2007; Shi et al. 2011). In cementitious materials, the diffusion coefficient of chloride is the most commonly studied parameter. The interpretation of the results obtained with electro-migration tests is not straightforward, and a vast literature has been dedicated to the modeling of these experimental tests (Krabbenhøft and Krabbenhøft 2008). When the diffusion parameters of a single species are sought, a common practice consists in applying the so-called single-species model, in which only the transport of the ion of interest is considered, and the electric field is approximated by a constant equal to the electric potential change over the sample divided by the sample length. More elaborate models rely on the solution of the Poisson–Nernst–Planck equation. According to Krabbenhøft and Krabbenhøft (2008), while the constant field approximation may be a very poor approximation, the effective diffusivities predicted by this model are often in agreement with natural diffusion tests, i.e., diffusion experiments carried out in the absence of an external electric field. The modeling of both the Cl⁻ breakthrough together and the current measured during an electro-migration experiment is however more challenging (Appelo 2017). Discrepancies between model predictions with the Poisson–Nernst–Planck equation and actual measurements were apparent, and the influence of the diffuse layer on the transport properties was put forward by various authors as a possible explanation of these discrepancies (Goto and Roy 1981; Friedmann et al. 2004, 2008a,b; Krabbenhøft and Krabbenhøft 2008). However, no modeling framework other than those based on phenomenological coefficients was available to decipher and quantify the possible mechanisms, until PHREEQC was adapted to consider the application of an external electrical field (Appelo 2017).

The "no external electric field" condition used to solve Equation (29) in reactive transport codes can be relaxed to model electro-migration experiments. In the presence of an external electrical field, an electrical flux J_c (A·m⁻² or C·m⁻²·s⁻¹) of ionic nature appears, which is related to the fluxes of aqueous species. The following method can be used to calculate the electrical flux in 1D (Appelo 2017):

$$J_c = \sum_j z_j F J_j = J_{c,d} + J_{c,e} \tag{39}$$

$J_{c,d}$ is the current contribution from the chemical activity gradient:

$$J_{c,d} = -\sum_j z_j F \phi \tau_j D_{0,j} c_{j,0} A_j \frac{\partial \ln(\gamma_j c_{j,0})}{\partial x} \tag{40}$$

and $J_{c,e}$ is the current contribution from the electrical potential gradient:

$$J_{c,e} = -\sum_j \frac{z_j^2 F^2 \phi \tau_j D_{0,j} c_j A_j}{RT} \frac{\partial \psi_e}{\partial x} = -\sum_j \phi \tau_j \Lambda_{m,j} c_j A_j \frac{\partial \psi_e}{\partial x} \tag{41}$$

where:

$$\Lambda_{m,j} = \frac{z_j^2 F^2 D_{0,j}}{RT} \tag{42}$$

is the molar conductivity of species i (A·V⁻¹·m²·mol⁻¹). If the subscript 1 is attributed to the driving force due to the presence of an electrical potential gradient and to the electrical flux, and the other subscripts are attributed to the driving forces due to the chemical activity gradients and the conjugated ionic fluxes, it follows:

$$L_{1,1} = -\sum_j \phi \tau_j \Lambda_{m,j} c_j A_j \tag{43}$$

$$L_{j \neq 1, j \neq 1} = -\phi \tau_j D_{0,j} c_{j,0} A_j \tag{44}$$

$$L_{1,j \neq 1} = -z_j F \phi \tau_j D_{0,j} c_{j,0} A_j = L_{j \neq 1,1} \tag{45}$$

$$L_{1,1} = -\sum_j \phi\tau_j\Lambda_{m,j}c_jA_j \qquad (46)$$

As a consequence, the model proposed by Appelo (2017) follows Onsager reciprocal relationships for bulk water as well as for diffuse layer water. Equation (46) means that there is no effect of the activity gradient of chemical species i on the flux of a chemical species j other than the influence mediated by the electrical potential gradient. Appelo (2017) introduced in his model the influence of the ionic strength on the self-diffusion coefficient of ionic species to model the change of molar conductivities with concentration and composition of the solution. This effect can be included in the $D_{0,j}$ terms present in Equations (43) to (45).

The numerical resolution of Equation (29) in the framework of a reactive transport approach requires that the local gradient of electrical potential $\dfrac{\partial\psi_e}{\partial x}$ is correctly evaluated. This evaluation should rely on the solution of the Nernst–Planck equation together with the Poisson equation (i.e., the Poisson–Nernst–Planck equation):

$$\frac{d^2\psi_e}{dx^2} - \frac{1}{\varepsilon}\sum_i(z_iFc_i + q) = 0 \qquad (47)$$

where $q = Q_{DL}$ in the diffuse layer and $q = 0$ in bulk water. However, the order of magnitude of F/ε in Equation (47) is 10^{15} V·m·mol^{-1}. Consequently, the second derivative of the potential (first term) in Equation (47) is negligible in comparison to the charge balance (second term) except for very small length scales or very large electric potentials (Truc et al. 2000b,c; Krabbenhøft and Krabbenhøft 2008). The length scale that is usually considered in reactive transport calculations, i.e., the size of a grid cell, is typically larger than a few micrometers. As a consequence, this approximation must be correct. In addition, Appelo (2017) pointed out that the precision of any numerical model for calculating chemical reactions is usually not better than 10^{-8} mol·m^{-3}, and a calculation of the second derivative of the potential from charge imbalance could not be more precise than 1.4×10^6 V·m^{-2}, a value that is usually not reached in electro-migration experiments (Appelo 2017). Therefore, the Poisson equation reduces to a local electro-neutrality condition, and there is no transient local charge build-up: J_c is thus constant throughout the sample.

Rearranging Equation (41) yields:

$$\frac{\partial\psi_e}{\partial x} = -\frac{J_{c,e}}{\sum_j \phi\tau_j\Lambda_{m,j}c_jA_j} = -R_cJ_{c,e} \qquad (48)$$

where R_c is the resistivity of the nanoporous media. After discretization on a grid cell, Equation (48) becomes:

$$\psi_e(x + \Delta x) - \psi_e(x) = -\Delta x R_c(x)J_{c,e}(x) \qquad (49)$$

and:

$$\psi_e(x_{n+1}) - \psi_e(x_0) = -\sum_{k=0}^{n}\Delta x_k R_c(x_k)J_{c,e}(x_k) \qquad (50)$$

where x_0 and x_{n+1} correspond to the positions where the potential $\psi_e(x_{n+1}) - \psi_e(x_0)$ is imposed. and $\Delta x_k R_c(x_k)$ are known quantities. Combining Equations (50) and (39), with J_c being constant throughout the sample, yields:

$$\psi_e\left(x_{n+1}\right) - \psi_e\left(x_0\right) = \sum_{k=0}^{n} -\Delta x_k R_c\left(x_k\right)\left(J_c - J_{c,d}\left(x_k\right)\right) \tag{51}$$

$J_{c,d}(x_k)$ is a known quantity from Equation (40) and the value of J_c can thus be computed.

The implementation of these equations in PHREEQC made it possible to unravel some of the diffusion properties of cementitious materials. In particular, the role of the diffuse layer and its average charge in the establishment of the measured current and ion breakthrough during electro-migration experiments was quantified, demonstrating that reactive transport can be used as an essential tool to understand and quantify highly coupled transport processes in nanoporous media (Appelo 2017).

BEYOND DIFFUSION, COUPLINGS WITH ADVECTIVE FLOW

Advective displacement method and reactive transport calculations

The "advective displacement" method consists in forcing flow through a confined sample by applying an hydraulic gradient in order to collect time series of small aliquots of displaced pore water (Fig. 9). It is usually assumed that early extracts closely resemble the *in situ* composition of the pore water. In addition, injection of tracers and the record of their breakthrough, or the elution of elements present initially in the sample, make it possible to derive a range of parameters for transport, adsorption/desorption, and dissolution/precipitation reactions using a reactive transport approach (Grambow et al. 2014; Montavon et al. 2014; Mäder and Waber 2017; Mäder 2018). Reactive transport modeling of advective displacement experiments have been carried out in a handful of studies with the consideration of the presence of a diffuse layer and the dual continuum model described in a previous section (Grambow et al. 2014; Alt-Epping et al. 2015, 2018). In all of these studies, a simplifying approximation was made: the porosity affected by the advective flow was set to the same exact value as the initial bulk porosity defined with the dual continuum model (Fig. 10A). While a significant portion of the diffuse layer porosity may remain indeed unaffected by the advective flow, it is unlikely that this portion is always equal to that of the diffuse layer porosity, which must be seen itself as a convenient simplified representation of the sample microstructure (see previous section and Fig. 6). Because the water density and viscosity in the diffuse layer is similar to that of bulk water, an increase of the diffuse layer porosity from $\phi_{DL,1}$ to $\phi_{DL,2}$, following e.g., a decrease of ionic strength or an increase of the ratio of monovalent counter-ions relative to multi-valent counter-ions in the bulk water, should

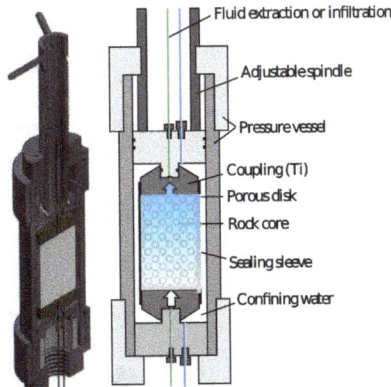

Figure 9. Advective displacement experimental setup. Reproduced from Mäder (2018) Advective displacement method for the characterisation of pore water chemistry and transport properties in claystone. Geofluids ID 8198762 CCBY.

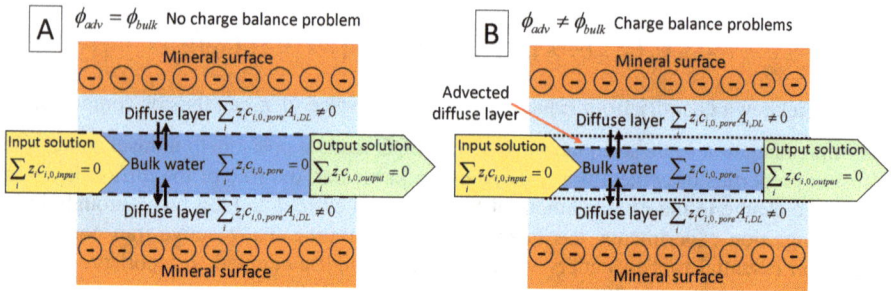

Figure 10. Advective flow in the presence of a diffuse layer and related problematics of charge balance. **A:** The advective flow is limited to the bulk water porosity. An electro-neutral solution displaces an electro-neutral solution and the charge balance is maintained throughout the sample. **B:** The advective flow displaces the bulk water porosity as well as a portion of the diffuse layer porosity. An electro-neutral solution displaces a charged solution: coupled processes must occur to maintain the local charge balance in the sample porosity and in the output solution.

lead to an advective displacement of a fraction of diffuse layer porosity equivalent to $\phi_{DL,2} - \phi_{DL,1}$. This displacement should result in the appearance of coupled flux because of charge imbalance (Fig. 10B). Yet, modeling results obtained with a fixed value of ϕ_{DL} led to satisfying agreement between experimental data and modeling predictions (Fig. 11). Several explanations can be put forward to explain this good agreement including: the change of the input solution may have been too limited to change significantly the size of the diffuse layer; or the concentrations of the elements of interest were mostly controlled by reaction processes (dissolution/precipitation, cation exchange in the absence of formation of a diffuse layer); or the fitted transport parameters may also lump together several coupled processes. In this last case, the parameters must be seen as being apparent rather than intrinsic to the material studied.

Advective displacement studies can be seen as the adaptation to intact clayey rocks of earlier experiments carried out on compacted bentonite or montmorillonite samples (McKelvey and Milne 1960; Kemper and Maasland 1964; Milne et al. 1964; Hanshaw and Coplen 1973). Ultrafiltration processes were clearly evidenced with steady state output solutions having a lower electrolyte concentration than input solutions. It is thus not possible to model these experiments with the model shown in Fig. 10A.

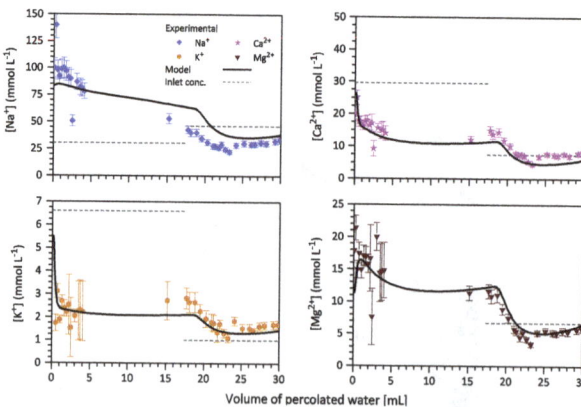

Figure 11. Example of experimental and modeling results obtained with an advective displacement experiment using a Callovian–Oxfordian argillite sample from the underground research laboratory in Bure (Meuse–Haute Marne, France). The type of model used in the study corresponds to the model shown in Fig. 10A. [Reproduced from Grambow et al. (2014) Nuclear waste disposal: I Laboratory simulation of repository properties Applied Geochemistry 49:237–246, Copyright 2014, with permission from Elsevier.]

Osmotic membrane efficiency

The influence of the diffuse layer on water and ion water transport processes in nanoporous media is most often taken into account implicitly using empirical phenomenological coefficients (Soler 2001; Malusis et al. 2012; Neuzil and Person 2017). As shown in a previous section, this modeling approach is restricted to systems in which a simple electrolyte such as NaCl is considered. In this approach, a key parameter is the coefficient of osmotic efficiency ω (–), which quantifies exclusion of ions from the total pore volume because of the presence of a diffuse layer. In the absence of thermal or electrical potential gradients, the ion flux of a single salt is modeled as the result of four contributions (Soler 2001; Sun et al. 2016; Neuzil and Person 2017):

$$J_i = J_{adv,i} + J_{diff,i} + J_{CO,i} + J_{HYP,i} \tag{52}$$

where $J_{CO,i}$ and $J_{HYP,i}$ are the chemical osmosis and hyperfiltration fluxes. Chemical osmosis corresponds to the viscous flow of water down the water activity gradient (or up the osmotic pressure gradient), which drags solutes along. The osmotic pressure of an aqueous solution, π, may be expressed in the Raoult-Lewis equation (Lewis 1908):

$$\pi - \frac{\alpha \pi^2}{2} = -\frac{RT}{V_o} \ln a_w \tag{53}$$

where α is the compressibility coefficient of water, V_o is its molecular volume, and a_w is its activity. Neglecting the water compressibility, the following equation is obtained for the osmotic pressure gradient:

$$\frac{\partial \pi}{\partial x} = RT \sum_j \frac{\partial c_j}{\partial x} \tag{54}$$

For aqueous solutions with molar concentration below 1 mol·L^{-1}, the osmotic pressure can be approximated with van't Hoff's Law:

$$\pi = RT \sum_j c_j \tag{55}$$

and the osmotic pressure gradient is:

$$\frac{\partial \pi}{\partial x} = RT \sum_j \frac{\partial c_j}{\partial x} \tag{56}$$

The corresponding chemical osmosis flux is:

$$J_{CO,i} = c_i \omega K \frac{\partial \Pi}{\partial x} \tag{57}$$

where $\Pi = \pi/\rho g$ is the osmotic pressure head (in m; ρ is the density of water and g the acceleration due to gravity in m·s^{-2}). The hyperfiltration flux corresponds to a correction to the advective and chemical osmosis flux due to the exclusion of solutes from the diffuse layer:

$$J_{HYP,i} = c_i \omega K \left(\frac{\partial h}{\partial x} - \omega \frac{\partial \Pi}{\partial x} \right) \tag{58}$$

The diffusion coefficient is usually taken at the same value for anions and cations (salt diffusion coefficient) and corrected from the membrane efficiency coefficient. It follows:

$$J_i = -D_i\left(1-\omega\right)\frac{\partial c_i}{\partial x} + \left(1-\omega\right)c_i\left(-K\frac{\partial h}{\partial x} + \omega K\frac{\partial \Pi}{\partial x}\right) \tag{59}$$

A value of $\omega = 1$ means that the membrane is ideal and there is no ionic flux through it. A value of $\omega = 0$ means that the media has no membrane properties.

Equation (59) or similar equations have been widely used in the literature to characterize membrane properties of clay materials (Soler 2001; Malusis and Shackelford 2002, 2004; Malusis et al. 2003, 2012; Neuzil and Person 2017). The coefficient of membrane efficiency is however dependent on the electrolyte nature and concentration, and as noted by Malusis and Shackelford (2012), the development of concentration gradients within the membrane preclude the calculation of intrinsic ω values from experimental data using Equation (59).

Nernst–Planck equation in the isothermal, no external electric field, no current condition, and in the presence of advective flow

In the presence of advective flow, the Nernst–Planck equation becomes:

$$J_i = -\phi\tau_i D_{0,i}c_{i,0}A_i\frac{\partial \ln\left(\gamma_i c_{i,0}\right)}{\partial x} - \frac{z_i F\phi\tau_i D_{0,i}c_{i,0}A_i}{RT}\frac{\partial \psi_e}{\partial x} + c_i A_i\phi v \tag{60}$$

where v again is the average pore fluid linear velocity (m·s^{-1}). The no current condition imposes:

$$\sum_i z_i J_i = 0 = -\sum_i z_i\phi\tau_i D_{0,i}c_{i,0}A_i\frac{\partial \ln\left(\gamma_i c_{i,0}\right)}{\partial x} - \sum_i \frac{z_i^2 F\phi\tau_i D_{0,i}c_{i,0}A_i}{RT}\frac{\partial \psi_e}{\partial x} + \sum_i z_i c_{i,0}A_i\phi v \tag{61}$$

By rearranging Equation (61), it is possible to express the value of $\frac{\partial \psi_e}{\partial x}$ as a function of chemical potential gradients, average linear pore fluid velocity, accumulation factors and concentrations in a multi-component system:

$$\frac{\partial \psi_e}{\partial x} = \frac{-\dfrac{Q_{DL}}{F}v - \sum_i z_i\tau_i D_{0,i}c_{i,0}A_i\dfrac{\partial \ln\left(\gamma_i c_{i,0}\right)}{\partial x}}{\sum_i \dfrac{z_i^2 F\tau_i D_{0,i}c_{i,0}A_i}{RT}} \tag{62}$$

The first term of the numerator of the right part in Equation (62) corresponds to a streaming potential, while the second term of the right part corresponds to a diffusion potential. Equation (62) is then reinserted into Equation (60):

$$J_i = -\phi\tau_i D_{0,i}c_{i,0}A_i\frac{\partial \ln\left(\gamma_i c_{i,0}\right)}{\partial x}$$
$$-z_i\phi\tau_i D_{0,i}c_{i,0}A_i\left(\frac{-\dfrac{Q_{DL}}{F}v - \sum_j z_j\tau_j D_{0,j}c_{j,0}A_j\dfrac{\partial \ln\left(\gamma_j c_{j,0}\right)}{\partial x}}{\sum_k z_k^2\tau_k D_{0,k}c_{k,0}A_k}\right) \tag{63}$$
$$+c_{i,0}A_i\phi v$$

Equation (63) can be further rearranged to yield:

$$J_i = -\phi \tau_i D_{0,i} c_{i,0} A_i \left(1 - \frac{z_i^2 \tau_i D_{0,i} c_{i,0}}{\sum_k z_k^2 \tau_k D_{0,k} c_{k,0} A_k} \right) \frac{\partial \ln(\gamma_i c_{i,0})}{\partial x}$$

$$+ \sum_{j \neq i} \frac{\phi \tau_i \tau_j D_{0,i} D_{0,j} z_i z_j c_{i,0} c_{j,0} A_i A_j \frac{\partial \ln(\gamma_j c_{j,0})}{\partial x}}{\sum_k z_k^2 \tau_k D_{0,k} c_{k,0} A_k} \qquad (64)$$

$$+ c_{i,0} \phi \left(A_i + \frac{z_i \tau_i D_{0,i} A_i \frac{Q_{DL}}{F}}{\sum_k z_k^2 \tau_k D_{0,k} c_{k,0} A_k} \right) v$$

which is a generalization of the single salt equation given in Revil and Leroy (2004). The species dependent coefficient of filtration efficiency, ω_i becomes:

$$\omega_i = 1 - A_i - \frac{z_i \tau_i D_{0,i} A_i \frac{Q_{DL}}{F}}{\sum_k z_k^2 \tau_k D_{0,k} c_{k,0} A_k} \qquad (65)$$

Equation (65) makes clear that, in multicomponent systems, the coefficient of osmotic efficiency is specific to each chemical species. As expected, Equation (65) indicates that the nanoporous medium is a perfect membrane with regards to the species i ($\omega_i = 1$) if $A_i = 0$, i.e., if the species i is completely excluded from the porosity. Equation (65) also indicates that the coefficient of osmotic efficiency is equal to 0 if $A_i = 1$ for all species, because $\sum_i z_i F c_{i,0} A_i = -Q_{DL} = 0$. In this last case, the porous media has no membrane properties. Equation (64) has not been implemented in a reactive transport code yet. To do this, it would be necessary to relate the linear fluid velocity to hydrostatic and osmotic pressure terms in a multi-component model framework.

From dual continuum to multi-continuum

If the entire porosity is affected by the advective fluid flow, the coefficients of osmotic efficiency can be calculated on the basis of the average pore concentrations computed in the previous section. Similarly, the osmotic pressure gradient can be calculated with Equation (56) combined with one of the models used to estimate the concentration in a charged pore. However, the concentration used for calculating the osmotic pressure may be different from the average concentration in the pore. In smectite osmotic swelling models, the reference concentration for this calculation is usually taken as the mid-pore concentration (Langmuir 1938; Warkentin and Schofield 1958; Blackmore and Miller 1961; Madsen and Müller-Vonmoos 1989; Liu 2013; Massat et al. 2016), i.e., a parameter the value of which cannot be easily computed from a mean electrostatic potential modeling approach. It would be surprising in any case if the advective flow affects uniformly the entire porosity, especially in clayey rocks and cementitious materials that exhibit a large range of pore sizes from nanometer size (interlayer porosity) to few micrometers (macropores) (Curtis et al. 2012; Kuila and Prasad 2013; Hemes et al. 2015, 2016; Gaboreau et al. 2016). It may be thus necessary to subdivide the porosity into different domains, in which specific surface charge and fluid flow are distributed heterogeneously. Multiple-porosity model for fluid flow in shale reservoirs are currently developed to capture the effect of the complexity of the shales pore structure on fluid flow (Yan et al. 2016). The consideration of ionic transport in modeling work adds one more level of complexity because of the presence of charged surfaces that make it necessary to adopt a multi-continuum approach. Even in the absence of fluid flow, a multi-continuum instead of a dual-continuum description may be necessary to model the diffusion behavior of some species such as Cs^+, where interlayer diffusion may be important compared to diffusion in bulk and diffuse layer water (Appelo et al. 2010).

CONCLUSIONS

The current implementations in reactive transport codes of coupled transport processes in the presence of a diffuse layer are in agreement with the theory of irreversible thermodynamics processes. These implementations available in only two codes—CrunchClay and PHREEQC—are currently limited to the modeling of purely diffusive systems, i.e., without advective flow where there is a need to consider flow with that portion of the pore structure affected by the diffuse layer. It is also possible to model electro-migration in PHREEQC. The multi-component capability of these codes makes it possible to overcome many limitations encountered with simplified approaches that are limited to single salt transport. In particular, the transport of electrolyte background ions and tracer ions can be simulated simultaneously, demonstrating the existence of highly coupled processes during their transport through nanoporous media. In addition, reactive transport codes are not limited to the simulation of systems with fixed surface charge and fixed diffuse layer dimensions, and can incorporate microstructural information in dual- to multi-continuum descriptions of the porosity. Applications of these capabilities to the simulation of clayey and cementitious material properties highlighted the prominent role of the diffuse layer in the transport properties of ions. This review provides some insights to advance reactive transport modeling in the direction of a full implementation of multi-component advective flux in the presence of a diffuse layer. In particular, the osmotic efficiency parameter must be revisited so that ion and neutral molecule specific properties can be simulated simultaneously. Holistic models of flow and ion transport in nanoporous charged media still face significant but ultimately energizing scientific and computational challenges.

ACKNOWLEDGMENTS

This work was supported by the Office of Science, Office of Basic Energy Sciences, of the U.S. Department of Energy (BES-DOE) under Contract No. DE-AC02-05CH11231, the TRANSREAC project from the Défi NEEDS – MIPOR, and by the TelluS Program of CNRS/INSU. Carl I. Steefel acknowledges funding from L'Institut Carnot for his visit to the BRGM. We gratefully acknowledge, Thomas Gimmi and Josep Soler for their constructive review of the manuscript.

REFERENCES

Alt-Epping P, Tournassat C, Rasouli P, Steefel CI, Mayer KU, Jenni A, Mäder U, Sengor SS, Fernandez R (2015) Benchmark reactive transport simulations of a column experiment in compacted bentonite with multispecies diffusion and explicit treatment of electrostatic effects. Comput Geosci 19:535–550
Alt-Epping P, Gimmi T, Wersin P, Jenni A (2018) Incorporating electrical double layers into reactive-transport simulations of processes in clays by using the Nernst–Planck equation: A benchmark revisited. Appl Geochem 89:1–10
Altmann S, Tournassat C, Goutelard F, Parneix J-C, Gimmi T, Maes N (2012) Diffusion-driven transport in clayrock formations. Appl Geochem 27:463–478
Andrietti F, Peres A, Pezzotta R (1976) Exact solution of the unidimensional Poisson–Boltzmann equation for a 1: 2 (2: 1) electrolyte. Biophys J 16:1121–1124
Appelo CAJ (2017) Solute transport solved with the Nernst–Planck equation for concrete pores with "free" water and a double layer. Cem Concr Res 101:102–113
Appelo CAJ, Wersin P (2007) Multicomponent diffusion modeling in clay systems with application to the diffusion of tritium, iodide, and sodium in Opalinus clay. Environ Sci Technol 41:5002–5007
Appelo CAJ, Vinsot A, Mettler S, Wechner S (2008) Obtaining the porewater composition of a clay rock by modeling the in- and out-diffusion of anions and cations from an in-situ experiment. J Contam Hydrol 101:67–76

Appelo CAJ, Van Loon LR, Wersin P (2010) Multicomponent diffusion of a suite of tracers (HTO, Cl, Br, I, Na, Sr, Cs) in a single sample of Opalinus clay. Geochim Cosmochim Acta 74:1201–1219

Babcock KL (1960) Some characteristics of a model Donnan system. Soil Sci 90:245–252

Babcock KL (1963) Theory of the chemical properties of soil colloidal systems at equilibrium. Hilgardia 34:417–542

Banin A, Lahav N (1968) Particle size and optical properties of montmorillonite in suspension. Israel J Chem 6:235–250

Bear J (1972) Dynamics of Fluids in Porous Media. Courier Dover Publications

Beaucaire C, Tertre E, Ferrage E, Grenut B, Pronier S, Madé B (2012) A thermodynamic model for the prediction of pore water composition of clayey rock at 25 and 80 °C—Comparison with results from hydrothermal alteration experiments. Chem Geol 334:62–76

Beauwens T, De Canniere P, Moors H, Wang L, Maes N (2005) Studying the migration behaviour of selenate in Boom Clay by electromigration. Eng Geol 77:285–293

Beinum W van, Hofmann A, Meeussen JC, Kretzschmar R (2005) Sorption kinetics of strontium in porous hydrous ferric oxide aggregates: I The Donnan diffusion model. J Colloid Interfac Sci 283:18–28

Bildstein O, Claret F, Frugier P (2019) RTM for waste repositories. Rev Mineral Geochem 85:419–457

Birgersson M (2017) A general framework for ion equilibrium calculations in compacted bentonite. Geochim Cosmochim Acta 200:186–(200)

Birgersson M, Karnland O (2009) Ion equilibrium between montmorillonite interlayer space and an external solution-Consequences for diffusional transport. Geochim Cosmochim Acta 73:1908–1923

Blackmore AV, Miller RD (1961) Tactoid size and osmotic swelling in calcium montmorillonite. Soil Sci Soc Proc 25:169–173

Borkovec M, Westall J (1983) Solution of the Poisson–Boltzmann equation for surface excesses of ions in the diffuse layer at the oxide electrolyte interface. J Electroanal Chem 150:325–337

Bourg IC (2015) Sealing shales versus brittle shales: A sharp threshold in the material properties and energy, technology uses of fine-grained sedimentary rocks. Environ Sci Technol Lett 2:255–259

Bourg IC, Sposito G (2011) Molecular dynamics simulations of the electrical double layer on smectite surfaces contacting concentrated mixed electrolyte (NaCl–CaCl$_2$) solutions. J Colloid Interfac Sci 360:701–715

Bourg IC, Steefel CI (2012) Molecular dynamics simulations of water structure and diffusion in silica nanopores. J Phys Chem C 116:11556–11564

Bourg IC, Lee SS, Fenter P, Tournassat C (2017) Stern layer structure and energetics at mica–water interfaces. J Phys Chem C 121:9402–9412

Bresler E (1973) Anion exclusion and coupling effects in nonsteady transport through unsaturated soils: I Theory. Soil Sci Soc Am J 37:663–669

Brigatti MF, Galán E, Theng BKG (2013) Chapter 2—Structure and mineralogy of clay minerals. *In:* Bergaya F, Lagaly G (Eds.), Handbook of Clay Science, Developments in Clay Science. Elsevier, p 21–81

Cama J, Soler JM, Ayora C (2019) Acid water–rock–cement interaction and multicomponent reactive transport modeling. Rev Mineral Geochem 85:459–498

Carnie SL, Torrie GM (1984) Advance in chemical physics. *In:* Prigogine I A RS (Eds.), Advances in Chemical Physics. John Wiley & Sons, Inc., p 141–253

Cey BD, Barbour S, Hendry MJ (2001) Osmotic flow through a Cretaceous clay in southern Saskatchewan, Canada. Can Geotech J 38:1025–1033

Chen Z, Singh RK (2002) General solution for Poisson–Boltzmann equation in semiinfinite planar symmetry. J Colloid Interfac Sci 245:301–306

Claret F, Marty N, Tournassat C (2018) Modeling the long-term stability of multi-barrier systems for nuclear waste disposal in geological clay formations. *In:* Xiao Y, Whitaker F, Xu T, Steefel C (Eds.), Reactive Transport Modeling: Applications in Subsurface Energy, and Environmental Problems. John Wiley & Sons, Ltd Chichester, UK, p 395–451

Collin M, Fournier M, Frugier P, Charpentier T, Moskura M, Deng L, Ren M, Du J, Gin S (2018a) Structure of international simple glass and properties of passivating layer formed in circumneutral pH conditions. npj Materials Degradation 2:4

Collin M, Gin S, Dazas B, Mahadevan T, Du J, Bourg IC (2018b) Molecular dynamics simulations of water structure and diffusion in a 1 nm diameter silica nanopore as a function of surface charge and alkali metal counterion identity. J Phys Chem C 122:17764–17776

Curtis ME, Sondergeld CH, Ambrose RJ, Rai CS (2012) Microstructural investigation of gas shales in two and three dimensions using nanometer-scale resolution imaging. AAPG Bull 96:665–677

Dagnelie R, Rasamimanana S, Blin V, Radwan J, Thory E, Robinet J-C, Lefèvre G (2018) Diffusion of organic anions in clay-rich media: Retardation and effect of anion exclusion. Chemosphere 213:472–480

Darcy H (1856) Note II Observation relative à l'écoulement de l'eau dans l'aqueduc du Rosoir. *In:* Les fontaines publiques de la ville de Dijon. Victor Dalmont, p 638–639

de Groot SR, Mazur P (1984) Non-equilibrium Thermodynamics. Courier Corporation

de Marsily G (1986) Quantitative Hydrogeology, Groundwater Hydrology for Engineers. Academic Press, New York

Delville A (2000) Electrostatic interparticle forces: from swelling to setting. *In:* PRO 13: 2nd International RILEM Symposium on Hydration and Setting—Why does cement set? An interdisciplinary approach. RILEM Publications, p 37

Descostes M, Blin V, Bazer-Bachi F, Meier P, Grenut B, Radwan J, Schlegel ML, Buschaert S, Coelho D, Tevissen E (2008) Diffusion of anionic species in Callovo-Oxfordian argillites and Oxfordian limestones (Meuse/Haute-Marne, France). Appl Geochem 23:655–677

Fernández AM, Sánchez-Ledesma DM, Tournassat C, Melón A, Gaucher EC, Astudillo J, Vinsot A (2014) Applying the squeezing technique to highly consolidated clayrocks for pore water characterisation: Lessons learned from experiments at the Mont Terri Rock Laboratory. Appl Geochem 49:2–21

Fick A (1855) V On liquid diffusion. The London, Edinburgh Dublin Phil Mag J Sci 10:30–39

Friedmann H, Amiri O, Ait-Mokhtar A, Dumargue P (2004) A direct method for determining chloride diffusion coefficient by using migration test. Cem Concr Res 34:1967–1973

Friedmann H, Amiri O, Ait-Mokhtar A (2008a) Shortcomings of geometrical approach in multi-species modelling of chloride migration in cement-based materials. Mag Concr Res 60:119–124

Friedmann H, Amiri O, Ait-Mokhtar A (2008b) Physical modeling of the electrical double layer effects on multispecies ions transport in cement-based materials. Cem Concr Res 38:1394–1400

Frugier P, Minet Y, Rajmohan N, Godon N, Gin S (2018) Modeling glass corrosion with GRAAL. npj Mater Degrad 2:35

Gaboreau S, Robinet J-C, Prêt D (2016) Optimization of pore network characterization of compacted clay materials by TEM and FIB/SEM imaging. Micropor Mesopor Mater 224:116–128

Gailhanou H, Lerouge C, Debure M, Gaboreau S, Gaucher EC, Grangeon S, Grenèche J-M, Kars M, Madé B, Marty NCM, Warmont F, Tournassat C (2017) Effects of a thermal perturbation on mineralogy and pore water composition in a clay-rock: an experimental and modeling study. Geochim Cosmochim Acta 197:193–214

Garavito AM, De Cannière P, Kooi H (2007) In situ chemical osmosis experiment in the Boom Clay at the Mol underground research laboratory. Phys Chem Earth, Parts A/B/C 32:421–433

Gates WP, Bouazza A, Churchman GJ (2009) Bentonite clay keeps pollutants at bay. Elements 5:105–110

Gaucher E, Robelin C, Matray J-M, Negrel G, Gros Y, Heitz JF, Vinsot A, Rebours H, Cassabagnere A, Bouchet A (2004) ANDRA underground research laboratory: Interpretation of the mineralogical and geochemical data acquired in the Callovian–Oxfordian Formation by investigative drilling. Phys Chem Earth, Parts A/B/C 29:55–77

Gaucher EC, Tournassat C, Pearson FJ, Blanc P, Crouzet C, Lerouge C, Altmann S (2009) A robust model for pore-water chemistry of clayrock. Geochim Cosmochim Acta 73:6470–6487

Gimmi T, Alt-Epping P (2018) Simulating Donnan equilibria based on the Nernst–Planck equation. Geochim Cosmochim Acta 232:1–13

Glaus MA, Birgersson M, Karnland O, Van Loon LR (2013) Seeming steady-state uphill diffusion of $^{22}Na^+$ in compacted montmorillonite. Environ Sci Technol 47:11522–11527

Glaus M, Aertsens M, Appelo C, Kupcik T, Maes N, Van Laer L, Van Loon L (2015) Cation diffusion in the electrical double layer enhances the mass transfer rates for Sr^{2+}, Co^{2+} and Zn^{2+} in compacted illite. Geochim Cosmochim Acta 165:376–388

Gonçalvès J, Rousseau-Gueutin P, Revil A (2007) Introducing interacting diffuse layers in TLM calculations: A reappraisal of the influence of the pore size on the swelling pressure and the osmotic efficiency of compacted bentonites. J Colloid Interfac Sci 316:92–99

Gonçalvès J, Adler PM, Cosenza P, Pazdniakou A, de Marsily G (2015) Chapter 8—Semipermeable membrane properties and chemomechanical coupling in clay barriers. In: Tournassat C, Steefel CI, Bourg IC, Bergaya F (Eds.), Natural and Engineered Clay Barriers, Developments in Clay Science. Elsevier, p 270–327

Goto S, Roy DM (1981) Diffusion of ions through hardened cement pastes. Cem Concr Res 11:751–757

Grambow B (2006) Nuclear waste glasses-How durable? Elements 2:357–364

Grambow B, Landesman C, Ribet S (2014) Nuclear waste disposal: I Laboratory simulation of repository properties. Appl Geochem 49:237–246

Grangeon S, Claret F, Lerouge C, Warmont F, Sato T, Anraku S, Numako C, Linard Y, Lanson B (2013) On the nature of structural disorder in calcium silicate hydrates with a calcium/silicon ratio similar to tobermorite. Cem Concr Res 52:31–37

Grangeon S, Claret F, Roosz C, Sato T, Gaboreau S, Linard Y (2016) Structure of nanocrystalline calcium silicate hydrates: insights from X-ray diffraction, synchrotron X-ray absorption and nuclear magnetic resonance. J Appl Crystallogr 49:771–783

Grangeon S, Fernandez-Martinez A, Baronnet A, Marty N, Poulain A, Elkaim E, Roosz C, Gaboreau S, Henocq P, Claret F (2017) Quantitative X-ray pair distribution function analysis of nanocrystalline calcium silicate hydrates: a contribution to the understanding of cement chemistry. J Appl Crystallogr 50:14–21

Gu X, Cole DR, Rother G, Mildner DF, Brantley SL (2015) Pores in Marcellus shale: a neutron scattering and FIB-SEM study. Energy Fuels 29:1295–1308

Guyonnet D, Touze-Foltz N, Norotte V, Pothier C, Didier G, Gailhanou H, Blanc P, Warmont F (2009) Performance-based indicators for controlling geosynthetic clay liners in landfill applications. Geotextiles Geomembranes 27:321–331

Haas J, Nonat A (2015) From C–S–H to C–A–S–H: Experimental study and thermodynamic modelling. Cem Concr Res 68:124–138

Hanshaw BB, Coplen TB (1973) Ultrafiltration by a compacted clay membrane—II Sodium ion exclusion at various ionic strengths. Geochim Cosmochim Acta 37:2311–2327

Hemes S, Desbois G, Urai JL, Schröppel B, Schwarz J-O (2015) Multi-scale characterization of porosity in Boom Clay (HADES-level, Mol, Belgium) using a combination of X-ray μ-CT, 2D BIB-SEM and FIB-SEM tomography. Micropor Mesopor Mater 208:1–20

Hemes S, Desbois G, Klaver J, Urai JL (2016) Microstructural characterisation of the Ypresian clays (Kallo−1) at nanometre resolution, using broad-ion beam milling and scanning electron microscopy. Netherlands J Geosci 95:293–313

Holland HD (1978) The Chemistry of the Atmosphere and Oceans. Wiley-interscience, New York

Holmboe M, Bourg IC (2014) Molecular dynamics simulations of water and sodium diffusion in smectite interlayer nanopores as a function of pore size and temperature. J Phys Chem C 118:1001–1013

Horseman S, Harrington J, Noy D (2007) Swelling and osmotic flow in a potential host rock. Phys Chem Earth, Parts A/B/C 32:408–420

Hunter RJ (2013) Zeta Potential in Colloid Science: Principles and Applications. Academic Press

Ioannidou K, Krakowiak KJ, Bauchy M, Hoover CG, Masoero E, Yip S, Ulm F-J, Levitz P, Pellenq RJ-M, Del Gado E (2016) Mesoscale texture of cement hydrates. PNAS 113:2029–(2034)

Jardat M, Dufreche JF, Marry V, Rotenberg B, Turq P (2009) Salt exclusion in charged porous media: a coarse-graining strategy in the case of montmorillonite clays. Phys Chem Chem Phys 11:2023–(2033)

Jougnot D, Revil A, Leroy P (2009) Diffusion of ionic tracers in the Callovo–Oxfordian clay-rock using the Donnan equilibrium model and the formation factor. Geochim Cosmochim Acta 73:2712–2726

Kemper WD (1960) Water and ion movement in thin films as influenced by the electrostatic charge and diffuse layer of cations associated with clay mineral surfaces. Soil Sci Soc Proc 10–16

Kemper W (1961a) Movement of water as effected by free energy and pressure gradients: II Experimental analysis of porous systems in which free energy and pressure gradients act in opposite directions. Soil Sci Soc Am J 25:260–265

Kemper W (1961b) Movement of water as effected by free energy and pressure gradients: I Application of classic equations for viscous and diffusive movements to the liquid phase in finely porous media. Soil Sci Soc Am J 25:255–260

Kemper W, Evans N (1963) Movement of water as effected by free energy and pressure gradients III Restriction of solutes by membranes. Soil Sci Soc Am J 27:485–490

Kemper WD, Maasland DEL (1964) Reduction in salt content of solution on passing through thin films adjacent to charged surfaces 1. Soil Sci Soc Am J 28:318–323

Kemper WD, Rollins JB (1966) Osmotic efficiency coefficients across compacted clays. Soil Sci Soc Am J 30:529–534

Keller LM, Holzer L, Wepf R, Gasser P (2011) 3D geometry and topology of pore pathways in Opalinus clay: Implications for mass transport. Appl Clay Sci 52:85–95

Keller LM, Holzer L, Schuetz P, Gasser P (2013) Pore space relevant for gas permeability in Opalinus clay: Statistical analysis of homogeneity, percolation, and representative volume element. J Geophys Res 118:2799–2812

Kjelstrup S, Bedeaux D (2008) Non-equilibrium Thermodynamics of Heterogeneous Systems. World Scientific

Krabbenhøft K, Krabbenhøft J (2008) Application of the Poisson–Nernst–Planck equations to the migration test. Cem Concr Res 38:77–88

Kuila U, Prasad M (2013) Specific surface area and pore-size distribution in clays and shales. Geophys Prospect 61:341–362

Lagneau V, Regnault O, Descostes M (2019) Industrial deployment of reactive transport simulation: An application to uranium in situ recovery. Rev Mineral Geochem 85:499–528

Langmuir I (1938) The role of attractive and repulsive forces in the formation of tactoids, thixotropic gels, protein crystals and coacervates. J Chem Phys 6:873–896

Lasaga AC (1998) Kinetic Theory in the Earth Sciences. Princeton University Press

Le Forestier L, Muller F, Villiéras F, Pelletier M (2010) Textural and hydration properties of a synthetic montmorillonite compared with a natural Na-exchanged clay analogue. Appl Clay Sci 48:18–25

Lee SS, Fenter P, Park C, Sturchio NC, Nagy KL (2010) Hydrated cation speciation at the muscovite (001)–water interface. Langmuir 26:16647–16651

Lee JW, Nilson RH, Templeton JA, Griffiths SK, Kung A, Wong BM (2012a) Comparison of molecular dynamics with classical density functional and Poisson–Boltzmann Theories of the electric double layer in nanochannels. J Chem Theory Comput 8:2012–2022

Lee SS, Fenter P, Nagy KL, Sturchio NC (2012b) Monovalent ion adsorption at the muscovite (001)–solution interface: Relationships among ion coverage and speciation, interfacial water structure, and substrate relaxation. Langmuir 28:8637–8650

Leroy P, Maineult A (2018) Exploring the electrical potential inside cylinders beyond the Debye–Hückel approximation: a computer code to solve the Poisson–Boltzmann equation for multivalent electrolytes. Geophys J Int 214:58–69

Leroy P, Revil A (2004) A triple-layer model of the surface electrochemical properties of clay minerals. J Colloid Interfac Sci 270:371–380

Leroy P, Revil A, Coelho D (2006) Diffusion of ionic species in bentonite. J Colloid Interfac Sci 296:248–255

Leroy P, Revil A, Altmann S, Tournassat C (2007) Modeling the composition of the pore water in a clay-rock geological formation (Callovo-Oxfordian, France). Geochim Cosmochim Acta 71:1087–1097

Lewis GN (1908) The osmotic pressure of concentrated solutions, and the laws of the perfect solution. J Am Chem Soc 30:668–683

Li Y-H, Gregory S (1974) Diffusion of ions in sea water and in deep-sea sediments. Geochim Cosmochim Acta 38:703–714

Lima E, Horinek D, Netz R, Biscaia E, Tavares F, Kunz W, Boström M (2008) Specific ion adsorption and surface forces in colloid science. J Phys Chem B 112:1580–1585

Liu L (2013) Prediction of swelling pressures of different types of bentonite in dilute solutions. Colloids Surf A 434:303–318

Loucaides S, Behrends T, Van Cappellen P (2010) Reactivity of biogenic silica: Surface versus bulk charge density. Geochim Cosmochim Acta 74:517–530

Luo G, Malkova S, Yoon J, Schultz DG, Lin B, Meron M, Benjamin I, Vansek P, Schlossman ML (2006) Ion distributions near a liquid-liquid interface. Science 311:216–218

Lynde C (1912) Osmosis in soils: Soils act as semi-permeable membranes. Agron J 4:102–108

Ma B, Fernandez-Martinez A, Grangeon S, Tournassat C, Findling N, Carrero S, Tisserand D, Bureau S, Elkaïm E, Marini C, Aquilanti G, Koishi A, Marty NCM, Charlet L (2018) Selenite uptake by Ca–Al LDH: a description of intercalated anion coordination geometries. Environ Sci Technol 52:1624–1632

Ma B, Fernandez-Martinez A, Grangeon S, Tournassat C, Findling N, Claret F, Koishi A, Marty NC, Tisserand D, Bureau S, Salas-Colera E (2017) Evidence of multiple sorption modes in layered double hydroxides using Mo as structural probe. Environ Sci Technol 51:5531–5540

Mäder U (2018) Advective displacement method for the characterisation of pore water chemistry and transport properties in claystone. Geofluids:8198762

Mäder UK, Waber HN (2017) Characterization of pore water, ion transport and water-rock interaction in claystone by advective displacement experiments. Procedia earth and planetary science 17:917–920

Madsen FT, Müller-Vonmoos M (1989) The swelling behaviour of clays. Appl Clay Sci 4:143–156

Maes N, Moors H, Dierckx A, De Cannière P, Put M (1999) The assessment of electromigration as a new technique to study diffusion of radionuclides in clayey soils. J Contam Hydrol 36:231–247

Maher K, Navarre-Stichler A (2019) Reactive transport processes that drive chemical weathering: From making space for water to dismantling continents. Rev Mineral Geochem 85:349–380

Malusis MA, Shackelford CD (2002) Theory for reactive solute transport through clay membrane barriers. J Contam Hydrol 59:291–316

Malusis MA, Shackelford CD (2004) Predicting solute flux through a clay membrane barrier. J Geotech Geoenviron Eng 130:477–487

Malusis MA, Shackelford CD, Olsen HW (2003) Flow and transport through clay membrane barriers. Eng Geol 70:235–248

Malusis MA, Shackelford CD, Maneval JE (2012) Critical review of coupled flux formulations for clay membranes based on nonequilibrium thermodynamics. J Contam Hydrol 138:40–59

Malusis MA, Kang JB, Shackelford CD (2015) Restricted salt diffusion in a geosynthetic clay liner. Environmental Geotechnics 2:68

Marry V, Turq P, Cartailler T, Levesque D (2002) Microscopic simulation for structure and dynamics of water and counterions in a monohydrated montmorillonite. J Chem Phys 117:3454–3463

Marry V, Rotenberg B, Turq P (2008) Structure and dynamics of water at a clay surface from molecular dynamics simulation. Phys Chem Chem Phys 10:4802–4813

Marty NCM, Cama J, Sato T, Chino D, Villiéras F, Razafitianamaharavo A, Brendlé J, Giffaut E, Soler JM, Gaucher EC, Tournassat C (2011) Dissolution kinetics of synthetic Na-smectite. An integrated experimental approach. Geochim Cosmochim Acta 75:5849–5864

Marty NC, Grangeon S, Elkaim E, Tournassat C, Fauchet C, Claret F (2018) Thermodynamic and crystallographic model for anion uptake by hydrated calcium aluminate (AFm): an example of molybdenum. Sci Rep 8:7943

Masoero E, Del Gado E, Pellenq R-M, Ulm F-J, Yip S (2012) Nanostructure and nanomechanics of cement: polydisperse colloidal packing. Phys Rev Lett 109:155503

Massat L, Cuisinier O, Bihannic I, Claret F, Pelletier M, Masrouri F, Gaboreau S (2016) Swelling pressure development and inter-aggregate porosity evolution upon hydration of a compacted swelling clay. Appl Clay Sci 124:197–210

Mazurek M, Alt-Epping P, Bath A, Gimmi T, Niklaus Waber H, Buschaert S, Cannière PD, De Craen M, Gautschi A, Savoye S, Vinsot A, Wernaere I, Wouters L (2011) Natural tracer profiles across argillaceous formations. Appl Geochem 26:1035–1064

Mazurek M, Oyama T, Wersin P, Alt-Epping P (2015) Pore-water squeezing from indurated shales. Chem Geol 400:106–121

McKelvey JG, Milne IH (1960) The flow of salt solutions through compacted clay. Clay Clay Mineral 9:248–259

Medved I, Cern R (2013) Osmosis in porous media: A review of recent studies. Micropor Mesopor Mater 170:299–317

Milne I, McKelvey J, Trump R (1964) Semi-permeability of bentonite membranes to brines. AAPG Bull 48:103–105

Molera M, Eriksen T (2002) Diffusion of $^{22}Na^+$, $^{85}Sr^{2+}$, $^{134}Cs^+$ and $^{57}Co^{2+}$ in bentonite clay compacted to different densities: experiments and modeling. Radiochim Acta 90:75–760

Montavon G, Sabatié-Gogova A, Ribet S, Bailly C, Bessaguet N, Durce D, Giffaut E, Landesman C, Grambow B (2014) Retention of iodide by the Callovo-Oxfordian formation: An experimental study. Appl Clay Sci 87:142–149

Narsilio G, Li R, Pivonka P, Smith D (2007) Comparative study of methods used to estimate ionic diffusion coefficients using migration tests. Cem Concr Res 37:1152–1163

Navarre-Sitchler A, Steefel CI, Sak PB, Brantley SL (2011) A reactive-transport model for weathering rind formation on basalt. Geochim Cosmochim Acta 75:7644–7667

Neuzil C (2000) Osmotic generation of "anomalous" fluid pressures in geological environments. Nature 403:182

Neuzil CE (2013) Can shale safely host US nuclear waste? Eos, Trans Am Geophys Union 94:261–262

Neuzil CE, Person M (2017) Reexamining ultrafiltration and solute transport in groundwater. Water Resour Res 53:4922–4941

Obliger A, Jardat M, Coelho D, Bekri S, Rotenberg B (2014) Pore network model of electrokinetic transport through charged porous media. Phys Rev E 89:043013

Olsen HW (1962) Hydraulic flow through saturated clays. Clays Clay Minerals 131–161

Onsager L (1931a) Reciprocal relations in irreversible processes. I Phys Rev 37:405

Onsager L (1931b) Reciprocal relations in irreversible processes. II Phys Rev 38:2265

Patriarche D, Ledoux E, Michelot J-L, Simon-Coinçon R, Savoye S (2004a) Diffusion as the main process for mass transport in very low water content argillites: 2. Fluid flow and mass transport modeling. Water Resour Res 40:W01517

Patriarche D, Michelot J-L, Ledoux E, Savoye S (2004b) Diffusion as the main process for mass transport in very low water content argillites: 1. Chloride as a natural tracer for mass transport—Diffusion coefficient and concentration measurements in interstitial water. Water Resour Res 40:W01516

Pearson FJ, Tournassat C, Gaucher EC (2011) Biogeochemical processes in a clay formation in situ experiment: Part E - Equilibrium controls on chemistry of pore water from the Opalinus Clay, Mont Terri Underground Research Laboratory, Switzerland. Appl Geochem 26:990–1008

Pellenq RJ-M, Van Damme H (2004) Why does concrete set?: The nature of cohesion forces in hardened cement-based materials. MRS Bull 29:319–323

Philipp T, Amann-Hildenbrand A, Laurich B, Desbois G, Littke R, Urai J (2017) The effect of microstructural heterogeneity on pore size distribution and permeability in Opalinus Clay (Mont Terri, Switzerland): insights from an integrated study of laboratory fluid flow and pore morphology from BIB-SEM images. Geol Soc, London Spec Publ 454:85–106

Poinssot C, Baeyens B, Bradbury MH (1999) Experimental and modelling studies of caesium sorption on illite. Geochim Cosmochim Acta 63:3217–3227

Pusch R (2001) The microstructure of MX–80 clay with respect to its bulk physical properties under different environmental conditions. SKB, TR-01–08

Revil A (1999) Ionic diffusivity, electrical conductivity, membrane and thermoelectric potentials in colloids and granular porous media: a unified model. J Colloid Interfac Sci 212:503–522

Revil A, Leroy P (2004) Constitutive equations for ionic transport in porous shales. J Geophys Res-Solid Earth 109

Revil A, Linde N (2006) Chemico-electromechanical coupling in microporous media. J Colloid Interfac Sci 302:682–694

Revil A, Woodruff W, Lu N (2011) Constitutive equations for coupled flows in clay materials. Water Resour Res 47:1–21

Richardson I (2008) The calcium silicate hydrates. Cem Concr Res 38:137–158

Richardson IG (2014) Model structures for C-(A)-SH (I). Acta Crystallogr Section B 70:903–923

Robinet J-C, Sardini P, Coelho D, Parneix J-C, Prêt D, Sammartino S, Boller E, Altmann S (2012) Effects of mineral distribution at mesoscopic scale on solute diffusion in a clay-rich rock: Example of the Callovo-Oxfordian mudstone (Bure, France). Water Resour Res 48:W05554

Rolle M, Le Borgne T (2019) Mixing and reactive fronts in the subsurface. Rev Mineral Geochem 85:111–142

Roosz C, Gaboreau S, Grangeon S, Prêt D, Montouillout V, Maubec N, Ory S, Blanc P, Vieillard P, Henocq P (2016) Distribution of water in synthetic calcium silicate hydrates. Langmuir 32:6794–6805

Rotenberg B, Marry V, Dufrêche J-F, Giffaut E, Turq P (2007) A multiscale approach to ion diffusion in clays: Building a two-state diffusion-reaction scheme from microscopic dynamics. J Colloid Interfac Sci 309:289–295

Rotenberg B, Marry V, Malikova N, Turq P (2010) Molecular simulation of aqueous solutions at clay surfaces. J Phys Condens Matter 22:284114

Rotenberg B, Marry V, Salanne M, Jardat M, Turq P (2014) Multiscale modelling of transport in clays from the molecular to the sample scale. C R Geosci 346:298–306

Rousseau-Gueutin P, De Greef V, Gonçalvès J, Violette S, Chanchole S (2009) Experimental device for chemical osmosis measurement on natural clay-rock samples maintained at in situ conditions: Implications for formation pressure interpretations. J Colloid Interfac Sci 337:106–116

Saiyouri N, Hicher PY, Tessier D (2000) Microstructural approach and transfer water modelling in highly compacted unsaturated swelling clays. Mech Cohesive-frictional Mater 5:41–60

Samson E, Marchand J, Snyder KA (2003) Calculation of ionic diffusion coefficients on the basis of migration test results. Mater Struct 36:156–165

Sayed Hassan M, Villieras F, Gaboriaud F, Razafitianamaharavo A (2006) AFM and low-pressure argon adsorption analysis of geometrical properties of phyllosilicates. J Colloid Interfac Sci 296:614–623

Schlegel ML, Nagy KL, Fenter P, Cheng L, Sturchio NC, Jacobsen SD (2006) Cation sorption on the muscovite (0 0 1) surface in chloride solutions using high-resolution X-ray reflectivity. Geochim Cosmochim Acta 70:3549–3565

Schramm LL, Kwak JCT (1982) Influence of exchangeable cation composition on the size and shape of montmorillonite particles in dilute suspension. Clay Clay Mineral 30:40–48

Shackelford CD (1991) Laboratory diffusion testing for waste disposal – A review. J Contam Hydrol 7:177–217

Shackelford CD, Daniel DE (1991) Diffusion in saturated soil. I: Background. J Geotech Eng 117:467–484

Shainberg I, Otoh H (1968) Size and shape of montmorillonite particles saturated with Na/Ca ions (inferred from viscosity and optical measurements). Israel J Chem 6:251–259

Shi X, Yang Z, Liu Y, Cross D (2011) Strength and corrosion properties of Portland cement mortar and concrete with mineral admixtures. Construct Building Mater 25:3245–3256

Siretanu I, Ebeling D, Andersson MP, Stipp SLS, Philipse A, Stuart MC, Ende D van den, Mugele F (2014) Direct observation of ionic structure at solid–liquid interfaces: a deep look into the Stern Layer. Sci Rep 4:4956–4956

Soler JM (2001) The effect of coupled transport phenomena in the Opalinus Clay and implications for radionuclide transport. J Contam Hydrol 53:63–84

Soler JM, Steefel CI, Gimmi T, Leupin OX, Cloet V (2019) Modeling the ionic strength effect on diffusion in clay. The DR-A experiment at Mont Terri. ACS Earth Space Chem 3:442–451

Soulaine C, Tchelepi HA (2016) Micro-continuum approach for pore-scale simulation of subsurface processes. Transp Porous Media 113:431–456

Soulaine C, Roman S, Kovscek A, Tchelepi HA (2018) Pore-scale modelling of multiphase reactive flow: application to mineral dissolution with production of CO_2. J Fluid Mech 855:616–645

Sposito G (2004) The Surface Chemistry of Natural Particles. Oxford University Press, New York

Steefel CI (2008) Geochemical kinetics and transport. *In:* Kinetics of Water–Rock Interaction. Springer, p 545–589

Steefel CI, Maher K (2009) Fluid-rock interaction: A reactive transport approach. Rev Mineral Geochem 70:485–532

Steefel CI, Appelo CAJ, Arora B, Jacques D, Kalbacher T, Kolditz O, Lagneau V, Lichtner PC, Mayer KU, Meeussen JCL, Molins S, Moulton D, Shao H, Šimunek J, Spycher N, Yabusaki SB, Yeh GT (2015a) Reactive transport codes for subsurface environmental simulation. Comput Geosci 19:445–478

Steefel CI, Beckingham LE, Landrot G (2015b) Micro-continuum approaches for modeling pore-scale geochemical processes. Rev Mineral Geochem 80:217–246

Sun X, Wu J, Shi X, Wu J (2016) Experimental and numerical modeling of chemical osmosis in the clay samples of the aquitard in the North China Plain. Environ Earth Sci 75:59

Tachi Y, Yotsuji K (2014) Diffusion and sorption of Cs^+, Na^+, I^- and HTO in compacted sodium montmorillonite as a function of porewater salinity: Integrated sorption and diffusion model. Geochim Cosmochim Acta 132:75–93

Tinnacher RM, Holmboe M, Tournassat C, Bourg IC, Davis JA (2016) Ion adsorption and diffusion in smectite: molecular, pore, and continuum scale views. Geochim Cosmochim Acta 177:130–149

Torrie GM, Valleau JP (1982) Electrical double layers. 4. Limitations of the Gouy–Chapman theory. J Phys Chem 86:3251–3257

Tournassat C, Appelo CAJ (2011) Modelling approaches for anion-exclusion in compacted Na-bentonite. Geochim Cosmochim Acta 75:3698–3710

Tournassat C, Steefel CI (2015) Ionic transport in nano-porous clays with consideration of electrostatic effects. Rev Mineral Geochem 80:287–330

Tournassat C, Steefel CI (2019) Modeling diffusion processes in the presence of a diffuse layer at charged mineral surfaces. A benchmark exercise. Comput Geosci, accepted

Tournassat C, Neaman A, Villiéras F, Bosbach D, Charlet L (2003) Nanomorphology of montmorillonite particles: Estimation of the clay edge sorption site density by low-pressure gas adsorption and AFM observations. Am Mineral 88:1989–1995

Tournassat C, Chapron Y, Leroy P, Boulahya F (2009a) Comparison of molecular dynamics simulations with Triple Layer and modified Gouy–Chapman models in a 0.1 M NaCl–montmorillonite system. J Colloid Interfac Sci 339:533–541

Tournassat C, Gailhanou H, Crouzet C, Braibant G, Gautier A, Gaucher EC (2009b) Cation exchange selectivity coefficient values on smectite and mixed-layer illite/smectite minerals. Soil Sci Soc Am J 73:928–942

Tournassat C, Alt-Epping P, Gaucher EC, Gimmi T, Leupin OX, Wersin P (2011) Biogeochemical processes in a clay formation in situ experiment: Part F - Reactive transport modelling. Appl Geochem 26:1009–1022

Tournassat C, Bourg IC, Steefel CI, Bergaya F (2015a) Chapter 1 - Surface properties of clay minerals. *In:* Tournassat C, Steefel CI, Bourg IC, Bergaya F (Eds.), Natural and Engineered Clay Barriers, Developments in Clay Science. Elsevier, p 5–31

Tournassat C, Vinsot A, Gaucher EC, Altmann S (2015b) Chapter 3 - Chemical conditions in clay-rocks. In: Tournassat C, Steefel CI, Bourg IC, Bergaya F (Eds.), Natural and Engineered Clay Barriers, Developments in Clay Science. Elsevier, p 71–100

Tournassat C, Bourg IC, Holmboe M, Sposito G, Steefel CI (2016a) Molecular dynamics simulations of anion exclusion in clay interlayer nanopores. Clay Clay Mineral 64:374–388

Tournassat C, Gaboreau S, Robinet J-C, Bourg IC, Steefel CI (2016b) Impact of microstructure on anion exclusion in compacted clay media. CMS Workshop lecture series 21:137–149

Truc O, Ollivier J, Carcassès M (2000a) A new way for determining the chloride diffusion coefficient in concrete from steady state migration test. Cem Concr Res 30:217–226

Truc O, Ollivier J-P, Nilsson L-O (2000b) Numerical simulation of multi-species transport through saturated concrete during a migration test—MsDiff code. Cem Concr Res 30:1581–1592

Truc O, Ollivier J-P, Nilsson L-O (2000c). Numerical simulation of multi-species diffusion. Mater Struct 33:566–573

Van Loon LR, Baeyens B, Bradbury MH (2005) Diffusion and retention of sodium and strontium in Opalinus clay: Comparison of sorption data from diffusion and batch sorption measurements, and geochemical calculations. Appl Geochem 20:2351–2363

Van Loon LR, Soler JM, Bradbury MH (2003) Diffusion of HTO, $^{36}Cl^-$ and $^{125}I^-$ in Opalinus Clay samples from Mont Terri: Effect of confining pressure. J Contam Hydrol 61:73–83

Van Loon LR, Glaus MA, Müller W (2007) Anion exclusion effects in compacted bentonites: Towards a better understanding of anion diffusion. Appl Geochem 22:2536–2552

Vinsot A, Appelo CAJ, Cailteau C, Wechner S, Pironon J, De Donato P, De Cannière P, Mettler S, Wersin P, Gäbler HE (2008a) CO_2 data on gas and pore water sampled in situ in the Opalinus Clay at the Mont Terri rock laboratory. Phys Chem Earth, Parts A/B/C 33, S54–S60

Vinsot A, Mettler S, Wechner S (2008b) In situ characterization of the Callovo-Oxfordian pore water composition. Phys Chem Earth, Parts A/B/C 33, S75–S86

Wang M, Chen S (2007) Electroosmosis in homogeneously charged micro- and nanoscale random porous media. J Colloid Interfac Sci 314:264–273

Warkentin BP, Schofield RK (1958) Swelling pressures of dilute Na-montmorillonite pastes. Clay Clay Mineral 7:343–349

Wersin P, Soler JM, Van Loon L, Eikenberg J, Baeyens B, Grolimund D, Gimmi T, Dewonck S (2008) Diffusion of HTO, Br^-, I^-, Cs^+, $^{85}Sr^{2+}$ and $^{60}Co^{2+}$ in a clay formation: Results and modelling from an in situ experiment in Opalinus Clay. Appl Geochem 23:678–691

Wigger C, Van Loon LR (2017) Importance of interlayer equivalent pores for anion diffusion in clay-rich sedimentary rocks. Environ Sci Technol 51:1998–2006

Yan B, Wang Y, Killough JE (2016) Beyond dual-porosity modeling for the simulation of complex flow mechanisms in shale reservoirs. Comput Geosci 20:69–91

Yokoyama S, Kuroda M, Sato T (2005) Atomic force microscopy study of montmorillonite dissolution under highly alkaline conditions. Clay Clay Mineral 53:147–154

Young A, Low PF (1965) Osmosis in argillaceous rocks. AAPG Bull 49:1004–1007

Zachara JM, Smith SC, McKinley JP, Resch CT (1993) Cadmium sorption on specimen and soil smectites in sodium and calcium electrolytes. Soil Sci Soc Am J 57:1491–1501

Zhao H, Bhattacharjee S, Chow R, Wallace D, Masliyah JH, Xu Z (2008) Probing surface charge potentials of clay basal planes and edges by direct force measurements. Langmuir 24:12899–12910

Reviews in Mineralogy & Geochemistry
Vol. 85 pp. 111-142, 2019
Copyright © Mineralogical Society of America

Mixing and Reactive Fronts in the Subsurface

Massimo Rolle

Department of Environmental Engineering
Technical University of Denmark
Lyngby
Denmark

masro@env.dtu.dk

Tanguy Le Borgne

Geosciences Rennes
University of Rennes 1
Rennes
France

tanguy.le-borgne@univ-rennes1.fr

INTRODUCTION

The interplay between mixing and reactive processes plays a pivotal role in a range of biogeochemical processes controlling the transport, transformation and turnover of chemical elements (McClain et al. 2003; Dentz et al. 2011; Valocchi et al. 2018). The study of these processes is of primary importance in many scientific disciplines and technical applications, including fluid mechanics, geochemistry, chemical engineering, reservoir engineering, subsurface and contaminant hydrology, water treatment, and remediation of contaminated sites. Mixing fronts often act as hot spots of biogeochemical reactions as they induce chemical disequilibrium and bring together reactants that may be otherwise segregated. By definition, reactive mixing fronts are characterized by strong and evolving chemical gradients; therefore, they cannot be in general understood and modeled under the conventional assumption of well-mixed chemical systems.

In this chapter, we provide an overview of some recent advances in the characterization, understanding and modeling of mixing and reactive fronts in the subsurface. We focus on incompressible water flow in porous media and we discuss two main types of mixing fronts, which play an important role in a range of applications from contaminant transport to geochemistry and eco-hydrology: (i) transient mixing fronts, where an advancing fluid progressively displaces a fluid of different chemical composition, (ii) steady-state mixing fronts, which develop around continuously released plumes. We present the basic mechanisms that govern the effective mixing and reaction rates of such fronts, and discuss recent findings that have changed our view on the deformation dynamics of these fronts in heterogeneous porous media and motivated the development of new modeling frameworks. We provide examples and discuss implications of mixing front dynamics for different biogeochemical reactions at scales ranging from the pore scale to the field scale.

Figure 1 illustrates a few examples of mixing and reaction fronts, which occur naturally (e.g., river/groundwater, sea/groundwater interactions) or as a result of contaminant release in groundwater from anthropogenic and/or natural sources. As they bring together reactants and create chemical disequilibrium, mixing fronts are often viewed as hot spots of biogeochemical reactions, where reaction rates can be much larger than in other areas (McMahon 2001;

1529-6466/19/0085-0005$05.00 (print)
1943-2666/19/0085-0005$05.00 (online)

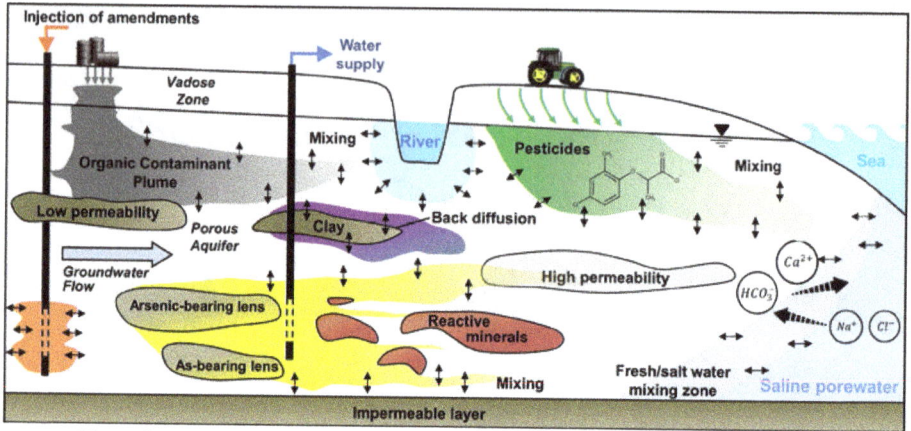

Figure 1. Schematic of groundwater contamination and mixing processes in the subsurface.

Martinez-Landa et al. 2012). They can thus control key geochemical processes at interfaces between different water bodies (groundwater, rivers, lakes, oceans) (McClain et al. 2003; McAllister et al. 2015; Stegen et al. 2015; Hester et al. 2017), determine the fate of released pollutants and the delivery of reactants and amendments (Kitanidis and McCarty 2012).

In the context of contaminant hydrology, hot spots of mixing occur at the fringes of contaminant plumes where the dissolved contaminants contact clean groundwater. The mixing of different waters triggers important reactions that may lead to contaminant degradation and to the alteration of the groundwater and/or porous medium chemistry through precipitation and dissolution reactions. For instance, the fate of organic contaminant plumes depends on microbially-mediated redox reactions in which the pollutants can be degraded through different electron acceptor processes (Chapelle et al. 1995; Christensen et al. 2001). In the mixing zones at the plume fringes favorable conditions occur since soluble electron donors and acceptors, such as the organic contaminants and dissolved oxygen, nitrate and sulfate present in the pristine groundwater, come into contact (Anneser et al. 2008; Rolle et al. 2008). This favors the degrading activity of indigenous microorganisms that can access the necessary substrates for their metabolic activity.

Subsurface porous media are poorly mixed environments in which water flow is dominated by viscous forces and mixing is hindered by the absence of turbulence. Under such conditions of creeping flow, dilution of conservative solutes (i.e., their distribution over increasingly larger volumes or water fluxes) and mixing of initially segregated reactants control many reactive processes. These processes often represent the bottleneck and the rate-limiting steps for the overall reaction rates since they are typically slow and characterized by long time scales. Aqueous diffusion of the solutes in the pore water is the ultimate control on mixing processes since diffusion is the only physical mechanism leading to the increase of entropy and to the decrease of the peak concentration of solute plumes (Kitanidis 1994; Chiogna et al. 2011a).

Scales of mixing in porous media

Mixing processes occur at different spatial and temporal scales. Figure 2 illustrates three important spatial scales of investigation of solute transport and mixing in porous media: the pore scale in solid–liquid domains with typical length of $\sim 10^{-2}$ m, the laboratory or Darcy scale with length of common laboratory experimental setups ($\sim 10^0$ m), and the field scale ($\sim 10^2$ m).

The study of flow, solute transport, and mixing at the pore scale has been instrumental to illuminate the coupling between different physical and reactive processes, to help gaining a fundamental understanding of these mechanisms, and to allow the interpretation of

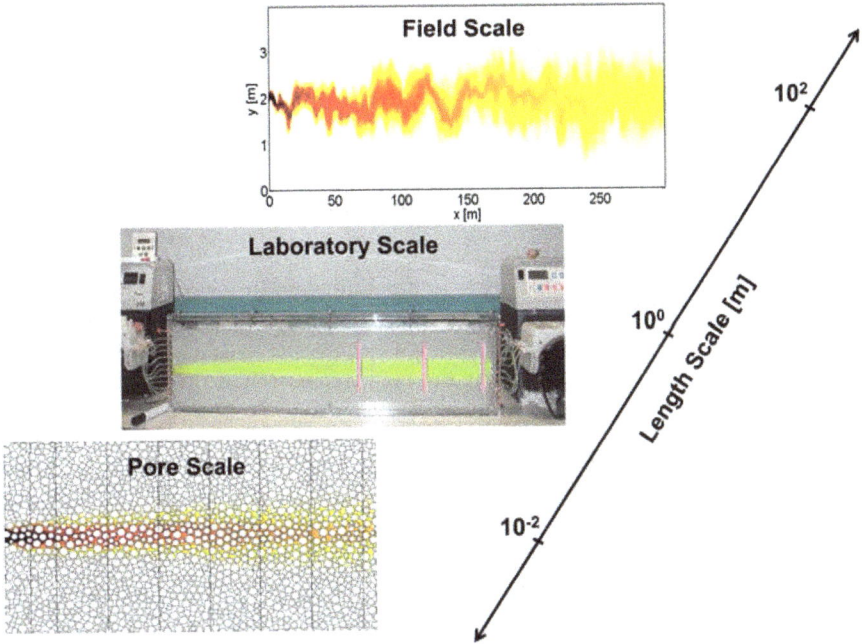

Figure 2. Scales of investigation of transport and mixing in the subsurface: examples showing continuously injected plumes at the pore, laboratory, and field scales (modified after Rolle et al. 2013a).

macroscopic experimental observations. At this scale the pore geometry is known and the flow problem is described by the Stokes equation and the continuity equation. The Stokes equation is a simplification of the Navier-Stokes equation for an incompressible fluid, disregarding inertia effects. Such momentum-flux terms are negligible in typical porous media flows, which are characterized by low Reynolds numbers. Considering steady-state conditions, the Stokes equation is a balance between potential energy and viscous dissipation terms:

$$-\rho g \nabla \varphi + \mu \nabla^2 \boldsymbol{u} = 0 \tag{1}$$

where ρ is the fluid density, g is the constant of acceleration due to gravity, φ is the hydraulic head ($p/(\rho g) + z$), p is the pressure, z is the elevation, μ is the dynamic viscosity, and $\boldsymbol{u}$ is the velocity vector. The continuity equation expresses the fluid mass conservation and can be written as: $\nabla \cdot \boldsymbol{u} = 0$.

Transport at the pore scale is governed by the fundamental processes of advection, diffusion, and chemical reactions. The governing transport equation expresses the mass balance for a given dissolved aqueous species with concentration C_i:

$$\frac{\partial C_i}{\partial t} + \nabla \cdot (\boldsymbol{u} C_i) - \nabla \cdot (D_{aq,i} \nabla C_i) = r_i \tag{2}$$

where $\boldsymbol{u}$ is the velocity field determined from the solution of the flow problem, $D_{aq,i}$ is the aqueous diffusion coefficients of species i, and r_i is the reactive term.

Microfluidics experiments and pore-scale modeling have been the main approaches to investigate flow and transport at the pore scale. The investigation has covered a wide variety of subsurface processes including conservative and reactive transport (Willingham et al. 2008;

Hochstetler and Kitanidis 2013; Crevacore et al. 2016), longitudinal and transverse mixing (Acharya et al. 2007; Rolle et al. 2012; Hochstetler et al. 2013; Rolle and Kitanidis 2014; de Anna et al. 2014b), displacement and dissolution of non-aqueous phases (Chomsurin and Werth 2003; Bandara et al. 2013), bioreactive transport and growth of microbial populations (Knutson et al. 2007; Zhang et al. 2010a; Tartakovsky et al. 2013), and precipitation /dissolution of reactive minerals (Li et al. 2008; Zhang et al. 2010b; Molins et al. 2012; Yoon et al. 2012).

Although pore-scale descriptions are intuitive and derived from first-principles, descriptions of flow and transport in porous media at larger scales require macroscopic representations based on upscaling. Already at the scale of a laboratory experiment, a detailed representation of a porous medium at the pore scale becomes infeasible or impractical due to the lack of information about the details of the geometry and structure of the grains and pores of the medium and to the large computational demand. Instead, the medium is treated as a "continuum" with properties and variables that change continuously from point to point. Upscaling techniques result in the macroscopic representation of fluid flow and solute transport in porous media from the detailed descriptions at the pore-scale (e.g., Whitaker 1999). Averaging over space and time scales in a representative elementary volume (REV) of the porous medium, containing a large number of grains and pores, allows the replacement of a solid–liquid domain with an equivalent continuum. Such coarse graining leads to gradually changing variables and to upscaled parameters, which are of key importance for describing flow and transport processes. The macroscopic, continuum description of fluid flow in a saturated porous medium can be written as:

$$S_s \frac{\partial \varphi}{\partial t} - \nabla \cdot (\mathbf{K}\nabla\varphi) = W \tag{3}$$

where S_s is the specific storage, φ is the hydraulic head, $\mathbf{K}$ is the hydraulic conductivity tensor, and W is a sink/source term.

The continuum-scale governing equation for solute transport is the advection-dispersion-reaction equation:

$$\frac{\partial C_i}{\partial t} + \nabla \cdot (vC_i) - \nabla \cdot (\mathbf{D_i}\nabla C_i) = r_i \tag{4}$$

This equation appears formally similar to the transport description at the pore scale, however here the concentration is defined over a larger support volume and varies more gradually compared to the concentration of an aqueous species at the pore scale. In Equation (4), v is the seepage velocity vector whose components are average values that, in homogeneous and mildly heterogeneous media, exhibit considerably lower fluctuations compared to the highly changing velocity in individual pore channels (u); $\mathbf{D_i}$ is the hydrodynamic dispersion tensor which accounts for the inherently coupled effects of aqueous diffusivity and pore velocity fluctuations in the pore channels; r_i is the reaction rate term that is now expressed as a function of the averaged continuum scale concentrations.

Equations (3) and (4) are the classical approaches adopted to describe flow and transport in the subsurface both at the laboratory and at the field scale. The key challenge under field conditions is the heterogeneity of geologic media. Such heterogeneity poses particular problems in upscaling solute transport that is more impacted by small scale fluctuations compared to fluid flow. If it would be possible to resolve the heterogeneity of geologic formations at a sufficiently small scale, a local scale continuum description of solute transport would allow capturing the main features controlling plume migration and mixing, as well as to represent the often observed non-Fickian behavior of solute breakthrough in subsurface systems (Kitanidis, 2017). This is not always possible due to the complexity of the subsurface and to the practical feasibility and costs of high resolution characterization of aquifer systems. Therefore, efforts have been dedicated to the development of effective descriptions of solute transport including macrodispersion approaches (e.g., Dagan 1989), dual and multicontinuum approaches (e.g., Haggerty and Gorelick 1995), and continuum time random walk (e.g., Berkowitz et al. 2006). Investigation performed at well-instrumented research sites has been fundamental for the development of new approaches and transport theories in hydrogeology. Notable examples are the long-term investigations at the Borden site (Canada) and at the Columbus site (MADE site, USA). Detailed investigation of the heterogeneity in these aquifers has been possible due to the extensive characterization and numerous studies performed at these sites over decades (Sudicky and Illman 2011; Zheng et al. 2011). The study of reactive transport and mixing at the field scale has focused on different aspects, including transport of organic compounds (e.g., Prommer et al. 2006; Amos et al. 2011), inorganic pollutants (e.g., Kocar et al. 2008; Ma et al. 2010; Fakhreddine et al. 2016), propagation of leachate plumes from leaking landfills (e.g., Rolle et al. 2008; Bjerg et al. 2011) and interpretation of isotopic signatures (e.g., Eckert et al. 2012; Van Breukelen and Rolle 2012; Druhan and Maher 2017).

MIXING AND MIXING-CONTROLLED REACTIONS

Dilution, or mixing of a non-reactive solute with the surrounding fluid, is the process by which the solute tends to occupy a larger volume. If a system is chemically or biologically reactive, mixing of initially segregated species brings these reactants in contact, thus allowing the reaction to proceed. The subsurface is typically a poorly mixed environment since the presence of creeping flow and slow diffusive transport in geologic formations strongly limits the effective distribution and contact of dissolved chemical species. Therefore, mass transfer limitations are ubiquitous in the subsurface and often determine the overall rate of degradation processes. The study of mass transfer-limiting processes and their coupling with chemical reactions is important to understand reactive transport and contaminant degradation since observations in porous media at different scales show that the reaction rates may be considerably smaller compared to well-mixed laboratory setups (Kapoor et al. 1997; Meile and Tuncay 2006; Li et al. 2008).

Dispersion in porous media

The investigation of hydrodynamic dispersion in porous media is a complex and fascinating topic that has attracted the interest of many researchers from various disciplines including chemical engineering, reservoir engineering and contaminant hydrology. Different theories, modeling approaches, and experimental methods have been proposed to study dispersion in porous media (Bear 1972; Gelhar 1987; Dagan 1989; Cushman 2013). In particular in the field of reactive transport, issues concerning solute dispersion have been recognized as major challenges and have fostered active research in the last decades. As recently discussed by Kitanidis (2017), the study and the communication of dispersion in hydrogeology is still an area posing important challenges among researchers and practitioners.

Dispersion terms are introduced at the continuum scale (Eqn. 4) by averaging the velocity and concentration fluctuations that cannot be resolved at the scale of observation (e.g., Kitanidis 1992; Whitaker 1999). In typical applications of solute transport in the subsurface, dispersion is described with the Fickian model considering the dispersive flux proportional to the gradient of the resolved concentration at the scale of interest. The dispersion term is strongly anisotropic since solute dispersion is inherently different along the direction of the flow (longitudinal dispersion) and perpendicular to the flow (transverse dispersion). Hydrodynamic dispersion depends on the coupled effects of the fundamental transport mechanisms of advection and diffusion. Although these mechanisms act together and are difficult to separate, hydrodynamic dispersion coefficients are typically parameterized as additive contributions of a velocity-independent pore diffusion term and a mechanical dispersion term that depends on the average flow velocity (Bear 1972).

In the classical description of dispersion commonly used in hydrogeology (Scheidegger 1961), the mechanical dispersion term is described based on a linear dependence on the seepage velocity and using dispersivity coefficients, assumed to be properties of the porous medium.

Dispersion processes have been studied over a wide range of scales. At the laboratory scale, experimental observations of solute transport in packed beds have provided important contributions to the understanding of hydrodynamic dispersion. Empirical correlations based on a wealth of experimental data (Guedes de Carvalho and Delgado 2005; Delgado 2006) and on results and upscaling of pore-scale simulations have been proposed to describe longitudinal and transverse dispersion. Although the relations proposed may differ in their dependencies and coefficients, many studies show that a linear or quasi-linear dependence on the velocity and the grain size describe well the mechanical dispersion term of the longitudinal dispersion coefficient (e.g., Bijeljic et al. 2004; Delgado 2006; Muniruzzaman and Rolle 2017). In the transverse direction, hydrodynamic dispersion is essentially "enhanced diffusion" through the velocity variability in the pore channels. Correlations that capture well the essence of local transverse dispersion have a mechanical dispersion term that is non-linearly related to the average flow velocity (Delgado 2006; Chiogna et al. 2010) and is dependent on the aqueous diffusion coefficient of the transported solutes (Rolle et al. 2012, 2013b; Ye et al. 2015a).

Fewer studies have investigated hydrodynamic dispersion at the block scale, which can correspond to the size of a grid cell in a numerically discretized domain or the scale of the sampling grid at a site (e.g., Rubin et al. 1999; Wang and Kitanidis 1999; de Barros and Dentz 2016). Heterogeneity cannot be resolved within the block and its effects on solute transport need to be factored in block-scale dispersion coefficients. The dispersion coefficients are essentially determined by averaging the velocity and concentration fluctuations within the block. As illustrated by a recent numerical study (Lee et al. 2018), the values of such coefficients depend on the interplay between advection and diffusion, the permeability contrast, and the structure of the heterogeneity within the block. At the field scale, dispersion has been studied intensively to describe solute transport in porous media. Macrodispersion theories have been developed to describe solute transport in heterogeneous flow fields and have contributed to form the research area of stochastic hydrogeology (e.g., Gelhar and Axness 1983; Dagan 1989; Rubin 2003). Macrodispersion coefficients expressed as half the rate of increase of the second central moment of the solute concentration can be linked to the heterogeneity of the formation and have been shown to describe well the evolution of conservative plumes after sufficient time from their release (e.g., Freyberg 1986; Garabedian et al. 1991). However, macrodispersion coefficients describing the spreading of solutes plumes would lead to substantial overestimation of the overall reactions if applied to describe mixing-controlled reactive transport (e.g., Semprini et al. 1990). Several studies have also emphasized the importance of the order of ensemble averaging and proposed the use of effective dispersion coefficients instead of ensemble (macrodispersion) coefficients to alleviate this problem (e.g., Dentz and de Barros 2015).

When dealing with solute transport in the subsurface it is necessary to separate the effects of spreading from mixing (Dentz et al. 2011). This task is not easy since these processes are closely coupled (Villermaux et al. 2019). To this end, stochastic-flux related approaches have been proposed to determine mixing relevant dispersion coefficients and to quantify the uncertainty of mixing and mixing-controlled reactions (Cirpka et al. 2011, 2012). In the following, we provide an overview of some recent approaches developed to study mixing and mixing-controlled reactions in porous media.

Mechanisms of mixing and mixing enhancement

The study of mixing processes in porous media is of primary importance to understand reactive solute transport in the subsurface. In particular, due to the poor mixing in subsurface flow systems it is crucial to identify the mechanisms that can lead to mixing enhancement. The heterogeneity of geologic formation can create complex flows that can substantially modify and enhance the mixing behavior of solute plumes. In the following, we briefly illustrate some key mechanisms for mixing in porous media that arise from the coupling between complex flows, causing stretching and deformation of solute plumes, and small scale processes such as diffusion and local scale dispersion. Figure 3 illustrates three important flow patterns causing deformation of solute plumes in porous media.

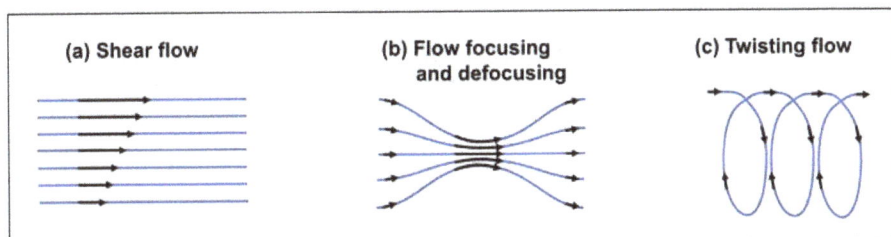

Figure 3. Example of flow patterns affecting mixing fronts and enhancing mixing and reaction rates in porous media: **(a)** shear flow results from differential velocities in neighboring streamlines; **(b)** flow focusing and defocusing results from variations of velocities along streamlines, which for incompressible fluids lead to contraction and divergence of streamlines; **(c)** twisting flow results from secondary motion in three-dimensional flow fields.

Front deformation and incomplete mixing. The elementary flow patterns illustrated in Fig. 3 are often combined to generate complex mixing dynamics at different scales. At the pore scale, the arrangement of solid grains and the pore space generates a broad velocity distribution (Bijeljic et al. 2011; de Anna et al. 2013; Kang et al. 2014; Holzner et al. 2015; Dentz et al. 2018) and complex streamline topologies (Lester et al. 2013), which leads to intricate mixing patterns (Fig. 4) (de Anna et al. 2014a,b; Jimenez-Martinez et al. 2015, 2017; Lester et al. 2016). At field scale, structural heterogeneities, such as sedimentary layers and fractures, as well as non-uniform boundary conditions, also control the patterns of mixing between different solutes (de Barros et al. 2012; Engdahl et al. 2014; Le Borgne et al. 2014; Chiogna et al. 2015; Cirpka et al. 2015; Bandopadhyay et al. 2018). Stretching of mixing fronts by velocity gradients and streamline topology (Fig. 3) can greatly enhance mixing rates compared to diffusive mixing rates by increasing the area available for diffusive mass transfer and steepening concentration gradients (Ottino 1989; Chiogna et al. 2011a, Le Borgne et al. 2014, Ye et al. 2015c).

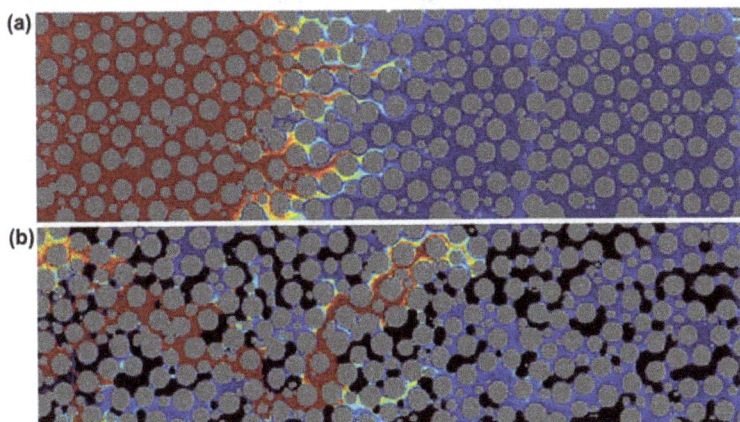

Figure 4. Conservative mixing experiments in 2D porous micromodels: **(a)** concentration field of a passive tracer front injected from left to right in a saturated porous medium (tracer concentration from red to yellow, water in blue and solid grains in grey); **(b)** concentration field of the same passive tracer injected in unsaturated conditions: the black zones represent air clusters (adapted from Jimenez-Martinez et al. 2015, 2017).

The fingering pattern that develops in mixing fronts can also leave poorly mixed areas within the front (Fig. 4). Such incomplete mixing is the main reason why mixing and related chemical reactions are not accurately predicted in general by conventional dispersion theories assuming locally well-mixed conditions (Gramling et al. 2002; Le Borgne et al. 2011; de Anna et al. 2014b; Wright et al. 2017). While new fingers are continuously formed under flow, aggregation of solute fingers (de Anna et al. 2014a; Jimenez-Martinez et al. 2015) and diffusive mass transfer into immobile zones (Haggerty and Gorelick 1995; de Dreuzy et al. 2013; Babey et al. 2015) progressively reduce incomplete mixing. For saturated flows, de Anna et al. (2014b) have shown that the effective reaction rate becomes consistent with classical dispersion theory when solute fingers have aggregated as discussed in the following section. However, for unsaturated flows, Jimenez-Martinez et al. (2015) have demonstrated that the incomplete mixing regime is much longer as clusters of air prevent full aggregation of fingers (Fig. 4).

Shear flows. A simple deformation mechanism for mixing fronts is shear (Fig. 3a), which is created by a gradient of velocity in the direction transverse to the local flow (Ottino 1989; Bolster et al. 2011a; Paster et al. 2015; Bandopadhyay et al. 2017; Souzy et al. 2018). In porous media, shear flows occur at multiple scales. At the pore scale, Poiseuille flow profiles, with zero velocity at the wall and maximum velocity at the pore center, act as shear flows in each pore. In Figure 4a, the deformation of the advancing front in finger patterns corresponds to a shear flow at pore scale where the velocity at the pore center is larger than at the walls. At larger scales, shear flows are created by permeability differences as illustrated in Figure 7 (Le Borgne et al. 2014, 2015). They can also be generated by non-uniform boundary conditions, even in homogeneous media. A hydraulic head gradient along the surface, leads to curved streamlines in the subsurface, on which the velocity decreases exponentially with depth, as occurs for instance in hyporheic flows and hillslopes (Toth 1963). This difference in velocities of neighboring streamlines acts effectively as a shear flow deforming advancing fronts of new water entering the subsurface (Bandopadhyay et al. 2018).

As the plumes advance in the domain, shear deformation increases the length of mixing fronts, thus enhancing mixing rates. In the simple shear flow scenario, where the velocity gradient is constant (Fig. 3a), the length of transient mixing fronts increases linearly in time (Bandopadhyay et al. 2017). In heterogeneous shear flows, where the shear rate is spatially distributed, the length of mixing fronts can follow a range of sub-exponential growth rates, including power law behaviors (Dentz et al. 2016). Under continuous solute injection, the differences in velocities in neighboring streamlines, driven by shear, affect the transient phase that leads ultimately to the steady concentration distribution and results in distinct solute residence times on each streamline, which may have important effects for different types of reactive processes.

Flow focusing. In heterogeneous porous media, streamlines converge and diverge depending on the spatial distribution of the hydraulic conductivity (Fig. 3b). In high permeability zones (e.g., permeable sand and gravel sediments embedded in matrices with lower permeability) the flow is focused and streamlines converge as most of the water flux occurs through the highly permeable material. The outcomes of the simulations shown in Figure 5 illustrate this effect in 2-D cross sections at the laboratory and field scales. In the laboratory-scale domain three rectangular high-K inclusions are embedded in a less permeable matrix (Fig. 5a). The flow is from left to right and the pattern of the computed streamlines clearly shows flow focusing into the high-K inclusions and defocusing downgradient of each inclusion. A similar but more complex pattern is observed in Figure 5b, where the streamlines are computed in a randomly generated binary field and the flow is focused into the irregularly shaped high permeability zones.

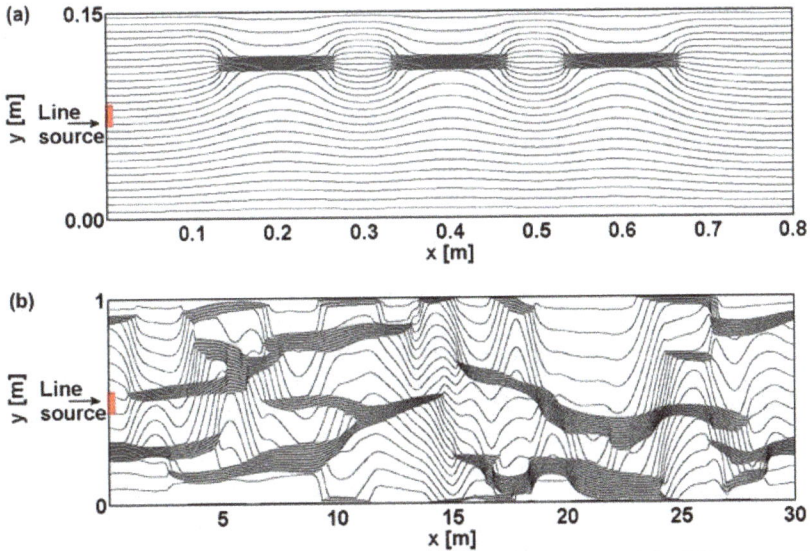

Figure 5. Flow focusing in heterogeneous porous media: streamlines computed for 2-D cross sectional domains at (a) laboratory scale and (b) field scale (from Chiogna et al. 2011a). [Used with permission from Chiogna G, Cirpka OA, Grathwohl P, Rolle M (2011a) Transverse mixing of conservative and reactive tracers in porous media: quantification through the concepts of flux-related and critical dilution indices. Water Resources Research, Vol 47, Fig. 2, p. 4.]

The focusing and defocusing of flow impacts solute transport and mixing in porous media since it affects: (i) the plume deformation, particularly relevant for transient fronts where the temporal dynamics of the front elongation can be related to the velocity distribution (Dentz et al. 2016); (ii) the residence time of a solute parcel within the inclusion is smaller than the time needed to cover the same distance in the low-K matrix; (iii) the distance between streamlines in the high-K zones is reduced and the transverse concentration gradients are increased; thus, transverse mass transfer between streamlines occurs over a smaller distance inside the inclusions; (iv) the transverse dispersion coefficient is higher in the high-K zones than in the surrounding matrix due to the higher velocity and to the larger grain size of the more permeable material.

In a study of continuously injected plumes, Werth et al. (2006) showed that the overall results of flow focusing and defocusing is an enhancement of transverse mixing due to the effects of increased mass transfer and dispersion coefficients within the high-permeability inclusions. The same authors also derived analytical expressions to quantify mixing enhancement factors in 2-D setups and showed that flow focusing in high-permeability inclusions leads to significant mixing and reaction enhancement. The combined effect of dimensionality and flow focusing on mixing of groundwater plumes has been investigated in successive studies. Ye et al. (2015b) performed laboratory-scale experiments directly comparing mixing enhancement through flow focusing in quasi 2-D and fully 3-D porous media. Expressions for mixing-enhancement factors in 3-D domains were also derived. The experiments and the derived analytical expressions showed that although plume dilution is larger in 3-D setups due to the additional degree of freedom, stemming from the third spatial dimension that ensures an increased contact area between the plume and the surrounding groundwater (Kitanidis 1994; Ye et al. 2015b), mixing enhancement is less effective in 3-D than in 2-D systems. Indeed, mixing enhancement in 2-D was shown to be the upper limiting case for fully 3-D cases. The study also demonstrated the importance of the parameterization of local transverse dispersion and showed that, in the hypothetical case of pure velocity-independent pore diffusion and identical flow focusing in both transverse directions, no enhancement of transverse mixing would occur. Finally, both experimental and numerical studies in 2-D and 3-D setups have highlighted the importance not only of the permeability contrast but also of the spatial distribution of the high-permeability inclusions. For instance, the effect of flow focusing on mixing enhancement is particularly effective when permeable inclusions are located close to plume sources. In such conditions flow focusing of an "undiluted plume" with sharp concentration gradients at the fringes results in very effective mixing enhancement.

Twisting of streamlines. Streamlines in heterogeneous formations can exhibit complex patterns that can significantly impact the extent of mixing of solute plumes both under steady-state and transient conditions (e.g., de Barros et al. 2012; Mays and Neupauer 2012; Trefry et al. 2012; Lester et al. 2013; Piscopo et al. 2013). Most studies have focused on two-dimensional setups probably due to the difficulty of running high-resolution three-dimensional numerical simulations and of performing three-dimensional flow-through experiments. However, 3-D systems are of great interest since in fully three-dimensional setups additional mechanisms can impact mixing dynamics. For instance, anisotropy of hydraulic conductivity can lead to groundwater whirls (Bakker and Hemker 2004; Stauffer 2007; Chiogna et al. 2014, 2015). In 3-D heterogeneous anisotropic media secondary motion involving streamlines twisting, folding and intertwining can overlay the primary velocity field (Fig. 3c). The impact of these processes on mixing enhancement for continuously emitted plumes has been studied in detailed 3-D numerical simulations at the field scale (Cirpka et al. 2015). At the pore scale, the topology of three-dimensional streamlines is expected to generate stretching and folding processes characteristic of chaotic mixing dynamics (Lester et al. 2013; Turuban et al. 2018). This leads to an exponential increase of mixing fronts, which is expected to dominate asymptotically the linear deformation created by shear (Lester et al. 2016).

Experimental evidence of the occurrence of streamline twisting and helical flow in porous media was provided by Ye et al. (2015c, 2016). Flow-through experiments were performed in a three-dimensional heterogeneous anisotropic porous medium consisting in alternating angled strips of coarse and fine materials in two different layers (Fig. 6a). Such architecture of the porous medium was chosen as a simplified representation of herringbone cross-stratification and entailed twisting of streamlines due to the alternating pattern and orientation of the high and low permeability zones (Fig. 6b). The twisting pattern of the streamlines strongly affects solute transport and mixing compared to an analogous homogeneous porous medium. Evidence of plume spiraling within the heterogeneous anisotropic medium was provided by using a dye tracer and by freezing and slicing the porous medium at different cross sections. The helical flow in the flow-through setup causes the deformation of the material surface of the plume. This results in an increase of the diffusive and transverse dispersive fluxes and, thus, of mass exchange between the streamlines leading to considerable mixing enhancement.

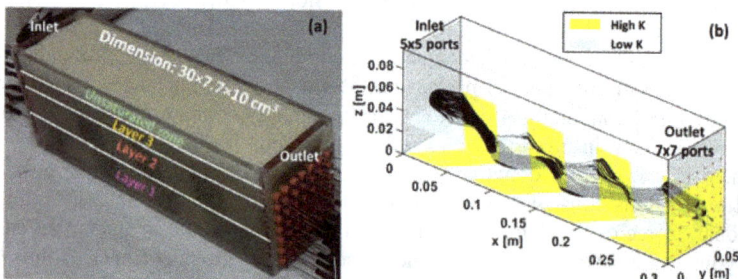

Figure 6. (a) 3-D experimental setup with a heterogeneous anisotropic porous medium and high-resolution sampling at the outlet; **(b)** twisting streamlines computed for the 3-D system (from Ye et al. 2016). [Used with permission from Ye Y, Chiogna G, Cirpka OA, Grathwohl P, Rolle M (2016) Experimental investigation of transverse mixing in porous media under helical flow conditions. Physical Review E, Vol 94, Fig. 1, p. 2]

Quantification of mixing

The quantification of mixing and mixing enhancement in porous media has attracted considerable interest particularly in recent years as it became apparent the need to distinguish between plume spreading and mixing for transport of solute plumes in the subsurface (Kitanidis 1994; Dentz et al. 2011; Chiogna et al. 2012; Le Borgne et al. 2015). Metrics used to quantify mixing for conservative and reactive transport include the dilution index (Kitanidis 1994; de Barros et al. 2012; Rolle and Kitanidis 2014; Di Dato et al. 2018), the scalar dissipation rate (De Simoni et al. 2005; Luo et al. 2008; Le Borgne et al. 2010; Engdahl et al. 2013), the concentration variance (Kapoor and Kitanidis 1996), and the gradient of the solute concentration square (Bolster et al. 2011b; Chiogna et al. 2011b). In principle, most of these metrics can be derived from the knowledge of the full concentration probability density function (Le Borgne et al. 2015). In this section, we illustrate two approaches for the quantification of mixing in porous media. We first discuss the case of a transient mixing front advancing into a domain using the lamella theory of mixing. Successively, we present an analysis of mixing for continuously injected plumes using the dilution index.

Lamella theory of mixing. The lamella theory of mixing is a Lagrangian framework that links the distribution of stretching rates along mixing interfaces to mixing and reaction rates. In this framework, the mixing interface is discretized into a series of elements, which deform advectively according to the local velocity field (Fig. 7). These elements are called lamellae in 2D and sheets in 3D. While we present here the theory for the 2D case, its extension to 3D is discussed in details by Martínez-Ruiz et al. (2018). The lamella representation is based on the formulation of the transport equations (equation 2 at the pore scale or equation 4 at the Darcy scale) in a reference frame attached to the interface between the two fluids that mix (Fig. 7b). The first axis of the reference frame, corresponding to coordinate σ, is aligned with the lamella, while the second axis, corresponding to coordinate n, is oriented in the direction perpendicular to the lamella. Since concentration gradients are maximum along coordinate n, diffusive mass transfer along σ can be neglected compared to diffusive mass transfer along n. Hence, the transport equation in this frame reduces to (Ranz 1979, Bandopadhyay et al. 2017, 2018):

$$\frac{\partial C_i}{\partial t} - \gamma n \frac{\partial C_i}{\partial n} - \frac{\partial}{\partial n}\left(D \frac{\partial C_i}{\partial n} \right) = r_i \qquad (5)$$

where γ is the stretching rate, D is the molecular diffusion coefficient at pore scale and the dispersion coefficient at Darcy scale, and r_i is the local reaction rate. The second term quantifies the effect of compression perpendicular to the direction of elongation in incompressible fluids and the third term quantifies the effect of diffusion.

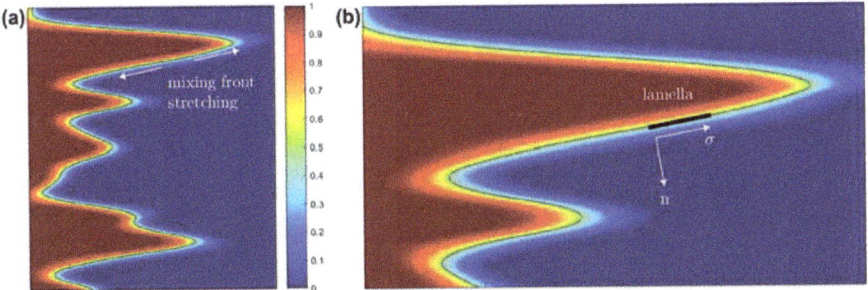

Figure 7. Illustration of the lamella description of mixing for a random shear flow. **(a)** Decomposition of a mixing front into a collection of lamellae, which quantify the local elongation (the color scale represents the concentration of the solute injected from the left, normalized to its maximum concentration); **(b)** zoom on a lamella showing the definition of the Lagrangian coordinate system attached to the lamella (adapted from Le Borgne et al. 2014).

Equation 5 can be used to derive analytical solutions for effective mixing and reactions rates in mixing fronts as a function of the stretching rate, which is defined for a lamella as:

$$\gamma = \frac{1}{\rho}\frac{d\rho}{dt} \tag{6}$$

where ρ is the elongation, defined as $\rho = L/L_0$, with L the current lamellae length and L_0 its initial length. The temporal evolution of the front elongation for the main types of deformation dynamics is given in Table 1. In uniform flows, there is no velocity gradient and the elongation is $\rho = 1$. In radial flows, as occur for instance around injection wells, the elongation of an injected front evolves as the square root of time due to streamline divergence (Le Borgne et al. 2014). Shear flows (Fig. 3a) lead to a linear increase of elongation. As discussed above, chaotic flows are generated by stretching and folding processes in three-dimensional porous media both at the pore scale and in anisotropic porous media (Lester et al. 2013, 2016; Turuban et al. 2018; Ye et al. 2016). They can also be engineered designing transient pumping and injection schemes to enhance mixing (Piscopo et al. 2013).

Elongation of mixing fronts increases the area available for diffusive mass transfer and enhances concentration gradients by compression in the direction perpendicular to the elongation direction. The mixing time is defined as the time at which compression balances with diffusion and concentration gradients cease to increase. It is therefore the time at which concentration gradients are maximum (de Anna et al. 2014a; Le Borgne et al. 2014, 2015; Bandopadhyay et al. 2018; Villermaux et al. 2019). Expressions for the mixing time are given in Table 1 for the different elongation dynamics.

Table 1. Temporal evolution of the elongation ρ for the main types of deformation and corresponding mixing times where compression and diffusion balance each other and concentration gradients are maximum.

	Uniform flow	Radial flows[a]	Shear flows[b]	Chaotic flows[c]
Elongation	$\rho = 1$	$\rho = \sqrt{\left(Q/\pi h \phi r_w^2\right)t}$	$\rho = \nabla v\, t$	$\rho = e^{\gamma t}$
Mixing time	N/A	$\tau = s_0 r_w \sqrt{\pi h \phi / DQ}$	$\tau = \nabla v^{-1}\left(s_0^2 \nabla v / D\right)^{1/3}$	$\tau = \gamma^{-1}\ln\left(D/s_0^2\gamma\right)$

Notes: [a] for an injection into an aquifer/fracture Q is the flow rate, h is the vertical size of the permeable domain, ϕ is the porosity and is the r_w well radius (Le Borgne et al. 2014), [b] ∇v is the velocity gradient (Bandopadhyay et al. 2017, 2018), [c] γ is the Lyapunov exponent (Lester et al. 2013). In all cases s_0 is the characteristic initial width of the front, and D is the diffusion coefficient at the pore scale and the dispersion coefficient at Darcy scale.

For the conservative case, $r_i = 0$, the solution of Equation (5) for a front is (Bandopadhyay et al. 2018):

$$C(n,t) = \frac{C_0}{2}\left(1 - \text{erf}\left(\frac{n\rho}{s_0\sqrt{1+4\theta}}\right)\right) \tag{7}$$

with s_0 the initial width of the front and θ the dimensionless time:

$$\theta = \int_0^t dt\,\frac{D\rho^2}{s_0^2} \tag{8}$$

To quantify the effect of front stretching on the enhancement of mixing rates, we use here the scalar dissipation rate as a metric of mixing (Le Borgne et al. 2010). In 2D, the scalar dissipation rate is defined as:

$$\chi = \iint dx\,dy\,D(\nabla c)^2 \tag{9}$$

with D the local diffusion/dispersion coefficient. As discussed in the next section, the interest of the scalar dissipation rate is that it is directly related to the reaction rate for fast equilibrium reactions (Le Borgne et al. 2010).

Using Equation (7), the effect of the front deformation on the effective mixing rate can be derived from the scalar dissipation rate as:

$$\chi = \iint dn\,dl\,D\big(\nabla c\big)^2 = \frac{Dl_0 c_0^2}{\sqrt{2\pi}s_0}\frac{\rho^2}{\sqrt{1+4\theta}} \tag{10}$$

with l_0 the initial length of the total front. The enhancement of mixing by front deformation is illustrated in Figure 8, where the mixing rate is computed from Equation (10) for the case of a shear flow (Fig. 7 and Table 1). While the mixing rate decreases in time in the absence of shear, due to the diffusive decay of concentration gradients, it increases in time with shear as concentration gradients are sustained by the front elongation. Therefore, front deformation by shear flow can increase the mixing rate by orders of magnitude compared to non-deformed fronts. This approach can be extended to random velocity fields both at the pore scale (de Anna et al. 2014a) and at the Darcy scale (Le Borgne et al. 2013, 2014, 2015; Bandopadhyay et al. 2018), and for continuously injected steady plumes (Lester et al. 2016). We discuss its application to reactive mixing in a latter section.

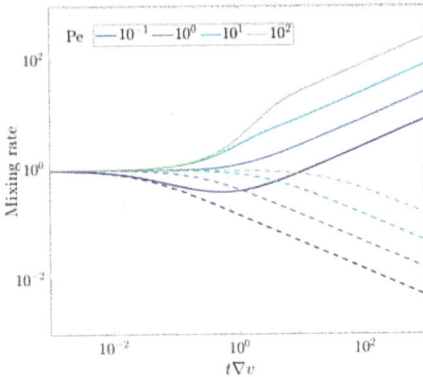

Figure 8. Temporal evolution of the mixing rate for different shear rates (continuous lines) compared to the mixing rates in the absence of shear (dashed lines). The mixing rate is defined as the scalar dissipation rate normalized by its initial value. These results were obtained assuming a constant diffusion coefficient (adapted from Bandopadhyay et al. 2018).

Analysis of continuously injected plumes with the flux-related dilution index. Here we illustrate a few examples of quantification of mixing and mixing enhancement for continuously emitted solute plumes using the concept of flux-related dilution index (Rolle et al. 2009). This metric is based on the volumetric dilution index introduced by Kitanidis (1994) but, instead of representing the distribution of a solute mass over an increasing volume, it quantifies dilution as the distribution of a given solute mass flux over a larger water flux. Physically, the flux-related dilution index represents the effective volumetric discharge transporting the solute flux at a given position along the main flow direction. In a bounded domain and for steady-state transport the flux-related dilution index is expressed as:

$$E_Q(x) = \exp\left(-\int_\Omega p_Q(\boldsymbol{x})\ln(p_Q(\boldsymbol{x}))q_x(\boldsymbol{x})d\Omega\right) \tag{11}$$

where q_x is the component of the specific discharge normal to the cross-sectional area Ω and $p_Q(x)$ is the flux-related probability density function:

$$p_Q(x) = \frac{C_i(x)}{\int\limits_{\Omega} C_i(x)q_x(x)d\Omega} \tag{12}$$

As the volumetric dilution index, $E_Q(x)$ is based on the Shannon entropy and, for conservative transport under steady-state conditions, monotonically increases with the travel distance since transverse dispersion distributes the solute mass flux over a larger water flux. The flux-related dilution index is particularly suited to quantify mixing for continuously emitted solutes, whereas the volumetric dilution index allows the quantification of mixing when compounds are introduced as pulses in a flow-through domain. A transient flux-related dilution index has been more recently introduced to analyze the dilution breakthrough of solute slugs and to complement the analysis based on the volumetric dilution index (Rolle and Kitanidis 2014). Figure 9 shows two examples of steady-state fluorescein plumes in a homogeneous and a heterogeneous porous media. In the heterogeneous setup two high-permeability inclusions cause focusing of the streamlines and of the dissolved plume and represent hot-spots for mixing enhancement.

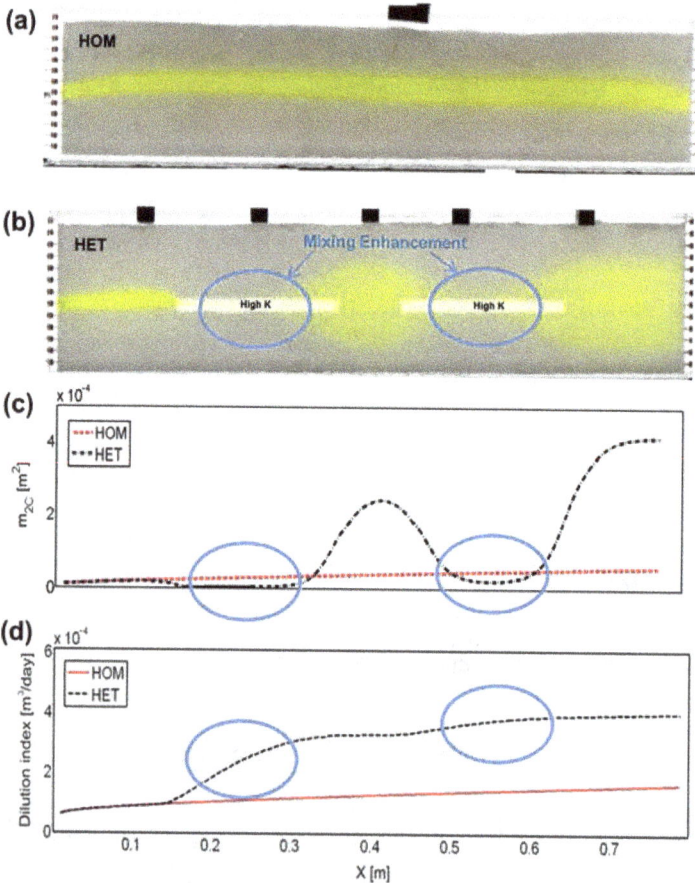

Figure 9. Solute plumes in homogeneous (**a**) and heterogeneous (**b**) laboratory scale setups, and computed second central moments (**c**) and flux-related dilution index (**d**) along the flow direction (modified after Rolle et al. 2009).

The computed second central moments of the solute distribution in the lateral direction cannot capture the enhancement of mixing due to flow focusing and show lower values in the high-permeability inclusions compared to the homogeneous setup. The flux-related dilution index, instead, allows the correct quantification of mixing in the two setups and adequately captures the monotonic increase of plume dilution and the enhancement due to flow focusing. In the homogeneous medium $E_Q(x)$ shows a slow increase throughout the length of the experimental setup. In the heterogeneous system the flux-related dilution index is the same in the low permeability matrix (same material as in the homogeneous setup) but suddenly increases as the plume is focused in the first high-permeability inclusion. Here, although the plume is less spread, mixing is significantly enhanced. The trend of $E_Q(x)$ is monotonically increasing also in the heterogeneous system, but the rate of increase is remarkably different at different spatial locations. For instance, the second high-permeability zone yields an increase of mixing compared to the matrix, but such increase is lower than in the first inclusion since the plume entering the downgradient high-K zone is already more well-mixed with the surrounding clean water.

The flux-related dilution index can be applied to quantify mixing at different scales. Figure 10 shows results of simulations performed at pore and field scales. The continuously released plumes in the two setups are illustrated in Figure 2 and the simulations were performed considering compounds with different aqueous diffusion coefficients spanning the typical range of variability for aqueous diffusivity in groundwater systems. The field scale simulations were performed in a stochastically generated permeability field with statistical properties consistent with the observations at the Columbus (MADE) field site (e.g., Rehfeldt et al. 1992; Zheng et al. 2011). The results show the increase of plume dilution with travel distance both at the pore and at the field scale. The trend at the pore scale is more regular since the pore scale domain, though heterogeneous, was constructed to reproduce macroscopically homogeneous media used in laboratory flow-through experiments (Hochstetler et al. 2013). At the field scale, the pattern of $E_Q(x)$ is more irregular with many flow focusing and defocusing events due to the high permeability contrast and heterogeneity. An important outcome of this multiscale analysis is the importance of aqueous diffusion for plume dilution at different scales and the capability of the proposed metric of mixing to capture this effect. Although diffusive processes occur at the small scale and the magnitude of aqueous diffusion coefficient may be perceived as small, their importance is pivotal for mixing in porous media and their effect propagate through scales and remain important for solute transport also at large spatial scales typical of field applications.

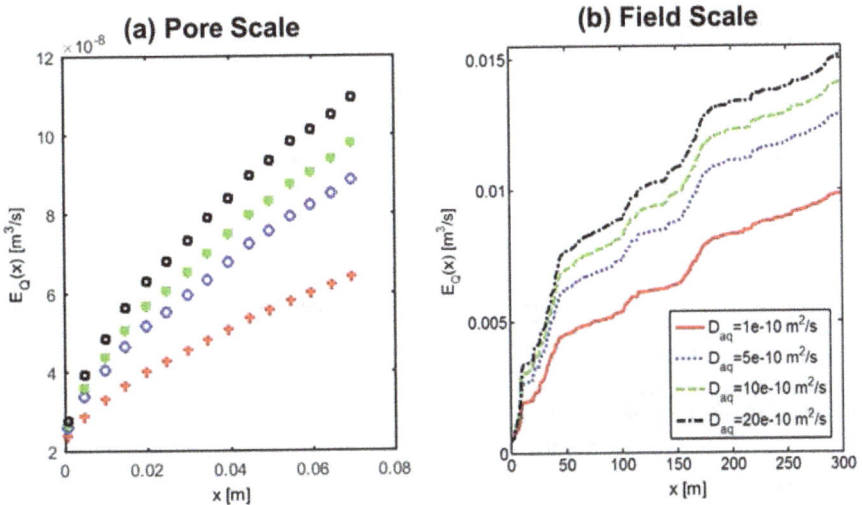

Figure 10. Flux-related dilution index for continuously injected plumes: simulations illustrating the impact of diffusion on plume dilution at the pore scale (**a**) and at the field scale (**b**) (modified after Rolle et al. 2013a).

Mixing dynamics and reactive processes

Mixing dynamics play a crucial role for triggering and sustaining many biogeochemical reactions in porous media (Valocchi et al. 2018). A number of relevant mixing-controlled reactive systems have been studied using both experimental and modeling approaches. Probably the most studied reactive system in the context of mixing-controlled reactive transport in porous media is the reaction between two dissolved reactants resulting in a dissolved product according to: A + B → C, and with kinetics described as an instantaneous or a bimolecular reaction rate (e.g., Cirpka and Valocchi 2007; Willingham et al. 2008). De Simoni et al. (2005, 2007) proposed approaches to solve mixing-controlled reactive transport problems involving multispecies transport and heterogeneous reactions with mineral precipitation and dissolution. Mixing processes were found to control equilibrium reaction rates, which were proportional to the rates of mixing. Therefore reactions can be induced by simply mixing waters under different equilibrium conditions as can happen, for instance, in carbonate systems. In analogy to the study of mixing for conservative solutes discussed above, in the following we provide an overview of the reactive lamella theory and the quantification of mixing with the dilution index in case of reactive transport.

Reactive lamella theory. In mixing fronts involving initially segregated reactants, reaction rates are enhanced by front deformation (de Anna et al. 2014a; Le Borgne et al. 2014). Using the lamella theory, Equation (5) may be used to quantify the effect of deformation on effective reaction rates. We present below the analytical expressions derived for the case of an irreversible reaction at arbitrary Damköhler numbers (Bandopadhyay et al. 2017) and that of a fast reversible reaction triggered by mixing of fluids at different chemical equilibrium (Le Borgne et al. 2014). The effective reaction rate is defined as:

$$R = \frac{dm_C}{dt} \tag{13}$$

where m_C is the total mass of product generated in the front.

We first consider the case of an irreversible reaction, involving two reactants on each side of the front, A and B that react to give a product C:

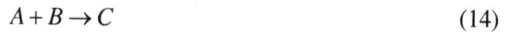

$$A + B \rightarrow C \tag{14}$$

with the local reaction rate:

$$\frac{dC_A}{dt} = -kC_A C_B \tag{15}$$

The effective reaction rates for a segment of the front of elongation subjected to a simple shear flow, $\rho = 1 + \nabla \nu t$ (Fig. 3a), are summarized in Table 2. For times smaller than the characteristic shear time $\tau_s = \nabla \nu^{-1}$, the elongation ρ is close to one and the reaction rate follows the effective kinetics expected for simple diffusion fronts (Larralde et al. 1992): for times smaller than the characteristic reaction time , $\tau_r = 1/kA_0$ the reaction rate increases as $R \sim t^{1/2}$, while for times larger than τ_r it decays as $R \sim t^{-1/2}$. This behavior is significantly modified by shear for $t > \nabla \nu^{-1}$, and the reaction rates scale as $R \sim t^{3/2}$ for $t < \tau_r$ and $R \sim t^{1/2}$ for $t > \tau_r$. These findings show that the effective kinetics of reaction fronts result from a close coupling between front deformation, diffusion and local reaction kinetics. This approach may be generalized to heterogeneous porous structures and permeability fields for which the estimated reaction kinetics differ significantly from those measured in well mixed batch reactors (Dentz et al. 2011).

We now consider a reactive front involving a fast reversible reaction:

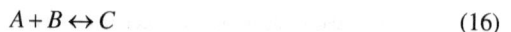

$$A + B \leftrightarrow C \tag{16}$$

Table 2. Synthesis of effective reaction rates for a mixing front subjected to a shear flow in the case of an irreversible reaction, where the Peclet number is $Pe = \tau_D/\tau_s$ the Damköhler number is $Da = \tau_D/\tau_r$, the characteristic diffusion time is $\tau_D = s_0^2/D$, and the characteristic shear time is $\tau_s = \nabla v^{-1}$ and the characteristic reaction rate is $\tau_r = 1/kA_0$.

weak stretching	$t < \tau_r$	$\tau_r < t < \tau_s$	$\tau_s < t$
$Pe < Da$	$R \sim Da(t/\tau_D)^{1/2}$	$R \sim t^{-1/2}$	$R \sim Pe(t/\tau_D)^{1/2}$
strong stretching	$t < \tau_s$	$\tau_s < t < \tau_r$	$\tau_r < t$
$Pe > Da$	$R \sim Da(t/\tau_D)^{1/2}$	$R \sim PeDa(t/\tau_D)^{3/2}$	$R \sim Pe(t/\tau_D)^{1/2}$

where C is a precipitate and the concentration of A and B are given by the mass action law (approximating the activities of the dissolved species by their concentrations):

$$K = C_A C_B \tag{17}$$

Each side of the front contains different concentrations of compounds A and B. Both fluids are at equilibrium with C, hence the concentrations of A and B follow Equation (17), although they are different on each side of the front. In this situation, mixing induces disequilibrium, which is balanced by a fast reaction in the front. While for the simple reaction given by Equation (16), this reaction is a precipitation, different chemical systems under similar hypotheses can lead to dissolution (de Simoni et al. 2007; Hidalgo et al. 2015). This framework has been used to model reactive mixing processes in freshwater/saline water fronts (Pool and Dentz 2018) and in CO_2 sequestration processes (Hidalgo et al. 2015).

For the chemical system defined by Equations (16) and (17), the local reaction rate in the mixing front, assuming same diffusive/dispersive properties of the reactants, can be obtained considering the conservative component:

$$u = C_A - C_B \tag{18}$$

and reads as (de Simoni et al. 2005; Le Borgne et al. 2014):

$$r = \frac{d^2C_i}{du^2} D(\nabla u)^2 \tag{19}$$

The first term of Equation (19) is linked to the chemical properties and is deduced from the relation:

$$C_A = \frac{u}{2} + \sqrt{\frac{u^2}{4} + K} \tag{20}$$

which results from solving the system of Equations given by (17) and (18). The second term of Equation (19) is the scalar dissipation rate (Eqn. 9), which quantifies the effect of mixing on reactions.

Expressions for the effective reaction rate derived from Equation (19) are detailed in Le Borgne et al. (2014) for different front deformation dynamics. They are similar to those derived for the scalar dissipation rate (Eqn. 10 and Fig. 8). Following this approach, the effect of front deformation on the enhancement of effective reaction rates can be derived for fast equilibrium reactions.

Continuously injected reactive plumes. Interesting insights on the interplay between mixing dynamics and reactive processes are provided by entropy balances that allow the connection of the concept of the dilution index with reactive mixing (Chiogna et al. 2012; Chiogna and Rolle 2017; Ye et al. 2018). Focusing on steady-state transport and continuous release of dissolved plumes, Chiogna et al. (2012) derived the transport equation for the entropy density of a reactive species:

$$\mathbf{v} \cdot \nabla \left(-p_Q \ln p_Q \right) - \nabla \cdot \left(\mathbf{D} \nabla \left(-p_Q \ln p_Q \right)\right) = \underbrace{-\left(1 + \ln p_Q \right) r_i^*}_{\text{reactive mixing term}} + \underbrace{\frac{1}{p_Q} \nabla p_Q^{\ T} \mathbf{D} \nabla p_Q}_{\text{dilution term}} \qquad (21)$$

The entropy balance for a reactant A (i.e., reactive species undergoing a homogeneous irreversible reaction as in Eqn. 14) involves two source/sink terms. The dilution term is a positive contribution and is the only term in the entropy balance for a conservative solute. The additional term is the reactive mixing contribution that acts as the only possible sink for the entropy. The interplay between these terms determines the mechanisms controlling transport and reactive mixing of a reactive species. For instance, Figure 11 illustrates the behavior for transport in heterogeneous porous media. Conservative and reactive plumes are continuously injected in a 2-D heterogeneous domain with statistics consistent with the Columbus aquifer (Rehfeldt et al. 1992). A conservative plume is shown in Figure 11a. Two scenarios were considered for reactive transport: (i) an instantaneous complete bimolecular reaction between the reactants A and B, and (ii) a slow degradation kinetics represented with a double Monod formulation dependent on the concentrations of the two reactants. The computed flux-related dilution index for the conservative and two reactive scenarios is shown in Figure 11b. In the conservative case the flux-related dilution index shows a monotonic increase consistent with the only presence of a source term in the entropy balance (Eqn. 21). Steeper increases of the flux-related dilution index correspond to zones of the domain where dilution is enhanced by flow focusing. The entropy of the reactive tracer, $E_Q(A)$, undergoing degradation with instantaneous and slow degradation rates is lower than the conservative case (Fig. 11b). However, for the first zone of the domain (i.e., gray area in Fig. 11) the trends for the two reactive cases is also increasing and the spatial derivatives have similar patterns and are positive. In this zone the

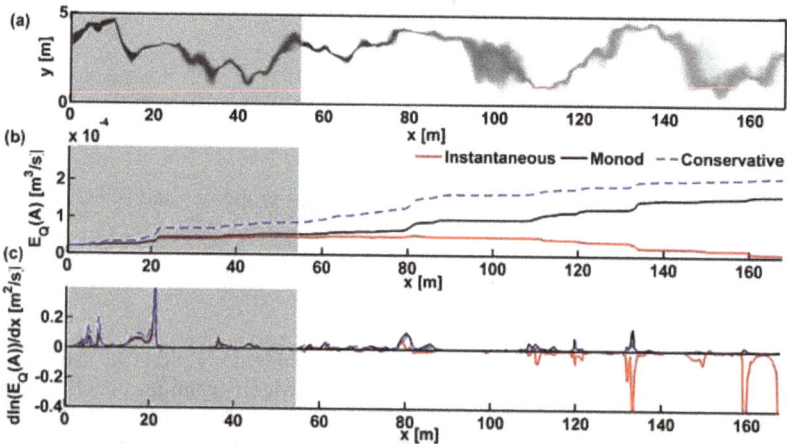

Figure 11. Entropy balance and field-scale dilution of a reactant: **(a)** conservative plume in a heterogeneous domain; **(b)** flux-related dilution index of solute A for the conservative and reactive scenarios; **(c)** spatial derivative of its natural logarithm (from Chiogna et al. 2012). [Used with permission from Chiogna G, Hochstetler DL, Bellin A, Kitanidis PK, Rolle M (2012) Mixing, entropy and reactive solute transport. Geophysical Research Letters, Vol 39, Fig. 3, p. 5.]

plumes of reactant A are mainly diluted as the dilution terms dominate the entropy balance. Farther downgradient, the trends differentiate remarkably as the importance of the reactive mixing terms increases. In the case of instantaneous reaction, the entropy of the reactant's plume decreases showing that the reactive term has become dominant in the entropy balance. In the case of slow double Monod kinetics, dilution is the dominant mechanism throughout the domain since lateral mixing distributes the reactant within the water flux more effectively than the mass-removal caused by the reaction. Nonetheless, the degradation of the reactant at the plume fringes causes significant difference in plume dilution with respect to the conservative case. This example is illustrative of the insights on the interplay between mixing and reactive processes provided by studying the entropy of solute plumes.

EXAMPLES OF MIXING AND REACTIONS AT DIFFERENT SCALES

In this chapter, we discuss a few selected examples of subsurface mixing and reaction processes at different scales. These examples cover the relevant scales of subsurface investigation, from pore to field scale, and are typically based on detailed experimental observations coupled with model-based interpretation.

Micromodel experiments

Micromodels and microfluidic experiments have been increasingly used for the detailed investigation of transport processes and reactions in porous media (e.g., Chomsurin and Werth 2003; Zhang et al. 2010a,b). Here we briefly illustrate two selected examples of studies that investigated reactive transport at the pore scale with focus on a homogeneous reaction (de Anna et al. 2014b) and on a heterogeneous reaction (Zhang et al. 2010b), respectively.

Chemiluminescence. To visualize and quantify the effect of mixing on chemical reactions at the pore scale, de Anna et al. (2014b) designed a micromodel experiment based on chemiluminescence (Fig. 12). In this reaction, photons are produced when two compounds A and B are mixed. In a two-dimensional micromodel composed of pillars, the light intensity quantifies the distribution of reaction rates in a mixing front, where A displaces B. The reactive front is composed of elongated filaments of high light intensity, i.e. high reaction rates, created by the effect of stretching due to velocity gradients at pore scale. As discussed above, this deformation process increases the area available for diffusive mass transfer and enhances concentration gradients. This increases the effective reaction rate, quantified here from the global light intensity produced in the mixing front (Fig. 12).

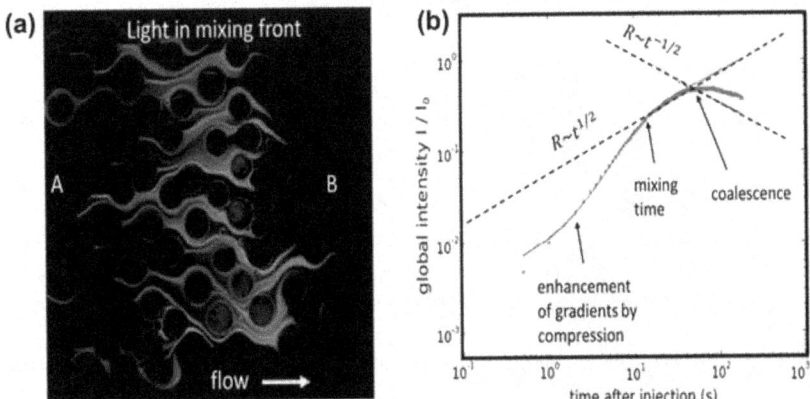

Figure 12. (a) Image of a light intensity in a 2D micromodel, where a compound A displaces a compound B and light is produced at the mixing front from a chemiluminescent reaction; **(b)** temporal evolution of the global intensity of light for different flow rates, adapted from de Anna et al. (2014b).

For a micromodel composed of random cylindrical pillars, stretching is approximately linear, since it is dominated by the shear induced by the velocity gradient between the no-slip boundary condition at the grain walls and the maximum velocity at the pore center (Table 1). The scaling that is expected for shear-enhanced mixing at large Damköhler number is $R \sim t^{\frac{1}{2}}$ (Table 2). This behavior is observed in this investigation, but only for part of the experiment (Fig. 12b). Two important phenomena change this scaling. Since chemical gradients were not initially sharp in the experiment, the initial compression of the plume enhances these gradients, which accelerates the reaction compared to the simple $R \sim t^{\frac{1}{2}}$ scaling. This behavior is observed until the mixing time (Table 1), when diffusion overcomes compression and concentration gradients are no longer enhanced. At later time, folding of the plume over itself promotes the coalescence of filaments. Hence, in this regime, the length of the area available to diffusive mass transfer ceases to increase. This leads to a $R \sim t^{-\frac{1}{2}}$ scaling, which corresponds to a diffusing front in the absence of shear (Larralde et al. 1992). In this late time regime, the effect of incomplete mixing tends to be negligible and the effective reaction rate may be evaluated by conventional dispersion theory. These observations hence provide a direct visualization and a quantitative measurement of the effect of the front deformation on the reaction kinetics at pore scale.

Calcite precipitation. Mixing induced mineral precipitation and dissolution have important implications for many subsurface applications including geological CO_2 sequestration, groundwater contamination and remediation, nuclear waste storage, and enhanced oil recovery. Here we briefly outline the study on mineral precipitation performed by Zhang et al. (2010b), where they carried out an experimental investigation of calcium carbonate precipitation induced by transverse mixing. Microfluidic experiments were performed in a micromodel pore structure in which solutions of calcium chloride and sodium carbonate at different saturation states where injected from two separate inlet channels (Fig. 13). Transverse mixing along the micromodel centerline allowed the contact between the reactants and the precipitation of $CaCO_3$. Calcium carbonate precipitates were observed primarily in the central mixing zone in all the experiments

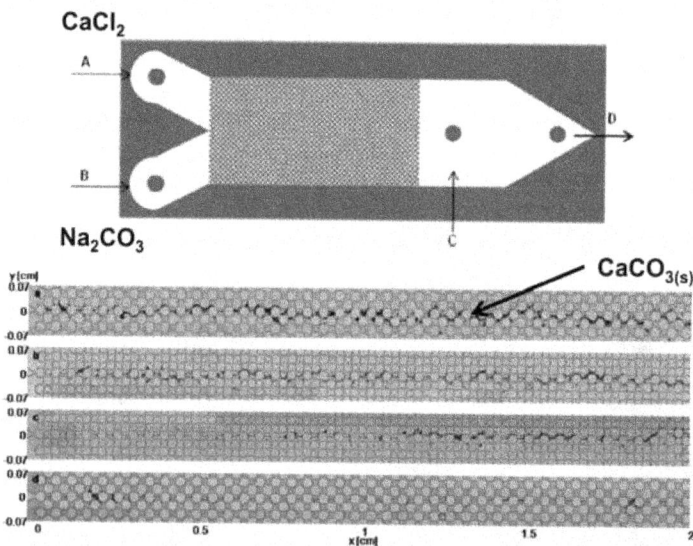

Figure 13. Micromodel setup and images of $CaCO_3$ precipitates formed along the mixing zone at different saturation states (from Zhang et al. 2010b). [Used with permission from Zhang C, Dehoff K, Hess N, Oostrom M, Wietsma TW, Valocchi AJ, Fouke BW, Werth C (2010b) Pore-scale study of transverse mixing induced CaCO3 precipitation and permeability reduction in a model subsurface sedimentary system. Environmental Science & Technology, Vol. 44, Fig. 1, 2 p. 7834, 7835.]

performed at different saturation states. Mineral characterization showed that vaterite and calcite were the predominant precipitate forms. Zhang et al. (2010b) observed that the amount of calcium carbonate precipitation decreased with increasing saturation state, showing that denser precipitates with lower porosity were formed at higher saturation. Such precipitates blocked the pore space and reduced the transverse mixing between calcium and carbonate. The reduction in porosity and mixing can lead to a reduction of the permeability of porous media and can limit mineral transformation towards more stable phases. This can have important consequences in flow-through systems where mixing of fluids is coupled with mineral precipitation reactions and can affect the performance of subsurface interventions causing, for instance, pressure buildup, decreased injectivity of fluids, and reduced remediation efficiency.

Intermediate laboratory scale experiments

Multidimensional laboratory flow-through experiments are instrumental for the investigation of mixing and mixing-controlled reactions in porous media. Different reactive systems have been investigated to study the interactions between mixing and reaction dynamics in porous media, including abiotic reactions between dissolved reactants (e.g., Gramling et al. 2002; Katz et al. 2010; Rolle et al. 2009), transport of pH fronts and electrostatic interactions (e.g., Loyaux-Lawniczak et al. 2012; Muniruzzaman et al. 2014; Muniruzzaman and Rolle 2015), precipitation/dissolution reactions (e.g., Tartakovsky et al. 2008; Poonoosami et al. 2015; Haberer et al. 2015; Battistel et al. 2019), and microbially-mediated reactions (e.g., Thullner al. 2002; Bauer et al. 2009a; Song et al. 2014). In the following, we provide two examples of mixing investigation in flow-through laboratory setups considering abiotic and biotic mixing-controlled reactions.

Abiotic mixing-controlled reactions. Reactive mixing and mixing enhancement by flow focusing and by transient flow fields was investigated by Rolle et al. (2009). Flow-through experiments were performed in a quasi two-dimensional flow-through chamber with dimensions: 77.9 cm × 15.0 cm × 1.1 cm. The setup was filled with a homogeneous porous medium with a grain size of 0.25–0.3 mm; high-permeability inclusions were embedded to realize a heterogeneous setup by using a coarser grain size (1–1.5 mm). The considered reactive system was an acid/base reaction between an alkaline (NaOH) and an acidic (HCl) solution. Such reaction occurs rapidly, thus, it is suitable to investigate reactive mixing in porous media. The alkaline solution (pH = 11.49) was continuously injected from the two central inlet ports of the flow-through chamber, whereas the acidic solution (pH = 2.03) was injected from the surrounding inlet ports. The solutions contained bromophenol blue as pH indicator that

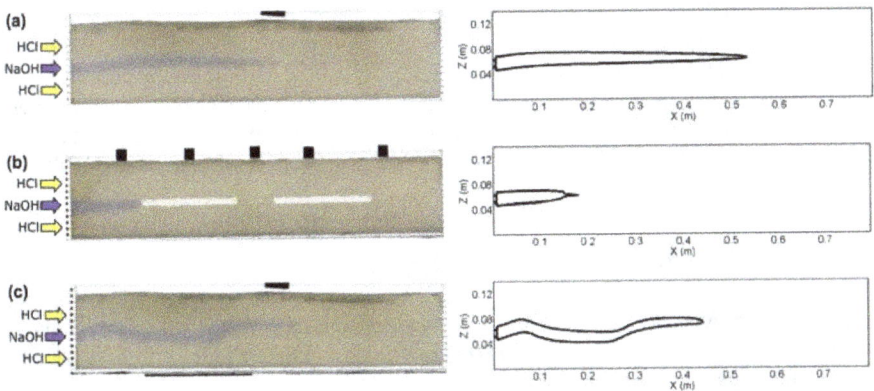

Figure 14. Abiotic mixing-controlled reactions: experimental observations and simulated results (contour at pH = 4.6). Steady-state plume in the homogeneous medium (a), steady-state plume in the heterogeneous medium (b), and plume under transient conditions in the homogeneous setup (c) (modified after Rolle et al. 2009).

allowed the visualization of the alkaline plume (Fig. 14). The acid/based reaction occurring at the plume fringes resulted in a steady-state alkaline plume within the porous media. In the homogeneous setup the plume had a length more than half the length of the flow-through chamber (Fig. 14a). The shape of the plume and its length were found to be considerably different in the experiments performed in the heterogeneous system. Here the alkaline plume disappeared at the location of the first high-permeability inclusion where mixing is enhanced by focusing of flow within the thin zone of coarser permeable material. Since the reaction was fast the enhancement of mixing directly translated into a reaction enhancement leading to a remarkably shorter steady-state plume. Considerably higher concentrations of NaOH in the same HCl background were necessary to establish longer plumes in the heterogeneous porous medium. A more than 7-fold increase of the NaOH concentration was required to obtain a steady-state alkaline plume reaching the outlet of the flow-through setup.

Experiments were also carried out to investigate the effect of transient flow conditions on reactive mixing. An oscillating flow field was established by using different flow rates and by pumping from alternating inlet and outlet ports. Such flow field caused the alkaline plume to bend and oscillate within the homogeneous porous medium (Fig. 14c). The experimental results showed transient alkaline plumes with lengths comparable to the one observed in the same porous medium under steady-state conditions.

Numerical simulations allowed the quantitative comparison between reactive mixing enhancement by flow focusing and transient flow. The results showed that both mechanisms contributed to increase reactive mixing. However the magnitude of reactive mixing enhancement was found to be considerably larger when the plume was focused in high-permeability inclusions. In fact, whereas the oscillating flow only caused moderate local mixing enhancement the focusing of the alkaline plume in the high-K inclusion resulted in a remarkable increase of reactive mixing.

Mixing and biodegradation. The coupling of mixing processes with microbially-mediated reactions is of utmost importance in subsurface porous media. For instance, biodegradation of organic contaminants often depends on the contact between substrates (e.g., soluble electron donors and electron acceptors) and nutrients at the fringes of pollutants' plumes. Laboratory flow-through microcosms (Bauer et al. 2009a) have been used to study biodegradation of common groundwater contaminants. Here we briefly illustrate some of the findings by Bauer et al. (2008) who performed flow-through biodegradation experiments to investigate microbially-mediated degradation of petroleum hydrocarbons under flow-through conditions. The setup was a quasi 2-D flow-through chamber similar to the one described above. A dissolved toluene plume was continuously injected from a central inlet whereas growth medium containing nutrients and electron acceptors was injected from the surrounding inlet ports. Experiments were performed under aerobic and anaerobic conditions. Two pure strains were considered: *Pseudomonas putida* mt-2 capable of aerobic degradation of toluene, and the *Aromatoleum aromaticum* strain EbN1 that degrades toluene using nitrate as electron acceptor. The degradation reactions under aerobic and anaerobic conditions can be written as:

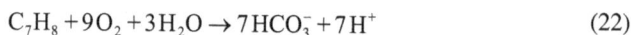

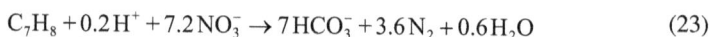

$$C_7H_8 + 9O_2 + 3H_2O \rightarrow 7HCO_3^- + 7H^+ \tag{22}$$

$$C_7H_8 + 0.2H^+ + 7.2NO_3^- \rightarrow 7HCO_3^- + 3.6N_2 + 0.6H_2O \tag{23}$$

Biodegradation takes place when the main reactants (i.e., the organic contaminant and the electron acceptors) come into contact. This occurs in thin (mm to cm scale) fringes surrounding the contaminant plumes where transverse dispersion allows the reactants to mix (Fig. 15). Such fringe zones represent biodegradation hotspots and are of fundamental importance to understand the transport and fate of contaminants in the subsurface. At the plume fringes reactants show steep concentration gradients that are the driving force for lateral mixing. The contact between initially segregated reactants (i.e., the organic contaminant in the plume and the electron acceptors in the background groundwater solution) provides optimal conditions for the biomass to grow. High-resolution subcoring of the porous medium and successive cell counts for the different samples allowed resolving the vertical distribution of the degrading biomass. The results show that highest numbers of degrading bacteria were found at the plume fringes (Fig. 15).

Successive studies investigated biodegradation of different monoaromatic petroleum hydrocarbons with focus on the effects of hydraulic heterogeneities on contaminant biodegradation (Bauer et al. 2009b) and on isotope fractionation of organic pollutants during fringe-controlled biodegradation (Rolle et al. 2010). Flow-through microcosms and their model-based interpretation have provided important insights on the coupling of biodegradation reactions with physical transport and mixing processes in porous media.

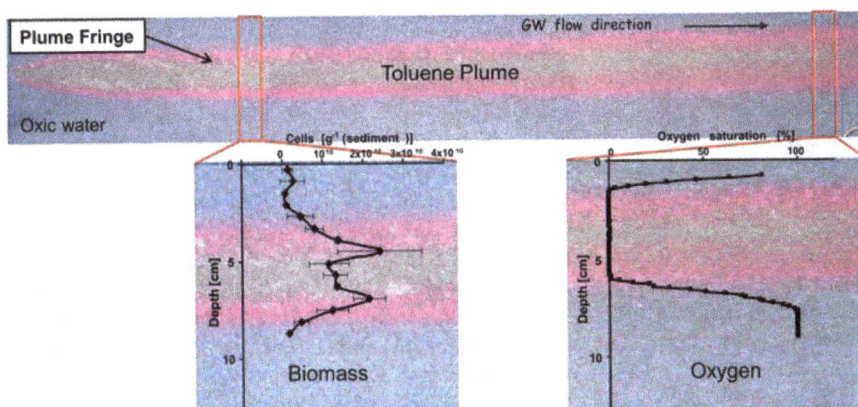

Figure 15. Toluene plume with bioactive fringes visualized with a redox indicator and results showing enhanced biomass growth and steep oxygen gradients at the plume fringes (modified after Bauer 2007 and Bauer et al. 2008).

Field-scale investigation

The investigation of mixing processes at the field scale is very challenging due to the heterogeneity of geologic media and to the complex interplay of physical, chemical, and biological processes that control mixing and reactive transport in the subsurface. Here we present two examples of investigation of mixing processes relevant for the fate of important organic and inorganic contaminants.

Biogeochemical gradients in a contaminated aquifer. The release of organic compounds in the subsurface triggers important redox processes that can result in a sequence of terminal electron acceptor processes and in a typical redox zonation of the contaminated aquifer (Chapelle et al. 1995; Christensen et al. 2001; Rolle et al. 2008). The redox processes can involve dissolved species and occur at the fringes of contaminant plumes, or solid electron acceptors such as reactive minerals (e.g., iron and manganese oxides and hydroxides) and, thus, take place in the core of the plume. Important electron acceptors that are replenished with the clean groundwater and come into contact with the dissolved organic contaminant plume through transverse mixing include dissolved oxygen, nitrate and sulfate. Plume fringes are typically very thin also at the field scale due to the limited extent of transverse mixing. To characterize these bioactive zones depth-specific and high-resolution sampling is required (e.g., Tuxen et al. 2006; Anneser et al. 2008; Rasa et al. 2011). Anneser et al. 2008 performed a detailed investigation of biogeochemical gradients in a tar-oil contaminated aquifer. The site was a former gasworks facility where the shallow Quaternary sand and gravel aquifer was contaminated by monoaromatic and polycyclic aromatic hydrocarbons. Dissolved plumes originated from the source zone and were transported downgradient. The main plume had a length of 200 m and was very shallow (~1 m). The groundwater geochemistry was considerably altered by the release of the organic contaminants and resulted in distinct biogeochemical gradients. A special multilevel monitoring well with very high spatial resolution (up to 2.5 cm) was installed with the aim of resolving the steep biogeochemical gradients at the site (Fig. 16).

The high-resolution sampling revealed very pronounced variation with depth of the contaminants' concentrations, groundwater species, isotopic signatures, and general parameters such as pH and redox potential. The results of the high-resolution investigation of the biogeochemical active plume fringes at the site were compared with a conventional multi-level well with spatial resolution of 0.5–1 m. The high-resolution investigation revealed cm-scale gradients of contaminants, electron acceptors, and metabolites that could not be resolved with the conventional multilevel sampling.

At the same field site, Winderl et al. (2008) used the sediment cores from the installed multilevel well to carry out a fine-scale study of the microbial community in the contaminated aquifer. Different microbial techniques allowed the qualitative and quantitative characterization of aquifer microorganisms with depth. In particular, the results showed the presence of a highly specialized community of anaerobic degraders at the lower fringe of the contaminant plume within the sulfidogenic gradient zone. This investigation established a clear link between the distribution of aquifer microbial community and biogeochemical gradients and confirmed that such biogeochemical gradients are hot spots of anaerobic toluene degradation. In a follow up investigation, Prommer et al. (2009) performed a reactive transport modeling study using the high-resolution dataset available at the field site. The model included flow, transport, contaminant degradation, microbial dynamics, and geochemical reactions in a two-dimensional transect along the flow direction. The simulation outcomes were compared with the high-resolution biogeochemical gradients measured in the aquifer (Fig. 16). Special focus of the model-based interpretation was the analysis of the isotopic signatures (carbon and sulfur isotopes). In particular for toluene, the most degradable compound in the plume, the simulations captured the evolution of the isotopic signature with a small enrichment in the plume core and a strong enrichment (up to 3.3‰) in the bioactive, sulfidogenic lower fringe of the plume. This finding provided additional evidence of enhanced biodegradation activity in the mixing zone between the organic contaminant plume and the surrounding groundwater rich of sulfate, the main electron acceptor at the site.

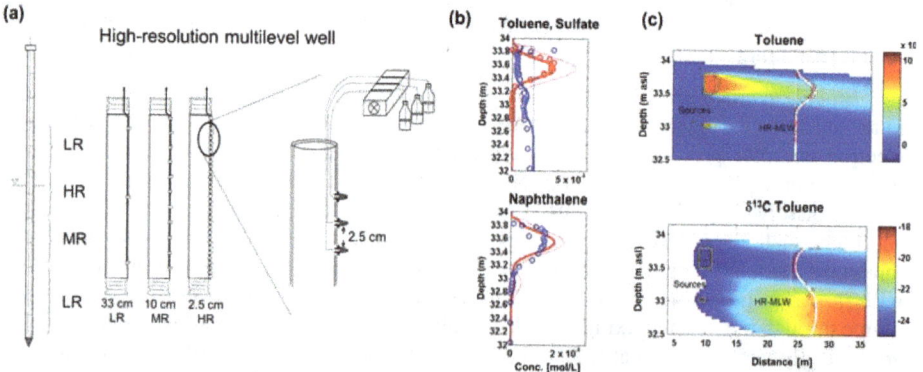

Figure 16. Plume fringes and biogeochemical gradients in a contaminated aquifer: (a) high-resolution multilevel well, (b) measured and simulated concentrations of two contaminants and of the main electron acceptor, (c) 2-D simulations and measurements of toluene concentrations and carbon isotopes (modified after Anneser et al. 2008 and Prommer et al. 2009).

Mixing fronts and arsenic attenuation in the hyporheic zone. Reactive mixing is also a key process in hyporheic zones, which represent hot spots of mixing and reaction involving surface water and groundwater (Hester et al. 2017). As demonstrated in the study of Datta et al. (2009), when anoxic groundwater discharges in rivers, mixing with oxic water leads to oxidation of dissolved iron and precipitation of iron oxides. This precipitation process also traps other dissolved elements, such as arsenic, that can be immobilized through sorption and incorporation in iron mineral phases (Kocar et al. 2006; Jessen et al. 2012; Stolze et al. 2019). Datta et al. (2009) measured large arsenic concentrations in shallow sediments along the Meghna river in Bangladesh (Fig. 17).

These large concentrations were found in the redox transition zone corresponding to the mixing front between groundwater and river water, where secondary minerals where formed upon the contact between the anoxic and oxic waters. Enrichment in arsenic in this mixing zone led to concentrations several orders of magnitude larger than the crustal abundance. Mixing fronts at the interface between anoxic and oxic water are thus suggested to act as reactive barriers, preventing large quantities of arsenic to flow into the river. However, depending on seasonal fluctuations in water levels, the location of the redox transition zone is expected to move, which may cause reducing conditions to reestablish thus leading to the dissolution of iron oxides and to the remobilization of the trapped arsenic. Such reactive barriers may be widespread in mixing fronts between surface water (rivers, lakes, oceans) and groundwater, thus playing an important role in the cycling of elements in hydrological systems.

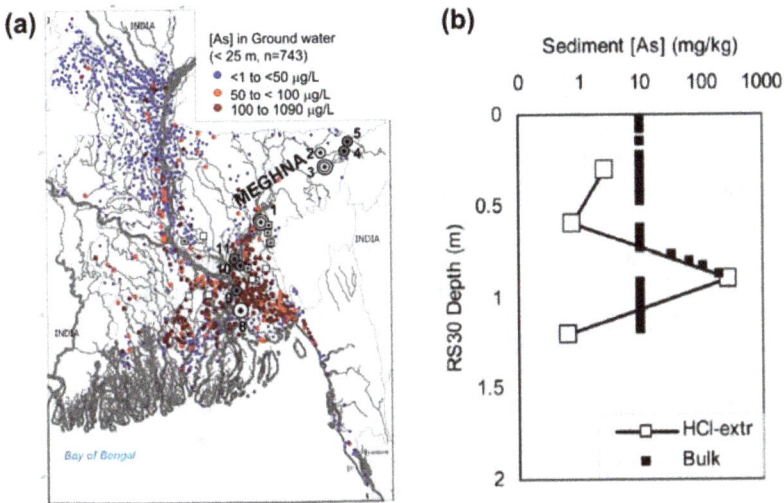

Figure 17. Arsenic concentration in shallow sediments along the Meghna river: **(a)** locations of sediment sampling sites and groundwater As concentrations; **(b)** depth-profile of sediment arsenic concentrations at one of the measurement location, showing a large increase in As concentration in the mixing zone (from Datta et al. 2009). [Used with permission from Datta S, Mailloux B, Jung H-B, Hoque MA, Stute M, Ahmed KM, Zheng Y (2009) Redox trapping of arsenic during groundwater discharge in sediments from the Meghna riverbank in Bangladesh. PNAS Vol. 106, Fig. 1, 3, p. 16931, 16933.]

CONCLUSIONS AND OUTLOOK

We have provided a brief overview of some recent conceptual and experimental advances shedding new light on the temporal and spatial dynamics of mixing and reaction fronts. We have selected illustrative examples of the coupling between physical mixing processes and biogeochemical reactions in highly non-uniform subsurface systems. These examples encompass both experimental and modeling applications at different scales. Mixing fronts, whether naturally occurring in hydrological systems or resulting from contaminant plumes, are often hot spots of biogeochemical reactions, where reaction rates are orders of magnitude larger than the background. Although the investigation of mixing in subsurface systems has advanced by leaps and bounds in recent years, formidable challenges remain to be addressed in order to be able to quantify and predict mixing-controlled reactions in natural and engineered systems. In the following, we present some thoughts about investigation directions for subsurface mixing and reactive processes that, in our view, represent upcoming challenges and may attract the attention of researcher in the next years.

Mixing and reactive processes in the subsurface have been the focus of different disciplines that have addressed these problems with very different approaches and often little degree of integration. This is, perhaps, due to the sharp boundary between physics and chemistry applied to subsurface processes. For instance, the literature of subsurface hydrology has focused on heterogeneity and scales with often simplified chemistry (e.g., the typical reaction: $A + B \rightarrow C$) and few attempts of including realistic complex kinetic and equilibrium reaction networks. On the other hand, most studies considering the complexity of biogeochemical processes have been carried out assuming simplifying flow and transport conditions, often in homogeneous domains. We foresee a "mixing of knowledge" between physical and chemical approaches to the investigation of reactive transport processes in subsurface formations. This will lead to more integrated studies that will tackle more complex and realistic mixing-controlled reactive systems. Important reactions and interactions that have been rarely considered in the context of the study of mixing processes will likely be investigated in greater detail. Such processes include surface reactions, solid/solution interactions of organic and inorganic compounds, and charge interactions both at the surface/ solution interface and within the pore water. Furthermore, advances in microbial, analytical and stable isotopes techniques will allow new possibilities for investigating microbially-mediated mixing-controlled reactions. This will lead to significant advances in understanding not only the impact of mixing processes on overall reaction rates but also on the structure and functioning of microbial communities in the subsurface (Stegen et al. 2015). Experimental and modeling studies will also be more integrated and will attempt to consider combined effects of physical and chemical heterogeneity on mixing processes in porous media.

Another major challenge is the one of dimensionality. Traditional investigations of transport processes in porous media have been largely based on column experiments. Such column investigations, and the associated one-dimensional perspective on contaminant transport, have many limitations and have contributed to hide the key importance of dimensionality for groundwater quality. The recent studies on mixing, some of them have been described above, have overcome such limitations and have often performed experimental and modeling investigation in 2-D setups. However, fully 3-D studies of mixing processes in porous media are still rare (e.g., Ye et al. 2015c; Rolle et al. 2018). Future experimental and modeling investigation is likely to focus on fully three-dimensional systems. The increased possibilities of high-resolution sampling and non-invasive sensor-based monitoring, as well as the advances in computational capabilities represent crucial advantages to study mixing processes in fully 3-D systems.

Issues of spatial scales and heterogeneity will still be persistent and will greatly influence the study of subsurface systems (Kitanidis 2015). Despite the great advances in describing the mechanisms of mixing and mixing-controlled reactions in porous media, the understanding and prediction of the impact of small-scale mixing processes at larger field scales remains a major theoretical and practical challenge. Furthermore, resolving coupled and dynamic mixing and reaction processes at the interface of different hydrological compartments, such as surface/ subsurface environments, is an outstanding challenge, which is attracting intense research focus (e.g., Hester et al. 2017). Key challenges include: (i) understanding and quantifying the effect of temporal fluctuations in surface water dynamics on hot moments of mixing and reaction at interfaces with subsurface environments; (ii) capturing the dynamics of mixing and reaction fronts by taking advantage of technological developments, such as distributed temperature sensing, high frequency chemistry monitoring and hydrogeophysics, to resolve mixing-relevant scales at higher spatial and temporal resolution; and (iii) representing the effect of mixing hot spots and hot moments in large scale hydrological models.

Reactive transport simulators (Steefel et al. 2015) are instrumental to study mixing processes in the subsurface. Advanced multiscale modeling tools (Molins and Knabner 2019, this volume) are needed to integrate the effects of physical, chemical, biological and electrostatic interactions (e.g., Thullner et al. 2005; Appelo and Rolle 2010; Tournassat and Steefel 2015; Rasouli et al. 2015; Muniruzzaman and Rolle 2016), to quantify mixing and mixing-controlled reactions at different scales using pore, continuum and hybrid approaches (e.g., Battiato and Tartakovsky 2011; Scheibe et al. 2015), and to describe reactive transport processes across different environmental compartments and critical zones (e.g., Li et al. 2017).

REFERENCES

Acharya RC, Valocchi AJ, Werth CJ, Willingham TW (2007) Pore-scale simulation of dispersion and reaction along a transverse mixing zone in two-dimensional porous media. Water Resour Resour 43:W10435

Amos RT, Bekins BA, Delin GN, Cozzarelli IM, Blowes DW, Kirshtein JD (2011) Methane oxidation in a crude oil contaminated aquifer: Delineation of aerobic reactions at the plume fringes. J Contam Hydrol 125:13–25

Anneser B, Einsiedl F, Meckenstock RU, Richters L, Wisotzky F, Griebler C (2008) High-resolution monitoring of biogeochemical gradients in a tar oil-contaminated aquifer. Appl Geochem 23:1715–1730

Appelo CAJ, Rolle M (2010) PHT3D: A reactive multicomponent transport model for saturated porous media. Groundwater 48:627–632

Babey T, de Dreuzy JR, Casenave C (2015) Multi-Rate Mass Transfer (MRMT) models for general diffusive porosity structures. Adv Water Resour 76:146–156

Bakker M, Hemker K (2004) Analytic solutions for groundwater whirls in box-shaped, layered anisotropic aquifers. Adv Water Res 27:1075–1086

Bandara UC, Tartakovsky AM, Oostrom M, Palmer BJ, Grate J, Zhang C (2013) Smoothed particle hydrodynamics pore-scale simulations of unstable immiscible flow in porous media. Adv Water Resour 62:356–369

Bandopadhyay A, Le Borgne T, Méheust Y, Dentz, M (2017). Enhanced reaction kinetics and reactive mixing scale dynamics in mixing fronts under shear flow for arbitrary Damköhler numbers. Adv Water Resour 100:78–95

Bandopadhyay A, Davy P, Le Borgne T (2018) Shear flows accelerate mixing dynamics in hyporheic zones and hillslopes. Geophys Res Lett 45:11–659

Battiato I, Tartakovsky DM (2011) Applicability regimes for macroscopic models of reactive transport in porous media. J Contam Hydrol 120–121:18–26

Battistel M, Muniruzzaman M, Onses F, Lee J, Rolle M (2019) Reactive fronts in chemically heterogeneous porous media: experimental and modeling investigation of pyrite oxidation. Appl Geochem 100:77–89

Bauer RD (2007) Control and limitations of microbial degradation in aromatic hydrocarbon plumes – experiments in 2-D model aquifers. PhD dissertation, University of Tuebingen, Tuebingen, Germany

Bauer RD, Maloszewski P, Zhang Y, Meckenstock RU, Griebler C (2008) Mixing-controlled aerobic and anaerobic biodegradation of toluene in porous media – results from two-dimensional laboratory experiments. J Contam Hydrol 96:150–168

Bauer RD, Rolle M, Kürzinger P, Grathwohl P, Meckenstock RU, Griebler C (2009a). Two-dimensional flow-through microcosms – Versatile test systems to study biodegradation processes in porous aquifers. J Hydrol 369:284–295

Bauer RD, Rolle M, Eberhardt C, Grathwohl P, Bauer S, Kolditz O, Meckenstock RU, Griebler C (2009b) Enhanced biodegradation by hydraulic heterogeneities in petroleum hydrocarbon plumes. J Contam Hydrol 105:56–68

Bear J (1972) Dynamics of Fluids in Porous Media, Elsevier, New York

Berkowitz B, Cortis A, Dentz M, Scher H (2006), Modeling non-Fickian transport in geological formations as a continuous time random walk. Rev Geophys 44: RG2003

Bijeljic B, Muggeridge AH, Blunt MJ (2004) Pore-scale modeling of longitudinal dispersion. Water Resour Res 40:W11501

Bijeljic B, Mostaghimi P, Blunt MJ (2011) Signature of non-Fickian solute transport in complex heterogeneous porous media. Phys Rev Lett: 107:204502

Bolster D, Dentz M, Le Borgne T (2011a) Hypermixing in linear shear flow. Water Resour Res 47:W09602

Bolster D, Valdés-Parada FJ, Le Borgne T, Dentz M, Carrera J (2011b) Mixing in confined stratified aquifers. J Contam Hydrol 120–121:198–212

Bjerg PL, Tuxen N, Reitzel LA, Albrechtsen H-J, Kjeldsen P (2011) Natural attenuation processes in landfill leachate plumes at three Danish sites. Ground Water 49:688–705

Chapelle FH, McMahon PB, Dubrovsky NM, Fujii RF, Oaksford ET, Vroblesky DA (1995) Deducing the distribution of terminal electron-accepting processes in hydrologically diverse groundwater systems. Water

Chiogna G, Rolle M (2017) Entropy-based critical reaction time for mixing-controlled reactive transport. Water Resour Res 53:7488–7498

Chiogna G, Eberhardt C, Cirpka OA, Grathwohl P, Rolle M (2010) Evidence of compound-dependent hydrodynamic and mechanical transverse dispersion by multitracer laboratory experiments. Environ Sci Technol 44:688–693

Chiogna G, Cirpka OA, Grathwohl P, Rolle M (2011a) Transverse mixing of conservative and reactive tracers in porous media: quantification through the concepts of flux-related and critical dilution indices. Water Resour Res 47:W02505

Chiogna G, Cirpka OA, Grathwohl P, Rolle M (2011b) Relevance of local compound-specific transverse dispersion for conservative and reactive mixing in heterogeneous porous media. Water Resour Res 47:W07540

Chiogna G, Hochstetler DL, Bellin A, Kitanidis PK, Rolle M (2012) Mixing, entropy and reactive solute transport. Geophys Res Lett 39: L20405

Chiogna G, Rolle M, Cirpka OA (2014) Helicity and flow topology in three-dimensional anisotropic porous media. Adv Water Resour 73:134–143

Chiogna G, Cirpka OA, Rolle M, Grathwohl P (2015) Helical flow in three-dimensional nonstationary anisotropic heterogeneous porous media. Water Resour Res 51:261–280

Chomsurin C, Werth CJ (2003) Analysis of pore-scale nonaqueous phase liquid dissolution in etched silicon pore networks. Water Resour Res 39:1265

Christensen TH, Bjerg PL, Banwart SA, Jakobsen R, Heron C, Albrechtsen HJ (2001) Characterization of redox conditions in groundwater contaminant plumes. J Contam Hydrol 45:165–241

Cirpka OA, Valocchi AJ (2007) Two-dimensional concentration distribution for mixing-controlled bioreactive transport in steady state. Adv Water Resour 30:1668–1679

Cirpka OA, de Barros FPJ, Chiogna G, Rolle M, Nowak W (2011) Stochastic flux-related analysis of transverse mixing in two-dimensional heterogeneous porous media. Water Resour Res 47:W06515

Cirpka OA, Rolle M, Chiogna G, de Barros FPJ, Nowak W (2012) Stochastic evaluation of mixing-controlled steady-state plume lengths in two-dimensional heterogeneous domains. J Contam Hydrol 138–139:22–39

Cirpka OA, Chiogna G, Rolle M, Bellin A (2015) Transverse mixing in three-dimensional nonstationary anisotropic heterogeneous porous media. Water Resour Res 51:241–260

Crevacore E, Tosco T, Sethi R, Boccardo G, Marchisio D (2016) Recirculation zones induce non-Fickian transport in three-dimensional periodic porous media. Phys Rev E 94:053118

Cushman JH (2013) The physics of fluids in hierarchical porous media: angstroms to miles (Vol. 10). Springer Science & Business Media

Dagan G (1989) Flow and transport in porous formations, Springer-Verlag, Berlin

Datta S, Mailloux B, Jung H-B, Hoque MA, Stute M, Ahmed KM, Zheng Y (2009) Redox trapping of arsenic during groundwater discharge in sediments from the Meghna riverbank in Bangladesh. PNAS 106:16930–16935

de Anna P, Le Borgne T, Dentz M, Tartakovsky AM, Bolster D, Davy P (2013) Flow intermittency, dispersion and correlated continuous time random walks in porous media. Phys Rev Lett: 110:184502

de Anna P, Dentz M, Tartakovsky A, Le Borgne T (2014a) The filamentary structure of mixing fronts and its control on reaction kinetics in porousmedia flows. Geophys Res Lett 41:4586–4593

de Anna P, Jimenez-Martinez J, Tabuteau H, Turuban R, Le Borgne T, Derrien M, Méheust Y (2014b) Mixing and reaction kinetics in porous media: An experimental pore scale quantification. Environ Sci Technol 48:508–516

de Barros FPJ, Dentz M (2016) Pictures of blockscale transport: effective versus ensemble dispersion and its uncertainty. Adv Water Resour 91:11–22

de Barros FPJ, Dentz M, Koch J, Nowak W (2012) Flow topology and scalar mixing in spatially heterogeneous flow fields. Geophys Res Lett 39: L08404

de Dreuzy JR, Rapaport A, Babey T, Harmand J (2013) Influence of porosity structures on mixing–induced reactivity at chemical equilibrium in mobile/immobile Multi-Rate Mass Transfer (MRMT) and Multiple INteracting Continua (MINC) models. Water Resour Res 49:8511–8530

Delgado JMPQ (2006) A critical review of dispersion in packed beds. Heat Mass Transfer 42:279–310

Dentz M, de Barros FPJ (2015) Mixing-scale dependent dispersion for transport in heterogeneous flows. J Fluid Mech 777:178–195

Dentz M, Le Borgne T, Englert A, Bijeljic B (2011) Mixing, spreading and reactions in heterogeneous media: A brief review. J Contam Hydrol 120–121:1–17

Dentz M, Lester DR, Le Borgne T, de Barros FPJ (2016) Coupled continuous-time random walks for fluid stretching in two-dimensional heterogeneous media. Phys Rev E 94:061102

Dentz M, Icardi M, Hidalgo JJ (2018) Mechanisms of dispersion in a porous medium. J Fluid Mech 841:851–882

De Simoni M, Carrera J, Sanchez-Vila X, Guadagnini A (2005) A procedure for the solution of multi-component reactive transport problems. Water Resour Res 41:W11410

De Simoni M, Sanchez-Vila X, Carrera J, Saaltink MW (2007) A mixing ratios-based formulation for multicomponent reactive transport. Water Resour Res 43:W07419

Di Dato M, de Barros FPJ, Fiori A, Bellin A (2018) Improving the efficiency of 3-D hydrogeological mixers: dilution enhancement via coupled engineering-induced transient flows and spatial heterogeneity. Water Resour Res 54:2095–2111

Druhan JL, Maher K (2017) The influence of mixing on stable isotope ratios in porous media: A revised Rayleigh model. Water Resour Res 53:1101–1124

Eckert D, Rolle M, Cirpka OA (2012) Numerical simulation of isotope fractionation in steady-state bioreactive transport controlled by transverse mixing. J Contam Hydrol 140:95–106

Engdahl NB, Ginn TR, Fogg GE (2013) Scalar dissipation rates in non-conservative transport systems. J Contam Hydrol 149:46–60

Engdahl NB, Benson DA, Bolster D (2014) Predicting the enhancement of mixing-driven reactions in nonuniform flows using measures of flow topology. Phys Rev E 90:051001

Fakhreddine S, Lee J, Kitanidis PK, Fendorf S, Rolle M (2016) Imaging geochemical heterogeneities using inverse reactive transport modeling: An example relevant for characterizing arsenic mobilization and distribution. Adv Water Resour 88:186–197

Freyberg DL (1986) A natural gradient experiment on solute transport in a sand aquifer: 2. Spatial moments and the advection and dispersion of nonreactive tracers. Water Resour Res 22:2031–2046

Garabedian SP, LeBlanc DR, Gelhar LW, Celia MA (1991) Large-scale natural gradient tracer test in sand and gravel, Cape Cod, Massachusetts: 2. Analysis of spatial moments for a nonreactive tracer. Water Resour Res 27:911–924

Gelhar LW (1987) Stochastic analysis of solute transport in saturated and unsaturated porous media. In: Advances in Transport Phenomena in Porous Media (pp. 657–700). Springer, Dordrecht

Gelhar LW, Axness CL (1983) Three-dimensional stochastic analysis of macrodispersion in aquifers. Water Resour Res 19:161–180

Gramling CM, Harvey CF, Meigs LC (2002) Reactive transport in porous media: A comparison of model prediction with laboratory visualization. Environ Sci Technol 36:2508–2514

Guedes de Carvalho JRF, Delgado JMPQ (2005) Overall map and correlation of dispersion data for flow through granular packed beds. Chem Eng Sci 60:365–75

Haberer CM, Muniruzzaman M, Grathwohl P, Rolle M (2015) Diffusive/Dispersive and reactive fronts in porous media: Fe (II)-oxidation at the unsaturated/saturated interface. Vadose Zone J 15:1–14

Haggerty R, Gorelick SM (1995) Multiple-rate mass transfer for modeling diffusion and surface reactions in media with pore-scale heterogeneity. Water Resour Res 31:2383–2400

Hester ET, Bayani Cardenas M, Haggerty R, Apte SV (2017) The importance and challenge of hyporheic mixing. Water Resour Res 53:3565–3575

Hidalgo JJ, Dentz M, Cabeza Y, Carrera J (2015) Dissolution patterns and mixing dynamics in unstable reactive flow. Geophys Res Lett 42:6357–6364

Hochstetler DL, Kitanidis PK (2013) The behavior of effective reaction rate constants for bimolecular reactions under physical equilibrium. J Contam Hydrol 144:88–98

Hochstetler DL, Rolle M, Chiogna G, Haberer CM, Grathwohl P, Kitanidis PK (2013) Effects of compound-specific transverse mixing on steady-state reactive plumes: Insights from pore-scale simulations and Darcy-scale experiments. Adv Water Resour 54:1–10

Holzner M, Willmann M, Morales V, Dentz M (2015) Intermittent lagrangian velocities and accelerations in three-dimensional porous medium flow. Phys Rev E 92:013015

Jessen S, Postma D, Larsen F, Nhan PQ, Hoa LQ, Trang PTK, Long TV, Viet PH, Jakobsen R (2012) Surface complexation modeling of groundwater arsenic mobility: Results of a forced gradient experiment in a Red River flood plain aquifer, Vietnam. Geochim Cosmochim Acta 98:186–201

Jimenez-Martinez J, de Anna P, Tabuteau H, Turuban R, Le Borgne T, Méheust Y (2015) Pore-scale mechanisms for the enhancement of mixing in unsaturated porous media and implications for chemical reactions. Geophys Res Lett 42:5316–5324

Jiménez-Martínez J, Le Borgne T, Tabuteau H, Méheust Y (2017) Impact of saturation on dispersion and mixing in porous media: Photobleaching pulse injection experiments and shear-enhanced mixing model. Water Resour Res 53:1457–1472

Kang PK, de Anna P, Nunes JP, Bijeljic B, Blunt MJ, Juanes R (2014) Pore-scale intermittent velocity structure underpinning anomalous transport through 3-D porous media. Geophys Res Lett 41:6184–6190

Katz GE, Berkowitz B, Guadagnini A, Saaltink MW (2011) Experimental and modeling investigation of multicomponent reactive transport in porous media. J Contam Hydrol 120–121:27–44

Kapoor V, Kitanidis PK (1996) Concentration fluctuations and dilution in two-dimensionally periodic heterogeneous porous media. Transport Porous Med 22:91–119

Kapoor V, Gelhar L, Miralles-Wilhelm F (1997) Bimolecular second-order reactions in spatially varying flows: segregation induced scale-dependent transformation rates. Water Resour Res 33:527–536

Kitanidis PK (1992) Analysis of macrodispersion through volume-averaging - moment equations. Stoch Hydrol Hydraul 6:5–25

Kitanidis PK (1994) The concept of the dilution index. Water Resour Res 30:2011–2026

Kitanidis PK (2015) Persistent questions of heterogeneity, uncertainty, and scales in subsurface flow and transport. Water Resour Res 51:5888–5904

Kitanidis PK (2017) Teaching and communicating dispersion in hydrogeology, with emphasis on the applicability of the Fickian model. Adv Water Resour 106:11–23

Kitanidis PK, McCarty PL (2012) Delivery and mixing in the subsurface: Processes and design principles for in situ remediation. Springer, New York, USA

Knutson C, Valocchi AJ, Werth CJ (2007) Comparison of continuum and pore-scale models of nutrient biodegradation under transverse mixing conditions. Adv Water Resour 30:1421–1431

Kocar BD, Polizzotto ML, Benner SG, Ying SC, Ung M, Ouch K, Samreth S, Suy B, Phan K, Sampson M, Fendorf S (2008) Integrated biogeochemical and hydrologic processes driving arsenic release from shallow sediments to groundwaters of the Mekong delta. Appl Geochem 23:3059–3071

Kocar B, Herbel M, Tufano K, Fendorf S (2006) Contrasting effects of dissimilatory iron (III) and arsenic (V) reduction on arsenic retention and transport. Environ Sci Technol 40:6715–6721

Larralde H, Araujo M, Havlin S, Stanley HE (1992) Reaction front for A + B $\rightarrow$ C diffusion-reaction systems with initially separated reactants. Phys Rev A 46:855–859

Le Borgne T, Dentz M, Bolster D, Carrera J, De Dreuzy J, Davy P (2010) Non-Fickian mixing: Temporal evolution of the scalar dissipation rate in heterogeneous porous media. Adv Water Resour 33:1468–1475

Le Borgne T, Dentz M, Davy P, Bolster D, de Dreuzy JR, Bour O (2011) Persistence of incomplete mixing: A key to anomalous transport. Phys Rev E 84:015301(R)

Le Borgne T, Ginn T, Dentz M (2014) Impact of fluid deformation on mixing-induced chemical reactions in heterogeneous flows. Geophys Res Lett 41:7898–7906

Le Borgne T, Dentz M, Villermaux E (2015) The lamellar description of mixing in porous media. J Fluid Mech 770:458–498

Lee J, Rolle M, Kitanidis PK (2018) Longitudinal dispersion coefficients for numerical modeling of solute transport in heterogeneous formations. J Contam Hydrol 212:41–54

Lester DR, Metcalfe G, Trefry MG (2013) Is chaotic advection inherent to porous media flow? Phys Rev Lett 111:174101

Lester DR, Dentz M, Le Borgne T (2016) Chaotic mixing in three-dimensional porous media. J Fluid Mech 803:144–174

Li L, Steefel CI, Yang L (2008) Scale dependence of mineral dissolution rates within single pores and fractures. Geochim Cosmochim Acta, 72:360–377

Li L, Maher K, Navarre-Sitchler A, Druhan J, Meile C, Lawrence C, Moore J, Perdrial J, Sullivan P, Thompson A, Jin L (2017) Expanding the role of reactive transport models in critical zone processes. Earth Sci Rev 165:280–301

Loyaux-Lawniczak S, Lehmann F, Ackerer P (2012) Acid/base front propagation in saturated porous media: 2D laboratory experiments and modeling. J Contam Hydrol 138–139:15–21

Luo J, Dentz M, Carrera J, Kitanidis PK (2008) Effective reaction parameters for mixing-controlled reactions in heterogeneous media. Water Resour Res 44:W02416

Ma R, Zheng C, Prommer H, Greskowiak J, Liu C, Zachara J, Rochold M (2010) A field-scale reactive transport model for U(VI) migration influenced by coupled multirate mass transfer and surface complexation reactions. Water Resour Res 46:W05509

Martinez-Landa L, Carrera J, Dentz M, Fernandez-Garcia D, Nardi A, Saaltink M (2012) Mixing induced reactive transport in fractured crystalline rocks. Appl Geochem 27:479–489

Martinez-Ruiz D, Meunier P, Favier B, Duchemin L, Villermaux E (2018) The diffusive sheet method for scalar mixing. J Fluid Mech 837:230–257

Mays DC, Neupauer R (2012) Plume spreading in groundwater by stretching and folding. Water Resour Res 48:W07501

McAllister SM, Barnett JM, Heiss JW, Findlay AJ, MacDonald DJ, Dow CL, Luther GW, Michael HA, Chan CS (2015) Dynamic hydrologic and biogeochemical processes drive microbially enhanced iron and sulfur cycling within the intertidal mixing zone of a beach aquifer. Limnol and Oceanogr 60:329–345

McClain ME, Boyer EW, Dent CL, Gergel SE, Grimm NB, Groffman PM, Hart SC, Harvey JW, Johnston CA, Mayorga E, McDowell WH (2003) Biogeochemical hot spots and hot moments at the interface of terrestrial and aquatic ecosystems. Ecosystems 6:301–312

McMahon P (2001) Aquifer/aquitard interfaces: Mixing zones that enhance biogeochemical reactions. Hydrogeol J 9:34–43

Meile C, Tuncay K (2006) Scale dependence of reaction rates in porous media. Adv Water Resour 29:62–71

Molins S, Knabner P (2019) Multiscale approaches in reactive transport modeling. Rev Mineral Geochem 85:27–48

Molins S, Trebotich D, Steefel C, Shen C (2012) An investigation of the effect of pore scale flow on average geochemical reaction rates using direct numerical simulation. Water Resour Res 48:W03527

Muniruzzaman M, Rolle M (2015) Impact of multicomponent ionic transport on pH fronts propagation in saturated porous media. Water Resour Res 51:6739–6755

Muniruzzaman M, Rolle M (2016) Modeling multicomponent ionic transport in groundwater with IPhreeqc coupling: Electrostatic interactions and geochemical reactions in homogeneous and heterogeneous domains. Adv Water Resour 98:1–15

Muniruzzaman M, Rolle M (2017) Experimental investigation of the impact of compound-specific dispersion and electrostatic interactions on transient transport and solute breakthrough. Water Resour Res 53:1189–1209

Muniruzzaman M, Haberer CM, Grathwohl P, Rolle M (2014) Multicomponent ionic dispersion during transport of electrolytes in heterogeneous porous media: Experiments and model-based interpretation. Geochim Cosmochim Acta 141:656–669

Ottino JM (1989) The kinematics of mixing: stretching, chaos, and transport. Cambridge University Press

Paster A, Aquino T, Bolster D (2015) Incomplete mixing and reactions in laminar shear flow. Phys Rev E 92:012922

Piscopo AN, Neupauer RM, Mays DC (2013) Engineered injection and extraction to enhance reaction for improved in situ remediation. Water Resour Res 49:3618–3625

Pool M, Dentz M (2018) Effects of heterogeneity, connectivity and density variations on mixing and chemical reactions under temporally fluctuating flow conditions and the formation of reaction patterns. Water Resour Res 54:186–204

Poonoosamy J, Kosakowski G, Van Loon L, Mäder U (2015) Dissolution-precipitation processes in tank experiments for testing numerical models for reactive transport calculations: Experiments and modelling. J Contam Hydrol 177–178:1–17

Prommer H, Tuxen N, Bjerg PL (2006) Fringe-controlled natural attenuation of phenoxy acids in a landfill plume: Integration of field-scale processes by reactive transport modeling. Environ Sci Technol 40:4732–4738

Prommer H, Anneser B, Rolle M, Einsiedl F, Griebler C (2009) Biogeochemical and isotopic gradients in a BTEX/PAH contaminant plume: model-based interpretation of a high-resolution field data set. Environ Sci Technol 43:8206–8212

Ranz WE (1979) Applications of a stretch model to mixing, diffusion, and reaction in laminar and turbulent flows. AIChE J 25:41–47

Rasa E, Chapman SW, Bekins BA, Fogg GE, Scow KM, Mackay DM (2011) Role of back diffusion and biodegradation reactions in sustaining an MTBE/TBA plume in alluvial media. J Contam Hydrol 126:235–247

Rasouli P, Steefel CI, Mayer UK, Rolle M (2015) Benchmarks for multicomponent diffusion and electrochemical migration. Computat Geosci 19:523–533

Rehfeldt KR, Boggs JM, Gelhar LW (1992) Field study of dispersion in an heterogeneous aquifer 3. geostatistical analysis of hydraulic conductivity. Water Resour Res 28:3309–3324

Rolle M, Kitanidis PK (2014) Effects of compound-specific dilution on transient transport and solute breakthrough: A pore-scale analysis. Adv Water Resour 71:186–199

Rolle M, Clement TP, Sethi R, Di Molfetta A (2008) A kinetic approach for simulating redox-controlled fringe and core biodegradation processes in groundwater: model development and application to a landfill site in Piedmont, Italy. Hydrol Process 22:4905–4921

Rolle M, Eberhardt C, Chiogna G, Cirpka OA, Grathwohl P (2009) Enhancement of dilution and transverse reactive mixing in porous media: Experiments and model-based interpretation. J Contam Hydrol 110:130–142

Rolle M, Chiogna G, Bauer R, Griebler C, Grathwohl P (2010) Isotopic fractionation by transverse dispersion: flow-through microcosms and reactive transport modeling study. Environ Sci Technol 44:6167–6173

Rolle M, Hochstetler DL, Chiogna G, Kitanidis PK, Grathwohl P (2012) Experimental investigation and pore-scale modeling interpretation of compound-specific transverse dispersion in porous media. Transport Porous Med 93:347–362

Rolle M, Chiogna G, Hochstetler DL, Kitanidis PK (2013a) On the importance of diffusion and compound-specific mixing for groundwater transport: An investigation from pore to field scale. J Contam Hydrol 153:51–68

Rolle M, Muniruzzaman M, Haberer CM, Grathwohl P (2013b) Coulombic effects in advection-dominated transport of electrolytes in porous media: Multicomponent ionic dispersion. Geochim Cosmochim Acta 120:195–205, doi: 10.1016/j.gca.2013.06.031

Rolle M, Sprocati R, Masi M, Jin B, Muniruzzaman M (2018) Nernst-Planck based description of transport, Coulombic interactions and geochemical reactions in porous media: Modeling approach and benchmark experiments. Water Resour Res 54:3176–3195

Rubin Y (2003) Applied Stochastic Hydrogeology, Oxford Univ. Press, Oxford, UK

Rubin Y, Sun A, Maxwell R, Bellin A (1999) The concept of block-effective macrodispersivity and a unified approach for grid-scale- and plume-scale-dependent transport. J Fluid Mech 395:161–180

Scheibe TD, Schuchardt K, Agarwal K, Chase J, Yang X, Palmer BJ, Tartakovsky AM, Elsethagen T, Redden G (2015) Hybrid multiscale simulation of a mixing-controlled reaction. Adv Water Resour 83:228–239

Scheidegger AE (1961) General theory of dispersion in porous media. J Geophys Res 66:3273–3278

Semprini L, Roberts PV, Hopkins GD, McCarty PL (1990) Field evaluation of in situ biodegradation of chlorinated ethenes: Part 2, Results of biostimulation and biotransformation experiments. Ground Water 28:715–727

Song X, Hong E, Seagren EA (2014) Laboratory-scale in situ bioremediation in heterogeneous porous media: Biokinetics-limited scenario. J Contam Hydrol 158:78–92

Souzy M, Zaier I, Lhuissier H, Le Borgne T, Metzger B (2018) Mixing lamellae in a shear flow. J Fluid Mech 838:R3-1-R3-12

Steefel CI, Appelo CAJ, Arora B, Jacques D, Kalbacher T, Kolditz O, Yeh GT (2015) Reactive transport codes for subsurface environmental simulation. Computat Geosci, 19:445–78

Stegen JC, Fredrickson JK, Wilkins MJ, Konopka AE, Nelson WC, Arntzen EV, Chrisler WB, Chu RK, Danczak RE, Fansler SJ, Kennedy DW (2015) Groundwater-surface water mixing shifts ecological assembly processes and stimulates organic carbon turnover. Nat Commun 7:11237

Stolze L, Zhang D, Guo H, Rolle M (2019) Surface complexation modeling of arsenic mobilization from goethite: Interpretation of an in-situ experiment. Geochim Cosmochim Acta 248:274–288

Sudicky EA, Illman WA (2011) Lessons learned from a suite of CFB Borden experiments. Ground Water 49:630–648

Stauffer F (2007) Impact of highly permeable sediment units with inclined bedding on solute transport in aquifers. Adv Water Res 30:2194–2201

Tartakovsky AM, Redden GD, Lichtner PC, Scheibe TC, Meakin P (2008) Mixing-induced precipitation: Experimental study and multi-scale numerical analysis. Water Resour Res 44:W06S04

Tartakovsky GD, Tartakovsky AM, Scheibe TD, Fang Y, Mahadevan R, Lovley DR (2013) Pore-scale simulation of microbial growth using a genome-scale metabolic model: Implications for Darcy-scale reactive transport. Adv Water Resour 59:256–270

Thullner M, Mauclaire L, Schroth MH, Kinzelbach W, Zeyer J (2002) Interaction between water flow and spatial distribution of microbial growth in a two-dimensional flow field in saturated porous media. J Contam Hydrol 58:169–189

Thullner M, Van Cappellen P, Regnier P (2005) Modeling the impact of microbial activity on redox dynamics in porous media. Geochim Cosmochim Acta 69:5005–5019

Toth J (1963) A theoretical analysis of groundwater ow in small drainage basins, J Geophys Res 68:4795–4812

Tournassat C, Steefel CI (2015) Ionic transport in nano-porous clays with consideration of electrostatic effects. Rev Mineral Geochem 80:287–329

Trefry MG, Lester DR, Metcalfe G, Ord A, Regenauer-Lieb K (2012) Toward enhanced subsurface intervention methods using chaotic advection. J Contam Hydrol 127:15–29

Turuban R, Lester DR, Le Borgne T, Méheust Y (2018) Space-group symmetries generate chaotic fluid advection in crystalline granular media. Phys Rev Lett 120:024501

Tuxen N, Albrechtsen H-J, Bjerg PL (2006) Identification of a reactive degradation zone at a landfill leachate plume fringe using high resolution sampling and incubation techniques. J Contam Hydrol 85:179–194

Valocchi AJ, Bolster D, Werth CJ (2018) Mixing-limited reactions in porous media. Transport Porous Med 126:1–26

Van Breukelen BM, Rolle M (2012) Transverse hydrodynamic dispersion effects on isotope signals in groundwater chlorinated solvents' plumes. Environ Sci Technol 46:7700–7708

Villermaux E (2019) Mixing versus stirring. Ann Rev Fluid Mech 51:245–273

Wang J, Kitanidis PK (1999) Analysis of macrodispersion through volume averaging: comparison with stochastic theory. Stochastic Environ Res Risk Assess 13:66–84

Werth CJ, Cirpka OA, Grathwohl P (2006) Enhanced mixing and reaction through flow focusing in heterogeneous porous media. Water Resour Res 42:W12414

Whitaker (1999) The Method of Volume Averaging. Springer

Willingham TW, Werth CJ, Valocchi AJ (2008) Evaluation of the effects of porous media structure on mixing-controlled reactions using pore-scale modeling and micromodel experiments. Environ Sci Technol 42:3185–3193

Winderl C, Anneser B, Griebler C, Meckenstock RU, Lueders T (2008) Depth-resolved quantification of anaerobic toluene degraders and aquifer microbial community patterns in distinct redox zones of a tar oil contaminant plume. Appl Environ Microbiol 74:792–801

Wright EE, Richter DH, Bolster D (2017) Effects of incomplete mixing on reactive transport in flows through heterogeneous porous media. Phys Rev Fluids 2:114501

Ye Y, Chiogna G, Cirpka OA, Grathwohl P, Rolle M (2015a) Experimental investigation of compound-specific dilution of solute plumes in saturated porous media: 2-D vs. 3-D flow-through systems. J Contam Hydrol 172:33–47

Ye Y, Chiogna G, Cirpka OA, Grathwohl P, Rolle M (2015b). Enhancement of plume dilution in three-dimensional porous media by flow-focusing in high-permeability inclusions. Water Resour Res 51:5582–5602

Ye Y, Chiogna G, Cirpka OA, Grathwohl P, Rolle M (2015c) Experimental evidence of helical flow in porous media. Phys Rev Lett 115:194502

Ye Y, Chiogna G, Cirpka OA, Grathwohl P, Rolle M (2016) Experimental investigation of transverse mixing in porous media under helical flow conditions. Phys Rev E 94:013113

Ye Y, Chiogna G, Lu C, Rolle M (2018) Effect of anisotropy on plume entropy and reactive mixing in helical flows. Transp Porous Med 121:315–332

Yoon H, Valocchi A, Werth C, Dewers T (2012) Pore-scale simulation of mixing-induced calcium carbonate precipitation and dissolution in a microfluidic pore network. Water Resour Res 48:W02524

Zhang C, Kang Q, Wang X, Zilles JL, Mueller RH, Werth CJ (2010a) Effects of pore-scale heterogeneity and transverse mixing on bacterial growth in porous media. Environ Sci Technol 44:3085–3092

Zhang C, Dehoff K, Hess N, Oostrom M, Wietsma TW, Valocchi AJ, Fouke BW, Werth C (2010b) Pore-scale study of transverse mixing induced CaCO$_3$ precipitation and permeability reduction in a model subsurface sedimentary system. Environ Sci Technol 44:7833–7838

Zheng C, Bianchi M, Gorelick S (2011) Lessons learned from 25 years of research at the MADE site. Ground Water 49:649–662

Reviews in Mineralogy & Geochemistry
Vol. 85 pp. 143-195, 2019
Copyright © Mineralogical Society of America

6

Multiphase Multicomponent Reactive Transport and Flow Modeling

Irina Sin and Jérôme Corvisier

MINES ParisTech - PSL University
Centre de Géosciences
35 rue Saint Honoré 77305
Fontainebleau Cedex
France

irina.sin@mines-paristech.fr; jerome.corvisier@mines-paristech.fr

INTRODUCTION

Upon injection into a reservoir, CO_2 migrates and interacts with the host rock and pore water (Benson et al. 2005, 2012). Supercritical or gaseous CO_2 dissolves in the pore aqueous solution as a function of the pressure, temperature, and salinity conditions. This process changes the local chemistry; in that it significantly decreases the pH and promotes geochemical reactions such as the dissolution of primary minerals and precipitation of secondary phases. These reactions then alter the porous network and can have a significant feedback on the migration of fluids (water and gas) in the reservoir. In addition, the dissolution of CO_2 in the aqueous phase changes the overall composition of both fluids, thus impacting the gravitational and viscous forces, as well as the motion of both phases. The importance of multiphase flow and reactive transport is thus critical to carbon capture and storage (CCS), and moreover, it is necessary for CCS operators to consider these induced chemical and physical processes in predictive simulations, thus ensuring that CO_2 does not leak into surrounding aquifers or the reservoir surface. Hence, accurate numerical multiphase flow and reactive transport modeling approaches are of paramount importance. Reactive transport simulations provide a unique means of quantitative and proper consideration of the coupling between the gas migration and its interactions with the pore water, in conjunction with the reactivity of the porous host rock (Garcia 2003; Ennis-King and Paterson 2005; Lu and Lichtner 2005; Nordbotten et al. 2005; Le Gallo et al. 2006; Trenty et al. 2006; Hassanzadeh et al. 2007; Pruess and Spycher 2007; Gaus et al. 2008; Nordbotten et al. 2008; Goerke et al. 2011; Sin 2015; Sin et al. 2017b). This chapter presents an overview of the current mathematical and numerical deterministic approaches to the modeling of such multiphase multicomponent reactive systems. Applications within the context of CO_2 storage, in addition to several other examples, are used to illustrate the concepts employed, their limitations, and the evolution of the simulation methods developed in the recent decades.

The presence of a gaseous phase, in addition to an aqueous and solid phase, adds complexity to the physical and chemical processes that should be considered in a reactive transport approach. Reactive transport models were developed initially to model saturated systems, e.g., only under consideration of the liquid-solid interactions. Several methods were developed accordingly for the modeling of single-phase reactive transport. For problems with large numbers of components, complex geochemical reaction paths, and a high degree of coupling between the transport and chemistry, the set of equations results in large nonlinear systems that require efficient mathematical and numerical techniques (Steefel 2019, this volume). Significant research efforts have been directed toward the reduction of this computational complexity, especially with respect to the dimensions of the resulting systems. In this context,

1529-6466/19/0085-0006$10.00 (print)
1943-2666/19/0085-0006$10.00 (online)   http://dx.doi.org/10.2138/rmg.2019.85.6

methods based on operator splitting between transport and reactions have typically been more advantageous for implementation due to their simplicity when compared with methods that collectively address the full set of equations (Yeh and Tripathi 1989, 1991; Xu and Pruess 1998; van der Lee et al. 2003; Molins et al. 2004; Kräutle and Knabner 2007; Molins and Mayer 2007; Steefel et al. 2014). The consideration of multiphase systems yields additional nonlinearities in flow- and geochemistry-related equations. The solution of the multiphase multicomponent flow and reactive transport (MMF&RT) increases in complexity in accordance with the evolution of the reactive system. The vapor–liquid equilibrium of multicomponent systems requires sophisticated models to handle vanishing phases and the related degeneration of equations that create numerical instabilities. Accurate modeling of the thermodynamic state for multicomponent fluids is therefore necessary for both the geochemistry and the flow, as it contributes to the phase equilibrium (transfers among phases and associated mass balance), in addition to the fluid and transport properties. Furthermore, it can have phenomenological effects, e.g., the changing of the flow regime, fluid dynamics, and matrix structure. Coupling methods for thermodynamic, hydrodynamic, and chemical modeling are therefore required, and they should ensure numerical stability and robustness.

In the modeling of multiphase systems, numerical issues can come both from the chemical heterogeneity and from the resolution of the set of nonlinear flow equations. The complexity of MMF&RT modeling can thus be attributed to both the complex physics and the high-dimensional nature of these problems. Moreover, several critical phenomena can be reproduced only by high-resolution simulations. Over the last four decades, numerous approaches have been developed and tested with global implicit or sequential iterative numerical methods (Peszynska and Sun 2002; Nghiem et al. 2004; Hao et al. 2012; Fan et al. 2012; Wei 2012; Wheeler et al. 2012; Xu et al. 2012; Nardi et al. 2014; Steefel et al. 2014; Hron et al. 2015; Parkhurst and Wissmeier 2015; Ahusborde and El Ossmani 2017; Sin et al. 2017a; Brunner and Knabner 2019). A range of multiphase multicomponent flow and transport solvers (e.g., White et al. 2012; Xu et al. 2014; Sin et al. 2017a; Voskov 2017) and geochemical solvers (e.g., Appelo et al. 2014; Corvisier et al. 2014; Leal et al. 2014) were recently developed to model CO_2 geological sequestration systems, among other applications. In this paper, a detailed state-of-the-art review focused on coupling issues is presented, and several recommendations are provided for future developments.

This chapter is divided into three main parts: (1) governing processes in multiphase multicomponent flow and reactive transport (MMF&RT), (2) mathematical and numerical approaches of MMF&RT modeling, and (3) applications with respect to CO_2 storage. Thereafter, discussion of other applications that illustrate the ability of existing methods to represent general non-ideal multicomponent gaseous systems within a wide range of temperature and pressure conditions, in addition to their possible interactions with water/brine/rock, are presented.

GOVERNING PROCESSES

The governing processes and present issues of multiphase flow and geochemical modeling are introduced in this section. The definition of wettability is first presented, which is difficult for several systems; followed by the definitions of capillary pressure, residual and capillary trapping, and heterogeneity. In addition, the relative permeability is introduced, which is a characteristic property of multiphase flow. Recently, novel experimental and numerical technologies have provided further insight into different concepts such as residual non-wetting saturation, maximum relative permeability, governing forces, hysteresis, the impact of heterogeneity, and upscaling from pore- and core-scales. The different regimes of gas diffusion are then discussed in conjunction with corresponding modeling methods. Thereafter, basic formulations of multiphase flow are presented, from immiscible to compositional flows.

After a brief presentation of geochemical modeling, e.g., mass balance and phase equilibria, a discussion of issues with respect to the modeling of multicomponent systems, such as the prediction of fugacity and activity coefficients, is given. Furthermore, thermodynamic properties such as density and viscosity are dependent on the composition, pressure, and temperature; and the challenge of describing appropriate equations of state (EOS) for multicomponent systems is significant. Hence, several methods used to predict these properties are discussed.

Capillary pressure and wettability

Porosity ϕ quantifies the space occupied by fluids or a void within a representative elementary volume (REV). When two fluids are separated by a distinct interface, these fluids are defined as immiscible, and the interfacial tension (IFT) results in a curved fluid–fluid surface within each pore. The cohesive and adhesive forces form a contact angle θ between the fluid–fluid interface and the solid face. If the contact angle θ is acute, the fluid is defined as wetting; otherwise, it is defined as non-wetting. For example, a gas and a liquid both in contact with the solid surfaces of the porous matrix create a gas–liquid interface with a curvature that results from balanced interfacial tensions.

In a system in which the number of fluids (N_f) is greater than one, each fluid represents a phase, and the system is referred to as multiphase. The saturation S_α is the ratio of the volume occupied by the phase α as a gas or liquid in a REV relative to the volume of pore space in this REV. Under the assumption that fluids occupy the entire pore space, the following relationship is satisfied:

$$\sum_{\alpha=1}^{N_f} S_\alpha = 1. \tag{1}$$

The gas and liquid fluids in contact with the solid phase apply different pressures on the curved interface that separates them. This variation in pressure from one side to the other is defined as the capillary pressure p_c:

$$p_c = p_g - p_l, \tag{2}$$

where p_g and p_l are the gas and liquid pressures, respectively. The capillary pressure at a point in the interface can be evaluated with the mean radius of curvature and a local interfacial tension γ using the Laplace equation, which takes the form of the Young–Laplace equation for a capillary tube of radius r:

$$p_c = \frac{2\gamma\cos\theta}{r}. \tag{3}$$

When a gas enters into a water-saturated porous medium, a certain pressure is required to displace the wetting fluid that occupies the porous space. This desaturation process is referred to as drainage. Displacement of the wetting fluid by the non-wetting fluid is described as the imbibition process (Fig. 1a). The relationship that links the capillary pressure and the liquid saturation is dependent on the direction of the displacement (e.g., drainage or imbibition) and thus hysteresis effects are commonly exhibited.

During the passage of a wetting phase (imbibition), a portion of non-wetting fluid can be trapped in the pore space. This capillary trapping process is highly dependent on the contact angle. Capillary pressure and wettability have been extensively investigated for oil–water systems (Leverett 1939; Mungan 1966; Batychy and McCaffery 1978). The contact angle is dependent on the mineralogy and surface roughness, and it is difficult to estimate. However, Armstrong et al. (2012) demonstrated that the contact angles of oil–water systems can be directly measured using X-ray computed tomography (CT). For the supercritical CO_2–water

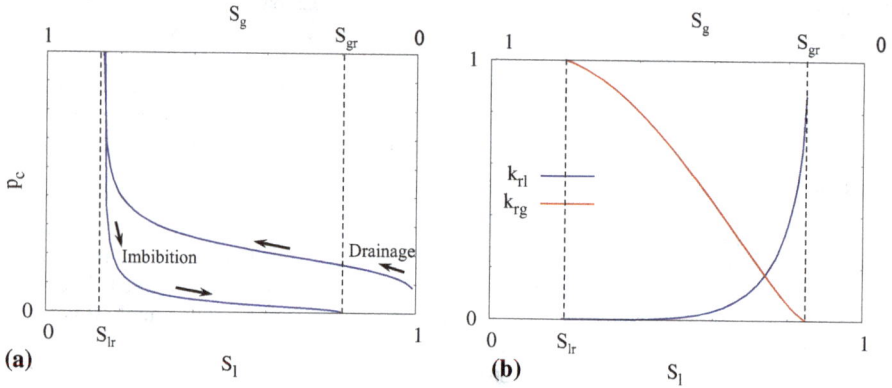

Figure 1. Characteristic properties of multiphase flow. **(a)** Capillary pressure drainage and imbibition curves. **(b)** Relative permeability curves.

system (scCO$_2$–water), most of the recent core-flooding experiments confirm that scCO$_2$ remains a non-wetting phase during water displacement (Krevor et al. 2012; Pentland et al. 2011; Pini et al. 2012; Andrew et al. 2014; Øren et al. 2019). Pini et al. (2012) observed contact angles of 20–45° at 25 °C and 0–5° at 50 °C for a CO$_2$–water system by fitting the contact angle for the IFT values obtained from Berea sandstone. Espinoza and Santamarina (2010) reported a contact angle of 30° for CO$_2$–brine–calcite at 0–10 MPa and 23.35 °C. Broseta et al. (2012) observed contact angles of 60–75° during imbibition at 0.5–14 MPa and 35 °C. Andrew et al. (2014) confirmed that scCO$_2$–brine–carbonate is a water-wet system at 10 MPa and 50 °C by measuring contact angles of 35–55° using micro-CT. Finally, Øren et al. (2019) observed angle ranges 20–30° and 30–40° for CO$_2$–brine analogue octane:brine within reservoir samples from the Paaratte sandstone formation in the Otway Basin, Australia, using a quasi-static pore-scale model based on in-situ three-dimensional (3D) pore-scale images.

Brooks and Corey (1964) measured the capillary pressure head ($p_l / \rho_l g$, where ρ_l is the liquid density and g is the gravity constant) for different types of soil. Similarities between the retention curves were observed, and the capillary pressure was found to be dependent on the effective saturation $\overline{S}_l$, which can be expressed as follows:

$$p_c\left(S_l\right) = p_e \overline{S}_l^{-\frac{1}{\lambda}}, \tag{4}$$

where p_e is the entry pressure (Fig. 1a); which is the pressure required to initiate a displacement of the wetting phase (e.g., water) by the non-wetting phase (e.g., air, oil, and CO$_2$). The effective saturation is defined by

$$\overline{S}_l = \frac{S_l - S_{lr}}{1 - S_{lr}}, \tag{5}$$

where S_{lr} is the residual or irreducible saturation of the wetting phase. The Brooks–Corey model for capillary pressure was designed for drainage processes, where λ is a fitting parameter that varies with respect to the soil texture, e.g., the pore size distribution, within the range 0.5–7.5 (Brooks and Corey 1964). Experimental studies conducted on Berea sandstone yielded λ in a range of 0.64–2.4 (Krevor et al. 2012; Pini and Benson 2013). Pini and Benson (2013) carried out core flooding experiments using different fluids: gN$_2$, gCO$_2$, and scCO$_2$ at 50 °C and 2.4 MPa, 2.4 MPa, and 9 MPa, respectively (where "g" represents gaseous, and "sc" represents supercritical). The entry pressure increased in accordance with an increase in the interfacial tension: 4.94 kPa (41 mN/m), 6.8 kPa (57 mN/m), and 8.13 kPa (65 mN/m) for scCO$_2$, gCO$_2$, and gN$_2$, respectively.

The trapping process was formulated by Land (1968) relating the initial and residual saturations as follows:

$$S_{gr} = \frac{S_{g,\text{init}}}{1+CS_{g,\text{init}}}, \tag{6}$$

where S_{gr} is the residual (trapped) saturation of the non-wetting phase; $S_{g,\text{init}}$ is the initial or turning point saturation; and C is the Land coefficient, which ranges from 0.9–2.5 for Paaratte sandstone (Øren et al. 2019). The Land coefficient has been found to range from 2–7 for Berea sandstone, and from 2–5 for other sandstone and carbonate rocks (Krevor et al. 2015). The collected data revealed that the trapped CO_2 saturation minimum is roughly 0.1, and most rocks can immobilize 30%–40% of CO_2. The Land model (Eqn. 6) is the most widely used, although there are other trapping models (Carlson 1981; Spiteri et al. 2008).

The Brooks–Corey model (Eqn. 4) commonly describes drainage processes. In contrast, multiple methods were developed for the imbibition process to capture the hysteresis behavior and capillary trapping depending on the initial saturation (Killough 1976; Ruprecht et al. 2014; Pini and Benson 2017). Pini and Benson (2017) proposed a novel imbibition model based on effective mobile non-wetting saturation $\overline{S}_{l,\text{mob}}$, such that $\overline{S}_{g,\text{mob}} + \overline{S}_{l,\text{mob}} = 1$:

$$p_{c,\text{imb}}(S_l) = p_{c,\text{init}}\left(\overline{S}_{l,\text{mob}}^{-\frac{1}{\lambda}} - 1\right) \text{for } \overline{S}_{l,\text{init}} \leq \overline{S}_{l,\text{mob}} \leq 1 \tag{7}$$

where $\overline{S}_{l,\text{mob}}$ is the effective mobile wetting phase saturation, and $\overline{S}_{l,\text{init}}$ is the turning-point saturation. Moreover, $p_{c,\text{init}}$ can be expressed based on the equality of imbibition and drainage capillary pressures at the turning point: $p_{c,\text{imb}} = p_c$ of a drainage curve (Eqn. 4). This Brooks–Corey based formulation (Eqns. 4 and 7) was used to analyze the results of capillary pressure heterogeneity and the hysteresis effect for a $scCO_2$–water system on Berea sandstone at the core and sub-core scales. A method of translation of physical properties and their heterogeneity for further numerical simulators was suggested by Pini and Benson (2017) using a capillary scaling method. Such studies (Pini et al. 2012; Pini and Benson 2013, 2017; Kuo and Benson 2015; Hejazi et al. 2019) describe a methodology that combines experiments and modeling for the analysis and mapping of capillary heterogeneity at the core and sub-core scales on sandstone and carbonate rocks.

The heterogeneity of rock samples is related to the grain shape and tortuosity, e.g., size and distribution of pore throats and bodies. This is characterized mostly by heterogeneous permeability K at a macroscopic scale. For multiphase flow, capillary pressure and relative permeability are major indicators of heterogeneity. The experimental data of capillary pressure are typically scaled by the Leverett J-function (Leverett 1941) $J(S) = \sqrt{K/\phi}\, p_c(S)/\gamma$; thus, they can be compared for similar textural properties (Pentland et al. 2011; Pini and Benson 2017). Using micro-CT imaging and numerical simulations, Andrew et al. (2014), Pini and Benson (2017), and Øren et al. (2019) demonstrated that the spatial distribution of heterogeneity at the micro-scale has a significant influence on the capillary pressure and relative permeability at the core scale. Recent experimental results for nine sandstone samples (Ni et al. 2019) confirmed that CO_2 trapping decreases in accordance with an increase in porosity, and increases in accordance with an increase in the degree of heterogeneity. Jackson and Krevor (2019) demonstrated that experimental observations of capillary heterogeneity, and therefore relative permeability, which includes the capillary heterogeneity at the core scale, can be translated to the meter scale. Thus core-averaged methods can be used to estimate the capillary pressure, relative permeability during drainage and imbibition, and residual trapping.

Several analytical and numerical studies have emphasized the effects of capillary trapping on CO_2 storage, the plume extent, and fate (Doughty 2007; Hesse et al. 2008;

Juanes et al. 2010; Nordbotten and Dahle 2011; Golding et al. 2013). The heterogeneity of porous media has a significant influence on the CO_2 dispersal and trapping due to local impermeable barriers that cannot be captured using simple upscaling techniques of vertical permeability reduction (Green and Ennis-King 2010; Hesse and Woods 2010). Trevisan et al. (2015) carried out reservoir scale experiments to investigate the impact of capillary heterogeneity on flow dynamics during drainage and imbibition. Moreover, heterogeneity, which is a variation of effective permeability, may yield a smaller onset time of convection than that in an equivalent homogeneous system (Green and Ennis-King 2010). Mouche et al. (2010) proposed a method for upscaling the vertical migration of CO_2 in a layered porous medium at capillary-dominant conditions. Rabinovich et al. (2015) developed a method for upscaling capillary pressure and rock relative permeability curves from the core to coarse scale, and from the sub-core to core scale. This approach allows for the modeling of drainage processes that consider high degrees of capillary heterogeneity at the fine scale, and do not consider buoyancy/gravity and the hysteresis effect. The modeling of heterogeneous media at the reservoir scale, in addition to obtaining and integrating information on the characteristic functions of different rocks at the core, sub-core, and pore scales, is thus necessary; although challenging.

Relative permeability

Darcy's law extended to multiphase flow is formulated as $u_\alpha = -k_{r\alpha}K/\mu_\alpha(\nabla p_\alpha - \rho_\alpha g)$, where the effective permeability $k_{r\alpha}K$ is a permeability tensor of phase α, and μ_α and ρ_α are the viscosity and density of phase α, respectively. The relative permeability $k_{r\alpha}$ is typically assumed to be a non-directional fluid–matrix property, although it can be a tensor. By construction, the relative permeability is dependent on the pore structure and saturation. Typical examples of non-wetting and wetting relative permeability curves are presented in Figure 1b. The relative permeability of the wetting phase decreases rapidly with the progress of drainage. The wetting fluid is immobile at significantly high saturations S_{lr} (Fig. 1b), whereas large pores are occupied by the non-wetting phase. This behavior is a characteristic of strongly water-wet systems. The relative permeability k_{rg} increases in accordance with an increase in saturation S_g up to its maximum value $k_{rg,max}$ at the maximum non-wetting phase saturation, e.g., the irreducible wetting phase saturation $S_{l,irr}$. These endpoint values are critical for rock characterization, e.g., the estimation of the storage capacity, residual trapping, and plume evolution in the CCS context.

Extensive measurements of relative permeability were carried out for CO_2–brine and H_2S–brine systems on sandstone, carbonate, shale, and anhydrite rocks from Alberta (Bachu and Bennion 2008; Bachu 2013), which yielded very low $k_{rg,max}$ values of drainage curves (approximately 0.1–0.6). Different values were obtained for $scCO_2$–brine/Berea by Krevor et al. (2012) and Pini and Benson (2013): $S_{l,irr}$ of approximately 0.45 and 0.35, and $k_{rg,max}$ of 0.38 and 0.9, respectively. Reported discrepancies between endpoint values can be attributed to the different measurement methods employed, in addition to the different theoretical assumptions of averaging. Krevor et al. (2012) suggested that endpoint values of gas saturation and permeability can be measured when sufficiently high capillary pressures are reached. This is attributed to the low viscosity of CO_2 in comparison with the viscosity of water (viscosity ratio μ_{water}/μ_{CO_2} is approximately 24–30 under high temperature and pressure conditions). Based on Darcy's law, changes in the capillary pressure are governed by changes in the viscous pressure during the experiments, as demonstrated by Pini and Benson (2013) using (steady-state) $scCO_2$, gCO_2, and gN_2 flooding experiments within a wide range of pressure, temperature, and salinity conditions. High capillary pressure conditions were obtained with large flooding rates. High values of relative permeabilities $k_{rg,max}$ and capillary pressures were simultaneously measured at the inlet face of the core. Moreover, the measurements of the saturation obtained using the X-ray CT data processing were directly related to the measured capillary pressure and relative permeability. The required high capillary pressure and high flow rate conditions may be difficult to achieve due to limited experimental capabilities (Krevor et al. 2012). This steady-state technique combined with the X-ray CT scanning described by Ramakrishnan and Cappiello (1991) and Pini and Benson (2013) was applied for measurements of the characteristic properties and allowed for the quantification of the endpoint behavior.

Sensitivity analysis experiments conducted on scCO$_2$, gCO$_2$, and gN$_2$–brine systems (Pini and Benson 2013) indicated that the IFT has a negligible influence on the relative permeability over the range of 40–65 mN/m, which corresponds to a wide range of reservoir pressure (P) and temperature (T) conditions. Consequently, analogue fluids can be used for the characterization of relative permeability. Al-Menhali et al. (2015) conducted a set of experiments to characterize the capillary pressure and relative permeability in CO$_2$–brine and N$_2$–water/Berea systems within a wide range of reservoir conditions (P = 5–20 MPa, T = 25–50 °C, and 0–5 mol/kg of NaCl). A slight difference between the N$_2$–brine and CO$_2$–brine relative permeability curves was observed, which may be due to small sample heterogeneities and other measurement uncertainties. However, in the case of the relative permeabilities measured for different gas viscosities at a fixed IFT, significant differences between the permeability curves under different viscosity conditions were demonstrated (Reynolds and Krevor 2015). Most importantly, the observed changes in the relative permeability were due to a shift in the balance between the capillary-dominated and viscous-dominated conditions, and not variations in the fluid properties.

The capillary number $N_{\text{capillary}}$, which is a ratio between the capillary to viscous forces, can be used to characterize the behavior of a system. There are different definitions of $N_{\text{capillary}}$, depending on the targeted application. For example, Virnovsky et al. (2004) proposed $N_{\text{capillary}} = H\Delta p / (L|\Delta p_c(f_w)|)$, where H and L are longitudinal and transverse length scales, respectively; the pressure drop Δp represents the viscous forces; and $|\Delta p_c(f_w)|$ is the magnitude of characteristic difference in the capillary pressure. Reynolds and Krevor (2015) demonstrated that if the system approaches a viscous limit with increasing capillary number $N_{\text{capillary}}$ (by increasing viscous pressure drop Δp), the measured relative permeability is actually the intrinsic relative permeability of the rock, and the saturation of the non-wetting phase is then insensitive to capillary heterogeneity and governed by the absolute permeability K. On the contrary, when the system is under capillary-limited conditions, e.g., $N_{\text{capillary}} \ll 1$, the effective relative permeability is measured, which is highly dependent on capillary heterogeneity, and therefore on the flow rate and fluid properties. Most relative permeability curves (Krevor et al. 2012; Pini and Benson 2013; Reynolds and Krevor 2015) are measured at the capillary limit and are then sensitive to capillary heterogeneity and reservoir conditions. Reynolds and Krevor (2015) proposed the use of scaled relative permeability for the characterization of heterogeneous cores, e.g., the measurement of the intrinsic and effective permeabilities of representative rock samples at the required reservoir conditions to better describe flow at the reservoir scale. The importance of sample heterogeneity was also highlighted by Zhang et al. (2013, 2017), who measured the relative permeabilities of Berea samples with bedding and lamination and of a heterogeneous core sample with a low permeability. Rabinovich et al. (2015) presented an upscaling technique of capillary pressure curves for heterogeneous samples at the capillary limit, not considering gravity and the hysteresis effect. Kuo and Benson (2015) proposed a two-dimensional (2D) semi-analytical method to predict the average saturation for 3D two-phase flow models of heterogeneous cores. A recent upscaling approach (Jackson et al. 2018; Jackson and Krevor 2019) was designed for the characterization of heterogeneous cores using combined experimental data and numerical modeling. The heterogeneity of a Bentheimer sandstone sample and that of a Bunter sandstone sample were defined with the viscous limit relative permeability and scaled capillary pressure curves. The characteristic properties were estimated within a wide range of $N_{\text{capillary}}$ values that can be used for flow simulations. In addition, the hysteretic curves were also obtained for a Bunter sandstone core.

Most of the experimental relative permeability data, especially the drainage curves, are generally fitted to the Brooks–Corey model (Brooks and Corey 1964; Krevor et al. 2012; Pini and Benson 2013; Reynolds and Krevor 2015; Jackson et al. 2018; Jackson and Krevor 2019):

$$k_{rl}(S_l) = \overline{S}_l^{\frac{2+3\lambda}{\lambda}} = \left(\frac{p_e}{p_c}\right)^{2+3\lambda}, \tag{8}$$

$$k_{rg}(S_l) = \left(1 - \overline{S}_l\right)^2 \left(1 - \overline{S}_l^{\frac{2+\lambda}{\lambda}}\right). \tag{9}$$

The relationships presented above correlate the permeabilities to the capillary pressure (Eqn. 4). Although data can easily be fitted to such a simple power model, relationships such as Equations (8) and (9) demonstrate the corresponding behavior of a capillary-controlled system. On the contrary when the capillary forces are absent, viscous forces govern the system, which leads to a uniform displacement of fluid. With respect to relative permeability curves, this is indicated by two diagonal lines. The models and frameworks commonly used for the construction of relative permeability curves and also of capillary pressure as well (Brooks and Corey 1964; Burdine 1953; van Genuchten 1980; Mualem 1976) can be generalized accordingly (Dury et al. 1999) as follows:

$$k_{rl}(S_l) = \overline{S}_l^{K_3}\left(1 - \left(1 - \overline{S}_l\right)^{K_1}\right)^{K_2}, \tag{10}$$

$$k_{rg}(S_l) = \left(1 - \overline{S}_l\right)^{K_3}\left(1 - \overline{S}_l^{K_1}\right)^{K_2}, \tag{11}$$

where K_i, $i = 1, 2, 3$ are fitting parameters. The terms on the right-hand side raised to the power K_3 in Equations (10) and (11) are correlated with the pore connectivity and tortuosity; the terms on the right-hand side raised to the power K_2 correspond to the capillarity effect. The same forms (Eqns. 10 and 11) can be employed for hysteretic relative permeability curves.

Several experimental and numerical studies have highlighted the hysteretic behavior of CO_2–brine systems, the shape of these curves, and residual trapping. An important difference between the drainage and imbibition processes was observed (Bachu 2013; Bachu and Bennion 2008; Ruprecht et al. 2014; Krevor et al. 2015; Jackson and Krevor 2019; Øren et al. 2019). Ruprecht et al. (2014) analyzed three core flooding cycles of alternated drainage and imbibition processes on a Berea sandstone and found that residual trapping and the CO_2 relative permeability significantly vary during the cycles. However, the water relative permeability did not exhibit hysteretic behavior. Øren et al. (2019) confirmed the hysteresis of the non-wetting relative permeability of a Paaratte sandstone using two different pore-scale modeling approaches. Interestingly, the observed non-wetting relative permeability curve of the imbibition process was higher than that of the drainage process, whereas it is typically opposite. This was attributed to the low aspect ratio of the Paaratte sample. The Killough model was employed to construct scanning curves for the capillary pressure and relative permeability, and the residual saturation was calculated by the Land equation (Eqn. 6). Moreover, Paterson et al. (2013) successfully carried out a test on residual trapping at the field scale, at the CO2CRC Otway site, Australia, during which 150 Mg of CO_2 was injected; followed by the injection of 454 Mg of formation water, which forced residual and dissolution trapping. Residual saturation values of about 0.15–0.23 were estimated using different observation techniques, which confirmed the importance of residual and dissolution trapping mechanisms. Despite these recent studies, our current knowledge of the behavior of hysteretic curves, scanning curves, and hysteresis of residual trapping is still limiting, and a comprehensive database for cyclic core flooding on representative rock samples is thus required.

Diffusion

Fick's law, which is expressed as $\boldsymbol{J}_{\alpha,k} = -\rho_\alpha \mathbf{D}_\alpha \nabla X_{\alpha,k}$, is generally used to describe molecular diffusion in liquid and gas phases; where $X_{\alpha,k}$ denotes the mass fraction of species k in fluid phase α, and $\mathbf{D}_\alpha$ is the diffusion tensor. Fick's law can also be expressed in terms of the concentration or mole fraction. The molecular diffusion $\mathbf{D}_\alpha$ can be formulated for each phase by an effective diffusion coefficient:

$$D_\alpha^{\text{eff}} = \tau \phi S_\alpha D_\alpha, \tag{12}$$

where τ is the tortuosity, which is generally represented by the Millington and Quirk model (Millington and Quirk 1961); and D_α is the molecular diffusion coefficient of fluid phase α. In this form, the diffusion coefficients in the aqueous phase are an average species-independent parameter required to maintain the electroneutrality of the solution (Tournassat and Steefel 2019, this volume). The molecular diffusion coefficient for the gas can be species-dependent, given that the gas species are neutral. Diffusion in the gas phase can be more complex than the molecular diffusion described by Fick's law. In particular, it may include Knudsen and nonequimolar diffusion. The Knudsen regime or free molecular flow occurs when the gas mean free path λ is significantly greater than the pore radius λ_p: $K_n = \lambda / \lambda_p > 10$. The Knudsen diffusion is dependent on the molecular weight, gas temperature, and pore radius. The kinetic theory of gases is then applied, and the dusty gas model of multicomponent systems can be employed (Cunningham and Williams 1980; Mason and Malinauskas 1983). Nonequimolar diffusion stems from a difference in molecular weight: lighter molecules are characterized by more rapid diffusion than heavier molecules. However, these diffusive slip fluxes do not separate gases.

For small Knudsen numbers, $10^{-3} < K_n < 10$, when the mean free path of gas molecules is approximately the same as the pore radius or less, transition or viscous slip flow occurs. Due to the slip effect, which is referred to as the Klinkenberg effect, Darcy's law typically underestimates gas flow. Klinkenberg (1941) proposed several models of gas slip permeability, which correct the gas permeability with respect to the mean pressure and a fitting parameter of slippage. The Klinkenberg first-order correction of permeability evolves into a modified gas velocity, as follows:

$$\mathbf{u}_g = -\left(1 + \frac{b_K}{p_g}\right)\frac{k_{rg}}{\mu_g}\mathbf{K}\left(\nabla p_g - \rho_g \mathbf{g}\right), \tag{13}$$

where b_K is the Klinkenberg first-order parameter. Gas flow in tight porous media can differ significantly from liquid flow, given that the effective permeability is pressure dependent and varies with respect to the gas composition. There are several forms of the Klinkenberg parameter. In particular, it can be presented with respect to the kinetic viscosity of the fluid, temperature, and molecular weight of the gas (Ertekin et al. 1986). Simplified forms with empirical parameters have been used to study the apparent permeability of gases such as hydrogen, helium, nitrogen, air, argon, and carbon dioxide in low permeable sands, sedimentary rocks, and shale formations (Heid et al. 1950; Florence et al. 2007; Ghanizadeh et al. 2014). Civan (2010) derived a correlation between the apparent gas permeability, Klinkenberg slippage parameter, and tortuosity. The apparent gas permeability can be greater than the initial intrinsic permeability by a factor of 10^3. Therefore, the Klinkenberg slippage effect is typically considered for the modeling of CO_2 sequestration in unconventional reservoirs such as coal bed methane (Liu et al. 2015; Fan et al. 2019) and shale gas reservoirs (Sun et al. 2013).

General formulations of multiphase flow

The formulations of multiphase flow briefly presented below are widely used for the modeling of hydrogeochemical and petroleum engineering problems. Mathematical and numerical methods of immiscible flow have been extensively investigated by the petroleum industry (Craig 1971; Barenblatt et al. 1972; Bear 1972; Peaceman 1977; Aziz and Settari 1979; Hassanizadeh and Gray 1979; Chavent and Jaffré 1986; de Marsily 1986; Whitaker 1986; Lake 1989; Chen et al. 2006). Several major formulations of immiscible flow are discussed in this section, as they are employed for compositional flow. The optional unsaturated flow is also derived, as it is applicable to a wide range of other environmental problems.

Immiscible multiphase flow. Mass conservation equations are required for each phase in a system of N_f fluid phases:

$$\frac{\partial m_\alpha}{\partial t} + \nabla \cdot \mathbf{F}_\alpha = q_\alpha, \tag{14}$$

where $m_\alpha = \phi S_\alpha \rho_\alpha$ is the mass of a fluid phase α, $\alpha \in \{0,...,N_f\}$, $\mathbf{F}_\alpha = \rho_\alpha \mathbf{u}_\alpha$ is the mass flux, and q_α is the source or sink term. The phase pressures p_α are related by the capillary pressure relationships:

$$p_{c\alpha} = p_{\alpha+1} - p_\alpha. \tag{15}$$

There are various formulations of the set of equations, and their structures modify the coupling between equations. Several examples are presented below, starting from a general formulation with respect to pressure and saturation.

The pressure-saturation formulation is derived from the substitution of the Darcy velocities in the mass conservation Equation (14):

$$\frac{\partial(\phi S_\alpha \rho_\alpha)}{\partial t} - \nabla \cdot \left(\rho_\alpha \frac{k_{r\alpha}}{\mu_\alpha} \mathbf{K}(\nabla p_\alpha - \rho_\alpha \mathbf{g}) \right) = q_\alpha. \tag{16}$$

For each phase α, a natural variable is selected, namely, S_α or p_α resulting in N_f nonlinear equations with N_f unknowns. The equations are strongly coupled through characteristic functions p_c and $k_{r\alpha}$. Primary variables should be carefully selected, as this leads to specific limitations. The modeling of a two-phase system allows for pairs (p_l, S_g) and (p_g, S_l). If (p_l, S_g) are primary variables, the conservation equation of the non-wetting phase contains a term $k_{rg}\nabla p_c$ with singularities at the endpoints. The application of the Brooks–Corey (BC) or the van Genuchten (vG) model yields the following:

$$\begin{array}{ccc} & \text{BC} & \text{vG} \\ \lim_{S_g \to 0} k_{rg} p'_c(S_g) = & 0 & 0 \end{array} \tag{17}$$

$$\lim_{S_g \to 1} k_{rg} p'_c(S_g) = \quad \infty \quad \infty$$

When selecting (p_g, S_l) as primary variables, the term $k_{rl}\nabla p_c$ has singularities at $S_l = 1$.

$$\begin{array}{ccc} & \text{BC} & \text{vG} \\ \lim_{S_l \to 0} k_{rl} p'_c(S_l) = & 0 & 0 \end{array} \tag{18}$$

$$\lim_{S_l \to 1} k_{rl} p'_c(S_l) = \quad <\infty \quad \infty$$

After the linearization of the resulting system of equations written in discrete residual form, numerical problems at the saturation endpoints then appear in the Jacobian matrix (Sin 2015). The pressure–pressure formulation can be used to prevent restrictions on saturation range, as discussed below. Alternatively, this issue can be solved by employing numerical methods, e.g., the use of spline curves.

The decoupled pressure and saturation formulations separate the pressure and saturation equations. Summarizing the conservation equations, the temporal derivative of saturation can be eliminated:

$$\frac{\partial \phi}{\partial t} + \sum_\alpha \frac{1}{\rho_\alpha}\left(\phi S_\alpha \frac{\partial \rho_\alpha}{\partial t} + \nabla \rho_\alpha \cdot \mathbf{u}_\alpha\right) + \nabla \cdot \mathbf{u} = \sum_\alpha q_\alpha, \quad (19)$$

where the total velocity $\mathbf{u} = \sum_\alpha \mathbf{u}_\alpha$. The Darcy velocities for two phases are defined as follows:

$$\mathbf{u}_l = f_l \mathbf{u} + f_l \lambda_g \mathbf{K}\left(\nabla p_c + (\rho_l - \rho_g)\mathbf{g}\right), \quad (20)$$

$$\mathbf{u}_g = f_g \mathbf{u} - f_g \lambda_l \mathbf{K}\left(\nabla p_c - (\rho_g - \rho_l)\mathbf{g}\right), \quad (21)$$

where $\lambda_\alpha = k_{r\alpha}/\mu_\alpha$ is the mobility, and $f_\alpha = \lambda_\alpha / \sum_\beta \lambda_\beta$ is the fractional flow.

Equation (19) is solved by considering a phase pressure or global pressure as a primary variable. If a phase pressure is selected, the saturation equation is then hyperbolic, and coupling of the system is strong. If the global pressure is a primary variable, the saturation equation is weakly coupled to the pressure equation.

When a phase pressure ρ_α is selected as a primary variable, Equation (16) is formulated for the Darcy velocity $\mathbf{u}_\alpha$ using Equations (20) or (21). This yields a pure hyperbolic equation for saturation. When using a phase pressure, the pressure equation contains a capillary term $f_\alpha \nabla p_c$. The result is that the phase pressure and saturation are decoupled.

Decoupled global pressure and saturation is also possible when the total velocity is given by $\mathbf{u} = -\lambda_{tot} \mathbf{K}\left(\nabla p - \sum_\alpha f_\alpha \rho_\alpha \mathbf{g}\right)$, where the total pressure $\nabla p = \sum_\alpha f_\alpha \nabla p_\alpha$. The phase velocities can then be expressed with respect to the total velocity and substituted into one of the mass balance equations. Hence, the pressure and saturation equations are weakly coupled through the mobilities and fractional flow. However, boundary and initial conditions have no physical meaning, and the pre- and post-treatment of variables is then required. These techniques for the immiscible multiphase flow allow for strong or weak mathematical decoupling, which can be employed for compositional flow.

Unsaturated flow. The modeling of unsaturated flow is critical in soil science, environmental engineering, and groundwater hydrology, e.g., to predict fluid motion in the vadose zone, changes in groundwater systems due to root water uptake, and the evaporation of water from soil. The vapor movement is significantly less important in soil science. The air pressure above the water table corresponds to the atmospheric pressure. Water is under tension and attracted upward by capillary forces. The total hydraulic head h_{total} consists of the pressure head h and elevation head z. When the pressure head is negative, water molecules are under tension, which indicates the capillary fringe. Air motion is neglected, which is an important hypothesis and differs from the multiphase flow formulation above: $p_{air} = 0$, $p_w = -p_c$. This yields the following:

$$u = -K_{cond}(h)\nabla(h+z) = -\left(K_{cond}(\theta)\frac{dh}{d\theta}\nabla\theta + K_{cond}(\theta)1_z\right) = -\left(D_c(\theta)\nabla\theta + K_{cond}(\theta)1_z\right), \quad (22)$$

where $\psi = p_c / \rho_w g = p_c / \gamma_w$ is the capillary pressure head.

The Darcy velocity can be expressed as follows:

$$u = -K_{cond}(h)\nabla(h+z) = -\frac{K_{cond}(h)}{\gamma_w}\nabla(p_w + \gamma_w z) = -\frac{K}{\mu_w}\nabla(p_w + \gamma_w z), \quad (23)$$

where $K_{cond}(h) = K\gamma_w/\mu_w$ is the hydraulic conductivity, which is related to the intrinsic permeability by the gravitational term γ_w and the water viscosity μ_w. Obviously, $K_{cond}(h)$ can be modeled using the Brooks–Corey model and other empirical models (Eqns. 8–11). Consequently, the Darcy velocity can be formulated with respect to the moisture $\theta = \phi S_w$:

$$u = -K_{cond}(h)\nabla(h+z) = -\left(K_{cond}(\theta)\frac{dh}{d\theta}\nabla\theta + K_{cond}(\theta)1_z\right) = -\left(D_c(\theta)\nabla\theta + K_{cond}(\theta)1_z\right), \quad (24)$$

where $D_c(\theta) = K_{cond}(\theta)/C(\theta)$ is the capillary diffusivity, and $C(h) = d\theta/dh$ is the moisture capacity. The water balance in the vadose zone is described by Richards equation, as follows:

$$\frac{\partial\theta}{\partial t} - \nabla\cdot\left(K_{cond}(\theta)\nabla(h+z)\right) = 0, \quad (25)$$

The equation above can be expressed in three different forms using Equation (24): mixed, h-based, and θ-based. The h-based formulation is not conservative, and the discretization of the accumulation term $C\partial h/\partial t$ can lead to numerical errors. Although the θ-based equation is conservative, it fails in saturated conditions. Therefore, the mixed formulation is generally advised. Given that Equation (25) can be coupled with multicomponent solute transport, it covers numerous applications in the vadose zone (Suarez and Simunek 1996). However, assumptions such as inert gasses and the incompressibility of fluids are limiting for a wide range of problems that involve gas dynamics, the heterogeneous composition of fluids, and important changes in their composition due to physical and chemical processes.

Compositional flow. When considering miscible fluids (i.e., no sharp interface between fluids and phase changes), the compositional flow should be solved. For a system of N_c components, the mass conservation equations and closure relationship for mass fractions can be expressed as follows:

$$\frac{\partial\left(\sum_{\alpha=1}^{N_f}\phi S_\alpha\rho_\alpha X_{\alpha,k}\right)}{\partial t} + \nabla\cdot\sum_{\alpha=1}^{N_f}\left(\rho_\alpha X_{\alpha,k}\mathbf{u}_\alpha + J_{\alpha,k}\right) = q_k, \quad (26)$$

$$\sum_{k=1}^{N_c}X_{\alpha,k} = 1. \quad (27)$$

The conservation equations can also be expressed with respect to mole fractions or concentrations.

The compositional flow problem consists of N_c equations and $2N_f$ constitutive relationships (Eqns. 1, 15, 26, and 27), and it describes a general multiphase multicomponent flow (MMF) with $2N_f + N_cN_f$ unknowns: p_α, S_α, and $X_{\alpha,k}$. The missing $N_c(N_f-1)$ equations correspond to the phase equilibrium relationship detailed below. Numerical and mathematical methods for MMF are presented in the section on multiphase multicomponent flow (MMF) approaches, followed by the coupling methods with reactive transport or geochemical reactions.

Geochemical reactions

In most geochemical or reactive transport codes, reservoirs are considered infinite with respect to gases, which is equivalent to the fixing of the corresponding dissolved concentrations in the aqueous solution. This assumption holds when modeling open or semi-open reactors, into which gases are continuously pumped to maintain the global pressure constant, or shallow aquifers in contact with the atmosphere. In particular, the mass balances that involve gaseous species are not considered, and the dissolved aqueous species just obey the corresponding mass action law. The system of equations required to be solved is then significantly simpler. Nevertheless, for closed reactors or for the general two-phase reactive-transport problem, this approach is not satisfactory, and codes intended for these purposes have been subsequently improved. The focus of this chapter is on the CO_2 geological storage application; hence, several aspects are therefore highlighted in this context. Many numerical studies were carried out using dissolved impurities and an artificial gas phase (Xu et al. 2007; Xiao et al. 2009; André et al. 2012). When CO_2 is dissolved, CO_2–water–rock interactions can be handled with a classical geochemical solver. However, to process changes in the pressure, temperature, and salinity conditions, accurate models are required to predict the gas–water equilibrium; especially when gasses other than CO_2 are considered. In addition, recent developments of gas–water–rock interactions allow for several other contexts (e.g., acidic–basic/oxidative–reducing conditions, high–low pressure/temperature/salinity, and kinetically-controlled biotic reactions involving gas); thus leading to various applications (natural gas reservoirs, hydrogen storage in salt caverns, enhance coal-bed methane, and soil gas migration).

Chemical equilibrium and mass balances. Geochemical solvers or reactive transport codes can manage chemical equilibrium calculations in two manners: based on a stoichiometric approach (PHREEQC, Parkhurst and Appelo 2013; CHESS, van der Lee 2009) or on a non-stoichiometric formulation (GEM-Selektor, Kulik et al. 2004).

On one hand, the stoichiometric approach requires the expression of all the chemical reactions and associated mass actions laws with respect to activities. On the other hand, the non-stoichiometric approach is based on the minimization of the total Gibbs energy of the considered systems, which is calculated from the chemical potential of all the species. In both cases, solvers also include mass or mole balance equations that are expressed with respect to the elements or total components, which are not presented here. They also generally comprise mineral precipitation/dissolution, solid solutions, kinetically controlled reactions, sorption, and isotopes.

Water–gas equilibrium. Many models have been developed since the early 1990s to predict the water–gas equilibrium; and in particular, mutual solubilities using various strategies. Studies are largely focused on hydrocarbons and CO_2, due to applications for the oil and gas industries (Michelsen and Mollerup 2007); or more recently, for the greenhouse gas geological storage and due to the quantity of experimental data available to fit or test models. Søreide and Whitson (1992) adapted the Peng–Robinson EOS (Peng and Robinson 1976); Duan and Sun (2003) derived a virial EOS; and Spycher et al. (2003) employed a cubic EOS (Redlich and Kwong 1949). Various geochemical solvers were recently developed with specific modules to incorporate the cubic EOS to address non-ideal gas phases at high pressures and temperatures (Appelo et al. 2014; Corvisier et al. 2014; Leal et al. 2014). The phase equilibrium is presented below.

The local equilibrium of a system requires a minimum of the Gibbs free energy at a constant temperature T, pressure p, and composition. However, the system could be monophasic or composed of multiple phases. In geochemical modeling, it is common to use the heterogeneous approach: an activity model describes the activity coefficient in the liquid phase, whereas the fugacity coefficient in the vapor is modeled by an equation of state. Given that not only pure water is considered but also various ionic species, the asymmetric convention (φ–γ) is used.

When considering non-stoichiometric geochemical solvers, the activities of the aqueous species (dissolved gases) and fugacities of the gases appear in the chemical potential functions:

$$\mu_i(T,p,n) = \mu_i^0(T,p) + RT\ln(C_i\gamma_i) \tag{28a}$$

$$\mu_i^g(T,p,n) = \mu_i^{0,g}(T,p) + RT\ln\left(y_i\varphi_i^g p\right) \tag{28b}$$

where C_i is a concentration quantity (e.g., molality), γ_i is an activity coefficient, y_i is the gaseous molar fraction, and φ_i^g is a fugacity coefficient. As stated by Gibbs, at local equilibrium, the chemical potentials of each component are equal in the multiphase system at constant temperature, pressure, and composition. This is a criterion, and the solution of the equilibrium conditions must then be coupled with a stability criterion. Lewis and Randall (1923) defined the fugacity of a component as follows:

$$f_i(T,p,x) = p\cdot\exp\left(\frac{g_i(T,p,x) - g_{i,\text{pure}}^{\text{Perfect Gas}}(T,p,x)}{RT}\right). \tag{29}$$

where g_i is the Gibb's energy and x is the composition. The fugacity therefore has the dimension of a pressure, and the fugacities of each component are equal at equilibrium:

$$f_i^{\text{liquid}}(T,p,x) = f_i^{\text{vapor}}(T,p,y) \tag{30}$$

and have to be determined for each component in each fluid phase. For stoichiometric solvers, considering a specific reference state via the Henry's law constant and the equality between fugacities produces a mass action law:

$$x_i\gamma_i K_i^h = y_i\varphi_i^g p, \tag{31}$$

where x_i and y_i are the mole fractions of species i in the aqueous and the gaseous phases respectively, and K_i^h is the Henry's law constant of species i. The Henry's constants could be expressed as a function of temperature, pressure, and even salinity. In addition, it can be derived from the constants expressed at the vapor saturation pressure p^{sat} and the molar volume at infinite dilution υ_i^∞:

$$K_i^h = K_i^h\left(T,p^{\text{sat}}\right)\exp\left(\frac{\upsilon_i^\infty\left(p-p^{\text{sat}}\right)}{RT}\right). \tag{32}$$

Detailed derivations of vapor–liquid equilibria, concepts, and techniques were reported by Firoozabadi (1999) and Michelsen and Mollerup (2007).

Equation of state. The EOS for the gas phase is solved with respect to the gas volume at fixed pressure or the pressure at constant volume. The fugacity coefficients can then be calculated. A perfect gas EOS is not sufficient when addressing relatively significant pressures and temperatures, and it is necessary to consider non-ideal gases. The cubic EOS is useful for reactive transport, given that the cubic equation can be directly solved, e.g., the use of an analytical Cardan method (Nickalls 1993, Spycher et al. 2003). When considering a mixture, a mixing rule can be employed, in addition to calibrated binary interaction parameters, thus rendering the cubic EOS general and well-suited for multicomponent systems.

The classic cubic equations are the Soave–Redlich–Kwong (SRK) (Soave 1972) and Peng–Robinson (PR) (Peng and Robinson 1976; Robinson and Peng 1978) equations. Various improvements of the EOS were proposed using different alpha functions adapted for precise compounds and more complex mixing rules. The predictive PR equation (Qian et al. 2013) improves the mixing rules proposing the temperature dependent binary parameters calculated via the global contribution method (GCM). Hence, the GCM is advantageous, given that the vapor–liquid equilibrium (VLE) can be calculated for various systems without introducing new empirical fitting parameters. There are accurate virial EOS models such as the Benedict–Webb–Rubin EOS (Benedict et al. 1940), which have been widely used for CO_2 (Duan and Sun 2003). However, they are typically different for each gaseous compound, and they require iterative solutions.

The classic cubic EOS formulation and mixing rule are general, i.e., applicable to every compound. Their ability to accurately represent the CO_2–H_2O and CO_2–H_2O–NaCl systems was demonstrated (Spycher et al. 2003; Spycher and Pruess 2005; Corvisier et al. 2014, 2017; Hajiw et al. 2018), in addition to other systems that include various compounds and gaseous mixtures. A comparison between experimental data and several other methods yielded good results (Hajiw et al. 2018).

Aqueous activity correction. Activity coefficients in solution also require models. The simplest approaches (Debye-Hückel 1923; Davies and Shedlovsky 1964; Colston et al. 1990; B-Dot, Helgeson 1969) allow for the modeling of diluted solutions, as they are based on long-distance electrostatic interactions, which are predominant in this case. For highly saline solutions, wherein the ionic strength is higher that 3 M, short-distance non-electrostatic interactions should be considered. The Pitzer (Pitzer 1973, 1991) and specific ion interaction (SIT) (Brønsted 1922; Grenthe and Puigdomenech 1997) models yield relatively good results for these solutions.

For neutral species such as dissolved gases, the Pitzer model yields the following expression:

$$\ln \gamma_i = 2C_i \lambda_{ii} + 2 \left(\sum_j \lambda_{ij} C_j \right) + 6C_i \left(\sum_j \mu_{iij} C_j \right) + \sum_j \sum_k C_j C_k \zeta_{ijk} \tag{33}$$

where λ_{ij}, μ_{ijk}, and ζ_{ijk} are the interaction parameters between i, j, and k species.

The SIT model limits to binary interactions and appears similar to first-order Pitzer models. Activity coefficients for such neutral species are expressed as follows:

$$\ln \gamma_i = \sum_j \varepsilon_{ij} C_j \tag{34}$$

where ε_{ij} are the binary interaction parameters between the i and j species.

Contrary to several activity models (Setchenow, Davies, or B-Dot), the two abovementioned models take into consideration and allow for the calibration of the influence of ions on the dissolved gas activity, and therefore, the gas solubility. In many CO_2 solubility models, the activity of dissolved CO_2 is calculated using specific models (Drummond 1982; Rumpf et al. 1994) based on available experimental data. Nevertheless, the SIT or Pitzer models allow for the use of the same model for various electrolytes and dissolved gases; thus simplifying the databases.

In dilute solutions, the activity of water is typically set as 1.0. However, its value deviates from 1 in saline solutions. The water activity can be expressed directly as a function of ionic strength or the osmotic coefficient (Helgeson 1969; Pitzer 1973, 1991), which gives reasonable results.

Parameterization. The quality of the representation of the gas–water equilibrium is significantly dependent on the parameters, and therefore on the database combined with the geochemical solver. Simulations that involve a poorly defined Henry's constant do not result in accurate predictions, and this is an important limit (mainly related to the current lack of sufficient and consistent experimental data) that modelers should consider.

With reference to the literature, the parameters associated with gases used in the EOS (critical temperature and pressure, and acentric factor) do not vary significantly among various sources, and they can be retrieved from the Yaws handbook (Yaws 1999).

The interaction parameters of gases, which are dependent on the selected EOS and mixing rule; and those of electrolytes, which are dependent on the selected aqueous activity model, come from experiments (binary mixtures PTy data and activity/solubility data, respectively), and are therefore significantly dependent on the quality of the selected data.

Henry constants or reference chemical potentials, and their dependence with respect to pressure are also drawn from solubility experimental data (Hajiw et al. 2018).

First 0D simulations. Geochemical simulators, as described here, allow for the simulation of many gas–water–brine systems, including CO_2–H_2O and CO_2–H_2O–$NaCl$. Significant research attention has been directed to these systems, and consistent data are increasingly available to help us select the appropriate EOS and activity models as demonstrated here below.

As an example, simulations were run using Chess on CO_2 solubility and co-solubility data at 50 °C, 100 °C, and 150 °C (Figs. 2 and 3), and the following modeling options were tested: the perfect gas EOS, PR EOS, and the pressure correction of thermodynamic equilibrium constants. These illustrate that it is necessary to use an accurate EOS, especially when the temperature and pressure increase; and highlight the importance of appropriately correcting the Henry's constants with respect to pressure. As an illustration, the absolute average deviation (AAD) between the

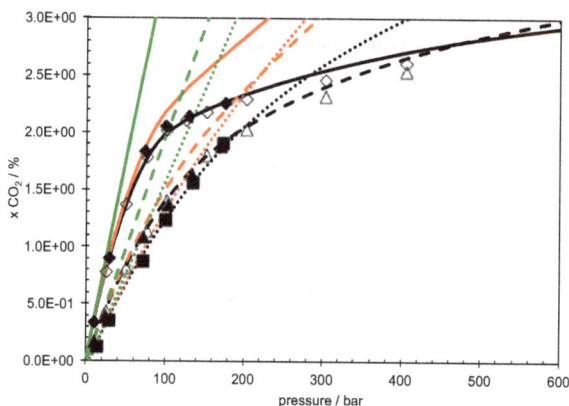

Figure 2. CO_2 molar fractions in water with respect to pressure. The symbols correspond to data from Hou et al. (2013) at 50 °C (black diamond), 100 °C (black triangle), and 150 °C (black square); and from Wiebe and Gaddy (1939) at 50 °C (white diamond) and 100 °C (white triangle). The green lines correspond to simulations using the perfect gas EOS, the red lines correspond to simulations using the Peng–Robinson EOS with no pressure corrections of thermodynamic constants, and the black lines correspond to simulations using the Peng–Robinson EOS with pressure corrections (the continuous, dashed, and dotted lines correspond to the three temperatures, respectively).

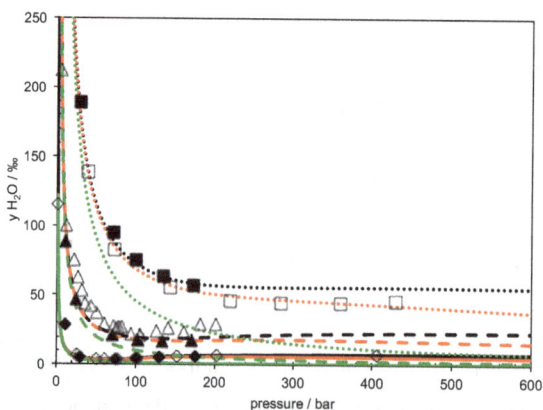

Figure 3. H_2O molar fractions in carbon dioxide with respect to pressure. The symbols correspond to data from Hou et al. (2013) at 50 °C (black diamond), 100 °C (black triangle), and 150 °C (black square); from Wiebe and Gaddy (1941) at 50 °C (white diamond); from Caumon et al. (2016) at 100 °C (white triangle); and from Taba-sinejad et al. (2011) at 150 °C (white squares). The green lines correspond to simulations using the perfect gas EOS, the red lines correspond to simulations using the Peng–Robinson EOS with no pressure corrections of thermodynamic constants, and the black lines correspond to simulations using the Peng–Robinson EOS with pressure corrections (the continuous, dashed, and dotted lines correspond to the three temperatures, respectively).

solubility experimental data and the model decreased from 85% using the perfect gas EOS and no pressure correction to 4% using the PR EOS and pressure correction.

Other simulations were run with Chess on CO_2 solubility data in NaCl brines at 50 °C, 100 °C, and 150 °C, and 150 bar (Fig. 4) using the B-Dot aqueous activity correction and the SIT model. These simulations highlight the importance of appropriately representing the electrolyte and its effects on the gas solubility (the salting-out effect). In this case, the AAD decreased from 7% with the B-Dot model to 3% with the SIT model. For NaCl brines, B-Dot simulations appropriately represent the data at significantly high salinities, but it may be noted that this is not the case for other brines (e.g., $CaCl_2$, $MgCl_2$).

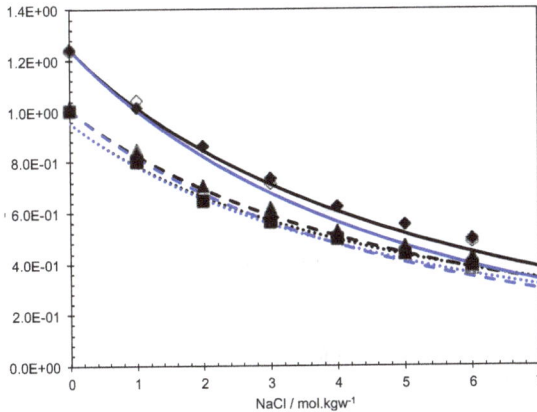

Figure 4. CO_2 molalities in NaCl brine. The symbols correspond to the data at 150 bar from Zhao et al. (2015), 50 °C (black diamond), 100 °C (black triangle), and 150 °C (black square); and the data from Messabeb et al. (2016) at 50 °C (white diamond), 100 °C (white triangle), and 150 °C (white square). The blue lines represent simulations using the B-Dot model, and the black lines represent simulations using the SIT model (the continuous, dashed, and dotted lines correspond to the three temperatures, respectively).

Density

Variations in the composition of phases have a direct influence on the saturation, solubilities and co-solubilities (e.g., the dew and bubble-point pressures and the *P–T* phase envelope). Other thermodynamic properties such as the density and viscosity are also dependent on the pressure, temperature, and composition; and they feature in the fluid dynamics calculations. The degree of influence varies with respect to the type of fluid. For example, the density of a CO_2 and SO_2 mixture can significantly increase with an increase in the SO_2 content. On the contrary, the density of a highly non-ideal mixture of CO_2 and H_2 can decrease with an increase in the H_2 content. The densities of the CO_2-rich mixture with 2–10 mol% of H_2 increases in accordance with an increase in the pressure, and decreases in accordance with an increase in the temperature and the H_2 concentration: at 313.15 K and 9 MPa, adding 2% of H_2 to the pure CO_2 decreases the molar density by 25% (Sanchez-Vicente et al. 2013). It is therefore necessary to consider these modifications and to accurately select a thermodynamic model.

The modeling of the density of pure bodies can be achieved by employing an EOS or empirical correlation; although there is no EOS that can universally predict the thermodynamic state and properties of any multicomponent system. For CO_2, CO_2–H_2O, H_2O–NaCl, and CO_2–H_2O–NaCl, several specific models were developed at certain temperature and pressure conditions. For example, the density and viscosity of formation waters in sedimentary basins (Adams and Bachu 2002; McCain Jr. 1991), the density of CO_2–H_2O (Garcia 2003) and the density of CO_2, CO_2–H_2O, and CO_2–H_2O–NaCl (Duan et al. 2008). However, these models are not applicable to other multicomponent systems.

The GCM is typically an advantageous method for modeling of multicomponent mixtures, given that the form of the EOS and mixing rule is general for any number of components. The SRK (Soave 1972) and PR (Robinson and Peng 1978) yield relatively accurate estimations of gas densities for pure bodies and gas mixtures (Sin 2015). Nazeri et al. (2017) obtained good results using the PR and SRK models for the gas, supercritical, and liquid densities of multicomponent systems that contained CO_2, CO, N_2, O_2, Ar, H_2, CH_4, C_2H_6, C_3H_8, SO_2, and H_2S. However, in the presence of water, the estimations would be significantly different. The predictions of the liquid densities for water systems made the PR or SRK EOS approaches inaccurate, as they are not adapted for highly polar systems.

Another GCM-based model is the volume translated PR (Ahlers and Gmehling 2001, 2002a,b; Ahlers et al. 2004), which is based on the pseudo-volume (Péneloux et al. 1982). The model therefore improves the prediction of volumetric properties. In addition, the model can be employed for the asymmetric VLE approach (Ahlers and Gmehling 2002a,b). The model demonstrates a good prediction of the liquid density of water systems with a rather wide range of components (Ahlers and Gmehling 2001; Sin 2015). Chapoy et al. (2013) developed a state viscosity model and density model with volume correction for the CO_2, O_2, Ar, and water system. The Cubic-Plus-Association EOS can also consider molecular structures with hydrogen bonding. Moreover, in combination with the PR model, a better modeling of both the VLE and liquid densities of water systems can be achieved. This was demonstrated for hydrocarbons and other compounds (Hajiw et al. 2015, 2018). The drawback is that it is computationally complex. Further development of GCM-based models is required, and more experimental data of multicomponent mixtures are necessary for validation.

Feedback on fluid and matrix properties

When CO_2 dissolves into reservoirs, it acidifies the water and initiates dissolution/precipitation reactions with rocks. Furthermore, if the injected gas is initially dry (as is typical for captured CO_2), water vapor is formed to satisfy the gas–water equilibrium conditions. Some of potential co-injected impurities modify and promote geochemical reactions. This indicates that gas and its geochemical reactions have a significant influence on the entire porous medium, e.g., host water/brine and rocks (Jung et al. 2013; Sterpenich et al. 2013; Lu et al. 2014; Pearce et al. 2015, 2016a,b; Lu et al. 2016) and engineered materials such as cement-based wells (Jacquemet et al. 2005). Models should be able to follow such modifications of the fluid and matrix and accurately quantify their effect on associated properties using a general or specific EOS.

Dissolution/precipitation reactions (not only induced by gas) modify hydrological properties such as the porosity, reactive surface area, and permeability at the pore- and core-scales (Gouze and Luquot 2011; Xu et al. 2017). Moreover, CO_2 injection into a heterogeneous medium may create wormholes, depending on the relationship between geochemical reactivity and fluid motion (Snippe et al. 2017). At the Darcy scale, the change in porosity due to the dissolution/precipitation of mineral m is typically given by $\phi = \phi^0 - \sum_{m=1}^{N_m}(c_m - c_m^0)/\rho_m$, where ϕ^0 is the initial porosity; c_m^0 and c_m are the initial and current concentrations of mineral m (unit: mol per volume); and ρ_m is the molar density of mineral m. In addition, the porosity varies with respect to pressure: $\phi = \phi^* e^{c_r(p-p^*)}$, where c_r is the rock compressibility; ϕ^* is the reference porosity; and p^* is the reference pressure. The intrinsic permeability–porosity relationship is generally represented by the Kozeny–Carman model. These relationships are actually more complex than this relatively simple description, especially at small scales. For example, Xu et al. (2017) demonstrated that mineral dissolution can enlarge pore throats, increase porosity and permeability, decrease capillary pressure, and influence relative permeability due to the injection of dissolved CO_2 in a Chaunoy sandstone core from the Paris Basin. A depressurization test revealed that gas exsolution and mineral precipitation may occur. However, this is completely different from injection of sc/gCO_2 (Jacquemet 2006; Jacquemet et al. 2008; Snippe et al. 2017), as experimental and numerical studies confirmed that the influence of scCO_2 injection on porosity and permeability is negligible. The modeling of such geochemically induced changes requires further experimental studies, more data, and an in-depth investigation to better understand these couplings and the evolving matrix structure, e.g., moving boundaries, reactive surfaces, tortuosity, diffusivity and the mass and volume balances/constraints. Further discussion of the modeling of evolving matrix properties and implications with respect to CCS are discussed in Seigneur et al. (2019, this volume) and Cama et al. (2019, this volume), respectively.

Changes in the gas composition have an influence on the gas volume and partial pressures, and therefore the compressibility and viscosity. This impacts the fluid motion: mass fluxes, gravity effects, and viscous forces. Molecular fluxes depend on a composition of the system, Hence, the interplay between these forces evolves into a heterogeneous composition of fluids, e.g., solubility-induced partitioning (see Applications section for further discussion). A notable example of the composition-dependent mass fluxes is density driven flow, which plays an important role in geological CO_2 storage. The formation water becomes denser with dissolved CO_2. Therefore, the accumulation of dissolved CO_2 at the top of an aquifer leads to instabilities. Upon reaching an onset time, denser fluid tends to sink. The convective motion created as a consequence enhances the dissolution rate. The density of fluids should then be accurately modeled to predict the onset time and convective fluxes, and to estimate convective dissolution trapping. Farajzadeh et al. (2007), Kneafsey and Pruess (2010), Neufeld et al. (2010), MacMinn et al. (2012), and Wang et al. (2017) experimentally confirmed the importance of convective fluxes. Several studies were carried out, which involved mathematical and numerical stability analyses of convective mixing for homogeneous and heterogeneous media, with and without capillary transition zones (Lindeberg and Wessel-Berg 1997; Xu et al. 2006; Hesse et al. 2008; Pau et al. 2010; Elenius et al. 2012, 2015; Li and Jiang 2014); for an anisotropic medium (Ennis-King and Paterson 2005); in open and closed systems (Riaz et al. 2006; Wen et al. 2018) and 3D heterogeneous domains (Green and Ennis-King 2018). At present, there is insufficient information as to the influence of gases other than CO_2 on convective mixing. Co-injected impurities can significantly change gas and liquid densities, affect the convection flux, and increase or decrease the onset time (see Applications section for further discussion). Coupled geochemical reactions and multiphase flow also influence gas exsolution, which can be induced during depressurization. Experimental studies have shown that the relative permeability for CO_2 exsolved from brine curves is significantly different from those of drainage process (Falta et al. 2013; Zuo et al. 2013, 2017; Xu et al. 2017).

MATHEMATICAL AND NUMERICAL APPROACHES OF MMF&RT MODELING

The previous section indicates that MMF&RT is a complex problem including mathematical and numerical issues with respect to the composition of flow and reactive transport. There are several formulations of MMF that attempt to accurately solve the evolving flow system and phase equilibrium. For MMF&RT, two coupling families exist: the global implicit and operator splitting approaches. Both families have advantages and drawbacks, and, the development of new coupling methods, formulations, and numerical techniques continues.

Multiphase multicomponent flow (MMF) approaches

The balance equations for multiphase multicomponent flow (Eqn. 26) can be formulated on mole fractions $x_{\alpha,k}$ of species k in phase α, as follows:

$$\frac{\partial\left(\sum_{\alpha=1}^{N_f}\phi S_\alpha \rho_{mol,\alpha} x_{\alpha,k}\right)}{\partial t} = -\nabla\cdot\sum_{\alpha=1}^{N_f}\left(\rho_{mol,\alpha} x_{\alpha,k}\mathbf{u}_\alpha - \rho_{mol,\alpha}\mathbf{D}_\alpha\nabla x_{\alpha,k}\right)+q_k. \tag{35}$$

where $\rho_{mol,\alpha}$ is the molar density of fluid α.

$$\sum_{k=1}^{N_c} x_{\alpha,k} = 1. \tag{36}$$

The balance system is equivalent to the set of mass conservation equations (26) and can be expressed with respect to concentrations, given the following:

$$\rho_{mol,\alpha} x_{\alpha,k} = c_{\alpha,k} = \frac{\rho_\alpha X_{\alpha,k}}{M_k}, \tag{37}$$

where M_k is the molar mass of species k.

For the system of N_c conservation equations (Eqn. 35), there are $2N_f + N_f N_c$ unknowns, which are referred to as natural variables (Coats 1980): S_α, p_α, and $x_{\alpha,k}$. Considering the closing relationships for pressure (Eqn. 15), saturation (Eqn. 1), and mole fractions (Eqn. 36) ($2N_f$ equations), the balance system (Eqn. 35) should be complemented by $N_c(N_f-1)$ phase equilibrium relationships. Using the Gibbs phase rule, an isothermal system with a locally known number of phases N_f can be described using a minimum of N_c-N_f+2 variables. For an isothermal two-phase system, N_c variables are required to define a thermodynamic state. There is a large selection of variables. The most evident is the pressure and N_c-1 variables of composition. The discretized nonlinear conservation equations should therefore be solved on the selected primary variables by employing an iterative method such as Newton's method. The residual is linearized, and the Jacobian is assembled for the iterative procedure. Approaches of MMF modeling are discussed further in the Primary variables section.

Phase equilibrium. The phase equilibrium of fluid phases can be solved using an EOS and/or activity models depending on the selected approach, (Water–gas equilibrium section). Another method, which is referred to as flash calculations, is also widely used. As it is based on K-values, which are defined as $K_i = y_i / x_i$, the data are generally available and tabulated for hydrocarbon mixtures; where x_i and y_i typically denote the liquid and gas mole fractions, respectively. The K-values can be derived from the $(\varphi-\varphi)$ and $(\varphi-\gamma)$ approaches:

$$K_i = \frac{\varphi_i^l}{\varphi_i^g} = \frac{K_i^h \gamma_i}{P\varphi_i^g}. \tag{38}$$

It can be noted that the K-values are therefore composition-independent under the hypothesis of an ideal mixture. The phase mole fractions V and L are defined such that $V+L=1$. This allows for the introduction of a new variable z_i, which is the overall mole fraction of species i:

$$z_i = x_i L + y_i V. \tag{39}$$

The mole fractions can then be expressed with respect to z_i, K_i, and V:

$$x_i = \frac{z_i}{1+V\left(K_i-1\right)}, y_i = x_i K_i = \frac{z_i K_i}{1+V\left(K_i-1\right)}. \tag{40}$$

The Rachford–Rice equation yields a constraint on the mole fractions of the system:

$$F(V) = \sum_{i=1}^{N_c} \frac{z_i(K_i - 1)}{1 + V(K_i - 1)} = 0, \tag{41}$$

which is solved with respect to V at a known pressure, temperature, and composition $\mathbf{z}$ using Newton's method. The mole fractions are then updated, and the phase equilibrium relationships are calculated to provide new K-values. The process is repeated until convergence occurs. The flash problem and stability analysis were extensively described by Firoozabadi (1999) and Michelsen and Mollerup (2007).

The phase equilibrium problem can be solved directly with the system of compositional flow (e.g., it is added to the Jacobian structure), or separately for each cell using any phase equilibrium method. An important advantage of using EOS models is that most of them can be solved analytically, e.g., for the cubic EOS. The flash problem requires an iterative method when a real mixture is modeled; otherwise, the K-values are constant, and the solution is significantly simplified.

Discretization. The balance equations (35) are nonlinear and strongly coupled due to the relative permeability, capillary pressure, and fluid properties. The equations are (near-)parabolic, which is similar to the system of immiscible two-phase flow (Eqn. 16). The fully implicit method (FIM) is generally employed for the time discretization, e.g., all the equations are solved simultaneously. The fluxes are treated using the two-point or multi-point flux approximation. The transmissibility approximation affects the numerical stability and accuracy. Implicit upstream ensures the stability and increases the risk of truncation errors at the same time (Blair and Weinaug 1969; Allen 1984). The relative permeabilities, viscosites, and densities at the interface between two cells follow the flow direction $\mathbf{u}_\alpha \mathbf{n}_{ij} > 0$, where $\mathbf{n}_{ij}$ is the outer normal to the interface. The phase densitites in the gravity term $\rho_\alpha \mathbf{g}$ can be weighted with respect to the effective phase volume (Coats 1980). When modeling a heterogeneous medium, the discontinuity can be handled with harmonic averaging applied to $Kk_{r\alpha}/\mu_\alpha$ or just using the intrinsic permeability. The system of nonlinear equations obtained from the time and space discretization should then be linearized with respect to the selected unknowns for Newton's method. The obtained linear system can be solved using a generalized minimal residual method (GMRES) algorithm (Saad and Schultz 1986) with a pre-conditioner such as ILU0 incomplete factorization (Saad 2003).

When the coupling between equations is weak, they can be solved separately. Similar to a decoupled pressure equation of immiscible flow (Eqn. 19), the equations in (35) are summed to derive a pressure equation. The total and phase velocities are defined using the fractional flow (Eqns. 20 and 21). This evolves into one pressure equation and a set of saturation/composition equations that are weakly coupled, and can therefore be solved semi-implicitly using implicit pressure explicit composition/saturation (IMPEC/IMPES). The pressure equation is first solved, followed by the composition equations with fixed fluxes. Under the assumption of incompressible flow, the pressure equation is elliptic (Eqn. 19). If the capillary pressure is neglected, the saturation/composition equations are purely hyperbolic and nonlinear in comparison with the linear transport equations. Given that this decreases the dependencies of the equations, they can be separated and efficiently solved using different schemes and preconditioners that are appropriate for each type of equation (Chen et al. 2006). However, if the equations are highly coupled, the solution of sequentially decoupled equations leads to the retardation of property calculations.

Despite the computational complexity of the implicit methods, they allow for a large time step, which is advantageous. The scheme is unconditionally stable; however, truncation errors may be introduced at large time steps. To reduce the numerical complexity, the flow system can be reformulated for the IMPEC. The explicitness indicates a restriction on the time step controlled by a Courant–Friedrichs–Lewy (CFL) criterion. Formulations of the

CFL condition were detailed by Settari and Aziz (1975), Peaceman (1977), and Coats (2003). The adaptive implicit method combines both methods using different levels of implicitness for each cell (Thomas and Thurnau 1983; Cao 2002).

In general, codes employ the finite volume method (FVM), as it is conservative and easy to implement. The main drawback of FVM schemes is that they are not adapted for modeling complex geometries and anisotropic porous medium. The FVMs were trevised to produce a new category of cell-centered finite volume methods (Hermeline 2007; Aavatsmark et al. 2008; Eymard et al. 2009). The methods are based on multi-point flux approximations, they ensure continuous pressure across the control volume interfaces, and are consistent on triangular meshes (Friis and Edwards 2011; Pal and Edwards 2011; Eymard et al. 2012). Finite element methods can result in violation of local conservation properties, which is important when modeling flow and reactive transport. Mixed finite element and discontinuous Galerkin methods can ensure local mass conservation (Raviart and Thomas 1977) and (Wheeler and Yotov 2006); the latter was extended to hexahedral meshes by Ingram et al. (2010).

Primary variables. Natural variable formulation (NVF) is widely used due to its direct implementation: S_α, p_α, and $x_{\alpha,k}$; although it requires variable switching when a phase appears or disappears (Coats 1980). When the primary variables are changed, the Jacobian should be reassembled to avoid a set that is numerically inefficient and may lead to inconsistencies. The NVF is not unique. Various formulations were developed by combining/ replacing variables/equations to solve the problem with respect to the mass or volume variables. Several alternatives to the NVF are reviewed below.

The pressure p_α and component masses m_α formulation (Acs et al. 1985) was designed for the compositional flow using N_c+1 primary variables: a pressure and the component masses in moles, and $m_i = \phi V \sum_\alpha \rho_{mol,\alpha} S_\alpha x_{\alpha,i}$. Replacing the mole fractions by the component masses implies an additional equation that constraints the volume. This results in N_c+1 equations.

The pressure p_α and overall composition z_i (Eqn. 39) formulation exploits the expression of the thermodynamic state and composition flow with respect to p_α and z_i. This can be achieved using the flash calculations and by re-writing the flow with respect to z_i. Other molar formulations with respect to pressure, phase fractions V and L, z_i, and the composition of N_f-1 phases or $\ln K_i$ are also possible.

The overall composition formulation with the flash calculations was compared to the NVF. The performance was tested on a problem related to CO_2 injection (Lu et al. 2010). A higher convergence rate of the overall composition formulation than the NVF with switching variables was observed. Voskov and Tchelepi (2012) compared the NVF, molar, and overall compositions using several problems that prioritize the NVF.

A tie-line based method was employed to parameterize the compositional space of the thermodynamic equilibrium problem (Entov et al. 2002). The technique was extended to an EOS-based two-phase multicomponent flow system, where the tie-lines are adaptively constructed, thus replacing or/and accelerating the phase stability calculations (Voskov and Tchelepi 2009).

A formulation with respect to the pressure and total concentrations using a concept of extended saturation was presented by Abadpour and Panfilov (2009). Voskov (2012) employed this approach for modeling of EOS-based compositional flow coupled with the negative flash method developed by Whitson and Michelsen (1989). The phase fraction and saturation can be negative or larger than one.

Extended NVFs were devised using Henry's law/solubility and the inverse capillary pressure function. First proposed by Ippisch (2003), the use of the solubility as a function of pressure was extended by Bourgeat et al. (2009): $\rho_{l,i}=\rho_l X_{l,i}=\rho_{l,i}(p_l+p_c(S_g))$, after which the saturation $S_g=S_g(\rho_{l,i},p_l)$. Jaffré and Sboui (2010) introduced three variables with complementary conditions for two-phase two-component flow: p_l, S_l, $X_{b,i}$. Angelini (2010) used extended pressures p_l, p_g (PPF). Neumann et al. (2013) considered p_c, p_g as primary variables using the inverse capillary pressure and nonlinear solubility functions.

The pressure p_α, saturation S_α, and fugacities $f_{\alpha,i}$ (PSF) formulation given by Lauser et al. (2011) proposes a fixed set of primary variables. The phase transition is controlled by complementarity conditions, and the flash calculations are excluded. The problem is solved using a semi-smooth Newton's method. Fugacity coefficients $\varphi_{\alpha,i}$ are calculated using EOS models.

Masson et al. (2014) presented a comparison between the NVF, PSF, and PPF for the modeling of different physical problems, namely, a drying process by suction, gas injection, and gas migration in a heterogeneous medium. The convergence test results revealed that the NVF and PSF demonstrated better performances than that of the PPF. Ben Gharbia et al. (2015) studied the NVF coupled with the negative flash and the PSF coupled with the cubic EOS. There were numerical difficulties with respect to the PSF during the phase state transition, and it was slightly less efficient than the NVF, which is still a reliable option. The practical usage and intended applications can serve a guide for the selection of a convenient set of primary variables.

Coupling MMF&RT

Coupling methods of MMF&RT should fully utilize the accumulated knowledge of reservoir engineering and reactive transport modeling. Their mathematical and numerical approaches were designed for the solution of different problems at different scales. Reservoir simulations are typically large-scale oriented and based on the global implicit approach (GIA) using flash calculations and EOS. Most reactive transport codes decouple the transport and the chemistry by the operator-splitting approach (OSA), thus rendering it flexible for extension. Coupling MMF&RT inherits all the complexity of both problems and the experience of both communities

Global implicit approach. The global implicit approach consists of the simultaneous solution of the nonlinear and linear partial differential equations (PDEs) and ordinary differential equations (ODEs) of transport/flow and the chemical system. Another method for solving the coupled system, which is referred to as the direct substitution approach (DSA), is derived from the RT modeling. The chemistry variables and the basis species concentrations $\mathbf{C}$ replace the total concentrations $\mathbf{T}$ in the transport equations. The total concentration of basis species is the mathematical sum of the basis species concentration over the entire system that can be constructed using the tableaux method described by Morel and Hering (1993). The chemistry is partly integrated into the transport equations, which are solved together with the heterogeneous reactions with respect to the basis species. Lichtner (1996) presented a general DSA for the solution of the RT problem, which can be applied to multiphase multicomponent systems. The concept can be expressed as follows:

$$\frac{\partial \mathbf{T}(\mathbf{C})}{\partial t} + \mathbf{F}\big(\mathbf{T}(\mathbf{C})\big) = \mathbf{S}_{eq}\mathbf{R}_{eq} + \mathbf{S}_{kin}\mathbf{R}_{kin}(\mathbf{C}), \tag{42}$$

where $\mathbf{F}$ is the linear transport or MMF operator; $\mathbf{S}_{eq}$ and $\mathbf{S}_{kin}$ typically denote the stoichiometric matrices; and $\mathbf{R}_{eq}$ and $\mathbf{R}_{kin}(\mathbf{C})$ are the reaction rates for the equilibrium and kinetic reactions. The formula was modified after Kräutle and Knabner (2007). The MMF&RT problem can then be solved with respect to the basis species $\mathbf{C}$ using Newton's method.

Global methods attempt to eliminate the equilibrium/kinetic rates. Sevougian et al. (1993), Lichtner (1996), Molins et al. (2004), Steefel and MacQuarrie (1996), and Saaltink et al. (1998) demonstrated that this can be achieved by the construction of an orthogonal matrix or linear combination of equations. For example, $\mathbf{ES}_{eq} = 0$, where $\mathbf{E}$ is the equilibrium rate annihilation matrix. These matrix transformations simplify and can reduce the number of differential equations; however, the equations are still coupled if there are heterogeneous reactions between the aqueous solution and solids. For example, a mineral can precipitate, dissolve, and re-precipitate depending on the aqueous composition and saturation. Kräutle and

Knabner (2007) found a method to decouple a subsystem of mobile species equations from the global problem. Two sets of variables were introduced: a linear combination of only mobile species, and a linear combination of only immobile species. Hence, there was no combination of mobile and immobile species. This substitution was computationally advantageous, and the PDE system was well structured. The method was then revised to model the disappearance or appearance of minerals, and it was recently extended to two-phase modeling (Kräutle 2011; Brunner and Knabner 2019). Issues due to the disappearance of minerals and gases were investigated for the geochemical solvers and reactive transport codes, and mathematical and numerical solutions were proposed (Kräutle 2011; Leal et al. 2013)

Reduction methods of the DSA have been used to couple the MMF and geochemistry. Nghiem et al. (2004) applied a classic DSA to the multiphase multicomponent problem under the consideration of modeling both equilibrium and kinetic reactions, dissolution and precipitation of minerals, and porosity changes due to changes in the geochemistry and permeability using the Kozeny–Carman model. The importance of efficient linear solvers was highlighted, as they are required for the solution of the resulting sparse matrix. The mass balance equations with respect to the basis species are completed with a volume constraint equation, phase equilibrium, and kinetic rate relationships. The DSA-based integration of RT in the code relies on the existing numerical methods, such as the adaptive implicit method (AIM) developed for a compositional simulator generalized equation-of-state model for greenhouse gas (GEM-GHG). Another example of a matrix reduction method was presented by Fan et al. (2012). The balance equations are transformed by the elimination of equilibrium rates. In addition, the resulting system is solved with respect to the elements rather than the primary species, as the composition of all the species can be expressed with respect to the elements/ atoms and corresponding number of atoms (Sevougian et al. 1993). The element-based method was developed in the automatic-differentiation-based general purpose research simulator (ADGPRS) framework using a two-stage pre-conditioner (Cao et al. 2005) and the AIM.

Despite the linear property of the transport operator, several reactive transport codes opted for the global implicit approach (GIA). MIN3P (Mayer et al. 2002); and PFLOTRAN (Lichtner et al. 2015) use this approach as well as CrunchFlow (Steefel 2009). Moreover, CrunchFlow and PFLOTRAN have an OSA option. Only PFLOTRAN has an MMF&RT coupling; and the GIA and OSA exist. Nevertheless, the reservoir modeling of large nonlinear systems using advanced solution techniques allows for the extension of fully implicit MMFs to chemical reactions. Although the computational complexity of the GIA is transparent, the increase in computational facilities allow for exploitation of the accuracy and robustness of the GIA. In addition to the abovementioned couplings, GEM-GHG (Nghiem et al. 2004) and ADGPRS (Fan et al. 2012) are coupled with GFLASH, which provides the equilibrium calculations based on the Gibbs free energy minimization (Voskov 2017). A fully implicit EOS-based compositional flow simulator, which is referred to as GPAS, was also extended for chemical flooding (Pope et al. 2005). However, the chemical reactions are modeled in the aqueous phase only. The coupling was achieved using a FIM and hybrid method. The hybrid option can be used to implicitly solve the subset of balance equations for hydrocarbons, and then explicitly solve the remainder. A concept of dominant and minor components is generally used to decouple flow and transport equations, as discussed below.

Operator splitting approach. Similar to the classic OSA used for RT coupling, the resulting system of MMF&RT can be decoupled by alternating the MMF and RT operators. The flow system is first solved, thus yielding the intermediate pressures, saturations, composition, velocities, fluid properties, and other secondary variables. The compositional formulation can be used, thus making it applicable to existing reservoir simulators and geochemical codes. For the extension of geochemical codes, it is sometimes assumed that there is a set of dominant components responsible for MMF, while other components are minor, and their impact on the flow can be neglected. This reduces the number of nonlinear equations, and only linear transport is solved for minor components. For example, two components, namely, CO_2 and H_2O, can be selected as dominant for the two-component two-phase flow. However, this assumption can limit the modeling of the complex dynamics of multicomponent systems (see Applications section for further discussion).

The reactive transport part $\mathfrak{R}$ may be sequentially linked and can be solved using a GIA or OSA (Yeh and Tripathi 1989). This allows for existing RT codes to be used as input/output modules. A general OSA has the following form:

$$\frac{\partial}{\partial t}\left(\mathbf{T}\left(\mathbf{C}^{n+1,2k}\right)-\mathbf{T}\left(\mathbf{C}^{n}\right)\right)+\mathbf{F}\left(\mathbf{T}\left(\mathbf{C}^{n+1,2k}\right)\right)=0, \tag{43a}$$

$$\mathfrak{R}\left(\mathbf{C}^{n+1,2k}\right)=\mathbf{C}^{n+1,2k+1}, \tag{43b}$$

where n is the time step number, and k corresponds to an iteration number. In the non-iterative coupling (SNIA), $k=0$, and the system proceeds to the next step after the RT. Alternatively, the objective of the sequential iterative approach (SIA) is to iteratively solve the flow for an updated chemical state, and then for the RT until convergence. Using the SNIA leads to persistent errors for reactive problems (Valocchi and Malmstead 1992). Another drawback is the time-step restriction, which is necessary to ensure accuracy of information transfer between the operators. The reliability of the OSA, and that of SIA in particular, was demonstrated for reactive transport and MMF&RT problems (Carrayrou et al. 2010; Saaltink et al. 2013). The chemistry has a significant influence on the fluid displacement, and vice versa. The OSA-based coupling depends on the applied internal methods, their connectivity, and on the modeled physical process.

Flexibility is a critical feature of the OSA, where new functionalities are easy to implement, and the operators/modules can be verified independently. The method is extensively applied in RT codes (Steefel et al. 2014). Several geochemical codes employed the method for modeling unsaturated flow, such as the bubble model developed in MIN3P (Mayer et al. 2012; Molins and Mayer 2007).

There are several examples of OSA-based couplings between MMF and RT:

- Code-Bright couples flow of water, air, and salt using the reactive transport code RETRASO (Olivella et al. 1996).

- DuMu^{X} employs an SNIA with a dominant and minor component concept (Ahusborde and El Ossmani 2017; Hron et al. 2015).

- HYDROGEOCHEM uses the global pressure formulation of multiphase flow, which is iteratively coupled with GIA/SIA-based RT (Yeh et al. 2004, 2012).

- HYTEC (Lagneau and van der Lee 2010; van der Lee et al. 2003) is an RT simulator incorporating the geochemical code CHEES. The RT is extended to a compressible flow module by employing a fully implicit EOS-based coupling method that allows treatment of real thermodynamic behavior of phase mixtures (Sin et al. 2017a).

- IPARS has several S(N)IA-based options with assumptions that decrease the coupling between flow and chemistry, e.g., slightly compressible (Peszynska and Sun 2002; Wheeler et al. 2012).

- NUFT sequentially couples compositional flow and GIA-based RT (Hao et al. 2012).

- Phreeqc was employed as a reactor for reservoir simulators. Parkhurst and Wissmeier (2015) presented an SNIA that alternates the transport and reaction steps, and the same approach is used in Phreeqc. In addition iCP is an SNIA-based coupling with COMSOL (Nardi et al. 2014); and MoReS coupled with Phreeqc uses an SNIA-based method, which is limited to slow chemical reaction rates (Farajzadeh et al. 2012; Wei 2012). PhreeqcRM is designed as a generic reaction module for OSA-based coupling.

- PFLOTRAN has an OSA coupling option for fully implicit flow and chemistry (Lichtner et al. 2015; Lu and Lichtner 2007).

- STOMP applies a tight SIA-based coupling of flow and chemistry alternating across all modules until convergence (White et al. 2012; White and Oostrom 2006).

- TOUGHREACT sequentially couples the MMF and linear transport with chemical reactions (Xu and Pruess 1998; Xu et al. 2012). TOUGHREACT is used within the current TOUGH3 simulator (Jung et al. 2017).

These examples illustrate that most MMF&RT couplings are not iterative between flow and RT. This may limit chemical processes, and therefore their influence on thermodynamics and hydrodynamics, similar to SNIA-based coupling for RT. The composition-dependent properties should be modeled implicitly to prevent a discrepancy between the MMF and RT operators. Consequently, the FIM is a favorable option. Implicitness provides codes with the unconditional stability property; thus, large time steps are allowed. This is of practical usage for modeling the rapid mechanical displacement of fluids. However, the reactive time step can be a restriction. When the range of physical processes occur over different characteristic scales, time-step splitting between operators can be a solution. The connecting properties should accurately be updated, and mass conservation should be strictly controlled. Another numerical alternative is the AIM that reduces the nonlinear flow system. Furthermore, the space domain decomposition is efficient when the computational intensity over the model domain is highly heterogeneous.

A tight SIA-based coupling, which couples flow and RT operators iteratively or with implicit methods, is not frequent in MMF&RT codes because of the large nonlinearity of the flow problem. However, it could be a good alternative, since simplified flow formulations typically have restricting assumptions; and they are not suitable for general-purpose simulations. The tight SIA-based approach with high implicitness is critical for the modeling of reaction-driven hydro- and thermodynamic processes.

Discussions

The modeling of complex thermodynamics is a necessary condition for MMF&RT. Multicomponent systems at supercritical conditions cannot be described using the ideal laws. Modeling mixtures requires implicitly updated fluid properties and relevant approximations of real fluid behavior. For example, an accurate modeling of compressible multicomponent real fluids can determine the evolution of fluid phases, and their displacement and influence on the matrix. The asymmetric approach of VLE allows for the modeling of various systems and gains from both constantly developing EOS and activity models. Based on the analysis of the MMF&RT methods and experience over the last four decades, the GIA or a tight SIA-based method with NPV or equivalent formulation can be considered a reliable option. The tight SIA can be preferable for reactive transport simulators. Due to the complexity of the MMF&RT structure, it is necessary to benchmark coupling methods. Although a few MMF benchmark problems exist (Pruess et al. 2004; Class et al. 2009; Nordbotten et al. 2012), they are CO_2-oriented and have only one gas species in the gas phase. Multicomponent benchmarks (Sin et al. 2017b) are therefore required to better understand the structures and behavior of different coupling methods.

Coupled physical and chemical processes at the continuum scale also occur at the pore scale, and they can be modeled by approaches such as the pore network model, Lattice Boltzmann model, direct numerical simulation, and particle methods. The advancement of computational capabilities allows for the application of these methods in complex physics and geometries. The Shan–Chen multicomponent lattice Boltzmann model was efficiently applied to fluid–fluid and fluid–solid interfacial reactions in miscible two-phase flow (Chen et al. 2018; Li et al. 2018) and reactive transport with evolving porosity (Gao et al. 2017). For example, Li et al. (2018) illustrated the influence of fluid–solid interface relations (e.g., adhesive forces, specific surface area) on the effective permeability and capillary pressure. Recent studies also show possibilities to model physics across scales integrating over a general framework. Battiato et al. (2011) proposed a hybrid method, wherein an iterative procedure couples two models and defines concentrations and fluxes through the boundaries of pore and continuum scale domains. A further investigation of hybrid methods is necessary to better understand the physiochemical processes that occur across scales and contribute one to each other (Molins and Knabner 2019 this volume).

APPLICATIONS

The increasing interest in the CCS field has initiated an intensive investigation of subsurface processes including numerous experiments and field studies, in addition to mathematical and numerical methods. Among various subjects, captured CO_2 gas contains small amounts of impurities such as N_2, O_2, Ar, CO, SO_x, NO_x, CH_4, H_2S, and H_2 (Kather 2009; IEAGHG 2011; Ussiri and Lal 2017). The type and concentration of co-injected impurities are dependent on the source of CO_2 and the selected capture method. The presence of impurities modifies the solubility of CO_2, and therefore influences the storage capacity (Li and Yan 2009a,b; Zirrahi et al. 2012; Ziabakhsh-Ganji and Kooi 2012, 2014). Moreover, The presence of impurities also affect the geochemistry and fluid dynamics of the system (Xu et al. 2007; Xiao et al. 2009; Corvisier et al. 2013, 2014, 2017; Jung et al. 2013; Sterpenich et al. 2013; Bacon et al. 2014; Lu et al. 2014; Ziabakhsh-Ganji and Kooi 2014; Pearce et al. 2015, 2016a,b; Wei et al. 2015; Lei et al. 2016; Lu et al. 2016; Waldmann and Rütters 2016; Todaka and Xu 2017).

Chromatographic effects have been observed in the field and laboratory following the injection of CO_2 gas with impurities. Wei et al. (2015) reported the injection of a CO_2-rich gas mixture with N_2 and O_2 during a Tongliao pilot experiment in China. Earlier arrivals of co-injected oxygen and nitrogen gases were observed when compared with the arrival of CO_2. Inversely, a gas mixture of 98% CO_2 and 2% H_2S was injected into a depleted gas reservoir in Canada. The earlier arrival of CO_2 and delayed breakthrough of H_2S were recorded at producing wells. The chromatographic effect is significant at large scales (Wei et al. 2015; Paterson et al. 2013). A series of dynamic solubility experiments was carried out by Bachu and Bennion (2009), which demonstrated the partition of a gas mixture because of the differential solubility of gases. Numerical models were able to reproduce these observations. Gas partition also occurs when other gases are already present in the porous medium.

Convective dissolution, which is an important trapping mechanism, is also significantly dependent on the presence of impurities because of their influence on the density of the plume. The interplay between gas partition and density-driven flow was demonstrated in three cases of CO_2 injection. and the modeling of MMF&RT allowed for the investigation of the critical effects of co-injected impurities on the evolution of the systems (Xu et al. 2007; Xiao et al. 2009; Lei et al. 2016; Todaka and Xu 2017).

Phase equilibrium and solubility

Batch geochemical simulations (0D simulations) offer a great opportunity to verify the ability of the abovementioned approaches to accurately simulate gas–water interactions in closed and homogeneous systems prior to the modeling of reactive transport processes at a larger scale. CO_2–H_2O and CO_2–H_2O–NaCl systems were studied intensively, and the consistent data obtained from the literature allow for the testing of the geochemical solvers and the associated databases. CO_2 solubility and co-solubility data at 50 °C, 100 °C, and 150 °C (Figs. 5–6) were modeled using three geochemical solvers and one thermodynamical model:

- Chess (van der Lee 2009; Corvisier et al. 2013, 2014, 2017) using the Peng–Robinson EOS, and the SIT model for the electrolyte.

- Phreeqc (Parkhurst and Appelo 2013; Appelo et al. 2014) using the Peng–Robinson EOS, and the Pitzer model for the electrolyte.

- Gem–Selektor (Kulik et al. 2004) using the SRK EOS and the extended Debye–Hückel model for the electrolyte.

- Duan thermodynamic model (Duan and Sun 2003) using a particular EOS and a fitted Pitzer model for the electrolyte.

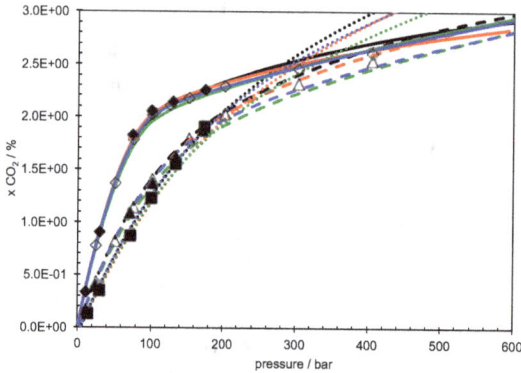

Figure 5. CO_2 molar fractions in water with respect to pressure. The symbols correspond to the data from Hou et al. (2013) at 50 °C (black diamond), 100 °C (black triangle), and 150 °C (black square); and that from Wiebe and Gaddy (1939) at 50 °C (white diamond) and 100 °C (white triangle). The green lines correspond to simulations using Gem–Selektor, the red lines correspond to simulations using phreeqc, the blue lines correspond to simulations using the model by Duan and Sun (2003), and the black lines correspond to simulations using Chess (the continuous, dashed, and dotted lines correspond to 50 °C, 100 °C, and 150 °C, respectively).

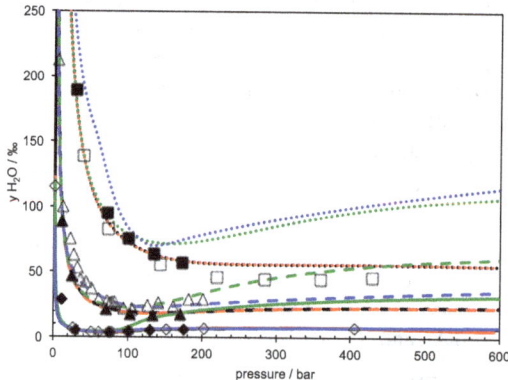

Figure 6. H_2O molar fractions in carbon dioxide with respect to pressure. The symbols correspond to the data from Hou et al. (2013) at 50 °C (black diamond), 100 °C (black triangle), and 150 °C (black square); that from Wiebe and Gaddy (1941) at 50 °C (white diamond); that from Caumon et al. (2016) at 100 °C (white triangle); and that from Tabasinejad et al. (2011) at 150 °C (white squares). The green lines correspond to simulations using Gem–Selektor, the red lines correspond to simulations using Phreeqc, the blue lines represent simulations using the model by Duan and Sun (2003), and the black lines correspond to simulations using Chess (the continuous, dashed, and dotted lines correspond to 50 °C, 100 °C, and 150 °C, respectively).

The four models were in relatively good agreement with the CO_2 solubility data for the entire range of the temperature and pressure.

For the co-solubility data, Chess and Phreeqc gave results in good agreement with the data, while results from Duan and Gem–Selektor showed significant discrepancies with the selected data (Figs 5 and 6). The absolute average deviation (AAD) between the experimental solubility data and the model decreased from 4% using Chess, Gem–Selektor, and Duan to 2% using Phreeqc. AAD for co-solubility data decreased from 43% using Gem-Selektor, 22% using Duan, 14% using Phreeqc to 13% using Chess.

Geochemical solvers can be used to solve gas–water equilibria and the entire speciation. Consequently, Chess, Phreeqc, and Gem–Selektor can be used calculate the pH of CO_2–H_2O systems. The extension of the approach by Duan and Sun (2003), which was developed by Li and Duan (2007) can be used to calculate the pH of such systems. These four models were used to simulate pH measurements at 50 °C, 100 °C, and 150 °C (Fig. 7). The AAD between the experimental pH values and the model predictions decreased from 2% using the Chess, Gem–Selektor, and Phreeqc models to 1% using the model by Li and Duan (2007). Various gas–water systems and gas mixtures–water were successfully simulated using Chess (Figs. 8 and 9). Hajiw et al. (2018) presented a comparison between the geochemical approach and process homogeneous approaches using the same PR EOS, and then illustrated the good performances of the geochemical solvers applied to these systems. The AAD for these systems are 3% for Ar, 10 % for CO, 6% for CH_4, 8% for H_2, 8% for H_2S, 4% for N_2, 3% for O_2, 11% for SO_2, 7% for CO_2–CH_4, and 3% for CO_2–N_2. Other simulations were run using the abovementioned models to reproduce the CO_2 solubility data measured in the presence in NaCl brines at 50 °C, 100 °C, and 150 °C at 150 bar (Fig. 10). The AAD was 12% using Gem–Selektor, 8% using Phreeqc, 4% using Duan, and 3% using Chess. The performances of the four models were good, although Gem–Selektor exhibited a slight overestimation of the CO_2 solubility at high salinities, which can be attributed to the employed aqueous activity model.

Several pH measurements of CO_2–H_2O–NaCl systems at 50 °C and 1 M, 3 M, and 5 M of NaCl were also simulated (Fig. 11) The AAD between the experimental pH and the model decreased from 27% using Gem–Selektor model to 3% using the Chess, Phreeqc, and Li and Duan model. The overestimated CO_2 solubility of Gem–Selektor probably leads to this rather important AAD for pH.

Overall, the benchmark presented here showed that that a variety of systems (gas mixtures, water, mixed brines, etc.) can be accurately simulated using geochemical solvers implemented in reactive-transport codes.

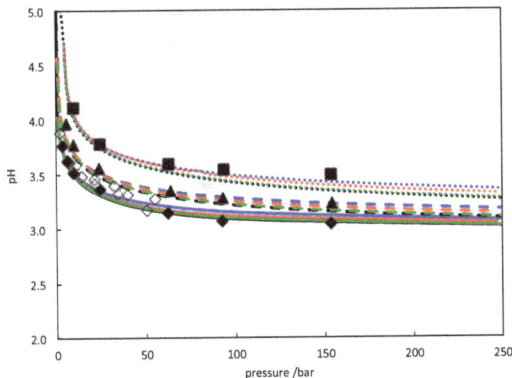

Figure 7. pH for CO_2–H_2O system with respect to pressure. The symbols correspond to the data from Peng et al. (2013) at 50 °C (black diamond), 100 °C (black triangle), and 150 °C (black square); and that from Haghi et al. (2017) at 50 °C (white diamond). The green lines correspond to simulations using Gem–Selektor, the red lines correspond to simulations using Phreeqc, the blue lines represent simulations using the model by Duan and Sun (2003), and the black lines correspond to simulations using Chess (the continuous, dashed, and dotted lines correspond to 50 °C, 100 °C, and 150 °C, respectively).

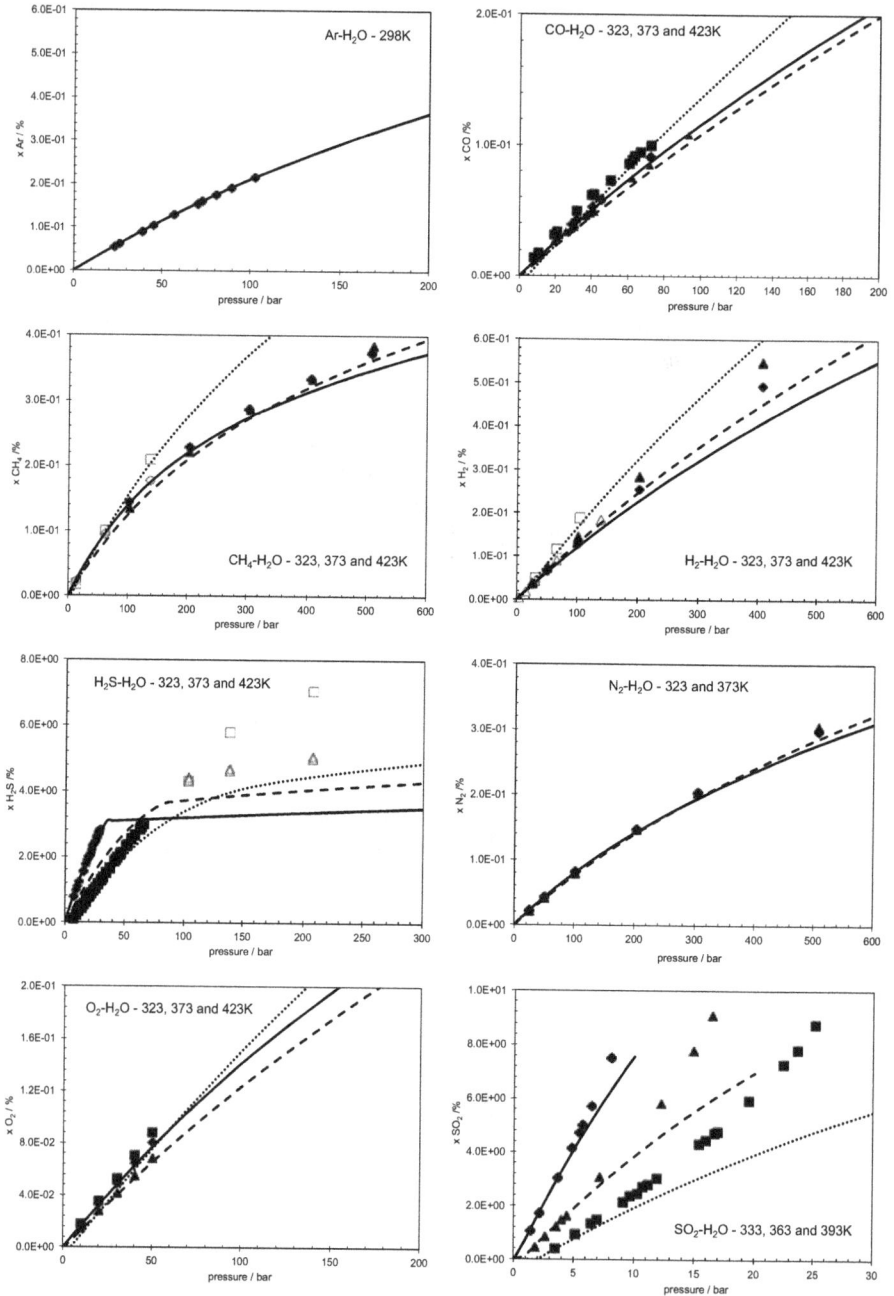

Figure 8. Molar fractions of dissolved gas in water for various gas–water systems (Ar, CO, CH$_4$, H$_2$, H$_2$S, N$_2$, O$_2$, and SO$_2$) with respect to pressure for different temperatures. The symbols correspond to data from various sources, and the lines correspond to simulations using Chess.

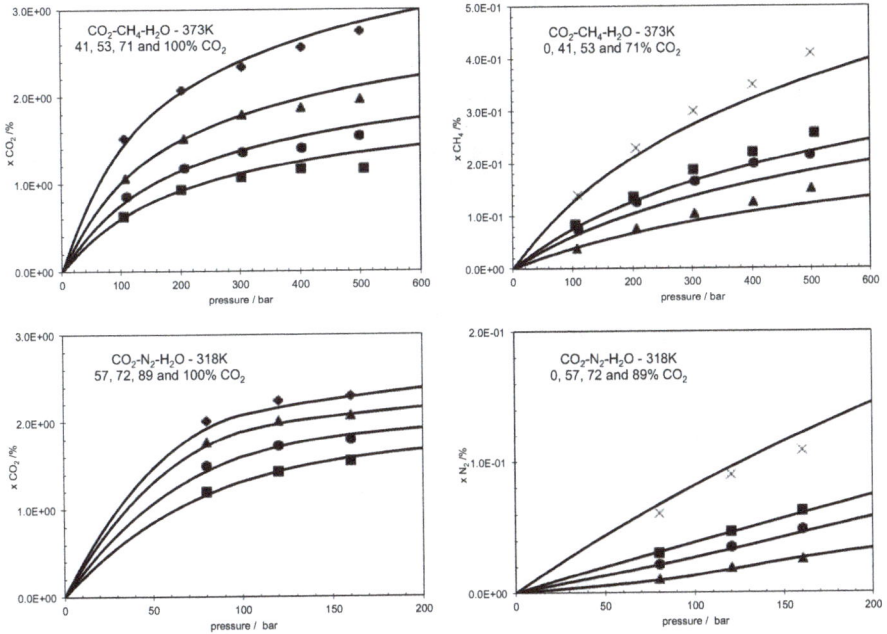

Figure 9. Molar fractions of dissolved gas in water for gas mixtures–water systems (CO_2–CH_4, and CO_2–N_2) with respect to pressure for different compositions. The symbols correspond to data from Qin et al. (2008) and Liu et al. (2012), and the lines correspond to simulations using Chess.

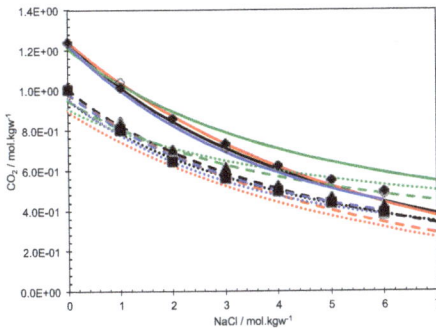

Figure 10. CO_2 molalities in NaCl brine with respect to the molality. The symbols correspond to data at 150 bar from Zhao et al. (2015) at 50 °C (black diamond), 100 °C (black triangle), and 150 °C (black square); and that from Messabeb et al. (2016) at 50 °C (white diamond), 100 °C (white triangle), and 150 °C (white square). The green lines correspond to simulations using Gem–Selektor, the red lines correspond to simulations using Phreeqc, the blue lines correspond to simulations using the Duan model, and the black lines correspond to simulations using Chess (the continuous, dashed, and dotted lines correspond to 50 °C, 100 °C, and 150 °C, respectively).

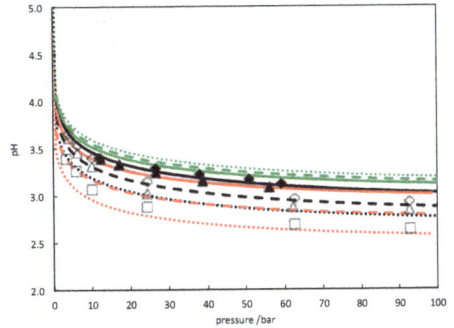

Figure 11. pH for CO_2–H_2O–NaCl brine with respect to the molality. The symbols correspond to data at 150 bar from Zhao et al. (2015) at 50 °C (black diamond), 100 °C (black triangle), and 150 °C; (black square) and that from Messabeb et al. (2016) at 50 °C (white diamond), 100 °C (white triangle), and 150 °C (white square). The green lines correspond to simulations using Gem–Selektor, the red lines correspond to simulations using Phreeqc, and the black lines correspond to simulations using Chess (the continuous, dashed, and dotted lines correspond to 50 °C, 100 °C, and 150 °C, respectively).

Gas chromatography

Dynamic solubility experiments were carried out by Bachu and Bennion (2009) to better understand the influence of a gas type (H_2S, N_2, SO_2, CH_4) and its concentration (2, 5, 30% of H_2S) on gas separation at different temperature, pressure, and salinity conditions (61 °C and 13.5 MPa; 25 °C and 6 MPa; fresh water and 119 ppm of salinity). Moreover, the breakthrough of gases was recorded and analyzed. The experimental results revealed that the breakthrough time and profiles of the gas components are dependent on the solubility ratio between gases. Less soluble gases arrive earlier with higher breakthrough concentrations.

In these experiments, mixtures of H_2S and CO_2 were injected in a tube homogeneously packed with silicate-rounded sand. Two feed gas cases were used: 2% H_2S and 98% CO_2, and 30% H_2S and 70% CO_2. The gas was injected at a constant rate of 7.5 cm³/h into the tube at a constant pressure of 13.5 MPa, temperature of 61 °C, and brine salinity of 118950 ppm (Bachu and Bennion 2009). The experiments were modeled using HYTEC (Sin and Corvisier 2018) and CMG-GEM (Bachu et al. 2009).

The modeled gas saturation profiles and aqueous composition of both codes for the co-injected 2 % H_2S and 98% CO_2 differed only slightly. The fluid displacement modeled by the HYTEC lags the curves of the CMG-GEM (Figs. 12a,b). This can be attributed to the different solubility models used in the codes. Model results illustrate the gradual evolution of the differential solubility of H_2S and CO_2 into an increasing difference between the H_2S and CO_2 profiles, as shown in Figure 12b (the mole fraction of dissolved H_2S is scaled). Since H_2S is more soluble than CO_2, the breakthrough of CO_2 was prior to that of H_2S (Fig. 13). The experimental and numerical results were in good agreement. They illustrated a later arrival of H_2S. After the breakthrough, the H_2S concentration increased and reached its concentration in the feed gas.

The same analysis of gas chromatography was applied for 95% CO_2 mixtures: H_2S, N_2, SO_2, and CH_4 (Bachu and Bennion 2009). Four numerical models were run using HYTEC. The numerical and experimental results of breakthrough of gases were compared, as shown in Figure 14a: the mole fraction of the impurity gas phases H_2S, N_2, SO_2, and CH_4 were normalized by the relevant initial mole fraction of the injected gas (5%).

As can be seen from the figure, the less soluble N_2 and CH_4 arrived first and exhibited higher peaks than that CO_2. The opposite was observed for H_2S and SO_2. The numerical results indicated the phenomenological behavior of fluids. The modeled height of the peaks was not precise. During the experiment with 5% SO_2, it was difficult to detect SO_2. Moreover, it was found in a very small amount in the gas phase, as it was almost completely dissolved in the brine.

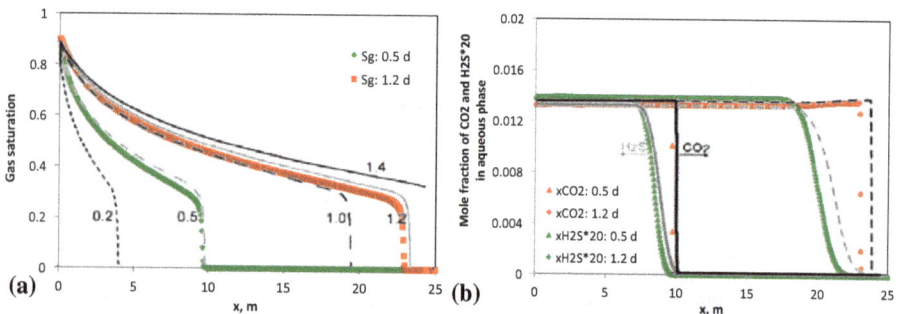

Figure 12. Injection of a feed gas of 2% H_2S and 98% CO_2 in a long sand-packed coil, as presented by Bachu and Bennion (2009). (a) Gas saturation evolution. Numerical results of CMG-GEM (gray and black lines) presented by Bachu et al. (2009) and HYTEC (colored lines). (b) Aqueous composition given by mole fractions of CO_2 and H_2S at 0.5 days and 1.2 days. Numerical results of CMG-GEM (gray and black lines) presented by Bachu et al. (2009) and HYTEC (CO_2 is represented by red triangles and diamonds, and H_2S is represented by green triangles and diamonds).

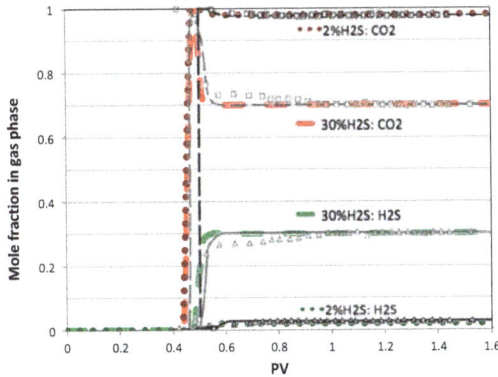

Figure 13. Injections of a feed gas of 2% H_{2S} and 98% CO_2, in addition to 30% H_2S and 70% CO_2, in a long sand-packed coil, as presented by Bachu and Bennion (2009): gas composition given by mole fractions of CO_2 and H_2S at the outlet of the used apparatus for the gas chromatography. The symbols correspond to experimental data from Bachu et al. (2009), and the lines correspond to the numerical results of CMG-GEM (gray and black lines) presented by Bachu et al. (2009) and HYTEC (CO_2 red and H_2S green dotted and dashed lines).

Although the numerical results were slightly different from the experimental data (Figs. 14a,b), and there were several uncertainties in the data, the numerical models present a good representation of the gas chromatography, e.g., the gas partition process of mixtures due to the differential solubility. It should be noted that the gases are displaced with the common velocity, which is the Darcy's velocity of the gas mixture. During the passage, several molecules of a substance are dissolved in the liquid phase, and other molecules are carried

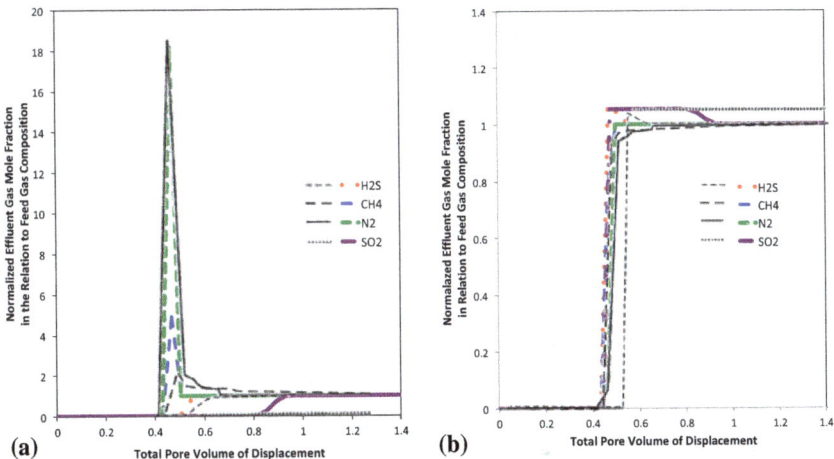

Figure 14. Injections of a feed gas of 5% impurity and 95% CO_2 in a long sand-packed coil for cases of H_2S, N_2, SO_2 and CH_4 carried out by Bachu and Bennion (2009). **(a)** Mole fraction of the impurity gases H_2S, N_2, SO_2, CH_4 at the outlet, normalized by the relevant mole fraction in the initially injected gas (5% for all cases), as a function of the pore volume of the total injection. Dashed and dotted black lines correspond to experimental data from Bachu and Bennion (2009), and colored lines correspond to numerical results of HYTEC. **(b)** Mole fraction of CO_2 at the outlet, normalized by the relevant mole fraction in the initially injected gas (95% for all cases), as a function of the pore volume of the total injection. Dashed and dotted black lines correspond to experimental data from Bachu and Bennion (2009), and colored lines correspond to numerical results of HYTEC.

by the general motion of the gas mixture. This is similar to the chromatography of solutes when the retardation is due to ion exchange. The accurate modelling of solubility is therefore necessary for the gas partition. The co-injected impurities can be classified in decreasing order of their solubilities (13.5 MPa, 61 °C), as follows:

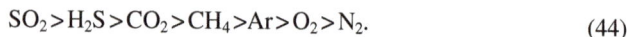

$$SO_2 > H_2S > CO_2 > CH_4 > Ar > O_2 > N_2. \tag{44}$$

The above corresponds to the breakthrough order of the gases (Fig. 15), a similar test with 5% Ar was performed.

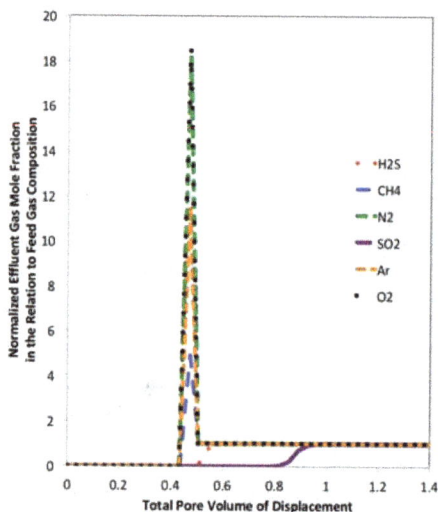

Figure 15. Injections of a feed gas of 5% impurity and 95% CO_2 in a long sand-packed coil for cases of H_2S, N_2, SO_2, CH_4, Ar, and O_2: mole fraction of the impurity gases at the outlet, normalized by the relevant mole fraction in the initially injected gas (5% for all cases), as a function of the pore volume of the total injection. Ar and O_2 were added. Numerical results of HYTEC only.

Gas chromatography and density-driven flow

Co-injected gases can accelerate or decelerate the plume displacement, affect the volume capacity, react with the host rock, or be used as tracers for monitoring. They are often modeled as dissolved in formation water (Xu et al. 2007; Xiao et al. 2009; Lei et al. 2016; Todaka and Xu 2017). This assumption changes the gas dynamics, mass/molar fluxes, and neglects the gas partition in the plume. Another important physical phenomenon is convective mixing. The impact of impurities on convective flow was studied by Li and Jiang (2014) and Raad and Hassanzadeh (2016), who analysed effects of impurities on the density driven flow, onset time, and flux rate. However, with reference to the literature, there is no model of impure CO_2 injection that considers the density of mixtures and convective dissolution at a large scale.

Impurities in the injected stream have an influence on the solubility of CO_2, the gas dynamics of the plume, and the gravitational forces of the entire system; since the densities of mixtures can vary significantly with respect to their compositions. For example, in the presence of N_2, O_2 and Ar, the CO_2-rich mixture is lighter than the pure CO_2 at high pressure and temperature conditions. In contrast, the $SO_2 + CO_2$ mixture is significantly heavier than the pure CO_2. Based on the analysis of the volumetric properties (Ahlers and Gmehling 2001; Al-Siyabi 2013; Li and Yan 2009b; Nazeri et al. 2017; Sin 2015), and under the consideration of previous observations of gas chromatography (Eqn. 44), three cases of gas injection were devised to study the gas partition and density driven flow. In the base case, the pure CO_2 was injected into a homogeneous long confined aquifer with a height of 100 m and depth of 1100 m. The fluid and matrix properties were obtained from Sin and Corvisier (2018). The injection rate was constant: 10 kg/s or 0.03 m³/s for the pure CO_2 scenario. Two models of impure CO_2 were then designed with the same volumetric rate (but different mass rates): injection of 95% CO_2, 4% N_2 and 1% O_2 mixture (air and CO_2 model), and 95% CO_2 and 5 % SO_2 mixture (SO_2 and CO_2 model). The length of the aquifer was 10 km in the present models. The grid resolution was 38400 cells, with a length that varied from 2–1000 m. The axisymmetric geometry allowed for an adequate representation of the convective mixing (Sin 2015). The number of gas components in the models varied from 2–4 (H_2O(g) included). Simulations were carried out using HYTEC (van der Lee et al. 2003; Lagneau and van der Lee 2010; Sin et al. 2017a).

The shape and general dynamics of the gas plume were similar in all cases, as expected (Fig. 16). The gas chromatography effect resulted in the accumulation of the less soluble N_2 and O_2 at the leading edge of the plume. In the SO_2+CO_2 case, the delayed arrival of a more soluble SO_2, which was concentrated near the injection well and lagging behind the CO_2, was expected (Fig. 17). The gas partition occurred throughout the plume. However, it was most pronounced at the top of the aquifer due to the advection forces that continuously supply less soluble gas compounds (compare Figs. 17, 18, 19). Heterogeneous gas distribution contributes significantly to the thermodynamic properties such as density. The gas density maps reveal the dependence of the volumetric properties on the gas composition (Fig. 20).

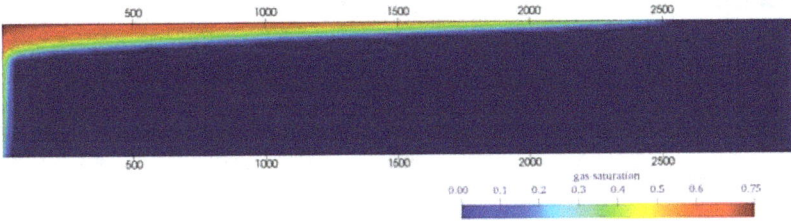

Figure 16. Injection of a feed gas of SO_2+CO_2 mixture in a 2D axisymmetric aquifer: gas saturation after 30 years of injection (modified after Sin and Corvisier 2018). The map is rescaled on the *X*-axis 5:1. See text for model details.

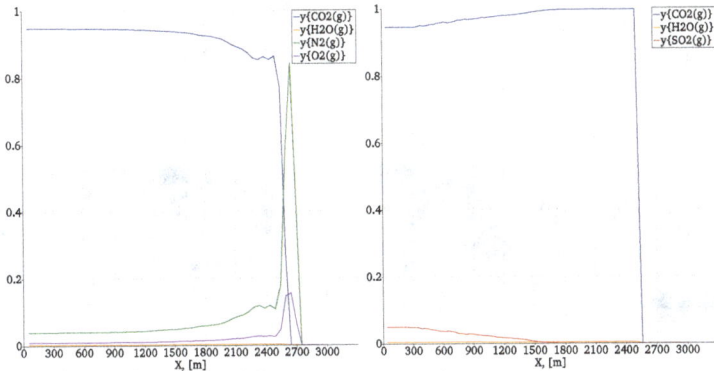

Figure 17. Injections of Air+CO_2 and SO_2+CO_2 mixtures in a 2D axisymmetric aquifer: gas composition at the top of the aquifer after 30 years of injection. See text for model details.

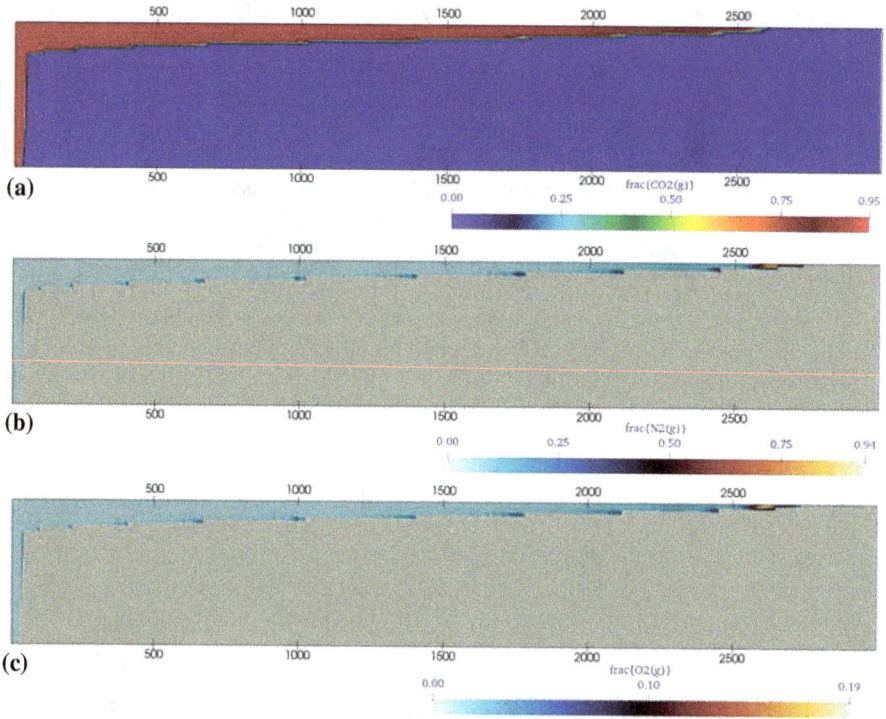

Figure 18. Interplay of the gas chromatography and density-driven flow. Injection of Air+CO_2 mixture in a 2D axisymmetric aquifer. Gas composition of the plume after 30 years of injection: mol fraction of (a) CO_2, (b) N_2, and (c) O_2 (modified after Sin and Corvisier 2018). The map is rescaled on the X-axis 5:1. See text for model details.

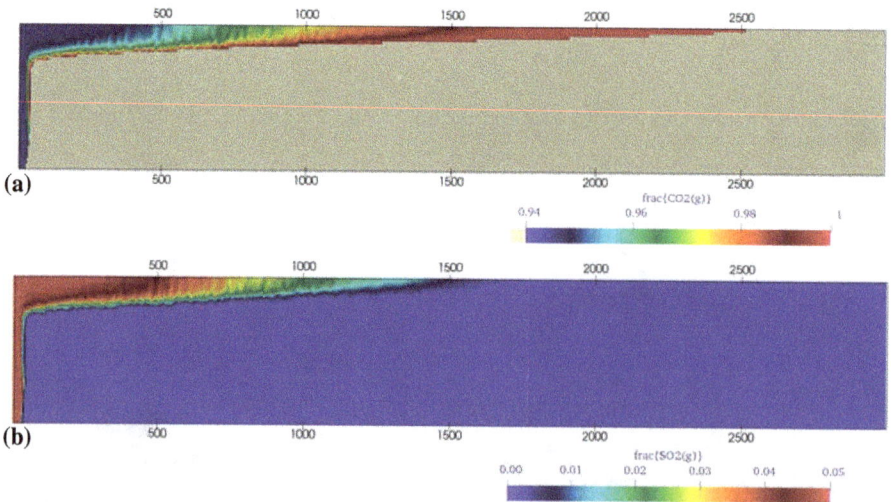

Figure 19. Interplay of the gas chromatography and density-driven flow. Injection of SO_2+CO_2 mixture in a 2D axisymmetric aquifer. Gas composition of the plume after 30 years of injection: mol fraction of (a) CO_2 and (b) SO_2 (modified after Sin and Corvisier 2018). The map is rescaled on the X-axis 5:1. See text for model details.

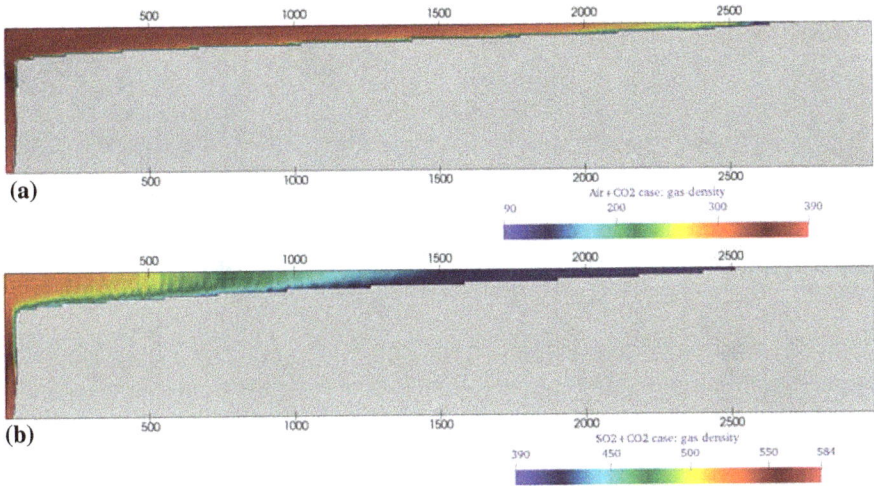

Figure 20. Interplay of the gas chromatography and density-driven flow. Injections of **(a)** Air+CO$_2$ and **(b)** SO$_2$+CO$_2$ mixtures in a 2D axisymmetric aquifer: gas density (in kg/m3) after 30 years of injection (modified after Sin and Corvisier 2018). The map is rescaled on the *X*-axis 5:1. See text for model details.

The formation water that contained dissolved CO$_2$ was heavier than the formation water without CO$_2$. The liquid density of the formed current increased with an increase in the content of CO$_2$, which was not the case for all the compounds, such as H$_2$S. The difference in density creates instabilities that balance the gravity forces that can finally initiate the convective motion under the current at the onset time. Several studies were conducted, which involved the experimental and numerical analysis of the nature of the process in closed and open systems for homogeneous and heterogeneous media (Lindeberg and Wessel-Berg 1997; Elenius et al. 2015; Emami-Meybodi et al. 2015; Green and Ennis-King 2018; Riaz et al. 2006; Thomas et al. 2018; Wen et al. 2018). The presence of impurities changes the densities of the gas and aqueous phases, and the convective fluxes then differs depending on a type and quantity of impurities. In the present models, the small amount of SO$_2$ (5%) resulted in a denser aqueous solution, and the increased convective motion accelerated the dissolution of gases.

Figures 21, 22, and 23 illustrate the effect of the gas partition on the distribution of the compounds dissolved in formation, which impacts the general evolution of the system. The liquid density of the SO$_2$+CO$_2$ model was almost greater than that of the Air+CO$_2$ model by a factor of 2, which indicates more CO$_2$ should be dissolved. In addition, the gas plume of the Air+CO$_2$ model was less dense than that of the SO$_2$+CO$_2$ model. Consequently, the gas current that contained air propagated faster than the gas current of the SO$_2$+CO$_2$ model (Fig. 24a). With respect to the CO$_2$ inventory (Fig. 24b), the ratio of CO$_2$(g) over the total CO$_2$ was very similar for the pure CO$_2$ and SO$_2$+CO$_2$ models. However, the quantity of injected CO$_2$ in the SO$_2$+CO$_2$ model was significantly greater than that in the pure CO$_2$ model due to the different densities of the injected streams (Fig. 24c).

The positive effect of SO$_2$ on the convective dissolution was evident. However, it should be noted that the convective rate decreased afterwards, as the formation water underneath the plume became homogeneously saturated in dissolved gases over time. The accepted limit of the SO$_2$ concentration in the injected stream should be defined with precautions with regard to its corrosive properties. SO$_2$ dissolves in brines and increases the acidity resulting in the dissolution of carbonate minerals. The evolution of porosity and a significant impact on the storage integrity can be expected. The effects of SO$_2$, CO$_2$, and N$_2$ on batch experiments of rock samples from the CO$_2$ pilot Heletz was investigated (Hedayati et al. 2018). Geochemical modeling revealed an increase in porosity due to SO$_2$ reactivity. With regard to N$_2$, and O$_2$, the partition of CO$_2$, N$_2$,

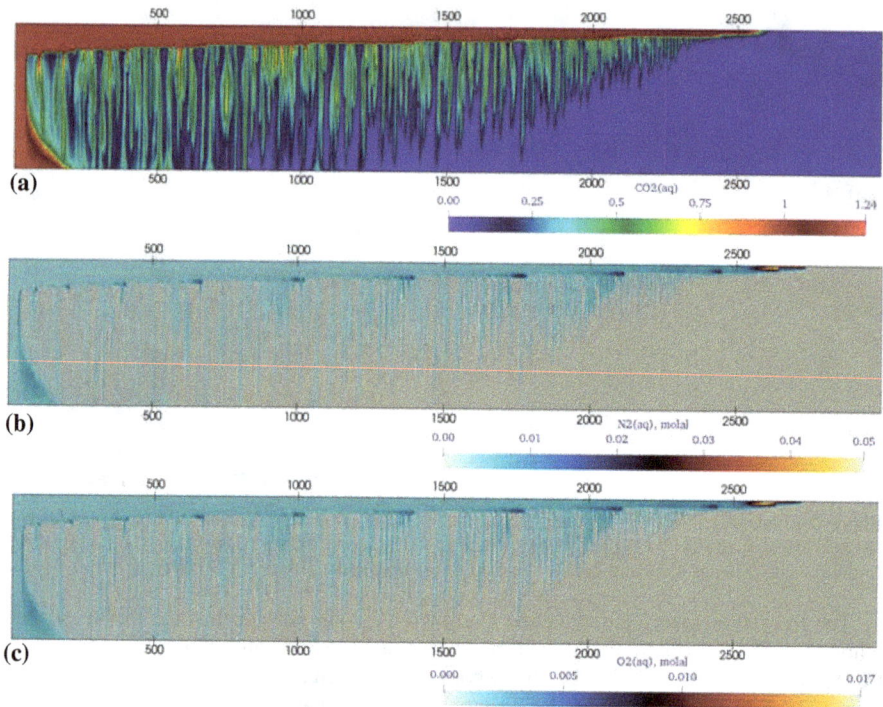

Figure 21. Interplay of the gas chromatography and density-driven flow. Injection of an Air+CO_2 mixture in a 2D axisymmetric aquifer. Composition of the aqueous phase (in molal) after 30 years of injection: (a) CO_2(aq), (b) N_2(aq), and (c) O_2 (modified after Sin and Corvisier 2018). The map is rescaled on the X-axis 5:1. See text for model details.

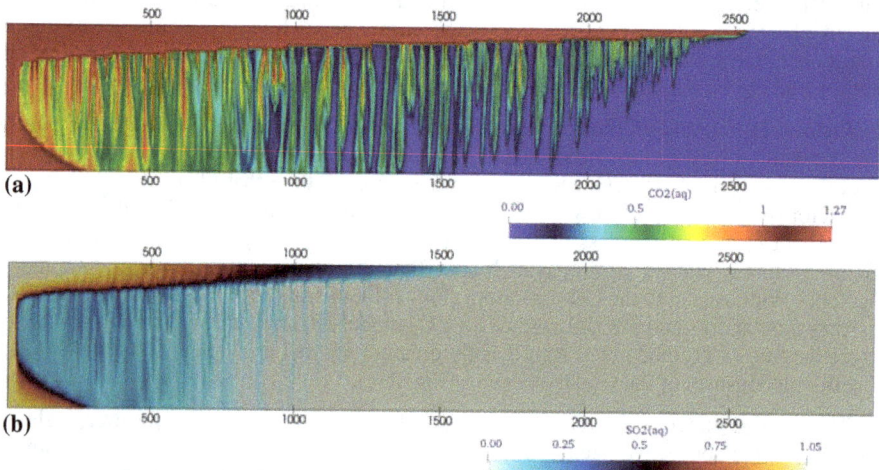

Figure 22. Interplay of the gas chromatography and density-driven flow. Injection of a SO_2+CO_2 mixture. Composition of the aqueous phase (in molal) after 30 years of injection: (a) CO_2(aq) and (b) SO_2(aq) (modified after Sin and Corvisier 2018). The map is rescaled on the X-axis 5:1. See text for model details.

and O_2 characterized by a N_2- and O_2-rich front of the plume edge makes them applicable for the monitoring as demonstrated by Wei et al. (2015). In addition, SO_2 is a redox reactive species that can transforms into sulfides, sulphate or precipitate in sulfide minerals.

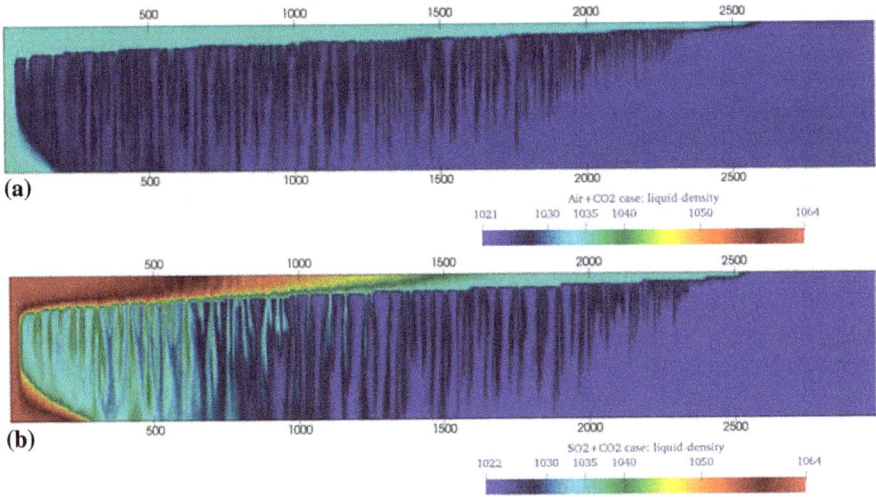

Figure 23. Interplay of the gas chromatography and density-driven flow. Injections of **(a)** Air+CO_2 and **(b)** SO_2+CO_2 mixtures in a 2D axisymmetric aquifer: liquid density (in kg/m^3) after 30 years of injection (modified after Sin and Corvisier 2018). The map is rescaled on the *X*-axis 5:1. See text for model details.

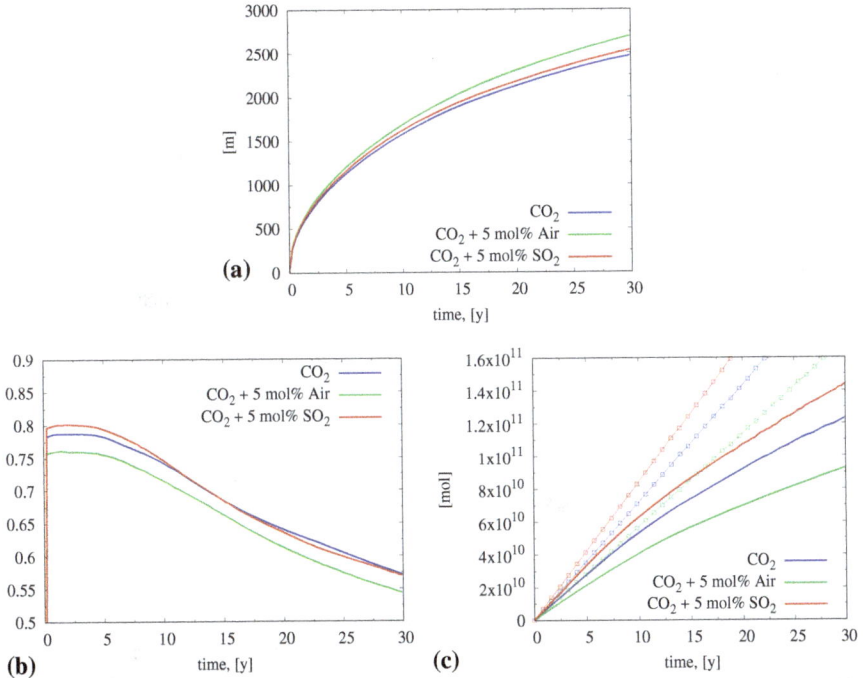

Figure 24. Injections of pure CO_2, and air+CO_2, and SO_2+CO_2 mixtures in a 2D axisymmetric aquifer: **(a)** radius of the plume, **(b)** ratio of CO_2(g) (in mol) to total CO_2 (in mol), and **(c)** CO_2(g) and total CO_2 in mol during 30 years of injection.

DISCUSSIONS AND PERSPECTIVES

The previous sections highlight efforts to model multiphase multicomponent reactive transport problems and several applications with respect to CCS and potential co-injected impurities. Until now, models required significant simplification for such applications. For example, as mentioned by Wolf et al. (2016), additional brine injections were considered in Toughreact, as impurities were dissolved and the gas phase was pure CO_2 (Xu et al. 2007; Xiao et al. 2009; Lei et al. 2016; Todaka and Xu 2017). Consequently, in such simulations, the mass balances, fluxes, and chemical reactions were misrepresented, especially in the vicinity of injection wells due to the artificial addition of water and ions. Wolf et al. (2016) reported on the ability of recent developments to consider trace gas impurities in TOUGHREACT (Xu et al. 2014) with the physical flow properties of a pure CO_2 plume with transported trace gas species. In the analysis of MMF&RT methods, a tight SIA-based coupling (such as STOMP) was highlighted. Bacon et al. (2014) conducted simulations of 99% $CO_2 + 1\%$ H_2S injection using STOMP, and the phase equilibrium was modeled using the asymmetric approach with the PR EOS similar to HYTEC (Sin et al. 2017a).

Despite all the intensive studies conducted on convective mixing, few studies considered co-injected impurities; and fewer considered geochemical reactions. For example, Thomas et al. (2016, 2018) recently carried out experiments on convective motion in alkaline and salt solutions. There are at least two major obstacles for this. First, a few simulators can model MMF&RT, which includes multicomponent gas mixtures and their properties, and carry out high-resolution simulations. Second, it is difficult to accurately predict thermodynamic properties, and experimental data for multicomponent systems are indeed critical. The modeling of such complex problems requires solid coupling methods. Recent MMF&RT developments offer large opportunities to model other systems than those previously presented. Several selected applications of MMF&RT and associated numerical difficulties are presented below.

Reservoir modeling is one of the major examples of MMF&RT applications. The modeling of multicomponent gas flow allows for a more in-depth understanding of long-term sour gas reservoir evolution. The knowledge of the compositional distribution within a gas reservoir has economical significance with respect to field development and resources production. Preferential leaching by underlying active aquifers is a significant factor that can cause a heterogeneous gas distribution. Soluble gases such as H_2S and CO_2 are dissolved and exported by the aquifer, thus developing diffusive gradients and a slow gas density-driven motion, which gradually develops over a long-term evolution. The impact of diffusive transport was studied only by Bonnaud et al (2012). Sin et al. (2017b) extended a problem and demonstrated the importance of modeling real fluids, the solubility of real gas mixtures, gas flows with composition-dependent thermodynamic properties at realistic temperature, and pressure conditions, which are characteristic for deep reservoirs. Based on this physical problem, a benchmark exercise was specifically designed to test multiphase multicomponent reactive transport codes (Sin et al. 2017b).

Carbon geo-sequestration also targets unconventional reservoirs such as shale and coalbed methane formations. Shale gas, which mainly consists of methane, is stored in reservoirs as free gas in fractures and as adsorbed gas on the surface of organic matter, which results in a nanoporous material. The CO_2 with its linear molecular structure enters into small pores thus leading to adsorption and further diffusion in the organic matrix. Moreover, CO_2 is preferentially adsorbed over CH_4 that makes the shale reservoirs suitable for CO_2 sequestration while enhancing gas recovery (EGR) (Kang et al. 2011). Godec et al. (2014b) estimated a potential storage of 740 Gt CO_2 in shale gas worldwide. A recent pilot study of the injection of 510 t CO_2 in the Chattanoga formation was successfully reported by Louk et al. (2017), which demonstrates the feasibility of CO_2-EGR. The modeling of multicomponent gas transport in shale formations comprises viscous Darcy flow in micropores and slip flow in nanopores. Competitive adsorption, gas compressibility, and the description of real gas are also required. Numerous concepts were proposed to better understand complex mechanism, such as a new formulation of apparent permeability with a derived definition of tortuosity (Civan 2010), a double porosity approach (Javadpour et al. 2007), and a triple porosity model (Yan et al. 2016).

The IEA Greenhouse Gas R&D Programme (IEAGHG) recently re-assessed the status of CO_2-enhanced coalbed methane (CO_2-ECBM) and evaluated the worldwide potential storage of nearly 488 $GtCO_2$ in "unmineable coal seams" (Godec et al. 2014a). Despite the lack of geological data on deep coal formation (field tests: USA, Godec et al. 2014a; New Mexico, Siriwardane et al. 2012; Jharia coalfields, India, Vishal et al. 2018), further investigations of the interactions between CO_2 and coals are required: in particular, a better understanding and development of comprehensive models of coal swelling/shrinkage and permeability decreases/increases due to gas adsorption and desorption. Intensive experimental and numerical studies on the adsorption of gases and coal were carried out, e.g., the adsorption of pure CO_2 and CH_4, their mixtures, their influence on the coal matrix and surface area, the verification of analytical models such as the commonly used Langmuir and Toth isotherms (Bae and Bhatia 2006; Ottiger et al. 2008), and the competitive adsorption of CO_2 and CH_4 mixtures on dry and wet coal (Lee et al. 2013). The injection of CO_2 and N_2 as a mixture or alternating gases has attracted significant research attention. Moreover, N_2 has a lower adsorption capacity than CO_2 and CH_4. Upon injection into coal, most of the N_2 is contained in cleats, whereas desorbed CH_4 migrates through the cleats. Given that only a small portion of N_2 is adsorbed when compared with CO_2, the matrix shrinks, and the permeability and porosity increase. Co-injecting N_2 can limit the swelling of the coal matrix and enhance methane recovery. This was confirmed in field studies (Godec et al. 2014a, Oudinot et al. 2017). Stevenson et al. (1991) presented modeling multicomponent adsorption equilibria on the basis of the real adsorbed solution theory, and then provided binary interaction parameters for CO_2, N_2, and CH_4. A recent study indicated a higher desorption of CH_4 when injecting a CO_2+SO_2 mixture (Luo et al. 2019). Another issue with respect to CO_2-ECBM is matrix swelling/shrinkage, which is dependent on the effective stress, in addition to the gas type and gas adsorption/desorption; and therefore the gas compressibility and temperature. Moreover, it is a reversible and anisotropic property. Pan and Connell (2012) presented a comprehensive review of the permeability models and experimental data. In addition to sorption-induced swelling, effective stress, pressure-dependent permeability, and a high gas compressibility, the Klinkenberg effect has a significant influence in the gas flow in cleats (Fan et al. 2019; Wang et al. 2014; Zhu et al. 2007). Moreover, few studies have been conducted on CO_2–brine–rock interactions (Wang et al. 2016).

Energy storage via gas storage in saline aquifers or salt caverns (called Power to Gas technology) is a potential application of the type of models presented in this contribution that take into account the influence of salinity on gas solubility. Hemme and van Berk (2017) proposed a numerical model using Phreeqc of CH_4 storage in a salt cavern with a complete geochemical approach, whereas the mesh and flow and transport were significantly simplified. With respect to hydrogen storage, we suggest that the Klinkenberg effect for gas motion should be considered.

Within the framework of radioactive waste storage, at least two subjects may require multiphase multicomponent models: the atmospheric carbonation of concrete and the corrosion of metallic components that is responsible for H_2 pressure build-up in the storage environment (see also Bildstein et al. 2019, this volume). The first topic is the subject of an ongoing benchmark exercise that combines the drying of concrete (i.e., desaturation), the flow and transport of CO_2 from the atmosphere, and the carbonation. The second one has been the subject of a study showing oxic and anoxic corrosion of steel and H_2 production, with an extent of oxidizing/reducing front depending on gas diffusion coefficients (De Windt et al. 2014).

The prediction of subsurface soil gas migration may also benefit from improvements in numerical modeling. Providing some developments to consider organic matter degradation and/or respiration from plants and microorganisms, such models could evaluate gas (major gases CO_2 and O_2, and inter gases as well) fluxes between the soil and atmosphere that could help monitor specific geochemical activities in soils (Alibert et al. 2018).

SUMMARY

In this chapter, the existing mathematical and numerical approaches to multiphase multicomponent reactive transport and flow were analyzed, in addition to the necessity for MMF&RT modeling, especially for CO_2-related problems.

Over the last two decades, there has been significant progress with respect to the understanding of the physics of multiphase flow. Novel measurement methods for capillary pressure and relative permeability combine experimental techniques and high-resolution numerical modeling. This allows for the accurate estimation of the characteristic properties of homogeneous and heterogeneous cores. Further studies are needed to obtain more experimental data for a wide range of representative rock types, to create a workflow for upscaling from the sub-core and core scales to the m-scale, and to accurately predict capillary trapping and hysteresis necessary for the evaluation of residual trapping during post-injection. Further investigations on relative permeability curves during CO_2 exsolution are required, as they are significantly different from those of drainage processes (Falta et al. 2013; Zuo et al. 2017).

Recent developments in modeling thermodynamic systems allow for a better representation of the phase equilibrium. Non-ideal solutions and gas mixtures can currently be considered using (cubic) EOS in conjuction with activity models, e.g., the asymmetric approach. Moreover, the modeling of fluid properties of multicomponent systems is challenging due to the lack of experimental data.

Modeling MMF&RT is a complex problem. The description started from the basics of multiphase flow and geochemical reactions, highlighting the complexity added by the presence of two or more fluid phases. The formulation of multiphase flow is not unique. The decoupled pressure formulation using the global pressure simplifies the resolution of the set of equations, but this approach is suitable only for loosely coupled problems. The method is employed in several couplings with the reactive transport. The flow calculations are then less intensive; however, the iterative loop between the flow and reactive transport is required to cover the gaps between steps. The classic form of conservation equations for two-phase or multiphase flow may exhibit stability issues when selecting the natural primary variables originated by Coats in 1980. To avoid changing the set of primary variables, numerous alternatives were proposed by combining variables, expressing them in different forms, and defining conservation equations with respect to the mole, mass, or volume. The extended set of primary variables with complementarity conditions and the overall composition formulation were tested on CCS related problems and demonstrated high efficiencies in these cases. However, the Coats form is competitive and generally used for coupling with reactive transport.

The GIA is reliable for practical use when applied with efficient linear solvers and reduction techniques. The OSA can be found in most codes due to its flexibility, but retardations of information circulated through the operators can evolve into convergence issues. An obvious remedy is to model implicitly flow and phase equilibrium, iterating between the flow and transport, or solving reactive transport by GIA. It is now clear that the flow and equations of state/phase equilibrium should be solved simultaneously. The stability of MMF&RT is improved for the modeling of coupled processes with an evolving matrix across spatial and temporal scales. Overall, the recommendation is to use a GIA or a tight SIA-based coupling with natural primary variables or equivalent formulation. We consider that the tight SIA is preferable for reactive transport simulators.

0D simulations demonstrated a high accuracy for various systems (e.g., various gaseous compounds, gas mixtures, brine); thus verifying the numerical representation of these systems and allowing for the validation of different approaches selected in the tested geochemical solvers. Additional comparisons with well-constrained experimental results for multi-dimensional problems involving multiphase transport and flow would be of great interest to validate the coupling. A few benchmark studies for MMF&RT were conducted; however, the gas phase was only represented by CO_2. Hence, MMF&RT benchmarks with multicomponent systems are still required.

REFERENCES

Aavatsmark I, Eigestad G, Mallison B, Nordbotten J (2008) A compact multipoint flux approximation method with improved robustness. Numer Methods Partial Differ Equations 24:1329–1360

Abadpour A, Panfilov M (2009) Method of negative saturations for modeling two-phase compositional flow with oversaturated zones. Transp Porous Media 79:197–214

Acs G, Doleschall S, Farkas E (1985) General purpose compositional model. SPE J 25:543–553

Adams J, Bachu S (2002) Equations of state for basin geofluids: algorithm review and intercomparison for brines. Geofluids 2:257–271

Ahlers J, Gmehling J (2001) Development of an universal group contribution equation of state: I. Prediction of liquid densities for pure compounds with a volume translated Peng–Robinson equation of state. Fluid Phase Equilib 191:177–188

Ahlers J, Gmehling J (2002a) Development of a universal group contribution equation of state. 2. Prediction of vapor–liquid equilibria for asymmetric systems. Ind Eng Chem Res 41:3489– 3498

Ahlers J, Gmehling J (2002b) Development of a universal group contribution equation of state 3. Prediction of vapor–liquid equilibria, excess enthalpies, and activity coefficients at infinite dilution with the VTPR model. Ind Eng Chem Res 41:5890–5899

Ahlers J, Yamaguchi T, Gmehling J (2004) Development of a universal group contribution equation of state. 5. Prediction of the solubility of high-boiling compounds in supercritical gases with the group contribution equation of state volume-translated Peng-Robinson. Ind Eng Chem Res 43:6569–6576

Ahusborde E, El Ossmani M (2017) A sequential approach for numerical simulation of two-phase multicomponent flow with reactive transport in porous media. Math Comput Simul 137:71–89

Al-Menhali A, Niu B, Krevor S (2015) Capillarity and wetting of carbon dioxide and brine during drainage in Berea sandstone at reservoir conditions. Water Resour Res 51:7895–7914

Al-Siyabi I (2013) Effect of impurities on CO_2 stream properties. PhD thesis, Heriot-Watt University

Alibert C, Pili E, Barré P, Massol F (2018) Influence of biogeochemical reactions on inert gas fluxes between soil and the atmosphere. *In:* EGU Gen Assembly Confer Abstr 20:2177

Allen MB (1984) Why upwinding is reasonable. *In:* Finite Elements in Water Resources, p 13–23

André L, Azaroual M, Bernstone C, Wittek A (2012) Modeling the geochemical impact of an injection of CO_2 and associated reactive impurities into a saline reservoir. *In:* TOUGH Symposium 2012, p 8

Andrew M, Bijeljic B, Blunt MJ (2014) Pore-scale contact angle measurements at reservoir conditions using X-ray microtomography. Adv Water Resour 68:24–31

Angelini O (2010) Étude de schémas numériques pour les écoulements diphasiques en milieu poreux déformable pour des maillages quelconques, Application au stockage de déchets radioactifs. PhD thesis, Université de Marne la Valée

Appelo C, Parkhurst DL, Post V (2014) Equations for calculating hydrogeochemical reactions of minerals and gases such as CO_2 at high pressures and temperatures. Geochim Cosmochim Acta 125:49–67

Armstrong RT, Porter ML, Wildenschild D (2012) Linking pore-scale interfacial curvature to column-scale capillary pressure. Adv Water Resour 46:55–62

Aziz K, Settari A (1979) Petroleum Reservoir Simulation. Elsevier

Bachu S (2013) Drainage and imbibition CO_2/brine relative permeability curves at in situ conditions for sandstone formations in western Canada. Energy Procedia 37:4428–4436

Bachu S, Bennion DB (2008) Effects of in-situ conditions on relative permeability characteristics of CO_2–brine systems. Environ Geol 54:1707–1722

Bachu S, Bennion DB (2009) Chromatographic partitioning of impurities contained in a CO_2 stream injected into a deep saline aquifer: Part 1. effects of gas composition and in situ conditions. Int J Greenhouse Gas Control 3:458–467

Bachu S, Pooladi-Darvish M, Hong H (2009) Chromatographic partitioning of impurities (H_2S) contained in a CO_2 stream injected into a deep saline aquifer: Part 2. effects of flow conditions. Int J Greenhouse Gas Control 3:468 – 473

Bacon DH, Ramanathan R, Schaef HT, McGrail BP (2014) Simulating geologic co-sequestration of carbon dioxide and hydrogen sulfide in a basalt formation. Int J Greenhouse Gas Control 21:165–176

Bae J-S, Bhatia SK (2006) High-pressure adsorption of methane and carbon dioxide on coal. Energy Fuels 20:2599–2607

Barenblatt G, Entov V, Ryzhik V (1972) Theory of unsteady filtration of fluids and gases. Nedra Publishing House, Moscow

Battiato I, Tartakovsky DM, Tartakovsky AM, Scheibe TD (2011) Hybrid models of reactive transport in porous and fractured media. Adv Water Resour 34:1140–1150

Batychy J, McCaffery FG (1978) Low interfacial tension displacement studies. *In:* Ann Tech Meet Pet Soc Can, Calgary, Canada

Bear J (1972) Dynamics of Fluids in Porous Media. Eisevier, New York

Ben Gharbia I, Flauraud E, Michel A (2015) Study of compositional multiphase flow formulations with cubic EOS. *In:* SPE Reservoir Simulation Symp

Benedict M, Webb GB, Rubin LC (1940) An empirical equation for thermodynamic properties of light hydrocarbons and their mixtures I: methane, ethane, propane and *n*-butane. J Chem Phys 8:334–345

Bennion B, Bachu S (2008) Drainage and imbibition relative permeability relationships for supercritical CO_2/brine and H_2S/brine systems in inter-granular sandstone, carbonate, shale, and anhydrite rocks. SPE Reservoir Eval Eng 11:487–496

Benson S, Cook P, Anderson J, Bachu S, Nimir H, Basu B, Bradshaw J, Deguchi G, Gale J, von Goerne G, Heidug W, Holloway S, Kamal R, Keith D, Lloyd P, Rocha P, Senior B, Thompson J, Torp T, Wildenborg T, Wilson M, Zarlenga F, Zhou D, Celia M, Gunter B, Ennis-King J, Lindeberg E, Lombardi S, Oldenburg C, Pruess K, Rigg A, Stevens S, Wilson E, Whittaker S (2005) Chapter 5—Underground geologic storage. *In:* IPCC Special Report on Carbon Dioxide Capture and Storage. Technical report, Cambridge University Press

Benson S, Bennaceur K, Cook P, Davison J, de Coninck H, Farhat K, Ramirez A, Simbeck D, Surles T, Verma P, Wright I (2012) Chapter 13—Carbon capture and storage. *In:* Global Energy Assessment—Toward a Sustainable Future. Technical report, Cambridge University Press

Bildstein O, Claret F, Frugier P (2019) RTM for waste repositories. Rev Mineral Geochem 85:419–457

Blair P, Weinaug C (1969) Solution of two-phase flow problems using implicit difference equations. SPE J 9:417–424

Bonnaud E, Dessort D, Lagneau V, Chiquet P (2012) A scenario for the creation of H_2S heterogeneities in acid gas reservoirs in contact with an active aquifer: a simulation study. *In:* SPE 161625:2123–2131

Bourgeat A, Jurak M, Smaï F (2009) Two-phase, partially miscible flow and transport modeling in porous media; application to gas migration in a nuclear waste repository. Comput Geosci 13:29–42

Brønsted J (1922) Studies on solubility. IV. The principle of the specific interaction of ions. J Am Chem Soc 44:877–898

Brooks R, Corey A (1964) Hydraulic Properties of Porous Media. Hydrology Papers, Colorado State University

Broseta D, Tonnet N, Shah V (2012) Are rocks still water-wet in the presence of dense CO_2 or H_2S? Geofluids 12:280–294

Brunner F, Knabner P (2019) A global implicit solver for miscible reactive multiphase multicomponent flow in porous media. Comput Geosci 23:127–148

Burdine N (1953) Relative permeability calculations from pore size distribution data. Trans. AIME 198:71–78

Cama J, Soler JM, Ayora C (2019) Acid water–rock–cement interaction and multicomponent reactive transport modeling. Rev Mineral Geochem 85:459–498

Cao H (2002) Development of techniques for general purpose simulators. PhD thesis, Stanford University

Cao H, Tchelepi HA, Wallis JR, Yardumian HE (2005) Parallel scalable unstructured CPR-type linear solver for reservoir simulation. *In:* SPE Ann Tech Confer Exhib

Carlson FM (1981) Simulation of relative permeability hysteresis to the nonwetting phase. *In:* SPE Ann Tech Confer Exhib

Carrayrou J, Hoffmann J, Knabner P, Kräutle S, De Dieuleveult C, Erhel J, Van Der Lee J, Lagneau V, Mayer KU, Macquarrie KT (2010) Comparison of numerical methods for simulating strongly nonlinear and heterogeneous reactive transport problems—the MoMaS benchmark case. Comput Geosci 14:483–502

Caumon M-C, Sterpenich J, Randi A, Pironon J (2016) Measuring mutual solubility in the H_2O–CO_2 system up to 200 bar and 100 °C by in situ Raman spectroscopy. Int J Greenhouse Gas Control 47:63–70

Chapoy A, Nazeri M, Kapateh M, Burgass R, Coquelet C, Tohidi B (2013) Effect of impurities on thermophysical properties and phase behaviour of a CO_2-rich system in CCS. Int J Greenhouse Gas Control 19:92–100

Chavent G, Jaffré J (1986) Mathematical Models and Finite Elements for Reservoir Simulation: Single Phase, Multiphase and Multicomponent Flows through Porous Media. Studies in Mathematics and its Applications Vol 17. North Holland, Amsterdam

Chen L, Wang M, Kang Q, Tao W (2018) Pore scale study of multiphase multicomponent reactive transport during CO_2 dissolution trapping. Adv Water Resour 116:208–218

Chen Z, Huan G, Ma Y (2006) Computational methods for multiphase flows in porous media, Volume 2. SIAM

Civan F (2010) Effective correlation of apparent gas permeability in tight porous media. Transp Porous Media 82:375–384

Class H, Ebigbo A, Helmig R, Dahle HK, Nordbotten JM, Celia MA, Audigane P, Darcis M, Ennis-King J, Fan Y, Flemisch B (2009) A benchmark study on problems related to CO_2 storage in geologic formations. Comput Geosci 13:409–434

Coats KH (1980) An equation of state compositional model. SPE J 20:363–376

Coats KH (2003) IMPES stability: selection of stable timesteps. SPE J 8:181–187

Colston BJ, Chandratillake MR, Robinson VJ (1990) Correction for ionic strength effects in modelling aqueous systems. Safety studies: NIREX radioactive waste disposal NSS/R204. Technical Report, University of Manchester, Department of Chemistry, M13 9PL

Corvisier J, Bonvalot A, Lagneau V, Chiquet P, Renard S, Sterpenich J, Pironon J (2013) Impact of co-injected gases on CO_2 storage sites: Geochemical modeling of experimental results. Energy Procedia 37:3699–3710

Corvisier J, El Ahmar E, Coquelet C, Sterpenich J, Privat R, Jaubert JN, Ballerat-Busserolles K, Coxam JY, Cézac P, Contamine F, Serin JP (2014) Simulations of the impact of co-injected gases on CO_2 storage, the SIGARR project: first results on water–gas interactions modeling. Energy Procedia 63:3160–3171

Corvisier J, Hajiw M, El Ahmar E, Coquelet C, Sterpenich J, Privat R, Jaubert JN, Ballerat-Busserolles K, Coxam JY, Cézac P, Contamine F (2017) Simulations of the impact of co-injected gases on CO_2 storage, the SIGARR project: processes and geochemical approaches for gas–water–salt interactions modeling. Energy Procedia 114:3322–3334

Craig FF (1971) The reservoir engineering aspects of waterflooding, Vol 3. HL Doherty Memorial Fund of AIME New York

Cunningham RE, Williams R (1980) Diffusion in Gases and Porous Media, Volume 1. Springer

Davies CW, Shedlovsky T (1964) Ion association. J Electrochem Soc 111:85C–86C

de Marsily G (1986) Quantitative Hydrogeology. Technical Report, Paris School of Mines, Fontainebleau

De Windt L, Marsal F, Corvisier J, Pellegrini D (2014) Modeling of oxygen gas diffusion and consumption during the oxic transient in a disposal cell of radioactive waste. Appl Geochem 41:115–127

Debye P, Hückel E (1923) The theory of the electrolyte II—The border law for electrical conductivity. Phys Z 24:305–325

Doughty C (2007) Modeling geologic storage of carbon dioxide: comparison of non-hysteretic and hysteretic characteristic curves. Energy Converse 48:1768–1781

Drummond S (1982) Boiling and mixing of hydrothermal fluids: Chemical effects on mineral precipitation, PhD Thesis, Penn State University

Duan Z, Sun R (2003) An improved model calculating CO_2 solubility in pure water, aqueous nacl solutions from 273 to 533 K and from 0 to 2000 bar. Chem Geol 193:257–271

Duan Z, Hu J, Li D, Mao S (2008) Densities of the CO_2–H_2O and CO_2–H_2O–NaCl systems up to 647 K and 100MPa. Energy Fuels 22:1666–1674

Dury O, Fischer U, Schulin R (1999) A comparison of relative nonwetting-phase permeability models. Water Resour Res 35:1481–1493

Elenius MT, Nordbotten JM, Kalisch H (2012) Effects of a capillary transition zone on the stability of a diffusive boundary layer. IMAJ Appl Math 77:771–787

Elenius M, Voskov D, Tchelepi H (2015) Interactions between gravity currents and convective dissolution. Adv Water Resour 83:77–88

Emami-Meybodi H, Hassanzadeh H, Green CP, Ennis-King J (2015) Convective dissolution of CO_2 in saline aquifers: Progress in modeling and experiments. Int J Greenhouse Gas Control 40:238–266

Ennis-King JP, Paterson L (2005) Role of convective mixing in the long-term storage of carbon dioxide in deep saline formations. SPE J 10:349–356

Entov V, Turetskaya F, Voskov D (2002) On approximation of phase equilibria of multicomponent hydrocarbon mixtures and prediction of oil displacement by gas injection. *In:* 8th Euro Confer Math Oil Recovery

Ertekin T, King GA, Schwerer FC (1986) Dynamic gas slippage: a unique dual-mechanism approach to the flow of gas in tight formations. SPE Form Eval 1:43–52

Espinoza DN, Santamarina JC (2010) Water–CO_2–mineral systems: Interfacial tension, contact angle, and diffusion—implications to CO_2 geological storage. Water Resour Res 46:W07537

Eymard R, Gallouët T, Herbin R (2009) Discretization of heterogeneous and anisotropic diffusion problems on general nonconforming meshes SUSHI: a scheme using stabilization and hybrid interfaces. IMA J Numer Anal 30:1009–1043

Eymard R, Guichard C, Herbin R (2012) Small-stencil 3D schemes for diffusive flows in porous media. ESAIM: Math Modell Numer Anal 46:265–290

Falta RW, Zuo L, Benson SM (2013) Migration of exsolved CO_2 following depressurization of saturated brines. Greenhouse Gases: Sci Technol 3:503–515

Fan Y, Durlofsky LJ, Tchelepi HA (2012) A fully-coupled flow-reactive-transport formulation based on element conservation, with application to CO_2 storage simulations. Adv Water Resour 42:47–61

Fan C, Elsworth D, Li S, Zhou L, Yang Z, Song Y (2019) Thermo-hydro-mechanical-chemical couplings controlling CH_4 production and CO_2 sequestration in enhanced coalbed methane recovery. Energy 173:1054–1077

Farajzadeh R, Barati A, Delil HA, Bruining J, Zitha PL (2007) Mass transfer of CO_2 into water and surfactant solutions. Pet Sci Technol 25:1493–1511

Farajzadeh R, Matsuura T, van Batenburg D, Dijk H (2012) Detailed modeling of the alkali/surfactant/polymer (asp) process by coupling a multipurpose reservoir simulator to the chemistry package PHREEQC. SPE Reservoir Eval Eng 15:423–435

Firoozabadi A (1999) Thermodynamics of Hydrocarbon Reservoirs. McGraw-Hill New York

Florence FA, Rushing J, Newsham KE, Blasingame TA (2007) Improved permeability prediction relations for low permeability sands. *In:* Rocky Mountain Oil & Gas Technology Symposium

Friis HA, Edwards MG (2011) A family of MPFA finite-volume schemes with full pressure support for the general tensor pressure equation on cell-centered triangular grids. J Comput Phys 230:205–231

Gao J, Xing H, Tian Z, Pearce JK, Sedek M, Golding SD, Rudolph V (2017) Reactive transport in porous media for CO_2 sequestration: Pore scale modeling using the lattice Boltzmann method. Comput Geosci 98:9–20

Garcia JE (2003) Fluid Dynamics of Carbon Dioxide Disposal into Saline Aquifers. PhD thesis, Lawrence Berkeley National Laboratory

Gaus I, Audigane P, André L, Lions J, Jacquemet N, Durst P, Czernichowski-Lauriol I, Azaroual M (2008) Geochemical and solute transport modelling for CO_2 storage, what to expect from it? Int J Greenhouse Gas Control 2:605–625

Ghanizadeh A, Amann-Hildenbrand A, Gasparik M, Gensterblum Y, Krooss BM, Littke R (2014) Experimental study of fluid transport processes in the matrix system of the European organic-rich shales: II. Posidonia Shale (lower Toarcian, Northern Germany) Int J Coal Geol 123:20–33

Godec M, Koperna G, Gale J (2014a) CO2–ECBM: a review of its status and global potential. Energy Procedia 63:5858–5869

Godec M, Koperna G, Petrusak R, Oudinot A (2014b) Enhanced gas recovery and CO2 storage in gas shales: a summary review of its status and potential. Energy Procedia 63:5849–5857

Goerke U-J, Park C-H, Wang W, Singh A, Kolditz O (2011) Numerical simulation of multiphase hydromechanical processes induced by CO2 injection into deep saline aquifers. Oil Gas Sci Technol 66:105–118

Golding MJ, Huppert HE, Neufeld JA (2013) The effects of capillary forces on the axisymmetric propagation of two-phase, constant-flux gravity currents in porous media. Physics Fluids 25:036602

Gouze P, Luquot L (2011) X-ray microtomography characterization of porosity, permeability and reactive surface changes during dissolution. J Contam Hydrol 120:45–55

Green CP, Ennis-King J (2010) Effect of vertical heterogeneity on long-term migration of CO2 in saline formations. Transp Porous Media 82:31–47

Green CP, Ennis-King J (2018) Steady flux regime during convective mixing in three-dimensional heterogeneous porous media. Fluids 3:58

Grenthe I, Puigdomenech I (1997) Modelling in Aquatic Chemistry. Nuclear Energy Agency, OECD, Paris, France

Haghi RK, Chapoy A, Peirera LM, Yang J, Tohidi B (2017) pH of CO2 saturated water and CO2 saturated brines: Experimental measurements and modelling. Int J Greenhouse Gas Control 66:190–203

Hajiw M, Chapoy A, Coquelet C (2015) Hydrocarbons–water phase equilibria using the CPA equation of state with a group contribution method. Can J Chem Eng 93:432–442

Hajiw M, Corvisier J, El Ahmar E, Coquelet C (2018) Impact of impurities on CO2 storage in saline aquifers: Modelling of gases solubility in water. Int J Greenhouse Gas Control 68:247–255

Hao Y, Sun Y, Nitao J (2012) Overview of NUFT: a versatile numerical model for simulating flow and reactive transport in porous media. Chapter 9 *In:* Groundwater Reactive Transport Models, p 212–239. Lawrence Livermore National Laboratory, USA

Hassanizadeh M, Gray WG (1979) General conservation equations for multi-phase systems: 2. Mass, momenta, energy, entropy equations. Adv Water Resour 2:191–203

Hassanzadeh H, Pooladi-Darvish M, Keith DW (2007) Scaling behavior of convective mixing, with application to geological storage of CO2. AIChE J 53:1121–1131

Hedayati M, Wigston A, Wolf JL, Rebscher D, Niemi A (2018) Impacts of SO2 gas impurity within a CO2 stream on reservoir rock of a CCS pilot site: Experimental and modelling approach. Int J Greenhouse Gas Control 70:32–44

Heid J, McMahon J, Nielsen R, Yuster S (1950) Study of the permeability of rocks to homogeneous fluids. *In:* Drilling and Production Practice. API

Hejazi SAH, Shah S, Pini R (2019) Dynamic measurements of drainage capillary pressure curves in carbonate rocks. Chem Eng Sci 200:268–284

Helgeson HC (1969) Thermodynamics of hydrothermal systems at elevated temperatures and pressures. Am J Sci 267:729–804

Hemme C, van Berk W (2017) Potential risk of H2S generation and release in salt cavern gas storage. J Nat Gas Sci Eng 47:114–123

Hermeline F (2007) Approximation of 2-D, 3-D diffusion operators with variable full tensor coefficients on arbitrary meshes. Comput Methods Appl Mech Eng 196:2497–2526

Hesse MA, Woods A (2010) Buoyant dispersal of CO2 during geological storage. Geophys Res Lett 37: L01403

Hesse MA, Orr FM, Tchelepi H (2008) Gravity currents with residual trapping. J Fluid Mech 611:35–60

Hou S-X, Maitland GC, Trusler JM (2013) Measurement, modeling of the phase behavior of the (carbon dioxide+water) mixture at temperatures from 298.15 K to 448.15 K. J Supercrit Fluids 73:87–96

Hron P, Jost D, Bastian P, Gallert C, Winter J, Ippisch O (2015) Application of reactive transport modeling to growth, transport of microorganisms in the capillary fringe. Vadose Zone J 14:vzj2014.07.0092

IEAGHG (2011) Effects of impurities on geological storage of CO2, 2011/04. Technical report

Ingram R, Wheeler MF, Yotov I (2010) A multipoint flux mixed finite element method on hexahedra. SIAM J Numer Anal 48:1281–1312

Ippisch O (2003) Coupled Transport in Natural Porous Media. PhD thesis, University of Heidelberg

Jackson SJ, Agada S, Reynolds CA, Krevor S (2018) Characterizing drainage multiphase flow in heterogeneous sandstones. Water Resour Res 54:3139–3161

Jackson SJ, Krevor S (2019) Characterization of hysteretic multiphase flow from the mm to m scale in heterogeneous rocks. *In:* E3S Web of Confer 89:02001. EDPSciences

Jacquemet N (2006) Durabilité des matériaux de puits pétroliers dans le cadre d'une séquestration géologique de dioxyde de carbone et d'hydrogène sulfuré. PhD thesis, Université Henri Poincaré-Nancy I

Jacquemet N, Pironon J, Caroli E (2005) A new experimental procedure for simulation of H2S+CO2 geological storage. application to well cement aging. Oil Gas Sci Technol 60:193–203

Jacquemet N, Pironon J, Saint-Marc J (2008) Mineralogical changes of a well cement in various H2S–CO2(–brine) fluids at high pressure, temperature. Environ Sci Technol 42:282–288

Jaffré J, Sboui A (2010) Henry's law and gas phase disappearance. Transp Porous Media 82:521–526

Javadpour F, Fisher D, Unsworth M (2007) Nanoscale gas flow in shale gas sediments. J Can Pet Technol 46:55–61

Juanes R, MacMinn CW, Szulczewski ML (2010) The footprint of the CO_2 plume during carbon dioxide storage in saline aquifers: storage efficiency for capillary trapping at the basin scale. Transp Porous Media 82:19–30

Jung HB, Um W, Cantrell KJ (2013) Effect of oxygen co-injected with carbon dioxide on gothic shale caprock– CO_2– brine interaction during geologic carbon sequestration. Chem Geol 354:1–14

Jung Y, Pau GSH, Finsterle S, Pollyea RM (2017) Tough3: A new efficient version of the tough suite of multiphase flow and transport simulators. Comput Geosci 108:2–7

Kang SM, Fathi E, Ambrose RJ, Akkutlu IY, Sigal RF (2011) Carbon dioxide storage capacity of organic-rich shales. SPE J 16:842–855

Kather A (2009) CO_2 quality and other relevant issues. *In:* 2nd Working Group Meeting on CO_2 Quality and Other Relevant Issues, Cottbus, Germany

Killough J (1976) Reservoir simulation with history-dependent saturation functions. SPE J 16:37–48

Klinkenberg L (1941) The permeability of porous media to liquids and gases. *In:* Drilling and Production Practice. API

Kneafsey TJ, Pruess K (2010) Laboratory flow experiments for visualizing carbon dioxide-induced, density-driven brine convection. Transp Porous Media 82:123–139

Kräutle S (2011) The semismooth newton method for multicomponent reactive transport with minerals. Adv Water Resour 34:137–151

Kräutle S, Knabner P (2007) A reduction scheme for coupled multicomponent transport-reaction problems in porous media: Generalization to problems with heterogeneous equilibrium reactions. Water Resour Res 43:W03429

Krevor SC, Pini R, Zuo L, Benson SM (2012) Relative permeability, trapping of CO_2 and water in sandstone rocks at reservoir conditions. Water Resour Res 48:W02532

Krevor S, Blunt MJ, Benson SM, Pentland CH, Reynolds C, Al-Menhali A, Niu B (2015) Capillary trapping for geologic carbon dioxide storage–from pore scale physics to field scale implications. Int J Greenhouse Gas Control 40:221–237

Kulik D, Berner U, Curti E (2004) Modelling chemical equilibrium partitioning with the GEMS-PSI code. PSI Scientific Report 2003 Volume IV, Nuclear Energy and Safety, Paul Scherer Institute

Kuo C-W, Benson SM (2015) Numerical and analytical study of effects of small scale heterogeneity on CO_2/brine multiphase flow system in horizontal corefloods. Adv Water Resour 79:1–17

Lagneau V, van der Lee J (2010) Operator-splitting-based reactive transport models in strong feedback of porosity change: The contribution of analytical solutions for accuracy validation and estimator improvement. J Contam Hydrol 112:118–129

Lake LW (1989) Enhanced Oil Recovery. Prentice-Hall Inc., Englewood Cliffs

Land CS (1968) Calculation of imbibition relative permeability for two- and three-phase flow from rock properties. SPE J 8:149–156

Lauser A, Hager C, Helmig R, Wohlmuth B (2011) A new approach for phase transitions in miscible multi-phase flow in porous media. Adv Water Resour 34:957–966

Le Gallo Y, Trenty L, Michel A, Vidal-Gilbert S, Parra T, Jeannin L (2006) Long-term flow simulations of CO_2 storage in saline aquifer. *In:* Proc GHGT8 Confer, Trondheim (Norway), p 18–22

Leal AM, Blunt MJ, LaForce TC (2013) A robust and efficient numerical method for multiphase equilibrium calculations: Application to CO_2–brine–rock systems at high temperatures, pressures and salinities. Adv Water Resour 62:409–430

Leal AM, Blunt MJ, LaForce TC (2014) Efficient chemical equilibrium calculations for geochemical speciation and reactive transport modelling. Geochim Cosmochim Acta 131:301–322

Lee H-H, Kim H-J, Shi Y, Keffer D, Lee C-H (2013) Competitive adsorption of CO_2/CH_4 mixture on dry and wet coal from subcritical to supercritical conditions. Chem Eng J 230:93–101

Lei H, Li J, Li X, Jiang Z (2016) Numerical modeling of co-injection of N_2 and O_2 with CO_2 into aquifers at the Tongliao CCS site. Int J Greenhouse Gas Control 54:228–241

Leverett MC (1939) Flow of oil–water mixtures through unconsolidated sands. Trans AIME 132:149–171

Leverett MC (1941) Capillary behavior in porous solids. Trans AIME 142:152–169

Lewis GN, Randall M (1923) Thermodynamics and the Free Energy of Chemical Substances. McGraw-Hill, New York

Li D, Duan Z (2007) The speciation equilibrium coupling with phase equilibrium in the H_2O–CO_2–NaCl system from 0 to 250 °C, from 0 to 1000 bar, from 0 to 5 molality of NaCl. Chem Geol, 244(3–4):730–751

Li D, Jiang Y (2014) A numerical study of the impurity effects of nitrogen and sulfur dioxide on the solubility trapping of carbon dioxide geological storage. Appl Energy 128:60–74

Li H, Yan J (2009a) Evaluating cubic equations of state for calculation of vapor–liquid equilibrium of CO_2 and CO_2-mixtures for CO_2 capture and storage processes. Appl Energy 86:826–836

Li H, Yan J (2009b) Impacts of equations of state (EOS), impurities on the volume calculation of CO_2 mixtures in the applications of CO_2 capture and storage (CCS) processes. Appl Energy 86:2760–2770

Li Z, Galindo-Torres S, Yan G, Scheuermann A, Li L (2018) A lattice Boltzmann investigation of steady-state fluid distribution, capillary pressure and relative permeability of a porous medium: Effects of fluid and geometrical properties. Adv Water Resour 116:153–166

Lichtner P (1996) Continuum formulation of multicomponent–multiphase reactive transport. Rev Mineral 34:1–81

Lichtner P, Hammond G, Lu C, Karra S, Bisht G, Andre B, Mills R, Kumar J (2015) PFLOTRANUser Manual: A massively parallel reactive flow and transport model for describing surface and subsurface processes. Technical Report LA-UR-15-20403, LANL, Los Alamos NM

Lindeberg E, Wessel-Berg D (1997) Vertical convection in an aquifer column under a gas cap of CO_2. Energy Convers Manage 38:S229– S234

Liu Q, Cheng Y, Zhou H, Guo P, An F, Chen H (2015) A mathematical model of coupled gas flow and coal deformation with gas diffusion and Klinkenberg effects. Rock Mech Rock Eng 48:1163–1180

Liu Y, Hou M, Ning H, Yang D, Yang G, Han B (2012) Phase equilibria of $CO_2 + N_2 + H_2O$ and $N_2 + CO_2 + H_2O + NaCl + KCl + CaCl_2$ systems at different temperatures and pressures. J Chem Eng Data 57:1928–1932

Louk K, Ripepi N, Luxbacher K, Gilliland E, Tang X, Keles C, Schlosser C, Diminick E, Keim S, Amante J, Michael K (2017) Monitoring CO_2 storage and enhanced gas recovery in unconventional shale reservoirs: Results from the Morgan county, Tennessee injection test. J Nat Gas Sci Eng 45:11–25

Lu C, Lichtner PC (2005) PFLOTRAN: Massively parallel 3-D simulator for CO_2 sequestration in geologic media. *In:* DOE-NETL Fourth Ann Confer Carbon Capture and Sequestration

Lu C, Lichtner PC (2007) High resolution numerical investigation on the effect of convective instability on long term CO_2 storage in saline aquifers. J Physics: Confer Ser 78:012042

Lu C, Lichtner PC, Hammond GE, Mills RT (2010) Evaluating variable switching and flash methods in modeling carbon sequestration in deep geologic formations using PFLOTRAN. Proc SciDAC, p 11–15

Lu J, Mickler PJ, Nicot J-P, Yang C, Romanak KD (2014) Geochemical impact of oxygen on siliciclastic carbon storage reservoirs. Int J Greenhouse Gas Control 21:214–231

Lu J, Mickler PJ, Nicot J-P, Yang C, Darvari R (2016) Geochemical impact of O_2 impurity in CO_2 stream on carbonate carbon-storage reservoirs. Int J Greenhouse Gas Control 47:159–175

Luo C, Zhang D, Lun Z, Zhao C, Wang H, Pan Z, Li Y, Zhang J, Jia S (2019) Displacement behaviors of adsorbed coalbed methane on coals by injection of SO_2/CO_2 binary mixture. Fuel 247:356–367

MacMinn CW, Neufeld JA, Hesse MA, Huppert HE (2012) Spreading, convective dissolution of carbon dioxide in vertically confined, horizontal aquifers. Water Resour Res 48:W11516

Mason EA, Malinauskas A (1983) Gas Transport in Porous Media: The Dusty-Gas Model. Chemical Engineering Monographs Volume 17. Elsevier Science

Masson R, Trenty L, Zhang Y (2014) Formulations of two phase liquid gas compositional Darcy flows with phase transitions. Int J Finite Volumes 11:34

Mayer KU, Frind EO, Blowes DW (2002) Multicomponent reactive transport modeling in variably saturated porous media using a generalized formulation for kinetically controlled reactions. Water Resour Res 38:13–1

Mayer K, Amos R, Molins S, Gérard F (2012) Reactive transport modeling in variably saturated media with MIN3P: Basic model formulation and model enhancements. Groundwater Reactive Transport Models, p 186–211

McCain Jr. W (1991) Reservoir-fluid property correlations—state of the art. SPE Reservoir Eng 6:266–272

Messabeb H, Contamine F, Cézac P, Serin JP, Gaucher EC (2016) Experimental measurement of CO_2 solubility in aqueous NaCl solution at temperature from 323.15 to 423.15 K and pressure of up to 20 MPa. J Chem Eng Data 61:3573–3584

Michelsen ML, Mollerup J (2007) Thermodynamic Models: Fundamentals & Computational Aspects. Tie-Line Publications, Denmark, 2 edition

Millington R, Quirk J (1961) Permeability of porous solids. Trans Faraday Soc 57:1200–1207

Molins S, Knabner P (2019) Multiscale approaches in reactive transport modeling. Rev Mineral Geochem 85:27–48

Molins S, Mayer K (2007) Coupling between geochemical reactions and multicomponent gas and solute transport in unsaturated media: A reactive transport modeling study. Water Resour Res 43:W05435

Molins S, Carrera J, Ayora C, Saaltink MW (2004) A formulation for decoupling components in reactive transport problems. Water Resour Res 40:W10301

Morel F, Hering J (1993) Principles and Applications of Aquatic Chemistry. New York: John Wiley & Sons

Mouche E, Hayek M, Mügler C (2010) Upscaling of CO_2 vertical migration through a periodic layered porous medium: The capillary-free and capillary-dominant cases. Adv Water Resour 33:1164–1175

Mualem Y (1976) A new model for predicting the hydraulic conductivity of unsaturated media. Water Resour Res 2:513–522

Mungan N (1966) Interfacial effects in immiscible liquid–liquid displacement in porous media. SPE J 6:247– 253

Nardi A, Idiart A, Trinchero P, de Vries LM, Molinero J (2014) Interface Comsol-PHREEQC (iCP), an efficient numerical framework for the solution of coupled multiphysics and geochemistry. Comput Geosci 69:10–21

Nazeri M, Chapoy A, Burgass R, Tohidi B (2017) Measured densities and derived thermodynamic properties of CO_2-rich mixtures in gas, liquid and supercritical phases from 273 K to 423 K and pressures up to 126 MPa. J Chem Thermodynam 111:157–172

Neufeld JA, Hesse MA, Riaz A, Hallworth MA, Tchelepi HA, Huppert HE (2010) Convective dissolution of carbon dioxide in saline aquifers. Geophys Res Lett 37:L22404

Neumann R, Bastian P, Ippisch O (2013) Modeling and simulation of two-phase two-component flow with disappearing nonwetting phase. Comput Geosci 17:139–149

Nghiem L, Sammon P, Grabenstetter J, Ohkuma H (2004) Modeling CO_2 storage in aquifers with a fully-coupled geochemical EOS compositional simulator. *In:* SPE /DOE symposium on improved oil recovery

Ni H, Boon M, Garing C, Benson SM (2019) Predicting CO_2 residual trapping ability based on experimental petrophysical properties for different sandstone types. Int J Greenhouse Gas Control 86:158–176

Nickalls RW (1993) A new approach to solving the cubic: Cardan's solution revealed. Math Gaz 77:354–359

Nordbotten JM, Celia MA, Bachu S, Dahle HK (2005) Semianalytical solution for CO_2 leakage through an abandoned well. Environ Sci Technol 39:602–611

Nordbotten JM, Kavetski D, Celia MA, Bachu S (2008) Model for CO_2 leakage including multiple geological layers and multiple leaky wells. Environ Sci Technol 43:743–749

Nordbotten JM, Dahle HK (2011) Impact of the capillary fringe in vertically integrated models for CO_2 storage. Water Resour Res 47:W02537

Nordbotten JM, Flemisch B, Gasda SE, Nilsen HM, Fan Y, Pickup GE, Wiese B, Celia MA, Dahle HK, Eigestad GT, Pruess K (2012) Uncertainties in practical simulation of CO_2 storage. Int J Greenhouse Gas Control 9:234–242

Olivella S, Gens A, Carrera J, Alonso E (1996) Numerical formulation for a simulator (CODEBRIGHT) for the coupled analysis of saline media. Eng Comput 13:87–112

Øren P, Ruspini L, Saadatfar M, Sok R, Knackstedt M, Herring A (2019) In-situ pore-scale imaging and image-based modelling of capillary trapping for geological storage of CO_2. Int J Greenhouse Gas Control 87:34–43

Ottiger S, Pini R, Storti G, Mazzotti M (2008) Competitive adsorption equilibria of CO_2, CH_4 on a dry coal. Adsorption 14:539–556

Oudinot AY, Riestenberg DE, Koperna, Jr GJ (2017) Enhanced gas recovery and CO_2 storage in coal bed methane reservoirs with N_2 co-injection. Energy Procedia 114:5356–5376

Pal M, Edwards MG (2011) Anisotropy favoring triangulation CVD (MPFA) finite-volume approximations. Int J Numer Methods Fluids 67:1247–1263

Pan Z, Connell LD (2012) Modelling permeability for coal reservoirs: a review of analytical models and testing data. Int J Coal Geol 92:1–44

Parkhurst DL, Appelo C (2013) Description of input, examples for PHREEQC version 3: a computer program for speciation, batch-reaction, one-dimensional transport, and inverse geochemical calculations. US Geological Survey, (No. 6-A43)

Parkhurst DL, Wissmeier L (2015) PhreeqcRM: A reaction module for transport simulators based on the geochemical model PHREEQC. Adv Water Resour 83:176–189

Paterson L, Boreham C, Bunch M, Dance T, Ennis-King J, Freifeld B, Haese R, Jenkins C, LaForce T, Raab M, Singh R (2013) Overview of the CO2CRC Otway residual saturation and dissolution test. Energy Procedia 37:6140–6148

Pau GS, Bell JB, Pruess K, Almgren AS, Lijewski MJ, Zhang K (2010) High-resolution simulation and characterization of density-driven flow in CO_2 storage in saline aquifers. Adv Water Resour 33:443–455

Peaceman DW (1977) Fundamentals of Numerical Reservoir Simulation. Developments in Petroleum Science Vol 6, Elsevier

Pearce JK, Kirste DM, Dawson GK, Farquhar SM, Biddle D, Golding SD, Rudolph V (2015) SO_2 impurity impacts on experimental and simulated CO_2–water–reservoir rock reactions at carbon storage conditions. Chem Geol 399:65–86

Pearce JK, Dawson GK, Law AC, Biddle D, Golding SD (2016a) Reactivity of micas and cap-rock in wet supercritical CO_2 with SO_2 and O_2 at CO_2 storage conditions. Appl Geochem 72:59–76

Pearce JK, Golab A, Dawson GK, Knuefing L, Goodwin C, Golding SD (2016b) Mineralogical controls on porosity and water chemistry during O_2–SO_2–CO_2 reaction of CO_2 storage reservoir and cap-rock core. Appl Geochem 75:152–168

Péneloux A, Rauzy E, Fréze R (1982) A consistent correction for Redlich-Kwong-Soave volumes. Fluid Phase Equilib 8:7–23

Peng C, Crawshaw JP, Maitland GC, Trusler JM, Vega-Maza D (2013) The pH of CO_2-saturated water at temperatures between 308 K and 423 K at pressures up to 15 MPa. J Supercrit Fluids 82:129–137

Peng D-Y, Robinson D (1976) A new two-constant equation of state. Ind Eng Chem Fundam 15:59–64

Pentland CH, El-Maghraby R, Iglauer S, Blunt MJ (2011) Measurements of the capillary trapping of supercritical carbon dioxide in Berea sandstone. Geophys Res Lett 38:L06401

Peszynska M, Sun S (2002) Reactive transport model coupled to multiphase flow models. *In:* Computational Methods in Water Resources SM Hassanizadeh RJ Schotting WG Gray, GF Pinder (Eds.) Elsevier, p. 923–930

Pini R, Benson SM (2013) Simultaneous determination of capillary pressure and relative permeability curves from core-flooding experiments with various fluid pairs. Water Resour Res 49:3516–3530

Pini R, Benson SM (2017) Capillary pressure heterogeneity and hysteresis for the supercritical CO_2/water system in a sandstone. Adv Water Resour 108:277–292

Pini R, Krevor SC, Benson SM (2012) Capillary pressure and heterogeneity for the CO_2/water system in sandstone rocks at reservoir conditions. Adv Water Resour 38:48–59

Pitzer KS (1973) Thermodynamics of electrolytes. I. Theoretical basis and general equations. J Phys Chem 77:268–277

Pitzer K (1991) Ion interaction approach: theory and data correlation. Chapter 3 *In:* Activity Coefficients in Electrolyte solutions 2nd Edition, p 75–153. CRC Revivals

Pope GA, Sepehrnoori K, Delshad M (2005) A new generation chemical flooding simulator. Technical report, Center for Petroleum and Geosystems Engineering, The University of Texas at Austin

Pruess K, Spycher N (2007) ECO2n–a fluid property module for the Tough2 code for studies of CO_2 storage in saline aquifers. Energy Convers Manage 48:1761–1767

Pruess K, Garcia J, Kovscek T, Oldenburg C, Rutqvist J, Steefel C, Xu T (2004) Code intercomparison builds confidence in numerical simulation models for geologic disposal of CO_2. Energy 29:1431–1444

Qian J-W, Privat R, Jaubert J-N (2013) Predicting the phase equilibria, critical phenomena, mixing enthalpies of binary aqueous systems containing alkanes, cycloalkanes, aromatics, alkenes, and gases (N_2, CO_2, H_2S, H_2) with the PPR78 equation of state. Ind Eng Chem Res 52:16457–16490

Qin J, Rosenbauer RJ, Duan Z (2008) Experimental measurements of vapor–liquid equilibria of the H_2O+ CO_2+CH_4 ternary system. J Chem Eng Data 53:1246–1249

Raad SMJ, Hassanzadeh H (2016) Does impure CO_2 impede or accelerate the onset of convective mixing in geological storage? Int J Greenhouse Gas Control 54:250–257

Rabinovich A, Itthisawatpan K, Durlofsky LJ 2015) Upscaling of CO_2 injection into brine with capillary heterogeneity effects. J Pet Sci Eng 134:60–75

Ramakrishnan T, Cappiello A (1991) A new technique to measure static and dynamic properties of a partially saturated porous medium. Chem Eng Sci 46:1157–1163

Raviart P-A, Thomas J-M (1977) A mixed finite element method for 2nd order elliptic problems. *In:* Mathematical aspects of finite element methods, p 292–315. Springer

Redlich O, Kwong JN 1949) On the thermodynamics of solutions. V. An equation of state. Fugacities of gaseous solutions. Chem Rev 44:233–244

Reynolds C, Krevor S (2015) Characterizing flow behavior for gas injection: Relative permeability of CO_2–brine and N_2–water in heterogeneous rocks. Water Resour Res 51:9464–9489

Riaz A, Hesse M, Tchelepi H, Orr F (2006) Onset of convection in a gravitationally unstable diffusive boundary layer in porous media. J Fluid Mech 548:87–111

Robinson DB, Peng D-Y (1978) The characterization of the heptanes and heavier fractions for the GPA Peng–Robinson programs. Gas Processors Assoc. Technical report

Rumpf B, Nicolaisen H, Öcal C, Maurer G (1994) Solubility of carbon dioxide in aqueous solutions of sodium chloride: experimental results and correlation. J Solution Chem 23:431–448

Ruprecht C, Pini R, Falta R, Benson S, Murdoch L (2014) Hysteretic trapping and relative permeability of CO_2 in sandstone at reservoir conditions. Int J Greenhouse Gas Control 27:15–27

Saad Y (2003) Iterative Methods for Sparse Linear Systems. SIAM, Philadelphia

Saad Y, Schultz MH (1986) GMRES: A generalized minimal residual algorithm for solving nonsymmetric linear systems. SIAM J Sci Stat Comput 7:856–869

Saaltink MW, Ayora C, Carrera J (1998) A mathematical formulation for reactive transport that eliminates mineral concentrations. Water Resour Res 34:1649–1656

Saaltink MW, Vilarrasa V, De Gaspari F, Silva O, Carrera J, Rötting TS (2013) A method for incorporating equilibrium chemical reactions into multiphase flow models for CO_2 storage. Adv Water Resour62:431–441

Sanchez-Vicente Y, Drage TC, Poliakoff M, Ke J, George MW (2013) Densities of the carbon dioxide+hydrogen, a system of relevance to carbon capture and storage. Int J Greenhouse Gas Control 13:78–86

Seigneur N, Mayer KU, Steefel CI (2019) Reactive transport in evolving porous media. Rev Mineral Geochem 85:197–238

Settari A, Aziz K (1975) Treatment of nonlinear terms in the numerical solution of partial differential equations for multiphase flow in porous media. Int J Multiphase Flow 1:817–844

Sevougian SD, Schechter RS, Lake LW (1993) Effect of partial local equilibrium on the propagation of precipitation/dissolution waves. Ind Eng Chem Res 32:2281–2304

Sin I (2015) Numerical simulation of compressible two-phase flow and reactive transport in porous media. Applications to the study of CO_2 storage and natural gas reservoirs. PhD thesis, MINES ParisTech

Sin I, Corvisier J (2018) Impact of co-injected impurities on hydrodynamics of CO_2 injection. Studying interplayed chromatographic partitioning and density driven flow and fate of the injected mixed gases: numerical and experimental results. Proc 14th Greenhouse Gas Control Technologies Conference, Melbourne

Sin I, Lagneau V, Corvisier J (2017a) Integrating a compressible multicomponent two-phase flow into an existing reactive transport simulator. Adv Water Resour 100:62–77

Sin I, Lagneau V, De Windt L, Corvisier J (2017b) 2D simulation of natural gas reservoir by two-phase multicomponent reactive flow and transport—description of a benchmarking exercise. Math Comput Simul 137:431–447

Siriwardane HJ, Bowes BD, Bromhal GS, Gondle RK, Wells AW, Strazisar BR (2012) Modeling of CBM production, CO_2 injection, and tracer movement at a field CO_2 sequestration site. Int J Coal Geol 96:120–136

Snippe J, Gdanski R, Ott H (2017) Multiphase modelling of wormhole formation in carbonates by the injection of CO_2. Energy Procedia 114:2972–2984

Soave G (1972) Equilibrium constants from a modified Redlich–Kwong equation of state. Chem Eng Sci 27:1197–1203

Søreide I, Whitson CH (1992) Peng-Robinson predictions for hydrocarbons, CO_2, N_2, and H_2S with pure water and NaCl brine. Fluid Phase Equilib 77:217–240

Spiteri EJ, Juanes R, Blunt MJ, Orr FM (2008) A new model of trapping and relative permeability hysteresis for all wettability characteristics. SPE J 13:277–288

Spycher N, Pruess K (2005) CO_2–H_2O mixtures in the geological sequestration of CO_2. II. Partitioning in chloride brines at 12–100 °C and up to 600 bar. Geochim Cosmochim Acta 69:3309–3320

Spycher N, Pruess K, Ennis-King J (2003) CO_2–H_2O mixtures in the geological sequestration of CO_2. I assessment and calculation of mutual solubilities from 12 to 100 °C and up to 600 bar. Geochim Cosmochim Acta 67:3015–3031

Steefel C (2009) CrunchFlow software for modeling multicomponent reactive flow and transport. User's manual. Earth Sciences Division. Lawrence Berkeley, National Laboratory, Berkeley, CA. October, p 12–91

Steefel CI (2019) Reactive transport at the crossroads. Rev Mineral Geochem 85:1–26

Steefel C, MacQuarrie KT (1996) Approaches to modeling of reactive transport in porous media. Rev Mineral 34:83–129

Steefel C, Appelo CAJ, Arora B, Jacques D, Kalbacher T, Kolditz O, Lagneau V, Lichtner P, Mayer K, Meeussen JCL (2014) Reactive transport codes for subsurface environmental simulation. Comput Geosci 19:445–478

Sterpenich J, Dubessy J, Pironon J, Renard S, Caumon MC, Randi A, Jaubert JN, Favre E, Roizard D, Parmentier M, Azaroual M (2013) Role of impurities on CO_2 injection: experimental, numerical simulations of thermodynamic properties of water-salt-gas mixtures (CO_2+co-injected gases) under geological storage conditions. Energy Procedia 37:3638–3645

Stevenson M, Pinczewski W, Somers M, Bagio S (1991) Adsorption/desorption of multicomponent gas mixtures at in-seam conditions. *In:* SPE Asia-Pacific Conference, 4–7 November, Perth. Australia

Suarez D, Simunek J (1996) Solute transport modeling under variably saturated water flow conditions. Rev Mineral Geochem 34:229–268

Sun H, Yao J, Gao S-H, Fan D-Y, Wang C-C., Sun Z-X (2013) Numerical study of CO_2 enhanced natural gas recovery and sequestration in shale gas reservoirs. Int J Greenhouse Gas Control 19:406–419

Tabasinejad F, Moore RG, Mehta SA, Van Fraassen KC, Barzin Y, Rushing JA, Newsham KE (2011) Water solubility in supercritical methane, nitrogen, and carbon dioxide: measurement and modeling from 422 to 483 K and pressures from 3.6 to 134 MPa. Ind Eng Chem Res 50:4029–4041

Thomas G, Thurnau D (1983) Reservoir simulation using an adaptive implicit method. SPE J 23:759–768

Thomas C, Loodts V, Rongy L, De Wit A (2016) Convective dissolution of CO_2 in reactive alkaline solutions: Active role of spectator ions. Int J Greenhouse Gas Control 53:230–242

Thomas C, Dehaeck S, De Wit A (2018) Convective dissolution of CO_2 in water and salt solutions. Int J Greenhouse Gas Control 72:105–116

Todaka N, Xu T (2017) Reactive transport simulation to assess geochemical impact of impurities on CO_2 injection into siliciclastic reservoir at the Otway site, Australia. Int J Greenhouse Gas Control 66:177–189

Tournassat C, Steefel CI (2019) Reactive transport modeling of coupled processes in nanoporous media. Rev Mineral Geochem 85:75–109

Trenty L, Michel A, Tillier E, Le Gallo Y (2006) A sequential splitting strategy for CO_2 storage modelling. *In:* ECMORX-10th European Conference on the Mathematics of Oil Recovery. Amsterdam, The Netherlands, 4–7 September 2006

Trevisan L, Pini R, Cihan A, Birkholzer JT, Zhou Q, Illangasekare TH (2015) Experimental analysis of spatial correlation effects on capillary trapping of supercritical CO_2 at the intermediate laboratory scale in heterogeneous porous media. Water Resour Res 51:8791–8805

Ussiri DA, Lal R (2017) Carbon Sequestration for Climate Change Mitigation and Adaptation. Springer

Valocchi AJ, Malmstead M (1992) Accuracy of operator splitting for advection-dispersion-reaction problems. Water Resour Res 28:1471–1476

van der Lee J (2009) Thermodynamic and mathematical concepts of CHESS Technical Report RT-20093103-JVDL, École des Mines de Paris, Centre de Géosciences, Fontainebleau, France

van der Lee J, de Windt L, Lagneau V, Goblet P (2003) Module-oriented modeling of reactive transport with HYTEC. Comput Geosci 29:265–275

van Genuchten M (1980) A closed form equation for predicting the hydraulic conductivity of unsaturated soils. Soil Sci Soc Am J 44:892–898

Virnovsky G, Friis H, Lohne A (2004) A steady-state upscaling approach for immiscible two-phase flow. Transp Porous Media 54:167–192

Vishal V, Mahanta B, Pradhan S, Singh T, Ranjith P (2018) Simulation of CO_2 enhanced coalbed methane recovery in Jharia coalfields, India. Energy 159:1185–1194

Voskov D (2012) An extended natural variable formulation for compositional simulation based on tie-line parameterization. Transp Porous Media 92:541–557

Voskov DV (2017) Operator-based linearization approach for modeling of multiphase multicomponent flow in porous media. J Comput Phys 337:275–288

Voskov DV, Tchelepi HA (2009) Compositional space parameterization: theory and application for immiscible displacements. SPE J 14:431–440

Voskov DV, Tchelepi HA (2012) Comparison of nonlinear formulations for two-phase multi-component EOS based simulation. J Pet Sci Eng 82:101–111

Waldmann S, Rütters H (2016) Geochemical effects of SO_2 during CO_2 storage in deep saline reservoir sandstones of Permian age (Rotliegend)–a modeling approach. Int J Greenhouse Gas Control 46:116–135

Wang G, Ren T, Wang K, Zhou A (2014) Improved apparent permeability models of gas flow in coal with Klinkenberg effect. Fuel 128:53–61

Wang K, Xu T, Wang F, Tian H (2016) Experimental study of CO_2–brine–rock interaction during CO_2 sequestration in deep coal seams. Int J Coal Geol 154:265–274

Wang L, Nakanishi Y, Hyodo A, Suekane T (2017) Three-dimensional finger structure of natural convection in homogeneous and heterogeneous porous medium. Energy Procedia 114:5048–5057

Wei L (2012) Sequential coupling of geochemical reactions with reservoir simulations for waterflood and EOR studies. SPE J 17:469–484

Wei N, Li X, Wang Y, Zhu Q, Liu S, Liu N, Su X (2015) Geochemical impact of aquifer storage for impure CO_2 containing O_2 and N_2: Tongliao field experiment. Appl Energy 145:198–210

Wen B, Ahkbari D, Zhang L, Hesse MA (2018) Dynamics of convective carbon dioxide dissolution in a closed porous media system. arXiv:1801.02537

Wheeler M, Sun S, Thomas S (2012) Modeling of Flow and Reactive Transport in IPARS. Bentham Science Publishers Ltd

Wheeler MF, Yotov I (2006) A multipoint flux mixed finite element method. SIAM J Numer Anal 44:2082–2106

Whitaker S (1986) Flow in porous media II: The governing equations for immiscible, two-phase flow. Transp Porous Media 1:105–125

White M, Bacon D, McGrail B, Watson D, White S, Zhang Z (2012) STOMPSubsurface Transport Over Multiple Phases: STOMP-CO2 and STOMP-CO2e Guide: Version 1.0. Pacific Northwest National Laboratory, Richland, WA, PNNL-21268 edition

White M, Oostrom M (2006) STOMP Subsurface Transport Over Multiple Phases, Version 4.0, User's Guide. Pacific Northwest National Laboratory, Richland, WA, PNNL-15782 edition

Whitson CH, Michelsen ML (1989) The negative flash. Fluid Phase Equilib 53:51–71

Wiebe R, Gaddy V (1939) The solubility in water of carbon dioxide at 50, 75 and 100, at pressures to 700 atmospheres. J Am Chem Soc 61:315–318

Wiebe R, Gaddy V (1941) Vapor phase composition of carbon dioxide–water mixtures at various temperatures and at pressures to 700 atmospheres. J Am Chem Soc 63:475–477

Wolf JL, Niemi A, Bensabat J, Rebscher D (2016) Benefits and restrictions of 2D reactive transport simulations of CO_2 and SO_2 co-injection into a saline aquifer using TOUGHREACT V3.0-OMP. Int J Greenhouse Gas Control 54:610–626

Xiao Y, Xu T, Pruess K (2009) The effects of gas-fluid-rock interactions on CO_2 injection and storage: Insights from reactive transport modeling. Energy Procedia 1:1783–1790

Xu R, Li R, Ma J, He D, Jiang P (2017) Effect of mineral dissolution/precipitation and CO_2 exsolution on CO_2 transport in geological carbon storage. Acc Chem Res 50:2056–2066

Xu T, Apps JA, Pruess K, Yamamoto H (2007) Numerical modeling of injection, mineral trapping of CO_2 with H_2S and SO_2 in a sandstone formation. Chem Geol 242:319–346

Xu T, Pruess K (1998) Coupled modeling of non-isothermal multiphase flow, solute transport and reactive chemistry in porous and fractured media: 1. Model development and validation. Lawrence Berkeley National Laboratory

Xu T, Sonnenthal E, Spycher N, Zheng L (2014) TOUGHREACT V3.0-OMP Reference Manual: AParallel Simulation Program for Non-isothermal Multiphase Geochemical Reactive Transport. Lawrence Berkeley National Laboratory, University of California, Berkeley, USA

Xu T, Spycher N, Sonnenthal E, Zheng L, Pruess K (2012) TOUGHREACT user's guide: A simulation program for non-isothermal multiphase reactive transport in variably saturated geologic media, version 2.0. Earth Sciences Division, Lawrence Berkeley National Laboratory, Berkeley, USA

Xu X, Chen S, Zhang D (2006) Convective stability analysis of the long-term storage of carbon dioxide in deep saline aquifers. Adv Water Resour 29:397–407

Yan B, Wang Y, Killough JE (2016) Beyond dual-porosity modeling for the simulation of complex flow mechanisms in shale reservoirs. Comput Geosci 20:69–91

Yaws CL (1999) Chemical Properties Handbook. McGraw-Hill

Yeh G-T, Sun J, Jardine P, Burgos W, Fang Y, Li M, Siegel M (2004) HYDROGEOCHEM 5.0: A three-dimensional model of coupled fluid flow, thermal transport, and hydrogeochemical transport through variably saturated conditions. Version 5.0. ORNL/TM-2004/107, Oak Ridge National Laboratory, Oak Ridge, TN

Yeh G-T, Tripathi VS (1989) A critical evaluation of recent developments in hydrogeochemical transport models of reactive multichemical components. Water Resour Res 25:93–108

Yeh G-T, Tripathi VS (1991) A model for simulating transport of reactive multi-species components: model development and demonstration. Water Resour Res 27:3075–3094

Yeh G-T, Tripathi VS, Gwo J, Cheng H, Cheng J, Salvage K, Li M, Fang Y, Li Y, Sun J, Zhang F, Siegel MD (2012) HYDROGEOCHEM: A coupled model of variably saturated flow, thermal transport, and reactive biogeochemical transport. Groundwater React Transp Models:3–41

Zhang Y, Kogure T, Chiyonobu S, Lei X, Xue Z (2013) Influence of heterogeneity on relative permeability for CO_2/brine: CT observations and numerical modeling. Energy Procedia 37:4647–4654

Zhang Y, Nishizawa O, Park H, Kiyama T, Xue Z (2017) Relative permeability of CO_2 in a low-permeability rock: Implications for CO_2 flow behavior in reservoirs with tight interlayers. Energy Procedia 114:4822–4831

Zhao H, Fedkin MV, Dilmore RM, Lvov SN (2015) Carbon dioxide solubility in aqueous solutions of sodium chloride at geological conditions: Experimental results at 323.15, 373.15, and 423.15 K and 150 bar and modeling up to 573.15 K and 2000 bar. Geochim Cosmochim Acta 149:165–189

Zhu W, Liu J, Sheng J, Elsworth D (2007) Analysis of coupled gas flow and deformation process with desorption and Klinkenberg effects in coal seams. Int J Rock Mech Min Sci 44:971–980

Ziabakhsh-Ganji Z, Kooi H (2012) An equation of state for thermodynamic equilibrium of gas mixtures and brines to allow simulation of the effects of impurities in subsurface CO_2 storage. Int J Greenhouse Gas Control 11:S21–S34

Ziabakhsh-Ganji Z, Kooi H (2014) Sensitivity of the CO_2 storage capacity of underground geological structures to the presence of SO_2 and other impurities. Appl Energy 135:43–52

Zirrahi M, Azin R, Hassanzadeh H, Moshfeghian M (2012) Mutual solubility of CH_4, CO_2, H_2S, and their mixtures in brine under subsurface disposal conditions. Fluid Phase Equilib 324:80–93

Zuo L, Ajo-Franklin JB, Voltolini M, Geller JT, Benson SM (2017) Pore-scale multiphase flow modeling and imaging of CO_2 exsolution in sandstone. J Pet Sci Eng 155:63–77

Zuo L, Zhang C, Falta RW, Benson SM (2013) Micromodel investigations of CO_2 exsolution from carbonated water in sedimentary rocks. Adv Water Res 53:188–197

Reviews in Mineralogy & Geochemistry
Vol. 85 pp. 197-238, 2019
Copyright © Mineralogical Society of America

7

Reactive Transport in Evolving Porous Media

Nicolas Seigneur, K. Ulrich Mayer

Department of Earth, Ocean and Atmospheric Sciences
University of British Columbia
2020-2207 Main Mall
Vancouver, British Columbia, V6T 1Z4
Canada

nseigneur@eoas.ubc.ca; umayer@eoas.ubc.ca

Carl I. Steefel

Energy Geosciences Division
Lawrence Berkeley National Laboratory
1 Cyclotron Road, Berkeley, CA 94720
USA

CISteefel@lbl.gov

INTRODUCTION

Reactive transport modeling is a process-based approach that accounts for advection, diffusion, dispersion and a multitude of biogeochemical reactions. The occurrence of these reactions, by nature, tends to affect the properties of porous media in many ways (Tenthorey and Gerald 2006). If these alteration reactions are significant, then feedback mechanisms could occur that influence the flow of groundwater as well as the migration of solutes and gases through porous media (Le Gallo et al. 1998; Kaszuba et al. 2005; Jin et al. 2013). In addition, changes induced by the reactions on the solid grains can also affect the rates of the reactions themselves (Hao et al. 2012; Harrison et al. 2017). A prime example for reactive transport in evolving porous media is the dissolution of mineral phases. If dissolution reactions are substantial, the porosity, i.e., the void space between grains or apertures of fractures in jointed rocks, will increase. Such an increase in porosity commonly has secondary effects, by altering the connectivity or larger scale pores in the porous medium under consideration (Navarre-Sitchler et al. 2009). Together, these changes in porosity and connectivity can substantially affect flow and transport processes by modifying the key transport parameters such as the medium's permeability and tortuosity, leading to alteration of the groundwater flow regime and modification of transport pathways. The impact of these changes can affect transport in the water phase as well as in the gas phase. In addition, because mineral dissolution reshapes the surface of the dissolving phases or leads to the complete dissolution of smaller particles, the system's reactivity can be affected as well, leading to a direct feedback on reaction progress and rates (Noiriel et al. 2009).

The dissolution of minerals is only one example for evolving porous media. Flow and transport properties can be affected by a multitude of other processes such as mineral precipitation, possibly leading to clogging (Gaucher and Blanc 2006; Brovelli et al. 2009; Chagneau et al. 2015) or modification of the pore size distribution (Emmanuel et al. 2010, 2015), microbial activity leading to the growth of biofilms (Kim and Fogler 2000; Ezeuko et al. 2011) or bioclogging (Thullner et al. 2002), swelling of clays (Herbert et al. 2008; Sedighi and Thomas 2014; Wang et al. 2014), and variations in surface loading and temperatures leading to the expansion or contraction of porous media and possibly the formation or closure

1529-6466/19/0085-0007$05.00 (print)
1943-2666/19/0085-0007$05.00 (online)

http://dx.doi.org/10.2138/rmg.2019.85.7

of fractures (Pfingsten 2002; MacQuarrie and Mayer 2005; Tian et al. 2014). In this chapter, we will focus on evolving porous media as affected by biogeochemical reactions with an emphasis on mineral dissolution and precipitation.

Specific examples where biogeochemical reactions result in the evolution of porous media include wellbore integrity studies. It has been shown that the productivity of wells can decrease due to clogging. Houben (2003) showed that chemical processes including corrosion and mineral precipitation account for more than 90% of ageing of wells, leading to a decrease in their productivity. In addition, geochemical interactions at interfaces between cementitious materials and clay can lead to the formation of an impermeable layer of calcite and other secondary minerals, effectively blocking solute transfer across the interfaces (Dauzères et al. 2010, 2019). These processes are important in the context of the safety assessment for the long-term storage of spent nuclear fuel (Atkinson et al. 1987; Spycher et al. 2003; Soler and Mäder 2005; Yang et al. 2008). An additional example where evolution of porous media is of importance is the long-term evolution of mine waste deposits. In these systems, the formation of alteration rims and secondary mineral formation has a strong impact on reactivity, leading to a decline of acid generation and metal release over time due to surface passivation (Blowes and Jambor 1990; Wunderly et al. 1996; Johnson et al. 2000). In extreme situations, extensive secondary mineral formation can lead to the formation of hard-pans (Blowes et al. 1991), with a pronounced effect on transport parameters. Furthermore, systems designed for the treatment of contaminated groundwater are also often affected by evolving transport and reactivity parameters. The precipitation of secondary mineral phases leads to passivation of carbonate minerals and limited pH-buffering capacity in limestone drains for the treatment of acid rock drainage (Hammarstrom et al. 2003; Rötting et al. 2008). Groundwater treatment by permeable reactive barriers (Blowes et al. 2000) or by bioremediation (Hazen and Fliermans 1995; Ritter and Scarborough 1995) is frequently affected by the growth of biofilms, which can have detrimental effects on the efficacy of the remediation effort and modifies permeability and transport parameters (Scherer et al. 2000; Li et al. 2006; Wantanaphong et al. 2006). The precipitation of secondary oxide phases in case of groundwater remediation via in-situ chemical oxidation with permanganate also may result in less effective contaminant treatment or contaminant encapsulation (Nelson et al. 2001; Henderson et al. 2009). Furthermore, evolving porous media plays a role in the clogging and passivation of permeable reactive barriers used for the treatment of chlorinated solvents and metals, which occurs due to pronounced changes in pH and redox conditions causing secondary mineral formation (Ouellet et al. 2006; Jeen et al. 2007).

Accounting for the evolution of porous media is not only of importance in man-made and engineered systems, but is equally relevant in natural environments, where transport and chemical properties can exhibit significant variation in space and time (Gouze and Coudrain-Ribstein 2002; White et al. 2005; Jin et al. 2013; Opolot and Finke 2015). A key example is the progress of weathering and soil formation. Also, karst formation caused by dissolution reactions of soluble minerals (predominantly carbonate minerals and gypsum) induced by the ingress of slightly acidic and dilute recharge. Birk et al. (2003, 2005) reported that the evolution of heterogeneity patterns strongly depends on flow and weathering rates. Karst development is obtained by positive feedback between reaction and flow (Hartmann et al. 2014): higher flow rates, lead to higher dissolution rates, resulting in higher porosity and permeability, in turn further increasing flow rates (Xiao et al. 2007). Kaufmann et al. (2010) showed that in karst systems permeability can vary by up to 9 orders of magnitude between fractures and the solid matrix. The morphology of the karst systems strongly depends on dissolution kinetics (Raines and Dewers 1997).

In addition, precipitation reactions in natural systems can have significant consequences for porous media evolution. Cementation (Giles et al. 2000; Putnis and Mauthe 2001; Pape et al. 2005) can reduce porosity and permeability, while strengthening the soil mechanical properties. It has also been shown that soil hydraulic properties can be decreased due to biogeochemical reactions causing bioclogging or biocementation (Ivanov and Chu 2008). Sedimentary rocks are often cemented due to the precipitation of clay minerals on the solid grain surfaces (Peters 2009). Such precipitation reactions can also have an impact on the reactivity of the primary minerals. Peters (2009) showed that while secondary minerals only occupied a volume fraction of 5-30%, they represented 65-86% of the reactive surface. Precipitation of secondary minerals can also lead to the sealing of fractures (Noiriel et al. 2010; Zhang 2011, 2013; Phillips et al. 2016). A decrease of permeability is often observed in certain hydrothermally altered rocks induced by mineral precipitation (Fontaine et al. 2001; Dobson et al. 2003a; Polak et al. 2003; Giger et al. 2007). Mineral precipitation can also lead to the formation of skarn (Meinert et al. 2003) or calcrete (Wang et al. 1994). In addition, several authors reported a correlation between chemical weathering and landslide occurrence (Jaboyedoff et al. 2004; Watanabe et al. 2005; Opfergelt et al. 2006; Regmi et al. 2013), implying that mechanical properties are evolving as chemical weathering progresses.

Despite the widespread relevance of reactive transport processes leading to changes in porous media properties, parameters such as porosity, permeability, effective diffusion coefficients and reactivity have often been considered constant in time in previous modeling efforts. The most common exception to this assumption has been in the consideration of the impact of changes in water content (Millington and Quirk 1961). Modeling studies of reaction induced porosity and permeability change date from the early 1990s, with Steefel and Lasaga (1992) presenting what appears to be the first numerical study of the reactive infiltration instability discussed by Ortoleva et al. (1987). In a later study of reactive transport in hydrothermal systems, Steefel and Lasaga (1994) analyzed the impact of quartz precipitation in fractures occurring as fluids cooled. They found that, unlike the reactive infiltration instability or "wormholing" effect due to dissolution of the rock matrix wherein reactive flow becomes increasingly focused in a smaller number of flow pathways, quartz precipitation had the effect of diffusing the flow paths over an increasingly wider volume, to the point where a convection cell might even reorganize completely. In another modeling study, Steefel and Lichtner (1994) showed how reactions in the rock matrix bordering a fracture could lead to "armoring" of the fracture, or isolation of the fracture volume from the surrounding rock matrix. Modeling was also used to show that the delicate balance between cementation of the rock matrix and the fracture itself can determine the subsequent evolution of the entire fractured rock system (Steefel and Lichtner 1998). The effect of evolving pore connectivity and tortuosity and their effect on diffusivity was considered in the modeling of chemical weathering of basaltic fragments (Navarre-Sitchler et al. 2011). Evolving reactivity has been less commonly treated in reactive transport models, except for the relatively straightforward case of dissolving (disappearing) phases where the reactive surface area necessarily must go to zero. However, one modeling study based on reactive flow experiments with calcite conducted in capillary tubes showed that it was necessary to consider the formation of high surface area precipitates in order to capture the evolving reactivity of the system (Noiriel et al. 2012). In a study of CO_2 injection into reactive volcanogenic sands, it was shown the use of a grain size distribution could capture at least part of the evolution of the reactivity of the system (Beckingham et al. 2017).

The simplification of the assumption of constant transport and reactivity parameters in modeling is due in part to the inherent difficulty of describing the temporal evolution of these parameters due to changes in porous media structure and composition. Traditionally, reactive transport has been simulated using upscaled parameters, i.e., on the continuum or Representative Elementary Volume (REV) scale. On this scale, relationships have been developed that describe interdependencies between transport parameters and porosity (e.g., Akanni et al. 1987; Hommel et al. 2018) as well as evolving surface area and reactivity. However, these relationships are commonly empirical in nature and have been developed based on fitting to experimental data (e.g., Low 1981; Tomadakis and Sotirchos 1993), or are based on theoretical considerations (e.g., Petersen 1958).

In reality, geochemical reactions and the processes controlling the evolution of porous media occur at the pore scale. Given the complexity and heterogeneity of the pore structure, the wide range of chemical perturbations and the coupling between these different processes, obtaining a representative evolution of the effective properties of the porous medium in a continuum approach can be a challenge. With the recent advent of improved imaging techniques such as X-ray microtomography (e.g., Werth et al. 2010), the development of improved pore-scale model formulations (e.g., Navarre-Sitchler et al. 2009; Dewanckele et al. 2012; Beckingham et al. 2013; Vilcáez et al. 2017), and substantial advances in computational power, the exploration of processes on the pore scale has become possible from both an experimental and modeling perspective. As a result, recent studies have focused more on describing these phenomena at the pore-scale (Molins et al. 2012, 2014; Steefel et al. 2013; Trebotich et al. 2014; Molins 2015; Deng et al. 2018). Although many advances have been made in pore scale investigations in recent years, it is difficult to simulate reactive transport on larger scales (on the orders of m's to km's) using pore scale approaches, although recently developed hybrid multi-scale approaches might handle this problem. For such studies, it is necessary to continue relying on continuum approaches. However, pore scale studies are providing deeper understanding and insights, which in turn can be incorporated to advance continuum approaches.

In this chapter, we will briefly review the governing equations of multiphase flow and reactive transport on the continuum scale, emphasizing the parameters that are affected by biogeochemical reactions in evolving porous media, and defining them from a pore scale perspective. We will then review recent studies carried out on the pore scale to provide insights and fundamental understanding on processes that affect and control porous media parameters. Subsequently, we will return to the continuum scale to provide a summary of the common approaches used to describe evolving flow, transport and reaction parameters such as porosity, permeability, tortuosity and reactive surface area. In this context, we review different approaches taken by the community to take these phenomena into considerations. We will then follow up with selected reactive transport modeling case studies, accounting for evolving properties in porous media. Finally, we will provide an outlook on future work and opportunities to work towards an integrated model for describing reactive transport in evolving porous media and challenges associated with filling the gap between pore scale and continuum approaches.

CONTINUUM SCALE GOVERNING EQUATIONS AND EVOLVING PARAMETERS

Governing equations

In porous media, variably saturated flow is modeled by the coupled two-phase flow equations (Xu et al. 2011), described by Sin et al. (2017) as:

$$\frac{\partial \phi S_\alpha \rho_\alpha}{\partial t} - \nabla \left[\frac{\rho_\alpha k_{r\alpha} \mathbf{k}}{\mu_\alpha} \left(\nabla p_\alpha - \rho_\alpha \mathbf{g} \right) \right] - \rho_\alpha Q_\alpha = 0 \qquad (\alpha = l, g) \tag{1}$$

where t [s] is time, ϕ is porosity [–], α denotes the fluid phase, either liquid (l) or gas (g), S_α is the saturation of the fluid phase α [–], $k_{r\alpha}$ is the phase relative permeability [–]. p_α is the fluid pressure [$kg\,m^{-1}\,s^{-2}$], and Q_α is a source-sink term for the fluid phase α [s^{-1}]. ρ_α is the fluid density [$kg\,m^{-3}$], g [$m\,s^{-2}$] is the gravitational acceleration, μ_α [$kg\,m^{-1}\,s^{-1}$] is the fluid dynamic viscosity and $\mathbf{k}$ is the intrinsic permeability tensor [m^2]. However, some of the reactive transport simulators do not model the coupled two-phase flow (Steefel et al. 2015a), but instead use a simplified approach through Richards' equation (Neuman 1973; Panday et al. 1993):

$$S_s S_l \frac{\partial h}{\partial t} + \phi \frac{\partial S_l}{\partial t} - \nabla \cdot \left[\frac{\rho_l g}{\mu_l} k_{rl} \mathbf{k} \nabla h \right] - Q_l = 0 \qquad (2)$$

where h is the hydraulic head [m] and S_s defines the specific storage coefficient [m^{-1}]. The latter approach does not consider gas advection.

In variably saturated media, the reactive transport equations in global implicit form can be written as (Lichtner 1996; Mayer et al. 2002)

$$\frac{\partial}{\partial t} \left[S_a \phi T_j^l \right] + \frac{\partial}{\partial t} \left[S_g \phi T_j^g \right] + \nabla \cdot \left[q_g T_j^g \right]$$
$$+ \nabla \cdot \left[q_l T_j^l \right] - \nabla \cdot \left[S_l \phi \mathbf{D}_l \nabla T_j^l \right] - \nabla \cdot \left[S_g \phi \mathbf{D}_g T_j^g \right] - Q_j^{ext} - Q_j^{int} = 0 \qquad j = 1, N_c \qquad (3)$$

where q_l and q_g are the Darcy flux vectors, representing the specific discharges in both phases [m s^{-1}]. For single phase flow formulation using Richards' equation (Eqn. 2), the gas specific discharge q_g is not considered. T_j^l [mol L^{-1} H$_2$O] and T_j^g [mol L^{-1} gas] define the total aqueous/gaseous component concentrations for the j-th component. Q_j^{int} [mol dm^{-3} porous medium] are internal source and sink terms due to all kinetically controlled reactions, including mineral dissolution–precipitation reactions. Q_j^{ext} [mol dm^{-3} porous medium] are external source and sink terms and define mass fluxes across the domain boundaries for the aqueous and gas phases and N_c defines the number of components. $\mathbf{D}_l$ and $\mathbf{D}_g$ are the hydrodynamic dispersion tensors [m^2 s^{-1}] for the aqueous and gas phases, respectively. The dispersion tensors include the contributions of molecular diffusion and mechanical dispersion, a process by which dissolved species migrate at different rates than the average linear groundwater velocity due the small-scale heterogeneities. In Equation (3), molecular diffusion is integrated into the hydrodynamic dispersion tensor in the form of the pore diffusion coefficient, defined as:

$$D_\alpha^p = \tau_\alpha D_\alpha^0 \qquad (4)$$

where D_α^p is the pore diffusion coefficient of phase α (aqueous or gas phase), τ_α is the tortuosity of phase α [–], and D_α^0 is the free phase diffusion coefficient in phase α. The tortuosity accounts for diffusion along tortuous pathways through the porous medium due to the presence of grains and partial phase saturation, leading to a reduction of the pore diffusion coefficient in comparison to the free phase diffusion coefficient. Following previous definitions, tortuosity is here defined as a value smaller than unity (as in Steefel and Maher 2009), capturing that the presence of tortuous pathways leads to a decline in the pore diffusion coefficient. However, it should be noted that in other studies, an inverse definition has been used, in which case tortuosity is proportional to the length of the tortuous diffusion pathway in relation to the direct pathway, therefore giving values greater than unity (Walsh and Brace 1984; Hommel et al. 2018). Using this alternative formulation, Equation (4) would have to be modified, since for any formulation, the pore diffusion coefficient is smaller than the free phase diffusion coefficient. For completeness, it must also be noted that the effective diffusion coefficient D_α^e, representative for the rate of solute or gas diffusion through a porous medium, is obtained by multiplying the pore diffusion coefficient (Eqn. 4) with the phase saturation and porosity, which is accounted for in Equation (3).

In addition to the mass balance equations for the aqueous and gas phase (Eqn. 3), it is also necessary to define mass balance equations for reactive minerals (Steefel and Lasaga 1994):

$$\frac{d\varphi_i}{dt} = V_i^m R_i^m \qquad i = 1, N_m \qquad (5)$$

where φ_i is the mineral volume fraction [dm^3 mineral dm^{-3} porous medium] of a mineral phase, V_i^m is the molar volume of the mineral [dm^3 mineral mol^{-1}], R_i^m is the mineral i overall dissolution–precipitation rate [mol dm^{-3} porous medium s^{-1}], and N_m is the number of mineral phases. The reaction rate R_i^m is commonly a function of solution composition including pH, but also the exposure of the mineral to the pore water, defining its intrinsic reactivity. Reactivity is commonly controlled by the reactive surface area. A mineral dissolution precipitation rate can therefore be expressed as:

$$R_i^m = S_i k_i f(C_j^l, pH) \qquad i = 1, Nm; \; j = 1, N_s \qquad (6)$$

where S_i [m^2 mineral dm^{-3} bulk] is the reactive surface area for the mineral i, k_i is the surface area normalized rate constant [mol m^{-2} mineral s^{-1}], C_j^l is the concentration of a chemical species in solution [mol L^{-1} H$_2$O] and N_s is the number of dissolved species.

GOVERNING PARAMETERS AND INTERDEPENDENCIES

Equations (1–6) include several parameters that can evolve in response to biogeochemical reactions, namely:

- porosity ϕ permeability tensor **k**;
- tortuosity τ_α;
- mineral reactive surface area S_i as a surrogate for reactivity.

Figure 1 depicts the dependencies of groundwater flow, transport processes and geochemical reactions on these parameters and illustrates how changes of these parameters lead to interdependencies and feedback mechanisms between the flow, transport and reaction domains. Since the dissolution of mineral phases will lead to a decline in the volume filled with the solid phase, an increase in porosity will result. The increase in porosity, in turn, will tend to cause changes in permeability and tortuosity. In addition, the reactive surface area may also be altered as a mineral progressively dissolves. Based on this analysis and considering Equations (1–6) together, it becomes clear that mineral dissolution has the potential to directly affect groundwater flow (both porosity ϕ and **k** are featured in Eqn. 1 and 2), advective solute/gas transport, since the specific discharges in Equation (3) depends on **k**, and diffusive solute and gas transport, since the diffusion coefficients are a function of τ_α (Eqn. 4). In addition, the fact that porosity is included in some terms of the multicomponent reactive transport Equation (4), but not in others, reveals the effect of changes in the water-rock ratios on the mass balance. Similar considerations are valid for precipitation reactions or the formation of biofilms. It becomes evident that biogeochemical processes have the potential to cause several feedback mechanisms, increasing

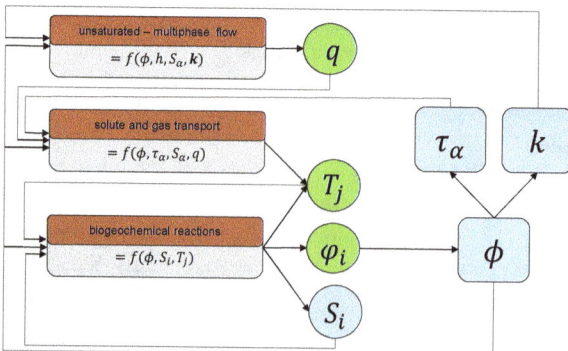

Figure 1. Schematic diagram outlining interactions between flow, transport and reaction processes, including feedback mechanisms.

the inter-dependencies of the equations governing flow and reactive transport (Fig. 1). In order to understand the evolution of these parameters at the continuum scale, it is instructive to review how biogeochemical processes control porous media evolution on the pore scale.

EVOLUTION OF POROUS MEDIA AT THE PORE SCALE

As most of the relevant physicochemical processes occur at the pore-scale, an understanding of processes that occur at this scale is essential for developing an appreciation on how flow, transport and reaction parameters evolve, how these processes and parameters are linked to each other, and how these parameters affect reactive transport both at the pore and continuum scales.

Porous media structure at the pore scale—Effective porosity and connectivity

With the advent of modern, non-invasive imaging techniques, it has become possible to visualize the structure of porous media on the pore scale. Werth et al. (2010) provide a review of these methods including X-Ray microtomography, Nuclear Magnetic Resonance Imaging (NMR) and optical imaging methods, as well as their applications in hydrogeology and reactive transport. Specifically, the use of 3D X-Ray microtomography has found increasing applications due to its high resolution. These imaging techniques allow visualizing pore structure at sub-micron resolution, providing a window into the complex nature of porous media.

Figure 2a depicts a 3D micro-CT image of the pore space of a limestone (Blunt et al. 2013). This limestone sample is characterized by large well-connected pores (Blunt et al. 2013). Based on such images, it is possible to construct pore network models (Fig. 2b), in which the pores are represented as spheres, while the pore throats are shown as cylinders, connecting the pores with each other (Blunt et al. 2013). In many situations, a fraction of the pores can be isolated, i.e., disconnected (trapped porosity), while other pores are difficult to reach (Navarre-Sitchler et al. 2009). Although connected to the pore-network, they are "dead-end" pores, and therefore do not contribute substantially to the "effective porosity" (Navarre-Sitchler et al. 2009). The connectivity, in principle, is defined by the ratio between the connected (i.e., effective) porosity and the overall porosity.

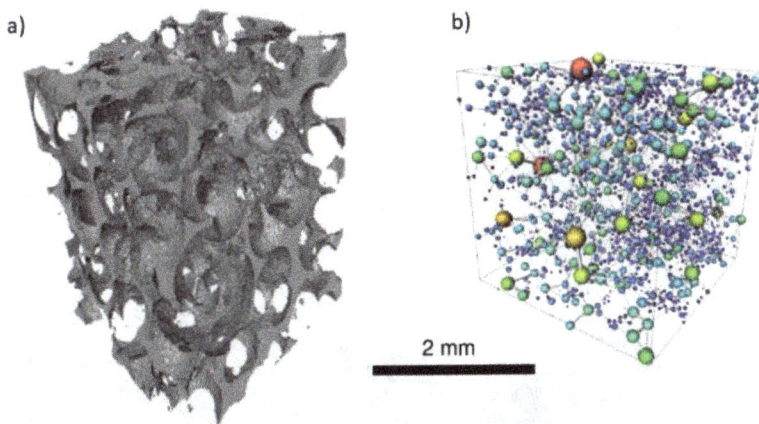

Figure 2. a) Pore-space images of a quarry carbonate. The image shows only the pore space. **b)** Pore networks extracted from the image shown in a). The pore space is represented as a lattice of wide pores (shown as spheres) connected by narrower throats (shown as cylinders). The size of the pore or throat indicates the inscribed radius. [Reproduced in modified form form MJ Blunt et al. (2013) Pore-scale imaging and modelling. Advances in Water Resources 51:187-216 with permission of Elsevier. Copyright 2013].

Evolution of flow and transport parameters at the pore scale

Examining the pore network shown in Figure 2 clarifies that flow and transport through such a material will heavily depend on the size of the pores, the diameter of the pore throats, and the connectivity between the pores. Large pores and direct connectivity in three dimensions implies a low degree of tortuosity, facilitating flow, and both diffusive as well as advective transport with relative ease.

In the case of an evolving porous medium affected by mineral dissolution–precipitation reactions, it is evident that both porosity and connectivity can be affected. The dissolution of a solid phase is directly linked to the increase of the overall porosity. However, dissolution does not necessarily increase effective porosity to the same degree. The effect can be substantial, if dissolution widens and opens up pore throats or opens up new diffusion pathways (Knackstedt and Zhang 1994; Moreira and Coury 2004), or it can be more limited, if dissolution is more equally distributed (Algive et al. 2010). Similarly, mineral precipitation might have a limited effect on connectivity, if minerals form on existing surfaces without clogging pore throats. On the other hand, precipitation may lead to the disconnection of existing solute migration pathways (Algive et al. 2010), with a pronounced impact on flow and transport parameters. These theoretical considerations imply that for the same change in total porosity, the dissolution and precipitation of minerals can lead to very different impacts on effective porosity and connectivity. As a consequence, flow and transport will also be affected in different ways due to the evolution of tortuosity and permeability as a function of effective porosity and connectivity, but not simply total porosity (Navarre-Sitchler et al. 2009).

Evolution of reactivity at the pore scale

In addition to changes in transport properties, the dissolution and precipitation of mineral phases, can result in changes in reactivity, leading to enhancement or reduction of reaction rates Normally, the dissolution of a mineral phase is controlled by mineral-specific surface-area-normalized reaction rates and by the reactive surface area of the mineral (Lasaga 1981, 2014).

The reactive surface area is controlled by a variety of factors, such as grain size (Emmanuel and Berkowitz 2007; Liu et al. 2008), exposure of the mineral and contact with the pore water (Landrot et al. 2012), the degree of occlusion (Peters 2009; Lai and Krevor 2014), surface roughness and presence of imperfections or etch-pits (Deng et al. 2018). Moreover, when multiple minerals are present, the surface fraction occupied by a certain mineral might significantly differ from its volume fraction (Lai and Krevor 2014). Also, the flow conditions can lead to more reactive zones, contributing to commonly observed deviations of reactive from geometric surface area. Together, these factors control the magnitude of the effective reactive surface area (Jung and Navarre-Sitchler 2018).

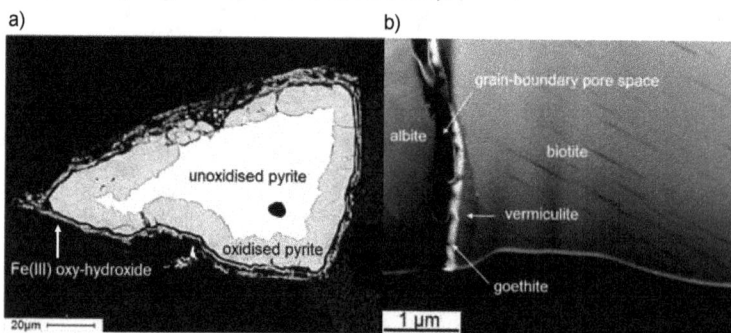

Figure 3. Evolution of reactivity and surface passivation: a) SEM image of partially oxidized pyrite grain with Fe(III) oxy-hydroxide surface coatings, b) STEM image showing goethite and vermiculite formation from incongruent dissolution of biotite (passivation) along a grain-boundary pore space. [Image credits: Tom Al, University of Ottawa, reproduced with permission.]

Any of these factors can change over time, as a porous medium progressively evolves, simply through shrinking of grain size, the enhancement of surface roughness and etch pit formation during dissolution (Teng 2004; Buss et al. 2007), incongruent dissolution as shown in Figure 3a for pyrite oxidation, or the formation of precipitates on the surface of a dissolving mineral. Secondary Fe(III)-oxy-hydroxides tend to precipitate on the surface of pyrite grains (Fig. 3a). Similarly, Figure 3b shows the precipitation of secondary oxide and clay minerals (goethite and vermiculite) on the surface of biotite, which is undergoing weathering. Incongruent dissolution and the formation of surface precipitates limit the interaction of the bulk solution with the surface of the minerals and in this way, tends to limit further mineral dissolution (Daval et al. 2009).

EFFECT OF MINERAL DISSOLUTION AND PRECIPITATION ON POROUS MEDIA

Considering that mineral-dissolution precipitation reactions have a pronounced impact on porous media evolution, we review these processes in more detail in the following sections, making a distinction between dissolution and precipitation reactions. In this context, we emphasize the evolution of porosity and transport parameters, as well as the interactions between the rate of flow and transport processes and reactions and the resulting effect on porous media evolution. Processes are discussed predominantly at the pore scale, keeping implications for the continuum scale in mind.

Mineral dissolution

In many situations, sedimentary deposits or fractured rocks are affected by weathering reactions involving the dissolution and alteration of their solid structure (Bjørlykke et al. 1989; Aharonov et al. 2004; Navarre-Sitchler et al. 2011; Jin et al. 2013). The oil and gas industry performed a substantial amount of research dedicated to understand and optimize the process of permeability enhancement via mineral dissolution with the goal of increasing production rates (Wang et al. 1993; Fredd and Fogler 1999; McDuff et al. 2010). Another example with relevance to the mining industry is the enhancement of in-situ uranium recovery (Simon et al. 2014; Regnault et al. 2015; Lagneau 2019, this volume) by using acidic solutions to dissolve uranium-containing minerals. In the case of construction materials, the dissolution of primary minerals is commonly a process that is unwanted. Indeed, the increase of porosity usually induces a decrease of the material's intended performance. Negative effects of leaching and acid attack on the properties of cement and concr have been demonstrated (Atkinson et al. 1987; Bentz and Garboczi 1992; Bertron et al. 2005).

As dissolution reactions can impact groundwater flow and solute transport through porous media, reactive transport modelling of the aforementioned problems has to account for the impact of dissolution on the relevant material properties. To predict how a porous medium's properties will evolve through dissolution reactions, it is necessary to understand the processes that control the development of dissolution patterns. Previous investigations of mineral dissolution have revealed that, under certain conditions, stable dissolution fronts develop (Barlet-Gouedard et al. 2006; Hesse et al. 2013; Walsh et al. 2013), while in other cases, such as karst development, dissolution fronts become unstable, leading to the development of fingering or wormholing (Hoefner and Fogler 1988; Daccord et al. 1989; Steefel and Lasaga 1990; Golfier et al. 2002).

Aharonov et al. 2004; Algive et al. 2012 and Luhmann et al. 2014 have shown that the dissolution regime depends on the characteristic times for advection, diffusion and reaction. Typically, relationships between characteristic times of transport and reaction processes are defined by the Peclet (Pe) and Damköhler (Da) numbers. The Peclet number addresses the relative importance of the advective and diffusive transport (Algive et al. 2010), while the Damköhler numbers reflect the ratio between the characteristic times of reaction compared to that of transport (Tartakovsky et al. 2007; Min et al. 2016). The Peclet number is defined as:

$$Pe = \frac{vL}{D} \tag{7}$$

where v represents the groundwater flow velocity, D the diffusion coefficient and L is a characteristic length of interest in the porous media. A high Peclet number reflects an advection-dominated transport regime. Two versions of the Damköhler number exist, which here will be referred to as the advective and diffusive Damköhler numbers, respectively. The Da numbers provide a relationship between the reaction rate and the transport (advection–diffusion) rates

$$Da_{adv} = \frac{\text{reaction rate}}{\text{advective transport rate}} \tag{8}$$

$$Da_{diff} = \frac{\text{reaction rate}}{\text{diffusive transport rate}} \tag{9}$$

Experimental and numerical research has focused on the prediction of wormhole formation in carbonate rocks (Fredd et al. 2000; Buijse et al. 2005; Quintard et al. 2007; Budek and Szymczak 2012).

Fredd et al. (2000) reviewed the response of carbonate rocks to ingress of acidic waters. Figure 4 depicts the different dissolution patterns for different acid injection rates (increasing from left to right). At very low injection rates (low Pe, high Da_{adv}), uniform face dissolution occurs near the injection point where all the acid is consumed. Increasing Pe facilitates deeper ingress of acidic waters along the main flow pathway where reactions take place, leading to a conical or dominant wormhole. The distinction between the conical/dominant wormholes and ramified or uniform dissolution was further evaluated by Detwiler et al. (2003). When the injection rate is further increased (to the point where Da_{adv} becomes $\ll 1$), acidic waters are forced through the entire porous medium without having time to react, leading to the formation of a ramified wormhole and ultimately uniform dissolution. The effects of Peclet and Damköhler number were analyzed with numerical modeling as early as 1990 (Steefel and Lasaga 1994), who showed the transition from dominant wormholes to ramified wormholes and uniform dissolution with decreasing Damköhler number.

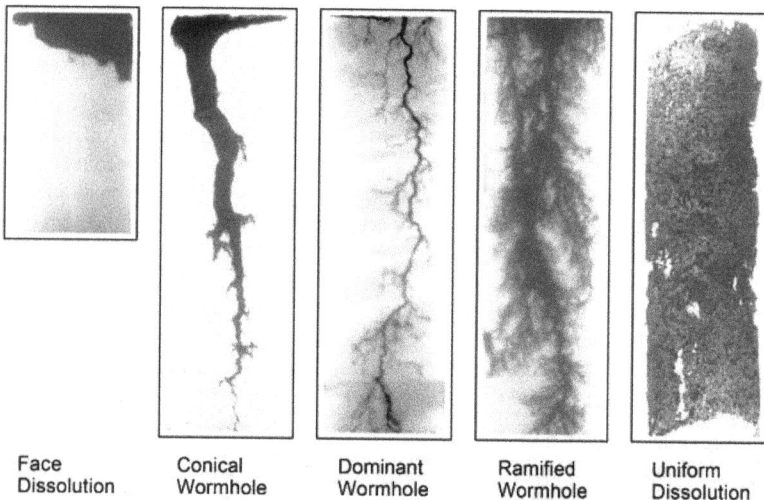

| Face Dissolution | Conical Wormhole | Dominant Wormhole | Ramified Wormhole | Uniform Dissolution |

Figure 4. Illustration of the different wormhole morphologies. Injection rates applied to the materials increase from the left to the right. [Republished with permission of the Society of Petroleum Engineers, from CN Fredd (2000) Dynamic model of wormhole formation demonstrates conditions for effective skin reduction during carbonate matrix acidizing. SPE Permian Basin Oil and Gas Recovery Conference:SPE-59537-MS. Copyright 2000.]

These diverse dissolution patterns lead to different evolution of transport properties. As uniform facial dissolution leads to a focused increased in permeability near the injection point, the macroscopic transport properties of the entire material do not change significantly. On the other hand, uniform dissolution tends to increase permeability throughout the entire sample. For situations between the two extremes cases, the development of a broad and relatively linear transport pathway ensues, through which most solutes can be transported. The latter case results in the formation of wormholes.

Min et al. (2016) performed a pore-scale study represented in Figure 5, which emphasizes the role of Da_{diff}. Figure 5a depicts a reaction-limited regime ($Da_{diff} \ll 1$) subject to diffusion-dominated transport ($Pe \ll 1$). In this case, dissolution is relatively slow, leading to solute release throughout the domain. However, since diffusion is the dominant transport process, the embedded solid inclusions dissolve rather uniformly. Figure 5b considers a diffusion-limited regime ($Da_{diff} > 1$) with low flow ($Pe \ll 1$). Reaction rates are faster than diffusion rates and one can see that dissolution reactions occur locally, leading to a steep concentration front. In Figure 5c, (Pe and $Da_{diff} > 1$), front instabilities are developed due to the advection dominated transport regime. Min et al. (2016) also simulated the evolution of permeability. The case depicted in Figure 5a corresponds to the largest increase in permeability, while the scenario shown in Figure 5b leads to the lowest increase in permeability.

The complexity of the dissolution regimes reported by Fredd et al. (2000) and Min et al. (2016) demonstrates that a quantitative assessment of dissolution patterns and rates, and feedback on porosity and permeability is difficult solely based on theoretical considerations. To properly take pore scale processes into account, pore-scale simulations of processes for relevant pore geometries provides a sensible path forward for deriving upscaled relationships for continuum scale modeling of evolving porous media. Recent developments in imaging and pore-scale modeling methods have made it possible to investigate these phenomena in more depth (Pereira Nunes et al. 2016; Soulaine and Tchelepi 2016; Tian and Wang 2018). Gao et al. (2017) and Gray et al. (2018) were able to approximately reproduce the experimentally observed dissolution patterns in microstructures obtained from CT scan images. Pre-and post-experimental X-ray

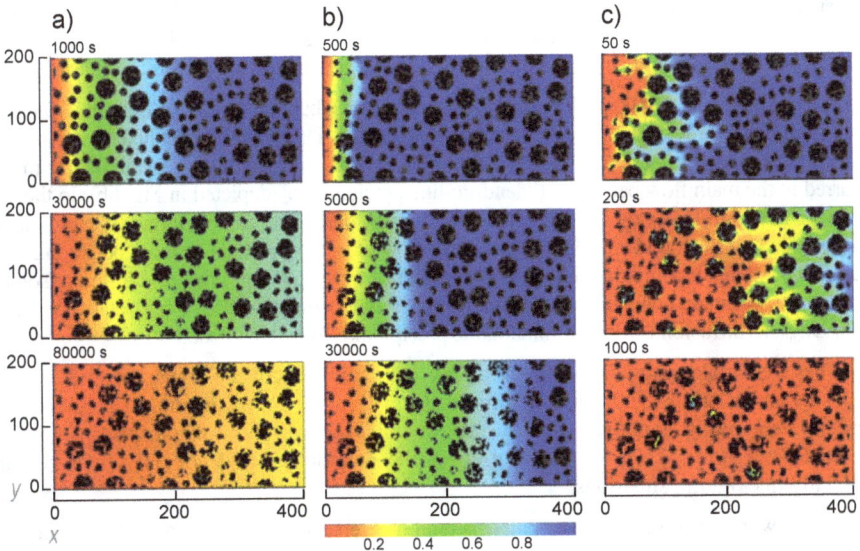

Figure 5. Mineral dissolution in a complex structure with randomly distributed insoluble material as a function of Peclet number and diffusive Damköhler number. **a)** slow reaction and slow flow, **b)** fast reaction and slow flow, **c)** fast reaction and fast flow. [Reproduced form T Min et al. (2016) Changes in porosity, permeability and surface area during rock dissolution: Effects of mineralogical heterogeneity. International Journal of Heat and Mass Transfer 103:900-913 with permission of Elsevier. Copyright 2016.]

synchrotron imaging of reactions in a fracture developed in Duperow Dolomite quantified the wormholing taking place due to the influx of acidic fluid, and this effect was captured in both reduced dimension models (Deng et al. 2016) and in pore scale models (Steefel et al. 2013).

Mineral precipitation

The precipitation of secondary minerals often impacts groundwater flow and solute transport. In some cases, precipitation reactions lead to the self-sealing of fractures (Steefel and Lasaga 1994; Dobson et al. 2003b; MacQuarrie and Mayer, 2005; Gherardi et al. 2007) or the reinforcement of the solid structure (Bickmore et al. 2001; Pfingsten 2002; Jacquemet et al. 2012). Similarly, carbonation of cementitious materials under water can lead to a calcite crust forming on the surface of the material (Dauzères et al. 2010; Seigneur et al. 2017), making the material almost impermeable. Crust formation has also been observed as a result of gypsum precipitation on historical monuments (Dewanckele et al. 2012). Furthermore, permeability decrease in sedimentary basins (Aharonov et al. 2004) or soils (White et al. 2005) due to mineral precipitation have been observed. The formation of secondary minerals may lead to fractures through generated internal mechanical stresses. This can lead to the deterioration of mechanical properties (Chagneau et al. 2015; Li et al. 2017). For cement materials, the delayed formation of ettringite (Taylor et al. 2001) or atmospheric carbonation (Auroy et al. 2013) are typical examples of precipitation processes that can induce fracturing, which can be detrimental in the context of CO_2 sequestration or radioactive waste management. In all the aforementioned cases, precipitation can lead to clogging, which significantly affects flow and solute transport. It is therefore important to understand the role of precipitation reactions on transport properties and to consider these evolving parameters in modeling analyses.

Kang et al. (2005) performed numerical studies of precipitation reactions involved in CO_2 sequestration. Working with 2D simplified images, they studied the influence of Pe and the Da_{diff} number on mineral precipitation reactions and subsequent permeability reduction. For a fixed high Pe, these simulations indicate that Da_{diff} controls the time required to plug the medium. Larger Da_{diff} lead to more localized precipitation near the inlet and more efficient clogging, reducing permeability to zero. Conversely, increasing Pe (for a fixed Da_{diff}) has the opposite effect. The efficient clogging of the medium relies on moderate Da_{diff} numbers and small Pe numbers. A small amount of precipitates can render the pore structure impermeable. For higher Pe, it takes more time (and more reactants) to clog the porous medium, as the precipitation front is wider.

Tartakovsky et al. (2007) performed a similar study and obtained comparable results, which are depicted in Figure 6. A high Pe number combined with a low Da_{diff} number resulted in precipitation throughout the porous medium (Fig. 6a). For high Pe and increased Da_{diff}, precipitation mostly occurred in the main flow pathways ("dendrite-like precipitation" depicted in Fig. 6b). In the case of low Pe and high Da_{diff}, most of the precipitation took place near the inlet, rapidly clogging the medium, leading to a rapid and sharp decrease of flowrates, as depicted in Figure 6c. Tartakovsky et al. (2008) obtained similar results for mixing-induced precipitation involving different solutes. For high Da_{diff}, the precipitation zone is thin, restricted to the region where both reactants meet. For lower Da_{diff}, as diffusion is more dominant, the precipitation zone is widened.

These examples emphasize the complex interactions between transport and reaction processes and relations between their time scales. Even if the total amount of precipitates is correctly captured, it is not straightforward to predict the evolution of transport parameters, which depend strongly on the distribution of the precipitates and not only their amount, especially in clogging situations.

While wormholing dissolution patterns tend to exert a positive feedback loop in terms of permeability enhancement, precipitation reactions can have the opposite effect by preferentially occurring in the high-flow zones (Steefel and Lasaga 1994; Aharonov et al. 1998; Mehmani et al. 2012), causing flow to diverge towards zones of lower permeability. This behavior can be described as a "dewormholing" process.

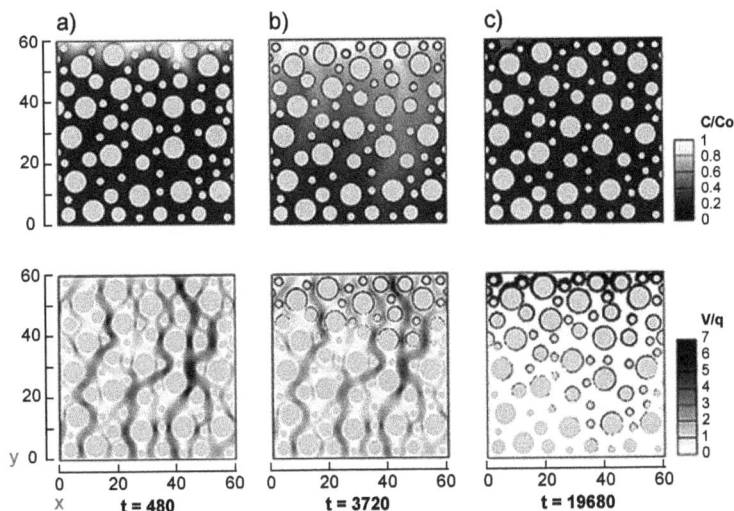

Figure 6. Distribution of precipitation, velocity and concentrations. Light gray particles represent primary mineral grains, black particles represent precipitated minerals. The grey scale in the upper panels denotes normalized concentrations C/C_o at 3 different output times (dimensionless model units), the bottom panels show normalized velocities at 3 times [Reproduced from Tartakovsky et al. (2007) A smoothed particle hydrodynamics model for reactive transport and mineral precipitation in porous and fractured porous media. Water Resour Res 43:W05437 with permission from John Wiley & Sons.]

However, other factors can also affect the location and distribution of precipitate formation. Rajyaguru et al. (2019) studied through-diffusion in chalk samples. Precipitation of barite and gypsum were investigated under a purely diffusive regime. Gypsum is characterized by a rather high solubility and kinetic rate compared to barite. It was observed through micro-CT analysis that the precipitation patterns were significantly different for barite and gypsum. While barite precipitated as a thin continuous disk at the center of the chalk sample (even though its intrinsic rate is lower), gypsum mostly precipitated as large isolated spheres. Even though the overall porosity decrease was similar (~2%), the diffusive properties were more substantially impacted in the barite experiment. The authors suggest that the barite precipitation was the result of both homogeneous and heterogeneous nucleation. Conversely, the heterogeneous precipitation of gypsum seemed to have been highly influenced by local pore structure heterogeneities. These experiments reveal the complexity of mineral precipitation within pore structures, even in simple diffusion-controlled transport regimes. Chagneau et al. (2015) showed that precipitation of celestite in a diffusion-reaction cell occurred almost exclusively in macropores based on the differential transport of anions (excluded from the smallest pores due to electrostatic effects) and cations.

Other studies also showed the importance of spatial heterogeneities for determining the distribution of precipitation, which in turn will control the evolution of tortuosity, permeability and reactivity of porous media (Andreani et al. 2009; Fox et al. 2016; Garcia-Rios et al. 2017).

INTEGRATED ASSESSMENT OF THE IMPACT OF MINERAL REACTIONS ON POROUS MEDIA

In many situations, mineral precipitation reactions occur in tandem or in response to dissolution reactions, leading to feedback between these reactions. Singurindy and Berkowitz (2003) provided images of the coupled dissolution/precipitation of carbonate and gypsum and showed how the dissolution patterns influenced the precipitation patterns. The evolution of a porous medium at the grain scale subject to dissolution and precipitation reactions is conceptually represented in two dimensions in Figure 7.

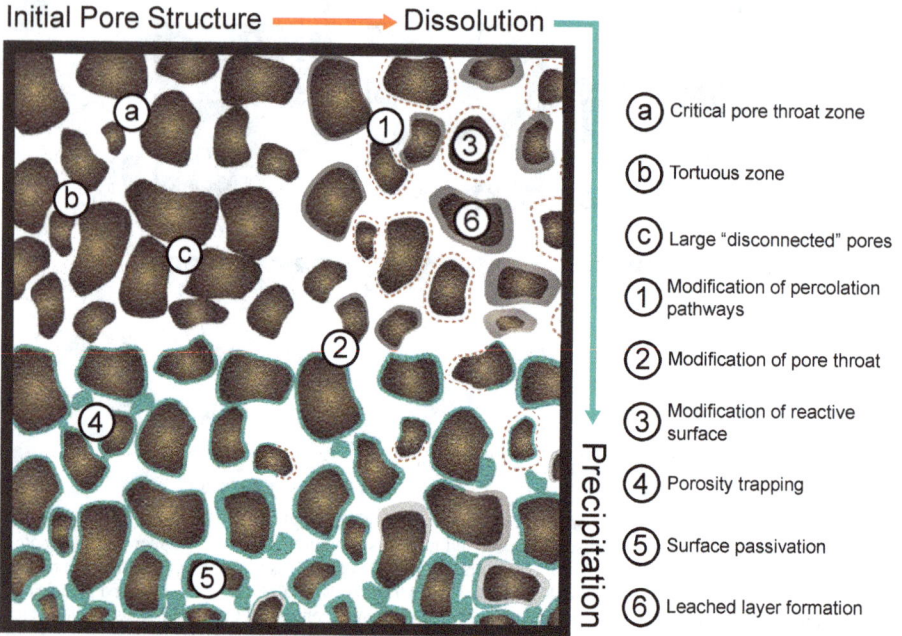

Figure 7. Schematic representation of the effect of dissolution–precipitation reactions on pore structure. Brown-colored grains depict primary phases, dashed red lines depict original grain size of dissolving primary phases, grey colored rims depict incongruent dissolution and leached layer formation, green-colored coatings and grains depict secondary mineral phases.

The initial pore structure depicted in Figure 7 (top left quadrant) exhibits several features. The region surrounding point (a) is critical, as it is well connected, providing a main pathway for water flow and solute migration. The region around point (b) represents a rather tortuous pathway for diffusion and advection. The region around (c) on the other hand, consists of large pores, which are poorly connected to the main transport pathways. Figure 7 also emphasizes that modifications to a pore structure can have varied impacts on macroscopic properties.

In a dissolution-dominated regime (top right quadrant of Fig. 7), dissolution of primary minerals can lead to the creation of new flow and transport pathways (1) or increase the size of critical pore throats (2). This evolution can have significant impacts on flow and transport processes. In addition, dissolution will impact the reactive surface of the initial solid structure (3) or can lead to the formation of leached layers in the case of incongruent dissolution (6).

In a precipitation-dominated regime (bottom left quadrant of Fig. 7), precipitation of secondary minerals can have opposite effects: decline of connectivity, leading to a reduction of flow and transport pathways (1), decrease of critical pore size (2), trapping of initially connected porosity (4). In addition, precipitation reactions can limit access to (4) or passivate (5) a fraction of the initial reactive surface.

In a regime that is both affected by dissolution and precipitation (depicted in the lower right quadrant of Fig. 7), coupled dissolution–precipitation reaction can lead to more complex effects. Figure 7 clarifies that the modification of the pore structure depends on the respective dissolution and precipitation kinetics and interactions between dissolution and precipitation reactions.

RELATIONSHIPS FOR MODELING EVOLUTION
OF CONTINUUM-SCALE PARAMETERS

Despite obvious challenges with the parameterization of evolving porous media, simulation tools are needed at the continuum scale to assess a variety of relevant environment and engineering problems. In the following, we provide a review of the most common approaches found in the literature to parameterize evolving parameters for reactive transport simulations at the continuum scale.

Evolution of porosity

As discussed above, the dissolution and precipitation of minerals lead to changes in porosity. From a modeling perspective, it is fairly straightforward to account for total porosity changes at the continuum scale. At any point in time, the porosity can be calculated based on the mineral volume fractions that constitute the grain matrix. In this context, porosity is defined as:

$$\phi = 1 - \sum_{i}^{N_m} \varphi_i \tag{10}$$

It must be kept in mind that porosity defines total porosity and does not take into consideration the fraction of disconnected porosity, dead-end pores and effective porosity. In many situations, the mineralogical composition of porous media is not well known and/or only a few mineral phases contribute significantly to porosity changes. Under such conditions, it is more practical to define porosity at a specific location in space and point in time as:

$$\phi = 1 - \varphi_{NR} - \sum_{i}^{N_{m,r}} \varphi_i \tag{11}$$

where φ_{NR} is the volume fraction occupied by non-reactive minerals and $N_{m,r}$ is the number of reactive minerals. Based on these considerations, the change in porosity can then be estimated as follows:

$$\frac{d\phi}{dt} = -\sum_{i=1}^{N_{m,r}} \frac{d\varphi_i}{dt} \tag{12}$$

Similar approaches can be taken to update effective porosity due to bioclogging, i.e., the formation of biofilms. While the previous equation accurately describes the physical processes at play, it does not provide any information about the structural evolution of the pore structure. Looking back at the few referenced studies of the previous section, this obviously constitutes a shortcoming. However, as we will discuss in the following sections, it is possible to attribute different continuum relationships for the porous media properties, in order to represent indirectly the pore structure evolution. This approach has been commonly used in reactive transport modeling and the evolving porosity has been used to update transport parameters such as tortuosity and permeability.

Tortuosity and diffusivity as a function of porosity

Relevance of purely diffusive regime. In the absence of significant advective fluxes (e.g.: in low permeability materials, low hydraulic head gradients), diffusion is the main driving process for solute transport. Diffusion-dominated transport is relevant for deep geologic repositories for long-term storage of radioactive waste. In this context, reactive transport modelling has been used to develop an improved understanding of the degradation of cementitious materials at cement/clay interfaces (Gaucher and Blanc 2006; Savage 2013; Jenni et al. 2017; Dauzères et al. 2019). Since these interfaces are affected by dissolution–precipitation reactions that induce porosity changes, diffusion properties of the materials are also affected, directly through changes of porosity, as well as through the evolution of tortuosity. Despite known limitations, it is therefore necessary to provide relationships able to describe the evolution of tortuosity as a function of changes in porosity.

Tortuosity–porosity relationships. A review of the literature reveals that many tortuosity–porosity relationships exist to describe the dependence of diffusion on porosity. These relationships have been developed to estimate effective diffusion coefficients based on porosity, a parameter that is easier to measure than the effective diffusion coefficient. These relationships are also commonly used in evolving porous media. Akanni et al. 1987; Boudreau 1996 and Ray et al. 2018 present a summary of the commonly used relationships. However, most reactive transport simulations in saturated porous media rely on Archie's law to account for the porosity dependence of the pore diffusion coefficient (Xie et al. 2015), which, in its simplest form can be written as:

$$\tau_l = \phi^n \tag{13}$$

with n being the cementation factor. Let us recall that, in this study, the effective diffusion coefficient for a fully saturated aqueous phase is defined as:

$$D_l^e = \phi \tau_l D_l^0 \tag{14}$$

Some confusion usually exists about these different relations. In some cases, studies relate a relationship between porosity and effective diffusion coefficient, instead of tortuosity. Other forms of Archie's law can be found, some of them describing a percolation threshold under which diffusion cannot occur (e.g., Bentz and Garboczi 1991 for cement materials). Another example is the relation developed for the effective diffusion coefficient in the liquid phase by Navarre-Sitchler et al. (2009), which will be further studied in a case study :

$$D_l^e = D_l^{\min} + D_l^0 (\phi_e)^2 \tag{15}$$

where $D_l^{\min}$ represents the effective diffusion coefficient of the matrix, and where the effective porosity, ϕ_e, is defined as:

$$\begin{aligned} \phi_e &= a(\phi - \phi_{\text{crit}})^n & \phi \geq \phi_{\text{crit}} \\ &= 0 & \phi < \phi_{\text{crit}} \end{aligned} \tag{16}$$

Another commonly used tortuosity–porosity relationship assumes an exponential relationship to porosity:

$$\tau_l = \frac{\exp(a\phi)}{\phi} \tag{17}$$

where a is an experimental fitting factor. Let us emphasize that these laws were usually developed to link the effective diffusion coefficient to porosity.

Under unsaturated conditions, it has been shown that tortuosity also depends on phase saturation. A formulation that is commonly used to estimate porosity was derived by Millington and Quirk (1961):

$$\tau_\alpha = S_\alpha^{7/3} \phi^{1/3} \tag{18}$$

This relationship, which shows a strong and non-linear dependency of tortuosity on phase saturation S_α, was derived for the gas diffusion in porous media (Millington 1959). However, most reactive transport codes adopt the Millington–Quirk relation (Millington and Quirk 1961) to model diffusion through the liquid phase as well. Considering the opposite wetting behavior of the gas and liquid phase, the use of similar relations to describe tortuosity in

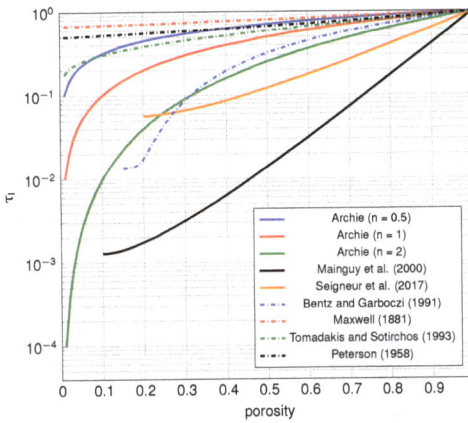

Figure 8. Graphical representation of common tortuosity–porosity relationships of the liquid phase.

these different phases remains questionable. Chou et al. 2012 provides a review of different tortuosity models for unsaturated conditions and concludes that the use of the Millington–Quirk relation overestimates experimental data.

Figure 8 depicts tortuosity as a function of porosity based on common relationships for saturated media, some of which are documented in the literature (Maxwell 1881; Petersen 1958; Tomadakis and Sotirchos 1993), while others are variations of Archie's law (Bentz and Garboczi 1991) or represent an exponential dependence on porosity (Mainguy et al. 2000; Seigneur et al. 2017). It can be observed that tortuosities obtained from the various relationships differ substantially, in particular for low porosities. The range between minimum and maximum tortuosities are approximately 2 orders of magnitude for a porosity of 0.5, while calculated tortuosities differ by more than 5 orders of magnitude for porosities less than 0.1. These pronounced differences suggest that tortuosity–porosity relationships are specific to porous media. The fact that some relationships do not yield a tortuosity of "zero" for a porosity of "zero" (even though the effective diffusion coefficient does go to 0) implies that these relationships are not applicable for conditions of complete pore clogging. However, there is experimental evidence that in some cases mineral precipitation will not completely clog the porous media. Li and co-workers observed that despite the formation of a distinct calcite cement band in cement subject to CO_2 attack, the porosity did not go completely to zero as evidenced by the continued progress of the primary portlandite dissolution front (Li et al. 2017). They attributed this to be due to either the lack of precipitation in nanopores that were still able to allow for diffusive transport of the uncharged H_2CO_3 molecule into the unaltered cement, or alternatively due to continued micro-fracturing. The lack of clogging of the nanopores is compatible with the interpretation of anion (and cation) transport in clay-rich material in which anion exclusion occurs (Chagneau et al. 2015).

Application to reactive transport modelling. Leaching of cementitious materials has been studied extensively, since dissolution reactions lead to increased diffusion into these materials, enhancing their deterioration. The prediction of degradation depths of these materials strongly depends on which cementation (or exponential) factor (Shao et al. 2013) or which feedback law is considered (Seigneur et al. 2017; Georget et al. 2018). This is due to the fact that dissolution rates of primary minerals are affected by the diffusion of calcium through the degraded layers. Several reactive transport studies managed to adequately reproduce the extent of measured degradation depths using either Archie's law (Galíndez et al. 2006; De Windt and Badreddine, 2007; Chagneau et al. 2015; Li et al. 2017; Georget et al. 2018) or an exponential relation (Mainguy et al. 2000; Seigneur et al. 2017) to describe tortuosity evolution as a function of changing porosity. However, the choice of the cementation factor n or the exponential factor a differs between the studies in a way that is not straightforward to predict, depending on the conditions and the materials considered.

Most of these modelling studies are based on an initial porosity and tortuosity representative for the investigated materials, and only use these empirical laws to compute the evolution of diffusion parameters. In these studies, the initial diffusive properties of the material are either known independently or were fitted to early time data. Since dissolution under a diffusion-controlled regime occurs in a fairly uniform manner and leads to a relatively mild increase in tortuosity, this approach was able to produce reasonable matches with observational data, albeit with different fitting factors.

However, when precipitation reactions occur simultaneously, it becomes more complicated to assess the relative dynamics between dissolution and precipitation reactions and their effect on tortuosity and diffusion properties. Several studies had to artificially increase the empirical factors when describing precipitation reactions (Huet et al. 2010; Brunet et al. 2013; Chagneau et al. 2015; Lalan 2016) or the reactivity (Jacquemet et al. 2012), or use other empirical assumptions (Walsh et al. 2013) to be able to reproduce breakthrough curves and/or degradation depths when simulating the carbonation of these materials.

These results are consistent with pore scale analyses (see above), implying that limited and localized mineral precipitation can significantly alter diffusive migration paths. Hence, the impact of precipitation reactions on diffusion is more difficult to assess than that of dissolution. These observations also imply that different materials are affected in distinctive ways, which is inherently related to the initial organization of their pore structure. Materials with a high degree of initial tortuosity, containing small pore throats, will be more sensitive to the effects of precipitation reactions, which was observed both experimentally and in numerical modelling studies (Kutchko et al. 2007; Brunet et al. 2013; Huber et al. 2014; Seigneur et al. 2017).

These results suggest that it might be necessary to use different tortuosity–porosity relationships for materials affected by dissolution or precipitation reactions. However, since these reactions are most commonly coupled and occur simultaneously, a proper assessment of the effects of these reactions on diffusivity remains a challenging issue.

Moreover, previous studies have shown that obtaining a better representation for the evolution of diffusion (sometimes through the use of unrealistic high cementation factors) can make it difficult to reproduce observed mineral precipitation and dissolution patterns. Jacquemet et al. 2012 indicated that the evolution of reactivity also plays a role in the degradation depth, although most of the research to date has focused on the evolution of the transport properties.

Permeability as a function of porosity

Relevance. Permeability and its evolution plays a significant role in advection-driven transport regimes, including the injection of fluids in disequilibrium with the surrounding rock formation (acidizing carbonate rocks or uranium recovery via sulfuric acid leaching). In these regimes, the modification of porosity can also impact permeability, and therefore groundwater flow patterns, and in turn the degree of chemical perturbations which alter the porous medium. It is therefore necessary to utilize relationships capable of capturing the impact of evolving porosity on permeability.

Review of Kozeny–Carman and power-laws relations. Like relationships between tortuosity and porosity, permeability–porosity relationships have been developed independently from the field of reactive transport modeling. The Kozeny–Carman relationship is the most commonly used permeability–porosity relationship in reactive transport simulations (Poonoosamy et al. 2018) for simulating the evolution of permeability affected by mineral dissolution–precipitation reactions. It is commonly used in the following form:

$$k = k_0 \frac{\phi^3}{(1-\phi)^2} \frac{(1-\phi_0)^2}{\phi_0^3} \tag{19}$$

where k_0 and ϕ_0 respectively represent the initial or reference permeability and porosity. Xu and Yu (2008) and Hommel et al. (2018) provide an extensive review of the most common permeability–porosity relationships. In general, power-laws as a function of porosity have been widely used to describe the evolution of permeability:

$$k = k_0 \left(\frac{\phi}{\phi_0} \right)^n \tag{20}$$

As an alternative, Verma and Pruess (1988) have introduced a critical porosity ϕ_{crit} in the power law:

$$k = k_0 \left(\frac{\phi - \phi_{crit}}{\phi_0 - \phi_{crit}} \right)^n \tag{21}$$

which corresponds to the porosity for which permeability becomes zero. The parameter n represents an empirical fitting parameter, and could be considered as similar in concept to thresholds for diffusion found in modifications of Archie's Law (Bentz and Garboczi 1992; Navarre-Sitchler et al. 2009). Values between 3 and 75 have been reported for the fitting parameter n (Brunet et al. 2013). Gouze and Luquot (2011) suggested that this wide range of power values can be explained by the fact that permeability is controlled by two parameters: the first parameter corresponds to the tortuosity of the pore space. The second parameter corresponds to the pore throat size, or the "hydraulic radius" (Knackstedt and Zhang 1994; Moreira and Coury 2004) representing the characteristic radius of the pore channels. Mercury intrusion porosimetry and the "ink-bottle effect" (Diamond 2000) provide useful experimental illustrations of the pore-throat size control on permeability. During the first stages of wormhole development, an increase of hydraulic radii occurs, while when channeling develops, tortuosity (as defined in this chapter) increases significantly. In more general terms, Hao et al. (2013) and Menke et al. (2016) suggest that the power value is representative of the heterogeneity of the pore structure.

Figure 9a depicts the common permeability–porosity relationships for an arbitrary reference porosity of 0.3. Similar to the dependence of tortuosity on porosity, it can be observed that the various relationships lead to widely differing results. A porosity increase, representing the case of mineral dissolution, leads to differences on the order of one order of magnitude between the relationships for a porosity enhancement of 0.3. On the other hand, much more significant differences are seen for the case of precipitation and porosity reduction.

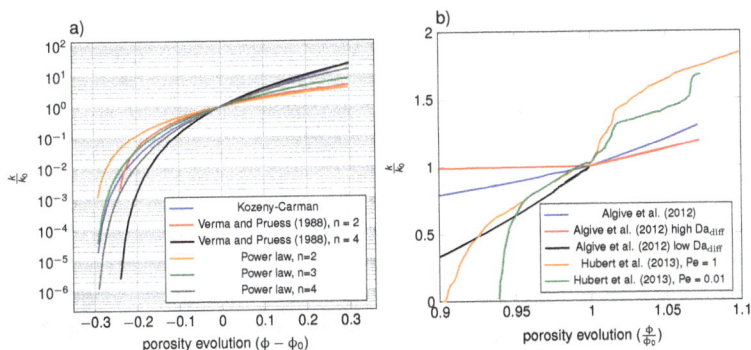

Figure 9. (a) Graphical representation of common permeability–porosity relationships. The reference porosity ϕ_0 was arbitrarily set to 0.3, (b) Permeability dependence on porosity based on pore network model simulations by Algive et al. (2012) and Huber et al. (2014). Presented results of Huber et al. (2014) correspond to a Da$_{diff}$ of 1, and an initial porosity of 0.3 was assumed.

It is worth pointing out that in certain cases, inverse correlations were found between permeability and porosity (Luquot et al. 2013, 2016; Garing et al. 2015). Such correlations usually arise when coupled dissolution and precipitation processes are occurring at the same time, but do not involve the same pore sizes (Luquot et al. 2014; Garing et al. 2015). None of the permeability–porosity relationships introduced above could represent these inverse relationships.

Applicability of power laws relations for porosity–permeability feedbacks. The evolution of permeability is more difficult to capture than the evolution of diffusivity. Indeed, in diffusion-dominated regimes, reaction fronts tend to be rather stable. Conversely, in advection-dominated regimes, pore-scale heterogeneities affect the distribution of flowrates and can induce unstable fronts and channeling.

Power laws are often used to describe permeability–porosity relationships. These relationships strongly depend on the initial material's pore structure and its heterogeneity (Rötting et al. 2015; Menke et al. 2016; Wolterbeek and Raoof 2018). Civan (2001) uses three different power laws to describe Fontainebleau sandstones that are similar in nature but have different initial porosities.

It has also been shown that the fitting parameter n depends on the Pe and Da numbers, since they influence the dissolution and precipitation patterns (Fig. 9b). Egermann et al. (2010) showed that the permeability evolution in five different dissolution experiments was well represented by power laws; however, each required a different fitting parameter. The values of the power increased from 0.5 (relatively uniform dissolution of grains) to 31 (dissolution focused on pore throats). Colón et al. (2004); Panga et al. (2005); Menke et al. (2016) provide additional examples with similar outcomes.

In addition, Smith et al. (2013) observed that the value of the fitting parameter n evolved with time as dissolution progressed. This was likely due to variations in heterogeneity caused by dissolution reactions in advective regimes. In other words, a constant fitting parameter n might not be suitable to describe the evolution for substantial modifications of porosity (Pape et al. 1998; Bernabé et al. 2003; Nogues et al. 2013; Ovaysi and Piri 2014; Vialle et al. 2014).

Considering that mineral dissolution reactions can lead to wormhole formations and precipitation reactions can lead to the closure of wormholes and fractures, the form of permeability–porosity relationships are typically different for precipitation and dissolution reactions (Bernabé et al. 2003; Li et al. 2010), as shown in Figure 9b. Furthermore, in cases of coupled dissolution and precipitation or dissolution–precipitation cycles, hysteresis was observed in the permeability–porosity curves (Algive et al. 2012; De Boever et al. 2012; Van der Land et al. 2013), whose importance depends on the Pe and Da numbers (Algive et al. 2012). In general, hysteresis is limited in the case of diffusion-dominated regimes with low kinetic rates (low Pe and high Da_{diff}), consistent with expected behavior, as discussed above in the section "*Effect of mineral dissolution and precipitation on porous media*" (Tartakovsky et al. 2008; Min et al. 2016).

These considerations explain why the permeability–porosity relationships have limited predictive capabilities in the context of evolving porous media. Indeed, for identical pore structure and otherwise identical conditions, considering diverse minerals with different kinetic rates (i.e., different Da numbers), experimental data can only be reproduced with different fitting parameters n. It seems unlikely that a single model, however complex, could predict changes in permeability in evolving porous media under different flow and transport conditions. These observations emphasize the need to choose an appropriate power-value, based on a priori knowledge or constrained by observational data.

Cubic law for porosity–permeability in fractures. In fractures, it is customary to use the cubic law for permeability, which is related to the local fracture aperture (and thus, local fracture porosity). This relation has a long history in the literature, with the most cited in the geosciences being perhaps the study by Witherspoon et al. (1980). Generalization to a local cubic law for rough fractures was discussed by Oron and Berkowitz (1998). The interested reader is referred to the chapter by Deng and Spycher (2019, this volume) for further discussion.

Evolving reactivity

In addition to the evolution of transport parameters, it is equally important to describe the evolution of reaction rates, i.e., the reactivity in evolving porous media. An adequate description of rates is necessary to capture long-term solute concentrations and loadings, pH-buffer reactions, and to adequately account for feedback with flow and transport processes.

Evolution of reactivity during mineral dissolution. Reactivity is often described via a reactive surface area of the minerals. Several models are commonly used to describe the evolution of the surface area S. The simplest formulation considers the uniform dissolution of spherical grains, in which case the surface area is inversely proportional to its radius:

$$S \propto \frac{3}{R} \tag{22}$$

Based on this dependence, a two-third power law relationship can be derived to describe the evolution of surface area with respect to a mineral's volume fraction (Steefel and Lichtner 1998):

$$S = S_0 \left(\frac{\varphi}{\varphi_0} \right)^{2/3} \tag{23}$$

where S_0 is a reference surface area and φ_0 defines the corresponding reference mineral volume fraction. This relationship is commonly used to describe surface controlled mineral dissolution reactions, accounting for a decrease in rate as dissolution proceeds (Fig. 10). Although this relationship is derived based on the dissolution of a single grain, it has been used frequently on the continuum scale in the form of Equation (23) (Steefel and Lichtner 1998). A decline in reactivity and surface area is intuitive; however, it does not hold under all circumstances. Dissolution reactions can also lead to an increase of reactive surface area, as illustrated by the sugar-lump model developed by Noiriel et al. (2009). Similarly, the formation of etch pits on the mineral surface may also lead to an increase in reactive surface area, although the reactive surface area (like the mineral volume fraction) must ultimately go to zero with continued dissolution.

Luquot and Gouze (2009) proposed a power-law relationship to describe the evolution of surface area as a function of porosity:

$$S = S_0 \left(\frac{\phi}{\phi_0} \right)^{-w} \tag{24}$$

This approach allowed Luhmann et al. (2014) to fit the decrease of reactivity as evidenced by decreasing concentrations in the breakthrough curves of their experiments.

However, surface reactions are also affected by pore scale flow dynamics. Li et al. (2008) considered the role of a diffusion boundary layer in the dissolution of calcite within a single cylindrical pore. Noiriel et al. (2012) also considered the role of a diffusion-controlled boundary layer on mineral dissolution rates, an effect that is expected in all but the most vigorously stirred system. Molins et al. (2014) performed pore-scale simulation of the injection of CO_2 into crushed calcite. While only a small amount of calcite dissolved in these simulations (porosity variation negligible), the results showed that mass transport limitation to the reactive surface also result in an important control on mineral dissolution. These conclusions were reinforced by a recent pore scale study on flow and transport in discrete reactive fractures (Molins et al. 2019). The shrinking sphere model depicted in Figure 10 assumes that concentrations in the bulk solution and in water directly in contact with the mineral surface are fully mixed and identical, which constitutes a simplification of the processes at work.

Minerals often do not undergo complete dissolution, but dissolve incongruently. Such dissolution reactions have been described by the shrinking core model (Levenspiel 1998), as depicted in Figure 11, in particular with applications to sulfide mineral oxidation (Ritchie 1994; Wunderly et al. 1996; Lefebvre et al. 2001; Mayer et al. 2002). In the case of sulfide oxidation, the rate-determining step is the diffusion of dissolved oxygen towards the mineral surface through a leached (oxidized) layer. As sulfide oxidation progresses, the thickness of the leached layer grows, and continued oxidative dissolution slows substantially, leading to a negative feedback on the grain scale. This model is not restricted to sulfide mineral oxidation, it can be applied to different minerals, also subject to incongruent dissolution and leached layer formation, or composite mineral grains, representing conditions when reactants diffuse through the porous leached layer generated by the dissolution of one of the minerals present in the grain (Steefel and Lichtner 1994, 1998; Deng et al. 2016). The shrinking core model can also be used to represent the effect of the formation of surface coatings associated with the precipitation of secondary minerals. In general, these considerations emphasize the mass transport limitations can affect reactivity and can significantly decrease the reaction rates (Dentz et al. 2011).

Assuming spherical particles, an effective first order rate constant capturing the effect of the formation of a leached layer or surface coating can be described as:

$$k_i^m = D_i^{e,m} \frac{r_i^p}{(r_i^p - r_i^r)r_i^r} \tag{25}$$

where $D_i^{e,m}$ represents the effective diffusion coefficient of the species driving mineral dissolution through the protective surface layer, r_i^p is the initial radius of the particle and r_i^r is the radius of

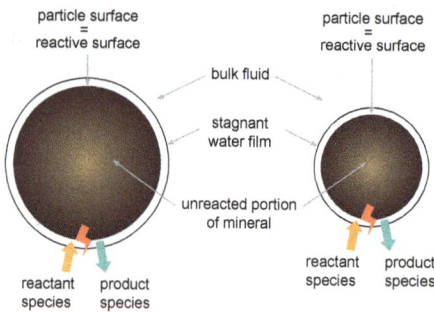

Figure 10. Schematic representation of the shrinking sphere model: the surface area and the volume fraction are related by a two-third power law.

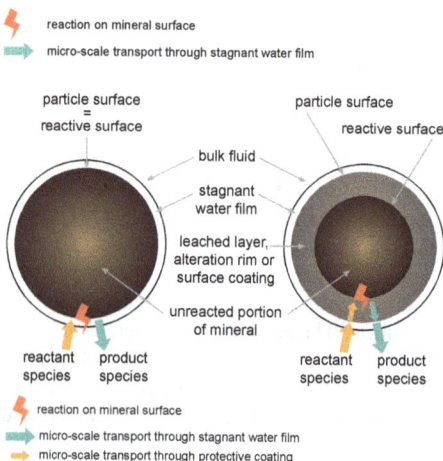

Figure 11. Schematic representation of the shrinking core model for transport-controlled reactions when an altered layer develops through an incongruent dissolution reaction.

the unreacted portion of the particle. Although the shrinking core model is derived for a single particle, this formulation makes it possible to extend its application to the continuum scale.

The application of the shrinking core model was shown to capture experimental observations when applied to the weathering of mine wastes (Wunderly et al. 1996; Brookfield et al. 2006; Demers et al. 2013; Soleimani et al. 2013; Doulati Ardejani et al. 2014; Smith et al. 2018), leaching of ore minerals (Ferrier et al. 2016), and for thermochemical sulfate reduction (Bildstein et al. 2001).

Although practical, the shrinking core model has several drawbacks: its formulation commonly does not account for a local mass balance during incongruent dissolution and is normally not accounting for the elements that are left behind in the leached layer and the elements that are released into the bulk solution. In addition, the formulation requires assumption of an initial leached layer thickness, which further adds to the empiricism of the model. An alternative approach would be to develop a formulation that is able to account for transition from surface-controlled dissolution to diffusion- (transport-) controlled dissolution (Fig. 12). However, such a formulation adds additional complications. For surface-controlled reactions at the continuum scale, it is commonly assumed that dissolved species concentrations on the reactive mineral surface are identical to species concentrations in the bulk solution (Fig. 12a), while for the shrinking core model, it is assumed that the species driving the dissolution of the mineral is consumed instantaneously when arriving at the mineral surface (Fig. 12c). However, for a transition regime, the concentration of the species driving the dissolution of the mineral at the reactive surface is unknown (Fig. 12b). Due to this additional complexity, this formulation has not been further developed to date.

Evolution of reactivity during mineral precipitation. The formation of secondary minerals may also affect the reactivity of primary mineral phases and their dissolution rates. The precipitation of these secondary phases on the surface of primary minerals can lead to surface passivation (e.g., Harrison et al. 2016) induced by heterogeneous nucleation (e.g., precipitation of carbonate minerals in permeable reactive barriers, precipitation of hydrated Mg-carbonates during carbon sequestration, precipitation of Fe and Mn oxides). Surface passivation is not straightforward to take into account, due to complex surface morphology of secondary phases and limited availability of data. Conceptually, the reactivity of primary mineral phases can be linked to the volume fraction of secondary precipitates. Jeen et al. (2007) used an exponential formulation to describe evolving reactivity of iron within a permeable reactive barrier as a decline in reactive surface area:

$$S = S_0 \exp(-a\varphi_p) \tag{26}$$

where φ_p represents the total volume fraction of precipitated secondary carbonate minerals. Jeen et al. (2007) showed a significantly better agreement with observed data using the surface passivation model than for using a two-third power law formulation (Eqn. 22). Similarly, Harrison et al. (2016) described the surface passivation using a volume-fraction threshold

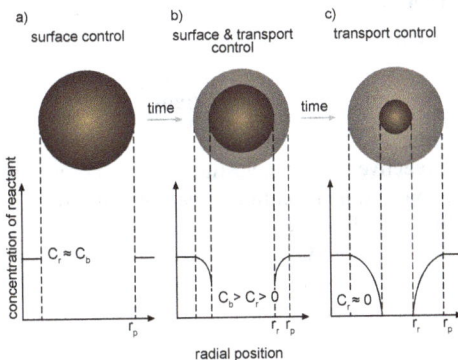

Figure 12. Concept of transitioning from surface-controlled mineral dissolution to transport-controlled conditions (shrinking core model). Cb and Cr respectively refer to the concentration of the reactant in the bulk solution and at the reactive surface.

model, implying that dissolution rates approached effectively zero above a threshold volume fraction for secondary precipitates. Also, Daval et al. (2009) described the passivation of wollastonite dissolution by the observed precipitation of calcite, by modeling a linear decrease in surface area with respect to calcite volume fraction.

Noiriel et al. (2012) studied the evolution of surface area during carbonate precipitation, based on imaging techniques, BET analysis and a micro-continuum model. The main conclusions were that the local increase in surface area arising from mineral precipitation had a significant impact on the integrated mineral concentrations and the breakthrough curves. Modelling the experiments with a constant surface area corresponding to the measured initial value failed to represent the observed data, indicating the necessity of taking into account the evolution of surface area during precipitation processes (Fig. 13). As discussed in more detail below (see case study by Beckingham et al. 2016, 2017), mineralogical heterogeneity and the evolution of reactive surface area can also substantially affect overall rock reactivity and solute release.

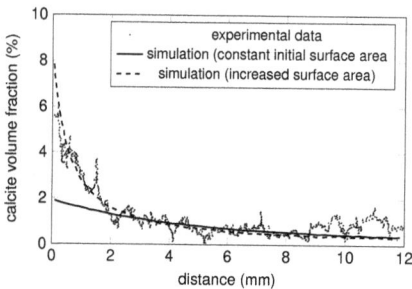

Figure 13. Reactive transport modeling of the accumulation of calcite in a capillary tube. Injection of calcite supersaturated solution results in precipitation of calcite on calcite seeds. In order to match the volume % calcite recorded by X-ray microtomography, it was necessary to use a higher specific surface area for neoformed calcite than was measured for the initial calcite grains in the capillary tube (reproduced from Noiriel et al. 2012)

CASE STUDIES OF REACTIVE TRANSPORT IN EVOLVING POROUS MEDIA

In the following, three case studies are presented covering the three main aspects of evolving porous media covered in this chapter, namely evolving permeability, diffusivity and reactivity. The first case study by Poonoosamy et al. (2015, 2016) focuses on permeability evolution in response to mineral-dissolution precipitation reactions. The second case study by Navarre-Sitchler et al. (2009) evaluates the effect of weathering on diffusivity on the pore scale and the continuum scale, while the third case study by Beckingham et al. (2016, 2017) focuses on the progress and evolution of mineral dissolution reactions in a heterogeneous rock matrix. These three case studies have provided successful comparison between experimental data and continuum approaches, incorporating details from pore-scale observations.

Effects of coupled dissolution–precipitation reactions on permeability and flow pattern

The experimental and modeling study by Poonoosamy et al. (2015, 2016) provides an example on the effect of coupled mineral dissolution–precipitation reactions, leading to a localized decrease of porosity with observable effects on flow and solute transport.

The 2D experimental setup, depicted in Figure 14, considers a layered porous medium composed on non-reactive quartz sand in which a reactive layer of celestite ($SrSO_4$), composed of two different particle sizes, was embedded. A concentrated solution of barium chloride was injected at a constant rate (100 μL/min) at the bottom left corner of the experimental reactor to induce celestite dissolution and barite precipitation ($BaSO_4$), associated with a porosity decrease. The effect of the porosity decrease on permeability was assessed by measuring fluid pressure upgradient and downgradient of the celestite layer over time.

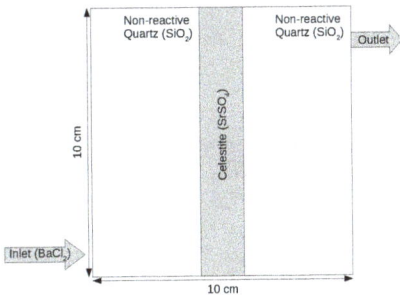

Figure 14. Schematic representation of the experimental setup by Poonoosamy et al. (2015). The 10cm×10cm domain contains non-reactive quartz near the inflow and outflow ports. Sandwiched in between is a reactive celestite layer. For reactive transport modeling, the celestite layer was subdivided into two regions containing a thin layer of finer grained material and a thicker layer of coarser-grained material.

Tracer tests were performed before and after completion of the experiment and showed significant discrepancies due to the porosity evolution of the reactive material (not shown). Barite precipitation was recorded throughout the reactive zones, with an enrichment near the left interface between the quartz sand and the region containing fine-grained celestite (Fig. 15a). High resolution µXRF imaging presented in Poonoosamy et al. (2016) provided insights into the dynamics of the dissolution/precipitation reactions. Dissolution of the small celestite grains was rapid and barite precipitation locally filled the open pore space. Larger celestite grains; however, were passivated by barite precipitation on the surface (Fig. 15b). Image analysis showed that barite precipitation led to partial surface passivation of the celestite grains as well as cementation of the solid matrix, disconnecting initially connected pores (Fig. 15b).

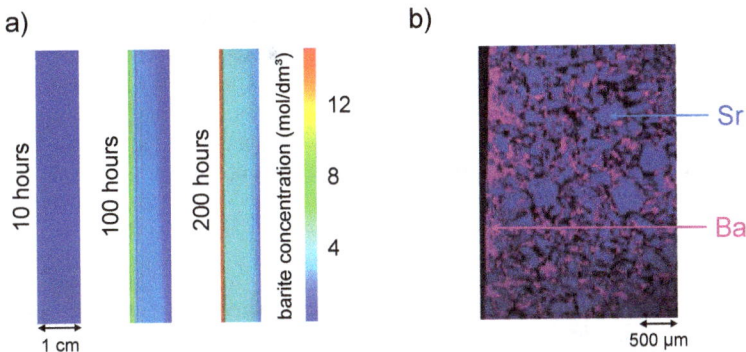

Figure 15. a) Barite precipitation patterns the in celestite-containing center of the domain after 10, 100 and 200 hours [Reproduced in modified form from J Poonoosamy et al. (2015) Dissolution–precipitation processes in tank experiments for testing numerical models for reactive transport calculations: Experiments and modelling. Journal of Contaminant Hydrology 177:1-17 with permission of Elsevier. Copyright 2015] **b)** EDS elemental mapping (pink: barium, dark blue: strontium) of a section of the celestite-containing zone at the end of the experiment, barite is seen to precipitate on the surface of celestite.

This zone of higher barite precipitation led to significant modifications of permeability and the flow patterns, as recorded by the measured pressure increase (Fig. 16a), despite maintaining constant flow rates. Figure 16b shows the dissolved quantities of celestite replaced by barite precipitation for the duration of the experiment.

These experimental results were analyzed with a reactive transport model to reproduce the observed mineral dissolution and precipitation patterns and the decrease in permeability. Reactive transport analysis required the ability to model density-dependent flow and transport, the evolution of porosity and the feedback of pore clogging on permeability and diffusion. Poonoosamy et al. (2015) presented simulations that reproduced the experimental results: evolution of the chloride concentration at the outlet (not shown), the increase in pressure due to pore clogging (Fig. 16a), and the amount of dissolved celestite and precipitated barite (Fig. 16b).

222

Seigneur et al.

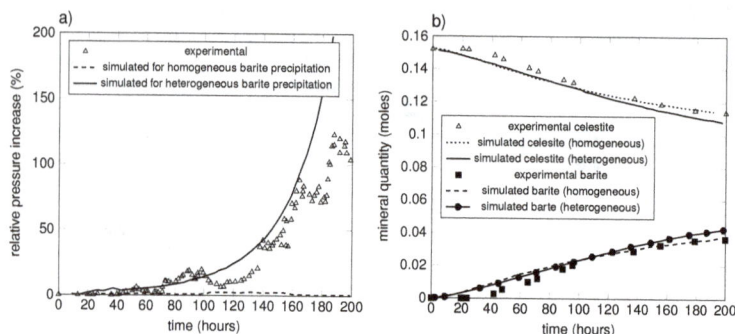

Figure 16. a) Relative pressure increase as a function of barite precipitation and pore clogging over the duration of the experiment **b)** Mineral quantities showing the dissolution of celestite and replacement by barite. [Reproduced in modified form from J Poonoosamy et al. (2015) Dissolution–precipitation processes in tank experiments for testing numerical models for reactive transport calculations: Experiments and modelling. Journal of Contaminant Hydrology 177:1-17 with permission of Elsevier. Copyright 2015.]

The impact of evolving porosity on tortuosity and permeability were modeled using Archie's law and the Kozeny–Carman relationship. To obtain agreement between the experimental results and the simulations, it was necessary to include a fine-grained celestite layer. Since a larger amount of small celestite grains are located in this zone, locally higher reaction rates were simulated which led to a more significant porosity and permeability reduction in a narrow region, which had a substantial impact on flow, matching the experimental values.

The benchmarking study of Poonoosamy et al. (2018) was based on these experiments. Several codes (CORE2D, MIN3P-THCm, OpenGeoSys-GEM, PFLOTRAN, TOUGHREACT), showed similar results regarding the evolution of porosity, mineral volume fraction, permeability and concentrations of the main elements. However, discrepancies in the spatial porosity and permeability patterns were observed. Therefore, the evolution of the flow patterns differ slightly between the codes, and discrepancies increase with time, and become more important when strong porosity changes occur (cases 3a and 3b from Poonoosamy et al. 2018).

Effects of dissolution on pore connectivity and diffusivity

To understand and predict the rates of chemical weathering under diffusion-controlled conditions, Navarre-Sitchler et al. (2009) carried out a combined X-ray synchrotron microtomography and tracer diffusion experiment on a sample of weathered basalt from a Costa Rican chronosequence. They used the resulting tortuosity-diffusivity model in the reactive transport code CrunchFlow (Steefel et al. 2015a) to simulate the evolution of the weathering rind over 340 thousand years (Navarre-Sitchler et al. 2011). Using a sample of basalt from a 125 thousand years Costa Rican alluvial terrace, they conducted a tracer experiment that was imaged with mXRF. Figure 17 shows the distribution of the bromide tracer (blue) after 7 days resulting from diffusion from left to right. Navarre-Sitchler et al. (2009) took the additional step of comparing these results to estimates of the diffusivity from pore-network modeling (see discussion below).

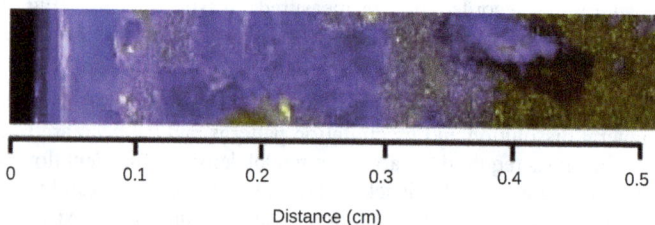

Figure 17. Bromide tracer (blue) diffusion profile after 7 days based on mXRF mapping at the Advanced Light Source, Lawrence Berkeley National Laboratory. Diffusion is from left to right. (Navarre-Sitchler et al. 2009)

Following the tracer experiment, Navarre-Sitchler et al. (2009) made use of X-ray synchrotron microtomography to map the pore structure of samples of the weathered basalt. Using a simple implementation of thresholding to map basalt versus pores, they were able to delineate chemical weathering related macroporous zones (> 4.4 µm voxel resolution) that were connected in 3-D. Navarre-Sitchler et al. (2009) then carried out numerical tracer diffusion experiments in 3-D cubes of weathered basalt by assuming a low diffusivity of 1.75×10^{-14} m^2 s^{-1} for the largely unconnected pore structure of unaltered basalt based on separate tracer diffusion experiments (not shown) and a free ion diffusivity of 10^{-9} m^2 s^{-1} (corresponding to a tortuosity of 1.0) for connected pores that can be fully resolved with the 4.4 µm discretization. Implemented in this way, the results of a 3-D numerical tracer diffusion experiment using CrunchFlow is similar to what could be obtained from a pore-network model. A 2-D slice through the skeletonized pore structure of one of the weathered zones is shown in Figure 18a (basalt in blue, pores in red). Results of the numerical tracer experiment are shown in Figure 18b, with diffusion of the tracer from the bottom of the figure towards the top. Note that in these 3-D tracer diffusion simulations, only two distinct tortuosities (or diffusion coefficients) are used and they are based on the segmented 3-D porosity map of the weathered sample.

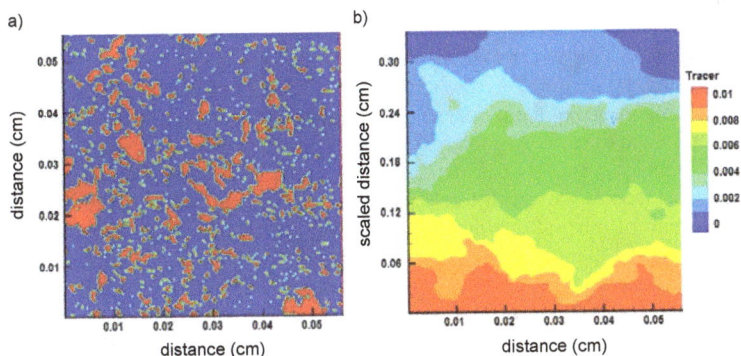

Figure 18. a) Segmented X-ray synchrotron microtomographic data collected at the Advanced Light Source at Lawrence Berkeley National Laboratory with a voxel resolution of 4.4 µm. Macropores developed as a result of chemical weathering in the basalt and are connected primarily in the third dimension (into the page), with red indicating pores, blue indicating basalt. **b)** Tracer diffusion simulation results using the pore structure shown in Figure 18a. Results are shown after 7 days of diffusion of the tracer from bottom to top. Simulation assumed a Dirichlet boundary condition at the bottom with a fixed tracer concentration of 0.01. See Navarre-Sitchler et al. (2009) for a more complete description of the experiment.

The segmented microtomography data on the weathered basalts was then used to quantify connected versus total porosity in the samples using the burning algorithm Percolate (Bentz et al. 2002). The data on all Costa Rican basalt samples analyzed indicate that connected porosity at > 4.4 µm increases markedly at about 9% total porosity (Fig. 19a).

The effective diffusion coefficient can be described with a modified Archie's Law (Eqn. 15) model that incorporates a critical porosity threshold ϕ_{crit} value of 9% (Eqn. 16), below which the porosity is considered to be largely unconnected. The parameter a in Equation (15) is taken as 1.3 while the parameter n is assumed to be 1.0 for 3-D volumes measuring 220 mm on a side (Navarre-Sitchler et al. 2009). Figure 19b compares experimental points to this relationship with the usual Archie's law described by Equation (13), which used the total porosity as mapped with the X-ray synchrotron microtomography with a 4.4 µm voxel resolution.

Based on this estimate of the effective diffusion coefficient shown in Figure 19b and captured in Equations (15) and (16), it was possible to compare the results of reactive transport simulations

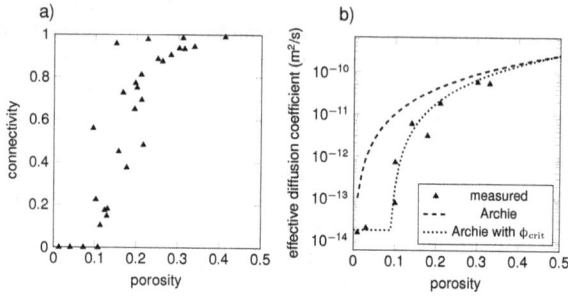

Figure 19. a) Connected versus total porosity for all sets of uCT data on Costa Ricanrasalts (Navarre-Sitchler et al. 2009). **b)** Values of D_l^e estimated using Archie's law (Eqn. 13 and 14) or from equation 15 using effective porosity (Eqn. 16) as the scaling parameter. These estimates are compared with measured liquid effective diffusion coefficients in basalt (combined data from Navarre-Sitchler et al. 2009 and Sato et al. 1997).

to observations of weathering rind thickness of basalt clasts in the Costa Rican alluvial terraces. As shown in the study, it is only possible to capture the four characteristic features of the weathered basalts, including (1) the mineralogy of weathering products, (2) weathering rind thickness, (3) the coincidence of plagioclase and pyroxene reaction fronts, and (4) the thickness of the weathering rinds, by including the modified Archie's Law formulation of effective diffusion coefficient. In particular, the coincidence of the pyroxene and plagioclase fronts, which have differing intrinsic solubilities and reactivities, is possible only with a threshold model of the kind implemented here (Navarre-Sitchler et al. 2011). In an unmodified Archie's Law formulation, the pyroxene and plagioclase fronts would move at different rates, so it is because of the generation of connected porosity as a result of mineral dissolution (weathering) that the mineral fronts advance in lock step. Figure 20 shows the position of the weathering fronts (data and solid lines model fit) as quantified using the mass transfer coefficients for calcium relative to immobile titanium at 240, 125 and 7 thousand years for quaternary terraces 1, 2 and 3, respectively.

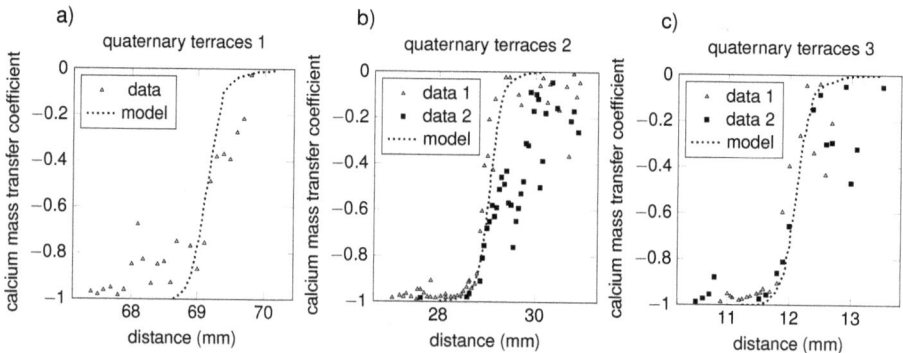

Figure 20. Model predictions compared to observed values of the calcium mass transfer coefficient. A positive value represents an enrichment of calcium relative to the immobile titanium. The three panels correspond to different quaternary terraces, and model predictions are compared to measured data. More information can be found in Navarre-Sitchler et al. (2011).

Evolving reactivity

A case study for evolving reactivity is provided by Beckingham et al. (2016, 2017) on highly reactive volcaniclastic sediments from the Nagaoka pilot CO_2 injection site in western Japan. The investigators carried out batch and coreflood experiments on a sample of sediment from the site to determine the time evolution of the sediment and fluid as a result of injection of CO_2-saturated fluid. They combined the experiments with multi-scale image analysis using synchrotron X-ray microCT, SEM QEMSCAN, XRD, SANS, and FIB-SEM to map the initial mineralogy and pore structure of the sediment (Fig. 21). The mineral mapping provided the initial conditions for continuum reactive transport modeling with the code CrunchFlow (Steefel et al. 2015a), and the simulation results were compared against effluent chemistry over the course of the batch and coreflood experiments.

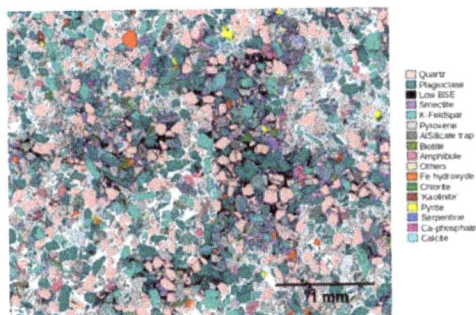

Figure 21. Registered SEM BSE and QEMSCAN image with multi-scale macro and micro pore connectivity analysis. Connected macro pores are depicted in white and connected micro-porous smectite regions are depicted in light blue [Reproduced form LE Beckingham et al. (2016) Evaluation of mineral reactive surface area estimates for prediction of reactivity of a multi-mineral sediment. Geochimica et Cosmochimica Acta188:310–329 with permission of Elsevier. Copyright 2016.].

The mapping with a multi-scale image analysis described provides both the modal mineralogy of the complex sediment and the co-location of connected porosity and reactive phases (Beckingham et al. 2016), thus defining more rigorously its bulk effective reactive surface area. In addition, the imaging was used to create a grain size distribution of the most reactive phases that could be used directly in the modeling. This was considered necessary because the fluid chemistry in both the batch and coreflood experiments evolved significantly over time, with early times showing very high cation (Ca^{2+} and Mg^{2+}) and silica concentrations that were as much as 10× higher than what they were at the end of the ~650 hours experiment.

The hypothesis was that the fine fraction of phases like pyroxene and plagioclase could account for the high reactivity early in the experiment because of their high specific surface area (Fig. 22). Using the mapping to constrain the volume fraction of the pyroxene and plagioclase (Table 1), it was possible to use the reactive transport simulations in both the batch and coreflood experiments to determine whether the pronounced evolved reactivity could be

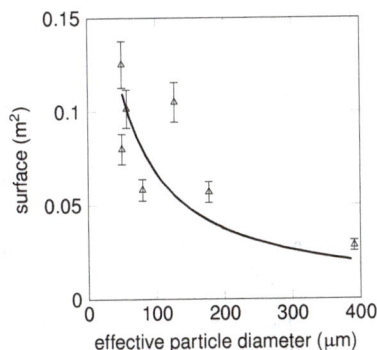

Figure 22: Literature values of BET surface area of pyroxene (Beckingham et al. 2016).

Table 1. Volume fractions and corresponding calculated surface areas for plagioclase minerals used in grain size distribution model (Beckingham et al. 2016)

Grain size (µm)	Weighted average diameter (µm)	Albite		An₂₅		Labradorite	
		Volume % total plagioclase	Surface area (m²/g)	Volume % total plagioclase	Surface area (m²/g)	Vol % total plagioclase	Surface area (m²/g)
1–20	0.025	0.0007	393.89	0.00016	375.27	0.0015	379.41
1–20	0.1	0.0007	98.47	0.00016	93.82	0.0015	94.85
1–20	0.5	0.0007	19.69	0.00016	18.76	0.0015	18.97
1–20	5	0.0007	1.97	0.00016	1.88	0.0015	1.89
20–40	23.38	0.32	0.42	0.07	0.40	0.66	0.41
40–60	43.02	1.26	0.23	0.29	0.22	2.63	0.22
60–80	66.99	1.79	0.15	0.41	0.14	3.72	0.14
80–100	90.22	1.67	0.11	0.38	0.10	3.47	0.11
100–120	108.94	2.04	0.091	0.47	0.086	4.24	0.087
120–140	131.07	2.19	0.075	0.51	0.072	4.57	0.072
140–160	150.12	2.53	0.066	0.58	0.062	5.27	0.063
160–180	166.74	3.47	0.059	0.80	0.056	7.22	0.057
180–200	190.33	4.13	0.052	0.95	0.049	8.60	0.049
200–220	212.11	1.42	0.046	0.33	0.044	2.97	0.045
220–240	227.50	2.64	0.043	0.61	0.041	5.49	0.042
240–260	247.05	3.38	0.039	0.78	0.038	7.04	0.038
260–280	274.33	3.02	0.036	0.71	0.034	6.42	0.034

attributed to the early reaction and then disappearance of the finest mineral fractions. While the trends were correct for pyroxene and plagioclase, it was found that it was also necessary to include a reactive glass phase to account for the high concentrations and reactivity (Fig. 23).

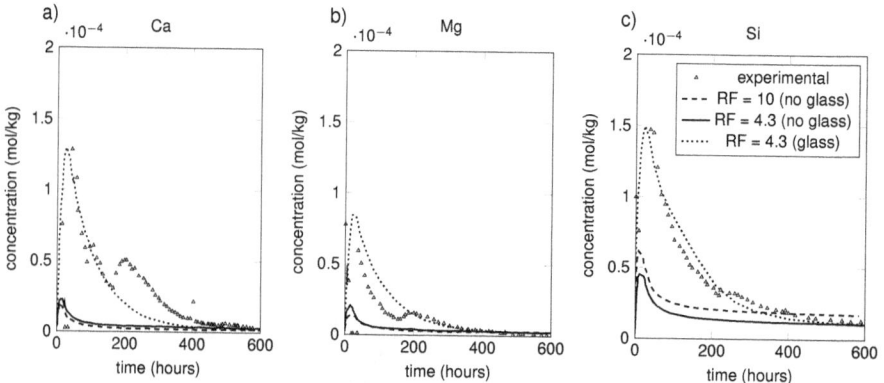

Figure 23. Observed (black circles) and model predicted (red lines) effluent concentrations incorporating grain size distributions for plagioclase and pyroxene minerals, with and without inclusion of glass phase. (Beckingham et al. 2016).

CHALLENGES OF AND OPPORTUNITIES
FOR REACTIVE TRANSPORT MODELING
IN EVOLVING POROUS MEDIA

State of the science for modeling reactive transport in evolving porous media

For the aforementioned reasons and given the large variety of pore structures and heterogeneities, it is evident that obtaining continuum scale relationships for effective transport properties and reactivities in porous media, in particular under dynamically evolving conditions, constitutes a major challenge.

Accurate simulation of breakthrough curves, mineral abundances and concentration distributions relies heavily on the adequate evaluation of the evolution of the porous medium's effective porosity, transport and reaction properties. As discussed above, a wide range of feedback relationships exists, relating transport parameters back to porosity; however, these relationships provide diverse predictions of the transport properties, in some cases differing by several orders of magnitude. In addition, it has been shown that identical systems might evolve differently as a function of transport regime, as well as the ratio between the time scales of reaction and transport. Furthermore, it is also common that relationships between transport parameters and porosity themselves evolve, as the pore structure is modified. Keeping in mind that the evolution of porous media is predominantly affected by processes occurring at the pore scale, and that mineralogical compositions and reactivities can vary widely, it becomes clear that the predictive capabilities of relationships describing the evolution of porosity, tortuosity, permeability and reactivity are limited. It is necessary to constrain these relationships by observational data on a case-by-case basis, although we are confident that some generalizations can eventually be drawn in terms of processes and materials. To this regard, the presented case studies show adequate approaches on how to incorporate pore-scale observations into the continuum approach. Although some theoretical underpinnings exist, relationships to describe the evolution of transport and reactivity parameters remain empirical in nature and are case specific. However, despite these shortcomings, empirical relationships, often in the form of power laws, have proven useful to provide a quantitative interpretation of observational data under dynamic and evolving conditions. Examples include problems involving diffusive mass transport (Seigneur et al. 2017; Georget et al. 2018), changing permeability (Mok et al. 2002; Luhmann et al. 2014) and evolving reactivity (Jeen et al. 2007; Noiriel et al. 2009). Although empirical in nature, these relationships provide insight and an opportunity for quantitative interpretation of feedback mechanisms.

Extreme cases: Complete clogging and solid structure collapse

It is relatively trivial to implement these empirical relationships in continuum scale reactive transport codes. However, the simulation of extreme cases of evolving porous media poses a challenge. On the one hand, complete clogging causes the disappearance of the water phase, which most geochemical solvers used in simulators are not able to deal with. Typically, reactive transport codes deal with the clogging problem by setting an arbitrary porosity threshold value, below which groundwater flow and solute migration does not occur. Although benchmarking exercises have been successful at verifying the correct implementation of permeability–porosity and tortuosity–porosity relationships, obtaining agreement between different reactive transport codes under clogging conditions has been more difficult (Xie et al. 2015). On the other hand, simulations can yield a substantial increase of porosity that, in fact, might not be physically representative, since the solid matrix would eventually collapse. This constitutes a mechanical issue which at this point in time is not commonly considered by reactive transport simulators.

Unsaturated flow relationships

Modelling of unsaturated cases, whether through Richards equation (Eqn. 2) or two-phase approaches (Eqn. 1), requires the use of relationships to take into account the material's hydraulic and transport properties. The models of Brooks Corey, van Genuchten provide the water retention characteristics (capillary-pressure), as well as the relative permeabilities of the fluid phases. These models involve several different empirical parameters. The latter have been assessed for different porous media and general understanding of their dependence on physical properties has been acquired (e.g., Zhang et al. 2017 for the dependence on grain sizes). Della Vecchia et al. (2015) also described a model linking pore size distribution to water retention models, specifically for clays. Similarly, Suazo et al. (2016) studied the evolution of the water retention curves during cement hydration. However, only a limited amount of research has been dedicated to the prediction of how these parameters evolve as the porous media undergoes chemical perturbations. Furthermore, no clear relationship between the unsaturated parameters and porosity has been described. For these reasons, most reactive transport simulations consider these parameters to be constant.

Numerical approaches for coupling of flow and reactive transport

The reactive transport problem is commonly solved by either the global implicit approach (e.g., Crunchflow, MIN3P), the sequential-iterative approach (Hytec), or the sequential non-iterative approach (e.g., PHT3D, HP1) (Steefel et al. 2015a). Independent of the numerical methods used, permeabilities and reactivities are typically updated in a time-lagged fashion (e.g., Mayer et al. 2002 for MIN3P). While this is not necessarily the case for porosity and diffusion coefficient (Lagneau and Van Der Lee 2010), this implies that at each time step, small mass balance errors are introduced.

An additional complication occurs due to the feedback of evolving porous media parameters between reactive transport and flow. In most models, groundwater flow and multicomponent reactive transport are solved sequentially. This approach is adequate if there is no feedback of the reactive transport solution on the flow problem. However, as depicted in Figure 1, this is not the case for evolving porous media, since a change in porosity affects the storage term in the flow equation (Eqn. 1 and Eqn. 2) and the flux term, by modifying the permeability. The decoupled treatment of groundwater flow and reactive transport makes it necessary to take relatively small time steps to avoid the introduction of mass balance errors. However, recent efforts to recouple multiphase flow and reactive transport in case of variable porosity have been implemented in Hytec (Seigneur et al. 2018).

Towards an integrated model for evolving porous media

As outlined above, the concepts of surface area, tortuosity and (effective) porosity play a significant role in evolving porous media. Usually, tortuosity and surface area are linked to diffusion and reactivity. Empirical relationships are generally used to update their evolution. Relationships for different evolving parameters are commonly independent from one another, despite the fact that they are inherently correlated, as suggested by numerical studies by Koponen et al. (1997). While general and universal correlations between these relationships are difficult to obtain, common trends can be estimated. Typically, as porosity increases, surface area tends to decrease (Koponen et al. 1997), while diffusivity (and hence, as defined in this chapter, tortuosity) increases (Knackstedt and Zhang 1994; Koponen et al. 1997). Saripalli et al. (2002) have also shown that tortuosity can be related to surface area (a higher surface area leads to a lower diffusivity/tortuosity).

Most of the efforts to estimate effective transport properties such as tortuosity and permeability have focused on developing relationships invoking porosity, tortuosity and/or surface area. While permeability and diffusivity represent fundamentally different transport processes, their dependence on tortuosity (hence surface area) and on the pore throat size are similar.

As was observed for the simulation of clogging reactions during cement carbonation, systematic errors are encountered when attempting to reproduce experimentally observed alteration depths. A failure to reproduce observed degradation depths, while managing to reproduce breakthrough curves, might be an indication of the need to work towards a more integrated framework of evolving parameters.

While the mathematical description of such an integrated approach is not yet determined, it is evident than an approach linked to porosity evolution alone is not suited to properly describe the processes which occur at the pore-scale and the continuum scale. Instead, geometrical considerations about the structural organization of the pore scale might be more appropriate. As an example, the surface area in itself carries more information about the potential interactions between the fluid and the solid matrix. On the one hand, high surface area indicates tortuous pathways which are linked with lower transport properties. In addition, reactivity is also intrinsically linked to the surface area.

Let us consider the observed inverse correlations between permeability and porosity. An integrated generalized approach based on surface area might be able to reproduce these results in a better way than one based on porosity alone.

Opportunities related to imaging techniques, pore scale modeling and upscaling

In recent years, imaging techniques have helped to improve our understanding of pore structure and its evolution, as well as the effective transport properties. Such techniques, combined with high-resolution of reactive transport at the pore scale (Molins et al. 2012; Steefel et al. 2013; Molins et al. 2014) are destined to be used increasingly, as they allow for a process-based assessment of the evolution of the pore structure. Imaging and pore scale modeling techniques have already provided outstanding results and insights. Molins et al. (2017) reproduced dissolution processes in a multi-mineral fracture zone by incorporating the heterogeneity of the spatial distribution of minerals. In another study, it was possible to reproduce the wormholing processes by direct simulation at the pore scale (Molins 2015). In addition, pore scale modeling has provided a means to document and quantitatively assess the effect of dissolution and precipitation reactions on the pore structure, as they occur within a porous material (Molins 2015; Noiriel 2015; Bultreys et al. 2016; Pereira Nunes et al. 2016). Future improvement in the resolution of imaging techniques might even make it possible to characterize porous media at even smaller scales and improve pore scale models built from these methods. However, it is unlikely that pore scale simulations will be able soon to deal with the large time and spatial scales required for engineering purposes.

Pore scale modelling is a promising tool and may eventually allow the derivation and upscaling of transport parameters for use in a continuum approach. This modeling technique may be able to help overcome the biggest gaps which still exist for the continuum approach, in particular when applied to evolving porous media. In the future, it is likely that combined approaches (micro-continuum) will provide a tool to bring the knowledge acquired from the pore space to the macroscale (Steefel et al. 2015b).

CONCLUSIONS

Biogeochemical reactions in porous media, involving mineral dissolution and precipitation, can lead to significant modifications of the pore structure with a potential dramatic evolution of the material's properties in a wide variety of natural and engineered systems. Reactive transport modelling is a useful deterministic tool to investigate these strongly coupled nonlinear effects. However, an accurate description of these phenomena requires a thorough understanding of their dynamics at the pore scale.

The last decade has provided major improvements with pore scale imaging and simulation. First, imaging techniques have allowed to reveal more detailed information about pore structures and their evolutions at a sub-micron level. Second, the parallel increase in computing capabilities opened the door to pore-scale simulations which can now be performed on 3D image samples. These tools, used together, provide precious data for developing an improved understanding of how porous media evolve under different conditions.

However, there are still major shortcomings to overcome. First, the resolution of imaging techniques is currently insufficient to provide adequate insights into the 3D-structure of fine grained materials. Indeed, porous media in which nanometric pores constitute the building blocks of the pore structure (clays, cement), cannot be accurately resolved by these imaging techniques at this point in time. The second problem, potentially more challenging, is that reactive transport simulations relevant to engineering problems usually involve scales (m–km) which are orders of magnitude larger than the scales at which processes relevant to porous media evolution occur (nm–μm). Continuum approaches for these engineering purposes likely still have a promising future, despite their inherent limitations.

Some significant limitations remain. The clogging phenomenon still constitutes a challenge due to the inherent complexity of capturing the effect on transport processes when near zero porosities are approached and the associated numerical issues. Furthermore, there is still a lack of experimental data to calibrate reactive transport models. Additionally, the development of fractures due to internal stresses (either due to heat generation, mineral dissolution or precipitation) have been observed in different experimental setups. Unfortunately, a full coupled problem involving heat and mechanics is not available.

Despite obvious challenges inherent to the continuum approach regarding the evolution of relevant parameters describing a porous medium's properties (tortuosity, permeability, reactivity and surface area), quantitative reactive transport analysis of evolving systems has been successful. Additionally, significant knowledge has been acquired through recently developed micro-continuum approaches. Empirical relations for transport (such as Archie's law, the Kozeny–Carman relationships or various power laws) or reactivity (threshold passivation or shrinking-core model) are not meant to be physically representative of pore-scale process. It must be acknowledged that it is not realistic to expect that using these relationships would lead to a fully process-based assessment of experimental data. Nevertheless, we now possess sufficient knowledge about how certain processes can be modelled through the use of these empirical relations.

ACKNOWLEGMENTS

The contributions by Nicolas Seigneur were supported by a postdoctoral fellowship through the TERRE-NET program, a strategic network funded by NSERC (Natural Science and Engineering Research Council of Canada).

K. Ulrich Mayer would like to acknowledge funding by NSERC through a Discovery Grant.

The contributions by Carl I. Steefel were supported by the Director, Office of Science, Basic Energy Sciences, Chemical Sciences, Geosciences, and Biosciences Division, of the U.S. Department of Energy under Contract No. DE-AC02-05CH11231 to Lawrence Berkeley National Laboratory.

We would like to acknowledge Jenna Poonoosamy for providing additional material for Figure 15b. We would like to thank Tom Al (UOttawa) for providing the images for Figure 3.

REFERENCES

Aharonov E, Tenthorey E CH S (1998) Precipitation sealing and diagenesis: 2. Theoretical analysis. J Geophys Res Solid Earth 10:23969–23981

Aharonov E, Spiegelman M, Kelemen P (2004) Three-dimensional flow and reaction in porous media: Implications for the Earth's mantle and sedimentary basins. J Geophys Res: Solid Earth 102:14821–14833

Akanni KA, Evans JW, Abramson IS (1987). Effective transport coefficients in heterogeneous media. Chem Eng Sci 42:1945–1954

Algive L, Bekri S, Vizika O (2010) Pore-network modeling dedicated to the determination of the petrophysical-property changes in the presence of reactive fluid. SPE J 15:618–633

Algive L, Békri S, Nader FH, Lerat O, Vizika O (2012) Impact oraf diagenetic alterations on the petrophysical and multiphase flow properties of carbonate rocks using a reactive pore network modeling approach. Oil & Gas Science and Technology–Revue d'IFP Energies nouvelles 67:147–160

Andreani M, Luquot L, Gouze P, Godard M, Hoise E, Gibert B (2009) Experimental study of carbon sequestration reactions controlled by the percolation of CO_2-rich brine through peridotites. Environ Sci Technol 43:1226-1231

Atkinson A, Everitt NM, Guppy R (1987) Evolution of pH in a radwaste repository: Experimental simulation of cement leaching. UKAEA Harwell Lab.(UK). Materials Development Div

Auroy M, Poyet S, Le Bescop P, Torrenti J-M (2013) Impact of carbonation on the durability of cementitious materials: water transport properties characterization. EPJ Web of Conferences 56:01008

Barlet-Gouedard V, Rimmele G, Goffe B, Porcherie O (2006) Mitigation strategies for the risk of CO_2 migration through wellbores. *In:* IADC/SPE Drilling Conference. Soc Petrol Eng

Beckingham LE, Peters CA, Um W, Jones KW, Lindquist W (2013) 2D and 3D imaging resolution trade-offs in quantifying pore throats for prediction of permeability. Adv Water Resour 62:1–12

Beckingham LE, Mitnick EH, Steefel CI, Zhang S, Voltolini M, Swift AM, Yang L, Cole DR, Sheets JM, Ajo-Franklin JB, DePaolo DJ (2016) Evaluation of mineral reactive surface area estimates for prediction of reactivity of a multi-mineral sediment. Geochim Cosmochim Acta 188:310–329

Beckingham LE, Steefel CI, Swift AM, Voltolini M, Yang L, Anovitz LM, Sheets JM, Cole DR, Kneafsey TJ, Mitnick EH, Zhang S (2017) Evaluation of accessible mineral surface areas for improved prediction of mineral reaction rates in porous media. Geochim Cosmochim Acta 205:31–49

Bentz DP, Garboczi EJ (1991) Percolation of phases in a three-dimensional cement paste microstructural model. Cem Concr Rese 21:325–344

Bentz DP, Garboczi EJ (1992) Modeling the leaching of calcium hydroxide from cement paste : effects on pore space percolation and diffusivity model description and technique microstructure Model Mater Struct 25:523–533

Bentz DP, Mizell S, Satterfield SG, Devaney JE, George WL, Ketcham PM, Graham JR, Porterfield JE, Quenard DA, Vallee FA, Sallee H (2002) The visible cement data set. J Res Nati Inst Stand Technol 107:137

Bernabé Y, Mok U, Evans B (2003) Permeability–porosity relationships in rocks subjected to various evolution processes. Pure Appl Geophys 160:937–960

Bertron A, Coutand M, Cameleyre X, Escadeillas G, Duchesne J (2005) Attaques chimique biologique des effluents agricoles et agroalimentaires sur les matériaux cimentaires. Matér Tech 93:111–121

Bickmore BR, Nagy KL, Young JS, Drexler JW (2001) Nitrate-cancrinite precipitation on quartz sand in simulated Hanford tank solutions. Environ Sci Technol 15:22

Bildstein O, Worden RH, Brosse E (2001) Assesment of anhydrite dissolution as the rate-limiting step during thermochemical sulfate reduction. Chem Geol 176:173–189

Birk S, Liedl R, Sauter M, Teutsch G (2003) Hydraulic boundary conditions as a controlling factor in karst genesis: A numerical modeling study on artesian conduit development in gypsum. Water Resour Res 39:1004

Birk S, Liedl R, Sauter M, Teutsch G (2005) Simulation of the development of gypsum maze caves. Environ Geol 48:296–306

Bjørlykke K, Ramm M, Saigal GC (1989) Sandstone diagenesis and porosity modification during basin evolution. Geologische Rundschau 78:243–268

Blowes DW, Jambor JL (1990) The pore-water geochemistry and the mineralogy of the vadose zone of sulfide tailings, Waite Amulet, Quebec, Canada. Appl Geochem 5:327–346

Blowes DW, Reardon EJ, Jambor JL, Cherry JA (1991) The formation and potential importance of cemented layers in inactive mine tailings. Geochim Cosmochim Acta 55:965–978

Blowes DW, Ptacek CJ, Benner SG, McRae CWT, Bennett TA, Puls RW (2000) Treatment of inorganic contaminants using permeable reactive barriers. J Contam Hydrol 45:123–137

Blunt MJ, Bijeljic B, Dong H, Gharbi O, Iglauer S, Mostaghimi P, Paluszny A, Pentland C (2013) Pore-scale imaging and modelling. Adv Water Resour 51:197–216

Boudreau BP (1996) The diffusive tortuosity of fine-grained unlithified sediments. Geochim Cosmochim Acta 60:3139–3142

Brookfield AE, Blowes DW, Mayer KU (2006) Integration of field measurements and reactive transport modelling to evaluate contaminant transport at a sulfide mine tailings impoundment. J Contam Hydrol 20:1–22

Brovelli A, Malaguerra F, Barry DA (2009) Bioclogging in porous media: Model development and sensitivity to initial conditions. Environ Modell Software 24:611–626

Brunet JP, Li L, Karpyn ZT, Kutchko BG, Strazisar B, Bromhal G (2013) Dynamic evolution of cement composition and transport properties under conditions relevant to geological carbon sequestration. Energy Fuels 21:8

Budek A, Szymczak P (2012) Network models of dissolution of porous media. Phys Rev E 86:4208–4220

Buijse MA, Glasbergen G (2005) A semi-empirical model to calculate wormhole growth in carbonate acidizing. *In:* SPE Annual Technical Conference and Exhibition. Soc Petrol Eng

Bultreys T, Boone MA, Boone MN, De Schryver T, Masschaele B, Van Hoorebeke L, Cnudde V (2016) Fast laboratory-based micro-computed tomography for pore-scale research: illustrative experiments and perspectives on the future. Adv Water Res 95:341–351

Buss HL, Lüttge A, Brantley SL (2007) Etch pit formation on iron silicate surfaces during siderophore-promoted dissolution. Chem Geol 240:326–342

Chagneau A, Claret F, Enzmann F, Kersten M, Heck S, Madé B, Schäfer T (2015) Mineral precipitation-induced porosity reduction and its effect on transport parameters in diffusion-controlled porous media. Geochem Trans 16:13

Chou H, Wu L, Zeng L, Chang A (2012) Evaluation of solute diffusion tortuosity factor models for variously saturated soils. Water Resour Res 48:W10539

Civan F (2001) Scale effect on porosity and permeability: Kinetics, model, and correlation. AIChE journal 47:271–87

Colón CF, Oelkers EH, Schott J (2004) Experimental investigation of the effect of dissolution on sandstone permeability, porosity, and reactive surface area1. Geochim Cosmochim Acta 68:805–817

Daccord G, Touboul E, Lenormand R (1989) Carbonate acidizing: toward a quantitative model of the wormholing phenomenon. SPE Prod Eng 4:63–68

Dauzères A, De Windt L, Sammaljärvi J, Bartier D, Techer I, Detilleux V, Siitari-Kauppi M (2019) Mineralogical and microstructural evolution of Portland cement paste/argillite interfaces at 70° C–Considerations for diffusion and porosity properties. Cem Concr Res 115:414–25

Dauzères A, Le Bescop P, Sardini P, Coumes CC (2010) Physico-chemical investigation of clayey/cement-based materials interaction in the context of geological waste disposal: Experimental approach and results. Cem Concr Res 40:1327–1340

Daval D, Martinez I, Corvisier J, Findling N, Goffé B, Guyot F (2009) Carbonation of Ca-bearing silicates, the case of wollastonite: Experimental investigations and kinetic modeling. Chem Geol 265:63–78

De Boever E, Varloteaux C, Nader FH, Foubert A, Békri S, Youssef S, Rosenberg E (2012) Quantification and prediction of the 3D pore network evolution in carbonate reservoir rocks. Oil Gas Sci Technol–Revue d'IFP Energies nouvelles 67:161–178

De Windt L, Badreddine R (2007) Modelling of long-term dynamic leaching tests applied to solidified/stabilised waste. Waste Manage 27:1638–1647

Della Vecchia G, Dieudonné A-C, Jommi C, Charlier R (2015) Accounting for evolving pore size distribution in water retention models for compacted clays. Int J Numer Anal Methods Geomech 39:702–723

Demers I, Molson J, Bussière B, Laflamme D (2013) Numerical modeling of contaminated neutral drainage from a waste-rock field test cell. Appl Geochem 33:346–356

Deng H, Spycher N (2019) Modeling reactive transport processes in fractures. Rev Mineral Geochem 85:49–74

Deng H, Molins S, Steefel C, DePaolo D, Voltolini M, Yang L, Ajo-Franklin JA (2016) 2.5 D reactive transport model for fracture alteration simulation. Environ Sci Technol 50:7564–7571

Deng H, Molins S, Trebotich D, Steefel CI, DePaolo D (2018) Pore-scale numerical investigation of the impacts of surface roughness: Upscaling of reaction rates in rough fractures. Geochim Cosmochim Acta 239:374–389

Dentz M, Gouze P, Carrera J (2011) Effective non-local reaction kinetics for transport in physically and chemically heterogeneous media. J Contam Hydrol 120–121:222–236

Detwiler RL, Glass RJ, Bourcier WL (2003) Experimental observations of fracture dissolution: The role of Peclet number on evolving aperture variability. Geophys Res Lett 30:1648

Dewanckele J, De Kock T, Boone MA, Cnudde V, Brabant L, Boone MN, Fronteau G, Van Hoorebeke L, Jacobs P (2012) 4D imaging and quantification of pore structure modifications inside natural building stones by means of high resolution X-ray CT Sci Total Environ 416:436–48

Diamond S (2000) Mercury porosimetry: an inappropriate method for the measurement of pore size distributions in cement-based materials. Cem Concr Res 30:1517–1525

Dobson PF, Kneafsey TJ, Hulen J, Simmons A (2003a) Porosity, permeability, and fluid flow in the Yellowstone geothermal system, Wyoming. J Volcanol Geothermal Res 123:313–324

Dobson PF, Kneafsey TJ, Sonnenthal EL, Spycher N, Apps JA (2003b) Experimental and numerical simulation of dissolution and precipitation: implications for fracture sealing at Yucca Mountain, Nevada. J Contam Hydrol 62:459–476

Doulati Ardejani F, Jannesar Malakooti S, Ziaedin Shafaei S, Shahhosseini M (2014) A numerical multi-component reactive model for pyrite oxidation and pollutant transportation in a pyritic, carbonate-rich coal waste pile in Northern Iran. Mine Water Environ 33:121–132

Egermann P, Bekri S, Vizika O (2010) An integrated approach to assess the petrophysical properties of rocks altered by rock-fluid interactions (CO_2 injection). Petrophysics 51: SPWLA-2010-v51n1a2

Emmanuel S, Berkowitz B (2007) Effects of pore-size controlled solubility on reactive transport in heterogeneous rock. Geophys Res Lett 34:L06404

Emmanuel S, Ague JJ, Walderhaug O (2010) Interfacial energy effects and the evolution of pore size distributions during quartz precipitation in sandstone. Geochim Cosmochim Acta 74:3539–3552

Emmanuel S, Anovitz LM, Day-Stirrat RJ (2015) Effects of coupled chemo-mechanical processes on the evolution of pore-size distributions in geological media. Rev Mineral Geochem 80:45–60

Ezeuko CC, Sen A, Grigoryan A, Gates ID (2011) Pore-network modeling of biofilm evolution in porous media. Biotechnology and bioengineering 108:2413–2423

Ferrier RJ, Cai L, Lin Q, Gorman GJ, Neethling SJ (2016) Models for apparent reaction kinetics in heap leaching: A new semi-empirical approach and its comparison to shrinking core and other particle-scale models. Hydrometallurgy 166:22–33

Fontaine FJH, Rabinowicz M, Boulègue J (2001) Permeability changes due to mineral diagenesis in fractured crust: implications for hydrothermal circulation at mid-ocean ridges. Earth Planet Sci Lett 184:407–425

Fox DT, Guo L, Fujita Y, Huang H, Redden G (2016) Experimental and numerical analysis of parallel reactant flow and transverse mixing with mineral precipitation in homogeneous and heterogeneous porous media. Transp Porous Media 111:605–626

Fredd CN (2000) Dynamic model of wormhole formation demonstrates conditions for effective skin reduction during carbonate matrix acidizing. *In:*SPE Permian Basin Oil and Gas Recovery Conference. Soc Petrol Eng

Fredd CN, Fogler HS (1999) Optimum conditions for wormhole formation in carbonate porous media: Influence of transport and reaction. SPE J 4:196–205

Galíndez JM, Molinero J, Samper J, Yang CB (2006) Simulating concr degradation processes by reactive transport models. J Phys (Paris) IV (Proceedings) 136:177–188

Gao J, Xing H, Tian Z, Pearce JK, Sedek M, Golding SD, Rudolph V (2017) Reactive transp porous media for CO_2 sequestration: Pore scale modeling using the lattice Boltzmann method. Comput Geosci 98:9–20

Garcia-Rios M, Luquot L, Soler JM, Cama J (2017) The role of mineral heterogeneity on the hydrogeochemical response of two fractured reservoir rocks in contact with dissolved CO_2. Appl Geochem 84:202–17

Garing C, Gouze P, Kassab M, Riva M, Guadagnini A (2015) Anti-correlated porosity–permeability changes during the dissolution of carbonate rocks: experimental evidences and modeling. Transp Porous Media 107:595–621

Gaucher EC, Blanc P (2006) Cement/clay interactions–a review: experiments, natural analogues, and modeling. Waste Manage 26:776–788

Georget F, Prévost JH, Huet B (2018) Impact of the microstructure model on coupled simulation of drying and accelerated carbonation. Cem Concr Res 104:1–2

Gherardi F, Xu T, Pruess K (2007) Numerical modeling of self-limiting and self-enhancing caprock alteration induced by CO_2 storage in a depleted gas reservoir. Chem Geol 244:103–129

Giger SB, Tenthorey E, Cox SF, Fitz Gerald JD (2007) Permeability evolution in quartz fault gouges under hydrothermal conditions. J Geophys Res: Solid Earth 112:B07202

Giles MR, Indrelid SL, Beynon GV, Amthor J (2000) The origin of large-scale quartz cementation: evidence from large data sets and coupled heat-fluid mass transport modelling. Spec Publ Int Assoc Sedimentol 29:21–38

Golfier F, Zarcone C, Bazin B, Lenormand R, Lasseux D, Quintard M (2002) On the ability of a Darcy-scale model to capture wormhole formation during the dissolution of a porous medium. J Fluid Mech 457:213–254

Gouze P, Coudrain-Ribstein A (2002) Chemical reactions and porosity changes during sedimentary diagenesis. Appl Geochem 17:39–47

Gouze P, Luquot L (2011) X-ray microtomography characterization of porosity, permeability and reactive surface changes during dissolution. J Contam Hydrol 120:45–55

Gray F, Anabaraonye B, Shah S, Boek E, Crawshaw J (2018) Chemical mechanisms of dissolution of calcite by HCl in porous media: Simulations and experiment. Adv Water Resour 121:369–387

Hammarstrom JM, Sibrell PL, Belkin HE (2003) Characterization of limestone reacted with acid-mine drainage in a pulsed limestone bed treatment system at the Friendship Hill National Historical Site, Pennsylvania, USA Appl Geochem 18:1705–1721

Hao L, Zhang S, Dong J, Ke W (2012) Evolution of corrosion of MnCuP weathering steel submitted to wet/dry cyclic tests in a simulated coastal atmosphere. Corrosion Sci 58:175–180

Hao Y, Smith M, Sholokhova Y, Carroll S (2013) CO_2-induced dissolution of low permeability carbonates. Part II: Numerical modeling of experiments. Adv Water Resour 62:388–408

Harrison AL, Dipple GM, Power IM, Mayer KU (2016) The impact of evolving mineral–water-gas interfacial areas on mineral–fluid reaction rates in unsaturated porous media. Chem Geol 421:65–80

Harrison AL, Dipple GM, Song W, Power IM, Mayer KU, Beinlich A, Sinton D (2017) Changes in mineral reactivity driven by pore fluid mobility in partially wetted porous media. Chem Geol 463:1–11

Hartmann A, Goldscheider N, Wagener T, Lange J, Weiler M (2014) Karst water resources in a changing world: Review of hydrological modeling approaches. Rev Geophys 52:218–242

Hazen TC, Fliermans CB (1995) Bioremediation of contaminated groundwater. Google Patents

Henderson TH, Mayer KU, Parker BL, Al TA (2009) Three-dimensional density-dependent flow and multicomponent reactive transport modeling of chlorinated solvent oxidation by potassium permanganate. J Contam Hydrol 106:195–211

Herbert H-J, Kasbohm J, Sprenger H, Fernández AM, Reichelt C (2008) Swelling pressures of MX-80 bentonite in solutions of different ionic strength. Phys Chem Earth, Parts A/B/C 33, S327–S342

Hesse MA, Strazisar BR, Bryant SL, Wenning QC, Huerta NJ, Lopano CL (2013) Development of reacted channel during flow of CO_2 Rich Water Along a Cement Fracture. Energy Procedia 37:5692–5701

Hoefner ML, Fogler HS (1988) Pore evolution and channel formation during flow and reaction in porous media. AIChE J 34:45–54

Hommel J, Coltman E, Class H (2018) Porosity–permeability relations for evolving pore space: a review with a focus on (bio-) geochemically altered porous media. Transp Porous Media 124:589–629

Houben GJ (2003) Iron oxide incrustations in wells. Part 1: genesis, mineralogy and geochemistry. Appl Geochem 18:927–939

Huber C, Shafei B, Parmigiani A (2014) A new pore-scale model for linear and non-linear heterogeneous dissolution and precipitation. Geochim Cosmochim Acta 124:109–130

Huet BM, Prevost JH, Scherer GW (2010) Quantitative reactive transport modeling of Portland cement in CO_2-saturated water. Int J Greenhouse Gas Control 4:561–574

Ivanov V, Chu J (2008) Applications of microorganisms to geotechnical engineering for bioclogging and biocementation of soil in situ. Rev Environ Sci Bio/Technol 7:139–153

Jaboyedoff M, Baillifard F, Bardou E, Girod F (2004) The effect of weathering on Alpine rock instability. Q J Eng Geol Hydrogeol 37:95–103

Jacquemet N, Pironon J, Lagneau V, Saint-Marc J (2012) Armouring of well cement in H_2S–CO_2 saturated brine by calcite coating–Experiments and numerical modelling. Appl Geochem 27:782–795

Jeen SW, Mayer KU, Gillham RW, Blowes DW (2007) Reactive transport modeling of trichloroethene treatment with declining reactivity of iron. Environ Sci Technol 41:1432–1438

Jenni A, Gimmi T, Alt-Epping P, Mäder U, Cloet V (2017) Interaction of ordinary Portland cement and Opalinus Clay: Dual porosity modelling compared to experimental data. Phys Chem Earth, Parts A/B/C 99:22–37

Jin L, Mathur R, Rother G, Cole D, Bazilevskaya E, Williams J, Carone A, Brantley S (2013) Evolution of porosity and geochemistry in Marcellus Formation black shale during weathering. Chem Geol 356:50–63

Johnson RH, Blowes DW, Robertson WD, Jambor JL (2000) The hydrogeochemistry of the Nickel Rim mine tailings impoundment, Sudbury, Ontario. J Contam Hydrol 41:49–80

Jung H, Navarre-Sitchler A (2018) Physical heterogeneity control on effective mineral dissolution rates. Geochim Cosmochim Acta 15:246–63

Kang Q, Tsimpanogiannis IN, Zhang D, Lichtner PC (2005) Numerical modeling of pore-scale phenomena during CO_2 sequestration in oceanic sediments. Fuel Process Technol 86:1647–1665

Kaszuba JP, Janecky DR, Snow MG (2005) Experimental evaluation of mixed fluid reactions between supercritical carbon dioxide and NaCl brine: Relevance to the integrity of a geologic carbon repository. Chem Geol 217:277–293

Kaufmann G, Romanov D, Hiller T (2010) Modeling three-dimensional karst aquifer evolution using different matrix-flow contributions. J Hydrol 388:241–250

Kim DS, Fogler HS (2000) Biomass evolution in porous media and its effects on permeability under starvation conditions. Biotechnol Bioeng 69:47–56

Knackstedt M, Zhang X (1994) Direct evaluation of length scales and structural parameters associated with flow in porous media. Phys Rev E 50:2134

Koponen A, Kataja M, Timonen J (1997) Permeability and effective porosity of porous media. Phys Rev E 56:3319

Kutchko BG, Strazisar BR, Dzombak DA, Lowry GV, Thaulow N (2007) Degradation of well cement by CO_2 under geologic sequestration conditions. Environ Sci Technol 41:4787–4792

Lagneau V, Van Der Lee J (2010) Operator-splitting-based reactive transport models in strong feedback of porosity change: The contribution of analytical solutions for accuracy validation and estimator improvement. J Contam Hydrol 112:118–129

Lai P, Krevor S (2014) Pore scale heterogeneity in the mineral distribution and surface area of Berea sandstone. Energy Procedia 63:3582–3588

Lalan P (2016) Influence d'une température de 70°C sur la géochimie, la microstructure et la diffusion aux interfaces béton/argile : expérimentations en laboratoire in situ et modélisation. (PhD Thesis). PSL Research University

Landrot G, Ajo-Franklin JB, Yang L, Cabrini S, Steefel CI (2012) Measurement of accessible reactive surface area in a sandstone, with application to CO_2 mineralization. Chem Geol 318:113–125

Lasaga AC (1981) Transition state theory. Rev Mineral 8:135–170

Lasaga AC (2014) Kinetic Theory in the Earth Sciences. Princeton University Press

Le Gallo Y, Bildstein O, Brosse E (1998) Coupled reaction-flow modeling of diagenetic changes in reservoir permeability, porosity and mineral compositions. J Hydrol 209:366–388

Lefebvre R, Hockley D, Smolensky J, Gélinas P (2001) Multiphase transfer processes in waste rock piles producing acid mine drainage: 1: Conceptual model and system characterization. J Contam Hydrol 52:137–164

Levenspiel O (1998) Chemical Reaction Engineering Book, 3rd edn., Chap. 9 and 10. John Wiley and Sons, New York

Li L, Benson CH, Lawson EM (2006) Modeling porosity reductions caused by mineral fouling in continuous-wall permeable reactive barriers. J Contam Hydrol 83:89–121

Li L, Steefel CI, Yang L (2008) Scale dependence of mineral dissolution rates within single pores and fractures. Geochim Cosmochim Acta 72:360–377

Li X, Huang H, Meakin P (2008) Level set simulation of coupled advection–diffusion and pore structure evolution due to mineral precipitation in porous media. Water Resour Res 44:W12407

Li X, Huang H, Meakin P (2010) A three-dimensional level set simulation of coupled reactive transport and precipitation/dissolution. Int J Heat Mass Transfer 53:2908–2923

Li Q, Steefel CI, Jun Y-S (2017) Incorporating nanoscale effects into a continuum-scale reactive transport model for CO_2-deteriorated cement. Environ Sci Technol 51:10861–10871

Lichtner PC (1996) Continuum formulation of multicomponent-multiphase reactive transport. Rev Mineral 34:1–82

Liu J, Aruguete DM, Jinschek JR, Rimstidt JD, Hochella Jr MF (2008) The non-oxidative dissolution of galena nanocrystals: Insights into mineral dissolution rates as a function of grain size, shape, and aggregation state. Geochim Cosmochim Acta 72:5984–5996

Low PF (1981) Principles of Ion Diffusion in Clays 1. *In:* Chemistry in the Soil Environment. Robert H. Dowdy (ed.) ASA Spec Publ 40, p 31–45

Luhmann AJ, Kong XZ, Tutolo BM, Garapati N, Bagley BC, Saar MO, Seyfried Jr WE (2014) Experimental dissolution of dolomite by CO_2-charged brine at 100 C and 150 bar: Evolution of porosity, permeability, and reactive surface area. Chem Geol 25:145–60

Luquot L, Gouze P (2009) Experimental determination of porosity and permeability changes induced by injection of CO2 into carbonate rocks. Chem Geol 265:148–159

Luquot L, Abdoulghafour H, Gouze P (2013) Hydro-dynamically controlled alteration of fractured Portland cements flowed by CO_2-rich brine. Int J Greenhouse Gas Control 16:167–79

Luquot L, Rodriguez O, Gouze P (2014) Experimental characterization of porosity structure and transport property changes in limestone undergoing different dissolution regimes. Transp Porous Media 101:507–532

Luquot L, Gouze P, Niemi A, Bensabat J, Carrera J (2016) CO_2-rich brine percolation experiments through Heletz reservoir rock samples (Israel): Role of the flow rate and brine composition. Int J Greenhouse Gas Control 48:44–58

MacQuarrie KTB, Mayer KU (2005) Reactive transport modeling in fractured rock: A state-of-the-science review. Earth Sci Rev 72:189–227

Mainguy M, Tognazzi C, Torrenti JM, Adenot F (2000) Modelling of leaching in pure cement paste and mortar. Cement Concr Res 30:83–90

Maxwell JC (1881) A Treatise on Electricity and Magnetism, Clarendon. Oxford

Mayer KU, Frind EO, Blowes DW (2002) Multicomponent reactive transport modeling in variably saturated porous media using a generalized formulation for kinetically controlled reactions. Water Resour Res 38:1174

McDuff D, Shuchart CE, Jackson S, Postl D, Brown JS (2010) Understanding wormholes in carbonates: Unprecedented experimental scale and 3-D visualization. *In:* SPE Ann Tech Conf Exhibition. Soc Petrol Eng

Mehmani Y, Sun T, Balhoff MT, Eichhubl P, Bryant S (2012) Multiblock pore-scale modeling and upscaling of reactive transport: application to carbon sequestration. Transp Porous Media 95:305–326

Meinert LD, Hedenquist JW, Satoh H, Matsuhisa Y (2003) Formation of anhydrous and hydrous skarn in Cu-Au ore deposits by magmatic fluids. Econ Geol 98:147–156

Menke HP, Andrew MG, Blunt MJ, Bijeljic B (2016) Reservoir condition imaging of reactive transport in heterogeneous carbonates using fast synchrotron tomography—Effect of initial pore structure and flow conditions. Chem Geol 15:15–26

Millington RJ (1959) Gas diffusion in porous media. Science 10:3367

Millington RJ, Quirk JP (1961) Permeability of porous solids. Trans Faraday Soc 57:1200–7

Min T, Gao Y, Chen L, Kang Q, Tao WW (2016) Changes in porosity, permeability and surface area during rock dissolution: Effects of mineralogical heterogeneity. Inter J Heat Mass Transfer 103:900–913

Mok U, Bernabé Y, Evans B (2002) Permeability, porosity and pore geometry of chemically altered porous silica glass. J Geophys Res: Solid Earth 107:ECV–4

Molins S (2015) Reactive interfaces in direct numerical simulation of pore-scale processes. Rev Mineral Geochem 80:461–481

Molins S, Trebotich D, Steefel CI, Shen C (2012). An investigation of the effect of pore scale flow on average geochemical reaction rates using direct numerical simulation. Water Resour Res 48:W03527

Molins S, Trebotich D, Yang L, Ajo-Franklin JB, Ligocki TJ, Shen C, Steefel CI (2014) Pore-scale controls on calcite dissolution rates from flow-through laboratory and numerical experiments. Environ Sci Technol 48:7453–7460

Molins S, Trebotich D, Miller GH, Steefel CI (2017) Mineralogical and transport controls on the evolution of porous media texture using direct numerical simulation. Water Resour Res 53:3645–3661

Molins S, Trebotich D, Arora B, Steefel CI, Deng H (2019) Multi-scale model of reactive transport in fractured media: diffusion limitations on rates. Transp Porous Media 128:701–721

Moreira EA, Coury JR (2004) The influence of structural parameters on the permeability of ceramic foams. Brazilian J Chem Eng 21:23–33

Navarre-Sitchler A, Steefel CI, Yang L, Tomutsa L, Brantley SL (2009) Evolution of porosity and diffusivity associated with chemical weathering of a basalt clast. J Geophys Res Earth Surf 114:F02016

Navarre-Sitchler A, Steefel CI, Sak PB, Brantley SL (2011) A reactive-transport model for weathering rind formation on basalt. Geochim Cosmochim Acta 75:7644–7667

Nelson MD, Parker BL, Al TA, Cherry JA, Loomer D (2001) Geochemical reactions resulting from in situ oxidation of PCE-DNAPL by $KMnO_4$ in a sandy aquifer. Environ Sci Technol 35:1266–1275

Neuman SP (1973) Saturated-unsaturated seepage by finite elements. J Hydraul Div 99:2233-2250

Nogues JP, Fitts JP, Celia MA, Peters CA (2013) Permeability evolution due to dissolution and precipitation of carbonates using reactive transport modelin in pore networks. Water Resour Res 49:6006-21

Noiriel C (2015) Resolving time-dependent evolution of pore-scale structure, permeability and reactivity using X-ray microtomography. Rev Mineral Geochem 80:247–285

Noiriel C, Luquot L, Madé B, Raimbault L, Gouze P, Van Der Lee J (2009) Changes in reactive surface area during limestone dissolution: An experimental and modelling study. Chem Geol 265:160–170

Noiriel C, Renard F, Doan M-L, Gratier J-P (2010) Intense fracturing and fracture sealing induced by mineral growth in porous rocks. Chem Geol 269:197–209

Noiriel C, Steefel CI, Yang L, Ajo-Franklin J (2012) Upscaling calcium carbonate precipitation rates from pore to continuum scale. Chem Geol 318:60–74

Opfergelt S, Delmelle P, Boivin P, Delvaux B (2006) The 1998 debris avalanche at Casita volcano, Nicaragua: Investigation of the role of hydrothermal smectite in promoting slope instability. Geophys Res Lett 33:L15305

Opolot E, Finke P (2015) Evaluating sensitivity of silicate mineral dissolution rates to physical weathering using a soil evolution model (SoilGen2. 25). BioGeosciences 12:6791–6808

Oron AP, Berkowitz B (1998) Flow in rock fractures: The local cubic law assumption reexamined. Water Resour Res 34:2811–2825

Ortoleva P, Chadam J, Merino E, Sen A (1987) Geochemical self-organization II; the reactive-infiltration instability. Am J Sci 287:1008–1040

Ouellet S, Bussière B, Mbonimpa M, Benzaazoua M, Aubertin M (2006) Reactivity and mineralogical evolution of an underground mine sulphidic cemented paste backfill. Mineral Eng 19:407–419

Ovaysi S, Piri M (2014) Pore-space alteration induced by brine acidification in subsurface geologic formations. Water Resour Res 50:440–452

Panday S, Huyakorn PS, Therrien R, Nichols RL (1993) Improved three-dimensional finite-element techniques for field simulation of variably saturated flow and transport. J Contam Hydrol 12:3–33

Panga MK, Ziauddin M, Balakotaiah V (2005) Two-scale continuum model for simulation of wormholes in carbonate acidization. AIChE J 51:3231–48

Pape H, Clauser C, Iffland J (1998) Permeability prediction for reservoir sandstones and basement rocks based on fractal pore space geometry. *In:* SEG Technical Program Expanded Abstracts (1998) Soc Explor Geophys, pp. 1032–1035

Pape H, Clauser C, Iffland J, Krug R, Wagner R (2005) Anhydrite cementation and compaction in geothermal reservoirs: Interaction of pore-space structure with flow, transport, P–T conditions, and chemical reactions. Int J Rock Mech Min Sci 42:1056–1069

Pereira Nunes JP, Bijeljic B, Blunt MJ (2016) Pore-space structure and average dissolution rates: A simulation study. Water Resour Res 52:7198–7212

Peters CA (2009) Accessibilities of reactive minerals in consolidated sedimentary rock: An imaging study of three sandstones. Chem Geol 265:198–208

Petersen EE (1958) Diffusion in a pore of varying cross section. AIChE Journal 4:343–5

Pfingsten W (2002) Experimental and modeling indications for self-sealing of a cementitious low-and intermediate-level waste repository by calcite precipitation. Nucl Technol 140:63–82

Phillips AJ, Cunningham AB, Gerlach R, Hiebert R, Hwang C, Lomans BP, Westrich J, Mantilla C, Kirksey J, Esposito R, Spangler L (2016) Fracture sealing with microbially-induced calcium carbonate precipitation: A field study. Environ Sci Technol 50:4111–4117

Polak A, Elsworth D, Yasuhara H, Grader AS, Halleck PM (2003) Permeability reduction of a natural fracture under net dissolution by hydrothermal fluids. Geophys Res Lett 30:2020

Poonoosamy J, Kosakowski G, Van Loon LR, Mäder U (2015) Dissolution–precipitation processes in tank experiments for testing numerical models for reactive transport calculations: Experiments and modelling. J Contam Hydrol 177:1–17

Poonoosamy J, Curti E, Kosakowski G, Grolimund D, Van Loon LR, Mäder U (2016) Barite precipitation following celestite dissolution in a porous medium: A SEM/BSE and μ-XRD/XRF study. Geochim Cosmochim Acta 182:131–144

Poonoosamy J, Wanner C, Epping PA, Águila JF, Samper J, Montenegro L, Xie M, Su D, Mayer KU, Mäder U, Van Loon LR (2018) Benchmarking of reactive transport codes for 2D simulations with mineral dissolution–precipitation reactions and feedback on transport parameters. Comput Geosci 1–22

Putnis A, Mauthe G (2001) The effect of pore size on cementation in porous rocks. Geofluids 1:37–41

Quintard M, Lasseux D, Zarcone C, Golfier F, Lernormand R, Bazin B (2007) Acidizing carbonate reservoirs: numerical modelling of wormhole propagation and comparison to experiments, *In:* SPE European Formation Damage Conference

Raines MA, Dewers TA (1997) Mixed transport/reaction control of gypsum dissolution kinetics in aqueous solutions and initiation of gypsum karst. Chem Geol 140:29–48

Rajyaguru A, L'Hôpital E, Savoye S, Wittebroodt C, Bildstein O, Arnoux P, Detilleux V, Fatnassi I, Gouze P, Lagneau V (2019) Experimental characterization of coupled diffusion reaction mechanisms in low permeability chalk. Chem Geol 5:29–39

Ray N, Rupp A, Schulz R, Knabner P (2018) Old and new approaches predicting the diffusion in porous media. Transp Porous Media 124:803–824

Regmi AD, Yoshida K, Dhital MR, Devkota K (2013) Effect of rock weathering, clay mineralogy, and geological structures in the formation of large landslide, a case study from Dumre Besei landslide, Lesser Himalaya Nepal. Landslides 10:1–13

Regnault O, Lagneau V, Fiet N (2015) 3D reactive transport simulations of uranium in situ leaching: Forecast and process optimization, *In:* Uranium-Past and Future Challenges, p 725–730

Ritchie AIM (1994) Sulfide oxidation mechanisms: controls and rates of oxygen transport, in short course handbook on environmental geochemistry of sulfide mine-wastes. Waterloo, Canada: Mineral Assoc Can

Ritter WF, Scarborough RW (1995) A review of bioremediation of contaminated soils and groundwater. J Environ Sci Health Part A 30:333–357

Rötting TS, Thomas RC, Ayora C, Carrera J (2008) Passive treatment of acid mine drainage with high metal concentrations using dispersed alkaline substrate. J Environ Qual 37:1741–1751

Rötting TS, Luquot L, Carrera J, Casalinuovo DJ (2015) Changes in porosity, permeability, water retention curve and reactive surface area during carbonate rock dissolution. Chem Geol 18:86–98

Saripalli KP, Serne RJ, Meyer PD, McGrail BP (2002) Prediction of diffusion coefficients in porous media using tortuosity factors based on interfacial areas. Groundwater 40:346–352

Sato H, Shibutani T, Yui M (1997) Experimental and modelling studies on diffusion of Cs, Ni and Sm in granodiorite, basalt and mudstone. J Contam Hydrol 26:119–133

Savage D (2013) Constraints on cement-clay interaction. Procedia Earth Planet Sci 7:770–773

Scherer MM, Richter S, Valentine RL, Alvarez PJJ (2000) Chemistry and microbiology of permeable reactive barriers for in situ groundwater clean up. Crit Rev Microbiol 26:221–264

Sedighi M, Thomas HR (2014) Micro porosity evolution in compacted swelling clays—A chemical approach. Appl Clay Sci 101:608–618

Seigneur N, L'Hôpital E, Dauzères A, Sammaljärvi J, Voutilainen M, Labeau PE, Dubus A, Detilleux V (2017) Transport properties evolution of cement model system under degradation—Incorporation of a pore-scale approach into reactive transport modelling. Phys Chem Earth 99:95–109

Seigneur N, Lagneau V, Corvisier J, Dauzères A (2018) Recoupling flow and chemistry in variably saturated reactive transport modelling-An algorithm to accurately couple the feedback of chemistry on water consumption, variable porosity and flow. Adv Water Resour 122:355-366

Shao H, Kosakowski G, Berner U, Kulik DA, Mäder U, Kolditz O (2013) Reactive transport modeling of the clogging process at Maqarin natural analogue site. Phys Chem Earth, Parts A/B/C 64:21–31

Simon RB, Thiry M, Schmitt JM, Lagneau V, Langlais V, Bélières M (2014) Kinetic reactive transport modelling of column tests for uranium In Situ Recovery (ISR) mining. Appl Geochem 51:116–129

Sin I, Lagneau V, Corvisier J (2017) Integrating a compressible multicomponent two-phase flow into an existing reactive transport simulator. Adv Water Resour 100:62–77

Singurindy O, Berkowitz B (2003) Evolution of hydraulic conductivity by precipitation and dissolution in carbonate rock. Water Resour Res 39:1016

Smith L, Sego DC, Langman JB, Blowes DW, Wilson D, Amos RT (2018) Diavik Waste Rock Project: Scale-up of a reactive transport model for temperature and sulfide-content dependent geochemical evolution of waste rock. Appl Geochem 96:177–190

Smith MM, Sholokhova Y, Hao Y, Carroll SA (2013) CO_2-induced dissolution of low permeability carbonates. Part I: Characterization and experiments. Adv Water Resour 62:370–387

Soleimani E, Moradzadeh A, Ansari Jafari M, Jodieri Shokri B, Doulati Ardejani F (2013) A combined mathematical geophysical model for prediction of pyrite oxidation and pollutant leaching associated with a coal washing waste dump. Int J Environ Sci Technol 5:517–526

Soler JM, Mäder UK (2005) Interaction between hyperalkaline fluids and rocks hosting repositories for radioactive waste: reactive transport simulations. Nucl Sci Eng 151:128–133

Soulaine C, Tchelepi HA (2016) Micro-continuum approach for pore-scale simulation of subsurface processes. Transp Porous Media 113:431–456

Spycher NF, Sonnenthal EL, Apps JA (2003) Fluid flow and reactive transport around potential nuclear waste emplacement tunnels at Yucca Mountain, Nevada. J Contam Hydrol 62:653–673

Steefel CI, Lasaga AC (1992) Putting transport into water-rock interaction models. Geology 20:680–684

Steefel CI, Lasaga AC (1994) A coupled model for transport of multiple chemical species and kinetic precipitation/dissolution reactions with application to reactive flow in single phase hydrothermal systems. Am J Sci 294:529–592

Steefel CI, Maher K (2009) Fluid-rock interaction: A reactive transport approach. Rev Mineral Geochem 70:485–532

Steefel CI, Lichtner PC (1994) Diffusion and reaction in rock matrix bordering a hyperalkaline fluid-filled fracture. Geochim Cosmochim Acta 58:3595–3612

Steefel CI, Lichtner PC (1998) Multicomponent reactive transport in discrete fractures: II: Infiltration of hyperalkaline groundwater at Maqarin, Jordan, a natural analogue site. J Hydrol 209:200–224

Steefel CI, Molins S, Trebotich D (2013) Pore scale processes associated with subsurface CO_2 injection and sequestration. Rev Mineral Geochem 77:259–303

Steefel CI, Appelo CA, Arora B, Jacques D, Kalbacher T, Kolditz O, Lagneau V, Lichtner PC, Mayer KU, Meeussen JC, Molins S (2015a) Reactive transport codes for subsurface environmental simulation. Comput Geosci 19:445–478

Steefel CI, Beckingham LE, Landrot G (2015b) Micro-continuum approaches for modeling pore-scale geochemical processes. Rev Mineral Geochem 80:217–246

Suazo G, Fourie A, Doherty J (2016) Experimental study of the evolution of the soil water retention curve for granular material undergoing cement hydration. J Geotech Geoenviron Eng 142:04016022

Tartakovsky AM, Meakin P, Scheibe TD, Wood BD (2007) A smoothed particle hydrodynamics model for reactive transport and mineral precipitation in porous and fractured porous media. Water Resour Res 43:W05437

Tartakovsky AM, Redden G, Lichtner PC, Scheibe TD, Meakin P (2008) Mixing-induced precipitation: Experimental study and multiscale numerical analysis. Water Resour Res 44:W06S04

Taylor HFW, Famy C, Scrivener KL (2001) Delayed ettringite formation. Cement and concr research 31:683–693

Teng HH (2004) Controls by saturation state on etch pit formation during calcite dissolution. Geochim Cosmochim Acta 68:253–262

Tenthorey E, Gerald JDF (2006) Feedbacks between deformation, hydrothermal reaction and permeability evolution in the crust: Experimental insights. Earth Planet Sci Lett 247:117–129

Thullner M, Zeyer J, Kinzelbach W (2002) Influence of microbial growth on hydraulic properties of pore networks. Transp Porous Media 49:99–122

Tian Z, Wang J (2018) Lattice Boltzmann simulation of dissolution-induced changes in permeability and porosity in 3D CO_2 reactive transport. J Hydrol 557:276–290

Tian H, Xu T, Wang F, Patil VV, Sun Y, Yue G (2014) A numerical study of mineral alteration and self-sealing efficiency of a caprock for CO_2 geological storage. Acta Geotechnica 9:87–100

Tomadakis MM, Sotirchos SV (1993) Transport properties of random arrays of freely overlapping cylinders with various orientation distributions. J Chem Phys 98:616–626

Trebotich D, Adams MF, Molins S, Steefel CI, Shen C (2014) High-resolution simulation of pore-scale reactive transport processes associated with carbon sequestration. Comput Sci Eng 16:22–31

Van der Land C, Wood R, Wu K, van Dijke MIJ, Jiang Z, Corbett PWM, Couples G (2013) Modelling the permeability evolution of carbonate rocks. Mar Petrol Geol 48:1–7

Verma A, Pruess K (1988) Thermohydrological conditions and silica redistribution near high-level nuclear wastes emplaced in saturated geological formations. J Geophys Res: Solid Earth 93:1159–1173

Vialle S, Contraires S, Zinzsner B, Clavaud JB, Mahiouz K, Zuddas P, Zamora M (2014) Percolation of CO_2-rich fluids in a limestone sample: Evolution of hydraulic, electrical, chemical, and structural properties. J Geophys Res Solid Earth 119:2828–2847

Vilcáez J, Morad S, Shikazono N (2017) Pore-scale simulation of transport properties of carbonate rocks using FIB-SEM 3D microstructure: Implications for field scale solute transport simulations. J Nat Gas Sci Eng 42:13–22

Walsh JB, Brace WF (1984) The effect of pressure on porosity and the transport properties of rock. J Geophys Res: Solid Earth 89:9425–9431

Walsh SDC, Du Frane WL, Mason HE, Carroll SA (2013) Permeability of wellbore-cement fractures following degradation by carbonated brine. Rock Mech Rock Eng 46:455–464

Wang Y, Hill AD, Schechter RS (1993) The optimum injection rate for matrix acidizing of carbonate formations. *In:* SPE Annual Technical Conference and Exhibition. Soc Petrol Eng

Wang Y, Nahon D, Merino E (1994) Dynamic model of the genesis of calcretes replacing silicate rocks in semi-arid regions. Geochim Cosmochim Acta 58:5131–5145

Wang Q, Cui Y-J, Tang AM, Delage P, Gatmiri B, Ye W-M (2014) Long-term effect of water chemistry on the swelling pressure of a bentonite-based material. Appl Clay Sci 87:157–162

Wantanaphong J, Mooney SJ, Bailey EH (2006) Quantification of pore clogging characteristics in potential permeable reactive barrier (PRB) substrates using image analysis. J Contam Hydrol 86:299–320

Watanabe N, Yonekura N, Sagara W, Cheibany OE, Marui H, Furuya G (2005) Chemical weathering and the occurrence of large-scale landslides in the Hime River Basin, Central Japan, *In:* Landslides. Springer, pp. 165–171

Werth CJ, Zhang C, Brusseau ML, Oostrom M, Baumann T (2010) A review of non-invasive imaging methods and applications in contaminant hydrogeology research. J Contam Hydrol 113:1–24

White AF, Schulz MS, Vivit DV, Blum AE, Stonestrom DA, Harden JW (2005) Chemical weathering rates of a soil chronosequence on granitic alluvium: III Hydrochemical evolution and contemporary solute fluxes and rates. Geochim Cosmochim Acta 69:1975–1996

Witherspoon PA, Wang JS, Iwai K, Gale JE (1980) Validity of cubic law for fluid flow in a deformable rock fracture. Water Resour Res 16:1016–1024

Wolterbeek TK, Raoof A (2018) Meter-scale reactive transport modeling of CO_2-rich fluid flow along debonded wellbore casing-cement interfaces. Environ Sci Technol 52:3786–3795

Wunderly MD, Blowes DW, Frind EO, Ptacek CJ (1996) Sulfide mineral oxidation and subsequent reactive transport of oxidation products in mine tailings impoundments: A numerical model. Water Resour Res 32:3173–87

Xiao Y, Jones GD (2007) Reactive transport models of limestone-dolomite transitions: implications for reservoir connectivity, *In:* Int Petrol Technol Conf

Xie M, Mayer KU, Claret F, Alt-Epping P, Jacques D, Steefel C, Chiaberge C, Simunek J (2015) Implementation and evaluation of permeability–porosity and tortuosity–porosity relationships linked to mineral dissolution–precipitation. Comput Geosci 19:655–671

Xu P, Yu B (2008) Developing a new form of permeability and Kozeny–Carman constant for homogeneous porous media by means of fractal geometry. Adv Water Resour 31:74–81

Xu T, Spycher N, Sonnenthal E, Zhang G, Zheng L, Pruess K (2011) TOUGHREACT Version 2.0: A simulator for subsurface reactive transport under non-isothermal multiphase flow conditions. Compu Geosci 37:763–774

Yang C, Samper J, Montenegro L (2008) A coupled non-isothermal reactive transport model for long-term geochemical evolution of a HLW repository in clay. Environ Geol 53:1627–1638

Zhang C-L (2011) Experimental evidence for self-sealing of fractures in claystone. Phys Chem Earth, Parts A/B/C 36:1972–1980

Zhang C-L (2013) Sealing of fractures in claystone. J Rock Mech Geotech Eng 5:214–220

Zhang YD, Park JS, Gao S, Sonta A, Horin B, Buscarnera G (2017) Effect of grain crushing and grain size on the evolution of water retention curves, *In:* PanAm Unsaturated Soils, p 268–278

Reviews in Mineralogy & Geochemistry
Vol. 85 pp. 239-264, 2019
Copyright © Mineralogical Society of America

8

Stable Isotope Fractionation
by Transport and Transformation

Jennifer L. Druhan

Department of Geology
University of Illinois Urbana Champaign
Urbana, Illinois, 61801
USA

jdruhan@illinois.edu

Matthew J. Winnick

Department of Geosciences
University of Massachusetts Amherst
Amherst, Massachusetts, 01003
USA

mwinnikc@geo.umass.edu

Martin Thullner

Department of Environmental Microbiology
UFZ—Helmholtz Centre for Environmental Research
04318 Leipzig
Germany

martin.thullner@ufz.de

INTRODUCTION

Of the 92 elements naturally present on modern Earth, only 21 are monotopic. The remainder are composed of multiple isotopes, many of which are either stable or decay over such extraordinarily long timescales that they may be considered effectively stable for appropriate applications. These isotopes of a given element are distinguished by the number of neutrons within their nucleus, resulting in subtle differences in mass. Though most fundamental characteristics of the isotopes of a given element are the same (e.g., charge, atomic number), their relative distributions in natural environments are altered as they are subjected to a range of mass-dependent transport and transformation processes. Urey and colleagues first demonstrated this mass-dependent partitioning within a thermodynamic framework (Urey 1947). Decades later, these small differences in isotopic abundances serve as the foundation for a suite of powerful diagnostic tools applied broadly across the Earth Sciences, constituting essential capabilities within the disciplines of hydrology, oceanography, petrology, paleoclimate, planetary science, paleontology, ecology, and microbiology, among others.

The mass-biased partitioning of isotopes, commonly referred to as mass-dependent fractionation, occurs through a variety of chemical reactions (e.g., complexation, redox, sorption, phase change) as well as some mechanisms of physical transport (e.g., molecular diffusion). In the isotope geochemistry literature (e.g., Hoefs 2004; Wolfsberg et al. 2010), fractionating pathways are commonly divided into two classes: equilibrium and kinetic. Equilibrium fractionation occurs when the presence of a 'heavier' isotope in a molecular structure causes a dampening of the bond energies associated with this element, such that the

1529-6466/19/0085-0008$05.00 (print)
1943-2666/19/0085-0008$05.00 (online)   http://dx.doi.org/10.2138/rmg.2019.85.8

distribution of these isotopically heavy molecules between two phases is skewed towards the lower energy state. Fractionation associated with unidirectional, time-dependent or otherwise non-equilibrium partitioning is classified as kinetic. This class of partitioning can occur as a result of both physical transport and chemical transformation, and thus encompasses all mass-dependent differences in rates or velocity between isotopologues (i.e., molecules of the same chemical formula and bonding structure that differ only in their isotopic composition).

Why are stable isotopes utilized so ubiquitously across the Earth Sciences? The simple answer is that measuring the isotopes of a given element often reduces the number of unknowns in a system. This allows a critical means of parsing contemporaneous processes that may be indistinguishable through measurements of chemical species alone. Consider a system with multiple input and/or output fluxes; a classic example of this is the flux of water vapor from land as a combination of evaporation and transpiration. These two pathways are indistinguishable in that they both contribute to increasing atmospheric water vapor pressure. However, bare-soil evaporation involves significant fractionation of both H and O isotopes in water, while plant transpiration, on steady-state timescales (days to weeks and longer), does not. Water isotopes measured in vapor fluxes from terrestrial environments thus uniquely allow for the partitioning of evaporation and plant transpiration (Gat and Matsui 1991).

This type of end-member mixing analysis is a common application of stable isotope geochemistry. However, many environmental systems involve more complex sources, transport, mixing and reaction networks. To expand on the previous example, partitioning evapotranspiration is complicated both by mixing with ambient atmospheric air and lateral flow, which must be accounted for through a model framework which distinguishes the contributions of reactivity and transport in order to provide quantitative estimates (e.g., Winnick et al. 2014). In the following chapter, we review multiple pathways of isotopic fractionation and their treatment within numerical reactive transport frameworks. We focus on near-surface hydrologic systems hosting a variety of water–rock–life interactions which reflect a fundamental balance between transport and transformation. Such conditions describe a wide variety of environments, ranging from the cycling of water through terrestrial ecosystems, to the burial of carbonates in marine sediments, to the metabolism of microbes. We begin with consideration of simple fractionation in the absence of transport, followed by sequential addition of common transport influences with an emphasis on the unique information contained within isotopic observations in these systems. Finally, we turn to more 'complex' reactive pathways that are not easily described by single unidirectional conversions.

Basic fractionation in a closed system

Changes in the distribution of stable isotopes among compounds or phases are tracked using their relative ratios, for example the ratio (R) of oxygen isotopes in water is given as $^{18}R_{H_2O} = (H_2^{18}O)/(H_2^{16}O)$. This value is not the same as the fractional abundance, in which the denominator would be the sum of the concentrations or mass of all isotopologues of the given compound or phase, but it is more commonly used. The natural distribution of most stable isotopes is such that the use of a rare isotope in the numerator of R leads to a small number, on the order of a few percent. Thus, it is often easier to reference this to a standard value and convert to per mil differences, leading to the common delta (δ) notation. For our example of oxygen $\delta^{18}O_{H_2O} = (^{18}R_{H_2O} / {}^{18}R_{std} - 1) \times 10^3$). Delta values which are greater than zero in this convention are then considered 'heavier' or 'enriched' relative to the standard (i.e., they contain more of the heavy ^{18}O isotope), and those that are negative are often referred to as 'light' or 'depleted'.

Closed vessels, or batch reactors, are common experimental designs used to quantify the fractionation factor ($\alpha = R_{product}/R_{reactant}$) of a given reaction (Fig. 1). Importantly, our discussion of a closed system entails reversible reactions in that there is no mechanism by which a product species may be removed or isolated from the reactant. This is not to say that a given reaction will operate reversibly, just that the potential is there. Closed system

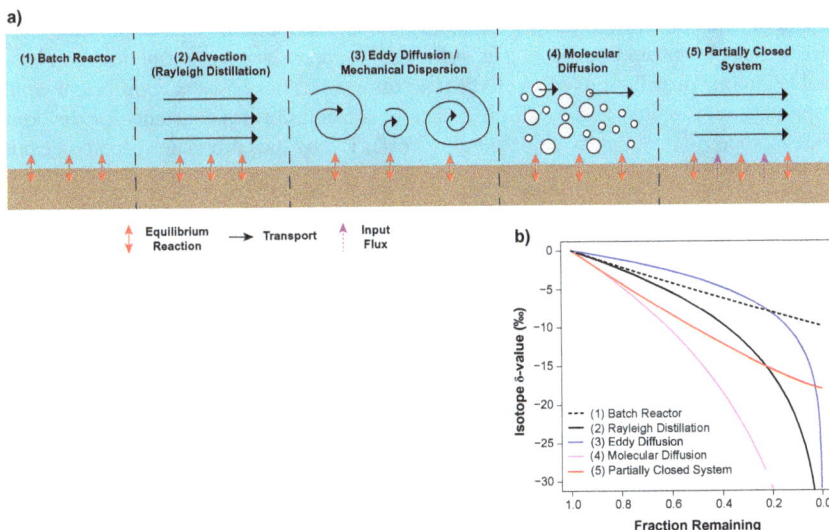

Figure 1. (a) Conceptual diagrams displaying different scenarios of transport for stable isotope reactive transport modelling. In these, a general chemical species in aqueous phase (blue background) is subject to transport (black arrows) and a reversible equilibrium reaction with a solid, stationary media (brown background) such as adsorption or mineral precipitation. Purple arrows represent a separate input flux of the chemical species with a constant isotopic composition. **(b)** Scaling relationships between the fraction of chemical species remaining in the aqueous phase during distillation and the isotopic composition of the chemical species for the transport scenarios depicted in **(a)**. All scenarios are modeled with the same equilibrium fractionation factor ($\alpha = 1.11$) such that the heavier isotope is preferentially removed from the aqueous phase. Molecular diffusion (4) is modelled assuming a diffusivity ratio between the two isotopologues of 1:0.65. The partially closed system (5) is modeled assuming an input flux equal to transport flux, with an isotopic composition of 0‰.

conditions further support the isolation of a single reactive transformation such that we may quantify the associated fractionation factor as a ratio between the isotopic composition of the (instantaneous) product species and the residual reactant (Mariotti et al. 1981; Criss 1991). In our example of oxygen this might be the distribution between liquid water instantaneously formed from a vapor and the residual vapor. By this convention an $\alpha = 1.0$ would mean that the isotope ratios of the two phases are equal and thus the reaction is non-fractionating. As with the isotope ratios above, typical values of α only deviate on order of a few percent from 1.0 and are thus often converted to a comparable epsilon (ε) notation using per mil differences (e.g., $\varepsilon = (\alpha - 1) \times 10^3$ or sometimes $\varepsilon = \ln(\alpha) \times 10^3$).

Fractionating reactions may evolve the distribution of initial reactants and products and their attendant isotope ratios either instantaneously (i.e., equilibrium is immediately achieved) or through time (i.e., kinetically controlled rate of progress towards equilibrium). For example, Mook et al. (1974) reported equilibrium fractionation factors between gas and dissolved phases of inorganic carbon species which are still widely used today. These experiments were conducted by allowing degassed water and a gaseous CO_2 of known isotopic composition to occupy the same closed vessel for several days before measurement, under the reasonable assumption that this was sufficient time for the new equilibrated distribution of carbonate species between and among phases to stabilize. In contrast the kinetics of many reactions are slow, thus behaving effectively unidirectionally, as frequently observed during redox transitions. A classic example is the fractionation of carbon isotopes during photosynthesis (Vogel 1980). These reactions are incomplete, and thus batch reactor experiments are often designed to monitor the reaction progress through time and the associated constraint of a kinetic fractionation factor.

Many excellent resources describe the intricacies and evolving quantitative treatment of these kinetic and equilibrium effects on stable isotope distributions (e.g., Sharp 2017; Teng et al. 2017). For our present purpose, it is sufficient to appreciate that a fractionation between the stable isotope distributions of an element across species or phases at equilibrium would be appropriately represented as a small difference in the equilibrium constants for the common and rare isotopes (Thorstenson and Parkhurst 2004). Taking the carbonate speciation example:

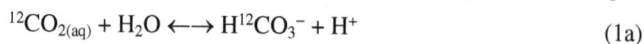

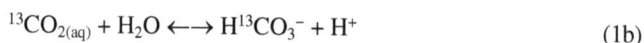

$$^{12}CO_{2(aq)} + H_2O \longleftrightarrow H^{12}CO_3^- + H^+ \tag{1a}$$

$$^{13}CO_{2(aq)} + H_2O \longleftrightarrow H^{13}CO_3^- + H^+ \tag{1b}$$

the equilibrium fractionation factor is easily represented between these carbon isotopologues as the ratio of the equilibrium constant (K_{eq}) for the rare ^{13}C relative to the common ^{12}C: $\alpha_{eq} = {}^{13}K_{eq}/{}^{12}K_{eq}$ for a given temperature. Similarly, taking the example of photosynthesis:

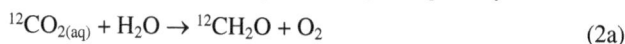

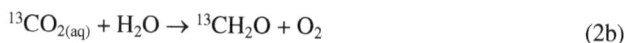

$$^{12}CO_{2(aq)} + H_2O \rightarrow {}^{12}CH_2O + O_2 \tag{2a}$$

$$^{13}CO_{2(aq)} + H_2O \rightarrow {}^{13}CH_2O + O_2 \tag{2b}$$

the ratio of the rates at which $^{13}CO_{2(aq)}$ and $^{12}CO_{2(aq)}$ are converted to their respective organic compounds taken over the ratio of the isotopes left in the residual CO_2 ($^{13}R_{CO2}$) yields the kinetic fractionation factor. For the simple case of a unidirectional first order reaction rate in which only one carbon atom is exchanged for each molecule reacted, this value reduces to the ratio of the rate constant (k) for the rare ^{13}C relative to the common ^{12}C: $\alpha_k = {}^{13}k/{}^{12}k$. It is worthwhile to note here that the definitions of equilibrium and kinetic fractionation factors we have described are the ratios of constants and are thus also constant in space and time. The observation of 'variable' or 'effective' fractionation factors commonly discussed in the literature thus suggests additional processes influencing these basic transformations.

From this standpoint two observations are apparent. First, such fractionating effects are easily implemented in virtually any modern reactive transport software by simply treating each isotopologue of a given compound as a unique 'species' (e.g., Thorstenson and Parkhurst 2004). The only requirements are that the sum of the individual isotopologues of a given species must be tracked to recover the concentration of that species, and each element divided into individual isotopes should be subject to precisely the same set of reactions which differ only in minor adjustments to the rate and equilibrium constants where α values are not equal to 1.0 (i.e., fractionating). Second, the potential for ambiguity in the implementation of these fractionations is clear. Many reactions are subject to both kinetic and equilibrium constraints, resulting in uncertainty as to how the fractionation should be imposed. Further, multiple fractionating pathways operating in tandem may result in complex signatures. We will return to these issues in the subsequent sections, but first it is necessary to move beyond a closed vessel and thus allow transport, isolation of products from reactants and mixing. Using the example of photosynthesis above, fractionation is commonly understood to represent a combination of diffusion through plant stomata as well as reaction driven by rubisco or PEP. Thus, natural systems are rarely devoid of transport effects.

FRACTIONATION IN OPEN SYSTEMS: THE INFLUENCE OF TRANSPORT

For our simple system featuring a single reversible fractionating pathway, allowing transport to move reactants and products relative to the other (e.g., rain condensing and falling out of a cloud) causes the isotopic ratios of these phases to evolve through space and time, and hence there is need for more sophisticated models (Fig. 1). Within this section, we will start by reviewing traditional Rayleigh distillation, and progressively incorporate complexity in terms of transport dynamics, multiple isotope sources (e.g., partially closed systems), and physical mixing to survey the continuum of processes that affect stable isotope distributions in nature. These processes are summarized in Figure 1 and are discussed in detail throughout the section.

Rayleigh distillation

The progressive removal of a phase from a reacting system and its effect on isotopic ratios may be described as a form of distillation (Rayleigh 1902). This framework was used as early as the 1930s to describe hydrogen isotope fractionation (Farkas and Farkas 1934). Dansgaard (1961) was the among the first to apply a Rayleigh distillation model to the isotopic dynamics of natural environments, describing the evolution of an air mass during rainout. As equilibrium fractionation causes the preferential incorporation of the heavier isotope (e.g., ^{18}O over ^{16}O) in the liquid phase, progressive rainout results in a shift towards lower isotopic values in both the remaining water vapor and subsequent rain following a power-law relationship. Quantitatively, the isotopic composition (R) varies from the initial value (R_0) with the fraction of vapor remaining (f) proportional to $\dfrac{R}{R_0} = f^{\alpha-1}$. While in the context of stable isotope analysis, Rayleigh distillation was originally introduced to describe temporal isotopic evolution, the same quantitative relationship has been applied spatially when transport is dominated by advection (Criss 1999; Allègre 2008). Thus, Rayleigh distillation has offered a simple framework which requires very little parameterization in order to apply across a wide diversity of systems, ranging from water vapor in clouds to groundwater contaminants in aquifers. There is no need, for example, to constrain flow or reaction rates. However, this model is strictly intended for application to single reactions in closed systems (or in a system featuring purely advective transport such that one is effectively moving a closed system through space), and any application to more complex conditions requires additional assumptions as well as the potential for misinterpretation or ambiguous results. Thus, the ease and versatility with which Rayleigh distillation may be applied often comes at the cost of an inability to parse between the influences of multiple transport and transformation pathways in such lumped parameter models.

Example: Equilibrium and kinetic effects

The limitations of a Rayleigh approach can be demonstrated for a simple system such as the reactive transformation of carbon compounds described above. Consider a 1D advective flow field in which we allow a simple, unidirectional kinetic reaction to oxidize organic carbon to CO_2 coupled to the reduction of O_2. This is essentially the reverse of the photosynthetic reaction described earlier (Eqns. 2a,b) and occurs ubiquitously throughout the terrestrial near-surface as a result of both autotrophic and heterotrophic respiratory pathways (Trumbore 2009; Dwivedi et al. 2019, this volume). This unidirectional kinetic reaction is associated with a slight mass-dependent partitioning in the stable isotopes of carbon. For the present simplified example, we choose a representative kinetic fractionation factor of $\alpha = 0.998$ (or equivalently $\varepsilon = 2‰$) between the reactant organic carbon and the product CO_2. In other words, $\alpha_k = 0.998$. In our simplified system, only a single organic carbon compound exists, with a fixed influent boundary concentration of 20 mM and $\delta^{13}C = 0‰$. Only one reactive pathway influences the progress of the reaction, as indicated by the fraction of organic carbon remaining (f), which decreases with distance down the flow path (Fig. 2A). The oxidation of this organic carbon correspondingly increases the total inorganic carbon with distance from the fixed inlet boundary concentration of 1.4 mM and $\delta^{13}C = +7.9‰$. The associated isotope ratio of the residual organic carbon evolves to increasingly fractionated values relative to the initial condition (Fig. 2B). In this example of a unidirectional reaction in which the product inorganic carbon is essentially 'removed from' (i.e., cannot react back to) the organic carbon phase, there is no difference between the behavior of this fractionating transformation of organic carbon along the 1D flow path (in space) vs. in a closed system (in time). We may simply replace distance along the direction of flow at steady state with time spent in a closed vessel, and the relationship between fractionation and reaction progress follows a simple Rayleigh distillation model (Fig. 2C).

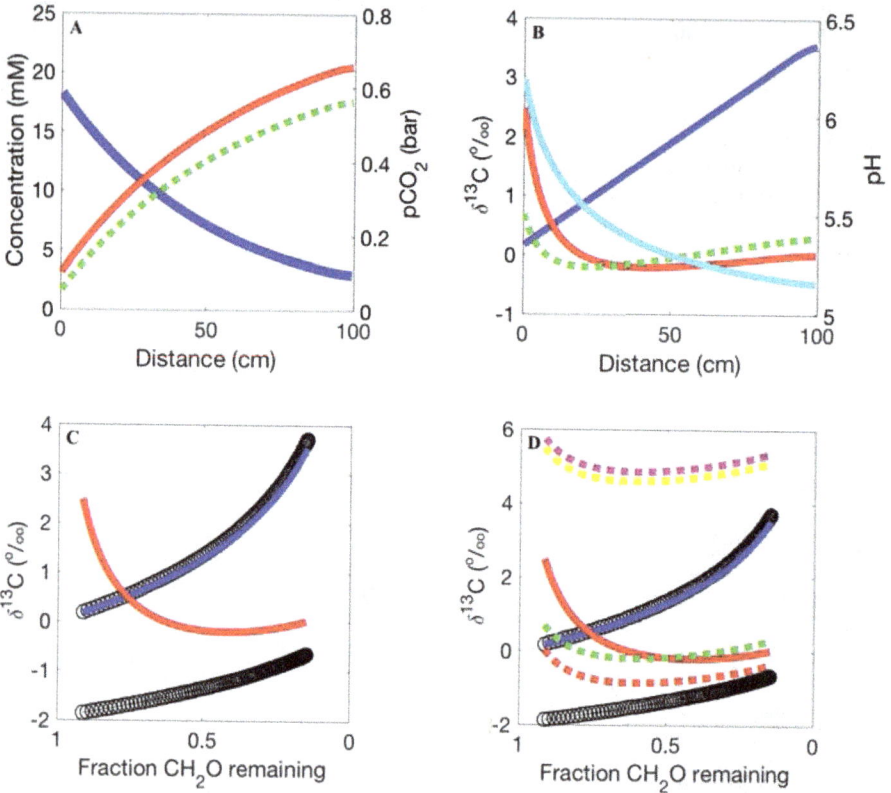

Figure 2. Organic carbon is oxidized to CO_2 following an irreversible first order reaction along a 1D flow path advecting from left to right at steady state. (**A**) Total organic carbon (solid blue line) and total inorganic carbon (solid red line) concentrations along with the speciated partial pressure of CO_2 (dashed green line) in equilibrium with the inorganic carbon as a function of distance. (**B**) The corresponding stable carbon isotope ratios of these three components along with the pH (light blue line) which is acidified as more CO_2 is produced. (**C**) The corresponding Rayleigh distillation model (black circles) accurately predicts the enrichment of residual organic carbon, but fails to produce the resulting accumulation of inorganic carbon due to mixing with the influent boundary condition. (**D**) The speciated inorganic carbon compounds (dashed lines) are illustrated for $CO_{2(g)}$ (green); $CO_{2(aq)}$ (red); HCO_3^- (yellow) and CO_3^{2-} (purple), all displaying fixed equilibrium partitioning.

Importantly, the total concentration of inorganic carbon, represented as a primary species in the reactive transport simulation (TOT_{CO_2}, Lichtner 1985; Steefel and MacQuarrie 1996) simply reflects the mass balance between the inorganic carbon content of the influent boundary condition and the inorganic carbon generated by the reaction (Fig. 2A). Similarly, the stable isotope composition of this total inorganic carbon reflects sequential mixing along the flow path between the influent +7.9‰ and contribution of an increasingly fractionated source from the organic carbon oxidation (Fig. 2B). As a result, the corresponding Rayleigh model (Fig. 2C) does not match the relationship between reaction progress and isotope ratio of the total inorganic carbon, because of the mixing between the influent and the fractionated product of the organic carbon oxidation. Some modifications could be made to the simple distillation equations used to generate these Rayleigh plots, including addition of a source term that would represent the influent contribution. This is similar to the partially closed system illustrated earlier (Fig. 1) in which the influx of a secondary source of the species contributes along the flow path, but here the contribution is only a result of the influent boundary condition.

However, such an approach fails to consider how the increasing CO_2 concentration with progress along the flow path (Fig. 2A) is partitioned into the range of co-existing carbonate species (CO_3^{2-}, HCO_3^-, $CO_{2(aq)}$ and $CO_{2(g)}$) with associated shifts in pH as a result of multi-component equilibria (e.g., Eqns. 1a and 1b) and attendant equilibrium fractionation factors. In our simple example, we allow this instantaneous equilibrium to occur along the flow path, such that pH drops due to the excess production of CO_2 (Fig. 2B). As a result, the majority of the total inorganic carbon is composed of $CO_{2(aq)}$ and a substantial increase in $pCO_{2(g)}$ also occurs (Fig. 2A). These speciated components of the total inorganic carbon content are differentiated in their isotope distributions through equilibrium fractionation, meaning that they are always offset from one another by fixed, instantaneous values (Fig. 2D), yet they always sum up to the isotopic ratio of the total inorganic pool. In this sense, despite the complexity of organic carbon oxidation and mixing with a unique influent composition, the isotopic partitioning between carbonate species also behaves as though they are in a closed system in which the reactants and products are always in communication.

Thus, out of a remarkably simple example, we have created behavior that at once displays both distillation (fractionation of organic carbon due to oxidation) and equilibrium (fractionation among inorganic carbon species) effects across a steady state flow path. Even in such a simple system subject to a single kinetic reaction and associated speciation, substantial complexity can result. If we were only interested in tracking the isotope partitioning of the organic carbon, then a Rayleigh model provides a powerful and versatile tool in that there is no need to constrain the flow rate, reaction rate or a variety of other factors impacting the system. However, the resulting isotope distribution of the inorganic carbon is already more complex due to the combination of multiple sources. Further, if we were to measure the isotopic composition of one of the co-existing inorganic carbon compounds, we would introduce the additional complexity of equilibrium speciation, which is temperature and pH dependent (e.g., Hansen et al. 2019). Finally, such generation of increased TIC commonly promotes the formation of carbonate solids with additional fractionating effects. Thus, when multiple sources and/or sinks of a compound are present even in a purely advective system the Rayleigh model provides an inaccurate prediction of observation.

From this perspective it becomes clear that substantial assumptions are made when isotopic ratios are used to infer past environmental conditions in the geologic record. For example, if we were to measure the carbon isotope ratio of a carbonate mineral formed from precipitation of inorganic carbon species in our example system, we would be attempting to deconvolve an isotope signature of that is evolving with distance due to the combination of mixing and fractionation (Fig. 1B) and potentially reflecting the effects of equilibrium with an associated gas phase pCO_2 which is also evolving with distance through the system (Fig. 1D). Yet such reconstructions are a common and widely employed method of understanding how the Earth system has changed through time. Thus, it becomes apparent that even simple representations of common reactive transport pathways require careful representation which benefit from multi-component, often numerical techniques.

Mixing

When advection is not the dominant mode of transport, the relationship between the fraction of a reactant remaining and its isotopic composition diverges from Rayleigh behavior. A common example is that of Fickian transport, in which diffusional fluxes are proportional to the concentration gradient, such as in molecular diffusion. A classic example is related to our earlier consideration of organic carbon oxidation and associated equilibrium speciation of carbonate ions. In soils, the generation of this oxidized CO_2 creates a concentration gradient, leading to diffusive transport upward towards the atmosphere, and an associated offset between the carbon isotope composition of soil gas and respired organic carbon (Cerling 1984; Davidson 1995). This diffusive fractionation must be considered when soil carbonate isotope signatures are used to reconstruct the pCO_2 at the time of formation.

Generally speaking, when molecular diffusion is combined with what would otherwise be a simple isotopic distillation, the observed or effective fractionation factor may appear to be either reduced or enhanced depending on the combination of transport and reaction fractionation factors. In Figure 1b, we demonstrate an example in which observed fractionation is enhanced when distillation preferentially reacts the heavy isotope (e.g., by precipitation or sorption) and molecular diffusion additionally decreases the rate at which the heavier isotope moves relative to its lighter counterpart, thus amplifying the overall apparent fractionation. Such predictions are relatively straightforward and versatile for many conditions of gas phase transport. Transport effects have also helped to identify where the underlying mechanisms governing observed fractionation are more complex than described above (Kopinke et al. 2018). During diffusion in liquid phases, for example, Richter et al. (2006) showed that the characteristic fractionation associated with solutes deviated from that predicted by a square root dependence on the ratio of the masses, as would be the case for an ideal gas. This observation has since been recognized as a result of the charged species and attendant solvation shells which compose a larger effective mass in solution (Bourg et al. 2010). As we move out to averages over larger volumes, both turbulent diffusion and mechanical dispersion also cause a departure from Rayleigh scaling. Turbulent diffusion occurs over such large spatiotemporal scales that molecular diffusion is often assumed insignificant, or at least non-fractionating, though this common assumption may be violated in aquifer systems (LaBolle et al. 2008). The result is that distillation during turbulent diffusion, where this mixing process is not considered fractionating, causes isotope values to scale proportionally with $f^{\sqrt{\alpha}-1}$ (Fig. 1b, Eriksson 1965) rather than the classic Rayleigh model ($f^{\alpha-1}$). In contrast, including the effects of both distillation and diffusive fractionation leads to the proportionality of $f^{\sqrt{\alpha/\alpha_{diff}}-1}$ where α is the fractionation factor associated with a reactive transformation as used above and a_{diff} is the separate fractionation factor associated with diffusion (Khan et al. 2018).

Fractionation of organic contaminants in groundwater has indicated significant complexity stemming from the combination of molecular diffusion, hydrodynamic dispersion and reactive transformation (Jin et al. 2014; Rolle and Jin 2017). Rolle and Jin (2017) even showed that fractionation can occur in the opposite direction suggested by the mass ratio for some deuterated isotopologues. Implementation of reactive transport simulations featuring the isotopic composition of individual compounds (largely hydrogen, carbon and chlorine) has proved highly beneficial. Often such models are calibrated against supporting laboratory experiments in order to constrain appropriate parameters in the absence of transport effects (van Breukelen et al. 2005; Hunkeler et al. 2009; Jin et al. 2013; Rodriguez-Escales et al. 2014) where application of simpler Rayleigh-type models have shown limited ability to appropriately describe observed spatiotemporal patterns (e.g., van Breukelen and Prommer 2008; Pooley et al. 2009; Thullner et al. 2012; Lutz and van Breukelen 2014; Khan et al. 2018). Such studies have largely centered on the utility of compound-specific stable isotope analysis (CSIA), analogous to the approach of tracking each speciated inorganic carbon compound in our prior example, but in application to the variety of reactants, reactive intermediaries and products which commonly occur during the biologically mediated degradation of organic contaminants and chlorinated solvents. Within this context, stable isotope patterns have highlighted the critical role of transverse dispersion in combining elevated contaminant concentrations with readily available reductive compounds along the fringe of the plume. Through direct simulation of the isotopic 'species' of the contaminants and their degradation biproducts, these studies have illustrated the role of the plume fringe not just as a mixing zone, but the process of dispersion as the rate limiting step which regulates the overall efficacy of remediation (Eckert et al. 2012; van Breukelen and Rolle 2012; Thullner et al. 2012). This observation applies in systems where transverse dispersion exhibits an appreciable dependence on diffusion and may thus generate fractionation (Rolle et al. 2010). Interestingly, it also holds true even when non-fractionating mixing effects outweigh actual diffusive fractionation due to differences in the mass of isotopologues (Xu et al. 2017; Rolle and Jin 2017).

A key result of these studies is that neglect of the influence of dispersive and diffusive mixing commonly results in the underestimation of fractionation factors associated with reactive transformations relative to the actual value in the absence of such transport effects (e.g., Kopinke et al. 2005). This underestimation can be further exacerbated by application of a Rayleigh—type model to a dataset comprised of fluid samples that are commonly representing a much larger depth-integrated section of the subsurface (e.g., the screen length of a well bore) than a single isolated representative elementary volume (Abe and Hunkeler 2006; Eckert et al. 2012; Green et al. 2010; La Bolle et al 2008; Van Breukelen 2007; Van Breukelen and Prommer 2008; Van Breukelen and Rolle 2012; Thullner et al. 2012). In this sense, the resulting coefficient describing the relationship between reaction progress and stable isotope enrichment, commonly termed the *effective fractionation factor*, is analogous to the observation of field-scale reaction rates which are routinely lower than what would be predicted from the same reactive process under laboratory conditions (e.g., Maher et al. 2006a). In leu of directly applying a Rayleigh model, often a laboratory determined fractionation factor is simply compared to field observations of isotope distribution as a prediction of reactive turnover. In this alternative approach, mixing leads to an overestimation of the fractionation pathway, such as biodegradation of a contaminant (e.g., Fischer et al. 2007; Thullner et al. 2012).

Example: Mixing of multiple flow paths during sampling

As in the example of evapotranspiration described above, the change in concentration of a contaminant with distance through the aquifer would be insufficient to deconvolve the relative contributions of dilution, mixing and reaction. We wish to use isotopes to assist in this analysis. Yet we now see that mixing (as a result of Fickian behavior), transport through heterogeneous media, or sampling across a large well screen, apparently compromises the log-linear relationship between reaction progress and isotope fractionation predicted by distillation or Rayleigh models. At the onset this observation might seem like a violation of mass balance. For example, it is easy to recognize that dilution will lower the concentration of a solute, but the ratio of the concentrations of two solutes should not be impacted. After all they are both diluted by the same factor. Similarly, how can mixing alter the ratio of two isotopes of the same element in a given compound in the absence of some additional source term or fractionating pathway? The basis for this behavior is that mixing of reactive solutes in open systems is not the same as adding the contents of one beaker into another. Aquifers are complex, non-uniform structures with preferential flow paths, low-permeability aquitards and a variety of scales at which non-uniform structure impacts fluid flow and solute mobility. Furthermore, collection of fluid samples is largely accomplished through the installation of well bores which often screen large sections of the vertical profile, integrating across multiple scales of this heterogeneity. The fluid that arrives at a well bore is thus a mixture of a variety of different flow paths, some of which are rapid, while others are slow. Not only does this influence the concentration of reactive solutes as a function of time spent in the system (i.e., the Damköhler number), but it also influences the percentage of the total volume of the sample composed of fluid that arrived quickly (the majority) and fluid that arrived slowly (the remainder). The result is that samples collected from these systems are partially mixed, or 'flux averaged', such that rapidly moving fluid which has been less reacted and fractionated exerts more influence on the overall sample composition. This non-uniform sample in turn undermines the assumptions inherent in Rayleigh distillation (Kopinke et al. 2005; Thullner et al. 2012; Druhan and Maher 2017).

Consider a simplified example based on the kinetically fractionating organic carbon oxidation used above (Fig. 2). Our organic carbon is now no longer a product of photosynthesis, but instead it represents an organic contaminant such as an aromatic hydrocarbon or chlorinated ethene, and it has been accidentally released into a shallow aquifer. The aquifer contains many naturally occurring electron acceptors (e.g., dissolved oxygen, nitrate) which may be sufficient to facilitate oxidative degradation of the organic contaminant. The fractionation factor for carbon isotopes implemented here is now $\alpha = 0.99$ (or equivalently $\varepsilon = 10‰$) such that the $\delta^{13}C$ of the residual organic contaminant remaining in the fluid enriches from an initial 0‰ to higher values. This reaction is effectively the same as the earlier example (Fig. 2) in terms of the reactive pathway from organic to inorganic carbon, but the fractionation factor is slightly larger. The 2D domain is bounded both above and below by impermeable layers, and all fluid flows from a high pressure head fixed across the left boundary condition to a low pressure head fixed across the right boundary. For the purposes of this simplified exercise we will remove the effects of dilution, such that all fluid entering the domain at the left-hand influent boundary contains a fixed, elevated concentration of the organic contaminant. This design alleviates the complexity of isotopic partitioning at the plume-fringe as discussed above, or in other words we are effectively looking inside of a much larger plume where surrounding uncontaminated water cannot reach. Thus, the only source of oxidative reactants capable of degrading this contaminant are in the influent fluid along the left boundary. It is as if our left boundary represents a line source of contaminant which is being picked up by the gradient of flow in the surrounding system. In this context we consider two physical structures. The first is a highly simplified flow field (Fig. 3A) consisting of horizontal layers of high and low permeability, where the low value is 60% of the high value. Differences in permeability within an order of magnitude are pervasive in shallow aquifer systems, and much larger ranges are common. For the 1 m flow field considered, a typical lateral dispersivity of $5\,cm^2$ is implemented, along with a transverse dispersivity of $0.5\,cm^2$, resulting in some mixing of the contaminant across the layers of high and low flow at steady state (Fig. 3B). This causes the illustrated organic carbon concentrations to appear smoother than they would for a purely advective stratified system. For the given flow domain and reaction rate, the organic contaminant concentration is diminished to approximately 30% of the influent value as fluid exits the right boundary. The second physical structure considered is a random realization of a permeability field which is based on a spatial covariance function with prescribed correlation coefficient and variance. The result is a system in which the permeability value of a given grid cell is likely to be similar to the values of its neighbors up to a certain distance, with an allowable variance in this correlation. The approach generates a completely random distribution of permeability in the limit as the correlation length goes to zero, and increasingly large zones of similar values as the correlation length increases. This method is a common means of generating an approximation to the size and distribution of structural features often observed in aquifers (Druhan and Maher 2014). The resulting velocity field for this arbitrary realization (Fig. 3C) has created a region of relatively high flow across the lower half of the domain, and lower flow across the top. In this simulation, no additional dispersivity is implemented, as the mixing process is largely generated through the highly variable distribution of velocities, referred to in Steefel and Maher (2009) as macrodispersion. The corresponding steady state distribution of the organic contaminant clearly illustrates mixing as a result of the non-uniform advective velocities across neighboring grid cells even for the use of a numerical scheme designed to minimize numerical dispersion (Steefel and Maher 2009). For this particular flow field realization, a generally higher concentration reaches the right boundary across the lower half of the domain (Fig. 3D). Despite the differences in structure, the overall change in concentration from the influent value to the flux-averaged value at the right boundary (i.e., the concentration that would be measured if the entire right boundary were screened and freely flowing into a well bore) of the two domains is quite similar.

At first it would seem that these two relatively simple 2D simulations of the degradation of a contaminant are easily predictable, both in terms of contaminant removal and associated isotopic fractionation. There is no influence of additional sources, and thus we would anticipate the classic Rayleigh relationship between reaction progress and fractionation. We would then be able to use departures from this relationship to discern additional effects, such as dilution, which might impact the concentration without altering the isotope ratio. Having intentionally removed the influence of dilution, we can directly compare the flux weighted average value of

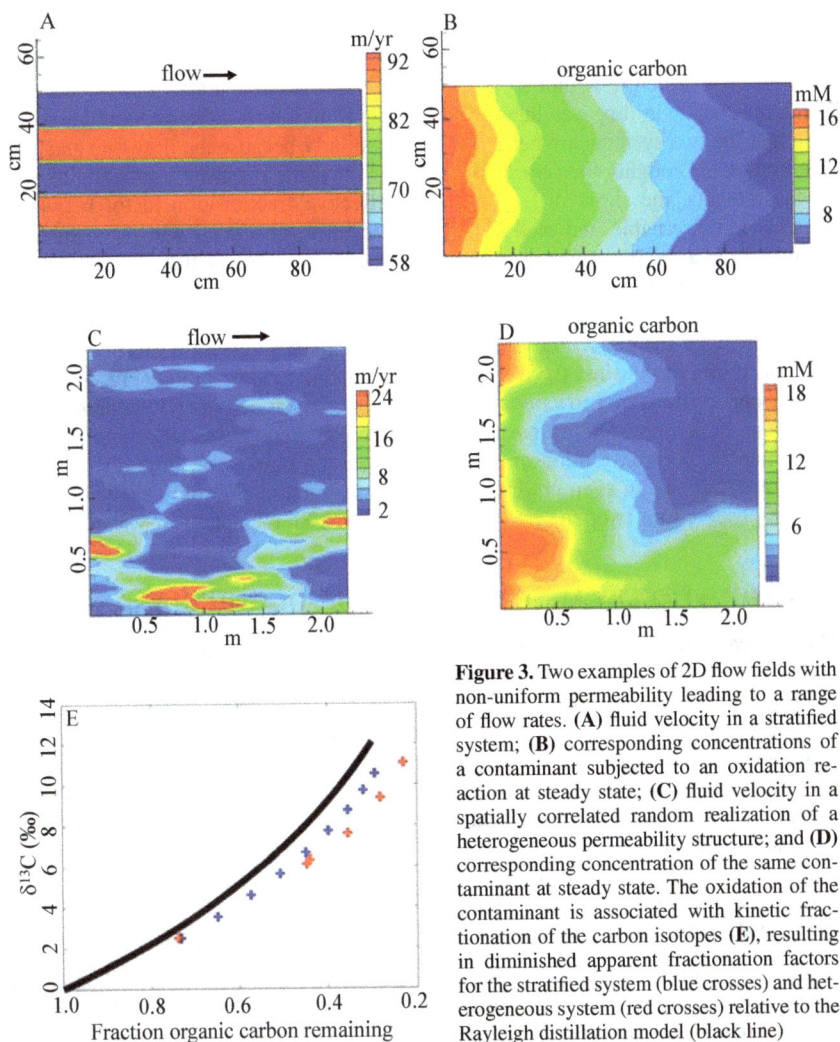

Figure 3. Two examples of 2D flow fields with non-uniform permeability leading to a range of flow rates. **(A)** fluid velocity in a stratified system; **(B)** corresponding concentrations of a contaminant subjected to an oxidation reaction at steady state; **(C)** fluid velocity in a spatially correlated random realization of a heterogeneous permeability structure; and **(D)** corresponding concentration of the same contaminant at steady state. The oxidation of the contaminant is associated with kinetic fractionation of the carbon isotopes **(E)**, resulting in diminished apparent fractionation factors for the stratified system (blue crosses) and heterogeneous system (red crosses) relative to the Rayleigh distillation model (black line)

concentration and isotope ratio at incremental distances from the left to the right boundary in comparison to the Rayleigh model, with the knowledge that we have assigned a fractionation factor of 10‰ in the forward simulations. The result is that both flow fields produce concentration-fractionation relationships that appear to have been subject to lower fractionation factors than the simple Rayleigh model would predict (Fig. 3E). In fact, the effective fractionation factors that we would infer from these data by assuming the Rayleigh model is still applicable are 8.5‰ for the stratified system and 7.5‰ for the heterogeneous system. The difference between these values and the 10‰ fractionation factor which we know is correct (because we used it in the model which generated these 'data') are well outside the range of error associated with most stable isotope measurements, and thus represent the potential for significant misunderstanding. For example, if we were attempting to characterize a natural system rather than our synthetic model, we might infer that roughly 10—15% of the decrease in concentration of the contaminant at the right boundary could be attributed to dilution, rather than degradation of the contaminant. This in turn would influence the quantitative estimate of remediation efficacy.

While this occurrence of diminished effective fractionation factors poses an engineering challenge in the application of isotope ratios as diagnostics of remediation, it is equally worthwhile to consider the potential utility of such behavior. For example, if we knew nothing about the physical structure within the two flow domains illustrated above, and were merely measuring the change in flux-weighted concentration from left to right, we would reasonably conclude that the two systems are behaving quite similarly because the flow rate and change in concentration are roughly the same. However, thanks to the exquisite sensitivity of isotope ratio measurements, the slight difference in the effective fractionation factor between the two systems is discernable and well outside of the error associated with such mass ratio measurements. As a result, the isotope ratios are the only diagnostic in this system that has the potential to suggest that the structures of the two domains, and thus the mixing of fluid flow paths within them, are distinct. Clearly the ability to discern these differences is still predicated on the sensitivity of the concentration measurements, as they remain a vital component to the appearance of a diminished fractionation factor relative to expectation (Fig. 3). Yet these preliminary observations hint towards the potential use of isotope ratios in tandem with bulk concentrations, as an improved means of quantifying reactivity in physically complex systems. In some instances, the convolution of functions describing fluid travel time distribution and Rayleigh fractionation can even produce closed form solutions, affording easy and versatile estimates of observations in open systems (Fischer et al. 2007; Druhan and Maher 2017).

FRACTIONATION BY MINERAL GROWTH: SOLUTIONS AND SOLIDS

Thus far, we have considered examples in which the combined effects of physical transport and chemical transformation remove the product of a fractionating reaction such that there is no opposing input flux or reversibility in the isotopic ratio. Many environmental systems in fact do not operate in this simple manner. Examples of partially closed-system behavior include the flux of water vapor via evaporation to clouds during rainout, the pooling of reactive intermediate species within a microbial cell, and the coupled dissolution of primary silicate minerals and precipitation of secondary clay phases during continental weathering. All of these systems involve a balance between mass input and output along with the rate of transport. For example, when we simply subject a normal Rayleigh distillation reaction to a constant return flux from the product, the modified behavior results in an effective fractionation which is dampened (Hendricks et al. 2000), as demonstrated by the example of a partially closed system in Figure 1.

Predictive models for many natural systems thus require capabilities beyond those afforded by the assumptions inherent in unidirectional distillation. Early examples of models separating the effects of reactivity and transport on isotopic distributions are found in application to the fractionation of carbon in soils and plants, as well as the spatiotemporal patterns of solutes in groundwater. Johnson and DePaolo (1994) developed coupled advection, mineral dissolution and isotopic exchange models to simulate strontium isotope ratios in groundwater. Years later, the same principles continue to be expanded in a variety of fields ranging from marine diagenesis to silicate weathering. A recent derivation by Huber et al (2017), for example, provides a closed form solution combining steady state recrystallization and sediment burial rates based on the partitioning of calcium isotopes between pore fluids and marine carbonates. Similar principles were used to develop a model for steady state stable isotope ratios of metals in weathering profiles which balances the dissolution of primary minerals and the loss of weathered mass due to erosion at the top of the soil profile (Bouchez et al. 2013).

Numerical reactive transport software packages offer the potential to generalize these approaches in application to a wide variety of systems (Steefel et al. 2015; Li et al. 2017). As discussed above, in principle, any software of this nature can simulate stable isotope partitioning by separating a species of interest into its respective isotopologues. In doing so, unique values may be assigned to the diffusion coefficients, kinetic rate constants, and/or equilibrium coefficients of the two isotopologues in the model, thus imposing a fractionation between them. In practice, this approach is suitable when simulating simple transport and reactive transformations, but more complex pathways such as phase changes and microbially catalyzed redox (discussed in the next section) often require specific rate expressions which may not be included in commonly available software.

The issue regarding formation and dissolution of solid phases may be considered by returning to our carbon isotope example. Carbonate minerals are some of the most prevalent rapidly reacting solids that commonly exist in the Earth's near surface environments. Many relevant conditions lead to both the dissolution (e.g., carbonate weathering fronts in soil profile development, sinkholes and caves typical of karst landscapes) and precipitation (e.g., travertine springs, marine foraminifera, speleothems) of this class of solids. The dissolved inorganic carbon either consumed or produced by these reactions is distributed through the contemporaneous speciation of CO_3^{2-}, HCO_3^-, $CO_{2(aq)}$ and $CO_{2(g)}$, and we may model the individual ^{12}C and ^{13}C 'species' of each primary and secondary compound in order to track the individual isotopes. Most modern numerical reactive transport software utilize thermodynamic databases such as EQ3/6 (Wolery 1992) which provides a set of instructions (i.e., the stoichiometry) to form each secondary species and solid phase out of the basis species present in solution. As described above, we can simply implement the isotopes of a given element by dividing each basis species containing this element into two (or more) 'species' representing the isotopologues of each compound. If these species are also present in a solid phase, then we can provide the stoichiometry of that solid, for example our carbonate solid may reflect the average 13-C natural abundance of approximately 1%. If the system is always in instantaneous equilibrium between the fluid and solid phase, then we need only treat a simple solid solution

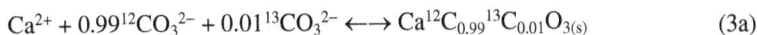

$$Ca^{2+} + 0.99\,^{12}CO_3^{2-} + 0.01\,^{13}CO_3^{2-} \longleftrightarrow Ca\,^{12}C_{0.99}\,^{13}C_{0.01}O_{3(s)} \qquad (3a)$$

which is aided by the near-ideality of the mixing between two isotopes of the same element. In this sense isotopes are a simplified example of solid solutions such as the range of composition bounded by the albite and anorthite endmembers, or the incorporation of trace elements substituting major ions (Helgeson et al. 1978; Kerrick and Darken 1975; Reed 1982; Glynn and Reardon 1990).

In a wide variety of environmental systems, isotopic signatures of minerals and surrounding fluids clearly demonstrate non-equilibrium effects. Lee and Bethke (1996) created a series of multi-component reaction models in which only the small component of the mineral participating in a reaction in the given timestep was allowed to interact with the fluid, while the rest was 'segregated', such that instantaneous equilibrium was not achieved. Using examples of limestone dolomitization, diagenetic sandstone alteration, and hydrothermal alteration of igneous rock, they illustrated improved agreement between their disequilibrium approach and common observations of hydrogen, carbon, oxygen and sulfur isotope distributions. Where kinetically controlled dissolution is necessary, this disequilibrium is not difficult to achieve. For example, a simple transition state theory (TST) rate expression of the form $R = kA(1 - Q/K_{eq})$ may be derived (Lasaga 1984) which allows the overall rate (R) to proceed as a product of the rate constant (k) the surface area of the solid phase (A) and the difference between the current ion activity product (Q) and the equilibrium value (K_{eq}). If the system is undersaturated, the value of Q/K_{eq} is less than 1 and the reaction dissolves the solid phase until these values are balanced. Critically, this rate law would work for the reaction above (Eqn. 3a) because we are simply delivering solute of a fixed isotope ratio equivalent to the solid phase value ($^{13}C/^{12}C \sim 0.01$) into solution. If the fluid contains carbonate ions with a different isotope ratio, then the resulting value would be a mixture as in the example of a partially closed system (Fig.1). There is no need to assign fractionation during dissolution, as mass-balance dictates no significant partitioning (examples and exceptions are reviewed in Druhan et al. 2015).

However, complexity arises when we attempt to simulate the formation of a new secondary carbonate solid, which would inherit the carbon isotope signature of the fluid from which it formed subject to any mass-dependent fractionation associated with the process of precipitation. Under this condition, the use of a single mineral composed of both isotopes (Eqn. 3a) is immediately problematic, as the model will be required to maintain this exact stoichiometry (and thus this exact isotope ratio) regardless of the isotope ratio of the fluid from which it was formed. As noted previously, we are always able to implement the individual isotopes of an element in each dissolved compound by creating individual primary and secondary 'species' which sum up to the concentration of each solute. However, if we are then required to provide a formula for how the mineral is created from these basis species, the same trick does not hold:

$$Ca^{2+} + {}^{12}CO_3^{2-} \longleftrightarrow Ca^{12}CO_{3(s)} \tag{3b}$$

$$Ca^{2+} + {}^{13}CO_3^{2-} \longleftrightarrow Ca^{13}CO_{3(s)} \tag{3c}$$

Here the single mineral phase $CaCO_{3(s)}$ is the sum of both solid phase isotopologues. They differ only in their isotopic composition, and therefore they should be subject to the same equilibrium constant (allowing for small differences due to equilibrium fractionation). However, the natural abundance of ${}^{13}C$ is approximately 1% that of ${}^{12}C$, meaning that for the same equilibrium constant and the same concentration of Ca^{2+}, Equation (3b) could easily become oversaturated (i.e., $Ca^{12}CO_{3(s)}$ precipitates) while Equation (3c) remains undersaturated, leading to the formation of an impossible pure ${}^{12}C$ solid.

A functional solution for this issue once again borrows from the rich history of solid solution modeling. Recognizing that the total activity of the solid phase must be the sum of the activities of the individual isotopologues, we can use their ideal nature to replace the individual activities (a_i) of each solid phase isotopologue with the equivalent mole fraction (x_i) of the total solid. This leads to an overall rate of mineral precipitation (R_{tot}) which is the sum of the rates of precipitation of each of the isotopologues (R_i):

$$R_{tot} = \sum_{i=1}^{n} x_i R_i \tag{4a}$$

where

$$R_i = k_i A_{tot} \left[1 - \left(\frac{Q_i}{K_i x_i} \right) \right]$$

Notably the parameters shown for the individual rates are the same as defined earlier for a simple TST rate law, where the values are specific to each isotope (with the exception of the surface area term which uses the total value). This formulation allows the assignment of both equilibrium fractionation through minor differences in K_i values and kinetic fractionation due to precipitation through differences in k_i values (Druhan et al. 2013). During dissolution, equivalent k_i values for all members of the solid solution creates stoichiometric addition to the fluid phase and avoids any spurious fractionation. This approach has been applied in a variety of recent isotopic reactive transport models, predominantly for systems in which the mineral phase precipitates without any subsequent re-dissolution or equilibration (Druhan et al 2012, 2014; Wanner et al. 2015; Hubbard et al. 2015; Druhan and Maher 2017).

A feature of this approach is that the value for the mole fraction x_i of each isotopologue of the solid must be defined. The implications for this designation are reviewed elsewhere (Druhan et al. 2015), but effectively revisit the inferences of Lee and Bethke (1996). At one extreme, we may use the isotopic composition of the entire solid phase to define each value of x_i, and at the other we may use such a vanishingly small component of the mineral surface that the solid is effectively isolated from further interaction (i.e., equilibration) with the fluid. In a sense this approach provides a means of circumventing the volume averaged representation of solids used in continuum scale models. From a different perspective, the isotope ratios of solids are challenging the frontiers of reactive transport simulations through the increasing need to represent isotopic structure (e.g., zoning) within pure mineral phases (Druhan and Winnick 2019). Finally, it is important to note that the formulations provided (Eqn. 4) are only strictly valid for first order rate expressions which do not incorporate any higher order exponents on the affinity term. This poses a second challenge for future model development in which we may generalize the formulation above to provide a simulation framework capable of treating the increasingly rich datasets of complementary isotopic and trace element ratios (e.g., $\delta^{30}Si$ and Ge/Si) and even the isotope ratios of contemporaneous major and trace elements (e.g., $\delta^{44}Ca$ and $\delta^{88}Sr$).

FRACTIONATION BY MICROBIAL POPULATIONS: THE MANY STEPS OF METABOLISM

From the discussion of mineral dissolution and formation above, it becomes apparent that the combination of two reactions functioning as a reversible system results in complex isotope signatures. This is one example of a broad variety of reactive transformations that are accomplished through multiple individual steps operating contemporaneously or sequentially in tandem or opposition to one another. For example, such multi-step transformations are common to microbially—mediated reactions which catalyze what would otherwise be sluggish electron transfer reactions. These pathways frequently impart large shifts in stable isotope ratios, and associated rate expressions commonly used to model microbially mediated reactions (e.g., Monod kinetics) have been developed to treat isotopic fractionation (e.g., Thullner et al. 2008), through the recognition of a common biomass population catalyzing each isotopologue of the same reactant. Application of such approaches to stable isotope signatures in reactive transport frameworks have largely focused on carbon isotopes of organics (e.g., Prommer et al. 2009, Thullner et al. 2012, Eckert et al. 2013) or sulfur isotopes of sulfate and similar oxyanion reduction pathways (e.g., Jamieson-Hanes et al. 2012; Druhan et al. 2012, 2014; Hubbard et al. 2014; Wanner et al. 2015). In what follows, we consider the general behavior of such multi-step reactive pathways that give rise to derivations such as Monod and Michaelis–Menten kinetics, and the behavior of stable isotope partitioning as a diagnostic of this sequential transformation. In these systems, the potential for macroscopic accumulation of intermediate species that are created and subsequently destroyed along a multi-step reactive chain strongly impacts the apparent or overall isotope fractionation observed between reactants and final products. This is exemplified by microbial catalysis of multi-step redox reactions, and the influence of mass transfer limitations which may precede these intracellular transformations

For the remainder of this subsection we will focus on kinetic stable isotope fractionation (driven by the extent in which the rates describing the breaking of chemical bonds containing a heavy isotope differ from rates for bonds containing only light isotopes). Within the context of this topic, it is common to express the inverse of our kinetic fractionation factor as the kinetic isotope effect (*KIE*) defined as $KIE = 1/\alpha_k$. For simple systems, α_k or *KIE* can be predicted from the involved elements, the type of the chemical bond(s) and the molecular structure of the reactant(s) (Elsner et al. 2005; González-Lafont and Lluch 2016). However, a chemical reaction—in particular those that are microbially driven—is typically composed of a sequence of reactive (and non-reactive) steps, the number and type depending on the nature of the overall reaction and on the scale of observation. Within such a sequence, the first irreversible step is most crucial for the stable isotope fractionation effect and in the simplest case (neglecting any prior reversible reactions) the *KIE* of this irreversible step would determine the observed overall fractionation factor. However, observed or effective fractionation (i.e., fractionation factors determined experimentally) may significantly differ from predictions based on the *KIE* only.

One major reason for such discrepancy is that any reversible steps preceding the first irreversible step may be rate limiting for the overall reaction, causing a 'masking' of the fractionation due to the irreversible step (Thullner et al. 2013) (see also section '*Mass transfer and bioavailability*' below). While the focus of this section is on microbially driven reactions, we note that some of the concepts presented here can be applied analogously to abiotic reactions, too.

As a simple example, we will depart from our previous usage of carbon isotope partitioning in favor of an equally rich and well-studied isotopic system: nitrogen cycling. Reactive transformation and attendant fractionation occur extensively among nitrogen species in response to changes in external forcing (e.g., shifting hydrologic regime, increased land surface temperatures, alteration in vegetation), and model predictions can vary drastically. Thus, shifts in the conditions under which a reaction occurs often lead to changes in the rate-limiting step of the overall reactive transformation and thus the associated expression of the stable isotope fractionation factor. If we simply lump this sequence of reactive steps into a single transformation from reactant to product, or even a simplification of the multi-step reactive chain (e.g., Monod kinetics) we may fail to adequately describe system response across a range of diverse and changing environmental conditions. To explore this behavior, we use a simple reactive transport model which explicitly treats the complete reaction network including all relevant intermediate species. The practical result of this approach is that reactive intermediates, even if they are short-lived, are allowed to influence the overall geochemical transformation of the system through direct simulation despite their often low concentrations. Using denitrification as the overall fractionating reactive transformation, a single step model would accomplish the complete 10 electron transfer pathway using one irreversible reductive transformation of nitrate (NO_3^-) to dinitrogen (N_2). In contrast we may break this process into four discrete steps:

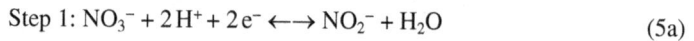

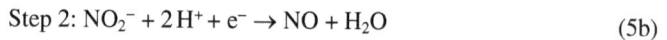

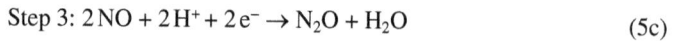

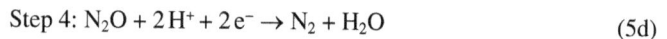

$$\text{Step 1: } NO_3^- + 2H^+ + 2e^- \longleftrightarrow NO_2^- + H_2O \tag{5a}$$

$$\text{Step 2: } NO_2^- + 2H^+ + e^- \rightarrow NO + H_2O \tag{5b}$$

$$\text{Step 3: } 2NO + 2H^+ + 2e^- \rightarrow N_2O + H_2O \tag{5c}$$

$$\text{Step 4: } N_2O + 2H^+ + 2e^- \rightarrow N_2 + H_2O \tag{5d}$$

where the overall transformation begins with a reduction of NO_3^- to NO_2^- through a reversible pathway, followed by three subsequent irreversible electron transfers, such that a total of five transformations are simulated in tandem (Fig. 4A). For the purposes of illustration, enrichment factors of $\varepsilon = 15‰$ for step 1, $\varepsilon = -10‰$ for the reverse of step 1, $\varepsilon = 12‰$ for step 2, $\varepsilon = 10‰$ for step 3 and $\varepsilon = 10‰$ for step 4 are all implemented as reasonable representative values. The model is run to steady state across an open, through-flowing 100 cm long 1D domain at a fixed flow rate and left-hand influent boundary condition containing 3 mM NO_3^- and negligible influent concentrations of the remaining nitrogen species, similar to the boundary conditions used in the organic carbon oxidation example but avoiding the complexity of influent product species (Fig. 2). With distance along the flow path (analogous to reaction time), NO_3^- is consumed and ultimately N_2 is produced. Clearly there is a small initial section of the domain in which N_2 is not the principle product species accumulating in solution, and thus we restrict further comparison to the section of the domain in which >50% of the influent NO_3^- has been consumed. This falls approximately 10 cm into the domain. Over the remaining 90 cm, NO_3^- decreases to effectively negligible concentration and N_2 accumulates, yet the corresponding Rayleigh space (Fig. 4B) indicates two important subtleties. First, the value of enrichment factor implemented in a Rayleigh distillation model that agrees with the fractionation of NO_3^- as a function of reaction progress is 13‰, a value that emerges as a result of the balance of rate limitation associated with the steps (Eqns. 5a–d) that exert unique influences at a given location along the steady state flow path. Second, the corresponding accumulated product N_2 isotope ratio is overestimated by the

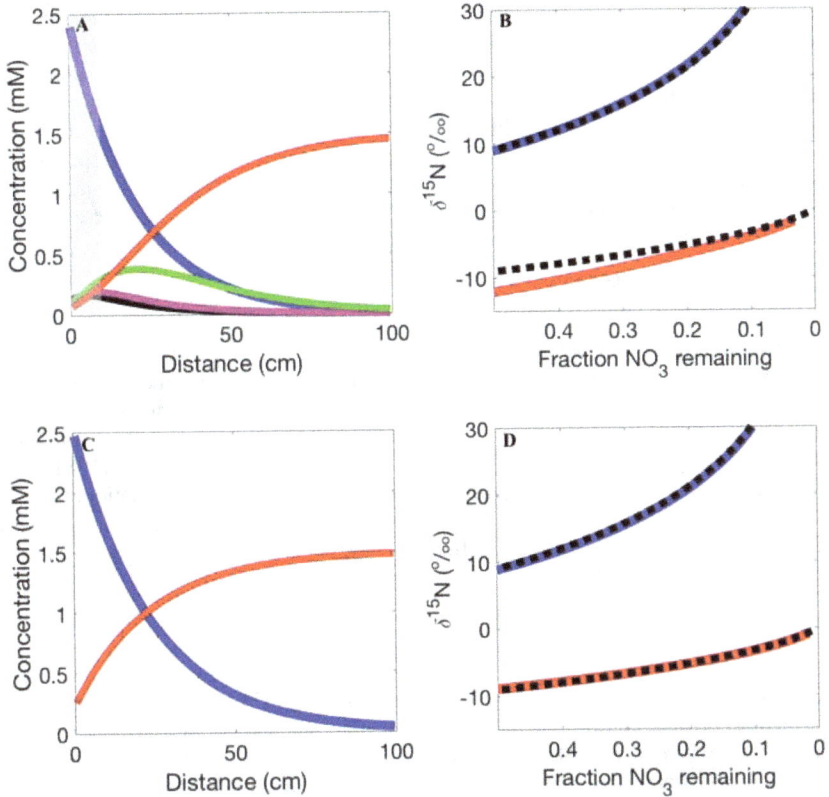

Figure 4. A multi-step, fractionating denitrification reaction at steady state along a 1D flow path advecting from left to right. (**A**) Concentrations of NO_3^- (blue line), intermediate species NO_2^- (black line), NO (purple line) and N_2O (green line) and final product species N_2 (red line) with distance along the flow path. Subsequent discussion is restricted to the component of the domain not shaded, in which >50% of the influent NO_3^- has been reacted and the final N_2 product is a substantial portion of the accumulated product. (**B**) The corresponding stable nitrogen isotope ratios of reactant NO_3^- (blue) and product N_2 (red) relative to a Rayleigh model for the instantaneous reactant and cumulative product (dashed line). (**C**) The same overall reaction simulated as a single unidirectional conversion of NO_3^- (blue) to N_2 (red) and (**D**) the corresponding distribution of isotopes as a function of reaction progress with the same Rayleigh model as (B).

associated Rayleigh model for the majority of the domain, despite the fact that we have restricted this analysis to exclude the initial period in which N_2 was not significantly accumulating.

As a further comparison, we may construct a corresponding single step model in which the overall transformation of NO_3^- to N_2 is accomplished in a single irreversible step (Fig. 4C) in which we fit an effective fractionation factor to recover the behavior of the multi-step simulation. In comparison to the multi-step model, the behavior of both reactant and product concentrations between 10 cm and 100 cm is essentially indistinguishable. Where NO_3^- reaches 50% of the influent value in the multi-step model, the corresponding single step concentration is 5% higher. At the effluent boundary, these values are within 1% of each other. The concentrations of the final product N_2 is similarly within 2% of one another at 100 cm, yet the accumulation of N_2 in the multi-step model is slightly slower at first, and effectively catches up to the single step value with distance along the flow path. As a result, the corresponding Rayleigh model for the single step reaction (Fig. 4D) requires the same $\varepsilon = 13‰$ effective enrichment factor as the

NO_3^- of the multi-step model, but in this case the corresponding N_2 isotope ratios precisely follow the cumulative Rayleigh model across the entire domain. Thus, while the overall conversion from NO_3^- to N_2 is effectively indistinguishable between the multi-step and single step simulations, the sequential occurrence of low concentration intermediates results in stable isotope partitioning that is distinct from the fractionation factor of any one step, and distinct in the accumulated product species by several per mil relative to a single-step framework. Such explicit incorporation of individual fractionating and non-fractionating steps within the simulation framework could be designed to distinguish the signatures of reaction intermediates, which govern the overall progress of the reaction, as expressed through differences in the enrichment of the stable isotopes (Aeppli et al. 2010; Hohener et al. 2015).

Enzymatic reactions: A comment on catalysis

The denitrification example illustrated above highlights the implications of microbially driven degradation reactions consisting of a pathway of enzymatically catalyzed biochemical transformations for stable isotope fractionation. Northrop (1981) described this behavior in detail, showing that observed stable isotope fractionation effects depend on the extent by which further reversible reaction steps precede the 'isotopically sensitive step' (or follow it in case the isotopically sensitive step is not irreversible). Although Northrop (1981) did not consider these further steps to directly cause any additional isotope fractionation, they could mask the observable fractionation. As a consequence, the 'apparent kinetic isotope effect' (*AKIE*), i.e., the effect observed for the entire pathway, is linked to—but not identical to—the *KIE* of the isotopically sensitive step:

$$AKIE = \frac{C + KIE}{C + 1} \tag{5}$$

with C describing the magnitude of the 'commitment to catalysis' as the tendency of the further reactive steps to proceed toward the isotopically sensitive step and not in the opposite direction (Northrop 1981; Elsner et al. 2005). The value of C depends on the kinetic parameters of the individual reaction steps along the transformative pathway, ranging from $C \approx 0$ (no masking, $AKIE \approx KIE$) to $C \gg \max(1, KIE)$ (strong masking, $AKIE \approx 1$). Quantitative relationships linking C to the rate parameters depend not only on the way the individual reactive steps interact along a degradation pathway but also on the concentration of the involved substrates (Northrop 1981). This implies that in the (common) case that a commitment to catalysis leads to a difference between *AKIE* and *KIE,* the resulting apparent stable isotope fractionation factor is not constant but a function of the concentration of the involved reactants—an effect not considered in most studies on microbially induced stable isotope fractionation. Brunner and Bernasconi (2005) analyzed the enzymatic degradation pathway of sulfate reduction and its implication for stable sulfur isotope fractionation. By expanding a previous mechanistic pathway model consisting of a sequence of reversible steps (Rees 1973) and considering more than one isotopically sensitive step, they showed that for the entire bacterial cell the overall sulfur isotope fractionation can vary from 3‰ for low sulfate concentrations (where no backward reaction takes place) to −70‰ for the opposite extreme (forward and backward reactions take place at equal fluxes). This corresponds to the range of values observed in natural environments. Mancini et al. (2006) subsequently showed that the concentration of trace elements (e.g., iron) can also influence the commitment to catalysis and thus the masking of stable isotope fractionation during aerobic toluene remediation. In a series of batch reactors, low iron concentrations were associated with toluene carbon and hydrogen fractionation factors of −2.5‰ and −159‰, respectively. In contrast, high iron concentrations led to less pronounced fractionation effects of −1.7‰ and −77‰, respectively. The authors suggested that this was a result of the iron concentrations affecting the relationship between the kinetic parameters of individual reaction steps along the toluene degradation pathway, which varied the magnitude of the commitment to catalysis.

Mass transfer and bioavailability

Any enzymatically catalyzed reaction requires the enzyme and the reactant to be microscopically close in vicinity to one another. With the exception of extracellular enzymes released by the microbial population, the abundance of enzymes is constrained to the interior of the individual cells, and substrates require some mass transfer to reach this location. Which mass transfer processes are involved depends on the participating microorganisms and the physical setting of the environment they reside in. At the smallest scale, a mass transfer across the cell membrane is needed and the kinetics of this substrate uptake step can influence the overall reaction rate (Button 1991, 1998). In addition, the concentration gradients in a system lead to differences between bioavailable substrate concentration (i.e., the concentration immediately outside of a cell) and the bulk concentration (defined by the spatial resolution of the observation) (Semple et al. 2004). In particular, the microbial substrate uptake itself can lead to a concentration depletion around the immediate vicinity of the cells, and slow mass transfer may become a rate-limiting process (Bosma et al. 1997; Johnsen et al. 2005; Heße et al. 2009; Gharasoo et al. 2012). Furthermore, substrate bioavailability can be affected by sorption to solids (Haws et al. 2006) or dissolution from a non-aqueous phase liquid (NAPL) (Amos et al. 2007). A number of experimental studies indicate that such mass transfer processes also affect the observed stable isotope fractionation effects in a system (i.e., the fractionation effects derived from measurable concentrations and observed stable isotope ratios). Differences between stable isotope fractionation factors obtained for free enzymes and for whole cell experiments demonstrated the influence of substrate uptake into the cell (Cichocka et al. 2007, 2008; Nijenhuis et al. 2005) and the magnitude of this effect depends on the properties of the cell membrane and of the substrate (Renpenning et al. 2015). For example Tobler et al. (2008) observed differences in stable isotope fractionation factors between cells attached to a solid surface and cells which are mobile in the water phase which they attributed to substrate mass transfer limitations for the stationary cells. Aeppli et al. (2009) showed that mass transfer from a NAPL to the aqueous phase masked stable isotope fractionation. Other authors have also showed that increasing concentrations of the microbial biomass can lead to decreasing observation of stable isotope fractionation (Templeton et al. 2006; Kampara et al. 2009) which is interpreted as a higher biomass concentration leading to greater substrate removal rates and thus more pronounced depletions of substrate in the immediate vicinity of the cells. This causes more severe limitation due to mass transfer creating masking of the stable isotope fractionation.

Approaches to describe the rate limitation of mass transfer can be linked to masking effects. A common method is the use of a linear exchange model linking bioavailable concentration c_{bio} to the observable bulk concentration c_{obs}:

$$r_{tr} = k_{tr}\left(c_{obs} - c_{bio}\right) \tag{6}$$

with r_{tr} is the mass transfer rate and k_{tr} is an effective 'rate' parameter describing the limitation. This effective parameter has been used to describe mass transfer due to transport processes (Baveye and Valocchi 1989; Bosma et al. 1997; Heße et al. 2010; Schmidt et al. 2018), dissolution processes (Barry et al. 2002) and sorption processes (Haws et al. 2006). Factors controlling the value of k_{tr} vary depending on the substrate, the microorganisms involved, the environmental setting as well as the relevant length scales, where the latter is also a result of the spatial observation scale. In particular, for the common case of diffusive mass transfer, k_{tr} linearly depends on the molecular diffusion coefficient of the substrate. While the implementation of such simple mass transfer into a numerical reactive transport model is straight forward, the combination of this mass transfer rate with a degradation rate given by Michaelis–Menten can also be achieved analytically. The closed form expression is known as the Best Equation (Best 1955) and describes mass transfer limited degradation processes (Bosma et al. 1997; Simoni et al. 2001; Heße et al. 2010). Expanding this approach to consider stable isotope fractionation leads to an equally simple and versatile expression for the observable stable isotope fractionation factor α* (Thullner et al. 2008, 2013; Heße et al. 2014):

$$\alpha^* = \alpha_{bio} \cdot \frac{1+A}{1+\dfrac{\alpha_{bio}}{\alpha_{tr}} \cdot A} \tag{7}$$

where α_{bio} is the intrinsic stable isotope fractionation factor of the microbial degradation reaction in the absence of transport effects. Notably, this expression is formally identical to Equation (5) describing the commitment to catalysis if $\alpha^* = 1/AKIE$ and $\alpha_0 = 1/KIE$. For degradation rates given by first-order kinetics, A is a fixed parameter given by $A = k_{bio}/k_{tr}$ and is thus a Damköhler number describing the relative magnitude of degradation vs. mass transfer. For fast mass transfer (compared to degradation) no masking of the stable isotope fractionation takes place, while slow mass transfer leads to severe masking ($\alpha^* \approx \alpha_{tr}$). The fractionation factor of the mass transfer process (α_{tr}) is commonly considered to be negligible (i.e., $\alpha_{tr} = 1$), but, as described earlier, molecular diffusion (LaBolle et al. 2008; Eggenkamp and Coleman 2009; Jin et al. 2014) and associated dispersion (Rolle et al. 2010), sorption (Kopinke et al. 2005) or phase exchange processes (Kuder et al. 2009; Horst et al. 2016; Kopinke et al. 2018) may be isotopologue-specific and can in fact cause an additional fractionating effect.

For a reaction which is reasonably described by Michaelis–Menten kinetics, i.e., $\partial c/\partial t = k_{max} \cdot c/(c+K_s)$, (and considering the common case of small stable isotope ratios R, i.e., $R + 1 \approx 1$) the value of A becomes a function of the observable bulk concentration (Thullner et al. 2008):

$$A = \frac{1}{2} \cdot \left(\frac{a}{k_{tr}} - \frac{c_{obs}}{K_S} - 1\right) + \sqrt{\frac{a}{k_{tr}} + \frac{1}{4} \cdot \left(\frac{a}{k_{tr}} - \frac{c_{obs}}{K_S} - 1\right)^2} \tag{8}$$

where $a = k_{max}/K_S$ is the specific affinity of the microorganisms and k_{max} and K_S are the maximum degradation rate and Michaelis–Menten constant, respectively. As a consequence, the magnitude of the masking also depends on substrate concentration. For high concentrations ($c_{obs}/K_S \gg 1$) the masking effect decreases and may eventually diminish ($\alpha^* \approx \alpha_0$), while for low concentrations ($c_{obs}/K_S < 1$) A approaches a fixed value of a/k_{tr} as for first-order reaction rates. This concept helped to explain observations by Kampara et al. (2008) of substrate concentration dependent stable isotope fractionation and has been implemented as an explicit representation of mass transfer and microbial degradation in reactive transport simulations of stable isotope fractionation in porous media (Thullner et al. 2008; Khan et al. 2018).

While such substrate concentration dependency of the stable isotope fractionation factor imposes some challenges in the application of CSIA for quantitative analyses of microbial degradation, as argued earlier in the context of structural heterogeneity, this sensitivity may provide an indicator for the influence of mass transfer in microbially mediated systems. A series of chemostat experiments (Ehrl et al. 2019) and supporting numerical models (Gharasoo et al. 2019) demonstrated dependencies of the observed stable isotope fractionation factor on the chemostat dilution rate. Thus, the residual substrate concentration in the chemostat could be explained using the above concept. Such data might be used to identify the importance of mass transfer across the cell membrane at different concentration levels as well as the magnitude of the trans-membrane diffusion coefficient. Kampara et al. (2008) introduced a theoretical concept of effective bioavailability $B_{eff} = r_{Best}/r_{MM}$ as a ratio between the actual degradation rate affected by mass transfer limitation (given by the Best equation) and the optimal degradation rate in the absence of any mass transfer limitations (given by Michaelis–Menten kinetics). Heße et al. (2014) introduced a relationship providing an approximate link between the effective bioavailability and the observed stable isotope fractionation, which allows estimates of B_{eff} for most conditions. In spite of the large influence masking may have on observed stable isotope fractionation factors, dual-isotope approaches (i.e., the combined analysis of stable isotope fractionation of two elements in the same compound(s) (Vogt et al. 2016)) are still readily applicable for the identification of degradation pathways in that masking of the fractionation takes place in an analogous manner for both elements (Thullner et al. 2013). Similarly, the commitment to catalysis has been observed to have analogous effects for stable isotope fractionation of different elements (Mancini et al. 2006).

In summary, masking of stable isotope fractionation can occur due to a variety of various processes taking place at different scales. Indicators for the occurrence of such masking processes include an offset between values of the stable isotope fractionation factor observed for a given system and values obtained for idealized conditions. Furthermore, masking effects can induce a concentration dependency in the observed stable isotope fractionation factor, where for sufficiently high concentrations masking may become negligible. The extent to which such processes can be explicitly resolved depends on the spatial resolution of the experimental observation scheme as well as on the spatial resolution of the applied reactive transport approach. For masking processes taking place at scales below observational resolution, the concepts presented here provide a means of constraining masking effects when modeling reactive processes and the associated stable isotope fractionation.

SUMMARY AND FUTURE OPPORTUNITIES

Reactive transport modeling quantitatively merges fluid flow and solute transport with a potentially limitless mixed suite of equilibrium (e.g aqueous complexation, ion exchange, surface complexation) and kinetic (e.g., mineral dissolution, precipitation, redox) reactive pathways between and among phases (solid, liquid and gas). Such an approach thus provides a unifying framework in which to describe and predict co-evolutional chemical, physical and biological dynamics in open, transient systems subject to the principles of mass, momentum and energy balance. Such diversity and flexibility have been pivotal in the quantitative analysis of low temperature biogeochemical environments, and many key processes are now well described within reactive transport frameworks. These include the conversion of rock to soil, the generation, movement and utilization of nutrients which sustain ecosystems, contaminant fate and transport, and watershed to regional elemental cycles (Steefel et al. 2015). In tandem with this diversity of applications, reactive transport software packages have been substantially improved and expanded to include coupled transport and reactive controls governing exchange reactions between fluid and mineral surfaces, weathering reactions, and biologically mediated reactivity (Steefel and Maher 2009; Lawrence et al. 2014; Li et al 2017; Maher and Mayer 2019). Applications of such coupled numerical descriptions of environmental transport and transformation serve as a platform supporting hypothesis-testing, experimental and field scale characterization of critical (e.g., rate limiting) processes, and as an integrator of datasets and quantitative framework for forecasting and uncertainty analysis.

The use of stable isotope partitioning within such modern reactive transport experiments and associated model architecture has still largely been focused in contaminant remediation (e.g., Berna et al. 2009; Sherwood Lollar et al. 1999, 2001; Jamieson-Hanes et al. 2012; Druhan et al. 2012, 2014). Recently, expanded applications of multi-component numerical reactive transport models treating the partitioning of both stable and radiogenic isotopes have begun to yield new insight into the (bio)geochemical cycling of elements through marine sediments (Dale et al. 2009; Maher et al. 2006a), secondary mineral growth (Druhan et al. 2013; Steefel et al. 2014), fluid transport and mechanisms of mixing (Prommer et al. 2009; Rolle et al.

2010; Eckert et al. 2012; Van Breukelen and Rolle 2012; Centler et al. 2013; Druhan and Maher 2014; 2017), terrestrial weathering rates and mechanisms (Maher et al. 2006b), vapor phase transport in soils (Khan et al. 2018) and increasingly complex contaminant fate and degradation pathways (Hunkeler et al. 2009; Gibson et al. 2011; Wanner et al. 2014).

As isotopic reactive transport simulations continue to advance, we are poised to gain critical new constraint of dynamic (bio)geochemical reaction networks such as those driven by surface-subsurface hydrologic coupling, including infiltration, land-surface alteration, hyporheic exchange, and meteorological forcing. In order to develop a predictive understanding of such integrated hydro-bio-geo-chemical reactors driven by structure, function, and transport, integrative models will need to be expanded to better represent the isotopic partitioning associated with (i) soil biogeochemistry including multi-state biomass fractions (i.e., living, dormant, and dead); (ii) hydraulic redistribution and effects of root–soil interaction, such as simultaneous root exudation and nutrient absorption; (iii) expression of three-dimensional fluxes to represent both vertical and lateral interconnected flows and processes; and (iv) the potential to include the influence of non-stationary external forcing and climatic factors. Each of these processes are described by a necessary parameter set, for which new advances in isotopic characterization can provide critical reduction in uncertainty through quantitative simulation frameworks. Similar advancements are now underway using isotope-enabled model representations of weathering environments, ore recovery, heavy metal mobility and metabolism both in modern and ancient environments (Druhan and Winnick 2019).

ACKNOWLEDGMENTS

JD would like to acknowledge support from the DOE Office of Science SBR program (DE-SC0019198) and BES program (DE-SC0019165). The authors wish to thank Shaun Brown and Noah Jemison for their careful reviews, as well as the tireless efforts of series editor Ian Swainson and managing editor Rachel Russell.

REFERENCES

Abe Y, Hunkeler D (2006) Does the Rayleigh equation apply to evaluate field isotope data in contaminant hydrogeology? Environ Sci Technol 40:1588–1596

Aeppli C, Berg M, Cirpka OA, Holliger C, Schwarzenbach RP, Hofstetter TB (2009) Influence of mass-transfer limitations on carbon isotope fractionation during microbial dechlorination of trichloroethene. Environ Sci Technol 43:8813–8820

Aeppli C, Hofstetter TB, Amaral HIF, Kipfer R, Schwarzenbach RP, Berg M (2010) Quantifying in situ transformation rates of chlorinated ethenes by combining compound-specific stable isotope analysis, groundwater dating, and carbon isotope mass balances. Environ Sci Technol 44:3705–3711

Allègre CJ (2008) Isotope Geology. Cambridge. University Press, Cambridge, 512 pp

Amos BK, Christ JA, Abriola LM, Pennell KD, Löffler FE (2007) Experimental evaluation and mathematical modeling of microbially enhanced tetracholorethene (PCE) dissolution, Environ Sci Technol 41:963–970

Barry DA, Prommer H, Miller CT, Engesgaard P, Brun A, Zheng C (2002) Modelling the fate of oxidisable organic contaminants in groundwater, Adv Water Resour 25:945–983

Baveye P Valocchi A (1989) An evaluation of mathematical models of the transport of biologically reacting solutes in saturated soils and aquifers. Water Resour Res 25:1413–1421

Berna EC, Johnson TM, Makdisi RS, Basui A (2010) Cr stable isotopes as indicators of Cr(VI) reduction in groundwater: A detailed time-series study of a point-source plume. Environ Sci Technol 44:1043–1048

Best JB (1955) The interference of intracellular enzymatic properties from kinetic data obtained on living cells. 1. Some kinetic considerations regarding an enzyme enclosed by a diffusion barrier., J Cell Comparat Physiol 46:1–27

Bosma TNP, Middeldorp PJM, Schraa G, Zehnder AJB (1997) Mass transfer limitation of biotransformation: Quantifying bioavailability. Environ Sci Technol 31:248–252

Bourg IC, Richter FM, Christensen JN, Sposito G (2010) Isotopic mass dependence of metal cation diffusion coefficients in liquid water. Geochim Cosmochim Acta 74:2249–2256

Bouchez J, Von Blanckenburg F, Schuessler JA (2013) Modeling novel stable isotope ratios in the weathering zone. Am J Sci 313:267–308

Brunner B, Bernasconi SM (2005) A revised isotope fractionation model for dissimilatory sulfate reduction in sulfate reducing bacteria. Geochim Cosmochim Acta 69:4759–4771

Button DK (1991) Biochemical basis for whole-cell uptake kinetics—Specific affinity, oligotrophic capacity, and the meaning of the michaelis constant. Appl Environ Microbiol 57:2033–2038

Button DK (1998) Nutrient uptake by microorganisms according to kinetic parameters from theory as related to cytoarchitecture. Microbiol Mol Biol Rev 62:636–645

Centler F, Heße F, Thullner M (2013) Estimating pathway-specific contributions to biodegradation in aquifers based on dual isotope analysis: Theoretical analysis and reactive transport simulations. J Contam Hydrol 152:97–116

Cerling TE (1984) The stable isotopic composition of modern soil carbonate and its relationship to climate. Earth Planet Sci Lett 71:229–240

Cichocka D, Siegert M, Imfeld G, Andert J, Beck K, Diekert G, Richnow HH, Nijenhuis I (2007) Factors controlling the carbon isotope fractionation of tetra- and trichloroethene during reductive dechlorination by *Sulfurospirillum* ssp and *Desulfitobacterium* sp strain PCE-S, FEMS Microbiol Ecol 62:98–107

Cichocka D, Imfeld G, Richnow HH, Nijenhuis I (2008) Variability in microbial carbon isotope fractionation of tetra- and trichloroethene upon reductive dechlorination. Chemosphere 71:639–648

Criss RE (1999) Principles of Stable Isotope Distribution, Oxford University Press, New York, Oxford

Dale AW, Brüchert V, Alperin M, Regnier P (2009) An integrated sulfur isotope model for Namibian shelf sediments. Geochim Cosmochim Acta 73:1924–1944

Davidson GR (1995) The stable isotopic composition and measurement of carbon in soil CO_2. Geochim Cosmochim Acta 59:2485–2489

Druhan JL, Maher K (2014) A model linking stable isotope fractionation to water flux and transit times in heterogeneous porous media. Proc Earth Planet Sci 10:179–188

Druhan JL, Maher K (2017) The influence of mixing on stable isotope ratios in porous media: A revised Rayleigh model. Water Resour Res 53:1101–1124

Druhan JL, Winnick MJ (2019) Reactive transport of stable isotopes. Elements 15:107–110

Druhan JL, Steefel CI, Molins S, Williams KH, Conrad ME, DePaolo DJ (2012) Timing the onset of sulfate reduction over multiple subsurface acetate amendments by measurement and modeling of sulfur isotope fractionation. Environ Sci Technol 46:8895–8902

Druhan JL, Steefel CI, Williams KH, DePaolo DJ (2013) Calcium isotope fractionation in groundwater: Molecular scale processes influencing field scale behavior. Geochim Cosmochim Acta 119:93–116

Druhan JL, Steefel CI, Conrad ME, DePaolo DJ (2014) A large column analog experiment of stable isotope variations during reactive transport: I A comprehensive model of sulfur cycling and $\delta^{34}S$ fractionation. Geochim Cosmochim Acta 124:366–393

Druhan JL, Brown ST, Huber C (2015) Isotopic gradients across fluid-mineral boundaries Rev Mineral Geochem 80:355–391

Dwivedi D, Tang J, Bouskill N, Georgiou K, Chacon SS, Riley WJ (2019) Abiotic and biotic controls on soil organo-mineral interactions: Developing model structures to analyze why soil organic matter persists. Rev Mineral Geochem 85:329–348

Eckert D, Rolle M, Cirpka OA (2012) Numerical simulation of isotope fractionation in steady-state bioreactive transport controlled by transverse mixing. J Contam Hydrol 140:95–106

Eckert D, Qiu S, Elsner M, Cirpka OA (2013) Model complexity needed for quantitative analysis of high resolution isotope and concentration data from a toluene-pulse experiment. Environ Sci Technol 47:6900–6907

Eggenkamp HGM, Coleman ML (2009) The effect of aqueous diffusion on the fractionation of chlorine and bromine stable isotopes. Geochim Cosmochim Acta 73:3539–3548

Ehrl BN, Kundu K, Gharasoo M, Marozava S, Elsner M (2019) Rate-limiting mass transfer in micropollutant degradation revealed by isotope fractionation in chemostat. Environmental Science and Technol 53:1197–1205

Elsner M, Zwank L, Hunkeler D, Schwarzenbach RP (2005) A new concept linking observable stable isotope fractionation to transformation pathways of organic pollutants. Environ Sci Technol 39:6896–6916

Farkas A, Farkas L (1934) Experiments on heavy hydrogen. Part I Proc R Soc London Ser A 144:467–480

Fischer A, Theuerkorn K, Stelzer N, Gehre M, Thullner M, Richnow HH (2007) Applicability of stable isotope fractionation analysis for the characterization of benzene biodegradation in a BTEX-contaminated aquifer. Environ Sci Technol 41:3689–3696

Gat JR, Matsui E (1991) Atmospheric water balance in the Amazon basin: An isotopic evapotranspiration model. J Geophys Res 96:0148–0227

Gharasoo M, Centler F, Regnier P, Harms H, Thullner M (2012) A reactive transport modeling approach to simulate biogeochemical processes in pore structures with pore-scale heterogeneities. Environ Model Softw 30:102–114

Gharasoo M, Ehrl BN, Cirpka OA, Elsner M (2019) Modeling of contaminant biodegradation and compound-specific isotope fractionation in chemostats at low dilution rates. Environ Sci Technol 53:1186–1196

Gibson BD, Amos RT, Blowes DW (2011) $^{34}S/^{32}S$ Fractionation during sulfate reduction in groundwater treatment systems: Reactive transport modeling. Environ Sci Technol 45:2863–2870

Glynn PD, Reardon EJ (1990) Solid-solution aqueous-solution equilibria; thermodynamic theory and representation. Am J Sci 290:164–201

González-Lafont À, Lluch JM (2016) Kinetic isotope effects in chemical and biochemical reactions: physical basis and theoretical methods of calculation, Wiley Interdisciplinary Rev: Comput Mol Sci 6:584–603

Green C T, Böhlke J K, Bekins B A, and Phillips S P (2010) Mixing effects on apparent reaction rates and isotope fractionation during denitrification in a heterogeneous aquifer. Water Resour Res 46: W08525

Hansen M, Scholz D, Schöne BR, Spötl C (2019) Simulating speleothem growth in the laboratory: Determination of the stable isotope fractionation ($\delta^{13}C$ and $\delta^{18}O$) between H_2O, DIC and $CaCO_3$. Chem Geol 509:20–44

Haws NW, Ball WP, Bouwer EJ (2006) Modeling and interpreting bioavailability of organic contaminant mixtures in subsurface environments, J Contam Hydrol 82:255–292

Helgeson H C, Delany J M, Nesbitt H W, Bird D K (1978) Summary and critique of the thermodynamic properties of rock-forming minerals. Am J Sci 278-A:1–229

Hendricks MB, DePaolo DJ, Cohen RC (2000) Space and time variation of $\delta^{18}O$ and δD in precipitation: Can paleotemperature be estimated from ice cores? Global Biogeochem Cycles 14:851–861

Heße F, Radu F A, Thullner M, Attinger S (2009) Upscaling of the advection–diffusion–reaction equation with Monod reaction. Adv Water Resour 32:1336–1351

Heße F, Harms H, Attinger S, Thullner M (2010) Linear exchange model for the description of mass transfer limited bioavailability at the pore scale. Environ Sci Technol 44:2064–2071

Heße F, Prykhodko V, Attinger S, Thullner M (2014) Assessment of the impact of pore-scale mass-transfer restrictions on microbially induced stable-isotope fractionation. Adv Water Resour 74:79–90

Hoefs J (2004) Stable Isotope Geochemistry, 5th ed. Springer, Berlin, Heidelberg, New York

Hohener P, Elsner M, Eisenmann H, Atteia O (2015) Improved constraints on in situ rates and on quantification of complete chloroethene degradation from stable carbon isotope mass balances in groundwater plumes. J Contam Hydrol 182:173–182

Horst A, Lacrampe-Couloume G, Sherwood Lollar B (2016) Vapor pressure isotope effects in halogenated organic compounds and alcohols dissolved in water. Anal Chem 88:12066–12071

Hubbard C G, Cheng Y, Engelbrekston A, Druhan J L, Li L, Ajo-Franklin J B, ... and Conrad M E (2014) Isotopic insights into microbial sulfur cycling in oil reservoirs. Frontiers Microbiol5:480

Huber C, Druhan JL, Fantle MS (2017) Perspectives on geochemical proxies: The impact of model and parameter selection on the quantification of carbonate recrystallization rates. Geochim Cosmochim Acta 217:171–192

Hunkeler D, Van Breukelen BM, Elsner M (2009) Modeling chlorine isotope trends during sequential transformation of chlorinated ethenes. Environ Sci Technol 43:6750–6756

Jamieson-Hanes JH, Amos RT, Blowes DW (2012) Reactive transport modeling of chromium isotope fractionation during Cr (VI) reduction. Environ Sci Technol 46:13311–13316

Jin B, Haderlein SB, Rolle M (2013) Integrated carbon and chlorine isotope modeling: applications to chlorinated aliphatic hydrocarbons dechlorination. Environ Sci Technol 47:1443–1451

Jin B, Rolle M, Li T, Haderlein SB (2014) Diffusive fractionation of BTEX and chlorinated ethenes in aqueous solution: quantification of spatial isotope gradients, Environ Sci Technol 48:6141–6150

Johnsen AR, Wick LY, Harms H (2015) Principles of microbial PAH-degradation in soil, Environ Pollut 133:71–84

Johnson TM, DePaolo DJ (1994) Interpretation of isotopic data in groundwater–rock systems: Model development and application to Sr isotope data from Yucca Mountain. Water Resour Res 30:1571–1587

Kampara M, Thullner M, Richnow HH, Harms H, Wick LY (2008) Impact of bioavailability restrictions on microbially induced stable isotope fractionation. 2. Experimental evidence, Environ Sci Technol 42:6552–6558

Kampara M, Thullner M, Harms H, Wick LY (2009) Impact of cell density on microbially induced stable isotope fractionation. Appl Microbiol Biotechnol 81:977–986

Kerrick DM, Darken LS (1975) Statistical thermodynamic models for ideal oxide and silicate solid solutions, with application to plagioclase. Geochim Cosmochim Acta 39:1431–1442

Khan AM, Wick LY, Thullner M (2018) Applying the Rayleigh approach for stable isotope-based analysis of VOC biodegradation in diffusion-dominated systems, Environ Sci Technol 52:7785–7795

Kopinke FD, Georgi A, Voskamp M, Richnow HH (2005) Carbon isotope fractionation of organic contaminants due to retardation on humic substances: Implications for natural attenuation studies in aquifers. Environ Sci Technol 39:6052–6062

Kopinke F D, Georgi A, Roland U (2018) Isotope fractionation in phase-transfer processes under thermodynamic and kinetic control—Implications for diffusive fractionation in aqueous solution. Sci Total Environ 610–611:495–502

Kuder T, Philp P, Allen J (2009) Effects of volatilization on carbon and hydrogen isotope ratios of MTBE. Environ Sci Technol 43:1763–1768

LaBolle EM, Fogg GE, Eweis JB, Gravner J, Leaist DG (2008) Isotopic fractionation by diffusion in groundwater. Water Resour Res 44:W07405

Lawrence C, Harden J, Maher K (2014) Modeling the influence of organic acids on soil weathering. Geochim Cosmochim Acta 139: 487–507

Lasaga AC (1984) Chemical kinetics of water-rock interactions. J Geophys Res 89:4009–4025

Lee M K, Bethke CM (1996) A model of isotope fractionation in reacting geochemical systems. Am J Sci 296:965

Li L, Maher K, Navarre-Sitchler A, Druhan JL, Meile C, Lawrence C, Moore J, Perdrial J, Sullivan P, Thompson A, Jin L, Bolton EW, Brantley SL, Dietrich W, Mayer KU, Steefel CI, Valocci A, Zachara J, Kocar B, Mcintoch J, Tutolo BM, Kumar M, Sonnenthal E, Bao C, Beisman J (2017) expanding the role of reactive transport models in critical zone processes. Earth Sci Rev 165:280–301

Lichtner PC (1985) Continuum model for simultaneous chemical reactions and mass transport in hydrothermal systems. Geochim Cosmochim Acta 49:779–800

Lutz SR, Van Breukelen B M (2014) Combined source apportionment and degradation quantification of organic pollutants with CSIA: 1. Model derivation. Environ Sci Technol, 48:6220–6228

Maher K, Mayer K U (2019) The art of reactive transport model building. Elements 15:117–118

Maher K, Steefel CI, DePaolo DJ, Viani BE (2006a) The mineral dissolution rate conundrum: Insights from reactive transport modeling of U isotopes and pore fluid chemistry in marine sediments. Geochim Cosmochim Acta 70:337–363

Maher K, DePaolo DJ, Christensen JN (2006b) U–Sr isotopic speedometer: fluid flow and chemical weathering rates in aquifers. Geochim Cosmochim Acta 70:4417–4435

Mariotti A, Germon J C, Hubert P, Kaiser P, Letolle R, Tardieux A, Tardieux P (1981) Experimental determination of nitrogen kinetic isotope fractionation: some principles; illustration for the denitrification and nitrification processes. Plant Soil. 162:413–430

Mancini SA, Hirschorn SK, Elsner M, Lacrampe-Couloume G, Sleep BE, Edwards EA, Sherwood Lollar B (2006) Effects of trace element concentration on enzyme controlled stable isotope fractionation during aerobic biodegradation of toluene, Environ Sci Technol 40:7675–7681

Mook WG, Bommerson JC, Staverman WH (1974) Carbon isotope fractionation between dissolved bicarbonate and gaseous carbon dioxide. Earth Planet Sci Lett 22:169–176

Nijenhuis I, Andert J, Beck K, Kästner M, Diekert G, Richnow HH (2005) Stable isotope fractionation of tetrachloroethene during reductive dechlorination by Sulfurospirillum multivorans and Desulfitobacterium sp strain PCE-S and abiotic reactions with cyanocobalamin, Appl Environ Microbiol 71:3413–3419

Northrop D B (1981) The expression of isotope effects on enzyme-catalyzed reactions, Ann Rev Biochem 50:103–131

Pooley KE, Blessing M, Schmidt TC, Haderlein SB, MacQuarrie KT, Prommer H (2009) Aerobic biodegradation of chlorinated ethenes in a fractured bedrock aquifer: quantitative assessment by compound-specific isotope analysis (CSIA) and reactive transport modeling. Environ Sci Technol 43:7458–7464

Prommer H, Anneser B, Rolle M, Einsiedl F, Griebler C. Biogeochemical and isotopic gradients in a BTEX/PAH contaminant plume: Model-based interpretation of a high-resolution field data set. Environ Sci Technol 43:8206-8612

Rayleigh L (1902) On the distillation of binary mixtures. Phil Mag 4:521–537

Reed MH (1982) Calculation of multicomponent chemical equilibria and reaction processes in systems involving minerals, gases and an aqueous phase. Geochim Cosmochim Acta 46:513–528

Rees CE (1973) A steady-state model for sulphur isotope fractionation in bacterial reduction processes. Geochim Cosmochim Acta 37:1141–1162

Renpenning J, Rapp I, Nijenhuis I (2015) Substrate hydrophobicity and cell composition influence the extent of rate limitation and masking of isotope fractionation during microbial reductive dehalogenation of chlorinated ethenes, Environ Sci Technol 49:4293–4301

Richter FM, RA Mendybaev, JN Christensen, ID Hutcheon, RW Williams, NC Sturchio, AD Beloso (2006), Kinetic isotope fractionation during diffusion of ionic species in water. Geochim Cosmochim Acta 70:277–289

Rodríguez-Escales P, van Breukelen B M, Vidal-Gavilan G, Soler A, Folch A (2014) Integrated modeling of biogeochemical reactions and associated isotope fractionations at batch scale: A tool to monitor enhanced biodenitrification applications. Chem Geol 365:20–29

Rolle M, Jin B (2017) Normal and inverse diffusive isotope fractionation of deuterated toluene and benzene in aqueous systems. Environ Sci Technol Lett 4:298–304

Rolle M, Chiogna G, Bauer R, Griebler C, Grathwohl P (2010) Isotopic fractionation by transverse dispersion: flow-through microcosms and reactive transport modeling study. Environ Sci Technol 44:6167–6173

Schmidt SI, Kreft J-U, Mackay R, Picioreanu C, Thullner M (2018) Elucidating the impact of micro-scale heterogeneous bacterial distribution on biodegradation. Adv Water Resour 116:67–76

Semple K T, Doick K J, Jones K C, Burauel P, Craven A, Harms H (2004) Defining bioavailabilty and bioaccesibility of contaminated soil and sediment is complicated. Environ Sci Technol 15:229A–231A

Simoni SF, Schäfer A, Harms H, Zehnder AJB (2001) Factors affecting mass transfer limited biodegradation in saturated porous media. J Contam Hydrol 50:99–120

Sherwood Lollar B, Slater G F, Ahad J, Sleep B, Spivack J, Brennan M, MacKenzie P (1999) Contrasting carbon isotope fractionation during biodegradation of trichloroethylene and toluene: implications for intrinsic bioremediation. Org Geochem 30:813–820

Sherwood Lollar B, Slater GF, Sleep B, Witt M, Klecka GM, Harkness M, Spivack J (2001) Stable carbon isotope evidence for intrinsic bioremediation of tetrachloroethene and trichloroethene at area 6, Dover Air Force Base. Environ Sci Technol 35:261–269

Steefel CI, MacQuarrie KTB (1996) Approaches to modeling reactive transport in porous media. Rev Mineral 34:83–125

Steefel CI, Maher K (2009) Fluid–rock interaction: A reactive transport approach. Rev Mineral Geochem 70:485–532

Steefel CI, Druhan JL, Maher K (2014) Modeling coupled chemical and isotopic equilibration rates. Proc Earth Planet Sci 10:208–217

Steefel C I, Appelo CAJ, Arora B, Jacques D, Kalbacher T, Kolditz O, Lagneau V, Lichtner PC, Mayer KU, Meeussen JCL, Molins S, Moulton D, Shao H, Šimunek J, Spycher N, Yabusaki SB, Yeh GT (2015) Reactive transport codes for subsurface environmental simulation. Comput Geosci 19:445–478

Sharp Z (2017) Principles of Stable Isotope Geochemistry, 2nd Edition (online) https://digitalrepository.unm.edu/cgi/viewcontent.cgi?article=1000&context=unm_oer

Templeton AS, Chu K-H, Alvarez-Cohen L, Conrad ME (2016) Variable carbon isotope fractionation expressed by aerobic CH_4-oxidizing bacteria, Geochim Cosmochim Acta 70:1739–1752

Teng F-Z, Dauphas N, Watkins JM (2017) Non-traditional stable isotopes: retrospective and prospective. Rev Mineral Geochem 82:1–26

Thorstenson DC, Parkhurst DL (2004) Calculation of individual isotope equilibrium constants for geochemical reactions. Geochim Cosmochim Acta 68:2449—2465

Thullner M, Kampara M, Richnow HH, Harms H, Wick LY (2008) Impact of bioavailability restrictions on microbially induced stable isotope fractionation. 1. Theoretical calculation. Environ Sci Technol 42:6544–6551

Thullner M, Centler F, Richnow HH, Fischer A (2012) Quantification of organic pollutant degradation in contaminated aquifers using compound specific stable isotope analysis–review of recent developments. Org Geochem 42:1440–1460

Thullner M, Fischer A, Richnow HH, Wick LY (2013) Influence of mass transfer on stable isotope fractionation. Appl Microbiol Biotechnol 97:441–452

Tobler NB, Hofstetter TB, Schwarzenbach RP (2008) Carbon and hydrogen isotope fractionation during anaerobic toluene oxidation by *Geobacter metallireducens* with different Fe(III) phases as terminal electron acceptors, Environ Sci Technol 42:7786–7792

Trumbore S (2009) Radiocarbon and soil carbon dynamics. Ann Rev Earth Planet Sci 37:47–66

Urey HC (1947) The thermodynamic properties of isotopic substances. J Chem Soc (Resumed) 1947:562–581

Van Breukelen BM, Hunkeler D, and Volkering F (2005) Quantification of sequential chlorinated ethene degradation by use of a reactive transport model incorporating isotope fractionation. Environ Sci Technol 39:4189–4197

Van Breukelen B M (2007) Extending the Rayleigh equation to allow competing isotope fractioning pathways to improve quantification of biodegradation. Environ Sci Technol 41:4004–4010

Van Breukelen BMV, Prommer H (2008) Beyond the Rayleigh equation: reactive transport modeling of isotope fractionation effects to improve quantification of biodegradation. Environ Sci Technol 42:2457–2463

Van Breukelen BM, Rolle M (2012) Transverse hydrodynamic dispersion effects on isotope signals in groundwater chlorinated solvents' plumes. Environ Sci Technol 46:7700–7708

Vogel JC (1980) Fractionation of the carbon isotopes during photosynthesis. Sitzungsber Heidelberger Akad Wiss 3:111–135

Vogt C, Dorer C, Musat F, Richnow HH (2016) Multi-element isotope fractionation concepts to characterize the biodegradation of hydrocarbons—from enzymes to the environment, Curr Opin Biotechnol 41:90–98

Wanner C, Druhan JL, Amos RT, Alt-Epping P, Steefel CI (2015) Benchmarking the simulation of Cr isotope fractionation. Comput Geosci 19:497–521

Wolery TJ (1992) EQ3NR, a computer program for geochemical aqueous speciation-solubility calculations: Theoretical manual, users guide, and related documentation (Version 7.0); Part 3 (No. UCRL-MA-110662-Pt. 3) Lawrence Livermore National Lab., CA (United States)

Wolfsberg M, van Hook WA, Paneth P, Rebelo LPN (2010) Isotope Effects in the Chemical, Geological, and Bio Sciences. Springer, Dodrecht, Heidelberg, London, New York

Winnick MJ, Chamberlain CP, Caves JK, Welker JM (2014) Quantifying the isotopic 'continental effect'. Earth Planet Sci Lett 406:123–133

Xu S, Lollar BS, Sleep BE (2017) Rethinking aqueous phase diffusion related isotope fractionation: Contrasting theoretical effects with observations at the field scale. Sci Total Environ 607:1085–1095

Reviews in Mineralogy & Geochemistry
Vol. 85 pp. 265-302, 2019
Copyright © Mineralogical Society of America

9

Microbial Controls on the Biogeochemical Dynamics in the Subsurface

Martin Thullner

UFZ—Helmholtz Centre for Environmental Research
Department of Environmental Microbiology
Leipzig
Germany

martin.thullner@ufz.de

Pierre Regnier

Université Libre de Bruxelles
Department Geoscience, Environment and Society
Brussels
Belgium

pregnier@ulb.ac.be

INTRODUCTION

Biogeochemical processes are of tremendous importance for determining the fate of many organic and inorganic compounds in the subsurface. Most global elemental cycles involve biogeochemical transformation, and the recycling of carbon and nutrients relies almost exclusively on biogeochemical processes. In particular, the majority of natural organic compounds are biogeochemically reactive, but also a large number of anthropogenic organic carbon compounds can be biogeochemically transformed, for instance, during the biodegradation of organic contaminants. Furthermore, inorganic compounds such as e.g., many nitrogen, phosphorus or sulfur compounds, metal compounds or minerals are directly or indirectly affected by biogeochemical reactions. To which extent and at which conditions a biogeochemical reaction takes place depends not only on the properties of the involved chemical reactants and products but also on the behavior of the microbial community (or communities) catalyzing the biogeochemical transformation. Porous media—in particular natural porous media—are complex and often heterogeneous structures, which imposes severe challenges in determining the exact physical, chemical and ecological conditions the microbial community is exposed to and to which extent it is able to provide any ecosystem service, such as the catalysis of a biogeochemical reaction.

Reactive transport models have become an established mean for the investigation and quantification of countless chemical transformations in porous media (e.g., Lichtner et al. (1996), Xiao et al. (2018), other chapters of this issue), and modeling approaches addressing the biogeochemical dynamics in natural porous media such as soil, aquifers and aquatic sediments exist already for decades (Berner 1980; Lichtner et al. 1996; Boudreau 1997; Murphy and Ginn 2000; Barry et al. 2002; Brun and Engesgaard 2002; Meysman et al. 2003; Thullner et al. 2007; McGuire and Treseder 2010; Meile and Scheibe 2018, 2019). Over the last years the improvement of experimental techniques has led to a better knowledge on the behavior of microorganisms that drive biogeochemical reactions and, more broadly, to an increased understanding of coupled reactive-transport processes in the subsurface. This has

1529-6466/19/0085-0009$05.00 (print)
1943-2666/19/0085-0009$05.00 (online)

http://dx.doi.org/10.2138/rmg.2019.85.9

led to increasingly sophisticated geomicrobial modeling approaches which have been used already in a reaction-transport framework or which provide the potential for being used to simulate biogeochemical processes and their dependency on the abundance and activity of microbial communities in subsurface environments.

The aim of this chapter is to build upon established modeling approaches for the simulation of biogeochemical processes in soils, aquifers and sediments across scales and to provide an overview of recent extensions of these established approaches towards increasingly complex representations of microbial dynamics in the subsurface. We thus start with section *'Traditional Approaches for Simulating Biogeochemical Processes in the Subsurface'* giving a brief summary of these established modeling approaches. This will be followed in section *'Thermodynamically Informed Models'* by a discussion of thermodynamic constraints on microbial activity and how to include such constraints into the simulation of biogeochemical reactions. Thereafter two sections will focus on the explicit representation of microbial dynamics. In recent years the increasing computational power as well as the increasing amount of experimental information on microbial systems has led to numerous new approaches for modeling the behavior of individual microbial species as well as of microbial communities (Song et al. 2014) and only a limited overview can be provided here. We separate these approaches into two (certainly overlapping) aspects: section *'Integration of Microbial Sequencing Data in Geomicrobial Models'* describes approaches of high complexity which consider a maximum of available data, mainly from various sequencing techniques, and section *'Considering the Ecological Behavior of Microorganisms'* describes approaches which put more emphasis on the ecological behavior of the microorganisms represented by certain traits and which are (implicitly) based on the assumption that the complexity of an ecological model should be limited (May 1974). Stable isotope approaches have become increasingly established for the investigation of biogeochemical processes and stable isotope data can be used to constrain the elemental fluxes between reactants and the microbial biomass or reaction products, and to identify which microbial pathway(s) has or have been involved into a degradation reaction. The section *'Considering Stable Isotope Signatures'* provides an overview on how such information can be embedded in reactive transport modeling approaches. Modeling microbially driven processes in a porous medium implicitly involves various spatial scales ranging from the micro scale on which microbial cells act to the observation scale which ranges from the cm scale in high resolution laboratory experiments up to the 100 km scale in global land and ocean models. In section *'Scale effects'* some consequences of linking microbial and transport scales are discussed and approaches on how to consider them within a modeling framework are presented.

TRADITIONAL APPROACHES FOR SIMULATING BIOGEOCHEMICAL PROCESSES IN THE SUBSURFACE

Reaction networks for biogeochemistry

Microorganisms generate the energy required for growth and maintenance by catalyzing the transfer of electrons from electron donor substrates (E_D) to terminal electron acceptors (E_A). As a result, they exert a major control on the redox chemistry of their immediate surroundings (Thullner et al. 2007). In surface and subsurface systems such as rivers and lakes, sediments, soils and groundwater, the vastly dominant source of electrons is most often reduced carbon bound in complex organic molecules of natural or anthropogenic origin. Their progressive degradation (typically oxidation) shapes the redox conditions of the surrounding environment and classically leads to a redox stratification. This zonation results from the sequential utilization of E_A's (O_2, NO_3^-, Mn(VI), Fe(III), $SO_4^=$) followed by methanogenesis, an order which is consistent with a progressive decrease in Gibbs energy yields released through these metabolic pathways (Claypool and Kaplan 1974; Froelich et al. 1979; Stumm and Morgan 1996).

In Reactive Transport Model (RTM) applications, redox reactions have traditionally been simulated using the well-known Michaelis–Menten model for enzymatically catalyzed reactions (e.g., Van Cappellen and Gaillard (1996); Boudreau (1997); Barry et al. (2002)) but other approaches have also been proposed, e.g., the 'reverse Michaelis–Menten' model (Schimel and Weintraub 2003) or the 'equilibrium chemistry approximation' (Tang and Riley 2013). The Michaelis–Menten model derives from theory of enzyme kinetics and is consistent with observations that show saturation behavior with increasing availability of E_D substrate (Thullner et al. 2007). Field and laboratory observations also reveal similar saturation behavior with respect to the concentrations of E_A's, and the reaction rate for a redox metabolic pathway is classically represented as:

$$\frac{dE_D}{dt} = -k \frac{E_D}{K^{E_D} + E_D} \cdot \frac{E_A}{K^{E_A} + E_A} \tag{1}$$

where k is the maximum rate and K^{E_D} and K^{E_A} denote half-saturation constants with respect to the electron donor and acceptor, respectively. Inhibition terms can be introduced in the rate law to account for the suppression of a specific metabolic pathway by the presence of higher energy-yielding E_A. This classical approach to simulate the redox zonation has been extensively used in RTM applications of natural and engineered environments, see, e.g., the reviews by Thullner et al. (2007), Arndt et al. (2013) and Paraska et al. (2014) for further details.

(Sub)-surface environmental conditions can often be traced back directly or indirectly to the oxidation of reduced carbon molecules. As a result, an accurate representation of these processes, including their temporal evolution, is often of the upmost importance for characterization of the biogeochemical dynamics in these systems. Redox processes control, among others, the recycling of inorganic carbon and nutrients, the precipitation/dissolution of carbonate and sulfide minerals, the pH of (pore)-waters, the sorption/desorption of toxic elements as well as the carbon sequestration in soils and sediments, all of which have been addressed in RTM applications (see, e.g., Paraska et al. 2014; Li et al. 2017 for recent reviews). Models relying on kinetic rate laws similar to Equation (1) or variations of it have been applied across all spatial scales, from pore networks to the assumed horizontally homogeneous plot or core scale (e.g., Gharasoo et al. 2012; De Biase et al. 2013; Druhan et al. 2014), and from catchment scales (e.g., Bao et al. 2014) to regional and global scales (e.g., Thullner et al. 2009; Wania et al. 2010; Krumins et al. 2013; Raivonen et al. 2017; Hülse et al. 2018). They remain popular, although microbial rate laws explicitly accounting for the dependence of the microbial reaction rate on the abundance of the microorganisms are increasingly applied in RTMs, and begin to be used in large scale Earth System Models (e.g., Wieder et al. 2013, 2014; Sulman et al. 2014; Wang et al. 2016; Huang et al. 2018).

Kinetics of geomicrobial reactions

The Monod model for microbial growth. Models in which microbial biomasses are explicitly included are typically applied to study the response of microbial communities to fluctuations in environmental conditions and the competition of different microbial groups for a common substrate (Boudreau 1999; Wirtz 2003; Thullner et al. 2005, 2007; Dale et al. 2008a; Regnier et al. 2011; Arndt et al. 2013). These biomass-explicit models have been coupled to RTMs to address environmental questions in a wide range of natural and engineered systems including water bodies (e.g., Vanderborght et al. 2002; Reed et al. 2014), groundwaters (e.g., Thullner and Schäfer 1999; Barry et al. 2002; Yabusaki et al. 2007; Li et al. 2009; Tartakovsky et al. 2009; Bao et al. 2014), lake sediments (e.g., Jin et al. 2013), marine sediments, (e.g., Dale et al. 2010; Regnier et al. 2011), and soils (e.g., Neill and Gignoux 2006; Wieder et al. 2013; Sierra et al. 2015; Huang et al. 2018).

Geomicrobial models rely on rate laws that explicitly account for the dependence of the microbial reaction rate on the biomass of the microorganisms, according to the Monod microbial growth equation (Monod 1949) or its derivatives (e.g., Soetaert and Herman 2009):

$$\frac{dB}{dt} = \mu B = \mu_{max} B \frac{S}{K_S + S} \tag{2}$$

where dB/dt is the rate of biomass production, B is the biomass, μ_{max} is the maximum specific growth rate, S is the concentration of the substrate, and K_S is the Monod half-saturation constant, that is, the substrate concentration when $\mu = 0.5\ \mu_{max}$. In a related model approach— the Contois model—K_S is considered to depend linearly on B (Contois 1959). About a decade ago, the functional dependency on substrate and biomass of the classical microbial growth model was challenged by Schimel and Weintraub (2003) and the reverse Michaelis– Menten kinetics was proposed as an alternative to simulate soil C decomposition. This model assumes that the rate is a nonlinear function of biomass while depending linearly on substrate concentration (Wang et al. 2016). Several later studies using microbial models showed that the classical Michaelis–Menten model may indeed fail to resolve the biogeochemical dynamics in both laboratory and natural settings (see Tang (2015) for an overview). This debate revolving around alternative nonlinear microbial models has received considerable attention in global scale applications of soil C decomposition and its response to global change. Comparative analysis of the model's emerging properties has also recently been carried out (Wang et al. 2016) and it has been demonstrated that both models are two special cases of the 'equilibrium chemistry approximation' (ECA) kinetics proposed by Tang (2015).

A particular biomass group i may rely on more than one substrate to sustain its metabolic needs. This dependency includes catabolic reactions involving the transfer of electrons from an external electron donor (E_D) to a terminal electron acceptor (E_A) as well as the possibility for a microbial group to rely on j electron donors to sustain its metabolic needs (Dale et al. 2006). The generalized form of Equation (2) thus reads:

$$\frac{dB_i}{dt} = \sum_{j=1}^{n} \mu_{max,i}^j B_i \frac{E_D^j}{K_i^j + E_D^j} \frac{E_A}{K_i^{E_A} + E_A} = \sum_{j=1}^{n} \mu_{max,i}^j B_i F_{K,i}^j \tag{3}$$

where n is the number of E_D's sustaining the growth of biomass group i, and $F_{K,i}^j$ are dimensionless kinetic limitation terms (0-1) for each of the j catabolic reaction that are function of the uptake efficiencies of the E_D's and E_A's by the microorganisms, as well as the local availabilities of the E_D and E_A. Note that Equation (3) can further be modified by including inhibition terms to account for the suppression of a specific metabolic pathway by the presence of higher energy-yielding E_A's or for the presence of toxic compounds. This is usually realized by introducing a series of hyperbolic functions, each characterized by an inhibition constant K_{in} (see, e.g., Van Cappellen et al. 1993; Boudreau 1997). In case where microorganisms are exposed to a new substrate an initial delay of their growth and degradation activity may be observed as the enzymatic apparatus of the cells needs to adapt to the new compound and a lag effect needs to be considered in the modeling approach (Wood et al. 1995; Ginn 1999; Nilsen et al. 2012).

The rates of biomass growth and substrate utilization are directly coupled through the growth yield, Y, expressed as the amount of biomass carbon produced per unit of mass electron donor consumed (Regnier et al. 2011). Values of Y depend on the Gibbs energy generated by the catabolic reaction j, the Gibbs energy needed for the formation of a new biomass i (anabolism), and the efficiency with which the organisms utilize energy (VanBriesen 2002). In other words, the growth yield partitions the flow of electrons between the catabolic pathway of energy generation and the anabolic pathway of microbial growth (Rittmann and McCarty 2001); accurately predicting Y is thus key to properly represent geomicrobial processes in

RTMs (see next section for further details). The change in electron donor concentration due to microbial processes can then be represented by:

$$\frac{dE_D^j}{dt} = -\sum_{i=1}^{m} \frac{1}{Y_i^j} \mu_{\max,i}^j \, B_i \, F_{K,i}^j \qquad (4)$$

where m is the number of microbial groups relying on the substrate (electron donor) j to sustain their metabolic energy needs and Y_i^j represents an observed growth yield, which is the efficiency of converting carbon into microbial products (Sinsabaugh et al. 2013; Bradley et al. 2018b). A similar equation can be written for the electron acceptor consumption, which can be catalyzed by multiple microbial groups relying on several electron donor substrates. The coupled Equations (3, 4) are at the foundation of geomicrobial models. An alternative approach to simulate microbially driven reactions is to constrain the substrate consumption through Michaelis–Menten kinetics and then quantify the microbial growth rate by multiplying the consumption rate by the yield (e.g., Thullner et al. (2007)). Both approaches are identical if Y_i^j is constant but if it varies, the choice defines if changes in yield imply unchanged growth rate but changing substrate consumption (Monod) or unchanged substrate consumption but changing growth rate (Michaelis–Menten).

As further discussed in section '*Thermodynamically Informed Models*' of this chapter, the computation of Y values depends on the Gibbs energies generated and consumed by the catabolic and anabolic (new biomass synthesis) reactions, respectively. The underlying principles to combine the catabolic and anabolic pathways in so-called macrochemical equations are described in detail in Dale et al. (2006) and Smeaton and Van Cappellen (2018). In short, electrons supplied by an E_D are used partially to generate energy and partially for biomass synthesis, the partitioning of this electron flow depending of the stoichiometry of the half-redox reactions involved in catabolism and anabolism. As an example, the macrochemical equation for anaerobic methane oxidation catalyzed by methane oxidizing archea (MOA),

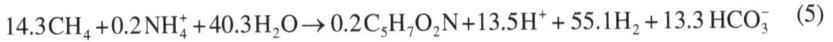

$$14.3\,CH_4 + 0.2\,NH_4^+ + 40.3\,H_2O \rightarrow 0.2\,C_5H_7O_2N + 13.5\,H^+ + 55.1\,H_2 + 13.3\,HCO_3^- \quad (5)$$

is obtained from the combination of the following two redox reactions :

$$0.2\,NH_4^+ + HCO_3^- + 4.8\,H^+ + 4e^- \rightarrow 0.2\,C_5H_7O_2N + 2.6\,H_2O \qquad (6)$$

$$14.3\,CH_4 + 42.9\,H_2O \rightarrow 55.1\,H_2 + 14.3\,HCO_3^- + 18.3\,H^+ + 4e^- \qquad (7)$$

for anabolism and catabolism, respectively. See Dale et al. (2006) for the determination of the stoichiometric coefficients in Equations (5–7). We thus note that only a fraction of the carbon source is fully oxidized to dissolved inorganic carbon (HCO_3^-), while the other fraction is only party oxidized into biomass ($C_5H_7O_2N$). Furthermore, in this reaction the MOAs produce a reactive intermediate (H_2) that can be used by other anaerobic microorganisms to sustain their metabolic needs, most notably sulfate reducing bacteria (SRB). The latter group uses H_2 as energy source according to the following macrochemical equation:

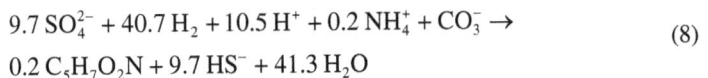

$$9.7\,SO_4^{2-} + 40.7\,H_2 + 10.5\,H^+ + 0.2\,NH_4^+ + CO_3^- \rightarrow$$
$$0.2\,C_5H_7O_2N + 9.7\,HS^- + 41.3\,H_2O \qquad (8)$$

Reaction 5 is endergonic under standard state conditions (see below for a definition) and thermodynamic calculations have in fact shown that CH_4 can only be oxidized to H_2 under a narrow range of *in-situ* pressure, temperature and solution composition (LaRowe et al. 2008). This thermodynamic constraint requires maintaining low H_2 concentrations, which is typically achieved through H_2 consumption by the SRB (Hoehler et al. 1994; Dale et al. 2008b). The syntrophic association of MOA and SRB thus catalyzes the net reaction of anaerobic oxidation of methane coupled to sulfate reduction, which can be obtained by combining Equations (5,8). If MOA and SRB biomasses are not explicitly simulated, the combination of the two catabolic pathways $CH_4 + 3\,H_2O \rightarrow 4\,H_2 + HCO_3^- + H^+$ and $4\,H_2 + SO_4^{2-} + H^+ \rightarrow HS^- + 4\,H_2O$ only leads to the much simpler net reaction stoichiometry for anaerobic oxidation of methane (assuming that the reactive intermediate (H_2) produced by the MOA is entirely consumed by the SRB):

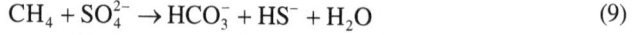

$$CH_4 + SO_4^{2-} \rightarrow HCO_3^- + HS^- + H_2O \tag{9}$$

In RTMs, the classical approach is to implement the net reaction Equation (9) for simulating the coupled methane-sulfur cycles (Regnier et al. 2011), but several authors have also explicitly modeled the dynamics of reactive intermediates (e.g., Dale et al. 2008b; Orcutt and Meile 2008; Alperin and Hoehler 2009b), thereby providing useful insights into the mechanisms and environmental controls of microbially mediated reactions. This metabolic modeling approach is straightforward to implement in RTMs as, for the above example, it only requires the addition of a new mass conservation equation for the intermediate species, H_2, while the rate expression for the net oxidation of methane by sulfate is replaced by two rate expressions, one for each of the individual reaction steps (Regnier et al. 2011).

Mortality and maintenance terms in geomicrobial models. The death of microorganisms and the contribution of dead biomass to substrate pools are included in geomicrobial models by assuming that mortality scales to the active biomass pool via a first-order decay rate constant μ_e (e.g., Dale et al. 2006). If the electron donor is an organic carbon substrate fed by the death of microbes, the conservation equations for biomass and substrates are modified according to:

$$\frac{dB_i}{dt} = \sum_{j=1}^{n}(\mu_{max,i}^j\, B_i\, F_{K,i}^j - \mu_{e,i}^j\, B_i) \tag{10}$$

$$\frac{dE_D^j}{dt} = -\sum_{i=1}^{m}(\frac{1}{Y_i^j}\mu_{max,i}^j\, B_i\, F_{K,i}^j - z_i^j\, \mu_{e,i}^j\, B_i) \tag{11}$$

with z_i^j as fraction of the dead biomass returning to the pool of substrate j.

The above two equations form the cornerstone of biomass-explicit modeling approaches in RTMs (e.g., Thullner et al. 2007). However, the observed growth yields in this model does not distinguish the energetic costs associated to growth from those due to maintenance (Bradley et al. 2018b). The classical models of Herbert (1958) and Pirt (1965) allow to separate the energetic costs required to generate new biomass (Lipson 2015) from the energetic costs to perform all maintenance functions. These classical models differ by the provenance of maintenance energy, Herbert considering maintenance costs as endogenous catabolism, i.e., the consumption of biomass while Pirt assumes that maintenance leads to an additional consumption of substrate. Both models have their inherent limitations (see, e.g., Wang and Post (2012); and references therein for details). In addition, observations show that the specific maintenance rate, and the provenance of maintenance energy, can vary under different environmental conditions (van Bodegom 2007). Together, these elements call for the development of an hybrid model that allows for alternative supplies of maintenance energy from biomass and/or substrate, as proposed by Wang and Post (2012) and Bradley et al. (2018b) for soils and sediments, respectively:

$$\frac{dB_i}{dt} = \sum_{j=1}^{n}\mu_{max,i}^j\, B_i\, F_{K,i}^j - m_{q,i}^j\, B_i\left(1 - F_{K,i}^j\right) - \mu_{e,i}^j\, B_i \tag{12}$$

where $m_{q,i}^j$ represents the specific maintenance rate, and the term in parenthesis tends to zero when substrates are abundant, consistent with the idea that microorganisms do not consume biomass for their maintenance requirements in this case. Electron donor substrates are consumed instead and calculated according to:

$$\frac{dE_D^j}{dt} = -\sum_{i=1}^{m} \frac{1}{Y_{G,i}^j} \mu_{max,i}^j B_i F_{K,i}^j + \frac{1}{Y_{G,i}^j} m_{q,i}^j B_i F_{K,i}^j - z_i^j \mu_{e,i}^j B_i \qquad (13)$$

where $Y_{G,i}^j$ is now the "true" growth yield, that is a parameter that only reflects the expenditure of energy to generate new biomass (Lipson 2015). Conversely, when substrates are scarce, the second term in Equation (13) tends to zero while the term in parenthesis in Equation (12) approaches 1.

THERMODYNAMICALLY INFORMED MODELS

In what follows, we review several quantitative approaches aiming at the integration of thermodynamic constraints in RTMs that explicitly simulate geomicrobial population dynamics. We restrict our analysis to concepts derived from equilibrium thermodynamics, which have been extensively applied within the framework of RTMs over the last decade. Appealing extensions to such classic, equilibrium thermodynamic approaches are models relying on the Maximum Entropy Production (MEP) concept (e.g., Vallino 2010). In these models, MEP is used either as an optimization goal or as a governing principle to understand and model biogeochemical processes (Meysman and Bruers 2010; Vallino 2010; Song et al. 2014). To our knowledge, however, application of these non-equilibrium thermodynamic approaches, including those relying on statistical thermodynamics (Song et al. 2014), have remained largely theoretical and conceptual. Despite their significant potential, the modeling of microbial processes based on MEP is thus not covered in this chapter.

Incorporating bioenergetics in models of microbial dynamics

Microbial dynamics is constrained by classical thermodynamics in two ways: (1) Growth yields (*Y*) are dependent on catabolic energy gains (Roden and Jin 2011) and (2) redox reactions can only proceed when the energy yield of the catabolic reaction exceeds a metabolic threshold. This bioenergetic limitation can be included in the Monod geomicrobial model for redox processes (section '*Traditional Approaches for Simulating Biogeochemical Processes in the Subsurface*') using a functional dependency on the thermodynamic driving force for the reaction, which depends on the Gibbs energy yield (Regnier et al. 2011):

$$\frac{dE_D^j}{dt} = -\sum_{i=1}^{m} \frac{1}{Y_i^j} \mu_{max,i}^j B_i F_{K,i}^j F_{T,i}^j \qquad (14)$$

where the thermodynamic limiting term, F_T, represents the limitation that energy yields have on catabolic reactions rates (Jin and Bethke 2002; Dale et al. 2008c; LaRowe et al. 2014). Therefore, the implementation of thermodynamic constraints in geomicrobial models require calculation of the energetics of catabolic reactions and subsequent quantification of the F_T term as well as derivation of a relation between *Y* and the energetics of cellular metabolism.

Energetics of catabolic reactions

Modeling studies on the energetics of redox reactions catalyzed by microorganisms have mostly concentrated on the relative Gibbs energy yields associated with different TEAs (Arndt et al. 2013), often to predict their classical sequential utilization (O_2, NO_3, MnO_2, $Fe(OH)_3$, SO_4) in subsurface environments (e.g., Thullner et al. 2007 and references therein). This well documented redox sequence may however no longer hold under non-standard state conditions prevailing in natural environments (Amend and Teske 2005; Bethke et al. 2011; LaRowe and Van Cappellen 2011) and sulfate may, for instance, become an energetically more favorable TEA than iron oxides (LaRowe and Van Cappellen 2011).

Recently, LaRowe and Van Cappellen (2011) have extended such classical thermodynamic analysis to a wide variety of natural electron donors, mainly organic matter compounds. The approach has the major advantage of not requiring structural information in order to estimate the energetic potential of complex, natural organic matter. Instead, it uses the average Nominal Oxidation State of the Carbon (*NOSC*) as a proxy to scale the bonding in organic compounds to their energetic content (Arndt et al. 2013). The *NOSC* is readily determined for a wide array of organic compounds as it only depends on the net charge Z and the stoichiometric numbers of the elements C, H, N, O, P and S in a given organic compound, respectively denoted by a, b, c, d, e and f:

$$NOSC = -\frac{-Z + 4a + b - 3c - 2d + 5e - 2f}{a} + 4 \tag{15}$$

Thermodynamic calculations for the oxidation half reactions of organic compounds show that Gibbs energies of reactions and *NOSC* values follow an inverse relationship (Fig. 1). This remarkable linear correlation can thus be used to estimate the energetics of catabolic reactions ($\Delta Gcat$) coupling organic matter oxidation with any TEA as long as the average *NOSC*, which is directly calculated from element ratios, is known. These findings are potentially of high value for reactive-transport modeling and, more specifically, geomicrobial modeling. Indeed, while much attention has been given to the simulation of the redox sequence in subsurface environments, much remains to be done regarding the quality and diversity of substrates available for microbial growth and how they shape the biogeochemistry of the subsurface. A better representation of the energetics of substrates in RTMs would thus help predict important patterns of organic matter degradation in natural environments. For instance, the degradation of labile organic carbon compounds (e.g., planktonic biomass, polysaccharides) leads to similarly high Gibbs energy yields in both oxic and anoxic settings (e.g., Henrichs (2005)). In stark contrast, the degradation rates of more refractory compounds such as lignins or lipids are much more sensitive to redox conditions (Canfield 1993; Canuel and Martens 1996; Hartnett et al. 1998; Henrichs 2005; Jin and Bethke 2009). In oxic environments, the degradation rates of refractory compounds remain high because of the very high oxidative potential and the resulting weak sensitivity towards the depletion of energy-rich organic compounds while in anoxic environments deprived of energy rich-organics and powerful TEAs, the degradation rate becomes thermodynamically limited (Arndt et al. 2013).

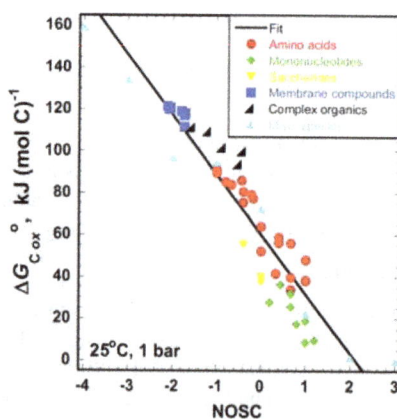

Figure 1. Standard molal Gibbs energies of the oxidation half reactions of organic compounds as a function of the average nominal oxidation state of carbon (NOSC) in the compounds, at 25 °C and 1 bar. The Gibbs energies are expressed in kJ per mole of carbon (Adapted from LaRowe and Van Cappellen 2011).

Bioenergetic theory: the F_T term

Microorganisms channel part of the catabolic energy released by redox reactions into metabolism and growth via intracellular synthesis of adenosine triphosphate, ATP (Regnier et al. 2011). In natural settings, the fraction of energy an organism uses to make ATP is essentially unknown (Harold 1986; Russell and Cook 1995). Therefore, the energetics of catabolic redox reactions only provides a quantitative estimate of the maximum amount of ATP that can be synthesized (LaRowe and Helgeson 2007; LaRowe et al. 2008; LaRowe et al. 2012). ATP synthesis implies that catabolic reactions must generate useable energy under the non-standard state conditions of natural environments (Smeaton and Van Cappellen 2018), that is, $\Delta G_{cat} < 0$:

$$\Delta G_{cat} = \Delta G^{\circ}_{cat} + RT \ln(Q_{cat}) \tag{16}$$

where ΔG_{cat} and ΔG°_{cat} are the Gibbs energy change of the catabolic reaction under actual and standard-state conditions, respectively, and Q_{cat} refers to the reaction quotient of the reaction. The notion of standard state results from the impossibility to define absolute values of some thermodynamic quantities. Only changes can be determined, which require definition of a baseline, or a standard state, for substances. Precise definitions for gases, pure liquids, solids and admixtures are provided in Cox (1982). For application of the concept of standard state to substances in solutions, which is particularly relevant to this chapter, the composition of the system (as well as the pressure) must be defined and is customarily taken as the standard state molality of 1 mol kg^{-1}. Note that temperature does not intervene in the definition of the standard states, but most thermodynamic tables report values at the recommended temperature of 298.15 K.

Values of Q_{cat}, are calculated using:

$$Q_{cat} = \prod_i a_i^{\upsilon_i} \tag{17}$$

where a_i designates the activity of the ith species and υ_i corresponds to the stoichiometric coefficient of the ith species in the given catabolic reaction (LaRowe et al. 2014).

It is thought that redox reactions can however only proceed when the Gibbs energy yield for the catabolic reaction ΔG_{cat} exceeds a minimum metabolic threshold. Such requirement can be added to the thermodynamic limiting term, F_T, in the Monod model. The standard formulation (Boudart 1976; Jin and Bethke 2002; Jin and Bethke 2007) relates the minimum excess catabolic energy production to the energy required to synthesize one mole of ATP from ADP and monophosphate, ΔG_{ATP}, in such a way that the F_T term reads:

$$F_T = 1 - \exp\left(\frac{\Delta G_{cat} + m\Delta G_{ATP}}{\chi RT} \right) \tag{18}$$

In Equation (18), m is the number of ATP molecules produced per catabolic formula reaction, R is the universal gas constant, T is the absolute temperature, and χ is the average stoichiometric number. The latter is equivalent to the number of times the rate limiting step occurs per mole of ATP made multiplied by the number of electrons transferred in the jth reaction (Jin and Bethke 2002; Jin and Bethke 2003; Jin and Bethke 2005). Thermodynamics therefore imposes that $\Delta G_{cat} + m\Delta G_{ATP}$ must be negative for a microbially mediated reaction to proceed. Because ΔG_{ATP} is defined as a positive (energy-requiring) value, ΔG_{cat} must be sufficiently negative to exceed $m\Delta G_{ATP}$ in absolute magnitude (Regnier et al. 2011). Equations (16, 17) show that the accumulation of reaction products (such as H_2 in Eqn. 5) may limit the thermodynamic drive of the catabolic reaction and thus reduce rates in the Monod model via the F_T term. For microbial reaction processes operating close to their thermodynamic limit, the reaction rates may become very sensitive to F_T (Thullner et al. 2007). This has been shown, for example, in the case of anaerobic oxidation of methane in marine sediments

(Regnier et al. 2005; Dale et al. 2006, 2008a) or for methanogenesis and sulfate reduction with H_2 in hydrothermal systems (LaRowe et al. 2014). When geomicrobial reactions become thermodynamically limited, most energy generated by the catabolic process is then diverted to maintenance functions, with little energy left to invest in growth.

Alternative expressions accounting for thermodynamic limitations of microbial reactions have been proposed in the literature (Cupples et al. 2004; Thullner et al. 2007). Recently, LaRowe et al. (2012) proposed a formulation for F_T that (1) relies on only one adjustable parameter rather than the three required by Equation (18) (ΔG_{ATP}, m and χ) and (2) circumvent the need to set *a priori* values of $m\Delta G_{ATP}$. Here, the approach relies on the amount of energy required to maintain a membrane potential ($\Delta G_{mp} = F\Delta\Psi$) as a proxy for the minimum amount of energy that a microbe needs to be considered active:

$$F_T = \begin{cases} \dfrac{1}{\exp\left(\dfrac{\Delta G_{cat} + F\Delta\psi}{RT}\right) + 1} & \text{for } \Delta G_{cat} \leq 0 \\[2em] 0 & \text{for } \Delta G_{cat} \geq 0 \end{cases} \qquad (19)$$

where ΔG_{cat} denotes the Gibbs energy of the catabolic reaction (per electron transferred), F is the Faraday constant and $\Delta\Psi$ is the electric potential (in volts) across an energy transducing membrane. Such expression has been used to simulate, e.g., microbial dynamics in hydrothermal vent systems (LaRowe et al. 2014) and coupled CH_4–SO_4 cycles in sediments on an active continental margin offshore New Zealand (Dale et al. 2010) where methane-rich fluids migrate upwards through the sediment (LaRowe et al. 2012). The sensitivity of a reaction rate to the F_T term in the above equation is shown for methanogenesis (MET) occurring in the wall of a hydrothermal vent chimney, as an example. Figure 2 shows that the computed MET rates are significantly reduced by the low energy yields in the inner part of the chimney wall, and become 0 towards the outer part of the chimney when the values of ΔG_{cat} are positive. Without the F_T term, the active zone of methanogenesis within the chimney wall would be larger.

The dependence of microbial reaction kinetics on bioenergetics is accounted for in the F_T term. However, catabolic energy gains also constrain the value of Y (Roden and Jin 2011) and thermodynamics is also needed to relate Y to the energetics of cellular metabolism (Smeaton and Van Cappellen 2018). Therefore, the growth yield, Y, provides another link between the kinetics and thermodynamics of microbially driven redox reactions.

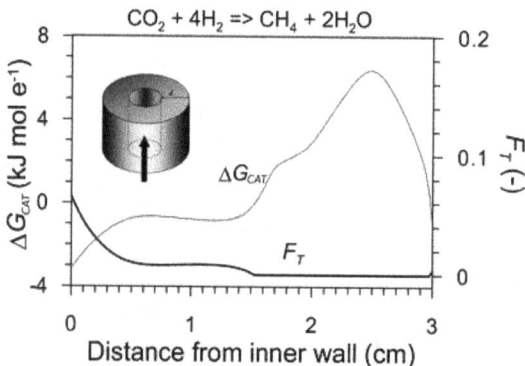

Figure 2. Values of the thermodynamic rate-limiting term, F_T and ΔG_{cat} across a hydrothermal vent chimney wall of thickness d (see inserted sketch) for methanogenesis. Adapted from LaRowe et al. (2014).

Quantifying growth yields

Growth yields (Y) are generally incorporated into biomass-explicit kinetic models using the Monod formulation, as summarized in section '*Traditional Approaches for Simulating Biogeochemical Processes in the Subsurface*' (e.g., Thullner et al. 2007). In simple terms, Y values depend on the Gibbs energy generated by the catabolic reaction, the Gibbs energy needed for the formation of new biomass, and the efficiency with which organisms utilize energy (VanBriesen 2002). Two distinct lines of approach that relate Y to the catabolic energy yield either through empirical relationships (e.g., Rittmann and McCarty 2001; Roden and Jin 2011) or through bioenergetically based models (Heijnen and Van Dijken 1992; Heijnen et al. 1992; McCarty 2007) such as the Gibbs Energy Dissipation model, have then been followed. In short, the latter approach accounts for the loss of energy due to entropy production and heat dissipation by cells (see Dale et al. 2006 for further details). As shown by VanBriesen (2002) and for a variety of organic carbon compounds and metabolic pathways (aerobic degradation, denitrification and methanogenesis), energy dissipation approaches are consistent with empirically based methods and predict quite similar yields with laboratory-determined values of Y. A slightly less complex theoretical approach has also recently been developed (the so-called Microbial Turnover to Biomass (MTB) method of Trapp et al. (2018)) and applied to the quantification of pesticide degradation and formation of biogenic non-extractable residues (Brock et al. 2017). Overall, the major drawback of these theoretical calculations is that they provide only maximum Y values, and do not consider energy changes due to variations in chemical composition of the system (Thullner et al. 2007). That is, Y values are calculated under biogeochemical standard state conditions that may deviate significantly from those encountered in natural settings (LaRowe and Amend 2015b).

To circumvent this important limitation, Smeaton and Van Cappellen (2018) recently developed a quantitative method for Y that accounts for changes in physical (e.g., temperature) and chemical conditions under which the metabolic processes are occurring. Their resulting semi-empirical model, the Gibbs Energy Dynamic Yield Method (GEDYM), is an extension of the one by Heijnen et al. (1992) and relies upon a much larger database of experimental Y values, most of which are relevant to low energy yielding catabolic processes such as methanogenesis and sulfate reduction. Briefly, the method computes for a given microorganism, Gibbs energy changes of metabolic reactions by linking Gibbs energy changes of their corresponding catabolic (ΔG_{cat}) and anabolic (ΔG_{syn}, see below) reactions through their growth yield. In their approach, Gibbs energy changes of metabolic, catabolic and anabolic reactions all depend on Q as shown by, e.g., Equation (16). Therefore, the resulting Y values account for changes in the chemical environment surrounding the cells (Smeaton and Van Cappellen 2018). As an example of application of the GEDYM model, Figure 3 reports Y values for sulfate reduction coupled to

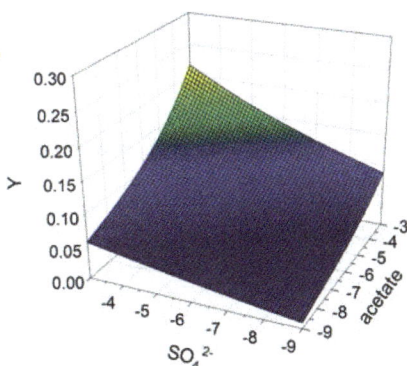

Figure 3. Predicted growth yields Y for sulfate reduction as a function of log acetate and sulfate activities. Y is expressed as C-mol biomass/mol acetate. Adapted from Smeaton and Van Cappellen (2018).

acetate oxidation for a broad range of environmental conditions, which in turn impact values of ΔG_{cat} and ΔG_{syn}. Y values applied in geochemical models generally fall within their predicted range but are assumed constant (Smeaton and Van Cappellen (2018) and references therein).

At this stage, it is important to note that the metabolic energy balance reported above only accounts for the catabolic energy invested in anabolism, that is, biomass synthesis. It therefore differs from the total cellular energy balance, which also includes the energy needed to sustain all maintenance functions. In a biomass-explicit kinetic model, maintenance energy requirements are implemented separately through an additional term, as shown in Equations (12, 13). The following section elaborates on the quantification of anabolic and maintenance energy requirements through classical thermodynamics.

Energetics of anabolism and maintenance

The calculation of the Gibbs energy required to synthesize biomolecules, ΔG_{syn}, from simple material such as CO_2 or acetate, involves three important steps: 1) definition of cellular biomass composition; 2) definition of the anabolic reaction stoichiometry; and 3) computation of the resulting Gibbs energy of the anabolic reaction.

In RTM applications, biomass synthesis generally assumes a unique biomass composition such as $C_5H_7O_2N$ (Rittmann and McCarty 2001; Dale et al. 2006, 2008a) or $C_5H_9O_{2.5}N$ which are based on the measured elemental concentrations of microbial species (Roels 1980; Smeaton and Van Cappellen 2018). A more complex approach consists in explicitly accounting for the variety of molecules making up the cell, that is, individual amino acids, nucleotides, lipids, saccharides, amines and other compounds (McCollom and Amend 2005) and subsequently calculate the energetics associated to each of these biomolecules. The polymerization of these various compounds in their respective biomacromolecules can also be accounted for in the energetics of anabolism, as performed in, e.g., Amend et al. (2013) and LaRowe and Amend (2016). To our knowledge, such approach as not yet been applied in a reactive-transport framework, despite being in principle compatible with RTM simulations.

Methods to constrain the anabolic reaction stoichiometry are presented in detail and illustrated in Rittmann and McCarty (2001), Dale et al. (2006), and Smeaton and Van Cappellen (2018). Briefly, establishing the full anabolic reaction requires combination of the half-redox reaction for biomass synthesis and the half-redox reaction providing the carbon source and electrons. If the carbon in microbial biomass is less oxidized than the carbon source, then the latter can also serve as electron donor in the full anabolic reaction. If not, the carbon source cannot be used to balance electrons during anabolism and an electron donor distinct from the carbon source is required in this case. Following LaRowe and Van Cappellen (2011), Smeaton and Van Cappellen (2018) have shown that the relative oxidation states of biomass ($ROS_B = 4 - NOSC_B$) and carbon source ($ROS_S = 4 - NOSC_S$) can directly be compared using the $NOSC$ of both compounds. Thus, only when $ROS_S < ROS_B$ the carbon source can also serve as electron donor.

As an example, consider biomass synthesis using ethanol as carbon substrate:

$$0.5\ C_2H_6O + 0.2\ NH_4^+ \rightarrow 0.2\ C_5H_7O_2N + 0.1\ H_2O + 2.2\ H^+ + 2.0\ e^- \qquad (20)$$

Here, $NOSC_S$ and ROS_S of ethanol are -2 and 6, respectively while $NOSC_B$ and ROS_B of biomass are 0 and 4, respectively. Therefore, since $ROS_S > ROS_B$, an electron acceptor other than the C source is needed to consume the electrons released by anabolism. For instance, the two electrons produced in Equation (20) can be consumed during sulfate reduction:

$$0.25\ SO_4^{2-} + 2.25\ H^+ + 2.0\ e^- \rightarrow 0.25\ HS^- + H_2O \qquad (21)$$

which leads to the full anabolic reaction:

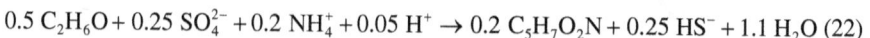

$$0.5\ C_2H_6O + 0.25\ SO_4^{2-} + 0.2\ NH_4^+ + 0.05\ H^+ \rightarrow 0.2\ C_5H_7O_2N + 0.25\ HS^- + 1.1\ H_2O \quad (22)$$

Importantly, the oxidation states of other elementary building blocks (e.g., N, S…) required for biomass synthesis also influence the stoichiometry of the full anabolic reaction (LaRowe and Amend 2016). For instance, in well-oxidized conditions NO_3^- will be used as nitrogen source during biosynthesis while under reducing condition, NH_4^+ will be used instead. In the latter case, the microorganisms will not need to invest a stream of electrons to reduce the nitrogen compounds during biosynthesis (LaRowe and Amend 2016). This example highlights that the prevailing environmental conditions not only affect the energetics of catabolism but may also influence the energetic of anabolism, an aspect not yet fully accounted for in RTM applications.

The full anabolic reaction formula derived above can then be used to compute the energetics of biomass synthesis. This step is relatively straightforward if a single biomass composition and standard state conditions are assumed, as shown by the RTM simulations of, e.g., Dale et al. (2008a). However, as already pointed out, similar computation using a range of biomacromolecular compounds synthesized under variable environmental conditions that depart from standard states has, to our knowledge, not yet been performed in the context of reactive-transport modeling. In principle, such coupling would nevertheless be achievable and ΔG_{syn} values would then be calculated according to LaRowe and Amend (2016):

$$\Delta G_{syn} = \sum_i \Delta G_{r,i} + \Delta G_{poly} \tag{23}$$

where $\Delta G_{r,i}$ stands for the Gibbs energy of the reaction describing the synthesis of the *i*th biomolecule and ΔG_{poly} denotes the Gibbs energy required to polymerize biomolecules into their respective biomacromolecules, a term neglected when simple non-polymerized biomass composition are assumed. Values of $\Delta G_{r,i}$ and ΔG_{poly} can then be calculated as a function of temperature and pressure using equations similar to Equations (16, 17).

The results by LaRowe and Amend (2016) clearly demonstrate that values of ΔG_{syn} depend strongly on the combination of the redox state of the precursor compounds and that of the environment, with values varying by up to about 40 kJ (g cell)$^{-1}$ depending on the different combinations. LaRowe and Amend also show that the contribution of ΔG_{poly} to ΔG_{syn} is significant when biomass synthesis occurs in the most conducive environmental conditions. Overall, the important control of environmental conditions on the energetics of anabolism calls for more direct linkages between spatio-temporal gradients in redox conditions simulated by RTMs and their effects on biomass synthesis.

In addition to the energy required to synthesize biomolecules, the total cellular energy balance also includes the maintenance energy ΔG_{main} which is needed to sustain all other functions in support of viability and that do not result in new biomass (LaRowe and Amend 2016). The total cellular energy requirement, ΔG_{cell}, thus reads:

$$\Delta G_{cell} = \Delta G_{syn} + \Delta G_{main} \tag{24}$$

where we note that the above formulation is certainly a simplification because energies required for biosynthesis and maintenance are not completely decoupled (van Bodegom 2007).

Thermodynamic models for the maintenance energy compute ΔG_{main} from typical doubling times of microbial populations and the power demand required to maintain cell integrity (LaRowe and Amend (2016) and references therein). The latter must nevertheless distinguish the power used while organisms are growing, from the so-called 'basal maintenance power' (Hoehler and Jørgensen 2013), which is the power required by microorganisms to remain viable only and is typically several orders of magnitude lower (LaRowe and Amend 2015a). Overall, maintenance energies can become important for populations that have long doubling/replacement times and large (cumulative) maintenance powers, ΔG_{main} exceeding ΔG_{syn} in this case, and may even become the dominant component of the total energy budget. Major avenues for future RTM research would thus be to calculate changes in power demands for maintenance as a function of environmental conditions and evolving states of the microorganisms as well as to better couple the energetics of biomolecule synthesis and maintenance into a unified modeling framework.

Carbon Use Efficiency as an alternative to Ys

Before proceeding, it is important to stress that thermodynamic constraints on microbial growth have been measured, reported and implemented in models using multiple currencies (Sinsabaugh et al. 2013). In particular, in the field of ecological modeling and especially in soil science, Y is typically not used and growth yields are constrained from rates of carbon transformation using the concepts of 'Carbon Use Efficiency' (CUE) and its equivalent Microbial Growth Efficiencies, MGE. The CUE is generally defined as the ratio of growth (μ) to assimilation, that is, $CUE = \mu/(\mu + R)$, where R includes any C losses to respiration (Sterner and Elser 2002; Manzoni et al. 2012b). Reported CUE values determined in the laboratory and in the field have been synthesized by del Giorgio and Cole (1998) for aquatic systems and by Six et al. (2006) for soil environments. These empirically determined CUEs span a broad range, from typical values as low as 0.01 up to values close to 0.8, partly reflecting the diversity of processes captured by the applied empirical methods (physiological, but also characteristic of community or ecosystem dynamics) that influence C metabolism across varying spatio-temporal scales (Manzoni et al. 2012b; Sinsabaugh et al. 2013; Geyer et al. 2016). In order to better organize and properly interpret CUEs, Geyer et al. (2016) recently proposed a conceptual framework that structures its definition according to increasingly temporal and spatial scales of investigation. In short, Geyer and co-workers distinguish the CUE of populations, which is governed by species-specific metabolic and thermodynamic constraints as described by the growth yield Ys discussed above, from CUEs of communities and ecosystems which include many additional controls such as substrate stoichiometry, external physico-chemical conditions (temperature, moisture, pH,...) or substrate recycling. These distinctions are important because many large scale models of soil carbon dynamics, especially those embedded in Earth System Models of the coupled biogeochemical cycles and climate such as the CENTURY model (Parton et al. 1988) rely on the CUE (or the MGE) concept, and mostly address the larger ecosystem scale (e.g., Allison (2014), Wieder et al. (2014), Huang et al. (2018)). The reader is referred to e.g., Geyer et al. (2016) for further discussion on the use of the CUE and its relation to Y across scales.

INTEGRATION OF MICROBIAL SEQUENCING DATA IN GEOMICROBIAL MODELS

In recent years the improvement of biomolecular methods has led to a number of powerful approaches providing sequencing data from e.g., genomics, proteomics, transcriptomics or metabolomics and the combination of them. Such 'omics' approaches provide information on the metabolic potential of individual microbial species as well as of microbial communities (Müller and Hiller 2013; Franzosa et al. 2015). Together with biochemical data this has led to the set-up of extensive biochemical reaction networks describing the entire metabolism of a cell (Fig. 4). The resulting 'genome-scale model' provides the basis for the prediction of cellular functions such as growth or the formation of specific metabolites (Durot et al. 2009; Kim et al. 2012; Monk and Palsson 2014; O'Brien et al. 2015). For this purpose the stoichiometric matrix describing the reaction network needs to be combined with a 'constrained-based modeling' approach making use of additional knowledge and assumptions e.g., regarding the occurrence and magnitude of individual reaction fluxes and the steady-state assumption (Bordbar et al. 2014) and of the available experimental data e.g., on gene expression, protein expression or metabolite concentration (Reed 2012). Among a variety of such approaches the so-called 'flux-balance analysis' is one of the most commonly used (Orth et al. 2010). This methodology

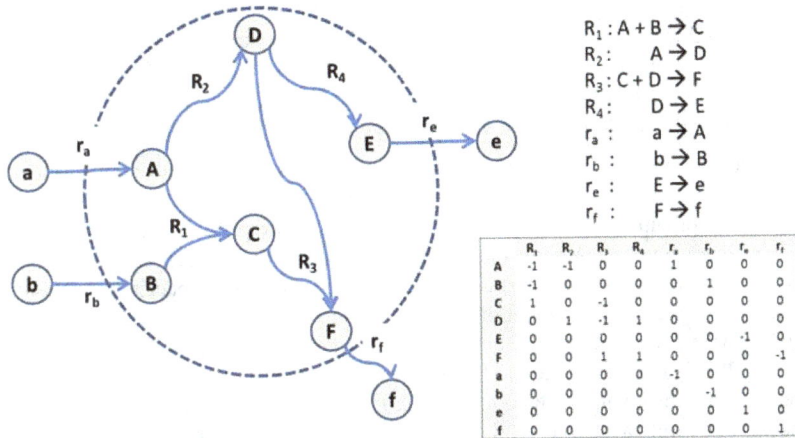

Figure 4. Left: Simplified generic example of a metabolic reaction network. $R_.$ denote metabolic transformations, $r_.$ denote uptake and release 'reactions', i.e. the exchange of chemical compounds between the cell interior (inside the dashed circle, upper case symbols) and its surrounding (lower case symbols). **Right:** On top the reactions of the network are provided with their stoichiometry which forms the basis for the stoichiometric matrix shown below, which is used for the calculation of mass fluxes along the network (adapted from (Durot et al. 2009).

allows considering dynamic changes of the chemical environment of the cells (Mahadevan et al. 2002), but several other methods have also been proposed (Lewis et al. 2012; Ramkrishna and Song 2012; Song et al. 2014). While 'omics'-based approaches have initially been used for modeling single microorganisms (or cell types) they have recently been expanded to describe metabolic fluxes in bulk microbial communities (Larsen et al. 2011; Biggs et al. 2015; Hanemaaijer et al. 2015; Gottstein et al. 2016; Perez-Garcia et al. 2016). The size of such communities is constrained by the required input data as well as the computational demands. So far, applications range from two-species systems (Klitgord and Segre 2010; Zomorrodi and Maranas 2012) to several hundreds of species (Magnúsdóttir et al. 2017).

Most metabolic flux approaches find their origin in the areas of medical and biotechnological research and have shown their potential for predicting the response of single cells to genetic modification or to drugs, or have been used to optimize the microbial production of specific compounds in industrial processes. However, they have also been used for the simulation of specific biochemical processes of relevance to the geosciences. Examples are methane oxidation by *Methylococcus capsulatus* (Lieven et al. 2018), methanogenesis by a community of *Desulfovibrio vulgaris* and *Methanococcus maripaludis* (Stolyar et al. 2007), and Fe(III) reduction by *Geobacter metallireducens* (Fang et al. 2012), by a community of *Geobacter sulfurreducens* and *Rhodoferax ferrireducens* (Zhuang et al. 2011) or by a community of the latter two species and *Shewanella oneidensis* (Zomorrodi et al. 2014). The Fe(III) reduction process is of particular interest in the context of uranium (U) reduction which is performed co-metabolically by some Fe(III) reducing microbes and leads to immobilization of uranium in contaminated aquifers.

The few applications of genome-scale descriptions of microorganisms in a RTM framework are also focusing on Fe-U cycling. King et al. (2009) coupled the dynamics of *Geobacter sulfurreducens* as described by Mahadevan et al. (2006) with a two-dimensional reactive transport solver and compared the resulting microbial growths and substrate consumptions with those obtained with more simplified description of microbial dynamics. Similarly Scheibe et al. (2009) coupled the same description of *Geobacter sulfurreducens* growth to the reactive transport model HYDROGEOCHEM to simulate acetate induced reduction of Fe(III) and U(VI) along a one-dimensional representation of an aquifer. In a follow-up study by Fang et al. (2011), a more integrated coupling of genome-scale data and reactive transport modeling was achieved. The same type of coupling was recently presented by Tartakovsky et al. (2013) using a "pore-scale smooth-particle hydrodynamics approach" (Tartakovsky et al. 2009). Their results were also compared to those obtained with a Monod-type model of bacterial growth and activity.

With the continuously increasing availability of sequencing data and computational power the inclusion of genome-scale models into reactive transport approaches has a clearly growing potential for applications focusing on the study of *in situ* processes occurring in porous media. There are however several important challenges limiting such applications. Natural microbial communities often consist of an unaccounted number of species and depending on the processes of interest the amount of data needed to set up a genome-scale model might be excessively large. An automated assembly of such data is supported by metabolic data bases (Karp et al. 2019) but results need to be checked for inconsistencies (Richter et al. 2015). In addition, a community-wide metabolic network implicitly considers all metabolites produced by a species to be fully available to all other species in the system, an approach which neglects limitations in availability due to transport into and out of a cell as well as to spatial separation of the cells. These limitations could be circumvented by considering the spatial distribution of the cells but existing approaches are limited to small number of species (Harcombe et al. 2014). Furthermore, substrate concentrations and environmental conditions may vary down to the pore-scale (Semple et al. 2004; Johnsen et al. 2005; Hesse et al. 2009; Schmidt et al. 2018),which challenges the genome-scale simulation of microbial dynamics in the same way as it does for other kinetic models of microbial activity.

Next to genome-scale models other less complex approaches that take available sequencing data into account exist. In these simpler methodologies, the data are used as proxies or biomarkers to constrain the dynamic behavior of a microbial community. While in the past, traditional microbial biomarkers such as cell counts, colony forming units, proteins or polysaccharides were used already as experimental references for the dynamics of microbial species or functional groups in the context of reactive transport simulations (Wirtz 2003; Thullner et al. 2004) more recent approaches combined these traditional approaches with marker genes to simulate the response of microorganism to fluctuating growth conditions (Stolpovsky et al. 2011) or carbon turnover and pesticide degradation in in a 1-D soil column (Pagel et al. 2016). Reed et al. (2014) and Louca et al. (2016) went further and applied a complex 'gene-centric' approach using marker genes as experimental references, the production of which was linked to biogeochemical reactions in the ocean.

CONSIDERING THE ECOLOGICAL BEHAVIOR OF MICROORGANISMS

The main reason for the simulation of microbial dynamics in the context of reactive transport modeling is their ability to catalyze biogeochemical reactions. This requires first of all modeling approaches describing the growth and metabolic activity of microbial species appropriately. However, the simulated systems represent an entire microbial ecosystem of varying complexity. In such an ecosystem the abundance of the microorganisms of interest can be affected not only by the concentration of substrates but also by a number of additional constraints, and the ability of the microorganisms to respond to these constraints also determines their abundance and their functional performance. There are numerous types of interactions between a (group of) microorganisms and their environment which for the sake of simplicity are separated here into two groups: microbe–microbe interactions, and the response of the microorganisms to abiotic constraints.

Microbe–microbe interactions

(Microbial) species may interact in different ways, with one species promoting or inhibiting the growth and activity of another species (Fig. 5, Lidicker 1979). While it can be challenging to determine such interactions in large communities (Faust and Raes 2012) specific interactions can be modeled using a generalized Lotka–Voltera approach describing the change of abundance B_i of a species i interacting with N other species according to:

$$\frac{\partial B_i}{\partial t} = \mu_i B_i + B_i \cdot \sum_{j=1}^{N} \gamma_{i,j} B_j \qquad (25)$$

with μ_i as specific growth rate and $\gamma_{i,j}$ as strength of the interaction between species i and j. γ values can be either positive or negative depending on the type of interaction, a value of 0 indicating no influence of species j on species i (Fig. 5). Applications of the Lotka–Voltera model in the context of microbial community dynamics range from single predator-prey interactions (Mauclaire et al. 2003) for which this approach was originally developed (Lotka 1925; Voltera 1926) to interactions in large communities found in the human gut (Stein et al. 2013; Kuntal et al. 2019) or in cheese (Mounier et al. 2008). Next to direct interactions such as protists predation (Mauclaire et al. 2003) or phage infections (Jover et al. 2013), indirect interactions have also been addressed by the generalized Lotka–Voltera approach. Indirect interactions may be mediated by a chemical compound produced or provided by a microbial species, which then promotes or inhibits the growth and activity of another species. Such interactions can be modeled by including a dependency of growth rates on the abundance of other species as described above but can of course also be addressed explicitly, e.g by linking the rate expressions describing the two species to the concentration of such mediator compounds or via a multi-species metabolic network simulation (Freilich et al. 2011). Similarly, the competition of species for a chemical compound can be addressed by a generalized Lotka–Voltera approach but can also be explicitly implemented in reactive transport simulations for, e.g., describing redox stratification associated with organic carbon degradation (Thullner et al. 2007).

A different type of microbe-microbe interaction occurs in the terrestrial mycosphere where bacteria can benefit from the presence of extensive networks of fungal mycelia (Harms et al. 2011; Worrich et al. 2018). These networks can act as 'fungal highways' which promote the mobility of bacterial cells (Kohlmeier et al. 2005) (see below) and as 'fungal pipelines' that facilitate the transport of water and nutrients to bacteria (Furuno et al. 2012; Schamfuss et al. 2013; Worrich et al. 2017). These effects have been simulated by considering a highly increased diffusion coefficient for transport of bacteria and/or nutrients along the fungal networks (Banitz et al. 2013).

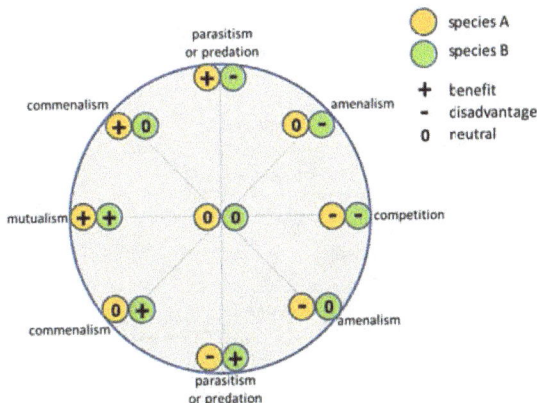

Figure 5. Overview of all possible pair-wise interactions between two species and of the resulting effect on each species (wheel display after Faust and Raes 2012 and Lidicker 1979). Shown signs also indicate the sign of the terms $\gamma_{i,j}$ in Equation (25).

Interactions between microbes and their physical environment

Subsurface environments impose a multitude of physical constraints on the abundance and activity of microorganisms. The solid matrix and its properties define the space and surface area available for the microorganisms while water flow (and distribution in unsaturated media) controls not only the transport of dissolved compounds but also the relocation of microorganisms. Due to the heterogeneity of the subsurface, these physical constraints can exhibit spatial variations at scales ranging from the pore scale to the field scale. Furthermore, flow and transport conditions can vary in time due to surface processes such as precipitation events, seasonal effects or anthropogenic activities. As a result, the microbial communities that live in the shallow subsurface need to adapt to changing environmental conditions. In particular such changes can involve periods of unfavorable conditions or disturbances (Fig. 6) (Allison and Martiny 2008; Shade et al. 2012) which can occur either as extended stress periods (Manzoni et al. 2012a; Worrich et al. 2016; Rocca et al. 2019) or as (series of) short disturbance events.

Microbial transport of bacteria by the flowing water is generally described by advective–dispersive transport combined with rate expressions for attachment and detachment of cells (Ginn et al. 2002; Tufenkji 2007). Attachment and detachment of cells are commonly simulated using first-order rate expressions with values of the rate parameters depending on the flow velocity and the resulting shear forces, the size and surface properties of the bacterial cells as well as the structure and surface properties of the solid matrix. Considering cells as solid particles allows application of the filtration theory (Tien and Payatakes 1979) to predict the attachment rate coefficient k_{att}:

$$k_{att} = \frac{3}{2} \cdot \frac{1-n}{d} \cdot v \cdot \eta \cdot \alpha \qquad (26)$$

with n as porosity, d as average grain size, v as transport velocity, η as collection efficiency (with values predicted from theory), and α as (empirically determined) collision efficiency. Limitations of the theoretical concepts of filtration theory to simulate the attachment process have been attributed e.g., to heterogeneities of the adhesion properties within the bacterial population (Simoni et al. 1998), to different microbial species affecting each other in their deposition behavior (Stumpp et al. 2011), or to repulsive interactions limiting the adhesion of cells on the matrix. While adhesion can alternatively be predicted by the (extended) DLVO theory (Hermansson 1999), there are also limitations to its application (Tufenkji 2007; Boks et al. 2008). Detachment rates depend on various factors (Peyton et al. 1995; Xavier et al. 2005) and are difficult to predict also as cells may increase their resistance to detachment producing

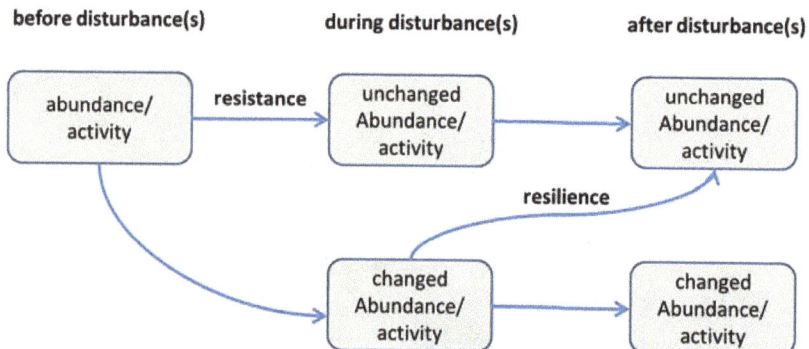

Figure 6. Schematic description of possible responses of a single microbial species to a (series of) disturbance(s) and of the meaning of resistance and resilience as ecological stability criteria (Grimm and Calabrese 2011). See e.g., Allison and Martiny (2008) for an analogous scheme for entire communities.

extracellular polymeric substances (EPS) (Tay et al. 2001) or may in turn actively detach from surfaces as response to nutrient availability or other favorable/unfavorable conditions (see Ginn et al. 2002 and literature cited therein). A general theory (such as filtration theory for attachment) providing a quantitative description of detachment processes has not been introduced, yet.

Besides being passively transported by the water flow or by the movement of solid particles they are attached to (Thullner et al. 2005) some bacterial species are motile and are thus actively able to swim in water (Blair 1995) or slide along a surface (Kearns 2010). This active movement may occur as random or chemotactic movement, that is to say, by sensing and following chemical concentration gradients (Berg 2000; Alexandre et al. 2004). This movement allows bacterial cells to relocate to more favorable conditions e.g., by moving towards higher nutrient concentrations or by moving towards lower concentrations of hazardous compounds—an ability which is however at the expense of significant metabolic costs. Furthermore, bacteria can produce chemoattractants triggering the chemotactic self-attraction of their own species and promoting their aggregation (Mittal et al. 2003; Park et al. 2003). In addition, bacterial motility may affect the attachment rates of cells in porous media (Nelson and Ginn 2001; de Kerchove and Elimelech 2008), their resilience towards disturbances (König et al. 2017) and their macroscopic transport behavior (Ford and Harvey 2007; Bai et al. 2016; Creppy et al. 2019), and it may promote the formation of pronounced spatial distribution patterns (Budrene and Berg 1991; Keymer et al. 2006). To model bacterial motility, the time evolution of bacterial abundance B in space can be described as

$$\frac{\partial B}{\partial t} = \nabla \left(D_b \cdot \nabla B - \sum_i \chi_{s,i} \cdot B \cdot \nabla s_i \right) \tag{27}$$

with D_b as random motility coefficient and $\chi_{s,i}$ as chemotactic sensitivities to the gradient of chemical compounds s_i. Sensitivities may be functions of the chemoattractant concentration (Keller and Segel 1971; Ford and Harvey 2007) or more simply considered as constants (Centler et al. 2011). In any case, positive $\chi_{s,i}$ values indicate attraction, while negative $\chi_{s,i}$ values indicate repulsion. Alternatively, the explicit link to the gradient of a chemical compound can be replaced by a dependency of the random motility coefficient D_b on the concentration of a chemical compound (Banitz et al. 2011a). Typically only a single compound (e.g., a nutrient) is considered to drive the chemotactic motility of bacteria, but two compounds might be considered if, in addition to a substrate, a bacterially emitted chemoattractant is also present (Saragosti et al. 2010; Centler et al. 2011).

Implementation of these motility concepts into reactive transport models have allowed to match high resolution data on bacterial chemotaxis (Pedit et al. 2002; Banitz et al. 2012) and predict traveling bands of bacteria (Hilpert 2005; Saragosti et al. 2010) as well as to simulate the formation of aggregated population patterns (Centler et al. 2011; Gharasoo et al. 2014; Centler and Thullner 2015) or the spreading of chemotactic bacteria at larger scales (Valdés-Parada et al. 2009). Microorganisms can also promote the dispersal of other microbial species (Ben-Jacob et al. 2016). Assuming fungal hyphae as pathway of increased bacterial motility ('fungal highways') in the simulation of microbial systems allowed describing the benefit of such high motility networks for biodegradation (Banitz et al. 2011a,b), for horizontal gene transfer (Berthold et al. 2016) or for the resistance of bacterial populations to disturbances (König et al. 2018a,b). A link between such reactive transport approaches with models describing fungal growth in porous media (Cazelles et al. 2013) has not been introduced, yet.

Dormancy is another microbial strategy to endure stress periods caused by natural effects (Lennon and Jones 2011; Joergensen and Wichern 2018) or anthropogenic perturbations (Balaban et al. 2004). When facing such unfavorable conditions microorganisms can switch from an active state into a dormant state which is characterized by a reduced metabolic activity, lower maintenance requirements and thus a better survival of the cells. If environmental conditions become favorable again microorganisms can reactivate and grow again (Dworkin and Shah 2010). For modeling microbial dormancy microorganisms are classically subdivided into an active and a dormant fraction (Fig. 7), but approaches considering a single fraction with transient changes of its dormancy degree exist, too (Resat et al. 2012). The transition between these two fractions is typically linked to environmental conditions by using a kinetic rate expression (depending on environmental variables) for the deactivation rate (transition into dormancy) and a complementary expression for the reactivation or resuscitation (transition from dormancy to active state) rate (Bär et al. 2002; Jones and Lennon 2010; Wang et al. 2014). Alternatively, a switch function that depends on environmental conditions can be used to determine the direction of the transition (Stolpovsky et al. 2011; Mellage et al. 2015; Bradley et al. 2018a). Other concepts consider both, deactivation (Chihara et al. 2015) and reactivation (Epstein 2009; Buerger et al. 2012) to be random processes, or rely on a growth rate dependent dormancy index to determine the fraction of dormant bacteria (Wirtz 2003). Furthermore, different degrees of dormancy may be considered assuming either distinct subgroups of dormant microorganisms or considering the depth of dormancy to increase with the duration of the unfavorable conditions (Stolpovsky et al. 2011). Models using such concepts have been applied to match data from specific experiments (Stolpovsky et al. 2011; Wang et al. 2014, 2015) and to analyze environmental samples ranging from dynamic systems such as the surface layer of lakes (Jones and Lennon 2010) to low activity systems like the deep subsurface of marine sediments or to soil organic matter composition at arbitrarily large scales (Huang et al. 2018). Furthermore, such simulations highlighted the relevance of dormancy for microbial competition (Stolpovsky et al. 2016) and long-term survival (Bär et al. 2002) in periodically changing environments. When implemented into RTMs dormancy approaches have allowed to match observations from shallow marine sediments (Wirtz 2003) and to study the influence of dormancy on the coexistence of competing species in heterogeneous porous media (Stolpovsky et al. 2012).

Noteworthy, yet not addressed this chapter, are feedbacks of the microorganisms on their physical environments. This includes changes of the hydraulic properties of the solid matrix (Hommel et al. 2018) due to biomass aggregation (Baveye et al. 1998; Yarwood et al. 2006; Thullner 2010), or due to microbially induced mineral dissolution, precipitation (Barkouki et al. 2011; Ebigbo et al. 2012) or gas formation (Mahabadi et al. 2018).

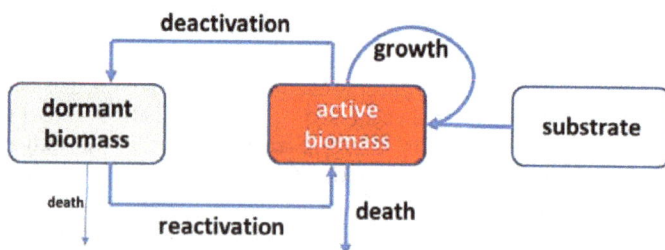

Figure 7. Schematic representation of active and dormant microorganisms and of the processes affecting them. Note that death rate of the dormant fraction is considered to be much smaller than for the active fraction (adapted from Stolpovsky et al. 2011).

CONSIDERING STABLE ISOTOPE SIGNATURES

Most elements, e.g., H, C, N, S, O associated in biogeochemically reactive compounds are found in more than one stable isotope form and the relative abundance of these isotopes can be determined with high accuracy. In recent years, isotopic methods have become increasingly used for the assessment of microbially driven processes in the subsurface. The most common approach is to use compound-specific stable isotope analysis (CSIA) to detect the stable isotope fractionation of reactants during (biogeo)chemical transformations (Brüchert 2004; Meckenstock et al. 2004; Elsner 2010; Thullner et al. 2012; Brunner et al. 2013; Blaser and Conrad 2016). The consideration of these fractionation effects in the context of reactive transport modeling is addressed in Druhan et al. (2019, this volume) and will thus not be discussed here. Instead, focus of this section is on the use of stable isotope signatures to trace the incorporation of inorganic compounds into microbial biomass, to identify the origin of microbial degradation products and to follow microbial degradation pathways.

Stable isotopes as tracers of matter fluxes into biomass

Since decades, the isotope signatures of animal biomass is known to be determined by the isotope signatures of their food sources (DeNiro and Epstein 1978). In the same way the isotopic composition of microbial biomass reflects the isotopic signatures of the used substrates (Coffin et al. 1989). Furthermore, stable isotope probing (SIP) approaches relying on the isotopic composition of specific biomarkers allow determining which (group of) microbial species pertaining to a complex community have taken up a specific substrate (Boschker and Middelburg 2002; Abraham 2014; Vogt et al. 2016b). Originally lipids were the commonly used biomarkers (Pancost and Sinninghe Damsté 2003; Yao et al. 2014; Wegener et al. 2016), which together with isotopically labeled substrates allowed for the detection of the substrate uptake into biomarkers (Pelz et al. 2001; Pombo et al. 2002; Stelzer et al. 2006). More recently, such SIP approaches were extended to the analysis of e.g., microbial DNA (Coyotzi et al. 2016), RNA (Lueders et al. 2016; Bradford et al. 2018) or proteins (Jehmlich et al. 2016) used as biomarkers for specific (groups of) microorganisms and the uptake of labeled substrate into their biomass.

The inclusion of stable isotope signatures into RTMs is well established. In short, heavy (h) and light isotope (l) fraction of a specific element in a reactive compound i are treated independently as two separate compounds with concentration $^h c_i$ and $^l c_i$, respectively. The isotopic signature for each compound is then given by the concentration ratio $R_i = {}^h c_i / {}^l c_i$ and commonly expressed as a relative deviation δ_i from a standard value (R_s):

$$\delta_i = (R_i - R_s)/R_s \qquad (28)$$

Note that δ-values are typically expressed in ‰. Analogously, for each reactive transformation rate j expressions $^h r_j$ and $^l r_j$ have to be defined for each fraction. The rate expressions have to fulfill the relation $^h r_j / {}^l r_j = \alpha_j R_i$ if compound i is consumed by process j. The fractionation factor α_j describes the magnitude of the stable isotope fractionation induced by the reactive transformation with $\alpha_j = 1$ indicating no fractionation. For further details—in particular regarding the simulation of stable isotope fractionation—see e.g., van Breukelen and Griffioen (2004), Thullner et al. (2008, 2012) and Druhan et al. (2019, this issue).

Although this RTM concept is well suited to describe the uptake of substrates and the resulting stable isotope signature into microbial biomass it has only rarely been applied for this purpose yet. Mauclaire et al. (2003) used stable isotope signatures to simulate carbon fluxes along a microbial food chain in a batch system (Fig. 8). ^{13}C-labeled toluene was used as primary substrate, which was taken up by a bacterial species that served as a prey for a protist predator. Isotopic signatures of the bacterial and protist biomasses were determined using biomarker fatty acids while total bacterial biomass was quantified using cell counting combined with image analysis techniques. Rate expressions based on Monod-type growth of the bacteria and Lotka–Voltera type predation for the protist were used to simulate changes of concentration and carbon isotope signatures in the different carbon pools (toluene, bacterial biomass, protist biomass). Using this approach, a carbon conversion efficiency of approximately 30% for the bacterial growth on toluene, and of approximately 10% for the protist predation on the bacteria was determined. Relying on ^{13}C-labeled *E. coli* cells added to a soil batch system Kindler et al. (2009)

A

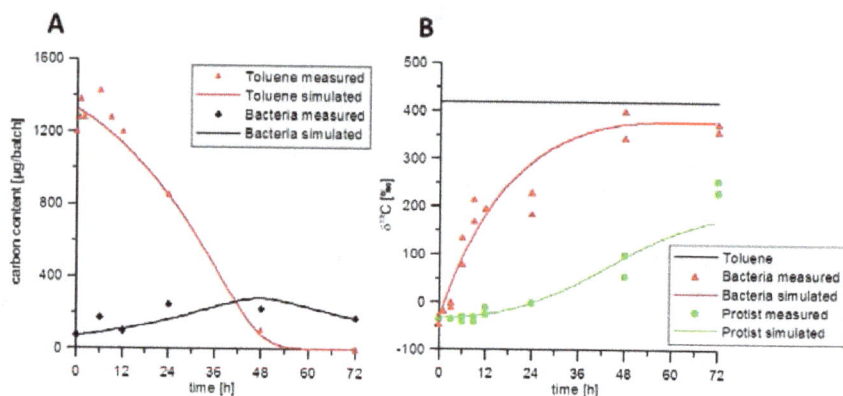

B

Figure 8. Simulation of carbon fluxes along a food chain consisting of (^{13}C-labeled) toluene, toluene feeding bacteria and a protist predator feeding on the bacteria (adapted from Mauclaire et al. 2003). **A:** measured and simulated concentrations; **B:** measured and simulated carbon isotope signatures.

quantified the transformation of the labeled carbon during recycling of the biomass in the microbial food chain. Fatty acids were used to distinguish between living and dead biomass, the latter considered as part of the soil organic matter. Instead of using a classical kinetic model approach the process was modeled by applying a series of feeding cycles to quantify the flux of carbon between living biomass, soil organic matter and inorganic carbon (CO_2). Results indicated that within each feeding cycle approximately 20% of the biomass food source was transferred into soil organic matter and that 33% of the added carbon source accumulated as soil organic matter. Carbon uptake by microbes and the associated changes of carbon isotope signatures in the biomass was simulated by Alperin and Hoehler (2009a) for anaerobic methane oxidation in marine sediments. To do so, they represented the network of biochemical reactions catalyzed by a consortium of methane oxidizing archea and sulfate reducing bacteria, and by organic matter oxidizing bacteria. In addition, various stable isotope fractionation effects were considered. Using this complex reaction network the authors showed that although the stable isotope signature of the biomass of the methane oxidizing consortia was similar to the stable isotope signature of methane, methane was not the only source of the biomass carbon, and the CO_2 released by the organic carbon oxidizing bacteria was another likely contributor.

Stable isotope signatures have also been used as tracers for the formation of reaction products during (microbially induced) reactive transformations (Avery et al. 1999; Bolliger et al. 1999; Kindler et al. 2009). The interpretation of such signatures can be done by mass balance approaches not requiring more elaborate model simulations as long as fractionation effects can be neglected.

Stable isotopes as indicators of degradation pathways

Next to using stable isotope fractionation as qualitative and quantitative indicators of biodegradation processes (not discussed in this chapter), the combined analysis of the fractionation of two elements provides additional information on the dominant fractionation pathway (Elsner 2010; Vogt et al. 2016a)—also known as dual or 2D isotope analysis. The basic principle of this approach takes advantage of the fact that during a reaction a stable isotope fractionation of different elements in the reactant can occur, each of which described by a pathway- and element-specific fractionation factor α. When analyzing the change of the stable isotope signatures of two of these elements a linear relation between these changes (expressed in δ-notation) is often observed as long as fractionation effects are not too strong. The ratio between these changes, that is the slope of the linear correlation when plotting the changes against each other, is specific for each degradation pathway and given by the ratio between the isotope enrichment factors $\varepsilon = 1 - \alpha$ (Fig. 9). Therefore, such plots can be used to identify a dominating degradation pathway, e.g., to distinguish between aerobic and anaerobic pathways (Fischer et al. 2007) or between biotic and abiotic transformations (Badin et al. 2016), or to elucidate the specific reaction mechanism (Meyer et al. 2009; Dorer et al. 2014). If fractionation effects are very strong as e.g., for H fractionation or if two pathways take place simultaneously (Van Breukelen 2007; Centler et al. 2013) a more complex analysis is needed.

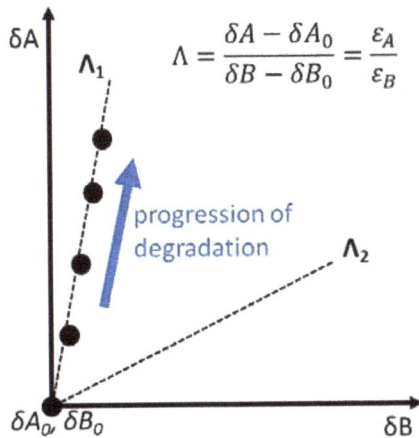

$$\Lambda = \frac{\delta A - \delta A_0}{\delta B - \delta B_0} = \frac{\varepsilon_A}{\varepsilon_B}$$

Figure 9. Schematic example for a dual isotope analysis of two arbitrary elements A and B. For a given pathway changes of the δ-values should plot along a line, the slope (Λ) of which given by the ratio of the stable isotope enrichment factors ε for each element. As values for ε are typically pathway specific, each pathway is characterized by a pathway specific slope (Λ_1 and Λ_2 in the present example; dashed lines). If measured δ-values (dot symbols) plot along a line with one of these slopes they indicate which degradation pathway has taken place.

SCALE EFFECTS

While the sections above focused on approaches describing the dynamics of microbial growth and activity two aspects have not been directly addressed: the measure by which the microbial abundance is quantified and the spatial representation of the microorganisms. Both of these interlinked aspects are explicitly or implicitly affected by the spatial resolution of the applied modeling approach.

Most reactive transport applications considering microbial abundance use a population-based concept in which biomass is quantified as concentration of microbial carbon mass, microbial cells or of specific parts (DNA, RNA, fatty acids or other biomarkers) of the microbial cells. Concentration may refer to individual species as sometimes implemented for simulating controlled laboratory experiments (Bauer et al. 2009; Dechesne et al. 2010; Banitz et al. 2011a; Monga et al. 2014) while for natural environments, the vast number of species in microbial communities often preclude the possibility to link the rate of a specific biogeochemical reaction to the abundance and activity of a restricted number of individual species. Instead the species are pooled into groups of specific traits, microbial guilds or functional groups e.g., catalysts of specific degradation pathways (Schäfer et al. 1998a; Brun et al. 2002; Wirtz 2003; Thullner et al. 2005; Yabusaki et al. 2011; Bouskill et al. 2012; Pagel et al. 2014; Yu and Zhuang 2019). In contrast, individual-based models consider the dynamic behavior of each single microbial cell in a system (Ferrer et al. 2008; Kreft et al. 2013; Hellweger et al. 2016; Jayathilake et al.

2017; Leveau et al. 2018), an approach that allows considering random variations of microbial behavior or intra-species variations of cell properties. Such individual-based models have been considered also within reactive transport approaches (Gras et al. 2011; Ebrahimi and Or 2014; Gharasoo et al. 2014; Centler and Thullner 2015; Ebrahimi and Or 2015; Kim and Or 2016) but applications are necessarily limited to small-scale (cm-scale or below) systems given typical abundances of 10^6 cells per cm^3 or more in most porous medium environments.

In most aqueous systems microorganisms tend to be evenly distributed in the water without exhibiting significant biomass concentration gradients at small spatial scales. In porous media microorganisms are primarily associated with the solid matrix (Lehman et al. 2001; Griebler et al. 2002; Mellage et al. 2015), which favors the development of small-scale spatial heterogeneities. Even along the surface of the solid matrix large variations in microbial abundance may occur (Dechesne et al. 2003; Nunan et al. 2003; Or et al. 2007; Iltis et al. 2011). As a consequence not all microorganisms at a macroscopic location (defined e.g., by the experimental sampling volume) are exposed to the same environmental conditions. In particular, small scale variations in substrate concentration may expose microorganisms to environmental conditions that differ from the mean state, that is to say, substrate bioavailability is not the same everywhere (Semple et al. 2004). The bioavailability of a substrate is thus no only limited by its interaction with solid and liquid phases (Johnsen et al. 2005; Haws et al. 2006; Amos et al. 2007) but also by a mass transfer process that is required to provide the substrate at the microscopic location of the microbial cells (Bosma et al. 1997). In most cases a limited bioavailability of a substrate leads to reduced degradation rates but in cases where high substrate concentrations become toxic to the degrading microorganisms, limited bioavailability may exceptionally promote higher degradation rates (Hanzel et al. 2012; Gharasoo et al. 2015).

Mass transfer processes can be explicitly included into reactive transport models of high spatial resolution. Such implementation requires a theoretical framework for simulating the distribution of the microbial biomass. The simpler approach represents the interface between pore water and solid matrix as a reactive surface of no volume homogeneously covered by microorganisms (Hesse et al. 2009; Gharasoo et al. 2012). A slightly more sophisticated approach is to consider that microorganisms form a biofilm covering the surface of the solid matrix. Biofilms are aggregations of microbial cells and extracellular polymeric substances (Costerton 1999; Watnick and Kolter 2000; Branda et al. 2005; Carrel et al. 2018), either forming a homogeneous layer of biomass or heterogeneously distributed biomass aggregates on the solid matrix. Since the introduction of the first biofilm models (Rittmann and McCarty 1980) increasingly complex approaches have been implemented (Kreft et al. 2001; van Loosdrecht et al. 2002; Picioreanu et al. 2004; Alpkvist et al. 2006; Cumsille et al. 2014) and combined with different pore-scale reactive transport models (Suchomel et al. 1998; Dupin et al. 2001; Thullner et al. 2002; Kapellos et al. 2007; Thullner and Baveye 2008; von der Schulenburg et al. 2009; Bottero et al. 2013; Qin and Hassanizadeh 2015; Tang and Liu 2017). These simulations have allowed constraining the distribution and thickness of the biofilm as a result of the mass transfer of substrates to and into the biofilm and its subsequent *in-situ* degradation by microorganisms. Results from pore-scale simulations of mass-transfer limited microbial degradation in porous media (see Fig. 10 for an example) show that the extent of the mass transfer limitation depends (among several other factors) on the size of the pores, on the heterogeneity of the size distribution and on the pore-scale distribution of the biomass along the solid matrix surface (Hesse et al. 2009; Gharasoo et al. 2012; Schmidt et al. 2018).

At the continuum scale, i.e. the scale at which porous media are described by representative elementary volumes (REV), pore scale structures and the pore-scale distribution of the biomass can't be resolved anymore (Fig. 11), and although often still called biofilm, model approaches typically consider only a bulk biophase of unknown spatial distribution (Schäfer et al. 1998b; Schäfer 2001; Ebigbo et al. 2010; Bozorg et al. 2011). Regardless of the distribution type (colonies, aggregates, biofilm etc.), a linear exchange term

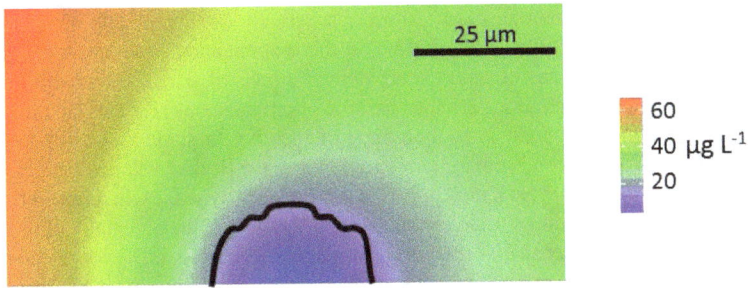

Figure 10. Simulated distribution of substrate (acetate) in and around a colony of 200 bacterial cells in a pore channel (adapted from Schmidt et al. 2018). Concentration inside the colony is determined by the distribution of the individual cells (no shown).

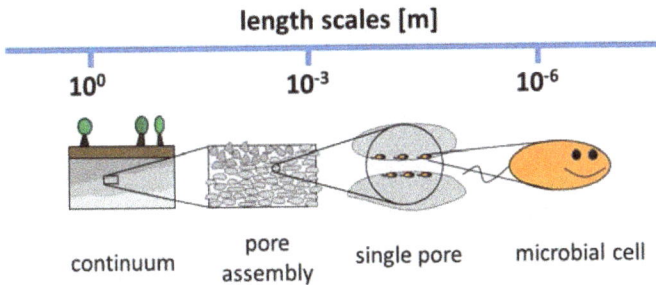

Figure 11. Overview of length scales involved when simulating microbially driven processes in porous media. The length scales of the simulated domains may go up to 10^4–10^5 m for catchment scale model and beyond for global scale models.

$$J_{tr} = k_{tr} \cdot \left(c_{bulk} - c_{bio} \right) \qquad (29)$$

has been suggested to describe the mass flux J_{tr} between the bulk aqueous phase and the biophase (Baveye and Valocchi 1989). This flux term links the bulk substrate concentration c_{bulk} (affected by the transport processes along the model domain) and the substrate concentration c_{bio} inside the biophase, the latter describing the bioavailable concentration. All other factors controlling the magnitude of J_{tr} are subsumed into the value of the mass-transfer rate coefficient k_{tr}. Predictions of the value of k_{tr} can be obtained from formal upscaling approaches (Hesse et al. 2010; Orgogozo et al. 2013), which do however require a series of simplifying assumptions. As a result, in many reactive transport applications the transfer rate is not well constrained in quantitative terms. In cases of a single rate-limiting chemical compound mass transfer (Eqn. 29) and a biodegradation rate following Michaelis–Menten kinetics (Eqn. 1), both rates can be combined into a single expression known as the Best equation (Best 1955), which has been shown to provide adequate descriptions of mass transfer limited microbial degradation rates (Bosma et al. 1997; Simoni et al. 2001). Alternatively, formal upscaling approaches have been used to derive effective rate expressions where small-scale mass transfer effects are affecting the values of the degradation rate or transport parameters at larger scales (Wood et al. 2007; Golfier et al. 2009; Hesse et al. 2009).

Interestingly, we are not aware of applications of microbial explicit RTMs at scales larger than the plot or the sediment core scale. For instance, the coupled microbial-geochemical dynamics has not yet been addressed at the catchment scale, at least with regard to processes associated with subsurface flow. This contrasts with catchment-based RTM applications simulating surface flow hydrodynamics in, e.g., rivers, lakes or estuaries where heterotrophic and autotrophic microbial community dynamics has already been explicitly simulated (e.g., Vanderborght et al. (2002), Billen et al. (2013)). In these applications, the interaction between substrates and microbial biomasses ignore the potential effects of spatial heterogeneities at the sub-grid scale and the concentrations are thus assumed homogeneous at scales smaller than the model resolution (typically hundreds to thousands of meters). In stark contrast to subsurface catchment scale models, several biomass-explicit models have been developed for global-scale applications over the last 5–10 years, mostly in relation to soil C dynamics. In terms of biological process representation, these large-scale microbial models are not fundamentally different from those applied at (much) smaller spatial scales. A major alteration, however, lies on the assumption that soil physics and microbiology are homogenous within the soil grid to which it is applied (e.g., Huang et al. (2018)). Taken that the typical resolution of land surface schemes of Earth System Model is on the order of 0.5–2° (i.e. from tens to hundreds of km), numerous physical–biogeochemical–microbial interactions occurring at smaller spatial scales are unresolved in these models and their parametrization remains a grand challenge for the future.

CONCLUSIONS

The advancement in experimental methods has provided an enormous amount of information on microbial community dynamics in the subsurface and the ongoing development of new methods for the analysis of microbial processes will further increase our knowledge in this field. As outlined in this chapter, this has triggered the development of a large number of modeling concepts that support the exploitation of this new knowledge in the broader context of reactive-transport modeling. Not all of these concepts have yet been embedded in reactive transport modeling approaches, but the existing combinations indicate the strong potential of improving the representation of microbial processes in such models. Next to technical developments there is however the need to establish effective modeling concepts that can be applied for the simulation of microbial community dynamics and their biogeochemical function in natural porous media. The complex behavior of microbial communities still imposes severe challenges in their experimental assessment as observations obtained from any location in the subsurface may integrate the effects of several micro-environments each with distinct communities. Furthermore, many microbial species in a natural community are still not known or characterized. The required effective model approaches thus imperatively need to reduce the real world complexity of a natural community into a less complex community represented in the model framework. However, this reduced community should still reproduce the dynamics of the real community adequately. How to meet this challenge in the best possible way depends not only on the available data but also on the microbial functions of interest and thus on the purpose of the specific reactive transport modeling effort. For many subsurface compartments there is a growing need for reactive transport approaches to consider the functional dynamics of the microbial community more adequately, from the pore scale to the global scale. This is in particular the case for the Critical Zone and its compartments (Griebler et al. 2014; Vereecken et al. 2016; Li et al. 2017; Baveye et al. 2018; Vogel et al. 2018) partly because surface processes (e.g., weather, climate, anthropogenic activities) lead to a dynamic variation in the environmental conditions. However, this can also be the case for marine sediments (van de Velde et al. 2018). The increasing global population will lead to further anthropogenic disturbances of the subsurface and global change is expected to lead to more dynamic weather conditions. This further increases the need for model approaches that can predict the response of the microbial functions to such perturbations together with the adaptation to a changing climate.

ACKNOWLEDGMENTS

MT was supported by funding from Helmholtz Centre for Environmental Research—UFZ via the integrated project Controlling Chemicals Fate (CCF) of the research topic Chemicals in the Environment (CITE) and by the German Research Foundation (DFG) CRC 1076 "AquaDiva". PR received funding from the VERIFY project from the European Union's Horizon 2020 research and innovation programme under grant agreement No. 776810 and from the F.R.S.-FNRS (sabbatical leave). The authors thank D. LaRowe for commenting an earlier version of this chapter and N. Roevros for her technical support.

REFERENCES

Abraham WR (2014) Applications and impacts of stable isotope probing for analysis of microbial interactions. Appl Microbiol Biotechnol 98:4817–4828

Alexandre G, Greer-Phillips S, Zhulin IB (2004) Ecological role of energy taxis in microorganisms. FEMS Microbiol Rev 28:113–126

Allison SD (2014) Modeling adaptation of carbon use efficiency in microbial communities. Front Microbiol 5:571

Allison SD, Martiny JBH (2008) Resistance, resilience, and redundancy in microbial communities. PNAS 105:11512–11519

Alperin MJ, Hoehler TM (2009a) Anaerobic methane oxidation by achea/sulfate-reducing bacteria aggregatres: 2. isotopic contraints. Am J Sci 309:958–984

Alperin MJ, Hoehler TM (2009b) Anaerobic methane oxidation by achea/sulfate-reducing bacteria aggregatres: 1. thermodynamic and physical constraints. Am J Sci 309:869–957

Alpkvist E, Picioreanu C, van Loosdrecht MCM, Heyden A (2006) Three-dimensional biofilm model with individual cells and continuum EPS matrix. Biotechnol Bioeng 94:961–979

Amend JP, Teske A (2005) Expanding frontiers in deep subsurface microbiology. Palaeogeogr Palaeoclimatol Palaeoecol 219:131–155

Amend JP, LaRowe DE, McCollom TM, Shock EL (2013) The energetics of organic synthesis inside and outside the cell. Philos Trans R Soc Lond B Biol Sci 368:20120255

Amos BK, Christ JA, Abriola LM, Pennell KD, Löffler FE (2007) Experimental evaluation and mathematical modeling of microbially enhanced tetracholorethene (PCE) dissolution. Environ Sci Technol 41:963–970

Arndt S, Jørgensen BB, LaRowe DE, Middelburg JJ, Pancost RD, Regnier P (2013) Quantifying the degradation of organic matter in marine sediments: A review and synthesis. Earth Sci Rev 123:53–86

Avery GB, Shannon RD, White JR, Martens CS, Alperin MJ (1999) Effect of seasonal changes in the pathways of methanogenesis on the $\delta^{13}C$ values of pore water methane in a Michigan peatland. Global Biogeochem Cycles 13:475–484

Badin A, Broholm MM, Jacobsen CS, Palau J, Dennis P, Hunkeler D (2016) Identification of abiotic and biotic reductive dechlorination in a chlorinated ethene plume after thermal source remediation by means of isotopic and molecular biology tools. J Contam Hydrol 192:1–19

Bai H, Cochet N, Drelich A, Pauss A, Lamy E (2016) Comparison of transport between two bacteria in saturated porous media with distinct pore size distribution. RSC Advances 6:14602–14614

Balaban NQ, Merrin J, Chait R, Kowalik L, Leibler S (2004) Bacterial persistence as a phenotypic switch. Science 305:1622–1625

Banitz T, Fetzer I, Johst K, Wick LY, Harms H, Frank K (2011a) Assessing biodegradation benefits from dispersal networks. Ecol Modell 222:2552–2560

Banitz T, Wick LY, Fetzer I, Frank K, Harms H, Johst K (2011b) Dispersal networks for enhancing bacterial degradation in heterogeneous environments. Environ Pollut 159:2781–2788

Banitz T, Johst K, Wick LY, Fetzer I, Harms H, Frank K (2012) The relevance of conditional dispersal for bacterial colony growth and biodegradation. Microbial Ecol 63:339–347

Banitz T, Johst K, Wick LY, Schamfuss S, Harms H, Frank K (2013) Highways versus pipelines: contributions of two fungal transport mechanisms to efficient bioremediation. Environ Microbiol Rep 5:211–218

Bao C, Wu H, Li L, Newcomer D, Long PE, Williams KH (2014) Uranium bioreduction rates across scales: biogeochemical hot moments and hot spots during a biostimulation experiment at Rifle, Colorado. Environ Sci Technol 48:10116–10127

Bär M, von Hardenberg J, Meron E, Provenzale A (2002) Modelling the survival of bacteria in drylands: the advantage of being dormant. Proc Biol Sci 269:937–942

Barkouki TH, Martinez BC, Mortensen BM, Weathers TS, De Jong JD, Ginn TR, Spycher NF, Smith RW, Fujita Y (2011) Forward and inverse bio-geochemical modeling of microbially induced calcite precipitation in half-meter column experiments. Transp Porous Media 90:23–39

Barry DA, Prommer H, Miller CT, Engesgaard P, Brun A, Zheng C (2002) Modelling the fate of oxidisable organic contaminants in groundwater. Adv Water Resour 25:945–983

Bauer RD, Rolle M, Bauer S, Eberhardt C, Grathwohl P, Kolditz O, Meckenstock RU, Griebler C (2009) Enhanced biodegradation by hydraulic heterogeneities in petroleum hydrocarbon plumes. J Contam Hydrol 105:56–68

Baveye P, Valocchi A (1989) An evaluation of mathematical-models of the transport of biologically reacting solutes in saturated soils and aquifers. Water Resour Res 25:1413–1421

Baveye P, Vandevivere P, Hoyle BL, DeLeo PC, de Lozada DS (1998) Environmental impact and mechanisms of the biological clogging of saturated soils and aquifer materials. Crit Rev Environ Sci Technol 28:123–191

Baveye PC, Otten W, Kravchenko A, Balseiro-Romero M, Beckers E, Chalhoub M, Darnault C, Eickhorst T, Garnier P, Hapca S, Kiranyaz S (2018) Emergent properties of microbial activity in heterogeneous soil microenvironments: different research approaches are slowly converging, yet major challenges remain. Frontiers Microbiol 9:1929

Ben-Jacob E, Finkelshtein A, Ariel G, Ingham C (2016) Multispecies swarms of social microorganisms as moving ecosystems. Trends Microbiol 24:257–269

Berg HC (2000) Motile behavior of bacteria. Phys Today 53:24–29

Berner RA (1980) Early Diagenesis. Princeton University Press, Princeton, NY, USA

Berthold T, Centler F, Hubschmann T, Remer R, Thullner M, Harms H, Wick LY (2016) Mycelia as a focal point for horizontal gene transfer among soil bacteria. Sci Rep 6:36390

Best JB (1955) The interference of intracellular enzymatic properties from kinetic data obtained on living cells. 1. Some kinetic considerations regarding an enzyme enclosed by a diffusion barrier. J Cell Comp Physiol 46:1–27

Bethke CM, Sanford RA, Kirk MF, Jin Q, Flynn TM (2011) The thermodynamic ladder in geomicrobiology. Am J Sci 311:183–210

Biggs MB, Medlock GL, Kolling GL, Papin JA (2015) Metabolic network modeling of microbial communities. Wiley Interdiscip Rev Syst Biol Med 7:317–334

Billen G, Garnier J, Lassaletta L (2013) The nitrogen cascade from agricultural soils to the sea: modelling nitrogen transfers at regional watershed and global scales. Philos Trans R Soc Lond B Biol Sci 368:20130123

Blair DF (1995) How bacteria sense and swim. Ann Rev Microbiol 49:489–522

Blaser M, Conrad R (2016) Stable carbon isotope fractionation as tracer of carbon cycling in anoxic soil ecosystems. Curr Opin Biotechnol 41:122–129

Boks NP, Norde W, van der Mei HC, Busscher HJ (2008) Forces involved in bacterial adhesion to hydrophilic and hydrophobic surfaces. Microbiology 154:3122–3133

Bolliger C, Höhener P, Hunkeler D, Haberli K, Zeyer J (1999) Intrinsic bioremediation of a petroleum hydrocarbon-contaminated aquifer and assessment of mineralization based on stable carbon isotopes. Biodegradation 10:201–217

Bordbar A, Monk JM, King ZA, Palsson BO (2014) Constraint-based models predict metabolic and associated cellular functions. Nat Rev Genet 15:107–120

Boschker HTS, Middelburg JJ (2002) Stable isotopes and biomarkers in microbial ecology. FEMS Microbiol Ecol 40:85–95

Bosma TNP, Middeldorp PJM, Schraa G, Zehnder AJB (1997) Mass transfer limitation of biotransformation: Quantifying bioavailability. Environ Sci Technol 31:248–252

Bottero S, Storck T, Heimovaara TJ, van Loosdrecht MC, Enzien MV, Picioreanu C (2013) Biofilm development and the dynamics of preferential flow paths in porous media. Biofouling 29:1069–1086

Boudart M (1976) Consistency between kinetics and thermodynamics. J Phys Chem 80:2869–2870

Boudreau BP (1997) Modelling Transport and Reactions in Aquatic Sediments. Springer

Boudreau BP (1999) A theoretical investigation of the organic carbon-microbial biomass relation in muddy sediments. Aquatic Microbial Ecol 17:181–189

Bouskill NJ, Tang J, Riley WJ, Brodie EL (2012) Trait-based representation of biological nitrification: model development, testing, and predicted community composition. Front Microbiol 3:364

Bozorg A, Sen A, Gates ID (2011) A new approach to model the spatiotemporal development of biofilm phase in porous media. Environ Microbiol 13:3010–3023

Bradford LM, Vestergaard G, Tancsics A, Zhu B, Schloter M, Lueders T (2018) Transcriptome-stable isotope probing provides targeted functional and taxonomic insights into microaerobic pollutant-degrading aquifer microbiota. Front Microbiol 9:2696

Bradley JA, Amend JP, LaRowe DE (2018a) Survival of the fewest: Microbial dormancy and maintenance in marine sediments through deep time. Geobiology 17:43–59

Bradley JA, Amend JP, LaRowe DE (2018b) Bioenergetic controls on microbial ecophysiology in marine sediments. Front Microbiol 9:180

Branda SS, Vik A, Friedman L, Kolter R (2005) Biofilms: the matrix revisited. Trends Microbiol 13:20–26

Brock AL, Kastner M, Trapp S (2017) Microbial growth yield estimates from thermodynamics and its importance for degradation of pesticides and formation of biogenic non-extractable residues. SAR QSAR Environ Res 28:629–650

Brüchert V (2004) Physiological and ecological aspects of sulfur isotope fractionation during bacterial sulfate reduction. *In*: Sulfur Biogeochemistry—Past and Present. Vol 379. Amend JP, Edwards KJ, Lyons TW, (eds). Geol Soc Am, p 1–16

Brun A, Engesgaard P (2002) Modelling of transport and biogeochemical processes in pollution plumes: literature review and model development. J Hydrol 256:211–227

Brun A, Engesgaard P, Christensen TH, Rosbjerg D (2002) Modelling of transport and biogeochemical processes in pollution plumes: Vejen landfill, Denmark. J Hydrol 256:228–247

Brunner B, Contreras S, Lehmann MF, Matantseva O, Rollog M, Kalvelage T, Klockgether G, Lavik G, Jetten MS, Kartal B, Kuypers MM.(2013) Nitrogen isotope effects induced by anammox bacteria. PNAS 110:18994–18999

Budrene EO, Berg HC (1991) Complex patterns formed by motile cells of *Escherichia-coli*. Nature 349:630–633

Buerger S, Spoering A, Gavrish E, Leslin C, Ling L, Epstein SS (2012) Microbial scout hypothesis, stochastic exit from dormancy, and the nature of slow growers. Appl Environ Microbiol 78:3221–3228

Canfield DE (1993) Organic matter oxidation in marine sediments. *In:* Organic Matter Oxidation in Marine Sediments. Springer Berlin Heidelberg, p 333–363

Canuel EA, Martens CS (1996) Reactivity of recently deposited organic matter: Degradation of lipid compounds near the sediment-water interface. Geochim Cosmochim Acta 60:1793–1806

Carrel M, Morales VL, Beltran MA, Derlon N, Kaufmann R, Morgenroth E, Holzner M (2018) Biofilms in 3D porous media: Delineating the influence of the pore network geometry, flow and mass transfer on biofilm development. Water Res 134:280–291

Cazelles K, Otten W, Baveye PC, Falconer RE (2013) Soil fungal dynamics: Parameterisation and sensitivity analysis of modelled physiological processes, soil architecture and carbon distribution. Ecol Modell 248:165–173

Centler F, Thullner M (2015) Chemotactic preferences govern competition and pattern formation in simulated two-strain microbial communities. Frontiers Microbiol 6:40

Centler F, Fetzer I, Thullner M (2011) Modeling population patterns of chemotactic bacteria in homogeneous porous media. J Theor Biol 287:82–91

Centler F, Hesse F, Thullner M (2013) Estimating pathway-specific contributions to biodegradation in aquifers based on dual isotope analysis: Theoretical analysis and reactive transport simulations. J Contam Hydrol 152C:97–116

Chihara K, Matsumoto S, Kagawa Y, Tsuneda S (2015) Mathematical modeling of dormant cell formation in growing biofilm. Front Microbiol 6:534

Claypool GE, Kaplan IR (1974) The origin and distribution of methane in marine sediments. *In:* Natural Gases in Marine Sediments. Kaplan IR, (ed) Plenum, p 99–139

Coffin RB, Fry B, Peterson BJ, Wrigth RT (1989) Carbon isotopic composition of estuarine bacteria. Limnol Oceanogr 34:1305–1310

Contois DE (1959) Kinetics of bacterial growth: Relationship between population density and specific growth rate of continuous cultures. J Gen Microbiol 21:40–50

Costerton JW (1999) Introduction to biofilm. Int J Antimicrob Agents 11:217–221

Cox JD (1982) Notation for states and processes, significance of the word standard in chemical thermodynamics, and remarks on commonly tabulated forms of thermodynamic functions. Pure Appl Chem 54:1239–1250

Coyotzi S, Pratscher J, Murrell JC, Neufeld JD (2016) Targeted metagenomics of active microbial populations with stable-isotope probing. Curr Opin Biotechnol 41:1–8

Creppy A, Clément E, Douarche C, Veronica DAM, Auradou H (2019) Effect of motility on the transport of bacterial populations through porous medium. Phys Rev Fluids 4:013102

Cumsille P, Asenjo JA, Conca C (2014) A novel model for biofilm growth and its resolution by using the hybrid immersed interface-level set method. Comput Math Appl 67:34–51

Cupples AM, Spormann AM, McCarty PL (2004) Vinyl chloride and cis-dichloroethene dechlorination kinetics and microorganism growth under substrate limiting conditions. Environ Sci Technol 38:1102–1107

Dale AW, Regnier P, Van Cappellen P (2006) Bioenergetic controls on anaerobic oxidation of methane (AOM) in coastal marine sediments: A theoretical analysis. Am J Sci 306:246–294

Dale AW, Van Cappellen P, Aguilera DR, Regnier P (2008a) Methane efflux from marine sediments in passive and active margins: Estimations from bioenergetic reaction–transport simulations. Earth Planet Sci Lett 265:329–344

Dale AW, Regnier P, Knab NJ, Jørgensen BB, Van Cappellen P (2008b) Anaerobic oxidation of methane (AOM) in marine sediments from the Skagerrak (Denmark): II. Reaction-transport modeling. Geochim Cosmochim Acta 72:2880–2894

Dale AW, Aguilera DR, Regnier P, Fossing H, Knab NJ, Jorgensen BB (2008c) Seasonal dynamics of the depth and rate of anaerobic oxidation of methane in Aarhus Bay (Denmark) sediments. J Mar Res 66:127–155

Dale AW, Sommer S, Haeckel M, Wallmann K, Linke P, Wegener G, Pfannkuche O (2010) Pathways and regulation of carbon, sulfur and energy transfer in marine sediments overlying methane gas hydrates on the Opouawe Bank (New Zealand). Geochim Cosmochim Acta 74:5763–5784

De Biase C, Carminati A, Oswald SE, Thullner M (2013) Numerical modeling analysis of VOC removal processes in different aerobic vertical flow systems for groundwater remediation. J Contam Hydrol 154:53–69

de Kerchove AJ, Elimelech M (2008) Bacterial swimming motility enhances cell deposition and suface coverage. Environ Sci Technol 42:4371–4377

Dechesne A, Owsianiak M, Bazire A, Grundmann GL, Binning PJ, Smets BF (2010) Biodegradation in a partially saturated sand matrix: compounding effects of water content, bacterial spatial distribution, and motility. Environ Sci Technol 44:2386–2392

Dechesne A, Pallud C, Debouzie D, Flandrois JP, Vogel TM, Gaudet JP, Grundmann GL (2003) A novel method for characterizing the microscale 3D spatial distribution of bacteria in soil. Soil Biol Biochem 35:1537–1546

del Giorgio PA, Cole JJ (1998) Bacterial growth efficiency in natural aquatic systems. Ann Rev Ecol System 29:503–541

DeNiro MJ, Epstein S (1978) Influence ofdiet on the distribution of carbon isotopes in animals. Geochim Cosmochim Acta 42:495–506

Dorer C, Vogt C, Kleinsteuber S, Stams AJ, Richnow HH (2014) Compound-specific isotope analysis as a tool to characterize biodegradation of ethylbenzene. Environ Sci Technol 48:9122–9132

Druhan JL, Steefel CI, Conrad ME, DePaolo DJ (2014) A large column analog experiment of stable isotope variations during reactive transport: I. A comprehensive model of sulfur cycling and δ^{34}S fractionation. Geochim Cosmochim Acta 124:366–393

Druhan JL, Winnick MJ, Thullner M (2019) Stable isotope fractionation by transport and transformation. Rev Mineral Geochem 85:239–264

Dupin HJ, Kitanidis PK, McCarty PL (2001) Simulations of two-dimensional modeling of biomass aggregate growth in network models. Water Resour Res 37:2981–2994

Durot M, Bourguignon PY, Schachter V (2009) Genome-scale models of bacterial metabolism: reconstruction and applications. FEMS Microbiol Rev 33:164–190

Dworkin J, Shah IM (2010) Exit from dormancy in microbial organisms. Nat Rev Microbiol 8:890–896

Ebigbo A, Helmig R, Cunningham AB, Class H, Gerlach R (2010) Modelling biofilm growth in the presence of carbon dioxide and water flow in the subsurface. Adv Water Resour 33:762–781

Ebigbo A, Phillips A, Gerlach R, Helmig R, Cunningham AB, Class H, Spangler LH (2012) Darcy-scale modeling of microbially induced carbonate mineral precipitation in sand columns. Water Resour Res 48:W07519

Ebrahimi AN, Or D (2014) Microbial dispersal in unsaturated porous media: Characteristics of motile bacterial cell motions in unsaturated angular pore networks. Water Resour Res 50:7406–7429

Ebrahimi A, Or D (2015) Hydration and diffusion processes shape microbial community organization and function in model soil aggregates. Water Resour Res 51:9804–9827

Elsner M (2010) Stable isotope fractionation to investigate natural transformation mechanisms of organic contaminants: principles, prospects and limitations. J Environ Monit 12:2005–2031

Epstein SS (2009) Microbial awakenings. Nature 457:1083

Fang Y, Wilkins MJ, Yabusaki SB, Lipton MS, Long PE (2012) Evaluation of a genome-scale in silico metabolic model for Geobacter metallireducens by using proteomic data from a field biostimulation experiment. Appl Environ Microbiol 78:8735–8742

Fang Y, Scheibe TD, Mahadevan R, Garg S, Long PE, Lovley DR (2011) Direct coupling of a genome-scale microbial in silico model and a groundwater reactive transport model. J Contam Hydrol 122:96–103

Faust K, Raes J (2012) Microbial interactions: from networks to models. Nat Rev Microbiol 10:538–550

Ferrer J, Prats C, Lopez D (2008) Individual-based modelling: an essential tool for microbiology. J Biol Phys 34:19–37

Fischer A, Theuerkorn K, Stelzer N, Gehre M, Thullner M, Richnow HH (2007) Applicability of stable isotope fractionation analysis for the characterization of benzene biodegradation in a BTEX-contaminated aquifer. Environ Sci Technol 41:3689–3696

Ford RM, Harvey RW (2007) Role of chemotaxis in the transport of bacteria through saturated porous media. Adv Water Resour 30:1608–1617

Franzosa EA, Hsu T, Sirota-Madi A, Shafquat A, Abu-Ali G, Morgan XC, Huttenhower C (2015) Sequencing and beyond: integrating molecular 'omics' for microbial community profiling. Nat Rev Microbiol 13:360–372

Freilich S, Zarecki R, Eilam O, Segal ES, Henry CS, Kupiec M, Gophna U, Sharan R, Ruppin E (2011) Competitive and cooperative metabolic interactions in bacterial communities. Nat Commun 2:589

Froelich PN, Klinkhammer GP, Bender ML, Luedtke NA, Heath GR, Cullen D, Dauphin P, Hammond D, Hartman B, Maynard V (1979) Early oxidation of organic matter in pelagic sediments of the eastern equatorial Atlantic: suboxic diagenesis. Geochim Cosmochim Acta 43:1075–1090

Furuno S, Foss S, Wild E, Jones KC, Semple KT, Harms H, Wick LY (2012) Mycelia promote active transport and spatial dispersion of polycyclic aromatic hydrocarbons. Environ Sci Technol 46:5463–5470

Geyer KM, Kyker-Snowman E, Grandy AS, Frey SD (2016) Microbial carbon use efficiency: accounting for population, community, and ecosystem-scale controls over the fate of metabolized organic matter. Biogeochemistry 127:173–188

Gharasoo M, Centler F, Fetzer I, Thullner M (2014) How the chemotactic characteristics of bacteria can determine their population patterns. Soil Biol Biochem 69:346–358

Gharasoo M, Centler F, Regnier P, Harms H, Thullner M (2012) A reactive transport modeling approach to simulate biogeochemical processes in pore structures with pore-scale heterogeneities. Environ Modell Software 30:102–114

Gharasoo M, Centler F, Van Cappellen P, Wick LY, Thullner M (2015) Kinetics of substrate biodegradation under the cumulative effects of bioavailability and self-inhibition. Environ Sci Technol 49:5529–5537

Ginn TR (1999) On the distribution of multicomponent mixtures over generalized exposure time in subsurface flow and reactive transport: Foundations, and formulations for groundwater age, chemical heterogeneity, and biodegradation. Water Resour Res 35:1395–1407

Ginn TR, Wood BD, Nelson KE, Scheibe TD, Murphy EM, Clement TP (2002) Processes in microbial transport in the natural subsurface. Adv Water Resour 25:1017–1042

Golfier F, Wood BD, Orgogozo L, Quintard M, Buès M (2009) Biofilms in porous media: Development of macroscopic transport equations via volume averaging with closure for local mass equilibrium conditions. Adv Water Resour 32:463–485

Gottstein W, Olivier BG, Bruggeman FJ, Teusink B (2016) Constraint-based stoichiometric modelling from single organisms to microbial communities. J R Soc Interface 13: 20160627

Gras A, Ginovart M, Valls J, Baveye PC (2011) Individual-based modelling of carbon and nitrogen dynamics in soils: Parameterization and sensitivity analysis of microbial components. Ecol Modell 222:1998–2010

Griebler C, Malard F, Lefebure T (2014) Current developments in groundwater ecology-from biodiversity to ecosystem function and services. Curr Opin Biotechnol 27:159–167

Griebler C, Mindl B, Slezak D, Geiger-Kaiser M (2002) Distribution patterns of attached and suspended bacteria in pristine and contaminated shallow aquifers studied with an *in situ* sediment exposure microcosm. Aquatic Microbial Ecology 28:117–129

Grimm V, Calabrese J (2011) What is resilience? A short introduction. *In:* Viability and Resilience of Complex Systems. Deffuant G, Gilbert N, (eds). Springer, Berlin, Germany, p 1-15

Hanemaaijer M, Röling WF, Olivier BG, Khandelwal RA, Teusink B, Bruggeman FJ (2015) Systems modeling approaches for microbial community studies: from metagenomics to inference of the community structure. Front Microbiol 6:213

Hanzel J, Thullner M, Harms H, Wick LY (2012) Walking the tightrope of bioavailability: growth dynamics of PAH degraders on vapour-phase PAH. Microbial Biotechnol 5:79–86

Harcombe WR, Riehl WJ, Dukovski I, Granger BR, Betts A, Lang AH, Bonilla G, Kar A, Leiby N, Mehta P, Marx CJ (2014) Metabolic resource allocation in individual microbes determines ecosystem interactions and spatial dynamics. Cell Rep 7:1104–1115

Harms H, Schlosser D, Wick LY (2011) Untapped potential: exploiting fungi in bioremediation of hazardous chemicals. Nat Rev Microbiol 9:177–192

Harold FM (1986) The Vital Force: A Study of Bioenergetics. W. H. Freeman, New York, NY, USA

Hartnett HE, Keil RG, Hedges JI, Devol AH (1998) Influence of oxygen exposure time on organic carbon preservation in continental margin sediments. Nature 391:572–574

Haws NW, Ball WP, Bouwer EJ (2006) Modeling and interpreting bioavailability of organic contaminant mixtures in subsurface environments. J Contam Hydrol 82:255–292

Heijnen JJ, Van Dijken JP (1992) In search of a thermodynamic description of biomass yields for the chemotropic growth of microorganisms. Biotechnol Bioeng 39:833–858

Heijnen JJ, van Loosdrecht MCM, Tijhuis L (1992) A black box mathematical model to calculate auto- and heterotrophic biomass yields based on Gibbs energy dissipation. Biotechnol Bioeng 40:1139–1154

Hellweger FL, Clegg RJ, Clark JM, Plugge CM, Kreft JU (2016) Advancing microbial sciences by individual-based modelling. Nat Rev Microbiol 14:461–471

Henrichs SM (2005) Organic matter in coastal marine sediments. *In:* The Global Coastal Ocean: Multiscale Interdisciplinary Processes. Vol 13. Robinson AR, Brink KH, (eds). Harvard University Press, Boston, MA, USA, p 129–162

Herbert D (1958) Some principles of continuous culture. *In:* Recent Progress in Microbiology. Tunevall E, (ed) Almqvist and Wiksell, Stockholm, Sweden, p 381–396

Hermansson M (1999) The DLVO theory in microbial adhesion. Colloids and Surfaces B-Biointerfaces 14:105–119

Hesse F, Radu FA, Thullner M, Attinger S (2009) Upscaling of the advection–diffusion–reaction equation with Monod reaction. Adv Water Resour 32:1336–1351

Hesse F, Harms H, Attinger S, Thullner M (2010) Linear Exchange Model for the Description of Mass Transfer Limited Bioavailability at the Pore Scale. Environ Sci Technol 44:2064–2071

Hilpert M (2005) Lattice-Boltzmann model for bacterial chemotaxis. J Math Biol 51:302–332

Hoehler TM, Jørgensen BB (2013) Microbial life under extreme energy limitation. Nat Rev Microbiol 11:83–94

Hoehler TM, Alperin MJ, Albert DB, Martens CS (1994) Field and laboratory studies of methane oxidation in an anoxic marine sediment—evidence for a methanogen—sulfate reducer consortium. Global Biogeochem Cycles 8:451–463

Hommel J, Coltman E, Class H (2018) porosity–permeability relations for evolving pore space: a review with a focus on (bio-)geochemically altered porous media. Transp Porous Media 124:589–629

Huang Y, Guenet B, Ciais P, Janssens IA, Soong JL, Wang Y, Goll D, Blagodatskaya E, Huang Y (2018) ORCHIMIC (v1.0), a microbe-mediated model for soil organic matter decomposition. Geosci Model Develop 11:2111–2138

Hülse D, Arndt S, Daines S, Regnier P, Ridgwell A (2018) OMEN-SED 1.0: a novel, numerically efficient organic matter sediment diagenesis module for coupling to Earth system models. Geosci Model Develop 11:2649–2689

Iltis GC, Armstrong RT, Jansik DP, Wood BD, Wildenschild D (2011) Imaging biofilm architecture within porous media using synchrotron-based X-ray computed microtomography. Water Resour Res 47: W02601

Jayathilake PG, Gupta P, Li B, Madsen C, Oyebamiji O, González-Cabaleiro R, Rushton S, Bridgens B, Swailes D, Allen B, McGough AS (2017) A mechanistic Individual-based Model of microbial communities. PloS one 12:e0181965

Jehmlich N, Vogt C, Lünsmann V, Richnow HH, von Bergen M (2016) Protein-SIP in environmental studies. Curr Opin Biotechnol 41:26–33

Jin QS, Bethke CM (2002) Kinetics of electron transfer through the respiratory chain. Biophys J 83:1797–1808

Jin Q, Bethke CM (2003) A new rate law describing microbial respiration. Appl Environ Microbiol 69:2340–2348

Jin Q, Bethke CM (2005) Predicting the rate of microbial respiration in geochemical environments. Geochim Cosmochim Acta 69:1133–1143

Jin Q, Bethke CM (2007) The thermodynamics and kinetics of microbial metabolism. Am J Sci 307:643–677

Jin Q, Bethke CM (2009) Cellular energy conservation and the rate of microbial sulfate reduction. Geology 37:1027–1030

Jin Q, Roden EE, Giska JR (2013) Geomicrobial kinetics: extrapolating laboratory studies to natural environments. Geomicrobiol J 30:173–185

Joergensen RG, Wichern F (2018) Alive and kicking: Why dormant soil microorganisms matter. Soil Biol Biochem 116:419–430

Johnsen AR, Wick LY, Harms H (2005) Principles of microbial PAH-degradation in soil. Environ Pollut 133:71–84

Jones SE, Lennon JT (2010) Dormancy contributes to the maintenance of microbial diversity. PNAS 107:5881–5886

Jover LF, Cortez MH, Weitz JS (2013) Mechanisms of multi-strain coexistence in host-phage systems with nested infection networks. J Theor Biol 332:65–77

Kapellos GE, Alexiou TS, Payatakes AC (2007) Hierarchical simulator of biofilm growth and dynamics in granular porous materials. Adv Water Resour 30:1648–1667

Karp PD, Ivanova N, Krummenacker M, Kyrpides N, Latendresse M, Midford P, Ong WK, Paley S, Seshadri R (2019) A comparison of microbial genome web portals. Frontiers Microbiol 10:208

Kearns DB (2010) A field guide to bacterial swarming motility. Nat Rev Microbiol 8:634–644

Keller EF, Segel LA (1971) Model for chemotaxis. J Theor Biol 30:225–234

Keymer JE, Galajda P, Muldoon C, Park S, Austin RH (2006) Bacterial metapopulations in nanofabricated landscapes. PNAS 103:17290–17295

Kim M, Or D (2016) Individual-based model of microbial life on hydrated rough soil surfaces. PloS one 11:e0147394

Kim TY, Sohn SB, Kim YB, Kim WJ, Lee SY (2012) Recent advances in reconstruction and applications of genome-scale metabolic models. Curr Opin Biotechnol 23:617–623

Kindler R, Miltner A, Thullner M, Richnow H-H, Kästner M (2009) Fate of bacterial biomass derived fatty acids in soil and their contribution to soil organic matter. Org Geochem 40:29–37

King EL, Tuncay K, Ortoleva P, Meile C (2009) In silico Geobacter sulfurreducens metabolism and its representation in reactive transport models. Appl Environ Microbiol 75:83–92

Klitgord N, Segre D (2010) Environments that induce synthetic microbial ecosystems. PLoS Comput Biol 6:e1001002

Kohlmeier S, Smits THM, Ford RM, Keel C, Harms H, Wick LY (2005) Taking the fungal highway: mobilization of pollutant-degrading bacteria by fungi. Environ Sci Technol 39:4640–4646

König S, Worrich A, Centler F, Wick LY, Miltner A, Kästner M, Thullner M, Frank K, Banitz T (2017) Modelling functional resilience of microbial ecosystems: Analysis of governing processes. Environ Modell Software 89:31–39

König S, Worrich A, Banitz T, Harms H, Kästner M, Miltner A, Wick LY, Frank K, Thullner M, Centler F (2018a) Functional resistance to recurrent spatially heterogeneous disturbances is facilitated by increased activity of surviving bacteria in a virtual ecosystem. Frontiers Microbiol 9:734

König S, Worrich A, Banitz T, Centler F, Harms H, Kästner M, Miltner A, Wick LY, Thullner M, Frank K (2018b) Spatiotemporal disturbance characteristics determine functional stability and collapse risk of simulated microbial ecosystems. Sci Rep 8:9488

Kreft JU, Picioreanu C, Wimpenny JWT, van Loosdrecht MCM (2001) Individual-based modelling of biofilms. Microbiology-Sgm 147:2897–2912

Kreft JU, Plugge CM, Grimm V, Prats C, Leveau JH, Banitz T, Baines S, Clark J, Ros A, Klapper I, Topping CJ (2013) Mighty small: Observing and modeling individual microbes becomes big science. PNAS 110:18027–18028

Krumins V, Gehlen M, Arndt S, Van Cappellen P, Regnier P (2013) Dissolved inorganic carbon and alkalinity fluxes from coastal marine sediments: model estimates for different shelf environments and sensitivity to global change. Biogeosciences 10:371–398

Kuntal BK, Gadgil C, Mande SS (2019) Web-gLV: A web based platform for Lotka–Volterra based modeling and simulation of microbial populations. Frontiers Microbiol 10:288

LaRowe DE, Amend JP (2015a) Power limits for microbial life. Frontiers Microbiol 6:718

LaRowe DE, Amend JP (2015b) Catabolic rates, population sizes and doubling/replacement times of microorganisms in natural settings. Am J Sci 315:167–203

LaRowe DE, Amend JP (2016) The energetics of anabolism in natural settings. ISME J 10:1285–1295

LaRowe DE, Helgeson HC (2007) Quantifying the energetics of metabolic reactions in diverse biogeochemical systems: electron flow and ATP synthesis. Geobiology 5:153–168

LaRowe DE, Van Cappellen P (2011) Degradation of natural organic matter: A thermodynamic analysis. Geochim Cosmochim Acta 75:2030–2042

LaRowe DE, Dale AW, Regnier P (2008) A thermodynamic analysis of the anaerobic oxidation of methane in marine sediments. Geobiology 6:436–449

LaRowe DE, Dale AW, Amend JP, Van Cappellen P (2012) Thermodynamic limitations on microbially catalyzed reaction rates. Geochim Cosmochim Acta 90:96–109

LaRowe DE, Dale AW, Aguilera DR, L'Heureux I, Amend JP, Regnier P (2014) Modeling microbial reaction rates in a submarine hydrothermal vent chimney wall. Geochim Cosmochim Acta 124:72–97

Larsen PE, Collart FR, Field D, Meyer F, Keegan KP, Henry CS, McGrath J, Quinn J, Gilbert JA (2011) Predicting Relative Metabolic Turnover (PRMT): determing metabolic turnover from a coastal marine metagenomic dataset. Microbial Inf Experiment 1:4

Lehman RM, Colwell FS, Bala GA (2001) Attached and unattached microbial communities in a simulated basalt aquifer under fracture- and porous-flow conditions. Appl Environ Microbiol 67:2799–2809

Lennon JT, Jones SE (2011) Microbial seed banks: the ecological and evolutionary implications of dormancy. Nat Rev Microbiol 9:119–130

Leveau JHJ, Hellweger FL, Kreft JU, Prats C, Zhang W (2018) Editorial: the individual microbe: single-cell analysis and agent-based modelling. Front Microbiol 9:2825

Lewis NE, Nagarajan H, Palsson BO (2012) Constraining the metabolic genotype-phenotype relationship using a phylogeny of in silico methods. Nat Rev Microbiol 10:291–305

Li L, Steefel CI, Williams KH, Wilkins MJ, Hubbard SS (2009) Mineral transformation and biomass accumulation associated with uranium bioremediation at Rifle, Colorado. Environ Sci Technol 43:5429–5435

Li L, Maher K, Navarre-Sitchler A, Druhan J, Meile C, Lawrence C, Moore J, Perdrial J, Sullivan P, Thompson A, Jin L (2017) Expanding the role of reactive transport models in critical zone processes. Earth Sci Rev 165:280–301

Lichtner PC, Steefell CI, Oelkers EH (eds) (1996) Reactive Transport in Porous Media. Rev Mineral Geochem Vol 34, MSA

Lidicker WZ (1979) A clarifiaction of interactions in ecological systems. Bioscience 29:475–477

Lieven C, Petersen LAH, Jorgensen SB, Gernaey KV, Herrgard MJ, Sonnenschein N (2018) A genome-scale metabolic model for *methylococcus capsulatus* (bath) suggests reduced efficiency electron transfer to the particulate methane monooxygenase. Front Microbiol 9:2947

Lipson DA (2015) The complex relationship between microbial growth rate and yield and its implications for ecosystem processes. Front Microbiol 6:615

Lotka AJ (1925) Elements of Physical Biology. Williams and Wilkins, Baltimore, MD, USA

Louca S, Hawley AK, Katsev S, Torres-Beltran M, Bhatia MP, Kheirandish S, Michiels CC, Capelle D, Lavik G, Doebeli M, Crowe SA (2016) Integrating biogeochemistry with multiomic sequence information in a model oxygen minimum zone. PNAS 113:E5925-E5933

Lueders T, Dumont MG, Bradford L, Manefield M (2016) RNA-stable isotope probing: from carbon flow within key microbiota to targeted transcriptomes. Curr Opin Biotechnol 41:83–89

Magnúsdóttir S, Heinken A, Kutt L, Ravcheev DA, Bauer E, Noronha A, Greenhalgh K, Jäger C, Baginska J, Wilmes P, Fleming RM (2017) Generation of genome-scale metabolic reconstructions for 773 members of the human gut microbiota. Nat Biotechnol 35:81–89

Mahabadi N, Zheng X, Yun TS, van Paassen L, Jang J (2018) Gas bubble migration and trapping in porous media: pore-scale simulation. J Geophys Res: Solid Earth 123:1060–1071

Mahadevan R, Bond DR, Butler JE, Esteve-Nunez A, Coppi MV, Palsson BO, Schilling CH, Lovley DR (2006) Characterization of metabolism in the Fe(III)-reducing organism *Geobacter sulfurreducens* by constraint-based modeling. Appl Environ Microbiol 72:1558–1568

Mahadevan R, Edwards JS, Doyle FJ, III (2002) Dynamic flux balance analysis of diauxic growth in *Escherichia coli*. Biophys J 83:1331–1340

Manzoni S, Schimel JP, Porporato A (2012a) Responses of soil microbial communities to water stress: results from a meta-analysis. Ecology 93:930–938

Manzoni S, Taylor P, Richter A, Porporato A, Agren GI (2012b) Environmental and stoichiometric controls on microbial carbon-use efficiency in soils. New Phytol 196:79–91

Mauclaire L, Pelz O, Thullner M, Abraham W-R, Zeyer J (2003) Assimilation of toluene carbon along a bacteria-protist food chain determined by ^{13}C-enrichment of biomarker fatty acids. J Microbiol Methods 55:635–649

May RM (1974) Stability and Complexity in Model Ecosystems. Princeton University Press, Princeton, NJ, USA

McCarty PL (2007) Thermodynamic electron equivalents model for bacterial yield prediction: modifications and comparative evaluations. Biotechnol Bioeng 97:377–388

McCollom TM, Amend JP (2005) A thermodynamic assessment of energy requirements for biomass synthesis by chemolithoautotrophic micro-organisms in oxic and anoxic environments. Geobiology 3:135–144

McGuire KL, Treseder KK (2010) Microbial communities and their relevance for ecosystem models: Decomposition as a case study. Soil Biol Biochem 42:529–535

Meckenstock RU, Morasch B, Griebler C, Richnow HH (2004) Stable isotope fractionation analysis as a tool to monitor biodegradation in contaminated aquifers. J Contam Hydrol 75:215–255

Meile C, Scheibe TD (2018) Reactive transport modeling and biogeochemical cycling. *In*: Reactive Transport Modeling: Applications in Subsurface Energy and Environmental Problems. Xiao Y, Whitaker F, Xu T, (eds). Wiley, Hoboken, NJ, USA, 2018, p 485–510

Meile C, Scheibe TD (2019) Reactive transport modeling of microbial dynamics. Elements 15:111–116

Mellage A, Eckert D, Grosbacher M, Inan AZ, Cirpka OA, Griebler C (2015) Dynamics of suspended and attached aerobic toluene degraders in small-scale flow-through sediment systems under growth and starvation conditions. Environ Sci Technol 49:7161–7169

Meyer AH, Penning H, Elsner M (2009) C and N isotope fractionation suggests similar mechanisms of microbial atrazine transformation despite involvement of different Enzymes (AtzA and TrzN). Environ Sci Technol 43:8079–8085

Meysman FJ, Bruers S (2010) Ecosystem functioning and maximum entropy production: a quantitative test of hypotheses. Philos Trans R Soc Lond B Biol Sci 365:1405–1416

Meysman FJR, Middelburg JJ, Herman PMJ, Heip CHR (2003) Reactive transport in surface sediments. I. Model complexity and software quality. Comput Geosci 29:291–300

Mittal N, Budrene EO, Brenner MP, van Oudenaarden A (2003) Motility of *Escherichia coli* cells in clusters formed by chemotactic aggregation. PNAS 100:13259–13263

Monga O, Garnier P, Pot V, Coucheney E, Nunan N, Otten W, Chenu C (2014) Simulating microbial degradation of organic matter in a simple porous system using the 3-D diffusion-based model MOSAIC. Biogeosciences 11:2201–2209

Monk J, Palsson BO (2014) Predicting microbial growth. Science 344:1448–1449

Monod J (1949) The growth of bacterial cultures. Ann Rev Microbiol 3:371–394

Mounier J, Monnet C, Vallaeys T, Arditi R, Sarthou AS, Helias A, Irlinger F (2008) Microbial interactions within a cheese microbial community. Appl Environ Microbiol 74:172–181

Müller S, Hiller K (2013) From multi-omics to basic structures of biological systems. Curr Opin Biotechnol 24:1–3

Murphy EM, Ginn TR (2000) Modeling microbial processes in porous media. Hydrogeol J 8:142–158

Neill C, Gignoux J (2006) Soil organic matter decomposition driven by microbial growth: A simple model for a complex network of interactions. Soil Biol Biochem 38:803–811

Nelson KE, Ginn TR (2001) Theoretical investigation of bacterial chemotaxis in porous media. Langmuir 17:5636–5645

Nilsen V, Wyller JA, Heistad A (2012) Efficient incorporation of microbial metabolic lag in subsurface transport modeling. Water Resour Res 48:W09519

Nunan N, Wu KJ, Young IM, Crawford JW, Ritz K (2003) Spatial distribution of bacterial communities and their relationships with the micro-architecture of soil. FEMS Microbiol Ecol 44:203–215

O'Brien EJ, Monk JM, Palsson BO (2015) Using Genome-scale models to predict biological capabilities. Cell 161:971–987

Or D, Smets BF, Wraith JM, Dechesne A, Friedman SP (2007) Physical constraints affecting bacterial habitats and activity in unsaturated porous media—a review. Adv Water Resour 30:1505–1527

Orcutt B, Meile C (2008) Constraints on mechanisms and rates of anaerobic oxidation of methane by microbial consortia: process-based modeling of ANME-2 archaea and sulfate reducing bacteria interactions. Biogeosciences 5:1587–1599

Orgogozo L, Golfier F, Buès MA, Quintard M, Koné T (2013) A dual-porosity theory for solute transport in biofilm-coated porous media. Adv Water Resour 62:266–279

Orth JD, Thiele I, Palsson BO (2010) What is flux balance analysis? Nat Biotechnol 28:245–248

Pagel H, Ingwersen J, Poll C, Kandeler E, Streck T (2014) Micro-scale modeling of pesticide degradation coupled to carbon turnover in the detritusphere: model description and sensitivity analysis. Biogeochemistry 117:185–204

Pagel H, Poll C, Ingwersen J, Kandeler E, Streck T (2016) Modeling coupled pesticide degradation and organic matter turnover: From gene abundance to process rates. Soil Biol Biochem 103:349–364

Pancost RD, Sinninghe Damsté JS (2003) Carbon isotopic compositions of prokaryotic lipids as tracers of carbon cycling in diverse settings. Chem Geol 195:29–58

Paraska DW, Hipsey MR, Salmon SU (2014) Sediment diagenesis models: Review approaches, challenges and opportunities. Environ Modell Software 61:297–325

Park S, Wolanin PM, Yuzbashyan EA, Lin H, Darnton NC, Stock JB, Silberzan P, Austin R (2003) Influence of topology on bacterial social interaction. PNAS 100:13910–13915

Parton WJ, Stewart JWB, Cole CV (1988) Dynamics of C, N, P and S in grassland soils: a model. Biogeochemistry 5:109–131

Pedit JA, Marx RB, Miller CT, Aitken MD (2002) Quantitative analysis of experiments on bacterial chemotaxis to naphthalene. Biotechnol Bioeng 78:626–634

Pelz O, Chatzinotas A, Zarda-Hess A, Abraham W-R, Zeyer J (2001) Tracing toluene-assimilating sulfate-reducing bacteria using ^{13}C-incorporation in fatty acids and whole-cell hybridization. FEMS Microbiol Ecol 38:123–131

Perez-Garcia O, Lear G, Singhal N (2016) Metabolic network modeling of microbial interactions in natural and engineered environmental systems. Front Microbiol 7:673

Peyton BM, Skeen RS, Hooker BS, Lundman RW, Cunningham AB (1995) Evaluation of bacterial detachment rates in porous-media. Appl Biochem Biotechnol 51–2:785–797

Picioreanu C, Kreft JU, van Loosdrecht MCM (2004) Particle-based multidimensional multispecies Biofilm model. Appl Environ Microbiol 70:3024–3040

Pirt SJ (1965) The maintenance energy of bacteria in growing cultures. Proc R Soc Ser B 163:224–231

Pombo S, Pelz O, Schroth MH, Zeyer J (2002) Field-scale ^{13}C-labeling of phospholipid fatty acids (PLFA) and dissolved inorganic carbon: tracing acetate assimilation and mineralization in a petroleum hydrocarbon-contaminated aquifer. FEMS Microbiol Ecol 41:259–267

Qin C-Z, Hassanizadeh SM (2015) Pore-network modeling of solute transport and biofilm growth in porous media. Transp Porous Media 110:345–367

Raivonen M, Smolander S, Backman L, Susiluoto J, Aalto T, Markkanen T, Mäkelä J, Rinne J, Peltola O, Aurela M, Lohila A (2017) HIMMELI v1.0: HelsinkI Model of MEthane buiLd-up and emIssion for peatlands. Geosci Model Develop 10:4665–4691

Ramkrishna D, Song H-S (2012) Dynamic models of metabolism: Review of the cybernetic approach. AIChE J 58:986–997

Reed JL (2012) Shrinking the metabolic solution space using experimental datasets. PLoS ComputBiol 8:e1002662

Reed DC, Algar CK, Huber JA, Dick GJ (2014) Gene-centric approach to integrating environmental genomics and biogeochemical models. PNAS 111:1879–1884

Regnier P, Dale AW, Pallud C, van Lith S, Bonneville S, Hyacinthe C, Thullner M, Laverman AM, Van Cappellen P (2005) Incorporating geomicrobial processes in reactive transport models of subsurface environments. In: Reactive Transport in Soil and Groundwater. Nuetzmann G, Viotti P, Aggaard P, (eds). Springer, Berlin, p 107–126

Regnier P, Dale AW, Arndt S, LaRowe DE, Mogollon J, Van Cappellen P (2011) Quantitative analysis of anaerobic oxidation of methane (AOM) in marine sediments: A modeling perspective. Earth Sci Rev 106:105–130

Resat H, Bailey V, McCue LA, Konopka A (2012) Modeling microbial dynamics in heterogeneous environments: growth on soil carbon sources. Microb Ecol 63:883–897

Richter S, Fetzer I, Thullner M, Centler F, Dittrich P (2015) Towards rule-based metabolic databases: a requirement analysis based on KEGG. Int J Data Min Bioinf 13:289–319

Rittmann BE, McCarty PL (1980) Model of steady-state-biofilm kinetics. Biotechnol Bioeng 22:2343–2357

Rittmann BE, McCarty PL (2001) Environmental Biotechnology: Principles and Applications. McGraw-Hill

Rocca JD, Simonin M, Blaszczak JR, Ernakovich JG, Gibbons SM, Midani FS, Washburne AD (2019) The microbiome stress project: toward a global meta-analysis of environmental stressors and their effects on microbial communities. Front Microbiol 9:3272

Roden EE, Jin Q (2011) Thermodynamics of microbial growth coupled to metabolism of glucose, ethanol, short-chain organic acids, and hydrogen. Appl Environ Microbiol 77:1907–1909

Roels JA (1980) Application of macroscopic principles to microbial metabolism. Biotechnol Bioeng 22:2457–2514

Russell JB, Cook GM (1995) Energetics of bacterial growth: balance of anabolic and catabolic reactions. Microbiol Rev 59:48–62

Saragosti J, Calvez V, Bournaveas N, Buguin A, Silberzan P, Perthame B (2010) Mathematical description of bacterial traveling pulses. PLoS Comput Biol 6

Schäfer D, Schäfer W, Kinzelbach W (1998a) Simulation of reactive processes related to biodegradation in aquifers - 2. Model application to a column study on organic carbon degradation. J Contam Hydrol 31:187–209

Schäfer D, Schäfer W, Kinzelbach W (1998b) Simulation of reactive processes related to biodegradation in aquifers - 1. Structure of the three-dimensional reactive transport model. J Contam Hydrol 31:167–186

Schäfer W (2001) Predicting natural attenuation of xylene in groundwater using a numerical model. J Contam Hydrol 52:57–83

Schamfuss S, Neu TR, van der Meer JR, Tecon R, Harms H, Wick LY (2013) Impact of mycelia on the accessibility of fluorene to PAH-degrading bacteria. Environ Sci Technol 47:6908–6915

Scheibe TD, Mahadevan R, Fang Y, Garg S, Long PE, Lovley DR (2009) Coupling a genome-scale metabolic model with a reactive transport model to describe in situ uranium bioremediation. Microb Biotechnol 2:274–286

Schimel JP, Weintraub MN (2003) The implications of exoenzyme activity on microbial carbon and nitrogen limitation in soil: a theoretical model. Soil Biol Biochem 35:549–563

Schmidt SI, Kreft J-U, Mackay R, Picioreanu C, Thullner M (2018) Elucidating the impact of micro-scale heterogeneous bacterial distribution on biodegradation. Adv Water Resour 116:67–76

Semple KT, Doick KJ, Jones KC, Burauel P, Craven A, Harms H (2004) Defining bioavailabilty and bioaccesibility of contaminated soil and sediment is complicated. Environ Sci Technol 15:229A-231A

Shade A, Peter H, Allison SD, Baho D, Berga M, Bürgmann H, Huber DH, Langenheder S, Lennon JT, Martiny JB, Matulich KL (2012) Fundamentals of microbial community resistance and resilience. Frontiers Microbiol Vol 3:417

Sierra CA, Melghani S, Müller M (2015) Model structure and parameter identification of soil organic matter models. Soil Biol Biochem 90:197-203

Simoni SF, Harms H, Bosma TNP, Zehnder AJB (1998) Population heterogeneity affects transport of bacteria through sand columns at low flow rates. Environ Sci Technol 32:2100–2105

Simoni SF, Schäfer A, Harms H, Zehnder AJB (2001) Factors affecting mass transfer limited biodegradation in saturated porous media. J Contam Hydrol 50:99–120

Sinsabaugh RL, Manzoni S, Moorhead DL, Richter A (2013) Carbon use efficiency of microbial communities: stoichiometry, methodology and modelling. Ecol Lett 16:930–939

Six J, Frey SD, Thiet RK, Batten KM (2006) Bacterial and fungal contributions to carbon sequestration in agroecosystems. Soil Sci Soc Am J 70

Smeaton CM, Van Cappellen P (2018) Gibbs Energy Dynamic Yield Method (GEDYM): Predicting microbial growth yields under energy-limiting conditions. Geochim Cosmochim Acta 241:1–16

Soetaert K, Herman P (2009) A Practical Guide to Ecological Modelling: Using R as a Simulation Platform. Springer, London, UK

Song H-S, Cannon W, Beliaev A, Konopka A (2014) Mathematical modeling of microbial community dynamics: a methodological review. Processes 2:711–752

Stein RR, Bucci V, Toussaint NC, Buffie CG, Ratsch G, Pamer EG, Sander C, Xavier JB (2013) Ecological modeling from time-series inference: insight into dynamics and stability of intestinal microbiota. PLoS Comput Biol 9:e1003388

Stelzer N, Büning C, Pfeifer F, Dohrmann AB, Tebbe CC, Nijenhuis I, Kästner M, Richnow HH (2006) In situ microcosms to evaluate natural attenuation potentials in contaminated aquifers. Org Geochem 37:1394–1410

Sterner RW, Elser JJ (2002) Ecological Stoichiometry: The Biology of Elements from Molecules to the Biosphere. In Book Ecological Stoichiometry: The Biology of Elements from Molecules to the Biosphere. Editor, (ed)^(eds). Princeton University Press, Princeton, NJ, USA

Stolpovsky K, Martinez-Lavanchy P, Heipieper HJ, Van Cappellen P, Thullner M (2011) Incorporating dormancy in dynamic microbial community models. Ecol Modell 222:3092–3102

Stolpovsky K, Gharasoo M, Thullner M (2012) The impact of pore-size heterogeneities on the spatiotemporal variation of microbial metabolic activity in porous media. Soil Sci 177:98–110

Stolpovsky K, Fetzer I, Van Cappellen P, Thullner M (2016) Influence of dormancy on microbial competition under intermittent substrate supply: insights from model simulations. FEMS Microbiol Ecol 92:fiw071

Stolyar S, Van Dien S, Hillesland KL, Pinel N, Lie TJ, Leigh JA, Stahl DA (2007) Metabolic modeling of a mutualistic microbial community. Mol Syst Biol 3:92

Stumm W, Morgan JJ (1996) Aquatic Chemistry. Wiley-Interscience, New York, NY, USA

Stumpp C, Lawrence JR, Hendry MJ, Maloszewski P (2011) Transport and bacterial interactions of three bacterial strains in saturated column experiments. Environ Sci Technol 45:2116–2123

Suchomel BJ, Chen BM, Allen MB (1998) Network model of flow, transport and biofilm effects in porous media. Transp Porous Media 30:1–23

Sulman BN, Phillips RP, Oishi AC, Shevliakova E, Pacala SW (2014) Microbe-driven turnover offsets mineral-mediated storage of soil carbon under elevated CO_2. Nat Clim Change 4:1099–1102

Tang JY (2015) On the relationship between Michaelis–Menten kinetics, equilibrium chemistry approximation kinetics, and quadratic kinetics. Geosci Model Develop 8:3823–3835

Tang JY, Riley WJ (2013) A total quasi-steady-state formulation of substrate uptake kinetics in complex networks and an example application to microbial litter decomposition. Biogeosciences 10:8329–8351

Tang Y, Liu H (2017) Modeling multidimensional and multispecies biofilms in porous media. Biotechnol Bioeng 114:1679–1687

Tartakovsky AM, Scheibe TD, Meakin P (2009) Pore-scale model for reactive transport and biomass growth. J Porous Media 12:417–434

Tartakovsky GD, Tartakovsky AM, Scheibe TD, Fang Y, Mahadevan R, Lovley DR (2013) Pore-scale simulation of microbial growth using a genome-scale metabolic model: Implications for Darcy-scale reactive transport. Adv Water Resour 59:256–270

Tay JH, Liu QS, Liu Y (2001) The effects of shear force on the formation, structure and metabolism of aerobic granules. Appl Microbiol Biotechnol 57:227–233

Thullner M (2010) Comparison of bioclogging effects in saturated porous media within one- and two-dimensional flow systems. Ecol Eng 36:176–196

Thullner M, Baveye P (2008) Computational pore network modeling of the influence of biofilm permeability on bioclogging in porous media. Biotechnol Bioeng 99:1337–1351 4

Thullner M, Schäfer W (1999) Modeling of a field experiment on bioremediation of cholobenzenes in groundwater. Biorem J 3:247–267

Thullner M, Zeyer J, Kinzelbach W (2002) Influence of microbial growth on hydraulic properties of pore networks. Transp Porous Media 49:99–122

Thullner M, Schroth MH, Zeyer J, Kinzelbach W (2004) Modeling of a microbial growth experiment with bioclogging in a two-dimensional saturated porous media flow field. J Contam Hydrol 70:37–62

Thullner M, Van Cappellen P, Regnier P (2005) Modeling the impact of microbial activity on redox dynamics in porous media. Geochim Cosmochim Acta 69:5005–5019

Thullner M, Regnier P, Van Cappellen P (2007) Modeling microbially induced carbon degradation in redox-stratified subsurface environments: Concepts and open questions. Geomicrobiol J 24:139–155

Thullner M, Kampara M, Richnow HH, Harms H, Wick LY (2008) Impact of bioavailability restrictions on microbially induced stable isotope fractionation. 1. Theoretical calculation. Environ Sci Technol 42:6544–6551

Thullner M, Dale AW, Regnier P (2009) Global-scale quantification of mineralization pathways in marine sediments: A reaction-transport modeling approach. Geochem Geophys Geosyst 10:Q10012

Thullner M, Centler F, Richnow H-H, Fischer A (2012) Quantification of organic pollutant degradation in contaminated aquifers using compound specific stable isotope analysis—Review of recent developments. Org Geochem 42:1440–1460

Tien C, Payatakes AC (1979) Advances in deep bed filtration. AIChE J 25:737–759

Trapp S, Brock AL, Nowak K, Kastner M (2018) Prediction of the formation of biogenic nonextractable residues during degradation of environmental chemicals from biomass yields. Environ Sci Technol 52:663–672

Tufenkji N (2007) Modeling microbial transport in porous media: Traditional approaches and recent developments. Adv Water Resour 30:1455–1469

Valdés-Parada FJ, Porter ML, Narayanaswamy K, Ford RM, Wood BD (2009) Upscaling microbial chemotaxis in porous media. Adv Water Resour 32:1413–1428

Vallino JJ (2010) Ecosystem biogeochemistry considered as a distributed metabolic network ordered by maximum entropy production. Phil Trans R Soc Lond B Biol Sci 365:1417–1427

van Bodegom P (2007) Microbial maintenance: a critical review on its quantification. Microb Ecol 53:513–523

Van Breukelen BM (2007) Extending the Rayleigh equation to allow competing isotope fractionating pathways to improve quantification of biodegradation. Environ Sci Technol 41:4004–4010

van Breukelen BM, Griffioen J (2004) Biogeochemical processes at the fringe of a landfill leachate pollution plume: potential for dissolved organic carbon, Fe(II), Mn(II), NH_4, and CH_4 oxidation. J Contam Hydrol 73:181–205

Van Cappellen P, Gaillard JF (1996) Biogeochemical dynamics in aquatic sediments. Rev Mineral Geochem 34:335–376

Van Cappellen P, Gaillard J-F, Rabouille C (1993) Biogeochemical transformations in sediments: Kinetic models of early diagenesis. *In:* Interactions of C, N, P and S Biogeochemical Cycles and Global Change. Springer Berlin, Heidelberg, p 401–445

van de Velde S, Van Lancker V, Hidalgo-Martinez S, Berelson WM, Meysman FJR (2018) Anthropogenic disturbance keeps the coastal seafloor biogeochemistry in a transient state. Sci Rep 8:5582

van Loosdrecht MCM, Heijnen JJ, Eberl H, Kreft J, Picioreanu C (2002) Mathematical modelling of biofilm structures. Antonie Van Leeuwenhoek Int J Gen Mol Microbiol 81:245–256

VanBriesen JM (2002) Evaluation of methods to predict bacterial yield using thermodynamics. Biodegradation 13:171–190

Vanderborght JP, Wollast R, Loijens M, Regnier P (2002) Application of a transport-reaction model to the estimation of biogas fluxes in the Scheldt estuary. Biogeochemistry 59:207–237

Vereecken H, Schnepf A, Hopmans JW, Javaux M, Or D, Roose T, Vanderborght J, Young MH, Amelung W, Aitkenhead M, Allison SD (2016) Modeling soil processes: review, key challenges, and new perspectives. Vadose Zone J 15:vzj2015.09.0131

Vogel HJ, Bartke S, Daedlow K, Helming K, Kögel-Knabner I, Lang B, Rabot E, Russell D, Stößel B, Weller U, Wiesmeier M (2018) A systemic approach for modeling soil functions. Soil 4:83–92

Vogt C, Dorer C, Musat F, Richnow HH (2016a) Multi-element isotope fractionation concepts to characterize the biodegradation of hydrocarbons - from enzymes to the environment. Curr Opin Biotechnol 41:90–98

Vogt C, Lueders T, Richnow HH, Kruger M, von Bergen M, Seifert J (2016b) Stable isotope probing approaches to study anaerobic hydrocarbon degradation and degraders. J Mol Microbiol Biotechnol 26:195–210

Voltera V (1926) Variazioni e fluttuazioni del numro d'individui specie animali conviventi. Memorie della Reale Accademia Nazionale dei Lincei 2:31–113

von der Schulenburg DAG, Pintelon TRR, Picioreanu C, Van Loosdrecht MCM, Johns ML (2009) Three-dimensional simulations of biofilm growth in porous media. AIChE J 55:494–504

Wang G, Post WM (2012) A theoretical reassessment of microbial maintenance and implications for microbial ecology modeling. FEMS Microbiol Ecol 81:610–617

Wang G, Mayes MA, Gu L, Schadt CW (2014) Representation of dormant and active microbial dynamics for ecosystem modeling. PloS one 9:e89252

Wang G, Jagadamma S, Mayes MA, Schadt CW, Steinweg JM, Gu L, Post WM (2015) Microbial dormancy improves development and experimental validation of ecosystem model. ISME J 9:226–237

Wang YP, Jiang J, Chen-Charpentier BM, Agusto FB, Hastings A, Hoffman FM, Rasmussen M, Smith MJ, Todd-Brown KE, Wang Y, Xu X (2016) Responses of two nonlinear microbial models to warming and increased carbon input. Biogeosciences 13:887–902

Wania R, Ross I, Prentice IC (2010) Implementation and evaluation of a new methane model within a dynamic global vegetation model: LPJ-WHyMe v1.3.1. Geosci Model Develop 3:565–584

Watnick P, Kolter R (2000) Biofilm, city of microbes. J Bacteriol 182:2675–2679

Wegener G, Kellermann MY, Elvert M (2016) Tracking activity and function of microorganisms by stable isotope probing of membrane lipids. Curr Opin Biotechnol 41:43–52

Wieder WR, Bonan GB, Allison SD (2013) Global soil carbon projections are improved by modelling microbial processes. Nat Clim Change 3:909–912

Wieder WR, Grandy AS, Kallenbach CM, Bonan GB (2014) Integrating microbial physiology and physio-chemical principles in soils with the MIcrobial-MIneral Carbon Stabilization (MIMICS) model. Biogeosciences 11:3899–3917

Wirtz KW (2003) Control of biogeochemical cycling by mobility and metabolic strategies of microbes in the sediments: an integrated model study. FEMS Microbiol Ecol 46:295–306

Wood BD, Ginn TR, Dawson CN (1995) Effects of microbial metabolic lag in contaminant transport and biodegradation modeling. Water Resour Res 31:553–563

Wood BD, Radakovich K, Golfier F (2007) Effective reaction at a fluid–solid interface: Applications to biotransformation in porous media. Adv Water Resour 30:1630–1647

Worrich A, Wick LY, Banitz T (2018) Ecology of contaminant biotransformation in the mycosphere: role of transport processes. Adv Appl Microbiol 104:93–133

Worrich A, König S, Banitz T, Centler F, Frank K, Thullner M, Harms H, Miltner A, Wick LY, Kästner M (2016) Bacterial dispersal promotes biodegradation in heterogeneous systems exposed to osmotic stress. Frontiers Microbiol 7:1214

Worrich A, Stryhanyuk H, Musat N, König S, Banitz T, Centler F, Frank K, Thullner M, Harms H, Richnow HH, Miltner A (2017) Mycelium-mediated transfer of water and nutrients stimulates bacterial activity in dry and oligotrophic environments. Nat Commun 8:15472

Xavier JD, Picioreanu C, van Loosdrecht MCM (2005) A general description of detachment for multidimensional modelling of biofilms. Biotechnol Bioeng 91:651–669

Xiao Y, Whitakeer F, Xu T, Steefel C (eds) (2018) Reactive Transport Modeling: Applications in Subsurface Energy and Environmental Problems. John Wiley and Sons, Hoboken, NJ, USA

Yabusaki SB, Fang Y, Williams KH, Murray CJ, Ward AL, Dayvault RD, Waichler SR, Newcomer DR, Spane FA, Long PE (2011) Variably saturated flow and multicomponent biogeochemical reactive transport modeling of a uranium bioremediation field experiment. J Contam Hydrol 126:271–290

Yabusaki SB, Fang Y, Long PE, Resch CT, Peacock AD, Komlos J, Jaffe PR, Morrison SJ, Dayvault RD, White DC, Anderson RT (2007) Uranium removal from groundwater via in situ biostimulation: Field-scale modeling of transport and biological processes. J Contam Hydrol 93:216–235

Yao H, Chapman SJ, Thornton B, Paterson E (2014) ^{13}C PLFAs: a key to open the soil microbial black box? Plant Soil 392:3–15

Yarwood RR, Rockhold ML, Niemet MR, Selker JS, Bottomley PJ (2006) Impact of microbial growth on water flow and solute transport in unsaturated porous media. Water Resour Res 42:W10405

Yu T, Zhuang Q (2019) Quantifying global N_2O emissions from natural ecosystem soils using trait-based biogeochemistry models. Biogeosciences 16:207–222

Zhuang K, Izallalen M, Mouser P, Richter H, Risso C, Mahadevan R, Lovley DR (2011) Genome-scale dynamic modeling of the competition between *Rhodoferax* and *Geobacter* in anoxic subsurface environments. ISME J 5:305–316

Zomorrodi AR, Maranas CD (2012) OptCom: a multi-level optimization framework for the metabolic modeling and analysis of microbial communities. PLoS Comput Biol 8:e1002363

Zomorrodi AR, Islam MM, Maranas CD (2014) d-OptCom: Dynamic multi-level and multi-objective metabolic modeling of microbial communities. ACS Synth Biol 3:247–257

Reviews in Mineralogy & Geochemistry
Vol. 85 pp. 303-328, 2019
Copyright © Mineralogical Society of America

10

Understanding and Predicting Vadose Zone Processes

Bhavna Arora[1], Dipankar Dwivedi[1], Boris Faybishenko[1], Raghavendra B. Jana[2], Haruko M. Wainwright[1]

[1]Earth and Environmental Sciences Area
Lawrence Berkeley National Laboratory
Berkeley, California 94720
USA

barora@lbl.gov; ddwivedi@lbl.gov; bafaybishenko@lbl.gov;hmwainwright@lbl.gov

[2]Center for Computational and Data-Intensive Science and Engineering
Skolkovo Institute of Science and Technology
Moscow 143026
Russia

r.jana@skoltech.ru

INTRODUCTION

The vadose zone (often called the unsaturated zone) is commonly defined as the geologic media spanning from the land surface to the groundwater table of the first unconfined aquifer (Stephens 2018). The vadose zone is known to play a critical role within the biosphere: (1) as a storage medium to supply water to the plants and atmosphere, and (2) as a controlling agent in the transmission of recharging water as well as contaminants from the land surface to groundwater (Nimmo 2005). Past interest in the vadose zone has largely been a result of public concern about the need to protect groundwater reserves for drinking and agricultural purposes. The main sources of groundwater contamination typically originate in the vadose zone as leaking underground storage tanks, municipal solids and hazardous waste landfills, waste management sites, unlined pits, ponds, and lagoons, household septic systems, pesticide application areas, or surface spills (LaGrega et al. 1994). More importantly, these contaminants are modified by interactions within the vadose zone. For example, Małecki and Matyjasik (2003) reported that the average total dissolved solids values for rainwater changed dramatically from 30.2 mg/L at the land surface to 318 mg/L in groundwater. They also noted a corresponding change in the water type from a SO_4–Cl–Ca–NH_4 to a HCO_3–SO_4–Ca–Mg type due to vadose zone hydrogeochemical interactions. This and other studies document the significant role of vadose zone in contaminant distribution and migration (e.g., Oostrom et al. 2016; Arora and Mohanty 2017; Wan et al. 2018). Research efforts have thus far focused on the vadose zone as a significant reservoir for the capture, storage, and release of contaminants.

Since vadose zone involves numerous coupled physical, geochemical, and microbial processes, the interplay of these processes is difficult to understand especially from a remediation perspective. Considerable attention has been given to characterize and quantify contaminant transport in subsurface heterogeneous media. For example, heterogeneity caused by macropores and fractures can lead to preferential flow and rapid infiltration, which, in turn, increases groundwater vulnerability to potential contamination. In contrast, clay lenses and layers can provide opportunities to sorb contaminants in place. More recently, a growing body of work has focused on the outsized impacts of nutrient cycling and higher reaction rates at

1529-6466/19/0085-0010$05.00 (print)
1943-2666/19/0085-0010$05.00 (online)

http://dx.doi.org/10.2138/rmg.2019.85.10

critical interfaces. Herein, critical interfaces are defined as the interacting boundaries between zones of distinct hydrological, biogeochemical and lithological properties (Li et al. 2017). Biogeochemical processes and biodiversity at these interfaces are often orders of magnitude higher than the surrounding matrix. For example, iron- and sulfate-reducing bacteria showed one to two orders of magnitude greater community numbers in a layered sand-over-loam column when compared to homogeneous sand and loam columns (Hansen et al. 2011). McGuire et al. (2005) studied the impact of a moderate-sized rainfall event on redox processes at a shallow, sandy aquifer contaminated with petroleum hydrocarbons and chlorinated solvents, and concluded that recharge effects on the progression of terminal electron accepting processes existed primarily at the interface between infiltrating water and the aquifer, and not at the average aquifer scale. Quantification of these critical interfaces and plume evolution are further complicated by transient flow and transport conditions in the vadose zone. Transient conditions, including seasonal, annual, and long-term variability in recharge as well as transient interactions between climatic, hydrologic and biogeochemical factors complicate the study of contaminant fate and transport in the vadose zone. For example, Arora et al. (2016) concluded that changes in concentrations of redox-sensitive chemicals in a shallow alluvial aquifer were directly related to rainfall and recharge events at bi-monthly and annual time scales. Han et al. (2001) determined that the wetting-drying moisture regime resulted in higher metal reactivity in arid regions, resulting in redistribution and fractionation of heavy metals such as Ni, Zn and Cu. Although these studies have clearly demonstrated that critical interfaces and transient conditions regulate biogeochemical reaction rates, it remains unclear how to characterize these interfaces and time periods, and quantitatively incorporate them in numerical models.

To capture complex vadose zone characteristics, significant advances have been made in subsurface characterization techniques which are non-invasive, have finer resolution and capture a large spatial extent (Wainwright et al. 2018). In particular, these include *in situ* water, solute and gas monitoring, geophysical and remotely sensed observations, as well as weather data, which all offer diverse sets of data. The data stream, however, comes with its

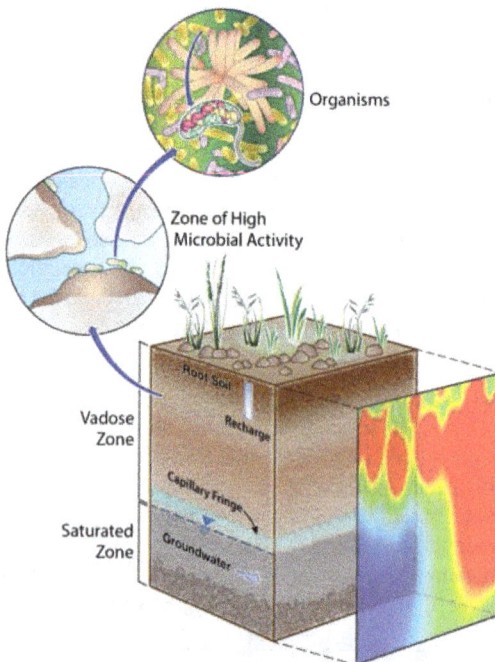

Figure 1. Schematic presentation of the vadose zone architecture and inherent complexities. For example, a solid-liquid interface acts as a zone of high biogeochemical activity.

own set of issues, such as sparse, heterogeneous and uncertain measurements, a variety of data sources, and differences in frequency of measurements. The major open questions related to these datasets include: (i) how to incorporate such diverse data types into reactive transport models; (ii) which methods to use to separate bulk properties from small scale heterogeneities and interfaces; and (ii) how to characterize error characteristics and incorporate this uncertainty into model predictions. The geoscience community is embracing the potential of additional data to improve model structure and therefore create opportunities for targeted sampling to better understand vadose zone dynamics (Mcmillan et al. 2012; Finsterle 2015).

To analyze complex environmental pollution problems, a quantitative study of water flow and contaminant transport in the vadose zone can be valuable and complementary to measurements of the heterogeneous and complex subsurface, as complicated by transient conditions. The past decade has seen a significant development of flow and reactive transport models as indispensable tools for studying vadose zone processes, including incorporating vadose zone complexities (e.g., sharp permeability interfaces) and transient conditions (e.g., climatic variability) to support decision making in such diverse areas as assessment of acid mine drainage, evaluation of the consequences of managed aquifer recharge, safety analysis of nuclear waste repositories, and analysis of global climate change. However, major gaps and challenges remain with regard to further conceptualization of vadose zone complexities, incorporation into and parameterization of models.

The focus of this chapter is to present a combination of modern approaches for vadose zone characterization and modeling. The key questions addressed here are how modeling activities can better serve quantification of vadose zone complexities, including heterogeneities and critical interfaces (e.g., Fig.1), while accounting for transient phenomena, and what areas and key challenges need to be addressed to improve the applicability and usefulness of current vadose zone models.

The chapter is organized as follows. The next section provides a review of field-based measurements to acquire diverse data streams relevant to the vadose zone. This is followed by a description of geochemical modeling of the vadose zone. At this point, the focus shifts to describing soil moisture studies from measurements to modeling. From here, the chapter proceeds to investigate some case studies which illuminate the power of geochemical modeling in deciphering vadose zone complexities and interactions. We then provide a brief discussion of current model limitations and future opportunities.

FIELD CONDITIONS, OBSERVATIONS AND TECHNIQUES

Vadose zone characterization and monitoring methods are commonly used for the development of a complete and accurate assessment of the inventory, distribution, and movement of contaminants in the unsaturated zone; development of improved predictive methods for liquid flow and contaminant transport; design of remediation systems (barrier systems, stabilization of buried wastes in situ, cover systems for waste isolation, in situ treatment barriers of dispersed contaminant plumes, bioreactive treatment methods of organic solvents in sediments and groundwater); design of chemical treatment technologies to destroy or immobilize highly concentrated contaminant sources (metals, radionuclides, explosive residues, and solvents) accumulated in the subsurface.

In the world of widely applied modern remote sensing and geophysical technologies for observations of vadose zone and groundwater processes, we need local scale measurements (Rubin et al. 1998; Faybishenko 2000), which is often called using a Latin word *in situ*—literally means *on site or at the original place*. Local hydro-geophysical methods are usually minimally invasive, conducted at small scales < 1 m, and are used to supplement and constrain larger scale geophysical and remote spatial-temporal measurements.

The goal of this section is to provide a synoptic view and a basic guide for common in situ vadose zone and groundwater measurements, as well as large scale remotely sensed observations, which are essential for hydro-geophysical characterization of the near-surface hydrological and biogeochemical properties.

Conceptualization

Conceptualization, or in other words—development of a conceptual model, is an approach to select *why, what, and how* to design and conduct measurements. Conceptualization of vadose zone systems is needed for the integrated qualitative and quantitative characterization of unsaturated flow and transport processes affected by natural behavior and man-induced changes. Development of appropriate conceptual models of water flow and chemical transport in the vadose zone is critical for developing adequate predictive modeling methods and designing cost-effective remediation techniques. Conceptual models of complex and heterogeneous vadose zone must take into account the processes of preferential and fast water seepage and contaminant transport toward the underlying aquifer. Such processes are enhanced under episodic natural precipitation, snowmelt, and extreme chemistry of waste leaks from tanks, cribs, and other surface sources. However, until recently, the effects of episodic infiltration and preferential flow on a field scale have not been taken into account when predicting flow and transport and developing remediation procedures. The pronounced temporal and spatial structure of water seepage and contaminant transport, which is difficult to detect, poses unique and difficult problems for characterization, monitoring, modeling, engineering of containment, and remediation of contaminants. Lack of understanding in this area has led to severely erroneous predictions of contaminant transport and incorrect remediation actions.

Because hydrological and biogeochemical cycles differ under conditions of the natural environment, for example, at the watershed scale, irrigation of agricultural lands, at contaminated sites, and urban areas, different types of conceptual models need to be developed, and different types of local scale measurements are needed for different environments.

A typical feature of the near-surface hydrological processes in different environments is preferential flow phenomenon (Allaire et al. 2009; Robinson et al. 2008). Preferential flow, often called fingering phenomena, is caused by the near-surface heterogeneity, cracks, and local surface depressions. Water moves faster through the preferential flow zones. Preferential flow is one of the main reasons of the spatial heterogeneity of the moisture content in soil. Due to the fingering effect, one would reasonably expect significant spatial and temporal variations of the soil moisture content and water potential in the near-surface zone. Note that more significant changes occur down to about 6–8 m depths.

To understand the vadose zone processes, one needs to measure both the soil moisture content and water pressure. The relationship between the moisture content and water pressure is fundamentally important to understanding soil water behavior. The graph of this relationship is commonly called the water retention curve.

In situ measurements in the vadose zone

Collecting soil samples. Soil investigations involve collecting soil samples—either disturbed or undisturbed soil samples, called monoliths, using special rings. Sensors are installed in boreholes—near-surface drilling can be done using hand held augers, usually to 2–3 m depths. Deeper drilling is carried out using mechanical augers, which are often integrated with electrical and gamma logging tools. The main requirement is to minimally disturb the soil during drilling of boreholes and sample collection.

Soil moisture measurements. Soil moisture is widely recognized as a key parameter in the mass and energy balance between the land surface and the atmosphere. The importance of soil moisture can be evident from the fact that researchers around the world are collecting soil moisture data, and nearly four decades of soil moisture data are now available from the International soil moisture network (Dorigo et al. 2011). Here, we describe only in situ measurements of soil moisture. A detailed description of remotely sensed observations to obtain soil moisture content and incorporate it into modeling is described in later sections.

Methods of measurements and collection of water samples are different depending on the moisture content. For example, one can easily collect water samples in the presence of so called gravitational water, but need to apply vacuum to withdraw water when the moisture content is below the residual water content. Deep lysimeters are used for collecting water samples below the depth of ~ 5 m.

The standard reference method for determining the soil moisture content is to oven dry soil samples at 105 °C. For organic soils and gypsiferous soils, the temperature should not exceed 70 °C (because of the problem with organic matter volatilization). Disturbed soil samples are commonly used to determine the gravimetric moisture content of soil. The moisture content is often expressed in volumetric units—gravimetric moisture content times the soil dry density will give you a volumetric moisture content.

Neutron moisture meter. A neutron moisture meter is based on the principle of utilizing neutron scattering around the borehole. Due to its hydrogen content, water is an effective neutron moderator, slowing high-energy neutrons. The technique is non-destructive, and is sensitive to moisture in the bulk of the target material.

Time Domain Reflectometry (TDR). TDR is a relatively new method for measurement of soil water content, first reported by Topp et al. (1980). It is based on the standard electromagnetic (EM) method for determining the moisture content. An electromagnetic pulse is generated by the TDR cable tester, and the propagation velocity the electromagnetic wave is measured along a transmission line (wave-guide) embedded in the soil. The support volume of TDR measurements depends on the probe design (Ferré et al. 1998; Skierucha et al. 2012). The main advantages of the TDR method over other methods for repetitive soil water content measurement (e.g., neutron probe) are: accuracy to within 1 or 2% of volumetric water content; calibration requirements are minimal (in many cases soil-specific calibration is not needed); averts radiation hazards associated with neutron probe or gamma-attenuation techniques; and capable of providing continuous soil water measurements through automation and multiplexing.

Capacitance/frequency domain technology. ECH$_2$O EC-5 is a simple sensor with excellent accuracy based on measuring the dielectric constant of soil based on the capacitance/frequency domain technology. Advantages are: minimum salinity and textural effects, factory calibration, small area of influence, and high resolution.

Figure 2 shows a schematic of the operational ranges of field and laboratory methods used in monitoring the matric suction in the vadose zone for soil water physical processes (Gee and Ward 1999).

Water pressure measurements. A tensiometer is used to directly measure the soil water pressure. The tensiometer consists of a porous tip/cup, a water-filled tube and a vacuum gauge. The porous cup of the tensiometer is buried in the soil. As water is pulled out of the tensiometer, the vacuum inside the tube increases. Soil suction creates the vacuum inside the tensiometer, which is measured by the gauge. When the vacuum created in the tensiometer corresponds to the soil suction, the water outflow from the tensiometer is ceased. The soil water suction is also called the soil water pressure. In conjunction with a water retention curve, tensiometers can be used to determine the soil moisture content. Tensiometers are used in irrigation scheduling to help farmers and other irrigation managers to determine when to water.

There are different types of porous cups and different designs. Theoretically, tensiometers with porous cups can be used in the range of water pressure from 0 (that is the reference value indicating the atmospheric pressure) to −1 bar, but practically to only −0.7–0.8 bars. There were recently developed the probes to measure the water pressure to −5 bars.

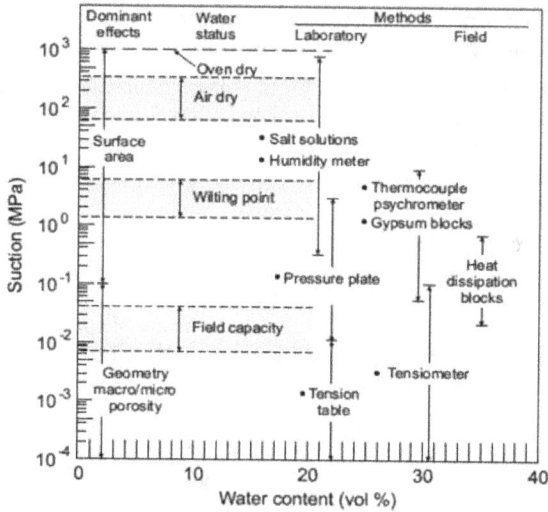

Figure 2. Schematic of the operational ranges of field and laboratory methods used in monitoring the matric suction in the vadose zone for soil water.

Water sampling. Water sampling is based on the idea of collecting soil water into the cup by applying vacuum inside the porous cup to exceed the capillary pressure holding water in soil. For shallow depths to ~ 5–7 m, you can bring the collected water to the surface using vacuum. For deeper depths, we need to first apply a vacuum to draw water inside the lysimeter and to push it to the surface by applying positive air/gas pressure.

Soil gas measurements. Soil gas is collected by applying vacuum through simple tubes installed in a borehole at different depths. Gas samples are collected into an impermeable container. Soil gas sampling is used to detect the source and locations of contaminated soils or groundwater. A soil gas probe can be connected to a preliminary vacuumed canister called Summa canisters (www.restek.com).

Evaporation and evapotranspiration. Evaporation is a vapor flux from the open water surface. Evaporometers for direct measurements of evaporation are installed at many meteorological stations around the world. Evapotranspiration is a term to define a cumulative vapor flux from both the surface and transpiration by plants. The most advanced apparatus to measure evapotranspiration is a weighing lysimeter (https://www.metergroup.com/environment/products/smart-field-lysimeter/#specifications). This is a tension controlled soil column with Tensiometers and soil moisture sensors to create the water and temperature conditions that are similar to those in the field. The lysimeter is installed on the scale to monitor the weight to calculate the total moisture content in the soil. For example, in Germany, there is a national lysimeter network for observing soil evapotranspiration called Tereno (Pütz et al. 2018). The observatory sites are located in different climatic conditions, characterized by different precipitation, elevation, and temperature (http://teodoor.icg.kfajuelich.de/ibg3searchportal2/index.jsp).

Eddy Covariance Stations. Measurements of gas exchange between terrain and the atmosphere are conducted at the Eddy Covariance Stations, which are part of the AmeriFlux monitoring network. There are 395 registered AmeriFlux sites. Measurements include monitoring of CO_2, vapor, temperature, wind, humidity, and precipitation. The main idea is measurements of high frequency wind and concentrations, which are then used to calculate fluxes. The detailed description of the Eddy Covariance stations can be found at https://fluxnet.ornl.gov/ or https://fluxnet.fluxdata.org/about/.

Surface runoff and soil loss due to erosion. One of the most important but difficult types of measurements is the evaluation of the surface runoff and associated soil loss due to erosion, for example, due to flooding. Measurements can be conducted using specially designed sites with different slope gradients and lengths, where water and sediments are collected at the exit from site.

Stable isotopes. One of the most effective methods to assess evapotranspiration is the use of measurements of stable isotopes—deuterium and oxygen-18 in soil water (Böhnke et al. 2002). The idea is that the stable hydrogen and oxygen isotopic compositions of fresh waters are highly correlated according to the 'global meteoric water line.' At some locations there is a small departure from this line, and the departure is even more increased due to evapotranspiration. So the departure from the local meteoritic water line can be used to assess evapotranspiration.

In situ measurements in the saturated zone

Groundwater level measurements. Groundwater level measurements are conducted in boreholes, which are usually equipped with the metal or plastic casing. Water enters the borehole through the screen being installed near the bottom of the well. Water level measurements are often accompanied by the barometric pressure measurements to make corrections on the effect of barometric pressure. The water level can be measured with a tape or using a water pressure transducer connected to the data acquisition system. A multi-parameter monitoring system, consisting of several sensors, can also be installed in a borehole to make measurements of different parameters.

Water sampling from boreholes. The chemical composition of a groundwater sample collected from a monitoring borehole is dependent on two factors: the length of the screened interval in the borehole, and the variability in permeability of the adjacent strata. During water sampling, the cone of depression of the water table is developed around the borehole. The radius of influence of pumping is largest at the depth of high permeability layer. In this sense collected samples can contain mixture of concentrations from different formations. Inflows from more permeable strata or fissures will dominate and bias the sample in the borehole and produce a flow-weighted average sample (FWA). In this case, an unrepresentative water sample is collected. Mixing mechanisms within boreholes can be very complex and research demonstrates that boreholes with longer screened intervals lead to greater bias, and therefore uncertainty, in the sampling result. Inflows from more permeable strata or fissures will dominate and bias the sample in the borehole and produce a flow-weighted average sample (FWA). Mixing mechanisms within boreholes can be very complex and research demonstrates that boreholes with longer screened intervals lead to greater bias, and therefore uncertainty, in the sampling result.

Remote sensing and geophysical observations

A suite of properties are expected to influence the vadose zone flow and biogeochemistry. Examples include the density, type, and distribution of plants and their dynamics; geomorphology and microtopography; the texture, density of near-surface soils; pore water chemistry and connectivity. Many of these properties can be measured through in situ sampling, but with the caveat that it is difficult to distribute these properties over space. Recently, there have been a variety of new technologies developed to capture the heterogeneity of the surface and subsurface properties—relevant to the vadose zone processes—by taking advantage of the state-of-the-art spatially integrating techniques—such as fiber optics sensing, geophysics, and unmanned aerial vehicle (UAVs).

Geophysical methods. Geophysical methods—including electrical resistivity, seismic, and radar—have been increasingly used to characterize the vadose zone in a non-invasive manner (e.g., Rubin and Hubbard 2005; Binley et al. 2015). They can be used to image soil moisture and biogeochemical properties (e.g., Dafflon et al. 2011, 2013;

Johnson et al. 2012, 2010; Wainwright et al. 2014). Autonomous electrical resistivity and phase tomography (ERT) monitoring, in particular, has the potential to achieve rapid and automated detection and identification of water infiltration within the vadose zone (Tran et al. 2016). In addition, recent developments in hydrogeophysical inversion methods allow for near-real-time characterization of vadose zone infiltration (e.g., Johnson et al. 2015). The inversion method also enable the estimation of soil hydraulic properties (Tran et al. 2016), which are critical for model predictions. These geophysical approaches can bridge the gap in sparse wellbore locations, by providing high-resolution and spatially and temporally extensive information in a minimally invasive manner (e.g., Johnson et al. 2015).

Remote sensing observations. Multiple platforms exist to remotely sense soil moisture and other landscape features for characterizing vadose zone processes. These include satellite, airplane-based and drone-based approaches. Among these, the UAV technologies are the latest growing area in the applications for environmental monitoring. The high-resolution surface elevation data from Light Detection and Ranging (LiDAR) or Photo Detection and Ranging (PhoDAR), for example, provides the surface elevation in a centimeter resolution which enables to map surface runoff patterns and infiltration. A recent study combined the UAV-based surface elevation mapping with isotopic analysis, and was successful in identifying surface flow accumulation and fast flow paths into the vadose zone (Christensen et al. 2018). In addition, thermal cameras are increasingly being used to map the locations of groundwater seep zones and river-groundwater interactions. Moreover, coupled spectral and structural information from UAV images can map the heterogeneity of plant species, photosynthetic activity as well as evapotranspiration, which are key components of vadose zone dynamics.

Fiber optics. Fiber optics technology uses optical fibers for measuring environmentally relevant parameters including temperature (Suarez et al. 2011), strain, soil moisture, acoustic waves (Bao and Chen 2012; Cox et al. 2012), and some chemical signatures (e.g., Potyrailo and Hieftje 1998). Each pulse of light samples the state of the fiber at all locations, yielding property measurements along the entire length at a fine lateral resolution (~ 0.25–1 m). In contrast to autonomous point sensors, the fiber optics technology can acquire real-time high-frequency datasets combining large extent and fine space/time sampling.

Field vadose zone characterization and monitoring methods are commonly used to constrain the results of remote geophysical and air-borne measurements, and to provide scientific and technical background to predict contaminant migration. By incorporating a combination of geophysical cross-borehole tomography and local-type measurements, numerical models can be developed to simulate the movement of organic and radioactive contaminants and their exchange between the zones of preferential flow and the soil or rock matrix. For example, the zonation approach (described below) serves as an important framework to include vadose zone complexities into models.

GEOCHEMICAL MODELING OF SUBSURFACE SYSTEMS

Over the last 20 years, there has been a growing reliance on reactive transport modeling (RTM) to address some of the most compelling issues facing our planet: nuclear waste management, contaminant remediation, and pollution prevention. While these issues are motivating the development of new and improved capabilities for subsurface environmental modeling using RTM, there remain longstanding challenges in characterizing the natural variability of hydrological, biological, and geochemical properties in subsurface environments and limited success in transferring models between sites and across scales. To achieve these ambitious objectives for subsurface reactive transport simulation, the subsurface science and engineering community is being driven to provide accurate assessments of engineering performance and risk for important issues with far-reaching consequences. In this regard, we present an overview of vadose zone complexities and how to incorporate these into models.

Conceptualization

To develop predictive capabilities of the vadose zone processes, a conceptual framework is necessary. Conceptual models typically describe the main physical processes taking place in the vadose zone, and how these processes interact or dominate the system. In general, the conceptual physical framework includes various processes and features such as a hydrologeologic setting describing the hydro-stratigraphic and structural details and relevant hydrological and biogeochemical processes. Relevant hydrogeological processes include surface infiltration rates (e.g., precipitation and evapotranspiration rates), flow rates (lateral and vertical) and details of the geologic formation. At the same time, relevant biogeochemical details include the presence of major reactive minerals, sediment characterization, pore water chemistry.

Because the presence of fast and slow flow paths significantly impact hydrological and biogeochemical processes in the vadose zone, it is important to capture the spatial and temporal signatures of vadose zone dynamics in the conceptual and consequently numerical frameworks. In particular, incorporation of preferential flow paths, clay lenses, and hydrostratigraphic layers with different properties need to be adequately represented in the numerical model. Several numerical solvers (e.g., TOUGHReact, PFLOTRAN, CrunchTope) exist that can be used to describe the hydrologic and biogeochemical implications of these vadose zone complexities. In the following section, we will briefly describe how to translate a conceptual framework into numerical models using appropriate parametrization depending on the level of complexity.

Description of the hydrological system

Variably saturated flow can be described using a set of equations for multiphase flow, explicitly considering liquid and air in the vadose zone (Steefel et al. 2015). Alternatively, the Richards equation can also be used for modeling of variably saturated flow in the vadose zone, which assumes the liquid phase is active while the gas phase is passive (Richards 1931). The Richards equation can be formulated either on the basis of aqueous phase saturation or fluid pressure. The Richards equation has been widely used in the literature for representing vadose zone processes (e.g., Yabusaki et al. 2017; Dwivedi et al. 2018a,b). However, a wide variety of problems such as nuclear waste disposal, acid rock drainage, and geological CO_2 sequestration may require multi-phase formulation. In the multi-phase formulation, the mass conservation equation requires the solution of each component present in different phases.

Simulations of hydrological and biogeochemical processes in heterogeneous soil and fractured-porous media require an application of the multi-continuum class models. In general, the multi-continuum approach includes different interacting regions with different hydrogeologic properties. For example, the dual-continuum approach includes two interacting regions—soil matrix and fracture domains; soil matrix is characterized with the less permeable intra-aggregate pore region, while the fracture domain is associated with the inter-aggregate pore region. Some examples of multi-continuum class of models include (1) the equivalent continuum model (ECM), which can incorporate the effects of fractures (Faybishenko et al. 2000); (2) dual permeability model (DPM) and dual, multi-porosity model, and multiple interacting continua approach (MINC)—representing fractures as regions of high permeability-low porosity in heterogeneous porous media (Barenblatt et al. 1960; Warren and Root 1963; Pruess 1985); and (3) discrete fracture and matrix model, which can represent a single fracture or an infinite number of equally spaced fractures in the porous media (Snow 1965). More details on different multi-continuum approaches including exchange functions and governing equations can be found in the literature (Faybishenko et al. 2000; Šimůnek et al. 2003; Arora et al. 2011).

One of the main parameters included in numerical models is the unsaturated hydraulic conductivity that is a nonlinear function of capillary pressure and liquid saturation; permeability values typically decrease as the water content and/or liquid pressure decrease (Mualem 1976;

van Genuchten 1980). Brooks-Corey and van Genuchten formulations have been extensively used in the literature to relate the relationship between the capillary pressure and aqueous phase saturation. Additionally, the Mualem and Burdine formulations are used for computing the relative permeability as a function of the effective aqueous saturation (Brooks and Corey 1964; Mualem 1976; van Genuchten 1980).

Description of the geochemical system

The key geochemical processes relevant in the vadose zone are aqueous chemistry, aqueous speciation, mineral precipitation and dissolution, sorption, ion exchange and redox reactions. To evaluate biogeochemical implications in the vadose zone, solute transport and reactions are coupled in the continuity equation (Dwivedi et al. 2016; Steefel et al. 2005). These reactions represent the transformation of reactant to product species. Reactions can either be classified as homogeneous or heterogeneous depending upon whether they occur within the same phase (e.g., aqueous) or across different phases (e.g., mineral reactions). Nevertheless, these reactions involve multiple geochemical species including major cations, anions, and trace elements. Correspondingly, a set of primary species is defined that determine the geochemical system. At the same time, a set of secondary species is defined on the basis of different combinations of primary species. The reactive transport codes can sweep through the database to pick secondary species, or secondary species can be explicitly defined in the input files as well. To compute the concentration of the secondary species, a geochemical equilibrium between aqueous species is assumed.

Biogeochemical reactions can either be kinetically controlled or equilibrium based. In general, mineral precipitation and dissolution, homogeneous reactions (aqueous phase), as well as microbially mediated reactions are kinetically controlled. On the contrary, sorption or surface complexation are equilibrium-based reactions. A kinetic treatment is more general; however, the type of treatment (kinetic vs. equilibrium) is contingent upon the relative time scales of the processes involved (Steefel and Lasaga 1994). For example, if the rate of mineral dissolution is significantly faster than the transport rate of a reactant, then it is reasonable to assume equilibrium-based mineral dissolution. An in-depth review on kinetically controlled and equilibrium based reactions is available elsewhere (Steefel et al. 2005; Steefel and Maher 2009; Li et al. 2010).

Model domain and discretization

Temporal discretization. Most of the modern reactive transport models have adaptive time steps. However, the maximum time step can be chosen for simulations based on the Courant–Friedrichs–Lewy (CFL) condition. The CFL condition ensures that the distance between mesh elements exceeds the distance that any particle travels during the time step (Sonnenthal et al. 2014).

Spatial discretization. At the watershed-, basin- or catchment-scales, the spatial discretization of the grid can be as fine as the digital elevation model (DEM). DEM can be derived from a high-resolution LiDAR dataset. DEM data are typically high-resolution (sub-meter) and can describe microtopographic features as well. However, it may be numerically intractable to run such simulations at these large scales. In that case, variable resolution meshes are preferred. In addition, mesh can be structured or unstructured. There are various tools that can create these meshes such as LaGrit, CUBIT, and Meshmaker (Hammond et al. 2014; Sonnenthal et al. 2014; Coon et al. 2016).

Material properties. As suggested above, hydrostratigraphic details are essential for appropriately representing the conceptual framework in numerical models. For example, details such as the thickness of the alluvial deposits and fluvial sediments including debris-flow can be spatially variable, and the soil properties can also vary in texture from sandy, silty to gravelly. These variations have a significant impact on flow and transport properties. Hence, different hydrostratigraphic layers need to be incorporated in the mesh by assigning different permeability values. The average hydrological properties (e.g., van Genuchten parameters) including permeability values can be estimated using pedotransfer functions from textural data of sediments from the site. Pedotransfer functions are described in more detail in the section.

An adequate representation of heterogeneity in model domain is also important to capture spatial and temporal signatures of vadose zone processes. In this regard, geophysical data and zonation approach are useful to map hydrogeological and geochemical properties such as permeability, percent of fines, and mineral ratio in the subsurface environment. In addition, reactive minerals can be characterized using XRD or digestion methods on soil samples. These properties can be assigned on the mesh by identifying appropriate regions. These regions include a set of control volumes having different material properties including soil types (e.g., van Genuchten parameters), permeability values, and minerals.

Finally, it is important to realize that the geochemical transformation of compounds in particular mineral reactions can alter the porosity and permeability of the porous media. Small changes in porosity can lead to substantial changes in permeability, and subsequently, these alteration in material properties are likely to modify the unsaturated flow properties of the porous media (Vaughan 1987). Most reactive transport codes can simulate these changes in material properties by computing volume changes of the matrix and fractures (Steefel et al. 2015; Xie et al. 2015).

Initial and boundary conditions

To evaluate vadose zone dynamics, reactive transport models require appropriate boundary conditions incorporating spatio-temporal variability of hydrologic and biogeochemical signatures. Hydrologic boundary conditions can be simulated using Dirichlet (or first-type) boundary condition, also known as hydrostatic or seepage face boundary conditions as shown in Figure 3.

For example, a river can be simulated using a dynamic seepage face boundary condition by applying observed transient river-stage measurements. A seepage face boundary is defined as an interface at atmospheric pressure (P_{atm}), where the water exits the porous medium (Bear 1975). On the contrary, the hydrostatic boundary condition is applied at the interface of the water table in the vadose zone, where negative pressure exists along the face of the model domain (Hammond et al. 2014; Sonnenthal et al. 2014).

It is essential to understand that redox or geochemical conditions change in response to perturbations such as precipitation events. Usually, recharge varies in both in space and time due to the topographic effects that cause substantial variability. Spatio-temporal recharge is typically applied as Neumann (or second-type) boundary condition (Hammond et al. 2014).

Pore water chemistry also changes with time; therefore, it is important to consider variations in the pore water composition to define geochemical boundary conditions in the model. It is possible to apply time-varying geochemical boundary conditions by assigning different pore water chemistry as a function of time in modern reactive transport codes. Moreover, it is important that models are appropriately initialized and run with proper meteorological, hydrological, and geochemical inputs. Typically, models should be run long enough so that model results are independent of initial conditions.

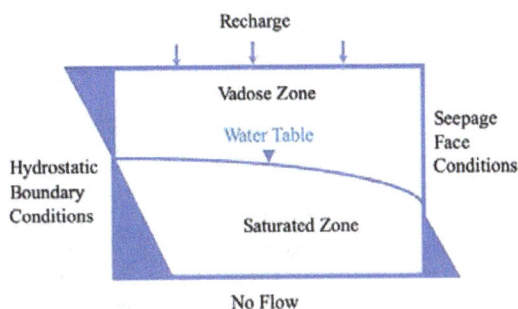

Figure 3. Schematic of boundary conditions in a reactive transport model.

SOIL MOISTURE STUDIES

As described above, soil moisture is a key variable controlling water, energy and nutrient fluxes from the land surface to the atmosphere. Although small in proportion (~0.15% of global freshwater), it is still an influential store of water in the hydrologic cycle (Dingman 1994; Western et al. 2002). Several studies have documented the role of soil moisture in modulating climate interactions, rainfall-runoff processes and plant growth. In the context of biogeochemical transport through the vadose zone, soil moisture is often the main variable controlling contaminant flow and transport. However, at the scale of the global hydrologic cycle, the processes governing water flow and solute transport through the vadose zone are the least understood (Wu and Li 2006). This is because a number of complexities such as, heterogeneity in soil texture, structure and composition, as well as transient environmental conditions can change soil moisture status in the vadose zone (Jana et al. 2007; Jana and Mohanty 2012a; Vereecken et al. 2014). Given the importance of the effect of soil moisture on the hydrologic cycle and related processes such as contaminant transport, agricultural runoff, crop production and flood control, it is obvious that concerted efforts must be made to measure, understand, and model soil moisture dynamics. Several studies have demonstrated that a better depiction of the spatio-temporal heterogeneity of soil moisture results in more accurate estimates from hydrologic and related process models (Koster et al. 2004; McCabe et al. 2005; van den Hurk et al. 1997). Here, we describe issues and complexities related to measurements and modeling of soil moisture within the vadose zone.

Soil moisture measurements

Soil moisture measurements, made in situ or by remote sensing, typically involve some trade-off between the scale triplet—support, spacing, and extent (Blöschl and Sivapalan 1995). Herein, support is the area (or time) over which the measurement is valid; spacing is the interval at which measurements are made; and extent is the overall coverage of the domain where measurements are made (Skøien and Blöschl 2006). In situ measurements offer support dimensions in the order of centimeters. However, it becomes impractical to have spacing of measurements in the same order, and to have such measurements over large extents. Intermediate support resolution (of the order of a few hundred meters to a kilometer) can be obtained using sensors mounted on airplanes or UAVs, but the extent of their coverage is typically limited to a few tens of kilometers. Satellite-based remote sensors, on the other hand, provide data with abundant (mostly global) coverage, but their support dimensions are of the order of hundreds and thousands of meters for soil moisture. Such poor resolution results in smoothing out much of the soil moisture heterogeneity in the footprint, which results in the omission of critical information from a remediation perspective. Recently, certain satellite-based sensors such as NASA's Soil Moisture Active Passive (SMAP) instrument were launched with the aim of providing soil moisture information at finer resolution (~3 km) (Entekhabi et al. 2010). However, these active (radar) sensors are no longer active due to technical issues. Other possible sources of relatively fine resolution remotely sensed soil moisture products can be obtained from Sentinel-1 and forthcoming missions such as NISAR and Tandem-L. These new efforts are described in more detail elsewhere (Mohanty et al. 2017).

In addition to spatial scales, temporal resolutions of the data also vary based on measurement platforms. The quickest return time of satellite remote sensors is of the order of 1-2 days. This means that information regarding intra-day variations in soil moisture are lost and only discrete snapshots of the moisture state are available. As a result, the immediate effects of certain low-intensity, short-duration precipitation events could go unobserved. In comparison, in situ sensors can provide almost real-time data as a continuous stream. Based on the type of application for soil moisture, different frequency of time series data may be preferred.

Time-stable locations. Large spatial extent, coarse resolution soil moisture measurements from satellite-based sensors are often calibrated and validated using in situ measurements. However, due to the mismatch in support scales and the inherent complexity of the vadose zone (such as, variations in topography, vegetation, soil physical properties and meteorology), a comprehensive validation becomes difficult to achieve (Cosh et al. 2004). Further, due to practical considerations, the number of in situ measurements is often limited in terms of both spatial as well as temporal spread. Such a situation necessitates the identification of locations within a remote-sensing footprint that can generally describe the mean soil moisture behavior within the footprint that is invariant over a sufficient period of time (Martínez-Fernández and Ceballos 2005). First proposed by Vachaud et al. (1985), such locations are referred to as time stable or temporally stable locations. Identification of such locations can help greatly reduce the number of locations to be sampled, and also enable permanent installation of in situ sensors for long-term studies relating measurements at the two disparate scales of measurement. It must be noted, however, that information regarding the soil moisture variability across the watershed is lost when measuring at only time stable locations. In order to overcome this drawback, newer non-linear model reduction techniques such as mode decomposition and discrete empirical interpolation method have been applied (Chaturantabut and Sorensen 2009; Hu and Si 2013). In particular, these newer approaches extend the concept of time stable locations to arrive at locations that provide not only the average soil moisture values for the area of interest, but also those that can help recover the dynamics across all locations in the domain. Using these techniques, it is possible to determine the least number of good candidate observation locations that would be required for a given accuracy of prediction across the footprint.

Soil moisture modeling

A major issue with quantifying soil moisture in the vadose zone is that the scale of measurement of soil moisture does not match that of the process/model (Jana and Mohanty 2012b). A further shortcoming of remotely sensed soil moisture data is that they are representative of only the near-surface layers (0-5cm). Due to the nature of the microwave remote sensing, deeper penetration of signals below this depth is not feasible. While the remote sensing footprint can provide excellent coverage (extent) and possibly intermediate level resolution in some cases (support and spacing), it is limited to only the skin (surface) and no vertical profile information is available. However, agricultural or contaminant transport studies may require and unfortunately rely on deeper soil moisture profiles. Consequently, extrapolation of remotely sensed soil moisture by modeling or scaling is necessary to obtain deeper layer moisture profiles. While we describe scaling in more detail below, assimilation of vadose zone model outputs with remotely sensed data has been performed in several studies. For example, González-Zamora et al. (2016) used a soil water index with a single parameter that was characteristic of the water travel time. This index served as a proxy for root zone soil moisture. Similarly, Pollacco and Mohanty (2012) used a joint parameter inversion scheme using remotely sensed soil moisture and evapotranspiration data to obtain root zone soil moisture profiles.

Scaling of soil moisture

The wide range of support and extent of available soil moisture data makes it necessary to bridge the gap between scales at which soil moisture data are available and the scale at which process representation is required. Two widely used approaches to bridge this disconnect between measurement and application scales are pedo-transfer functions (PTFs) and scaling. PTFs involve obtaining the required soil hydraulic parameters from available or easily measurable soil properties such as the texture and structure (Pachepsky and Rawls 2010). Scaling involves conversion of available measured data at one resolution to another resolution by aggregation (upscaling) or disaggregation (downscaling).

Pedo-transfer functions. The pedo-transfer function (PTF) approach is used to estimate the soil water retention relationship based on several types of basic soil properties, such as particle size distribution, i.e., sand, silt, and clay content, as well as organic carbon) (Tietje and Tapkenhinrichs 1993). The point regression PTF approach is used to predict the soil water content values at specific soil water potentials using empirically derived regression equations (e.g., Rawls et al. 1982; Ahuja et al. 1985). Function parameter PTFs such as those developed by Vereecken et al. (1989) predict the parameters for the entire water retention and/or hydraulic conductivity functions. Both approaches have been widely used for various soil databases. However, it has been demonstrated that point regression PTFs generally perform better (Tomasella et al. 2003). As the relationship between basic soil properties and soil water retention parameters is rather complicated, the variability in the retention parameters is often driven by different subsets of soil physical properties at distinct ranges of soil water pressure (Jana et al. 2007). Some attempts have been made in the past to apply PTFs across spatial resolutions in order to obtain scaled soil hydraulic parameters (e.g., Jana et al. 2007, 2008, 2012).

Scaling algorithms. The essential aim of scaling is to describe vadose zone complexity and heterogeneous features, and more importantly, the integrated effect of this complexity on flow or transport processes (Western et al. 2002). Scaling algorithms as applied in the vadose zone literature can be broadly classified into two categories. The first, and most common, approach is to derive parameters of the constituent flow equations (e.g., the Richards' equation) that effectively produce the same fluxes at the scale of interest as the scale of their measurement. This approach is, logically, termed as parametric scaling. The other approach, termed process-based scaling, attempts to modify the forms of the equations describing the soil moisture dynamics based on key governing variables at each scale of interest. The reasoning behind this approach is that different heterogeneous features and vadose zone complexities will become relevant at different scales, so the crucial question is to determine the key governing variables (Vereecken et al. 2016). Because there is significant knowledge gap about the nature of the relationships between these governing variables and the soil moisture dynamics, process scaling is less favored in practice.

Upscaling is the process of supplanting a heterogeneous domain of soil properties with an effective, homogeneous one. Most upscaling efforts for soil hydraulic parameters are based on the assumption that the movement of water occurs only in the vertical plane, and that no lateral movements take place (Khaleel et al. 2002; Vereecken et al. 2007). However, as would be expected, this assumption is valid only at relatively finer scales of around a few hundred meters (Vereecken et al. 2014). At larger footprints, it is no longer safe to assume that complex soil features such as topography, land cover, and meteorology do not exert significant influence on soil moisture dynamics. Hence, it becomes necessary to incorporate these complex features into the scaling methodology. One of the early steps in this direction was taken by Jana and Mohanty (2012a) who incorporated the influence of topography in an upscaling algorithm. A power averaging approach was used where the upscaled parameter, P^*, was defined as

$$P^* = \frac{\sum_{i=1}^{n}\left(1+T\left(p_i\right)\right)p_i}{\sum_{i=1}^{n}\left(1+T\left(p_i\right)\right)} \tag{1}$$

where

$$T\left(P_i\right) = \sum_{j=1, j\neq i}^{i} Sup\left(p_i, p_j\right) \tag{2}$$

$$Sup\left(p_i, p_j\right) = e^{-\eta\left(p_i-p_j\right)^2} \tag{3}$$

and the scale parameter, η, is given by

$$\eta = \left(\frac{z_{j_{max}} - z_{j_{min}}}{z_i - z_j} \right)^2 \frac{\sqrt{\left(x_i - x_j\right)^2 + \left(y_i - y_j\right)^2 + \left(z_i - z_j\right)^2}}{S} \tag{4}$$

In the above system of equations, $\mathrm{Sup}(p_i; p_j)$ is the support from parameter pi to parameter p_j, i and j are indices for locations where the parameters are measured, x, y, and z are Cartesian coordinates of the measurement location, and S is the scale to which the parameters are being aggregated. As Eq. (4) suggests, the scale parameter is a product of the normalized difference in elevation between the two locations i and j, and the linear distance between measurement values, normalized by the scale dimension. When upscaling soil hydraulic parameters to resolutions of a kilometer or beyond, topography-based scaling outperformed other algorithms which ignored the effect of such complex features (Jana and Mohanty 2012a).

Downscaling is the process of disaggregating large homogeneous footprints of soil moisture into a heterogeneous field of finer resolution values. Several downscaling approaches have been applied in the past including data fusion (e.g., Das et al. 2011; Montzka et al. 2016), statistical modeling (e.g., Ines et al. 2013; Verhoest et al. 2015) and data assimilation (e.g., Sahoo et al. 2013). A comprehensive review of a number of spatial downscaling algorithms is provided by Peng et al. (2017).

Remarks

A recent survey of the hydrological sciences community showed the need for better integration of field work and modeling across a range of spatial and temporal scales in order to advance understanding in this area (Blume et al. 2016). The study pointed out that one of the greatest challenges to this advancement is the lack of improved hydrological theories. In research related to soil moisture, it has been argued that concepts such as the Richards' equation, valid at the continuum scale, are routinely applied across much larger extents wherein assumptions made at the continuum scale are no longer valid (e.g., Blöschl and Sivapalan 1995). More than two decades after this observation, little has changed.

We suggest that future efforts be focused on developing comprehensive scaling algorithms that are capable of refactoring the role of vadose zone complexities, especially under varied saturation conditions and transient phenomenon (Fig. 4). Further, a holistic approach to characterize and quantify soil moisture, beyond efforts from the hydrological community, will be useful. For example, statistical data mining approaches can prescribe important approaches like identification of time stable locations.

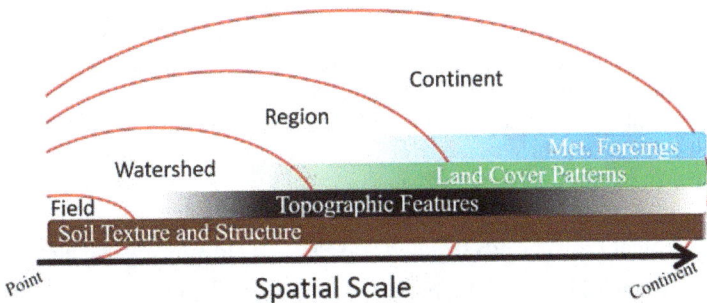

Figure 4. Dominant controls of soil moisture variability as a function of spatial scale (modified from Jana 2010).

APPLICATION OF GEOCHEMICAL MODELING
TO VADOSE ZONE STUDIES

To gain predictive understanding of heterogeneous and temporally variable vadose zone processes, it is critical to integrate multi-type multi-scale data encompassing both surface and the subsurface. In particular, spatially extensive characterization technologies do not often directly measure the properties of interest; such as permeability, or they have a large uncertainty (such as petrophysical relationships for geophysics). It is cost prohibitive to measure all the properties across many locations. Similarly, temporal vadose zone patterns can be difficult to interpret because their variability can depend on several factors, including temperature, precipitation, vegetation and stream discharge. To tackle this challenge, statistical approaches combined with geophysical and geochemical data have been developed recently to capture spatial and temporal variability in field scale environments (e.g., Arora et al. 2013, 2019; Sassen et al. 2012; Wainwright et al. 2014). This section describes these zonation-based and time variable approaches and its application at two field studies—the Rifle floodplain and the Savannah River site. A brief description of these sites is also included in the following section.

Identifying critical interfaces through zonation

The zonation approach is based on the suppositions that: (1) many subsurface and surface hydrogeomorphlogical, vegetation, and physiochemical properties are likely to be correlated with each other; (2) some properties may exert a dominant control on the vadose zone processes (such as clay content, soil moisture, microtopography); and (3) some spatially extensive measurements (such as geophysical and remote sensing) may provide reasonable proxies for controlling variables, thus offering an approach to characterize key properties over large regions and in a minimally invasive manner. This is the extension of lithofacies and hydrofacies concepts (e.g., Fogg et al. 1998; Klingbei et al. 1999; Weissmann et al. 2002; Heinz et al. 2003; Yabusaki et al. 2011), in which geological units represents a set of common hydrological or geochemical properties. The zonation approach made these concepts more general and quantitative such that the zones or units are discovered and defined by a suite of spatially extensive datasets such as geophysics and remote sensing as well as point measurements. The zonation approach works by using a statistical data-mining approach to integrate spatially extensive datasets for identifying zones that have distinct characteristics relative to neighboring zones. For example, Sassen et al. (2012) and Wainwright et al. (2014) used geophysical data and zonation approach to map hydrogeological and geochemical properties (i.e., permeability, percent of fines, mineral ratio) in the subsurface at a uranium contaminated site. Sassen et al. (2012) showed that the correlation between hydraulic and geochemical properties would need to be taken into account to describe the flow processes. Hubbard et al. (2013) and Wainwright et al. (2015) applied the zonation-based approach to characterize the spatial organization of the Arctic tundra ecosystem and permafrost environment. They showed that the zonation approach could capture the heterogeneity of near-surface soil characteristics and greenhouse gas fluxes, and that it enables mapping of soil properties and fluxes across the site based on sparse point-scale measurements. In this manner, the zonation approach provides a tractable and effective way to capture the spatial heterogeneity of co-varied properties and employs varied datasets across scales. We further describe the zonation approach through two case studies, wherein the approach was successful in describing and quantifying contaminant fate and transport.

Identifying critical time periods through wavelet and entropy approaches

It is important to address time invariance in vadose zone dynamics because biogeochemical interactions are inherently non-linear and complex, so much so that changes to measured water chemistry parameters (such as pH, SO_4) and gas fluxes (e.g., CO_2) can indicate the influence of multiple processes simultaneously. This is problematic when developing conceptual models and identifying governing controls on vadose zone processes that need to be incorporated into numerical models. Wavelet and entropy-based approaches offer the opportunity to systematically interrogate complex, multivariate datasets to characterize natural variability and identify the governing processes that exert control over vadose zone patterns at different time scales. For example, Arora et al. (2013) used wavelets to associate variations in geochemical concentrations at a landfill site to water table and precipitation dynamics, which showed dominance at 8 month scales. They further documented that temporal variability in sulfate concentrations was significantly impacted by FeS cycling and uptake by vegetation, which would need to be incorporated into models to accurately represent redox dynamics at the site. Similarly, Arora et al. (2019) applied the entropy-based approach to characterize temporal variability of the Arctic tundra ecosystem and permafrost environment. They showed that the entropy approach could capture the variable nature of relationships between gas fluxes and near surface soil characteristics (e.g., soil moisture, temperature), and thereby document the impact of changing environmental conditions (e.g., warming) on these patterns in a future climate. In this manner, wavelet and entropy-based techniques provide a simplistic yet tractable approach to identify temporal patterns and capture the influence of co-variability of hydrological and biogeochemical processes, as well as climatic controls on vadose zone dynamics and conditions.

Case studies

We describe the zonation approach through two case studies, wherein spatially heterogeneous clusters were successful in describing and quantifying contaminant fate and transport. We further describe a wavelet-entropy approach, whose characterization of temporal patterns was successfully embedded into a reactive transport model of the Rifle site.

Rifle case study. The Rifle site is located in the town of Rifle, CO, adjacent to the Colorado River. The Department of Energy (DOE) Rifle site is a former uranium mill tailings site, where soil and groundwater have been contaminated with uranium above the background level. Extensive characterization of the site has occurred over the past ten years to identify and quantify biogeochemical cycling in the Rifle floodplain. In particular, studies have found regions near riverbanks—known as naturally reduced zones (NRZs)—that have relatively high organic carbon, reducing conditions and elevated uranium concentration (Campbell et al. 2012). Historically, NRZs have been considered important as diffusion-limited interfaces rich in uranium and pyrite. A zonation-based approach was used for the probabilistic mapping of NRZs within a three-dimensional subsurface domain using a geophysical method of induced polarization imaging, since it identifies the electrical chargeability of subsurface materials and hence is sensitive to certain minerals (i.e., reduced sulfides) associated with NRZs (Wainwright et al. 2016).

Furthermore, a novel wavelet-entropy approach was applied to identify temporal variability in biogeochemical processes (Arora et al. 2016a). In particular, this approach works by interrogating complex, multivariate geochemical datasets to identify transient phenomenon within the floodplain environment. Results indicated that seasonal and annual hydrologic perturbations associated with rainfall and snowmelt events drove biogeochemical cycling at the site.

A 2-D reactive transport model was developed to investigate the impact of these NRZs and transient phenomenon—i.e., water table variations and temperature gradients—on subsurface carbon fluxes in the Rifle floodplain (Arora et al. 2016b). Results from model simulations ignoring NRZs underpredicted atmospheric CO_2 fluxes by almost 230% compared to simulations with NRZ relevant microbial reaction pathways. Results further indicated that subsurface carbon exports from the Rifle site were underestimated by almost 170% (to $3.3\,g\,m^{-2}\,d^{-1}$) when transient conditions were ignored in the simulations. This modeling study concluded that spatially (e.g., naturally reduced zones) and temporally discrete conditions (e.g., temperature fluctuations) can significantly contribute to carbon cycling in floodplains and need to be appropriately represented in model simulations. Other modeling efforts at the site have therefore explicitly used 3D map of NRZs to quantify oxygen and nitrogen fluxes (Fig. 5) (e.g., Yabusaki et al. 2017; Dwivedi et al. 2018a).

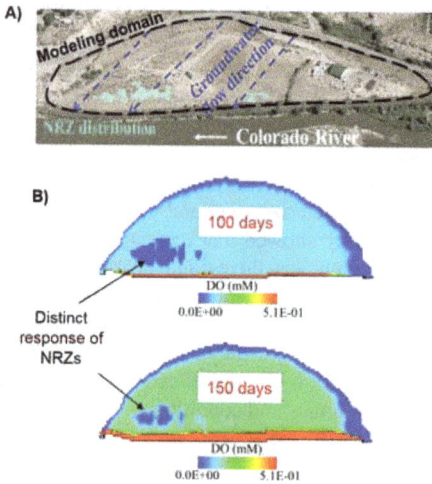

Figure 5. A) Distribution of naturally reduced zones (NRZs) at the Rifle site determined using induced polarization imaging, and **B)** impact of NRZs on simulated dissolved oxygen (DO) profiles at 5.0 m depth in groundwater on day 100 (i.e., before the flow reversal from the Colorado River) and 150 (i.e., during the flow reversal) of a 365 day simulation (modified from Dwivedi et al. 2018a).

Savannah case study. The DOE's Savannah River Site F-Area includes three unlined seepage basins, where low-level radioactive waste solutions were disposed (Bea et al. 2013; Wainwright et al. 2014). The site located near Aiken, South Carolina is a former nuclear weapons production site. Groundwater at the site has been contaminated with uranium and other radionuclides associated with plutonium separation for over 30 years. The site has been extensively characterized in terms of subsurface structure (Sassen et al. 2012; Wainwright et al. 2014) and uranium geochemistry (Dong et al. 2012). These studies found that the hydrogeological and geochemical parameters (such as permeability, percent of fines, mineral ratio) are tightly correlated to each other, and distinct geological units—reactive facies—exist, each of which has a unique distribution of co-varied properties. Thus, reactive facies based zonation approach has been found to provide a relevant description of physiochemical controls at plume relevant scales. In particular, a seismic technique was used to identify the spatial distribution of reactive facies along the uranium plume center line (Wainwright et al. 2014).

A modeling study found that incorporating reactive facies based properties and parameters improved the accuracy of model predictions, which are important to understand long-term uranium mobility and plume evolution at the site (Sassen et al. 2012). Moreover, a reactive transport model including the saturated and unsaturated (vadose) zones, U(VI) and H^+ adsorption (surface complexation) onto sediments, dissolution and precipitation of Al and Fe minerals, and key hydrodynamic processes was developed for the site (Bea et al. 2013). Herein, uncertainty quantification analyses were conducted to assess the sensitivity of model results to various input parameters affecting the migration of the acidic-U(VI) plume at the F-Area. This study concluded that model (and parameter) sensitivity evolves in space and time, and recommended that uncertainty-based analyses be conducted to assess the temporal efficiency of remediation strategies in contaminated sites. A recent study further compared two surface complexation models in describing U(VI) plume evolution at two different time periods of the site operation—the basin infiltration and post-closure times (Arora et al. 2018). The study concluded that there was uncertainty in describing the pre-closure behavior of U(VI) from the F-Area because of the different model approaches. This suggests that a careful evaluation and buildup of the reactive transport models is necessary for use in evaluating remediation strategies, nuclear cleanup, and other opportunities.

CURRENT MODEL LIMITATIONS AND FUTURE OPPORTUNITIES

Advancement in computational methods and measurement techniques have expanded the role of RTMs to address a wide range of spatial and temporal scales—from pore to catchment scale and across 10^0 to 10^3 years (Vereecken et al. 2016 ; Li et al. 2017). With this advancement, however, there is a growing need to evaluate the robustness and performance of RTMs and describe model limitations. In this regard, we present the hindrances to model application, formal treatment of uncertainty, the role of benchmarking activities, and future opportunities therein.

Model validation

Geochemical model validation is important if model results are being used to provide accurate assessments of engineering performance, risk analysis, and support policy formulations, which have far-reaching consequences. However, in contrast to hydrologic models, the extent to which geochemical models can be validated has been questioned (Tsang 1987). In this regard, model validation has been primarily related to careful evaluation of the conceptual framework and support for better model calibration (e.g., through the incorporation of spatial and temporal heterogeneities) (Arora et al. 2018).

Although geochemical models have been successful in describing processes within the heterogeneous vadose zone, as shown through case studies above, several aspects need to be determined before developing such a model. In particular, a careful evaluation of the input data for the species of interest, determination of the accuracy required of the modeling results, and whether the conditions of the system being modeled match the range of conditions for which the model is valid is needed (Herbert 1996). For example, Um et al. (2008) concluded that a non-electrostatic surface complexation model developed using variable pH measurements on synthetic ferrihydrite resulted in inaccurate predictions of U(VI) mobility in subsurface environments. This and other studies demonstrate the importance of obtaining experimental data over a sufficiently wide range of chemical conditions when developing a surface complexation model (Fox et al. 2006, 2012; Hyun et al. 2009). Similarly, a critique of the thermodynamic database would be necessary to evaluate to what extent input parameters affect model predictions.

Inverse estimation. A reactive transport model generally consists of a number of flow, transport and kinetic parameters, whose values are not necessarily known a priori. Many of these parameters are determined through experimental or laboratory studies, and yet others need to be estimated through inverse modeling. Further, the choice of data types for the inverse analyses has a significant impact on the resulting model predictions and model performance. For example, Arora et al. (2012) concluded that lumped macropore-matrix observations of water content, cumulative outflow, and effluent concentration increased errors in inversely estimating model parameters representing preferential flow and bromide transport at the column scale in comparison to using separate measurements of macorpore and soil matrix.

Uncertainty quantification

Uncertainty in vadose zone models can arise from a number of different sources including inappropriate conceptual framework and assumptions, uncertainty in the input and thermodynamic parameters, and the inherent stochastic nature of the subsurface system. Conceptual model uncertainties for vadose zone models have been quantified using Bayesian model averaging (e.g., Wöhling et al. 2008; Gupta et al. 2012). This Bayesian approach evaluates conceptual uncertainty by using multimodel ensemble simulations.

Uncertainties in input model parameters can arise from measurement errors and inverse estimation. Inverse parameter estimation is challenging because of nonidentifiability of the solution set, ill-posedness of the inverse problem, and non-unique parameter set (Carrera and Neuman 1986). Moreover, heterogeneous vadose zone models that incorporate chemical and mineral heterogeneities or preferential flow typically suffer from increased interdependence of model parameters, which increases the risk of reaching local minima in the parameter set (e.g., Arora et al. 2012). Thus, uncertainty estimation seems important from a geochemical perspective. Bayesian methods present a powerful solution to the challenges of model nonlinearity and identifying the maximum likelihood of parameters.

Development of benchmarks

Despite the availability of a large number of models available to simulate vadose zone flow and reactive transport, substantial differences exist among them—some that are related to the conceptual framework, process couplings, and yet others that ascribe numerical differences. Considering these differences, benchmarking activities are critical to building confidence in models and providing measures of model performance. Several benchmarking studies have been conducted in the past that compared thermodynamic database and speciation (e.g., Nordstrom and Archer 2003), and others that compared conceptual and numerical capabilities of models (e.g., Mayer et al. 2015). These are described in more detail elsewhere (Arora et al. 2015 ; Steefel et al. 2015; Dwivedi et al. 2016).

Future opportunities

There is a significant need for the further development of monitoring and predictive tools for allocation and management of water resources, decision making for environmental protection, and increasing cost saving for remediation efforts. New tools and approaches are needed that are open source and easy to use, with capabilities ranging from inverse parameter estimation and uncertainty quantification, to launching and monitoring simulations. Herein, we describe some opportunities for future growth.

Data-worth analysis. Addressing remediation issues in a data-driven, cost-effective manner is of increasing significance, especially because of their economic and environmental implications. It is therefore essential to determine the influential vadose zone properties with sufficient accuracy so that relevant predictions can be made with confidence. Data-worth analysis is one such metric that measures the extent to which the collected data can reduce the uncertainty of target predictions that are critical for decision-making. The data-worth analysis works by ranking the contribution that each (potential or existing) data point makes to the solution of an inverse problem (used for model calibration and parameter estimation) and a subsequent predictive simulation (Finsterle 2015).

Scaling. One of the biggest challenges for vadose zone modeling is to apply laboratory experimental results to natural field systems. Geoscientists have often delved into the rate conundrum particularly dealing with significant discrepancies between rate constants derived from experimental studies to application at the field scale (e.g., Maher et al. 2006). While several scaling algorithms have focused on soil moisture estimation in the vadose zone, relatively fewer approaches have considered scaling that have biogeochemical implications.

Comprehensive vs parsimonious approach. Some authors have argued that rather than developing a geochemical model with a large number of parameters that are poorly defined at a larger extent, a parsimonious approach may work better (e.g., Basu et al. 2010). The latter approach can still describe contaminant distribution as a function of landscape characteristics while being more numerically efficient than a reactive transport model. However, the former approach has several advantages in teasing out individual components and processes especially to describe the stochastic and non-linear nature of the vadose zone (e.g., Fatichi et al. 2016). In particular, reactive transport models may be better poised to forecast the spatiotemporal evolution of plume or non-linear feedbacks with climate than empirically based models.

A recent review paper (Vereecken et al. 2016) suggests that a major challenge in developing a numerical model for the vadose zone is the coherent integration of all the information on (i) the multiscale architecture (including the heterogeneous vadose zone architecture); (ii) the process formulation for the chosen range of scales; (iii) the system's coupling to the environment, which is typically represented as an external forcing but should also include the feedbacks to the atmosphere and/or groundwater; and (iv) the available data, which often need to be transferred into the chosen range of scales. Thus, future opportunities in the development of vadose zone models should focus on holistic approaches that honor the new data types and streams, but also include all aspects of the critical zone—from the bedrock to the top of the canopy.

SUMMARY

Vadose zone hydrologic and biogeochemical processes play a significant role in the capture, storage and distribution of contaminants between the land surface and groundwater. One major issue facing geoscientists in dealing with investigations of the unsaturated zone flow and transport processes is the evaluation of heterogeneity of subsurface media. This chapter presents a summary of approaches for monitoring and modeling of vadose zone dynamics in the presence of heterogeneities and complex features, as well as incorporating transient conditions. Modeling results can then be used to provide early warning of soil and groundwater contamination before problems arise, provide scientific and regulatory credibility to environmental management decision-making process to enhance protection of human health and the environment.

We recommend that future studies target the use of RTMs to identify and quantify critical interfaces that control large-scale biogeochemical reaction rates and ecosystem functioning. Improvements also need to be made in devising scaling approaches to reduce the disconnect between measured data and the scale at which processes occur.

ACKNOWLEDGMENTS

This material is based upon work supported by the U.S. Department of Energy, Office of Science, Office of Biological and Environmental Research (as part of the Watershed Function Scientific Focus Area), and Office of Science, Office of Advanced Scientific Computing (as part of the project "Deduce: Distributed Dynamic Data Analytics Infrastructure for Collaborative Environments") under Contract No. DE-AC02-05CH11231.

REFERENCES

Ahuja LR, Naney JW, Williams RD (1985) Estimating soil water characteristics from simpler properties or limited data1. Soil Sci Soc Am J 49:1100

Allaire SE, Roulier S, Cessna AJ (2009) Quantifying preferential flow in soils: A review of different techniques. J Hydrol 378:179–204

Arora B, Davis JA, Spycher NF, Dong W, Wainwright HM (2018) comparison of electrostatic and non-electrostatic models for u(vi) sorption on aquifer sediments. Groundwater 56:73–86

Arora B, Dwivedi D, Hubbard SS, Steefel CI, Williams KH (2016a) Identifying geochemical hot moments and their controls on a contaminated river floodplain system using wavelet and entropy approaches. Environ Model Softw 85:27–41

Arora B, Mohanty BP (2017) Influence of spatial heterogeneity and hydrological perturbations on redox dynamics: a column study. Procedia Earth Planet Sci 17:869–872

Arora B, Mohanty BP, McGuire JT (2012) Uncertainty in dual permeability model parameters for structured soils. Water Resour Res 48:W01524

Arora B, Mohanty BP, McGuire JT (2011) Inverse estimation of parameters for multidomain flow models in soil columns with different macropore densities. Water Resour Res 47:W04512

Arora B, Mohanty BP, McGuire JT, Cozzarelli IM (2013) Temporal dynamics of biogeochemical processes at the Norman Landfill site. Water Resour Res 49:6909–6926

Arora B, Şengör SS, Spycher NF, Steefel CI (2015) A reactive transport benchmark on heavy metal cycling in lake sediments. Comput Geosci 19:613–633

Arora B, Spycher NF, Steefel CI, Molins S, Bill M, Conrad ME, Dong W, Faybishenko B, Tokunaga TK, Wan J, Williams KH, Yabusaki SB 2016b. Influence of hydrological, biogeochemical and temperature transients on subsurface carbon fluxes in a flood plain environment. Biogeochemistry 127:367–396

Arora B, Wainwright HM, Dwivedi D, Vaughn LJS, Curtis JB, Torn MS, Dafflon B, Hubbard SS (2019) Evaluating temporal controls on greenhouse gas (GHG) fluxes in an Arctic tundra environment: An entropy-based approach. Sci Total Environ 649:284–299

Bao X, Chen L (2012) Recent progress in distributed fiber optic sensors. Sensors 12:8601–8639

Barenblatt G, Zheltov I, Kochina I (1960) Basic concepts in the theory of seepage of homogeneous liquids in fissured rocks [strata]. J Appl Math Mech 24:1286–1303

Basu NB, Rao PSC, Winzeler HE, Kumar S, Owens P, Merwade V (2010) Parsimonious modeling of hydrologic responses in engineered watersheds: Structural heterogeneity versus functional homogeneity. Water Resour Res 46:1–16

Bea SA, Wainwright H, Spycher N, Faybishenko B, Hubbard SS, Denham ME (2013) Identifying key controls on the behavior of an acidic-U(VI) plume in the Savannah River Site using reactive transport modeling. J Contam Hydrol 151:34–54

Bear J (1975) Dynamics of fluids in porous media. Soil Sci 120:162–163

Binley A, Hubbard SS, Huisman JA, Revil A, Robinson DA, Singha K, Slater LD (2015) The emergence of hydrogeophysics for improved understanding of subsurface processes over multiple scales. Water Resour Res 51:3837–3866

Blöschl G, Sivapalan M (1995) Scale issues in hydrological modelling: A review. Hydrol Process 3–4:251–290

Blume T, van Meerveld I, Weiler M (2016) The role of experimental work in hydrological sciences—insights from a community survey. Hydrol Sci J 1–4: 334–337

Böhnke R, Geyer S, Kowski P (2002) Using environmental isotopes ^{2}H and ^{18}O for identification of infiltration processes in floodplain ecosystems of the River Elbe. Isotopes Environ Health Stud. 38:1–13

Brooks RH, Corey, a T (1964) Hydraulic properties of porous media. Hydrology Papers, Colorado State University. Pap. 3. Fort Collins CO

Campbell KM, Kukkadapu RK, Qafoku NP, Peacock AD, Lesher E, Williams KH, Bargar JR, Wilkins MJ, Figueroa L, Ranville J, Davis JA, Long PE (2012) Geochemical, mineralogical and microbiological characteristics of sediment from a naturally reduced zone in a uranium-contaminated aquifer. Appl Geochem 27:1499–1511

Carrera J, Neuman SP (1986) Estimation of aquifer parameters under transient and steady state conditions: 2. uniqueness, stability, and solution algorithms. Water Resour Res 22:211–227

Chaturantabut S, Sorensen DC (2009) Discrete Empirical Interpolation for nonlinear model reduction. *In:* Proceedings of the 48h IEEE Conference on Decision and Control (CDC) Held Jointly with 2009 28th Chinese Control Conference. IEEE, pp. 4316–4321

Christensen JN, Dafflon B, Shiel AE, Tokunaga TK, Wan J, Faybishenko B, Dong W, Williams KH, Hobson C, Brown ST, Hubbard SS (2018) Using strontium isotopes to evaluate the spatial variation of groundwater recharge. Sci Total Environ 637–638:672–685

Coon ET, David Moulton J, Painter SL (2016) Managing complexity in simulations of land surface and near-surface processes. Environ Model Softw 78:134–149

Cosh MH, Jackson TJ, Bindlish R, Prueger JH (2004) Watershed scale temporal and spatial stability of soil moisture and its role in validating satellite estimates. Remote Sens Environ 92:427–435

Cox B, Wills P, Kiyashchenko D, Mestayer J, Lopez J, Bourne S, Lupton R, Solano G, Henderson N, Hill D, Roy J (2012) Distributed acoustic sensing for geophysical measurement, monitoring and verification | Feb. 2012 |. CSEG Rec 37:7–13

Dafflon B, Irving J, Barrash W (2011) Inversion of multiple intersecting high-resolution crosshole GPR profiles for hydrological characterization at the Boise hydrogeophysical research site. J Appl Geophys 73:305–314

Dafflon B, Wu Y, Hubbard SS, Birkholzer JT, Daley TM, Pugh JD, Peterson JE, Trautz RC (2013) Monitoring CO_2 intrusion and associated geochemical transformations in a shallow groundwater system using complex electrical methods. Environ Sci Technol. 47:314–321

Das NN, Entekhabi D, Njoku EG (2011) An algorithm for merging SMAP radiometer and radar data for high-resolution soil-moisture retrieval. IEEE Trans. Geosci Remote Sens 49:1504–1512

Dingman SL (1994) Physical Hydrology. Macmillan, New York

Dong W, Tokunaga TK, Davis JA, Wan J (2012) Uranium(VI) adsorption and surface complexation modeling onto background sediments from the F-Area Savannah River site. Environ Sci Technol 46:1565–1571

Dorigo WA, Wagner W, Hohensinn R, Hahn S, Paulik C, Xaver A, Gruber A, Drusch M, Mecklenburg S, Van Oevelen P, Robock A, Jackson T (2011) The International Soil Moisture Network: A data hosting facility for global in situ soil moisture measurements. Hydrol Earth Syst Sci 8:1609–1663

Dwivedi D, Arora B, Molins S, Steefel CI (2016) Benchmarking reactive transport codes for subsurface environmental problems, *In:* Thangarajan D, Singh VP (Eds.), Groundwater Research on Exploration, Assessment, Modelling and Management of Groundwater Resources and Pollution. CRC Taylor and Francis Group, pp. 301–318

Dwivedi D, Arora B, Steefel CI, Dafflon B, Versteeg R (2018a) Hot spots and hot moments of nitrogen in a riparian corridor. Water Resour Res 54:205–222

Dwivedi D, Steefel CI, Arora B, Newcomer M, Moulton JD, Dafflon B, Faybishenko B, Fox P, Nico P, Spycher N, Carroll R, Williams KH (2018b) Geochemical exports to river from the intrameander hyporheic zone under transient hydrologic conditions: East River Mountainous Watershed, Colorado. Water Resour Res 54:8456–8477

Entekhabi D, Njoku EG, O'Neill PE, Kellogg KH, Crow WT, Edelstein WN, Entin JK, Goodman SD, Jackson TJ, Johnson J, Kimball J, Piepmeier JR, Koster RD, Martin N, McDonald KC, Moghaddam M, Moran S, Reichle R, Shi JC, Spencer MW, Thurman SW, Tsang L, Van Zyl J (2010) The Soil Moisture Active Passive (SMAP) Mission. Proc IEEE 98:704–716

Fatichi S, Vivoni ER, Ogden FL, Ivanov VY, Mirus B, Gochis D, Downer CW, Camporese M, Davison JH, Ebel B, Jones N, Kim J, Mascaro G, Niswonger R, Restrepo P, Rigon R, Shen C, Sulis M, Tarboton D (2016) An overview of current applications, challenges, and future trends in distributed process-based models in hydrology. J Hydrol 537:45–60

Faybishenko B (2000) Vadose Zone characterization and monitoring: current technologies, applications, and future developments. *In:* Looney B, Falta R (Eds.), Vadose Zone Science and Technology Solutions. Battelle Press, Ohio, pp. 133–396

Faybishenko B, Witherspoon PA, Benson SM (Eds.) (2000) Dynamics of Fluids in Fractured Rock, Geophysical Monograph Series. American Geophysical Union, Washington D C

Ferré PA, Knight JH, Rudolph DL, Kachanoski RG (1998) The sample areas of conventional and alternative time domain reflectometry probes. Water Resour Res 34:2971–2979

Finsterle S (2015) Practical notes on local data-worth analysis. Water Resour Res 51:9904–9924

Fogg GE, Noyes CD, Carle SF (1998) Geologically based model of heterogeneous hydraulic conductivity in an alluvial setting, Hydrogeol J 6:131–143

Fox PM, Davis JA, Hay MB, Conrad ME, Campbell KM, Williams KH, Long PE (2012) Rate-limited U(VI) desorption during a small-scale tracer test in a heterogeneous uranium-contaminated aquifer. Water Resour Res 48:1–18

Fox PM, Davis JA, Zachara JM (2006) The effect of calcium on aqueous uranium(VI) speciation and adsorption to ferrihydrite and quartz. Geochim Cosmochim Acta 70:1379–1387

Gee GW, Ward AL (1999) Innovations in two-phase measurements of soil hydraulic properties. *In:* van Genuchten MT, Leij FJ, Wu L (Eds.), Proceedings of the International Workshop Characterization and Measurement of the Hydraulic Properties of Unsaturated Porous Media. University of California, Riverside, p. 241–269

González-Zamora, Á., Sánchez N, Martínez-Fernández J, Wagner W (2016) Root-zone plant available water estimation using the SMOS-derived soil water index. Adv Water Resour 96:339–353

Gupta H V, Clark MP, Vrugt JA, Abramowitz G, Ye M (2012) Towards a comprehensive assessment of model structural adequacy. Water Resour Res 48:1–16

Hammond GE, Lichtner PC, Mills RT (2014) Evaluating the performance of parallel subsurface simulators: An illustrative example with PFLOTRAN Water Resour Res 50:208–228

Han FX, Banin A, Triplett GB (2001) Redistribution of heavy metals in arid-zone soils under a wetting-drying cycle soil moisture regime. Soil Sci 166:18–28

Hansen DJ, McGuire JT, Mohanty BP (2011) Enhanced biogeochemical cycling and subsequent reduction of hydraulic conductivity associated with soil-layer interfaces in the Vadose Zone. J Environ Qual 40:1941–1954

Heinz J, Kleineidam S, Teutsch G, Aigner T (2003) Heterogeneity patterns of Quaternary glaciofluvial gravel bodies (SW-Germany): Application to hydrogeology. Sediment Geol 158:1–23

Herbert BE (1996) Application of geochemical speciation models for groundwater chemistry modeling and evaluation of remediation technologies. Geochemical Speciat Model p. 20

Hu W, Si BC (2013) Soil water prediction based on its scale-specific control using multivariate empirical mode decomposition. Geoderma 193–194:180–188

Hubbard SS, Gangodagamage C, Dafflon B, Wainwright H, Peterson J, Gusmeroli A, Ulrich C, Wu Y, Wilson C, Rowland J, Tweedie C, Wullschleger SD (2013) Quantifying and relating land-surface and subsurface variability in permafrost environments using LiDAR and surface geophysical datasets. Hydrogeol J 21:149–169

Hyun SP, Fox PM, Davis JA, Campbell KM, Hayes KF, Long PE (2009) Surface complexation modeling of U(VI) adsorption by aquifer sediments from a former mill tailings site at Rifle, Colorado. Environ Sci Technol 43:9368–9373

Ines AVM, Mohanty BP, Shin Y (2013) An unmixing algorithm for remotely sensed soil moisture. Water Resour Res 49:408–425

Jana RB (2010) Scaling Characteristics of Soil Hydraulic Parameters at Varying Spatial Resolutions. Texas A&M University

Jana RB, Mohanty BP (2012a) A comparative study of multiple approaches to soil hydraulic parameter scaling applied at the hillslope scale. Water Resour Res 48:W02520

Jana RB, Mohanty BP (2012b) A topography-based scaling algorithm for soil hydraulic parameters at hillslope scales: Field testing. Water Resour Res 48:W02519

Jana RB, Mohanty BP, Sheng Z (2012) Upscaling soil hydraulic parameters in the Picacho Mountain region using Bayesian neural networks. Trans ASABE 55:463–473

Jana RB, Mohanty BP, Springer EP (2008) Multiscale Bayesian neural networks for soil water content estimation. Water Resour Res 44: W08408

Jana RB, Mohanty BP, Springer EP (2007) Multiscale pedotransfer functions for soil water retention. Vadose Zo J 6:868

Johnson T, Versteeg R, Thomle J, Hammond G, Chen X, Zachara J (2015) Four-dimensional electrical conductivity monitoring of stage-driven river water intrusion: Accounting for water table effects using a transient mesh boundary and conditional inversion constraints. Water Resour Res 51:6177–6196

Johnson TC, Versteeg RJ, Rockhold M, Slater LD, Ntarlagiannis D, Greenwood WJ, Zachara J (2012) Characterization of a contaminated wellfield using 3D electrical resistivity tomography implemented with geostatistical, discontinuous boundary, and known conductivity constraints. GEOPHYSICS 77, EN85-EN96

Johnson TC, Versteeg RJ, Ward A, Day-Lewis FD, Revil A (2010) Improved hydrogeophysical characterization and monitoring through parallel modeling and inversion of time-domain resistivity andinduced-polarization data. GEOPHYSICS 75, WA27-WA41

Khaleel R, Yeh T-CJ, Lu Z (2002) Upscaled flow and transport properties for heterogeneous unsaturated media. Water Resour Res 38:1053

Klingbeil R Kleineidam S Asprion U Aigner T, Teutsch G (1999), Relating lithofacies to hydrofacies: Outcrop-based hydrogeological characterization of Quaternary gravel deposits, Sediment Geol 129:299–310

Koster RD, Dirmeyer PA, Guo Z, Bonan G, Chan E, Cox P, Gordon CT, Kanae S, Kowalczyk E, Lawrence D, Liu P, Lu C-H, Malyshev S, McAvaney B, Mitchell K, Mocko D, Oki T, Oleson K, Pitman A, Sud YC, Taylor CM, Verseghy D, Vasic R, Xue Y, Yamada T, GLACE Team (2004) Regions of strong coupling between soil moisture and precipitation. Science 305:1138–40

LaGrega MD, Buckingham PL, Evans JC (1994) Hazardous Waste Management (Second ed). Waveland Press, Inc., Long Grove, IL

Li L, Maher K, Navarre-Sitchler A, Druhan J, Meile C, Lawrence C, Moore J, Perdrial J, Sullivan P, Thompson A, Jin L, Bolton EW, Brantley SL, Dietrich WE, Mayer KU, Steefel CI, Valocchi A, Zachara J, Kocar B, Mcintosh J, Tutolo BM, Kumar M, Sonnenthal E, Bao C, Beisman J (2017) Expanding the role of reactive transport models in critical zone processes. Earth-Science Rev 165:280–301

Li L, Steefel CI, Kowalsky MB, Englert A, Hubbard SS (2010) Effects of physical and geochemical heterogeneities on mineral transformation and biomass accumulation during biostimulation experiments at Rifle, Colorado. J Contam Hydrol 112:45–63

Maher K, Steefel CI, DePaolo DJ, Viani BE (2006) The mineral dissolution rate conundrum: Insights from reactive transport modeling of U isotopes and pore fluid chemistry in marine sediments. Geochim Cosmochim Acta 70:337–363

Małecki J, Matyjasik M (2003) Vadose zone—challenges in hydrochemistry. Acta Geol Pol 52:449–458

Martínez-Fernández J, Ceballos A (2005) Mean soil moisture estimation using temporal stability analysis. J Hydrol 312:28–38

Mayer KU, Alt-Epping P, Jacques D, Arora B, Steefel CI (2015) Benchmark problems for reactive transport modeling of the generation and attenuation of acid rock drainage. Comput Geosci 19:599–611

McCabe MF, Franks SW, Kalma JD (2005) Calibration of a land surface model using multiple data sets. J Hydrol 302:209–222

McGuire JT, Long DT, Hyndman DW (2005) Analysis of recharge-induced geochemical change in a contaminated aquifer. Ground Water 43:518–530

Mcmillan H, Krueger T, Freer J (2012) Benchmarking observational uncertainties for hydrology: Rainfall, river discharge and water quality. Hydrol Process 26:4078–4111

Mohanty BP, Cosh MH, Lakshmi V, Montzka C (2017) Soil moisture remote sensing: state-of-the-science. Vadose Zo J 16:/vzj2016.10.0105

Montzka C, Jagdhuber T, Horn R, Bogena HR, Hajnsek I, Reigber A, Vereecken H (2016) Investigation of SMAP fusion algorithms with airborne active and passive L-Band microwave remote sensing. IEEE Trans Geosci Remote Sens 54:3878–3889

Mualem Y (1976) A new model for predicting the hydraulic conductivity of unsaturated porous media. Water Resour Res 12:513–522

Nimmo JR (2005) Unsaturated Zone Flow Processes, In: Anderson MG, Bear J (Eds.), Encyclopedia of Hydrological Sciences. Wiley, Chichester, UK, pp. 2299–2322

Nordstrom DK, Archer DG (2003) Arsenic thermodynamic data and environmental geochemistry. In:Arsenic in Ground Water. Kluwer Academic Publishers, Boston, pp. 1–25

Oostrom M, Truex MJ, Last GV, Strickland CE, Tartakovsky GD (2016) Evaluation of deep vadose zone contaminant flux into groundwater: Approach and case study. J Contam Hydrol 189:27–43

Pachepsky YA, Rawls WJ (2010) Accuracy and reliability of pedotransfer functions as affected by grouping soils. Soil Sci Soc Am J 63:1748–1757

Peng J, Loew A, Merlin O, Verhoest NEC (2017) A review of spatial downscaling of satellite remotely sensed soil moisture. Rev Geophys 55:341–366

Pollacco JAP, Mohanty BP (2012) Uncertainties of water fluxes in soil–vegetation–atmosphere transfer models: inverting surface soil moisture and evapotranspiration retrieved from remote sensing. Vadose Zo J 11: vzj2011.0167

Potyrailo RA, Hieftje GM (1998) Advanced strategies for spatially resolved analyte mapping with distributed fiber-optic sensors for environmental and process applications, Proc Conf Environ and Remediation Technol. SPIE Vol. 3534:49–63. International Society for Optics and Photonics

Pruess K (1985) A practical method for modeling fluid and heat flow in fractured porous media. Soc Pet Eng J 25:14–26

Pütz T, Fank J, Flury M (2018) Lysimeters in Vadose Zone research. Vadose Zo J 17:1–21

Rawls WJ, Brakensiek DL, Saxtonn KE (1982) Estimation of soil water properties. Trans ASAE 25:1316–1320

Richards LA (1931) Capillary conduction of liquids through porous mediums. J Appl Phys 1:318–333

Robinson DA, Campbell CS, Hopmans JW, Hornbuckle BK, Jones SB, Knight R, Ogden F, Selker J, Wendroth O (2008) Soil moisture measurement for ecological and hydrological watershed-scale observatories: A review. Vadose Zo J 7:358–389

Rubin Y, Hubbard S, Wilson A, Cushey M (1998) Aquifer characterization. *In:* Delleur J (Ed.), The Handbook of Groundwater Engineering. CRC Press, New York, NY

Rubin Y, Hubbard SS (2005) Hydrogeophysics. Springer, New York, NY

Sahoo AK, De Lannoy GJM, Reichle RH, Houser PR (2013) Assimilation and downscaling of satellite observed soil moisture over the Little River Experimental Watershed in Georgia, USA Adv. Water Resour 52:19–33

Sassen DS, Hubbard SS, Bea SA, Chen J, Spycher N, Denham ME (2012) Reactive facies: An approach for parameterizing field-scale reactive transport models using geophysical methods. Water Resour Res 48:W10526

Šimůnek J, Jarvis NJ, van Genuchten MT, Gärdenäs A (2003) Review and comparison of models for describing non-equilibrium and preferential flow and transport in the vadose zone. J Hydrol 272:14–35

Skierucha W, Wilczek A, Szypłowska A, Sławiński C, Lamorski K, Skierucha W, Wilczek A, Szypłowska A, Sławiński C, Lamorski K (2012) A TDR-based soil moisture monitoring system with simultaneous measurement of soil temperature and electrical conductivity. Sensors 12:13545–13566

Skøien JO, Blöschl G (2006) Sampling scale effects in random fields and implications for environmental monitoring. Environ Monit Assess 114:521–552

Snow DT (1965) A parallel plate model of fractured permeable media. University of California Berkeley

Sonnenthal E, Spycher N, Xu T, Zheng L, Miller N, Pruess K (2014) TOUGHREACT V3.0-OMP reference manual: A parallel simulation program for non-isothermal multiphase geochemical reactive transport. Berkeley, CA

Steefel CI, Appelo CAJ, Arora B, Jacques D, Kalbacher T, Kolditz O, Lagneau V, Lichtner PC, Mayer KU, Meussen JCL, Molins S, Moulten D, Shao H, Šimůnek J, Spycher N, Yabusaki SB, Yeh GT (2015) Reactive transport codes for subsurface environmental simulation. Comput Geosci 19:445–478

Steefel CI, DePaolo DJ, Lichtner PC (2005) Reactive transport modeling: An essential tool and a new research approach for the Earth sciences. Earth Planet Sci Lett 240:539–558

Steefel CI, Lasaga AC (1994) A coupled model for transport of multiple chemical species and kinetic precipitation/dissolution reactions with application to reactive flow in single phase hydrothermal systems. Am J Sci 294:529–592

Steefel CI, Maher K (2009) Fluid–rock interaction: a reactive transport approach. Rev Mineral Geochem 70:485–532

Stephens DB (2018) Vadose Zone Hydrology. CRC Press.

Suarez F, Hausner MB, Dozier J, Selker JS, Tyler SW (2011) Heat transfer in the environmeny : development and use of fiber-optic distributed temperature sensing, *In:* Developments in Heat Transfer. InTechOpen.

Tietje O, Tapkenhinrichs M (1993) Evaluation of pedo-transfer functions. Soil Sci Soc Am J 57:1088–1095

Tomasella J, Pachepsky Y, Crestana S, Rawls WJ (2003) Comparison of two techniques to develop pedotransfer functions for water retention. Soil Sci Soc Am J 67:1085–1092

Topp GC, Davis JL, Annan AP (1980) Electromagnetic determination of soil water content: Measurements in coaxial transmission lines. Water Resour Res 16:574–582

Tran AP, Dafflon B, Hubbard SS, Kowalsky MB, Long P, Tokunaga TK, Williams KH (2016) Quantifying shallow subsurface water and heat dynamics using coupled hydrological-thermal-geophysical inversion. Hydrol Earth Syst Sci 20:3477–3491

Tsang C-F (1987) Comments on model validation. Transp Porous Media 2:623–629

Um W, Serne RJ, Brown CF, Rod KA (2008) Uranium(VI) sorption on iron oxides in Hanford Site sediment: Application of a surface complexation model. Appl Geochem 23:2649–2657

Vachaud G, Passerat De Silans A, Balabanis P, Vauclin M (1985) Temporal stability of spatially measured soil water probability density function 1. Soil Sci Soc Am J 49:822–828

van den Hurk BJJM, Bastiaanssen WGM, Pelgrum H, van Meijgaard E, Hurk BJJM van den, Bastiaanssen WGM, Pelgrum H, Meijgaard E van (1997) A New methodology for assimilation of initial soil moisture fields in weather prediction models using Meteosat and NOAA Data. J Appl Meteorol 36:1271–1283

van Genuchten MT (1980) A closed-form equation for predicting the hydraulic conductivity of unsaturated soils1. Soil Sci Soc Am J 44:892–898

Vaughan PJ (1987) Analysis of permeability reduction during flow of heated, aqueous fluid through westerly granite. *In:* Tsang C-F (Ed.), Coupled Processes Associated with Nuclear Waste Repositories. Academic Press, New York

Vereecken H, Huisman JA, Pachepsky Y, Montzka C, van der Kruk J, Bogena H, Weihermu˙ller L, Herbst M, Martinez G, Vanderborght J (2014) On the spatio-temporal dynamics of soil moisture at the field scale. J Hydrol 516:76–96

Vereecken H, Kasteel R, Vanderborght J, Harter T (2007) Upscaling hydraulic properties and soil water flow processes in heterogeneous soils. Vadose Zo J 6:1–28

Vereecken H, Maes J, Feyen J, Darius P (1989) Estimating the soil moisture retention characteristic from texture, bulk density, and carbon content. Soil Sci 64:155–165

Vereecken H, Schnepf A, Hopmans JW, Javaux M, Or D, Roose T, Vanderborght J, Young MH, Amelung W, Aitkenhead M, Allison SD, Assouline S, Baveye P, Berli M, Brüggemann N, Finke P, Flury M, Gaiser T, Govers G, Ghezzehei T, Hallett P, Hendricks Franssen HJ, Heppell J, Horn R, Huisman JA, Jacques D, Jonard F, Kollet S, Lafolie F, Lamorski K, Leitner D, McBratney A, Minasny B, Montzka C, Nowak W, Pachepsky Y, Padarian J, Romano N, Roth K, Rothfuss Y, Rowe EC, Schwen A, Šimůnek J, Tiktak A, Van Dam J, van der Zee SEATM, Vogel HJ, Vrugt JA, Wöhling T, Young IM (2016) Modeling soil processes: review, key challenges, and new perspectives. Vadose Zo J 15:vzj2015.09.0131

Verhoest NEC, Van Den Berg MJ, Martens B, Lievens H, Wood EF, Pan M, Kerr YH, Al Bitar A, Tomer SK, Drusch M, Vernieuwe H, De Baets B, Walker JP, Dumedah G, Pauwels VRN (2015) Copula-based downscaling of coarse-scale soil moisture observations with implicit bias correction. IEEE Trans. Geosci Remote Sens 53:3507–3521

Wainwright H, Arora B, Hubbard S, Lipnikov K, Moulton D, Flach G, Eddy-Dilek C, Denham M (2018) Sustainable remediation in complex geologic systems. *In:* Encyclopedia of Inorganic and Bioinorganic Chemistry. John Wiley & Sons, Ltd, Chichester, UK, pp. 1–12

Wainwright HM, Chen J, Sassen DS, Hubbard SS (2014) Bayesian hierarchical approach and geophysical data sets for estimation of reactive facies over plume scales. Water Resour Res 50:4564–4584

Wainwright HM, Dafflon B, Smith LJ, Hahn MS, Curtis JB, Wu Y, Ulrich C, Peterson JE, Torn MS, Hubbard SS (2015) Identifying multiscale zonation and assessing the relative importance of polygon geomorphology on carbon fluxes in an Arctic tundra ecosystem. J Geophys Res Biogeosci 120:788–808

Wainwright HM, Flores Orozco A, Bücker M, Dafflon B, Chen J, Hubbard SS, Williams KH (2016) Hierarchical Bayesian method for mapping biogeochemical hot spots using induced polarization imaging. Water Resour Res 52:533–551

Wan J, Tokunaga TK, Dong W, Williams KH, Kim Y, Conrad ME, Bill M, Riley WJ, Hubbard SS (2018) Deep unsaturated zone contributions to carbon cycling in semiarid environments. J Geophys Res Biogeosci 123:3045–3054

Warren JE, Root PJ (1963) The behavior of naturally fractured reservoirs. Soc Pet Eng J 3:245–255

Weissmann G S Y Zhang E M LaBolle, and G E Fogg (2002) Dispersion of groundwater age in an alluvial aquifer system, Water Resour Res 38:1198

Western AW, Grayson RB, Blöschl G (2002) Scaling of soil moisture: a hydrologic perspective. Ann Rev Earth Planet Sci 30:149–180

Wilson LH, Johnson PC, Rocco JR (2005) Collecting and Interpreting Soil Gas Samples from the Vadose Zone: A Practical Strategy for Assessing the Subsurface Vapor-to-Indoor Air Migration Pathway at Petroleum Hydrocarbon Sites (2005). Washington D C

Wöhling T, Vrugt JA, Barkle GF (2008) Comparison of three multiobjective optimization algorithms for inverse modeling of Vadose zone hydraulic properties. Soil Sci Soc Am J 72:305–319

Wu J, Li H (2006) Concepts of scale and scaling *In:* Scaling and Uncertainty Analysis in Ecology: Methods and Applications. Wu J, Jones B, Li H, Loucks OL. (Eds) Springer

Xie M, Mayer KU, Claret F, Alt-Epping P, Jacques D, Steefel C, Chiaberge C, Simunek J (2015) Implementation and evaluation of permeability–porosity and tortuosity–porosity relationships linked to mineral dissolution–precipitation. Comput Geosci 19:655–671

Yabusaki SB, Fang Y, Williams KH, Murray CJ, Ward AL, Dayvault R, Waichler SR, Newcomer DR, Spane FA, Long PE (2011) Variably saturated flow and multicomponent biogeochemical reactive transport modeling of a uranium bioremediation field experiment. J Contam Hydrol 126:271–290

Yabusaki SB, Wilkins MJ, Fang Y, Williams KH, Arora B, Bargar J, Beller HR, Bouskill NJ, Brodie EL, Christensen JN, Conrad ME, Danczak RE, King E, Soltanian MR, Spycher NF, Steefel CI, Tokunaga TK, Versteeg R, Waichler SR, Wainwright HM (2017) Water table dynamics and biogeochemical cycling in a shallow, variably-saturated floodplain. Environ Sci Technol 51:3307–3317

Reviews in Mineralogy & Geochemistry
Vol. 85 pp. 329-348, 2019
Copyright © Mineralogical Society of America

11

Abiotic and Biotic Controls on Soil Organo–Mineral Interactions: Developing Model Structures to Analyze Why Soil Organic Matter Persists

Dipankar Dwivedi[1], Jinyun Tang[2], Nicholas Bouskill[2], Katerina Georgiou[3], Stephany S. Chacon[2], William J. Riley[2]

*[1]Geosciences Division
Lawrence Berkeley National Laboratory
1 Cyclotron Road, M.S. 74R316C
Berkeley, CA 94720
USA*

*[2]Climate and Environmental Sciences Division
Lawrence Berkeley National Laboratory
1 Cyclotron Road, M.S. 74R316C
Berkeley, CA 94720
USA*

*[3]Department of Earth System Science
Stanford University
Stanford, CA 94305
USA*

DDwivedi@lbl.gov; jinyuntang@lbl.gov; njbouskill@lbl.gov;

georgiou@stanford.edu; SSChacon@lbl.gov; wjriley@lbl.gov

INTRODUCTION

Soil organic matter (SOM) represents the single largest actively cycling reservoir of terrestrial organic carbon, accounting for more than three times as much carbon as that present in the atmosphere or terrestrial vegetation (Schmidt et al. 2011; Lehmann and Kleber 2015). SOM is vulnerable to decomposition to either CO_2 or CH_4, which can increase atmospheric greenhouse gas concentrations (GHGs) and serve as a positive feedback to climate change. Conversely, the formation and stabilization of SOM within aggregates or associated with soil minerals can lead to carbon sequestration, representing a negative feedback to climate change. However, the conundrum as to why some SOM decomposes rapidly, while other thermodynamically unstable SOM can persist on centennial time scales (Hedges et al. 2000), leads to substantial uncertainty in model structures, as well as uncertainty in the predictability of the land carbon sink trajectory.

This chapter aims to tackle a part of this problem through generating recommendations for improved model structure in the representation of SOM cycling within global Earth System Models. The chapter is organized as follows: We first review the recent work contributing to the understanding of SOM stability. Then, we discuss current model structures and how these model structures deal with various processes, including data-model integration, microbial modeling, organo–mineral interactions, and SOM persistence. Finally, we conclude with recommendations for the future development of soil carbon models and the required relevant processes that determine the stability of SOM.

1529-6466/19/0085-0011$05.00 (print)
1943-2666/19/0085-0011$05.00 (online)   http://dx.doi.org/10.2138/rmg.2019.85.11

THE EMERGENT PICTURE OF SOIL ORGANIC MATTER STABILITY

Traditional conceptual models of SOM formation and stability have focused on the chemical structure of SOM compounds. These models consider the recalcitrance of certain SOM compounds to biological decomposition as the primary determinant of persistence and long-term stability (Schmidt et al. 2011; Lehmann and Kleber 2015). Within this perspective, plant-derived polymers are the primary source of organic matter (Fig. 1a). These polymers were thought to form through a suite of reactions, via biotic (Fig. 1a.i) or abiotic mechanisms, leaving large, dark-colored (humic) compounds rich in carbon and nitrogen (N) (Lehmann and Kleber 2015). It was thought that the formation of these compounds renders SOM resistant to further decomposition. However, the hypothesis that chemical recalcitrance (and humification) explains SOM stability has gradually fallen out of favor, due in part to a failure to demonstrate that the secondary synthesis of humic compounds is relevant in natural systems (Keiluweit et al. 2015; Kleber et al. 2015), but also because chemically complex compounds (e.g., lignin) that should theoretically persist in soil can have shorter residence times than supposedly labile molecules (Schmidt et al. 2011). In response, new theories of SOM formation and stability have begun to emerge (Schmidt et al. 2011; Riley et al. 2014; Cotrufo et al. 2015; Lehmann and Kleber 2015; Castellano et al. 2015; Dwivedi et al. 2017a).

Many of these emerging theories put interactions between physical and geochemical conditions and biology at the forefront of SOM stability. Schmidt et al. (2011), for example, concluded that SOM persistence depended on ecosystem properties; they described an array

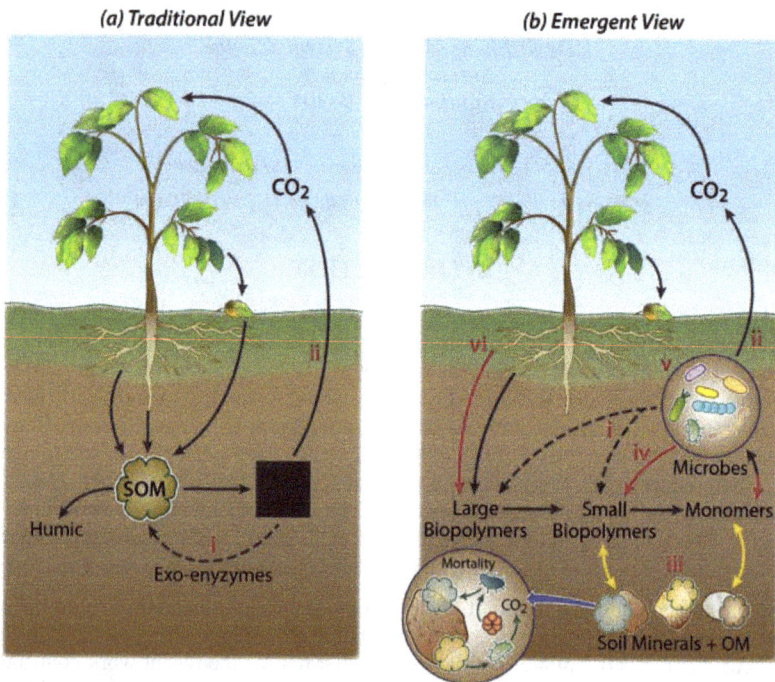

Figure 1. Schematic depicting the **(a)** traditional and **(b)** emergent conceptual understanding of soil organic matter (SOM) cycling. The traditional view does not explicitly represent the primary underlying agents and processes (e.g., microbes, aggregation) that drive SOM cycling. In contrast, the emergent view suggests that SOM decomposition is a function of a wide range of ecosystem properties and mechanisms. The key processes depicted in the figure include: (i) exoenzyme production, (ii) decomposition, (iii) organo-mineral interactions, (iv) microbial necromass, (v) ecology of belowground biota, and (vi) root contributions and mobilization of nutrients. Additional details are provided in the text.

of characteristic environmental traits that contribute to the stability of SOM and sequestration of carbon in soils. These properties include climate, physical properties (e.g., the formation of preferential flow paths within soils that determine the degree of connectivity between microbes and substrates—Hunt and Manzoni 2016), chemical properties (e.g., soil pH or the chemical quality of plant exudates and litter—Castellano et al. 2015), physical protection mechanisms (e.g., the sorption of organic matter onto mineral surfaces, Fig. 1b.iii—Kleber et al. 2015), soil redox status, the ecology of belowground biota (including bacteria, archaea, fungi and soil macrofauna, Fig. 1b.v—Bardgett and Van Der Putten 2014), and the metabolic constraints on OM mineralization. In such a network of factors, the diverse biotic components interact chemically and physically with each other, and with the surrounding heterogeneous matrix of minerals and soil aggregates, resulting in the emergence of SOM.

Microbial and plant processes

The majority of SOM is decomposed by microbes and respired into the atmosphere over short (< 50 year) time scales. However, a significant portion of this SOM can become chemically or physically protected—before or after microbial processing—and persist for much longer (100 to 1,000 year) time scales. Several studies have recognized the importance of diverse microorganisms in the formation of SOM (Allison et al. 2010; Knicker 2011; Cotrufo et al. 2013). The Microbial Efficiency-Matrix Stabilization framework, for example, describes SOM dynamics resulting from individual physiologies of diverse microorganisms (Cotrufo et al. 2013). The concept behind the "framework" asserts that plant exudates and litter, while being significant components of SOM themselves (Castellano et al. 2015), are decomposed (Fig. 1a/b.ii), assimilated (Fig. 1b.v), and processed by the microbial community.

Plants play an active role in soil carbon cycling through litterfall and root exudation, thereby influencing microbial growth (Fig. 1b.iv). For instance, roots excrete small-molecular-weight compounds to aid plant acquisition of nutrients and water. In addition, when plants supply small molecular-weight compounds such as sugars or reducible compounds, they prime microbial growth in soil. In general, the 'priming' of the soil microbial community can stimulate the production of exoenzymes near the root zone, increasing depolymerization of large biopolymers to monomers, which can become accessible to plant roots (Fig. 1b.i) (Kuzyakov et al. 2000; Gunina and Kuzyakov 2015). Moreover, roots also play a significant role in SOM stability through activities aimed at liberating and mobilizing nutrients (Fig. 1b.vi). Organic acids and chelators can disrupt the ligand bonds between mineral surfaces and organic matter or nutrients. Such actions can subsequently release previously immobilized phosphorus and other inorganic nutrients (Hocking 2001; Wu et al. 2018), and destabilize mineral-associated organic matter (MAOM) (Keiluweit et al. 2015; Jilling et al. 2018).

Overall, microbial processing of decomposed plant exudates and litter results in the microbial "exudation" of different metabolic byproducts, including enzymes, lipids, glycoproteins, extracellular polysaccharides (EPS), and the eventual accumulation of chemically heterogeneous necromass (Fig. 1b.iv) that are the major precursors to stable SOM (Miltner et al. 2012; Cotrufo et al. 2013, 2015; Kögel-Knabner 2017). Indeed, recent evidence suggests that the chemical composition of persistent SOM more closely resembles that of microbial cells and their byproducts than that of plant materials (Kögel-Knabner 2017). Decomposition-resistant SOM often exhibits a narrow carbon-to-nitrogen (carbon:nitrogen) ratio (between 8 to 10) that is more representative of microbial residues than the wide C:N ratio (upwards of 30) characteristic of plant materials (Baisden 2002; Conen et al. 2008). These microbial cells and excretions are often stabilized via organo–mineral and organo–metal–oxide interactions.

Organo–mineral interactions

Physicochemical associations between soil minerals and plant or microbial-derived organic matter can result in the protection of theoretically thermodynamically unstable (or labile) compounds within soil for decades. Soil minerals often constitute a mixture of phyllosilicates, oxides, and hydroxides, as well as short-range order and noncrystalline minerals in some soil types. These particles are generally found in the clay (< 2 µm) size-fraction of soils. The accessibility of SOM to microorganisms can be hindered through mineral interactions by adsorbing or occluding organic compounds and immobilizing exoenzymes responsible for the depolymerization reactions. Recent studies have demonstrated that different classes of microbial biochemicals show variability in the degree to which they adsorb onto mineral surfaces (Jagadamma et al. 2012, 2014). For example, Swenson et al. (2015) and Newcomb et al. (2017) have demonstrated generalizable patterns in the competitive sorption of different functional groups to the surface of ferrihydrite, an amorphous iron oxide. Phosphate-, N-, and aromatic-containing metabolites and carboxylates showed strong adsorptive behavior onto mineral surfaces (Swenson et al. 2015). Conversely, differential biochemical adsorption shapes the composition, stoichiometry (e.g., carbon:nitrogen:phosphorous ratio), and accessibility of the dissolved organic matter (DOM) pool. For example, the immobilization of exoenzymes on mineral surfaces can alter catalytic function by blocking the active site or changing enzyme structure (Quiquampoix and Ratcliffe 1992; Baron et al. 1999; Norde and Giacomelli 2000; Servagent-Noinville et al. 2000). Although adsorbed exoenzymes generally decrease in catalytic capacity, their residence times are higher than free exoenzymes (Yan et al. 2010). These mechanisms, in turn, can shape the structure, function, and activity of the microbial community, which can promote the production of enzymes or organic acids that mineralize available polymers and desorb stable SOM compounds (Kaiser et al. 2014).

Finally, empirical observations and modeling efforts have typically concentrated on surface soils (< 20 cm) for the estimation of carbon storage and turnover rates (Trumbore et al. 1995; Hicks Pries et al. 2018). However, over half of global SOM is stored in subsoils (> 20 cm), and this SOM is significantly older than surface SOM (Jobbágy and Jackson 2000; Mathieu et al. 2015). These patterns suggest a distinct balance of decomposition and preservation processes at depth relative to that at the surface. In all, this emerging picture of MAOM places a premium on understanding the interdependent impacts of soil hydrology, geology, geochemistry, and ecology across the complete soil profile. In particular, it calls for an improved understanding of the feedback between soil properties and the metabolisms of microbes (fungi, bacteria, and archaea). Insight into these emergent properties of soil carbon cycling has in particular benefited from improvements in approaches to study fine-scale MAOM dynamics.

EMERGING TECHNOLOGIES

Emerging technologies are shaping the picture of MAOM dynamics as affected by biology, chemistry, and soil structure. Advances in analytical techniques are revealing the chemistry of minerals and the composition of soil organic matter. Extremely detailed information on the composition of extractable soil organic matter can be obtained by using Fourier Transform Ion Cyclotron Resonance Mass spectrometry (FT-ICR-MS) and quantified by ^{1}H and ^{13}C solution-state Nuclear Magnetic Resonance (Simpson et al. 2001; Tfaily et al. 2015). For non-extractable organic matter, solid-state NMR can characterize ^{13}C labeled organic matter sorbed on surfaces of non-paramagnetic minerals (Clemente et al. 2012). FT-Infrared spectrometry is compatible with living systems and can detect functional group characteristics of MAOM, even when associated with paramagnetic minerals such as iron oxides (Omoike and Chorover 2004; Lehmann et al. 2007). The capacity for imaging soil in nondestructive ways has grown. Advancements in X-ray computed tomography of intact soils yields information regarding the soil structure, pore network and even roots, which can be processed into three-dimensional images (Taina et al. 2011; Mairhofer et al. 2012). More importantly, when combining imaging techniques with chemical analysis, components of MAOM can be spatially resolved. Techniques such as synchrotron radiation scanning transmission X-ray microscopy (STXM) coupled with near edge X-ray absorption fine structure (NEXAFS) has been used to determine the spatial heterogeneity of MAOM in two dimensions (Lehmann et al. 2008). A more detailed literature review on the benefits and limitations of advanced analytical techniques used for MAOM has been presented previously (Kleber et al. 2015).

Microbial traits that affect SOM stability

The emerging understanding of SOM dynamics also calls for greater efforts to identify and quantify the role of microbial functional traits in determining the fate of organic compounds within soils. Microbial communities are complex adaptive systems (Levin 1999), in which heterogeneous collections of individual organisms that coexist, co-operate, and compete across wide-ranging environmental gradients shape the evolution and emergence of community structure. The fitness of individuals within a community is governed by their inherent physiological traits and the energetic costs of intracellular trade-offs (Maharjan et al. 2013). Microbial traits can be defined as physiological or morphological characteristics closely linked with ecosystem function and performance (Viollea et al. 2014). These traits can be broadly separated into response (determining how organisms respond to environmental change, such as drought) and effect (determining an organism's effect on ecosystem processes, such as nitrification) traits (Allison 2012). The expression of, and resource allocation to, different traits (e.g., exoenzyme production) is regulated by intracellular energetic trade-offs (Beardmore et al. 2011). These factors determine an individual microbe's ability to compete for resources within a wider community, and to tolerate and thrive under extraneous selective pressures, including fluctuating environmental conditions. Traits and trade-offs that promote optimal relative fitness will subsequently be selected for and fixed (at least temporarily) within a population (Razinkov et al. 2013).

Many relevant traits (e.g., osmolyte production or exoenzyme production) show deep evolutionary divergence and nonrandom distributions (Schimel and Schaeffer 2012), meaning that environmental selection for specific ecological strategies can lead to concomitant changes in a suite of genome-linked traits related to SOM formation and degradation. Reorganization of microbial community composition can also change anabolic trait distribution and select for traits that convert plant-derived carbon to metabolites (e.g., osmolytes), or necromass that may be preferentially stabilized on soil minerals (Liang et al. 2017). On a broader scale, community changes that result in an increase in the soil fungal:bacterial ratio can also result in higher SOM due to the higher biomass associated with the proliferation of fungi (Strickland and Rousk 2010). The extent of bacterial secondary metabolism (e.g., antibiotics, non-structural carbohydrates, siderophores, etc.), which plays a large role in shaping the chemical composition of microbial exudates and necromass, is beginning to be quantified through advances in genomic and metabolomic technologies. For example, Crits-Christoph et al. (2018) recently used large-scale metagenomic sequencing to reconstruct the genomes of a number of common soil bacteria. They uncovered the potential for extensive secondary metabolite production, suggesting a largely underestimated capacity of these bacteria to produce and exude biosynthetic products and contribute to the composition of SOM. However, while "omic" techniques provide greater understanding of the complexity of the microbial community and its metabolic diversity, the need for large amounts of material for nucleic acid extraction (covering multiple niches) and the destructive nature of that extraction, limit understanding of temporal dynamics. These dynamics include how community-composition interactions necessitate the development and application of microbial centric models that permit testing of hypotheses related to cell-cell interactions, response to environmental fluctuations, and the microbial feedback to carbon dynamics (Riley et al. 2014; Georgiou et al. 2017).

While these small-scale processes undoubtedly interact, giving rise to larger-scale phenomena, a tractable balance of simplicity and complexity must be sought in the development of parsimonious, yet sufficiently accurate, macro-scale models of SOM cycling. To this end, it is critical to thoroughly understand key SOM decomposition and stabilization mechanisms, where and when they are important, and how nonlinear interactions between components may compound. Changes in climate and vegetation may also cause important shifts in the relative role of key mechanisms, with broad implications for the terrestrial carbon cycle.

MODELING SOM DYNAMICS

There is a suite of SOM dynamics models incorporated into land models ranging from simple pool-based CENTURY-like representations (Parton et al. 1987, 1998, 2010; Jenkinson et al. 2008) to more complex models such as the Biotic and Abiotic Model of SOM (BAMS; Riley et al. 2014; Dwivedi et al. 2017; Tang et al. 2019) and the COntinuous representation of SOC in the organic layer and the mineral soil, Microbial Interactions and Sorptive StabilizatION (COMISSION; Ahrens et al. 2015), which explicitly represents physicochemical processes and microbially mediated reactions. Here we briefly describe the traditional and some of the emerging modeling approaches. More importantly, this section aims to describe differences in the traditional and emerging modeling approaches, and the need for a reactive transport modeling framework representing relevant mechanisms.

Traditional modeling approaches

Traditionally, mean residence times (MRT) are used to categorize soil organic carbon into different "pools". The decomposition reaction rate is the inverse of MRT, and therefore pools with shorter MRT have higher inherent decomposition reactivity (Parton et al. 1987; Davidson and Janssens 2006; Kleber 2010). These pools have been extensively used to investigate SOM decomposition in models such as CENTURY, ROTH-C, and their progeny models. These pools are intended to represent from the most labile to the most recalcitrant carbon in CENTURY-Like models (Parton et al. 1987, 2010), and from humified organic matter to inert in ROTH-C models (Jenkinson and Coleman 2008; Jenkinson et al. 2008). Similarly, SOM decomposition rates are also empirically referred to as fast, slow, and passive for pools ranging from the most labile, or humified, to the most recalcitrant, or inert. These traditional approaches have been very widely applied, but lack representation of the underlying microbial and abiotic processes described above. They have also been shown to have very high parameter equifinality, in spite of being calibratable in many systems (Luo et al. 2015, 2017).

The traditional CENTURY-like models are formulated by assuming sufficient microbial activity to support organic matter decomposition, because litter bag and soil incubation experiments often observe consistent patterns of exponential organic matter decomposition over time (Davidson et al. 1995; Berg and McClaugherty 2014). Soil minerals are considered to exert only a multiplicative effect on the turnover rates of the different components of SOM. Other multiplicative factors are similarly applied to represent how temperature and moisture regulate SOM decomposition. However, spectrum analysis of field measurements such as concentration-gradient-based soil surface CO_2 efflux time series often show that there are many dynamic signals that cannot be resolved with traditional models (e.g., Vargas et al. 2010; Manzoni et al. 2018). Notably, the traditional models fail to capture the priming effect in soil, which involves strong interactions between microbes and their bioproducts, substrates, and soil minerals (e.g., Fontaine and Barot 2005). Furthermore, when organo–mineral interactions are involved, substrate bio-availability usually decreases. This observation suggests that in soils where soil minerals are plentiful, oligotrophic microbes are favored. However, model representation of oligotrophic microbes may not be necessary, considering that microbes can live in a dormant state, making it difficult to determine whether oligotrophic microbes are the major contributors to observed microbial activity or simply whether a smaller population of more copiotrophic organisms are active (Joergensen and Wichern 2018).

Although these simplified pool-based models have been important for estimating SOM stocks, they are unable to capture SOM dynamics beyond first order effects. For example, most site- and global-scale terrestrial SOM models apply first-order relationships, and therefore cannot predict explicit turnover times for different compounds. Further, a major obstacle remains in matching the model with data in terms of understanding which components of SOM are decomposed and which are stabilized, because the pools in the traditional models are not structurally well defined (i.e., not observable; Abramoff et al. 2018). Most importantly, these traditional modeling approaches cannot answer why organic matter that is thermodynamically unstable persists in soils, sometimes for a millennium (Schmidt et al. 2011). In view of these modeling inconsistencies, new models have been developed and applied to understand the effects of microbial processes and organo–mineral interactions on SOM dynamics.

Transitional modeling approaches

Interactions between microbes and soil minerals can be explicitly represented in more mechanistic models—some of the earlier attempts can be found in Smith (1979) and Grant et al. (1993). The recently published batch (i.e., no vertical resolution) models (e.g., MEND; Wang et al. 2013, RESOM; Tang and Riley 2014, CORPSE; Sulman et al. 2014, and MIMICS; Wieder et al. 2014) use diverse approaches to formulate organo–mineral interactions. There are also several new models (described below in Section `Current use of reactive transport modeling of SOM dynamics`) that explicitly represent microbes, soil mineral interactions, and vertical structure.

The MEND model incorporates a mineral-adsorbed soil carbon pool affected by separate adsorption and desorption kinetics that result in a Langmuir isotherm at equilibrium. However, Wang et al. (2013) argued that by incorporating microbial dormancy into the MEND modeling framework, only a single microbial functional group is required to reproduce laboratory aerobic incubation measurements of biomass and CO_2 respiration rates. A similar approach is adopted by the CORPSE model. RESOM uses the equilibrium form of the Langmuir isotherm, which is then integrated with the Equilibrium Chemistry Approximation kinetics (Tang and Riley 2013; Tang 2015; Zhu et al. 2017) to represent interactions between dissolvable carbon, mineral surfaces, exoenzymes, and microbes. The MIMICS model avoids an explicit formulation of organo–mineral interactions, and instead lumps these interactions by changing the half saturation constants used to drive Michaelis-Menten type microbial kinetics. These models were evaluated together in a recent model intercomparison (Sulman et al. 2018); that study concluded that the observational data were insufficient to invalidate any of the model structures. The authors suggested that long-term observations of soil carbon input (from plants) and perturbation experiments (such as warming) are needed to better evaluate and constrain these newly developed models.

Next-generation process representation approaches

Soil is a complex system and considering its physical and chemical evolution is critical for accurately capturing SOM dynamics (Riley et al. 2019). The following discussion is intended to review modeling needs for the following important topics: (1) microbial dynamics, (2) mineral associated organic matter (MAOM) interactions, and (3) the effects of SOM molecular structure. We subsequently make recommendations as to how these processes can be integrated into a reactive transport model.

Microbial Processes. The structure, function, and response to perturbation of the soil microbial community plays an important role in shaping soil biogeochemistry and SOM dynamics (see Section `Microbial and plant processes` above), motivating the active representation of the belowground community in next-generation model structures. Allison (2012), using the DEMENT model to explore how microbial traits would covary with the evolution of decomposing litter, found that incorporating microbes with different growth traits enabled the model to explain more than 60% of the variance in decomposition rates of 15 Hawaii litter types. Using a similar modeling strategy, Kaiser et al. (2014) found that microbial diversity is essential in explaining the stoichiometric evolution in decomposing litters. However, relatively few studies explored the influence of microbial diversity on carbon formation and decomposition in the mineral soil. When this is done, microbial community structure is often aggregated into the *r* versus *K* strategists that are adopted from the MacArthur and Wilson (1967) classification scheme. For instance, the MIMICS model (Wieder et al. 2014) used a copiotrophic group to represent the r-strategists and an oligotrophic group to represent the *K*-strategists. This *r* vs *K* representation of microbial community structure has been argued to be critical to accurately simulating the priming effect (Fontaine et al. 2003). However, models like MEND, CORPSE, and RESOM are able to simulate priming with only one microbial group, as long as substrate diversity is included in the model. In addition, Lawrence et al. (2009) showed that short-term SOM dynamics resulting from interactions between microbes and environmental conditions (e.g., drying–rewetting events) could not be adequately represented using a first-order representation of decomposition (as in Century or RothC). They suggested that while first-order models can capture the bulk response of SOM dynamics under steady-state conditions, the inclusion of exoenzyme and microbial controls on decomposition is necessary to appropriately simulate pulsed rewetting dynamics.

Mineral Associated Organic Matter (MAOM). Traditional models that employ a function of soil texture to represent soil mineral effects on soil carbon dynamics, as a multiplicative modifier to decomposition rates, are likely underestimating the roles played by soil minerals. For example, soil minerals shift between different oxidation status due to particular electron orbital configurations, which enable them to act as either reductants or oxidants of organic matter. A number of microorganisms are able to take advantage of this chemical flexibility and couple these soil minerals with organic carbon to sequester energy when the most favorable oxidant, i.e., oxygen, is in short supply. Likewise, none of the models mentioned above sufficiently accounts for organo–mineral interactions by only considering sorption–desorption dynamics.

Soil minerals are important factors in determining soil aqueous chemistry. Minerals, microbial metabolic byproducts (e.g., organic acids), and inorganic compounds exert strong controls on soil pH and redox potential. However, few contemporary soil carbon models have taken such organo–mineral interactions into account. Rather, when they are needed, such as in modeling methane dynamics, pH and redox potential are formulated as multiplier functions without accounting for their mechanistic evolutions (Zhang et al. 2002; Li et al. 2004; Zhuang et al. 2004; Riley et al. 2011). A slightly more mechanistically based model is attempted in Tang et al. (2016), where they merged the Windermere Humic Aqueous Model (WHAM; Tipping 1994) with the CLM-CN model to simulate the effect of iron cycling on CO_2 and CH_4 production in anoxic Arctic soil microcosms. Their model also simulated the evolution of pH using the aqueous chemistry capability provided by WHAM. However, the decomposition dynamics follow the traditional century-like formulation, so the model is therefore still not mechanistically representing organo–mineral interactions. Zheng et al. (2019) recently updated the framework from Tang et al (2016) with the PHREEQC 3.0 model (Charlton and Parkhurst 2011) and applied the resultant code to incubated organic soils from an Arctic tundra site. Although their simulations indicated some favorable comparisons with CO_2 and CH_4 fluxes, the representation of organo–mineral interactions was not improved.

The most comprehensive model integrating soil pH, redox, and sorption-desorption aspects of organo–mineral interaction is likely the *ecosys* model (e.g., Grant et al. 2003, 2017a, b), which includes a comprehensive parameterization of aqueous electrolyte dynamics and their relations with soil pH and ion-exchange capacity (Grant and Heaney 2010). However, this parameterization was primarily used to model inorganic phosphorus dynamics. As such, the role of soil minerals (e.g., ferrihydrite) as alternative electron acceptors for organic carbon decomposition is not modeled.

In summary, although much of the necessary chemistry knowledge exists, a modeling framework that systematically evaluates how organo–mineral interactions will play out in various aspects of soil biogeochemistry has not been developed (but see Riley et al. 2019 for a possible path forward in this regard). Further, more knowledge on how organo–mineral interactions affect biological aspects of microbial dynamics is needed. For instance, there is not a good method applicable to large-scale models to represent how organo–mineral interactions contribute to the formation of biofilms at different places within the soil, even though such structures are of great significance in determining how microbes deal with fluctuations in soil water (Or et al. 2007).

Molecular Structures. Most land models use CENTURY-Like SOM pools to represent SOM dynamics by applying pseudo-first-order approximations (Jenkinson et al. 2008; Parton et al. 2010). Although these simplified representations of SOM compounds are informative, some investigators have argued that the structures of diverse organic matter compounds may need to be represented to capture SOM dynamics, for several reasons. First, the Century-like pools are largely qualitative and vague, thereby leading to large uncertainty when confronting predictions of SOM dynamics with observations. In general, it is not possible to precisely categorize pools of SOM by turnover times that have such a broad range (e.g., a few days to thousands of years). In nature, the formation or decomposition of SOM compounds are properties of the ecosystem (Schmidt et al. 2011), and the traditional qualitative structural classification is a major obstacle toward providing a mechanistic basis for SOM dynamics. For instance, a group of SOM compounds may have a completely different response to soil moisture and temperature than another group of SOM compounds, even though these distinct groups may have comparable turnover times. Moreover, organo–mineral interactions may also show contrasting behaviors; for example, a group of compounds falling into comparable intrinsic decomposition rates may have different affinities to soil minerals or responses to environmental conditions (Gordon and Millero 1985; Gu et al. 1994; Mikutta et al. 2007; Kleber et al. 2011). Overall, we argue that there is a need to represent SOM compounds using their molecular structures in reactive transport models along with physical protection mechanisms (e.g., MAOM) and different functional groups of microbes (e.g., Riley et al. 2014; Dwivedi et al. 2017; Tang et al. 2019.)

As discussed in the Introduction and the sction `*Current Use of Reactive Transport Modeling of SOM Dynamic*', SOM decomposition is governed by environmental conditions such as physical heterogeneity, physical disconnection, plant inputs, microbial diversity, and microbial activity, as well as the molecular structure of organic matter (Schmidt et al. 2011; Riley et al. 2014; Dwivedi et al. 2017a). To capture such complex physico-chemical and biological interactions, we propose that a vertically resolved reactive transport modeling framework is required.

CURRENT USE OF REACTIVE TRANSPORT MODELING OF SOM DYNAMICS

Flow and reactive transport processes influence SOM formation and decomposition. In particular, geochemical processes can affect the mobility of chemical species through different mechanisms of dissolution-precipitation, sorption-desorption, ion-exchange, redox-reactions, complexation, and colloidal interactions (Arora et al. 2016, 2018; Yabusaki et al. 2017; Dwivedi et al. 2018b, a). There are a variety of RTMs and numerical codes—such as TOUGHReact (Xu et al. 2006; Maggi et al. 2008; Gu et al. 2009), PFLOTRAN (Hammond et al. 2014), CRUNCH (Steefel et al. 2015), *ecosys* (Grant 2013), BAMS (Riley et al. 2014; Dwivedi et al. 2017a; Tang et al. 2019), and BeTR (Tang et al. 2013; Tang and Riley 2018)—that are available to describe and can represent the interaction of various complex and competing biogeochemical processes across spatial and time scales. These numerical codes can represent aqueous, gaseous, and sorbed phases, along with a large suite of processes—including hydrology, soil energy dynamics, advection and diffusion, aqueous speciation, minerals precipitation and dissolution, microbial dynamics, adsorption and desorption, and equilibrium and nonequilibrium chemical reactions.

Several studies have demonstrated the use of reactive transport models to address SOM decomposition and stabilization (Grant et al. 2003; Ahrens et al. 2015; Dwivedi et al. 2017a; Georgiou et al. 2018). These models explicitly represent physical protection mechanisms and different functional groups of microbes. BAMS also represented a reduced number of SOM molecular structures (Riley et al. 2014; Dwivedi et al. 2017; Tang et al. 2019). Here, we provide a brief description of how these processes are represented in reactive transport models.

Organo–mineral interactions are represented in a fashion similar to the isotherm-based approach in the single layer batch model. For instance, BAMS1 (Riley et al. 2014) adopted linear kinetics to model sorption and desorption processes, whereas in BAMS2 (Dwivedi et al. 2017a) the formulation was replaced with the Surface Complexation Model (SCM), allowing variable biogeochemical conditions within a thermodynamic framework. Both BAMS1 and BAMS2 were able to reproduce the almost exponentially decreasing profile of soil carbon content and increasing age of soil radiocarbon with depth. The COMISSION model (Ahrens et al. 2015) similarly considers interactions between microbes, minerals, and substrates, using Langmuir kinetics and reactive transport modeling, and was able to obtain analogous results. In contrast, *ecosys* uses Freundlich kinetics to model the sorption dynamics of soluble carbon to soil mineral surfaces and unhydrolyzed organic matter (Grant et al. 2003). Of all these models, *ecosys* is the only one that also considers a full range of terrestrial ecosystem properties; it has been shown in many studies across various ecosystems to be able to capture much of the observed temporal variability in net ecosystem carbon exchanges (e.g., Grant et al. 2003, 2006, 2017a,b). Overall, these modeling exercises suggest transport combined with soil–mineral-regulated SOM stabilization and microbial decomposition help explain observed soil carbon content and age profiles. Furthermore, the specific mathematical form used to represent organo–mineral interaction plays a minor role, given the limited data available to evaluate the models.

Riley et al. (2014) made the first attempt to explicitly represent molecular structures of SOM compounds and integrate processes into a reactive transport model. Because SOM compounds consist of several plant-synthesized, degraded compounds, and microbial biomass as well as necromass, it is difficult to include a full representation of each species. Therefore, they grouped SOM compounds based on different metrics such as O:C ratio, charges (positive or negative), and degree of polarity. Riley et al. (2014) include above- and belowground litter, as well as root exudates for carbon inputs. The litter was degraded into several simpler organic compound groups such as cellulose, hemicellulose, lignin, monosaccharides, phenols, amino acids, lipids, and nucleotides. Root exudates were considered to be simpler organics, and were included among monomers (i.e., monosaccharides, amino acids, organic acids, and lipids pools). Subsequently, Dwivedi et al. (2017) modified this reaction network by collapsing all the monomers into a single pool. Although observations of individual SOM compounds were not available for comparison with predictions, individual SOM compounds highlighted the processes predicted to control SOM stocks and cycling.

To a certain extent, these reactive transport models demonstrated the value of including different microbial functional groups. BAMS included fungi and aerobic bacteria functional groups by assigning affinities to decompose plant litter and SOM (Neely et al. 1991; Romaní et al. 2006; Thevenot et al. 2010; DeAngelis et al. 2013). Subsequently, dissolved organic carbon (DOC) and peptidoglycan (i.e., necromass or dead cell wall material (DCWM); Frostegard and Baath 1996) were produced upon death of these microbial groups. Even though substantially underrepresenting functional diversity in soils, these models indicated key processes and pulse responses, and the value of examining SOM dynamics with this type of approach. It is important to realize that the governing ecological functions in the ecosystem are based upon biodiversity, and it is very data-intensive to decipher linkages between ecosystem function and biodiversity (Goldfarb et al. 2011).

There are some models that have gone beyond representing microbial diversity by considering soil fauna in the decomposition dynamics of SOM. For instance, Hunt et al. (1987) modeled the detrital food-web in a grassland by representing 15 groups of organisms, ranging from bacteria and fungi to predacious mites. The simulations show that trophic interactions are essential in maintaining the ecosystem wide nitrogen, echoing the message by Kaiser et al. (2014) that biodiversity is important for enabling nutrient-based regulation of the carbon cycle. Considering that SOM decomposition always involves a spectrum of bioreactions that include nitrification, denitrification, methanogenesis, and methanotrophy, it may be important to consider microbial diversity when all these different bioreaction pathways and different gas fluxes are to be modeled. For a carbon-only model, before more comprehensive data are available to benchmark model evaluations, it is hard to justify whether microbial community structure should be included in the model formulation.

Below, we describe the most important governing equations for simulating reactive flow and transport in the continuum subsurface environment.

Variably saturated flow

Multiphase flow (e.g., liquid and gas) or the Richards equation can be used to describe variably saturated flow. The Richards equation assumes only one active phase (e.g., liquid), while the gas phase is assumed to be passive.

Richards equation. The Richards equation (Richards 1931) has been widely used in the literature for describing partially saturated flow assuming a passive air phase present at atmospheric pressure (e.g., Neuman 1973; Panday et al. 1993; Dwivedi et al. 2016, 2017b). The saturation-based formulation of the Richards equation is given as:

$$S_a S_s \frac{\partial h}{\partial t} + \phi \frac{\partial S_a}{\partial t} = \nabla[k_{ra} K \nabla h] + Q_a \qquad (1)$$

where ∇ denotes the divergence operator; h [m] is the hydraulic head; K [m s^{-1}] is the hydraulic conductivity tensor; t [s] is the time; ϕ [m^3 void m^{-3} medium] is a dimensionless quantity, which represents the porosity of the porous media; S_a [m^3 H$_2$O m^{-3} void] and dimensionless quantity k_{ra} [–] denote the saturation of the aqueous phase and the relative permeability, respectively; S_s [m^{-1}] is the specific storage coefficient; Q_a [m^3 H$_2$O m^{-3} medium s^{-1}] is the volumetric source or sink term.

The corresponding pressure-based formulation is given as:

$$\frac{\partial[S_a \phi \rho_f]}{\partial t} = \nabla\left[-\rho_f \frac{k_{ra} k_{sat}}{\mu}(\nabla P - \rho_f g e_z)\right] + \rho_f Q_a \qquad (2)$$

ρ_f [kg m^{-3} water] represents the fluid density; ∇P [kg m^{-1} s^{-2}, or Pa m^{-1}] represents the gradient of the fluid pressure; g [m s^{-2}] denotes the acceleration due to gravity; e_z is a unit vector in the vertical direction; k_{sat} [m^2] is the permeability tensor for fully saturated conditions.

Multiphase Flow. The occurrence of components like CO$_2$ or H$_2$O in various phases (gas, liquid, solid) can be described using multiphase flow. Multiphase flow is described by solving the mass conservation equation of multiphase transport of each component j in phase α:

$$\frac{\partial\left[\phi \sum_\alpha \rho_\alpha S_\alpha Y_{j\alpha}\right]}{\partial t} = \nabla\left[-\sum \rho_\alpha Y_{j\alpha} \frac{k_{ra} k_{sat}}{\mu}(\nabla P - \rho_\alpha g e_z)\right] + \rho_\alpha Q_j \qquad (3)$$

where $Y_{j\alpha}$ is the mass fraction of component j in phase α; μ [kg s^{-1} m^{-1}] is the dynamic viscosity; and all other variables (e.g., k_{ra}, etc.) are same as defined earlier.

There are several models in the literature that can be used to describe the relationship between the aqueous phase saturation and capillary pressure. However, Brooks-Corey and van Genuchten formulations are the most well-known. In addition, the Mualem and Burdine formulations are typically used to describe relative permeability functions (Brooks and Corey 1964; Mualem 1976; van Genuchten 1980). We encourage readers to explore user's manuals of numerical codes such as TOUGHReact, PFLOTRAN, or Crunch for more details on these models.

Reactive transport equations

To simulate reactive transport, the transport and biogeochemical reactions are combined in the continuity equation as follows:

$$\frac{\partial\left(\phi S_L C_i\right)}{\partial t} = \nabla\left(\phi S_L D_i^* \nabla C_i\right) - \nabla\left(q C_i\right) - \sum_{r=1}^{N_r} v_{ir} R_r - \sum_{m=1}^{N_m} v_{im} R_m - \sum_{g=1}^{N_g} v_{il} R_l \qquad (4)$$

The left-hand side of the Equation (4) denotes the accumulation term (mol m^{-3} medium s^{-1}), which is the product of the porosity (ϕ) and liquid saturation (S_L), and the concentration (C_i). On the right-hand side, the first and second term together describe advective-diffusive transport; the third, fourth, and fifth terms represent various reactions that are partitioned between aqueous-phase reactions (R_r), mineral reactions (R_m), and gas reactions (R_l), respectively. Furthermore, a stoichiometric relationship is used to compute transformation from reactant to product species. In Equation (4) D_i^* denotes the diffusion coefficient, which is specific to chemical species considered as indicated by the subscript i; v_{ij} are the stoichiometric coefficients of reactant j in reaction i, whereas there are N_r, N_m, N_g reactants in aqueous, mineral, and gas phases, respectively.

These reactions can further be classified into equilibrium-based or kinetic. Equilibrium-based reactions typically require the assumption that there exists a local equilibrium between reactants and products. However, rate-based reactions (kinetic) are always the more general (Steefel and Lasaga 1994). SOM decomposition involves microbial reactions, which we describe below.

Biological reaction rates

The microbial decomposition rate of substrate C_i can be described using Michaelis-Menten (Michaelis and Menten 1913) kinetics:

$$\frac{dC_i}{dt} = -\mu_i \frac{C_i}{K_i + C_i} \frac{O_2}{K_{O_2} + O_2} \frac{B}{Y} \qquad (5)$$

where μ_i (s^{-1}) is the maximum specific consumption rate of substrate i (s^{-1}); O_2 (mol O_2 m^{-3}) is the aqueous O_2 concentration; and B (mg C-wet-biomass L^{-1}) is the wet biomass carbon of microbial functional groups considered in the reaction network (e.g., bacteria and fungi). The effects of pH, saturation levels of soil, and temperature (T) can be important for decomposition rates (e.g., Schimel et al. 2011). Equation (5) can be modified to using functions $f_1(\theta)$, $f_2(pH)$, and $f_3(T)$ to describe these environmental effects on microbial activity (Maggi et al. 2008), as follows:

$$\frac{dC_i}{dt} = -\mu_i \frac{C_i}{K_i + C_i} \frac{O_2}{K_{O_2} + O_2} \frac{B}{Y} f_1(\theta) f_2(pH) f_3(T) \qquad (6)$$

While the Monod-Michaelis-Menten kinetics, including the related multi-substrate progeny, have enjoyed a wide range of applications, Tang and Riley (2013) showed that the Monod kinetics fail to account for the substrate limitation in their approximation to the law of mass action, and misplace such limitation as a linear competition by juxtaposing many Monod terms (e.g., for Eqn. (5), the Monod representation of competition may lead to simple addition of more Monod terms without accounting for one microbe's influence on another microbe's K parameter (Tang 2015; Tang and Riley 2017). This shortcoming will result in poor model performances under conditions of substrate limitation, a situation that is characteristic of soil. Further, the Monod kinetics will oversimplify the connections between substrates and consumers when describing a multi-substrate—multi-consumer network, resulting in the modeled fluxes being oversensitive to the kinetic parameters that would be formulated in the corresponding law of action model. The Equilibrium Chemistry Approximation kinetics (ECA) and its extended SUPECA (synthesizing unit plus ECA) kinetics are both able to adequately address these two shortcomings of the Monod kinetics, and also provide a very straightforward way to link the kinetic parameters with thermodynamics and biological traits (Tang and Riley 2013, 2017; Zhu et al. 2016a, b).

In most formulations of microbial growth in reactive transport models, substrate use efficiency is the only parameter to describe the partition of an assimilated substrate into either biomass or respired product. However, the catabolic and anabolic separation is more complex than this simple representation. When one takes into account building up metabolic reserves, cell maintenance and growth, and the enzyme and polymer exudation involved in microbial physiology, the emergent CUE can have strong nonlinear variations with respect to the environmental conditions, such as temperature or moisture (Tang and Riley 2014; Allison and Goulden 2017). CUE may be further modified by regulations of nutrient availability (Manzoni et al. 2018). All these conclusions suggest a more nuanced parameterization of substrate use is needed in future modeling of microbial-activity-regulated SOM dynamics.

Mineral reactions

SOM dynamics are influenced by minerals present in the geochemical system. For example, pH affects SOM reactions, and the presence or absence of calcite minerals may therefore affect SOM dynamics. Similarly, iron minerals (e.g., ferrihydrite) can be important for SOM dynamics, as a significant fraction of the aqueous phase of SOM can undergo sorption and stay protected from microbial reactions (Dzombak and Morel 1996; Dwivedi et al. 2017a). Mineral precipitation and dissolution reactions can be described using Transition State Theory (or TST) type rate laws:

$$R_m = (k_{neutral} + k_{H^+}[a_{H^+}] + \Sigma k_j \prod_i [a_{ij}])(1 - Q_m / K_{eq,m}) \qquad (7)$$

where rate constants for neutral, acid or additional (j_{th}) reaction mechanisms are denoted by $k_{neutral}$, k_{H^+} and k_j, respectively; a_{ij} indicates the activity of the i_{th} aqueous species in the j_{th} reaction; finally, Q_m denotes the ion activity product of the m_{th} mineral phase and $K_{eq,m}$ denotes its corresponding equilibrium constant.

Sorption reactions

Sorption is a complex process, one that depends upon organic molecule characteristics, surface area, and site density of minerals, as well as aqueous chemistry (Dudal and Gérard 2004). There are several approaches that can be used to describe sorption processes and MAOM. However, kinetic and equilibrium adsorption isotherms (e.g., linear, Langmuir, Freundlich; Davis and Kent 1990) have been widely used in the literature to describe SOM dynamics (Grant et al. 2011; Mayes et al. 2012; Riley et al. 2014; Ahrens et al. 2015). Typically, sorption isotherms describe a functional relationship between adsorbate and adsorbent for a constant-temperature condition. Correspondingly, empirical sorption isotherms have been developed to describe both kinetic (i.e., sorption site is slow and assumed to achieve adsorption equilibrium described using forward and reverse rates) and equilibrium (i.e., sorption site is assumed to reach adsorption equilibrium fast) systems. Because these empirical formulations cannot account for the out-of-sample effects of chemical conditions for which they are derived, thermodynamic surface complexation models (SCM) were also used (Dwivedi et al., 2017). A simple linear sorption relationship can be imposed using forward (adsorption; k (s^{-1})) and reverse (desorption; k_r (s^{-1})) rates. In the absence of any competing source or sink of a specific SOM species, an effective equilibrium linear sorption relationship will become as follows:

$$K_d = \frac{k_f}{k_r} \qquad (8)$$

Nonlinear kinetic models require estimates of these rate constants (forward and reverse) and an additional exponential parameter (Goldfarb et al. 2011). On the other hand, SCMs offer a mechanistic approach for representing sorption processes and are generally more robust and applicable over variable geochemical conditions (Dzombak 1990; Goldfarb et al. 2011). The SCM models describe the sorption of solutes on solid surfaces as a chemical reaction between aqueous species and surface sites (surface complexation) (Dzombak 1990).

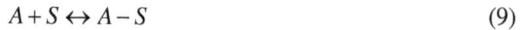

$$A + S \leftrightarrow A - S \qquad (9)$$

where A and S are the sorbate molecule and sorbent site (surface), respectively; and $A - S$ is the sorbed complex. The sorption reactions are described by a mass action law equation assuming a local equilibrium:

$$K_{Eq} = \frac{[A - S]}{[A][S]} \qquad (10)$$

where K_{Eq} denotes the equilibrium sorption constant; $[A - S]$ is the concentration of surface complexes (mol kg^{-1}water); $[A]$ denotes the concentration of sorbate molecule (mol kg^{-1}water), and $[S]$ denotes the sorbent site (mol kg^{-1}water).

Although we are not aware of any study considering rate-limited and equilibrium-based models together to describe sorption processes, we argue that rate-limited and equilibrium-based sorption do co-occur in soils. We suggest future modeling approaches consider rate-limited and equilibrium-based sorption to determine the relative importance of these processes for SOM dynamics.

SUMMARY

Soils store vast amounts of terrestrial organic carbon, more than the atmosphere and terrestrial vegetation combined. This carbon is vulnerable to release to the atmosphere under a changing climate. Although the mechanisms leading to the decomposition of SOM are well documented, uncertainties persist regarding the stability of SOM. To address this critical gap, a robust predictive understanding and modeling of SOM dynamics is essential for examining short- and long-term changes in soil carbon storage and feedbacks with climate. In this chapter, we reviewed recent research that improves emergent understanding of the important factors contributing to SOM stability. While there currently exists a suite of models representing SOM dynamics that span a range of complexity, some recent mechanistic models are more consistent with this emergent understanding of SOM persistence. Yet even those recent models do not represent several processes that can be important for SOM dynamics. We conclude that the next-generation models need to represent the full spectrum of quantitatively important mechanisms for determining SOM persistence—including rate-limited and equilibrium-based sorption, the formation of soil aggregates, representative soil minerals, microbial community dynamics, and vegetation interactions—to accurately predict short- and long-term SOM dynamics. Because this recommendation is obviously challenging, we have assembled an open-source, reactive-transport based SOM model that can be used to robustly integrate many of these processes (called BeTR-S; https://github.com/BeTR-biogeochemistry-modeling/sbetr) and invite the community to experiment with it (Riley et al. 2019).

Overall, it is important to understand SOM dynamics because SOM decomposition gives rise to potential greenhouse gases. Microbial dynamics, MAOM, and the molecular structure of SOM compounds have implications for understanding and modeling the short- and long-term responses of soil carbon stocks under local and regional climatic perturbations. Finally, this chapter illustrates the need to evaluate SOM dynamics in a reactive transport modeling framework that includes ecosystem properties.

ACKNOWLEDGMENT

This material is based upon work supported by the U.S. Department of Energy, Office of Science, Office of Biological and Environmental Research (as part of the Watershed Function Scientific Focus Area), and Office of Science, Office of Advanced Scientific Computing (as part of the project "Deduce: Distributed Dynamic Data Analytics Infrastructure for Collaborative Environments") under Contract No. DE-AC02-05CH11231. Jinyun Tang acknowledges support from the Next Generation Ecosystem Experiment (NGEE-Arctic). The USDA NIFA Postdoctoral Fellowship program supported Katerina Georgiou. Nicholas Bouskill acknowledges support from a DOE Early Career Research Project (# FP00005182). William Riley acknowledges support from the Terrestrial Ecosystem Science Scientific Focus Area of Berkeley Lab. We thank Diana Swantek of Berkeley Lab for assistance with preparing Figure 1.

REFERENCES

Abramoff R, Xu X, Hartman M, O'Brien S, Feng W, Davidson E, Finzi A, Moorhead D, Schimel J, Tom M, Mayes MA (2018) The Millennial model: in search of measurable pools and transformations for modeling soil carbon in the new century. Biogeochemistry137:51–71

Ahrens B, Braakhekke M, Guggenberger G, Schrumpf M, Reichstein M (2015) Contribution of sorption, DOC transport and microbial interactions to the ^{14}C age of a soil organic carbon profile: Insights from a calibrated process model. Soil Biol Biochem 88:390–402

Allison SD (2012) A trait-based approach for modelling microbial litter decomposition. Ecol Lett 15:1058–1070

Allison SD, Goulden ML (2017) Consequences of drought tolerance traits for microbial decomposition in the DEMENT model. Soil Biol Biochem 7:104–113

Allison SD, Wallenstein MD, Bradford MA (2010) Soil-carbon response to warming dependent on microbial physiology. Nat Geosci 3:336–340

Arora B, Spycher NF, Steefel CI, Molins S, Bill M, Conrad ME, Dong W, Faybishenko B, Tokunaga TK, Wan J, Williams KH, Yabusaki SB (2016) Influence of hydrological, biogeochemical and temperature transients on subsurface carbon fluxes in a flood plain environment. Biogeochemistry 127:367–396

Arora B, Davis JA, Spycher NF, Dong W, Wainwright HM (2018) Comparison of electrostatic and non-electrostatic models for U (VI) sorption on aquifer sediments. Ground Water 56:73–86

Baisden WT (2002) Turnover and storage of C and N in five density fractions from California annual grassland surface soils. Global Biogeochem. Cycles 16:1117

Bardgett RD, Van Der Putten WH (2014) Belowground biodiversity and ecosystem functioning. Nature 515:505–511

Baron MH, Revault M, Servagent-Noinville S, Abadie J, Quiquampoix H (1999) *Chymotrypsin* adsorption on montmorillonite: Enzymatic activity and kinetic FTIR structural analysis. J Colloid Interface Sci 214:319–332

Beardmore RE, Gudelj I, Lipson DA, Hurst LD (2011) Metabolic trade-offs and the maintenance of the fittest and the flattest. Nature 472:342–346

Berg B, McClaugherty C (2014) Plant litter: Decomposition, Humus Formation, Carbon Sequestration. Springer-Verlag Berlin Heidelberg

Brooks R, Corey A (1964) Hydraulic properties of porous media. Hydrol Pap Color State Univ 3:37 pp

Castellano MJ, Mueller KE, Olk DC, Sawyer JE, Six J (2015) Integrating plant litter quality, soil organic matter stabilization and the carbon saturation concept. Glob Change Biol 21:3200–3209

Charlton SR, Parkhurst DL (2011) Modules based on the geochemical model PHREEQC for use in scripting and programming languages. Comput Geosci 37:1653–1663

Clemente JS, Gregorich EG, Simpson AJ, Kumar R, Courtier-Murias D, Simpson MJ (2012) Comparison of nuclear magnetic resonance methods for the analysis of organic matter composition from soil density and particle fractions. Environ Chem 9:97–107

Conen F, Zimmermann M, Leifeld J, Seth B, Alewell C (2008) Relative stability of soil carbon revealed by shifts in δ^{15}N and C:N ratio. Biogeosciences 5:123–128

Cotrufo MF, Soong JL, Horton AJ, Campbell EE, Haddix ML, Wall DH, Parton WJ (2015) Formation of soil organic matter via biochemical and physical pathways of litter mass loss. Nat Geosci 8:776–780

Cotrufo MF, Wallenstein MD, Boot CM, Denef K, Paul E (2013) The Microbial Efficiency-Matrix Stabilization (MEMS) framework integrates plant litter decomposition with soil organic matter stabilization: do labile plant inputs form stable soil organic matter? Glob Change Biol 19:988–995

Crits-Christoph A, Diamond S, Butterfield CN, Thomas BC, Banfield JF (2018) Novel soil bacteria possess diverse genes for secondary metabolite biosynthesis. Nature 558:440–444

Davidson EA, Agren G, Daniel O, Emeis K, Largeau C, Lee C, Mopper K, Oades JM, Reeburgh WS, Schimel DS, Zepp RG (1995) What are the physical, chemical and biological processes that control the formation and degradation of nonliving organic matter? *In:* The Role of Nonliving Organic Matter in the Earth's Carbon Cycle. John Wiley & Sons Ltd

Davis JA, Kent DB (1990) Surface complexation mode+ling in aqueous geochemistry. Rev Mineral 23:176–260

DeAngelis KM, Chivian D, Fortney JL, Arkin AP, Simmons B, Hazen TC, Silver WL (2013) Changes in microbial dynamics during long-term decomposition in tropical forests. Soil Biol Biochem 66:60–68

Dudal Y, Gérard F (2004) Accounting for natural organic matter in aqueous chemical equilibrium models: A review of the theories and applications. Earth Sci Rev 66:199–216

Dwivedi D, Mohanty BP, Lesikar BJ (2016) Impact of the linked surface water-soil water-groundwater system on transport of *E. coli* in the Subsurface. Water Air Soil Pollut 227:351

Dwivedi D, Riley WJ, Torn MS, Spycher N, Maggi F, Tang JY (2017a) Mineral properties, microbes, transport, and plant-input profiles control vertical distribution and age of soil carbon stocks. Soil Biol Biochem 107:244–259

Dwivedi D, Steefel CI, Arora B, Bisht G (2017b) Impact of intra-meander hyporheic flow on nitrogen cycling. Procedia Earth Planet Sci 17:404–407

Dwivedi D, Arora B, Steefel CI, Dafflon B, Versteeg R (2018a) Hot spots and hot moments of nitrogen in a riparian corridor. Water Resour Res 54:205–222

Dwivedi D, Steefel CI, Arora B, Newcomer M, Moulton JD, Dafflon B, Faybishenko B, Fox P, Nico P, Spycher N, Carroll R, Williams KH (2018b) Geochemical exports to river from the intrameander hyporheic zone under transient hydrologic conditions: East River mountainous watershed, Colorado. Water Resour Res 4:205–222

Dzombak DA (1990) Surface Complexation Modeling: Hydrous Ferric Oxide. John Wiley & Sons

Dzombak DA, Morel FMM (1996) Surface Complexation Modeling: Hydrous Ferric Oxide. John Wiley & Sons

Fontaine S, Barot S (2005) Size and functional diversity of microbe populations control plant persistence and long-term soil carbon accumulation. Ecol Lett 8:1075–1087

Fontaine S, Mariotti A, Abbadie L (2003) The priming effect of organic matter: A question of microbial competition? Soil Biol Biochem 35:837–843

Frostegard A, Baath E (1996) The use of phospholipid fatty acid analysis to estimate bacterial and fungal biomass in soil. Biol Fertil Soils 22:59–65

Georgiou K, Abramoff RZ, Harte J, Riley WJ, Torn MS (2017) Microbial community-level regulation explains soil carbon responses to long-term litter manipulations. Nat Commun 8:1223

Georgiou K, Harte J, Mesbah A, Riley WJ (2018) A method of alternating characteristics with application to advection-dominated environmental systems. Comput Geosci 2018 22:851–865

Goldfarb KC, Karaoz U, Hanson CA, Santee CA, Bradford MA, Treseder KK, Wallenstein MD, Brodie EL (2011) Differential growth responses of soil bacterial taxa to carbon substrates of varying chemical recalcitrance. Front Microbiol 2:94

Gordon AS, Millero FJ (1985) Adsorption mediated decrease in the biodegradation rate of organic compounds. Microbiol Ecol 11:289

Grant RF (2013) Modelling changes in nitrogen cycling to sustain increases in forest productivity under elevated atmospheric CO_2 and contrasting site conditions. Biogeosciences 10:7703–7721

Grant RF, Heaney DJ (2010) Inorganic phosphorus transformation and transport in soils: mathematical modeling in Ecosys. Soil Sci Soc Am J 61:752–764

Grant RF, Juma NG, McGill WB (1993) Simulation of carbon and nitrogen transformations in soil: Mineralization. Soil Biol Biochem

Grant RF, Oechel WC, Ping CL (2003) Modelling carbon balances of coastal arctic tundra under changing climate. Glob Change Biol 9:16–36

Grant RF, Black TA, Gaumont-Guay D, Klujn N, Barr AG, Morgenstern K, Nesic Z (2006) Net ecosystem productivity of boreal aspen forests under drought and climate change: Mathematical modelling with Ecosys. Agric For Meteorol 140:152–170

Grant RF, Humphreys ER, Lafleur PM, Dimitrov DD (2011) Ecological controls on net ecosystem productivity of a mesic arctic tundra under current and future climates. J Geophys Res Biogeosci 116:G01031

Grant RF, Mekonnen ZA, Riley WJ, Arora B, Torn MS (2017a) Mathematical modelling of arctic polygonal tundra with ecosys: 2. microtopography determines how CO_2 and CH_4 exchange responds to changes in temperature and precipitation. J Geophys Res Biogeosci 122:3174–3187

Grant RF, Mekonnen ZA, Riley WJ, Wainwright HM, Graham D, Torn MS (2017b) Mathematical modelling of arctic polygonal tundra with ecosys: 1. Microtopography determines how active layer depths respond to changes in temperature and precipitation. J Geophys Res Biogeosci 122:3161–3173

Gu BH, Schmitt J, Chen ZH, Liang LY, McCarthy JF (1994) Adsorption and desorption of natural organic matter on iron oxide: mechanisms and models. Environ Sci Technol 8:38–46

Gu CH, Maggi F, Riley WJ, Hornberger GM, Xu T, Oldenburg CM, Spycher N, Miller NL, Venterea RT, Steefel C (2009) Aqueous and gaseous nitrogen losses induced by fertilizer application, J Geophys Res-Biogeo 114:G01006

Gunina A, Kuzyakov Y (2015) Sugars in soil and sweets for microorganisms: Review of origin, content, composition and fate. Soil Biol Biochem 90:87–100

Hammond GE, Lichtner PC, Mills RT (2014) Evaluating the performance of parallel subsurface simulators: An illustrative example with PFLOTRAN. Water Resour Res 50:208–228

Hedges JI, Eglinton G, Hatcher PG, Kirchman DL, Arnosti C, Derenne S, Evershed RP, Kögel-Knabner I, De Leeuw JW, Littke R, Michaelis W, Rullkötter J (2000) The molecularly-uncharacterized component of nonliving organic matter in natural environments. Org Geochem 31:945–958

Hicks Pries CE, Sulman BN, West C, O'Neill C, Poppleton E, Porras RC, Castanha C, Zhu B, Wiedemeier DB, Torn MS (2018) Root litter decomposition slows with soil depth. Soil Biol Biochem 25:103–114

Hocking PJ (2004) Organic acids exuded from roots in phosphorus uptake and aluminum tolerance of plants in acid soils. Adv Agron 74:63–97

Hunt AG, Manzoni S (2016) Networks on Networks The Physics of Geobiology and Geochemistry. Morgan & Claypool Publishers

Hunt HW, Coleman DC, Ingham ER, Ingham RE, Elliott ET, Moore JC, Rose SL, Reid CPP, Morley CR (1987) The detrital food web in a shortgrass prairie. Biol Fertil Soils 3:57

Jagadamma S, Mayes MA, Phillips JR (2012) Selective sorption of dissolved organic carbon compounds by temperate soils. PLoS One 7:e50434

Jagadamma S, Mayes MA, Zinn YL, Gísladóttir G, Russell AE (2014) Sorption of organic carbon compounds to the fine fraction of surface and subsurface soils. Geoderma 213:79–86

Jenkinson DS, Coleman K (2008) The turnover of organic carbon in subsoils. Part 2. Modelling carbon turnover, The turnover of organic carbon in subsoils. Part 2. Modelling carbon turnover. Eur J Soil Sci 59:400–413

Jenkinson DS, Poulton PR, Bryant C (2008) The turnover of organic carbon in subsoils. Part 1. Natural and bomb radiocarbon in soil profiles from the Rothamsted long-term field experiments. Eur J Soil Sci 59:391–399

Jilling A, Keiluweit M, Contosta AR, Frey S, Schimel J, Schnecker J, Smith RG, Tiemann L, Grandy AS (2018) Minerals in the rhizosphere: overlooked mediators of soil nitrogen availability to plants and microbes. Biogeochemistry 139:103

Jobbágy EG, Jackson RB (2000) The vertical distribution of soil organic carbon and its relation to climate and vegetation. Ecol Appl 10:423– 436

Joergensen RG, Wichern F (2018) Alive and kicking: Why dormant soil microorganisms matter. Soil Biol Biochem 116:419–430

Kaiser C, Franklin O, Dieckmann U, Richter A (2014) Microbial community dynamics alleviate stoichiometric constraints during litter decay. Ecol Lett 17:680–690

Keiluweit M, Bougoure JJ, Nico PS, Pett-Ridge J, Weber PK, Kleber M (2015) Mineral protection of soil carbon counteracted by root exudates. Nat Clim Change 5:588–595

Kleber M, Nico PS, Plante A, Filley T, Kramer M, Swanston C, Sollins P (2011) Old and stable soil organic matter is not necessarily chemically recalcitrant: Implications for modeling concepts and temperature sensitivity. Glob Change Biol 17:1097–1107

Kleber M, Eusterhues K, Keiluweit M, Mikutta C, Mikutta R, Nico PS (2015) Mineral–organic associations: formation, properties, and relevance in soil environments. Adv Agron 130:1–140

Knicker H (2011) Soil organic N—An under-rated player for C sequestration in soils? Soil Biol Biochem 3:1118–1129

Kögel-Knabner I (2017) The macromolecular organic composition of plant and microbial residues as inputs to soil organic matter: Fourteen years on. Soil Biol Biochem 105:A3–A8

Kuzyakov Y, Friedel JK, Stahr K (2000) Review of mechanisms and quantification of priming effects. Soil Biol Biochem 32:1485–1498

Lawrence CR, Neff JC, Schimel JP (2009) Does adding microbial mechanisms of decomposition improve soil organic matter models? A comparison of four models using data from a pulsed rewetting experiment. Soil Biol Biochem 41:1923–1934

Lehmann J, Kleber M (2015) The contentious nature of soil organic matter. Nature 528:60

Lehmann J, Kinyangi J, Solomon D (2007) Organic matter stabilization in soil microaggregates: Implications from spatial heterogeneity of soil organic carbon contents and carbon forms. Biogeochemistry 85:45–57

Lehmann J, Solomon D, Kinyangi J, Dathe L, Wirick S, Jacobsen C (2008) Spatial complexity of soil organic matter forms at nanometre scales. Nat Geosci 1:238–242

Levin SA (1999) Fragile Dominion: Complexity and the Commons. Basic Books. Perseus Books

Li C, Cui J, Sun G, Trettin C (2004) Modeling impacts of management on carbon sequestration and trace gas emissions in forested wetland ecosystems. Environ Manage 33(Suppl 1):S176–S186

Liang C, Schimel JP, Jastrow JD (2017) The importance of anabolism in microbial control over soil carbon storage. Nat. Microbiol 2:17105

Luo Z, Wang E, Zheng H, Baldock JA, Sun OJ, Shao Q (2015) Convergent modelling of past soil organic carbon stocks but divergent projections. Biogeosciences 12:4373–4383

Luo Z, Wang E, Sun OJ (2017) Uncertain future soil carbon dynamics under global change predicted by models constrained by total carbon measurements. Ecol Appl 27:1001–1009

MacArthur RH, Wilson EO (1967) The Theory of Island Biogeography. Princeton Univ. Press, Princeton, NJ

Maggi F, Gu C, Riley WJ, Hornberger GM, Venterea RT, Xu T, Spycher N, Steefel C, Miller NL, Oldenburg CM (2008) A mechanistic treatment of the dominant soil nitrogen cycling processes: Model development, testing, and application. J Geophys Res 113:G02016

Maharjan R, Nilsson S, Sung J, Haynes K, Beardmore RE, Hurst LD, Ferenci T, Gudelj I (2013) The form of a trade-off determines the response to competition. Ecol Lett 16:1267–1276

Mairhofer S, Zappala S, Tracy SR, Sturrock C, Bennett M, Mooney SJ, Pridmore T (2012) RooTrak: automated recovery of three-dimensional plant root architecture in soil from X-ray microcomputed tomography images using visual tracking. Plant Physiol 158:561–569

Manzoni S, Capek P, Porada P, Thurner M, Winterdahl M, Beer C, Bruchert V, Frouz J, Herrmann AM, Lindahl BD, Lyon SW (2018) Reviews and syntheses: Carbon use efficiency from organisms to ecosystems—Definitions, theories, and empirical evidence. Biogeosciences 15:5929–5949

Mathieu JA, Hatté C, Balesdent J, Parent É (2015) Deep soil carbon dynamics are driven more by soil type than by climate: A worldwide meta-analysis of radiocarbon profiles. Glob Change Biol 1:4278–4292

Mayes MA, Heal KR, Brandt CC, Phillips JR, Jardine PM (2012) Relation between soil order and sorption of dissolved organic carbon in temperate subsoils. Soil Sci Soc Am J 76:1027

Michaelis L, Menten ML (1913) Die Kinetik der Invertinwirkung. Biochem Z 49:333–369

Mikutta R, Mikutta C, Kalbitz K, Scheel T, Kaiser K, Jahn R (2007) Biodegradation of forest floor organic matter bound to minerals via different binding mechanisms. Geochim Cosmochim Acta 71:2569–2590

Miltner A, Bombach P, Schmidt-Brücken B, Kästner M (2012) SOM genesis: Microbial biomass as a significant source. Biogeochemistry 111:41

Mualem Y (1976) A new model for predicting the hydraulic conductivity of unsaturated porous media. Water Resour Res 12:513–522

Neely CL, Beare MH, Hargrove WL, Coleman DC (1991) Relationships between fungal and bacterial substrate-induced respiration, biomass and plant residue decomposition. Soil Biol. Biochem. 23:947–954

Neuman SP (1973) Saturated-unsaturated seepage. J Hydraul Div Am Soc Civ Eng 99(HY12):2233–2250

Newcomb CJ, Qafoku NP, Grate JW, Bailey VL, De Yoreo JJ (2017) Developing a molecular picture of soil organic matter-mineral interactions by quantifying organo–mineral binding. Nat Commun 8:396

Norde W, Giacomelli CE (2000) BSA structural changes during homomolecular exchange between the adsorbed and the dissolved states. J Biotechnol 79:259–268

Omoike A, Chorover J (2004) Spectroscopic study of extracellular polymeric substances from *Bacillus subtilis*: Aqueous chemistry and adsorption effects. Biomacromolecules 5:1219–1230

Or D, Smets BF, Wraith JM, Dechesne A, Friedman SP (2007) Physical constraints affecting bacterial habitats and activity in unsaturated porous media—a review. Adv Water Resour 30:1505–1527

Panday S, Huyakorn PS, Therrien R, Nichols RL (1993) Improved three-dimensional finite-element techniques for field simulation of variably saturated flow and transport. J Contam Hydrol 12:3–33

Parton WJ, Schimel DS, Cole CV, Ojima DS (1987) Analysis of factors controlling soil organic matter levels in great plains grasslands 1. Soil Sci Soc Am J 51:1173

Parton WJ, Hartman M, Ojima D, Schimel D (1998) DAYCENT and its land surface submodel: Description and testing. Glob Planet Change 19:35–48

Parton WJ, Hanson PJ, Swanston C, Torn M, Trumbore SE, Riley W, Kelly R (2010) ForCent model development and testing using the enriched background isotope study experiment. J Geophys Res Biogeosci 115:G04001

Quiquampoix H, Ratcliffe RG (1992) A ^{31}P NMR study of the adsorption of bovine serum albumin on montmorillonite using phosphate and the paramagnetic cation Mn^{2+}: modification of conformation with pH. J Colloid Interface Sci 148:343–352

Razinkov IA, Baumgartner BL, Bennett MR, Tsimring LS, Hasty J (2013) Measuring competitive fitness in dynamic environments. J Phys Chem B 117:13175–13181

Richards LA (1931) Capillary conduction of liquids through porous mediums. J Appl Phys 1:318–333

Riley WJ, Subin ZM, Lawrence DM, Swenson SC, Torn MS, Meng L, Mahowald NM, Hess P (2011) Barriers to predicting changes in global terrestrial methane fluxes : analyses using CLM4Me , a methane biogeochemistry model integrated in CESM. Biogeosciences 8:1925–1953

Riley WJ, Maggi F, Kleber M, Torn MS, Tang JY, Dwivedi D, Guerry N (2014) Long residence times of rapidly decomposable soil organic matter: application of a multi-phase, multi-component, and vertically resolved model (BAMS1) to soil carbon dynamics. Geosci Model Dev 7:1335–1355

Riley WJ, Sierra C, Tang JY, Bouskill NJ, Zhu Q, Abramoff R (2019) Next generation soil biogeochemistry model representations: A proposed community open source model farm (BeTR-S). *In:* Multi-scale Biogeochemical Processes in Soil Ecosystems: Critical Reactions and Resilience to Climate Changes, Y. Yang, M. Keiluweit, N. Senesi, B. Xing (eds) In Press

Romaní AM, Fischer H, Mille-Lindblom C, Tranvik LJ (2006) Interactions of bacteria and fungi on decomposing litter: Differential extracellular enzyme activities. Ecology 87:2559–2569

Schimel JP, Schaeffer SM (2012) Microbial control over carbon cycling in soil. Front Microbiol 3:348

Schimel JP, Wetterstedt JÅM, Holden PA, Trumbore SE (2011) Drying/rewetting cycles mobilize old C from deep soils from a California annual grassland. Soil Biol Biochem 43:1101–1103

Schmidt MWI, Torn MS, Abiven S, Dittmar T, Guggenberger G, Janssens IA, Kleber M, Kögel-Knabner I, Lehmann J, Manning DAC, Nannipieri P, Rasse DP, Weiner S, Trumbore SE (2011) Persistence of soil organic matter as an ecosystem property. Nature 478:49–56

Servagent-Noinville S, Revault M, Quiquampoix H, Baron MH (2000) Conformational changes of bovine serum albumin induced by adsorption on different clay surfaces: FTIR analysis. J Colloid Interface Sci 221:273–283

Simpson AJ, Kingery WL, Shaw DR, Spraul M, Humpfer E, Dvortsak P (2001) The application of 1H HR-MAS NMR spectroscopy for the study of structures and associations of organic components at the solid—Aqueous interface of a whole soil. Environ Sci Technol 35:3321–3332

Smith OL (1979) An analytical model of the decomposition. J Biomech 9:397–405

Steefel CI, Lasaga AC (1994) A coupled model for transport of multiple chemical species and kinetic precipitation/dissolution reactions with application to reactive flow in single phase hydrothermal systems. Am J Sci 294:529–592

Steefel CI, Appelo CAJJ, Arora B, Jacques D, Kalbacher T, Kolditz O, Lagneau V, Lichtner PC, Mayer KU, Meeussen JCLL, Molins S, Moulton D, Shao H, Šimůnek J, Spycher N, Yabusaki SB, Yeh GT (2015) Reactive transport codes for subsurface environmental simulation. Comput Geosci 19:445–478

Strickland MS, Rousk J (2010) Considering fungal: Bacterial dominance in soils—Methods, controls, and ecosystem implications. Soil Biol Biochem 42:1385–1395

Sulman BN, Phillips RP, Oishi AC, Shevliakova E, Pacala SW (2014) Microbe-driven turnover offsets mineral-mediated storage of soil carbon under elevated CO_2. Nat Clim Change 4:1099–1102

Sulman BN, Moore JAM, Abramoff R, Averill C, Kivlin S, Georgiou K, Sridhar B, Hartman MD, Wang G, Wieder WR, Bradford MA, Luo Y, Mayes MA, Morrison E, Riley WJ, Salazar A, Schimel JP, Tang J, Classen AT (2018) Multiple models and experiments underscore large uncertainty in soil carbon dynamics. Biogeochemistry 141:109

Swenson TL, Bowen BP, Nico PS, Northen TR (2015) Competitive sorption of microbial metabolites on an iron oxide mineral. Soil Biol Biochem 90:34–41

Taina IA, Heck RJ, Elliot TR (2011) Application of X-ray computed tomography to soil science: A literature review. Can J Soil Sci 88:1–19

Tang JY (2015) On the relationships between the Michaelis-Menten kinetics, reverse Michaelis-Menten kinetics, equilibrium chemistry approximation kinetics, and quadratic kinetics. Geosci Model Dev 8:3823–3835

Tang JY, Riley WJ (2013) A total quasi-steady-state formulation of substrate uptake kinetics in complex networks and an example application to microbial litter decomposition. Biogeosciences 10:8329–8351

Tang J, Riley WJ (2014) Weaker carbon–climate feedbacks resulting from microbial and abiotic interactions. Nature Clim Change:2438

Tang JY, Riley WJ (2017) SUPECA kinetics for scaling redox reactions in networks of mixed substrates and consumers and an example application to aerobic soil respiration. Geosci Model Dev 10:3277–3295

Tang JY, Riley WJ, Koven CD, Subin ZM (2013) CLM4-BeTR, a generic biogeochemical transport and reaction module for CLM4: model development, evaluation, and application. Geosci Model Dev 6:127–140

Tang G, Zheng J, Xu X, Yang Z, Graham DE, Gu B, Painter SL, Thornton PE (2016) Biogeochemical modeling of CO_2 and CH_4 production in anoxic Arctic soil microcosms. Biogeosciences 13:5021–5041

Tang J, Riley WJ (2018) Predicted land carbon dynamics are strongly dependent on the numerical coupling of nitrogen mobilizing and immobilizing processes : A Demonstration with the E3SM Land Model. Earth Interact 22:1–8

Tang FHM, Riley WJ, Maggi F (2019), Hourly and daily rainfall intensification causes opposing effects on C and N emissions, storage, and leaching in dry and wet grasslands. Biogeochemistry. In Press.

Tfaily MM, Chu RK, Tolić N, Roscioli KM, Anderton CR, Paša-Tolić L, Robinson EW, Hess NJ (2015) Advanced solvent based methods for molecular characterization of soil organic matter by high-resolution mass spectrometry. Anal Chem 87:5206–5215

Thevenot M, Dignac MF, Rumpel C (2010) Fate of lignins in soils: A review. Soil Biol Biochem 42:1200–1211

Tipping E (1994) WHAMC-A chemical equilibrium model and computer code for waters, sediments, and soils incorporating a discrete site/electrostatic model of ion-binding by humic substances. Comput Geosci 20:973–1023

Trumbore SE, Davidson EA, Barbosa De Camargo P, Nepstad DC, Martinelli LA (1995) Belowground cycling of carbon in forests and pastures of eastern Amazonia. Global Biogeochem Cycles 9:515–528

van Genuchten MT (1980) A closed-form equation for predicting the hydraulic conductivity of unsaturated soils 1. Soil Sci Soc Am J 44:892

Vargas R, Detto M, Baldocchi DD, Allen MF (2010) Multiscale analysis of temporal variability of soil CO_2 production as influenced by weather and vegetation. Glob Change Biol 16:1589–1605

Viollea C, Reich PB, Pacala SW, Enquist BJ, Kattge J (2014) The emergence and promise of functional biogeography. PNAS 111:13690–13696

Wang G, Post WM, Mayes MA (2013) Development of microbial-enzyme-mediated decomposition model parameters through steady-state and dynamic analyses. Ecol Appl 3:255–272

Wieder WR, Grandy AS, Kallenbach CM, Bonan GB (2014) Integrating microbial physiology and physio-chemical principles in soils with the MIcrobial-MIneral Carbon Stabilization (MIMICS) model. Biogeosciences 11:3899–3917

Wu L, Kobayashi Y, Wasaki J, Koyama H (2018) Organic acid excretion from roots: a plant mechanism for enhancing phosphorus acquisition, enhancing aluminum tolerance, and recruiting beneficial rhizobacteria. Soil Sci Plant Nutr 64:697–704

Xu T, Sonnenthal E, Spycher N, Pruess K (2006) TOUGHREACT—A simulation program for non-isothermal multiphase reactive geochemical transport in variably saturated geologic media: Applications to geothermal injectivity and CO_2 geological sequestration. Comput Geosci 32:145–165

Yabusaki SB, Wilkins MJ, Yilin F, Williams KH, Arora B, Bargar JR, Beller HR, Bouskill NJ, Brodie EL, Christensen JN, Conrad ME, Danczak RE, King E, Spycher NF, Steefel CI, Tokunaga TK, Versteeg RJ, Waichler SR, Wainwright HM (2017) Water table dynamics and biogeochemical cycling in a shallow, variably-saturated floodplain. Environ Sci Technol 51:3307–3317

Yan J, Pan G, Li L, Quan G, Ding C, Luo A (2010) Adsorption, immobilization, and activity of β-glucosidase on different soil colloids. J Colloid Interface Sci 348:565–570

Zhang Y, Li C, Trettin CC, Li H, Sun G (2002) An integrated model of soil, hydrology, and vegetation for carbon dynamics in wetland ecosystems. Global Biogeochem Cycles 16:1–17

Zheng J, Thornton PE, Painter SL, Gu B, Wullschleger SD, and Graham DE (2019) Modeling anaerobic soil organic carbon decomposition in Arctic polygon tundra: Insights into soil geochemical influences on carbon mineralization. Biogeosciences 16:663–680

Zhu Q, Iversen CM, Riley WJ, Slette IJ (2016a) A new theory of plant–microbe nutrient competition resolves inconsistencies between observations and model predictions. Ecol Appl 27:875–886

Zhu Q, Riley WJ, Tang J, Koven CD (2016b) Multiple soil nutrient competition between plants, microbes, and mineral surfaces: model development, parameterization, and example applications in several tropical forests. Biogeosciences 13:341–363

Zhu Q, Riley WJ, Tang JY (2017), A new theory of plant and microbe nutrient competition resolves inconsistencies between observations and models. Ecol Appl 27:875–888

Zhuang Q, Melillo JM, Kicklighter DW, Prinn RG, McGuire AD, Steudler PA, Felzer BS, Hu S (2004) Methane fluxes between terrestrial ecosystems and the atmosphere at northern high latitudes during the past century: A retrospective analysis with a process-based biogeochemistry model. Global Biogeochem Cycles 16:663–680

https://github.com/BeTR-biogeochemistry-modeling/sbetr; accessed on 06 May 2019.

Reviews in Mineralogy & Geochemistry
Vol. 85 pp. 349-380, 2019
Copyright © Mineralogical Society of America

12

Reactive Transport Processes that Drive Chemical Weathering: From Making Space for Water to Dismantling Continents

Kate Maher

Department of Earth System Science
Stanford University
Stanford, CA 943205
USA

kmaher@stanford.edu

Alexis Navarre-Sitchler

Department of Geology and Geological Engineering
and
Hydrologic Sciences and Engineering Program
Colorado School of Mines
Golden, CO 80401
USA

asitchle@mines.edu

INTRODUCTION

Chemical weathering, or the breakdown of rock to form regolith, occurs within an interface between the highly dynamic atmosphere and the comparably quiescent bedrock boundary known as the Critical Zone (Amundson et al. 2007; Anderson et al. 2007). The Critical Zone hosts a complex biosphere that redistributes water and nutrients while injecting carbon into the regolith as gas, dissolved compounds and less soluble organic matter. Most of this injected carbon reacts with mineral surfaces and is either exported as bicarbonate to rivers, and ultimately the ocean, or stored in newly formed secondary minerals. Simultaneously, tectonic forces push bedrock upward, driving erosion that ultimately transfers both weathered and unweathered material into floodplains and deltas, creating secondary weathering zones (West et al. 2002; Bouchez et al. 2012). Given the juxtaposition of climatic, biological and tectonic drivers that influence the geochemistry and hydrology, reactive transport has been central to advancing our understanding of chemical weathering processes. For more extensive reviews on weathering, we refer the reader to several excellent reviews (e.g., Brantley and Lebedeva 2011; Riebe et al. 2017). Here, we focus primarily on the application of reactive transport models to advance our understanding of how the regolith is sculpted over time by various geological and biological agents. Specifically, we review the ways in which reactive transport approaches have helped to understand the initiation of weathering via the opening of porosity at the bedrock interface, the formation of weathering zones on hillslopes that extend from the bedrock interface to the land-surface, and the role that climate, tectonics and organisms may play both in shaping the subsurface architecture that regulates water and nutrient availability and the export of solute via rivers.

1529-6466/19/0085-0012$05.00 (print)
1943-2666/19/0085-0012$05.00 (online)

http://dx.doi.org/10.2138/rmg.2018.85.12

Models for chemical weathering now range from process-based numerical algorithms that solve the advection-dispersion equation in tandem with a system of geochemical constraints, or reactive transport models (RTMs), to simplified expressions that honor reactive transport principles, to more empirical (or zero dimensional) representations. A number of different reactive transport models are used to study weathering processes, and the key features of the majority of them are summarized in Steefel et al. (2015) and thus we refrain from a deeper discussion of the models. In general, all of them contain a flexible provision for describing mineral dissolution and precipitation, cation exchange and sorption, redox reactions and microbial metabolisms (Steefel et al. 2005, 2015). Treatment of flow in the models is calculated using Darcy's law for groundwater, whereas the treatment of unsaturated flow is highly variable, from constant flow at a prescribed water saturation to full treatment of multi-phase flow (Steefel et al. 2015). By extension, transport of species is then by advection and diffusion. Current models do not include geomorphic rules to treat erosion and sedimentation, limiting models to shorter time-scale problems or requiring additional coupling to account for sediment transport. However, some models do allow for prescribed erosion and sedimentation (e.g., Giambalvo et al. 2002; Maher et al. 2006). As the studies we describe here demonstrate, reactive transport models with full kinetic rate laws and transport descriptions are likely to be more accurate for examining field weathering rates compared to mass balance approaches, but nevertheless have some limitations that are currently circumvented by using intermediate complexity models that honor reactive transport principles but contain more simplistic representations of the geochemical processes.

This chapter begins with an introduction to the scales and processes that influence reactive transport in the weathering zone and how they are conceptualized as reactive transport processes, then proceeds to discuss the scales at which models allow us to examine the processes that control weathering zone evolution, starting with one-dimensional (1-D) models of weathering fronts, their extension into two-dimensional (2-D) models, finally building up to models of the weathering zone as it influences catchment-scale transfer and export of dissolved rock. By using scale and dimensionality to organize this review, we emphasize both (1) the importance of combining well-constrained field systems to test our understanding of field observations and their representation in models and (2) the application of RTMs to guide simplified models of larger-scale systems that would otherwise be computationally prohibitive. We conclude with a summary of the frontiers for application of reactive transport models to weathering processes.

OVERVIEW OF TRANSFERS WITHIN THE WEATHERING ZONE

The development of models for chemical weathering in the Critical Zone requires careful consideration of the many modes and vectors of mass transport (Fig. 1A). For example, surface lowering by erosion effectively acts to advect solid material upward through the weathering zone relative to the land surface (Waldbauer and Chamberlain 2005; Hilley and Porder 2008), fluids may travel vertically during infiltration through the unsaturated zone and laterally once they encounter the water table (McGlynn et al. 2002; Germann and Zimmermann 2005), and gaseous species diffuse according to concentration gradients that are be controlled by plants, microbial communities and geochemical reactions (Cerling 1991; Bazilevskaya et al. 2015). Different types of heterogeneity, from roots that allow preferential flow of water (Noguchi et al. 1999) to fracture networks and weathering zones that influence hillslope drainage (Fletcher et al. 2006; Lebedeva and Brantley 2013; St. Clair et al. 2015; Pandey and Rajaram 2016; Lebedeva and Brantley 2017), create non-uniform distributions of hydraulic properties that result in an array of different flow paths and flow rates (Maxwell and Kollet 2008b; Meyerhoff and Maxwell 2011). In light of the myriad mass transfer mechanisms, scale has always been an important consideration in the design of weathering studies (e.g., Navarre-Sitchler and Brantley 2007; Li et al. 2017) and, by extension, in the application of reactive transport approaches to interrogate them.

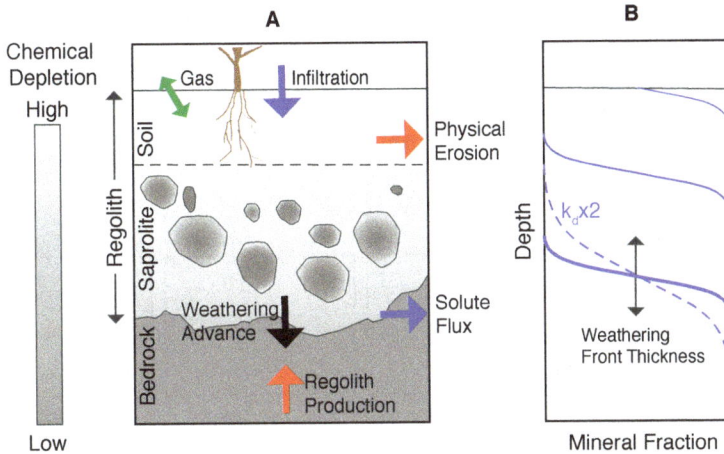

Figure 1. (A) Definitions and modes and vectors of mass transfer in the weathering zone that are commonly addressed through reactive transport models. The arrows show the predominant direction of transfer. For steady state regolith thickness, the weathering advance rate must be balanced by physical erosion and the weathering flux. The color scaling of the regolith reflects the degree of chemical depletion. **(B)** Conceptualization of the weathering front. The bold line shows the steady-state weathering profile, whereas the lighter lines show the evolution of weathering front over time. The dashed line shows the effect of a change in the intrinsic dissolution rate of the mineral, where a slower (faster) kinetic rate constant results in a steeper(shallower) gradient and thicker (thinner) weathering front compared to the base case but no change in weathering advance rate.

All of these modes of transport in turn influence the mass loss by weathering. Weathering occurs when water containing dissolved reactive gasses infiltrates into the subsurface coming into contact with bedrock minerals. Through time, as more precipitation infiltrates, the bedrock minerals dissolve and secondary minerals precipitate resulting in progressive chemical depletion, especially for base cations such as Ca^{2+}, Na^+, K^+ and Mg^{2+}. The location in the subsurface where weathering occurs (i.e., the transition between weathered and unweathered bedrock) is termed the weathering front (Fig. 1B). The mass loss by weathering is commonly measured by the relative depletion of the regolith compared to the unweathered protolith (for a full discussion of different methods for calculating chemical depletion, we refer the reader to several recent reviews (e.g., Brantley and Lebedeva 2011; Riebe et al. 2017)). At the profile scale, weathering fronts generally migrate downward over time, but when observed at smaller scales they may advance inward from fractures or discontinuities (Fig. 1A) that allow water and reactants to preferentially access the fresh bedrock (Fletcher et al. 2006; Buss et al. 2008; Navarre-Sitchler et al. 2011). Thus, the weathering zone may be comprised of nested weathering fronts that operate in tandem to control the formation of regolith. A major goal of reactive transport modeling of weathering systems has been to mechanistically interrogate the processes that control the advance of weathering fronts.

At the smallest scale, dissolution at mineral surfaces sets the flux of solute from the solid reservoir to the dissolved (Fig. 2A). Decades of experimental work have focused on the factors that control the dissolution rate of most rock forming minerals, revealing considerable complexity along the way (Oelkers et al. 1994; Banfield et al. 1995; Hellmann et al. 2012; Maher et al. 2016; Daval et al. 2018). In general, reactive transport models allow for a flexible treatment of mineral dissolution rates (R_d, mol yr^{-1}) through the assignment of kinetic rate laws. For the majority of rock-forming minerals this is handled through a modified version of Transition State Theory (TST) that assumes reversibility of the reaction in terms of the departure from thermodynamic equilibrium (Aagaard and Helgeson 1982; Lasaga 1984, 1998):

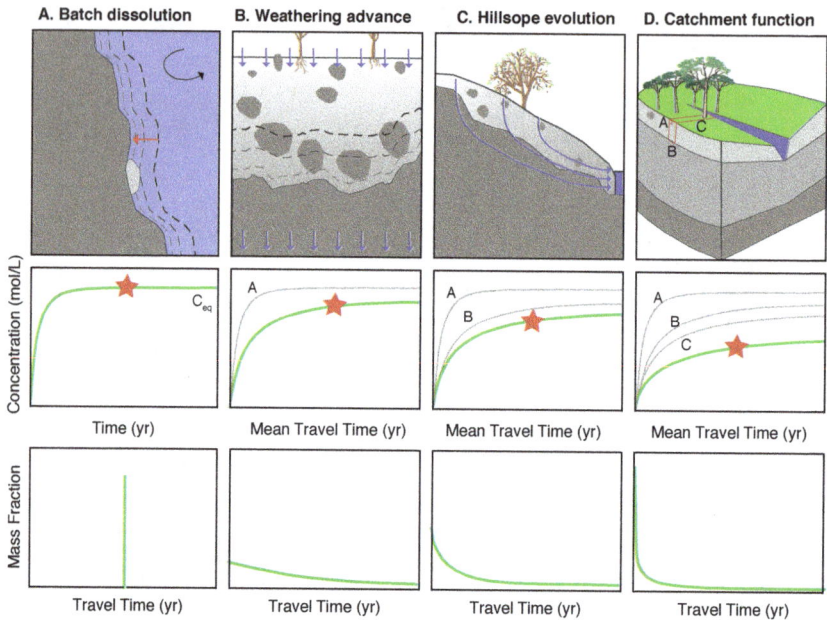

Figure 2. Schematic of scales of consideration for reactive transport modeling of weathering processes. Each panel shows a visual conceptualization (top row), followed by the schematic relationship between dissolved concentration and mean travel time (middle row). The red stars indicate the mean travel time for the system. Assuming the same mean travel time for all scales, the mass fraction of travel times dictate the concentration–mean travel time curves (bottom row). **(A)** Dissolution at the mineral surface and the change in concentration with time predicted by a linear dissolution rate law. The dashed lines show the retreat of the mineral surface and the precipitation of a secondary phase (light material) at the surface. In a homogeneous system, the change in concentration with time is equivalent to the change in concentration with fluid travel time. **(B)** Profile-scale weathering processes where the dashed lines show the advance of the weathering front over time. As fluid flow becomes more heterogeneous, the mixing of short and long travel times can result in a slower approach to equilibrium. The degree of chemical depletion is shown by the color scale as in Fig. 1. **(C)** Hillslope-scale reactive transport incorporates an even greater degree of heterogeneity and more pronounced distribution of travel times. As shown by comparison to the other scales, even greater dilution can arise from the variability in fluid flow paths. **(D)** At the catchment-scale, reactive transport models must capture finer-scale processes while managing an ever-greater degree of heterogeneity. Assuming the same mean travel time for all scales, the increasing heterogeneity results in a lower concentration for the same mean travel time. In reality, as scale increases, mean travel time is expected to increase. Travel time distributions are computed using an exponential distribution in **(B)** and using a gamma distribution in **(C)** and **(D)**.

$$R_d = A_s k_d \prod_{i=1} a_i^p \left[1 - \left(\frac{Q}{K_{eq}} \right)^m \right]^n \tag{1}$$

where A_s (m^2) is the total surface area of the mineral, k_d (mol m^{-2} yr^{-1}) is the far-from-equilibrium dissolution rate constant, a_i^p (-) gives the exponential dependence (p) on the activity, a, of aqueous species i, K_{eq} is the equilibrium constant for the reaction and Q is the ion activity product that defines the saturation state of the bulk solution. The exponents n and m allow for non-linear dependence on the saturation state or reaction affinity term, respectively, and are commonly determined empirically through experimental studies. The assignment of exponents on the affinity terms and/or parallel rate laws, or "sigmoidal rate laws", can result in highly non-linear rates as a function of the reaction affinity. In addition, because some rock forming minerals, such as feldspar, will not precipitate at earth surface conditions, the rate-laws are often defined to only allow for dissolution (or only precipitation, as appropriate). Kinetic rate laws compiled in Palandri and Kharaka (2004) and Bandstra et al. (2008) are often used as a starting point, but as discussed below, more complex representations have often yielded compelling agreement with field data. Figure 2A shows that over time in a closed system the concentration in solution reaches a concentration plateau where $Q = K_{eq}$ and the dissolution rate goes to zero, sometimes defined as the thermodynamic limit (Lebedeva et al. 2010; Maher 2010; Maher and Chamberlain 2014) and in weathering systems where it is attained, the local equilibrium approximation is a valid assumption (Lichtner 1988, 1993; Lebedeva et al. 2007). In well-mixed or homogeneous flow systems, the time scale for reactions is set by the fluid residence time (i.e., the system volume divided by the volumetric flow rate) or the fluid travel time (i.e., the length of the system divided by the advective velocity), respectively (for a comparison see Maloszewski and Zuber 1982; Maher and Mayer 2019). Thus, the transport timescales can be considered interchangeable measures of the time available for a reaction to proceed. Many reduced-order models of weathering take advantage of this assumption (Maher and Chamberlain 2014; von Blanckenburg et al. 2015; Anderson et al. 2019; Harman and Cosans 2019).

Weathering initiates when fluid accesses small (ca. 100 nm) pores in minerals and rocks and reacts with exposed surface area (Navarre-Sitchler et al. 2009). As weathering progresses porosity opens up through dissolution or fracturing and weathering advances into the rock. As discussed in a recent review by Navarre-Sitchler et al. (2015), a variety of measurement techniques, including electron microprobe analysis (EMPA) and small-angle neutron scattering (SANS), can now capture the pores and pore networks within clasts to provide insights into the initial stages of weathering. These studies, often combined with tracer studies or numerical approaches, also reveal that only a fraction of the porosity is accessible to the fluid (Navarre-Sitchler et al. 2009; Zahasky et al. 2018). When the "effective" porosity, or porosity accessible to fluid, is low, diffusion-limited transport can allow pore fluids to reach equilibrium. Hence, depending on the flow rate and intrinsic kinetic rate, the average surface area and saturation state terms in Equation (1) derived from the pore scale distributions can differ from a continuum representation that does not account for the sub-grid (pore-scale) effects (Li et al. 2004, 2008). Yet, these pore scale effects ultimately determine the evolution of the pore network and the potential for weathering-induced fracturing to further generate porosity (Navarre-Sitchler et al. 2015). Thus, approaches are needed beyond the continuum or volume-averaged representation described in Figure 1 and commonly used in RTMs because the continuum assumption may no longer apply at the pore scale (Molins et al. 2012). Direct numerical simulation at the pore scale, which resolves fluid velocity gradients within individual pores, is one approach that is being used to address this problem (Molins 2015; Molins and Knabner 2019, this volume). The development of pore-scale models and multi-scale approaches are needed to derive the appropriate upscaled representation for a particular porous media (e.g., Molins and Knabner 2019, this volume). Although we do not focus on pore scale here, a similar problem occurs at the profile and catchment scales, as discussed in several subsequent case studies.

Beyond the pore scale, to understand the evolution of weathering profiles over time and in response to erosion and fluid flow, studies have focused on slices of the Critical Zone that are predominantly characterized by one-dimensional transport of fluids and solids (White et al. 2001; Brantley and White 2009). These have included chronosequences, or soils of different ages, and ridge-top weathering, where input of upslope material is minimized (Fig. 2B). As we discuss below, the ability to rigorously constrain the mass balance at this scale, including flow and resulting transport along with solid phase compositions, enables modeling studies to examine a number of hypotheses about controls on the weathering front. The weathering front thickness (L_w, m) is the portion of a depth profile over which the concentration of a component decreases relative to the parent material, resulting in a concentration gradient (Fig. 1B).

Mathematically, L_w is usually proportional to the ratio of the advective velocity (v, m yr^{-1}) to the product of the rate constant and bulk surface area (A_{bulk}, m^2 m^{-3}) in the absence of erosion (Ortoleva et al. 1987; Lichtner 1988; Murphy et al. 1989):

$$L_w = \frac{vC_{eq}\rho_{fl-s}}{k_d A_{bulk}} \quad (2)$$

C_{eq} (mol L^{-1}) is the equilibrium concentration ($= K_{eq}$) for a single component monomineralic system and r_{fl-s} (L m^{-3}) is the ratio of fluid volume to total solid and accounts for the fluid to rock ratio. Relatedly, as the reacting mineral is depleted, the reaction front propagates downward over time (lighter lines in Fig. 1B) relative to the rate at which erosion removes material from the top. When the rates of the two processes balance, this is referred to as steady-state regolith. In the absence of erosion and for a developed profile where mineral is entirely depleted at the top, the weathering advance rate (ω) is the rate at which the mid-point of the reaction front moves down over time and is a function of both the mineral solubility and advective velocity (Lichtner 1988):

$$\omega = \frac{v(C_{eq} - C_0)\overline{V}_m\rho_{fl-s}}{V_m^0} \quad (3)$$

where C_0 (mol L^{-1}) is the initial concentration, V_m^0 is the volume fraction of mineral in the parent material and $\overline{V}_m$ (m^3 mol^{-1}) is the molar volume. The relationships described above, as well as other approaches that rely on depletions in regolith primary minerals (Brantley and Lebedeva 2011) reflect the time-integrated weathering front advance and profile evolution. In contrast, contemporary weathering rates can also be defined as the product of the Darcy flux, q (where $q = v\phi$, and ϕ is the water-filled porosity) and the dissolved concentration, C, exiting the regolith (or measured in a stream). Relative to the geometric approximations above, no assumptions are made about steady-state profile development or attainment of equilibrium when calculating a contemporary weathering rate from the solute flux. However, if the assumptions about profile development are valid, the weathering advance rate should be consistent with the contemporary weathering rate via rearrangement of Equation (3).

Ultimately, weathering rates are determined by how fast a mineral dissolves and how long fluid stays in contact with the mineral. This can be formalized in the non-dimensional Damköhler number (Da), specifically accounting for fluid transport or residence time (Damköhler 1936; Steefel and Maher 2009; Maher 2010):

$$Da = \frac{\tau_f}{\tau_r} = \left[\frac{L}{v}\right]\left[\frac{k_d A_{bulk}}{C_{eq}\rho_{fl-s}}\right] \quad (4)$$

where τ_f is the characteristic fluid travel time calculated as the characteristic length scale (L, m), divided by v in left-hand brackets and τ_r is the characteristic timescale of the weathering reactions, represented in terms of the ratio of C_{eq} (mol L^{-1}) to the net weathering rate ($k_d A_{bulk}$, mol m^{-3} yr^{-1}) which includes the specific surface and mass fraction of reacting mineral. This equation, when applied to a homogenous system where there are no distributions of fluid transport times or reaction rates, provides an evaluation of controls on overall mass transfer rates (Fig. 2A). Where Da is large, reaction rates are fast relative to fluid transport (i.e., fluid travel times are long) and fluids approach saturation with respect to dissolving minerals along the flow path reaching saturation prior to sampling. In these cases, reaction rates are termed transport-limited as increases in fluid velocity (decreases in residence time) remove solutes and allow minerals to dissolve at faster mass transfer rates. Where Da is small, fluid transport is fast relative to reaction rates and fluids do not approach saturation allowing minerals to continue to dissolve at far-from-equilibrium rates. Because, systems characterized by high Da generate solute more efficiently and will thus have the highest weathering rate for a given flow rate (Maher and Chamberlain 2014). These simple quasi one-component definitions allow one to interpret weathering profiles and processes, but to fully address the controls on weathering advance and weathering fluxes, additional factors such as secondary mineral precipitation, reactive gases, pH changes, exchangeable cations and plant inputs necessitate the use of reactive transport models.

Although we can observe weathering fronts at the profile scale, an important assumption typically made at this scale is that fluid travel time increases with depth into the regolith and thus the accumulation of solute from the dissolution of rock progresses along the flow path according to the appropriate kinetic rate law (Fig. 2B). This assumption allows modeling studies to examine mechanistic controls on mineral dissolution. However, even if flow is dominantly in one-direction, heterogeneity imposed by fracturing and feedbacks between chemical composition and reaction extent can result in a distribution of flow paths and fluid travel times, as well as diffusion-limited domains that play an important role in governing solute export at this scale (compare the thin grey line for homogenous case to the green line that accounts for an exponential distribution of travel times in Fig. 2B).

The knowledge gained from weathering profiles has further inspired exploration of hillslope-scale systems, where solute concentrations may vary according to both local heterogeneity and the mixing of fluids with different compositions or travel times (Fig. 2C). Here, the presence of a distribution of fluid travel times becomes even more important. Because of the variable flow paths and residence times, fluids associated with shorter fluid travel times may have acquired less solute from dissolution of minerals (low Da). Thus, if the effect of hillslope geometry is to shift the travel time distribution such that a greater fraction of the water is characterized by shorter travel times for the same mean travel time, the effective concentration will be lower (Maher 2011). When these highly evolved fluids mix with more dilute fluids, the averaging of the concentrations according to the fluxes results in a new relationship between solute concentration and mean fluid travel time (Maher and Druhan 2014; Druhan and Maher 2017). Such flux averaging generally results in fluids that are more dilute compared to the uniform travel time—reaction progress case of Figure 2A. In addition, differences in chemical depletion, which also affect the Da, can reduce the efficiency of solute production. These additional complexities make it difficult to relate groundwater or stream fluxes directly to mineral dissolution rates. Nevertheless, at this scale, hypotheses about physical controls on the evolution of regolith, such as the interplay between climate, weathering, rock fracturing, and drainage (e.g., Rempe and Dietrich 2014; St. Clair et al. 2015; Brantley et al. 2017; Maffre et al. 2018; Anderson et al. 2019), can start to be addressed. Finally, at the catchment scale, we see the aggregation of these subunits and their heterogeneity in the form of flux-averaged concentrations exiting via the stream as the dissolved load (Fig. 2D). Again, assuming that an even more heterogeneous distribution of fluid travel times emerges, the flux averaged concentration in the stream may be even lower than for a hillslope, even with the same mean travel time. A similar mixing occurs for sediment, where variable residence times in the regolith as a function of landscape position can result in mixtures of fresh and highly weathered sediment that appear distinct from point measurements on individual hillslopes. At this scale, models often require more extensive simplifications that must honor the processes we observe at hillslopes, while at the same time such models can inform about the landscape scale patterns, including aspect, lithology, vegetation distributions, elevation that shape the relevance of any given hillslope or regolith profile to the system as a whole (Pelletier et al. 2013).

As discussed below, reactive transport modeling studies are now revealing the ways in which heterogeneity plays a role in field observations. Below we first review the process-oriented studies focused at the profile scale followed by the ways in which these studies have advanced hillslope-scale and catchment-scale models. We conclude with a synthesis of promising directions using RTMs to guide our growing knowledge of weathering.

CONTROLS ON THE MIGRATION OF WEATHERING FRONTS

Chronosequences as a bridge between the laboratory and the field

Chronosequences have long been an important tool for not only understanding the geologic and climatic history of a region, but for interrogating weathering processes under relatively simple conditions (Birkeland and Burke 1988; Blum and Erel 1995; White et al. 1996, 2005, 2008, 2009; Bullen et al. 1997; Vidic 1998; White and Brantley 2003). Most chronosequences are a series of soils of different ages formed from sediment derived from glaciers, lakes, oceans, and rivers, such that the end of deposition marks the onset of soil development providing a clock for weathering processes. Larger rock clasts within these chronosequences can also weather in place, developing weathering rinds that preserve a separate record of weathering through time (e.g., Sak et al. 2004; Navarre-Sitchler et al. 2009, 2011). Chronosequences can also develop from weathering of different aged lava flows, such as the well-known chronosequences in Hawaii (e.g., Chadwick et al. 1999). Depending on the nature of sediment transport, well-sorted and relatively homogeneous sediments of an array of ages can be found. Sedimentary chronosequences are also amenable to age dating using cosmogenic nuclides (or other methods) and do not suffer from the additional complications that arise from conversion of bedrock to saprolite and then mobile regolith, although compositional heterogeneity may be greater and thus it can be more difficult to determine the composition of the parent material.

Given the simplifications that arise from examining chronosequences, several studies have leveraged detailed field studies to examine mineral dissolution and its relationship to weathering front advance rates using RTMs. In particular, one series of studies focused on a well-characterized series of marine terraces near Santa Cruz, CA, using mass balance approaches based on the age of the terraces and either the solid mass depletion or solute gradients to calculate weathering rates (White et al. 2008, 2009). Building on this work, Maher et al. (2009) used a forward modeling approach (i.e., model simulations start from the assumed initial conditions and run forward to the soil age) to examine why the rates of mineral dissolution calculated using both solute and solid mass balance were several orders of magnitude slower than the corresponding laboratory rates. The mass balance calculations had not explicitly considered the dependence of the rates on saturation state (i.e., Eqn. 1) whereas laboratory studies have shown that feldspar dissolution rates can depend in a non-linear way on the activity of Al^{3+} in solution (Oelkers et al. 1994), or on the bulk saturation state due to opening of etch pits far-from-equilibrium compared to more uniform surface removal close-to-equilibrium (Hellmann and Tisserand 2006). Although different mechanisms are assumed to dominate, both dependencies result in lower rate laws close to equilibrium compared to a "linear" rate law that is often assumed (Fig. 2a). Previous studies had also pointed out that slow clay precipitation allows for accumulation of weathering products in solution, which may in turn suppress the dissolution rate (Zhu 2005; Maher et al. 2006). The wealth of field data, including porewater data that constrained the saturation state, allowed these hypotheses to be explored using an RTM.

Results from the modeling are shown in Figure 3 for the plagioclase abundance, dissolution rate and saturation state after 226 kyr of weathering. The model could reproduce both the gradient in mineral abundance across the weathering front and the weathering front advance rate with laboratory determined rate laws only if the sigmoidal rate laws were introduced and clay precipitation rates were decreased to allow dissolution products to accumulate in solution (see Fig. 9 in Maher et al. 2009). As the panels in Figure 3 demonstrate, the advantage of an RTM is that both the solid profiles (mineral abundance and major element concentrations) and solute profiles can be used as constraints. As components appear on different sides of the mass action equations, this greatly reduces the number of free parameters. Nevertheless, application of inverse modeling

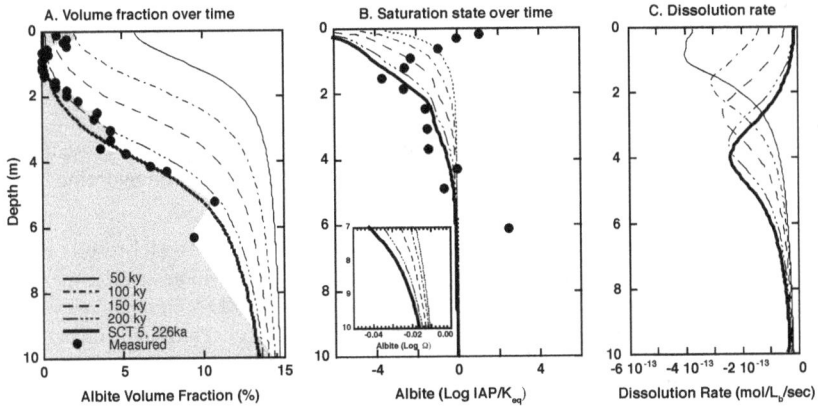

Figure 3. Results of RTM simulations of the Santa Cruz Marine Terrace chronosequence. Model profiles after 226 kyr are compared to data for the volume fraction of mineral in (**A**). (**B**) Saturation state for the plagioclase computed from pore water data at the site. (**C**) The corresponding rate profile. The maximum rates correspond to and optimum dictated by the blance between total mineral surface area and saturation state. Modified from Maher et al. (2009).

using the RTM revealed that several of the parameters fit by the model were strongly correlated, especially the flow rate and kaolinite precipitation rate constant. To understand this correlation, sensitivity tests suggested that increasing the clay precipitation increased the weathering advance rate but did not affect the weathering gradient, in a manner that is similar to the effect of changes in advective velocity (or time) as suggested by Equations (2) and (3) above.

Additional modeling work at the Santa Cruz site has also examined the role of organic acids in chemical weathering (Lawrence et al. 2014). The presence of plant exudates and decomposition products, such as low-molecular weight organic acids in the weathering zone, is thought to accelerate weathering in response to biota (Berner 1992; Drever 1994; Jones 1998; Ganor et al. 2009). In laboratory studies, organic acids accelerate the dissolution rate of feldspars by weakening the Al–O bonds at the mineral surface (Stillings et al. 1996) and increase the effective solubility by forming aqueous complexes that reduce the activity of Al^{3+} in solution (Huang and Keller 1970; Drever and Stillings 1997). However, because increasing the dissolution rate constant (or solubilization rate) changes the weathering gradient, the increased kinetics have little effect on the overall weathering advance rate under transport-controlled conditions (Lawrence et al. 2014). In addition, modeling showed that most organic acids are decomposed fairly rapidly in the upper layers of the soil, such that the weathering front over time moves away from the zone of biological influence. Here, the application of RTM provided a unique means of assessing the implications of experimentally established frameworks within a field system where they would otherwise be difficult to observe.

Another case study at a proximal chronosequence located in Merced, CA applied similar approaches to evaluate weathering over even longer timescales, using extensive field data to parameterize and evaluate the RTM. Here, Moore et al. (2012) used the RTM FLOTRAN to model mineral dissolution and precipitation at another granitic chronosequence at Merced, CA (White et al. 1996, 2005). Relative to Santa Cruz, the Merced chronosequence is characterized by much older surfaces, with the oldest approaching 3,000 ka, and considerably drier conditions, although still within a Mediterranean climate regime. In contrast to the Santa Cruz, CA study, in order to model the observed profiles, even using sigmoidal rate laws, required an adjustment relative to laboratory rates of between 470–1560 and 47 for the traditional TST and sigmoidal rate laws, respectively. This was attributed to variations in hydrologically assessable surface area, which was substantiated by tracer studies at the site (Green et al. 2005). Thus, a key conclusion of Moore et al. (2012) is the need to quantify and model the impacts of heterogeneous flow over geological time scales. As shown in Figure 4, sensitivity analysis can expand the conclusions by considering a broader range of variables, for example the value of k_d and v. Here, Moore et al. (2012) show the transition in flow rate where the weathering front advance changes from transport controlled to kinetically controlled weathering varies also as a function of k_d. This threshold, or non-linear behavior, would not emerge from the simple equations presented above but suggests that the scaling of reaction fronts will be highly dependent on hydrologic regime, as well as the reactivity of the dissolving minerals.

Both chronosequence studies above also demonstrated that the model results depended on the amount of CO_2 in the weathering zone (Maher et al. 2009; Moore et al. 2012). Carbon dioxide, which can range from atmospheric levels to ca. 10–20,000 ppm in soils, impacts the pH, and by extension, the saturation state. Thus, increasing CO_2 concentrations, increasing clay precipitation rates and increasing flow rates can all accelerate the weathering advance rate in the same way: by influencing the departure from equilibrium. Using an RTM-based sensitivity analysis, Winnick and Maher (2018) demonstrate that increasing CO_2 in the weathering zone increases C_{eq} and thus alters the Da and the resulting concentration-travel time curves in Figure 2. The high sensitivity of weathering fluxes to belowground CO_2 provides a direct link between the biosphere and weathering processes that may partly explain the variability in the Da of major rivers (Ibarra et al. 2016) and some of the global trends in atmospheric CO_2 throughout Earth history as the biosphere has evolved (Berner 1992, 1995; Berner and Kothavala 2001; Berner et al. 2003; Beerling and Berner 2005; Pagani et al. 2009).

Although much of the work summarized above focused on sedimentary chronosequences that were granitic in nature, recent work has also addressed different lithologies. For example, weathering of shale bedrock over 10 kyr indicated that weathering was initiated by pyrite oxidation and dissolution in the first 1 kyr, leading to dissolution of chlorite and precipitation of Fe(III)-hydroxides in response to the lower pH (Heidari et al. 2017). In agreement with prior studies from granitic chronosequences, the presence of CO_2, the assignment of specific surface area, and flow rates exerted an important control on the weathering fronts. However, in contrast to previous work, the presence of O_2 in the profile was critical for controlling the oxidative dissolution of pyrite. This study further highlights the value of the stoichiometry imposed by the components of the system, as well as the complexity that RTMs can afford in examining multi-component weathering systems.

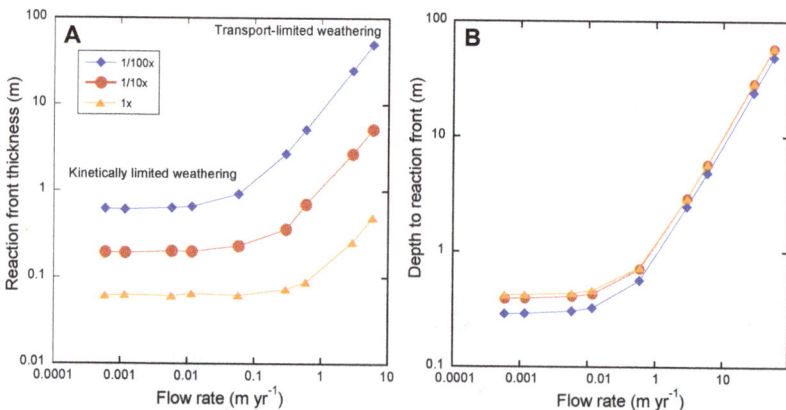

Figure 4. Sensitivity analysis of RTM simulations of the Merced, CA chronosequence showing the relationships between weathering front thickness (**A**) and depth to weathering front (**B**) as a function of flow rate using a simplified version of the Merced 250 ka soil for different values of the dissolution rate constant, k_d (Eqns. 1 and 2). The flow rate where the weathering front transitions to a mixed kinetic-equilibrium regime to kinetic-limited regime varies with the assumed rate constant. Modified from Moore et al. (2012).

Evaluating the assumptions of isotopic proxies for weathering using RTM approaches

Given the complexity of weathering processes, isotopic proxies or tracers have been extensively used to track exchanges between the water, soil and plant reservoirs. As such, the incorporation of isotopes into reactive transport models has been increasingly used to test the assumptions of various isotopic proxies (Maher et al. 2006; Druhan et al. 2013, 2014; Steefel et al. 2014; Wanner et al. 2014; Maher and von Blanckenburg 2016; Druhan and Maher 2017). In a more theoretical application of RTM to chronosequences, an RTM model was developed to look at the application of meteoric ^{10}Be to dating sedimentary profiles and by extension, the use of the ratio of ^{10}Be/^{9}Be as a proxy for weathering rates (Willenbring and von Blanckenburg 2010; von Blanckenburg et al. 2012, 2015; von Blanckenburg and Bouchez 2014). Meteoric ^{10}Be is deposited via the atmosphere and sorbs strongly to mineral surfaces within the regolith as long as pH is around circumneutral. Beryllium-9 is derived from weathering of primary minerals, mostly feldspars, but is similarly retained by sorption and co-precipitation into secondary minerals. Thus, poor retention of ^{10}Be can reduce the inventory and compromise the dating method. Methods for correcting for ^{10}Be loss can include the use of ^{9}Be based on the fraction of weathered ^{9}Be that is retained by the secondary minerals. However, ^{9}Be is released to solution at the weathering front, whereas ^{10}Be is introduced through the surface of the profile such that incomplete retention may differentially impact the two isotopes. To reflect the processes controlling Be, a surface complexation model (Dzombak and Morel 1990) was developed for each isotopologue. Using this treatment, the amount of sorbed Be is a function of the amount of secondary mineral that has formed, which is in turn a function of soil age. The modeling results show that the ^{9}Be loss correction only introduces substantial error after significant weathering has occurred. Furthermore, because the model enforces a rigorous mass balance it was possible to test the assumptions contained in the weathering proxy. Even though depth gradients in ^{10}Be/^{9}Be develop in tandem with the weathering gradients, the inventory of ^{10}Be/^{9}Be still reflects the relative input and losses of both isotopes such that the ^{10}Be accumulation provides a clock for the weathering of ^{9}Be (Fig. 5). Specifically, the modeling showed that the inventory of ^{9}Be/^{10}Be tracks the gross weathering rate, supporting the use of the proxy as an alternative approach to calculate weathering rates. Reactive transport modeling of Li isotopes during weathering (Wanner et al. 2014), potentially in combination with comprehensive differential mass balance approaches (Bouchez et al. 2013), is another promising approach for assessing an important marine proxy for weathering.

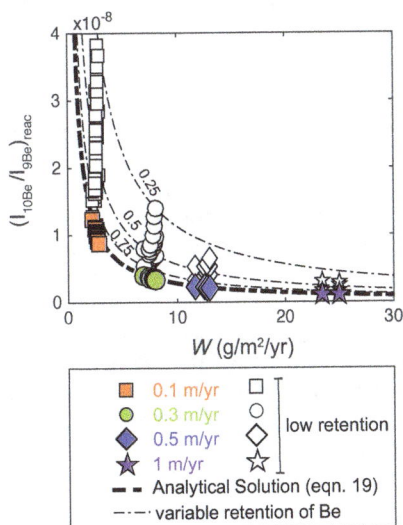

Figure 5. The sorbed Be inventory (I ^{10}Be / I ^{9}Be)reac as a function of the weathering flux (W) for four different flow rates. The open symbols show the results for lower retention simulations. The heavy dashed line corresponds to Equation (18) in Maher and von Blanckenbug (2016) assuming the fraction of Be retained is 1, implying full retention; light stippled lines correspond to less efficient retention as indicated.

The effect of heterogeneity on reaction fronts and overall chemical weathering rates

The one-dimensional profile studies above all assume uniform mineral distribution and fluid flow, thus neglecting chemical and physical heterogeneity that is almost certainly present in all weathering systems. Chemical heterogeneity ranges from defects in crystal structure at angstrom scale (Arvidson et al. 2003; Luttge et al. 2013) to heterogeneous mineral distribution at mm to cm scale (Beckingham et al. 2016, 2017), and lithologic variation from cm to km scale (Gaillardet et al. 2003; Ma et al. 2011; Cai et al. 2018). These chemical heterogeneities can limit mineral surface area in contact with reacting fluids and control opening of pore space through preferential dissolution of a fast-reacting mineral (Navarre-Sitchler et al. 2008; Navarre-Sitchler et al. 2015). Heterogeneity in physical properties ranges from irregularities in pore network distribution and connectivity at nm to μm scale (Navarre-Sitchler et al. 2008), macropore formation in soils at μm to cm scale (Beven and Germann 1982), and fractures in bedrock at cm to m scale (Fletcher et al. 2006; Jamtveit and Hammer 2012). These physical heterogeneities generate variations in fluid flow velocity such that residence times of fluids are variable (e.g., Dagan and Indelman 1999; Werth et al. 2006; Sanchez-Vila et al. 2007; Steefel 2008; Steefel and Maher 2009; Porta et al. 2012, 2013).

The impact of heterogeneity can be conceptualized in terms of the impact on the Damköhler numbers. In heterogeneous systems, local saturation conditions and fluid velocities produce local, or grid-scale, Damköhler numbers that describe locally the controls on reaction rates (e.g., Siirila and Maxwell 2012; Jung and Navarre-Sitchler 2018a). In highly heterogeneous systems where residence times of fluids have a wide distribution, flux weighted fluid concentrations and average residence times can produce domain-average Damköhler numbers <1 that give the appearance of far from equilibrium reaction conditions when in fact most reactions are controlled locally by transport conditions (Jung and Navarre-Sitchler 2018a). Additionally, as minerals dissolve and rocks weather, pore space is created (Navarre-Sitchler et al. 2008; Navarre-Sitchler et al. 2015) and the chemical and physical properties of the rocks evolve inducing changes in local Damköhler conditions though time. The time- and spatial-scales over which these properties evolve can range from mm to m and months to millennia depending on the rates of fluid flow and geochemical reactions. These dynamic conditions in field systems, where past conditions must usually be inferred, combined with limitations in data related to sample collection from the subsurface, mapping of heterogeneity, and flux-weighted averaging of water chemistry, make it difficult to parse out the influence of heterogeneity on long-term weathering rates. Reactive transport models, where heterogeneity and mineral composition can be prescribed, allow for exploration and elucidation of the coupling of physical heterogeneity and geochemical rates across space and time.

In order to use reactive transport models to evaluate controls of physical heterogeneity on weathering, models must move from simplified 1-D domains to either 2-D or 3-D domains, which increases computational complexity. Additionally, simulations of heterogeneous field systems can be limited by current ability to characterize physical and chemical heterogeneity in the subsurface. Collection of rock samples from the surface to constrain mineralogy, porosity, and surface area can bias characterization toward rock samples that are potentially less resistant to weathering. Drill cores can provide a valuable window into the subsurface, but data points are often separated by 10s of meters or more and extrapolation between wells is often very uncertain. Shallow geophysical techniques are rapidly enhancing our ability to quantify physical heterogeneity in the critical zone (Holbrook et al. 2014; Parsekian et al. 2015; St. Clair et al. 2015), but there exists no technique or suite of techniques to date that would allow for complete characterization of physical heterogeneity at mm to km scale in-situ. While advances in characterization are being made, three recent studies of weathering in 2-D domains have systematically quantified the role of physical heterogeneity on local- and domain-scale mineral dissolution rates for feldspar minerals. Results from these studies provide valuable insights into the controls of physical heterogeneity on weathering that will aid in developing simulations of heterogeneous field systems in the future.

In two studies, Jung and Navarre-Sitchler (2018a,b) simulated anorthite ($CaAl_2Si_2O_8$) weathering to kaolinite in $1\,m \times 1\,m$ and $4\,m \times 4\,m$ domains. Simulations were performed in CrunchFlow with gaussian random field distributions of permeability (Fig. 6A–C). When heterogeneity is represented by variation in permeability in a porous media, increased correlation lengths and variance lead to highly localized fast reactions (near far-from-equilibrium rates) in fast fluid velocity pathways with low residence times early in the development of a weathering front. As anorthite is depleted in these fast fluid velocity pathways, anorthite dissolution dominantly occurs at the interface between high and low permeability zones, where fluids have longer residence times. Thus, over time reaction rates slow as reacting fluids increase in saturation relative to anorthite. Eventually, reactions become transport limited with all weathering occurring within blocks of low permeability and diffusion of solutes away from the reacting phases becomes a controlling process. This temporal evolution of reaction rates, over long-time periods (1000s of years) can reduce overall mass transfer rates by orders of magnitude relative to early time far-from-equilibrium rates (Fig. 7). Despite the transport limitation locally, flux weighted concentrations in fluid sampled at the outlet of the domain are low and suggest reactive fluids undersaturated with respect to anorthite, thus masking the signal of transport limitation on weathering rates, as described conceptually in Figure 2D.

The decrease in weathering rates in time in heterogeneous systems becomes even more pronounced in fractured systems relative to gaussian random fields in porous media. Pandey and Rajaram (2016) simulated orthoclase weathering to kaolinite in a $10\,m \times 10\,m$ domain over

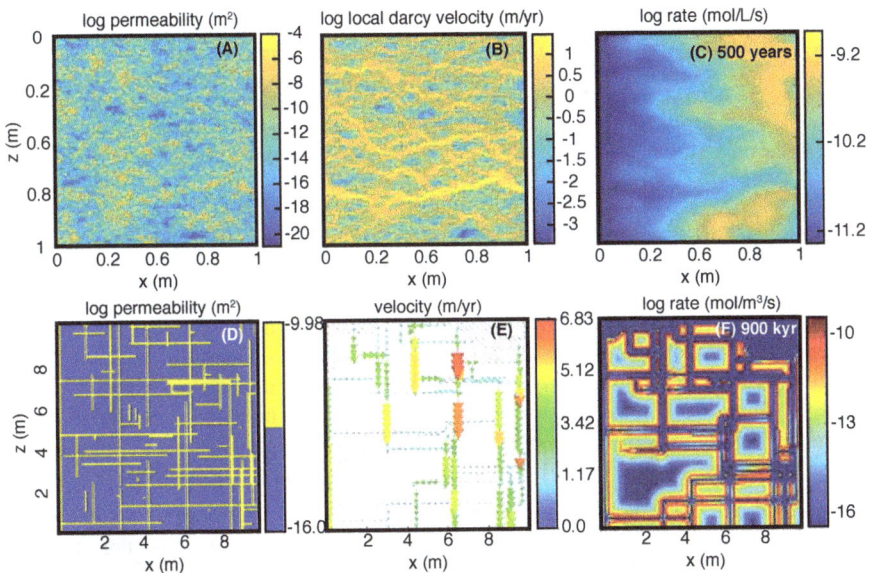

Figure 6. Heterogeneously distributed permeability in a Gaussian random field (**A**) and a discrete fracture network (**D**) with corresponding local velocity fields at steady state (**B** and **E**). The spatially variable velocity leads to spatially variable reaction rates locally in both gaussian random fields and discrete fracture networks where fast reaction occurs in fast fluid flow paths relative to areas with low permeability (**C** and **F**). Parts of this figure modified from, Jung and Navarre-Sitchler (2018a), Figs. 1,4; Jung and Navarre-Sitchler (2018b) Fig. 6; and Pandey and Rajaram (2016), Figs. 1,2.

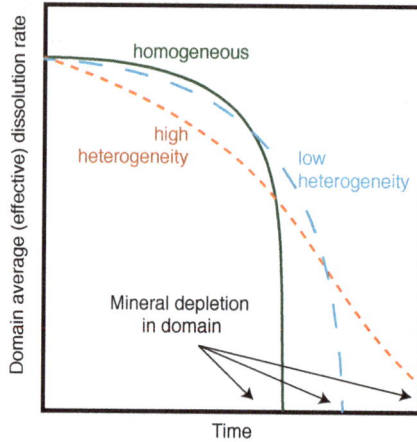

Figure 7. Conceptual relationship derived from results of Jung and Navarre-Sitchler (2018a,b) and Pandey and Rajaram (2016) between domain averaged (or effective) mineral dissolution rates in heterogeneous systems where heterogeneity is represented by variations in permeability. The fastest domain average rates occur in homogeneous simulations (solid line) compared to simulations with heterogeneity. For low heterogeneity systems (low permeability variance or small correlation lengths in gaussian random fields) early dissolution rates are similar to homogeneous domains but as minerals start to deplete in the high permeability pathways, the domain average dissolution rate slows. In high heterogeneous domains (e.g., high permeability variance, long correlation lengths in gaussian random fields, or in discrete fracture networks) domain average dissolution rates are lower than both low heterogeneity and homogeneous domains. However, at late time, the mineral in the homogenous domain depletes fastest, reducing reaction rates to zero at times earlier than either heterogeneous domain. Reaction happens over the longest period of time in the high heterogeneous domains due to a shift from kinetically limited dissolution rates in early time when there is reactive mineral in the high permeability pathways to transport-limited rates in low permeability zones as reactive mineral is depleted. The specific timing of rate reductions and relationship between homogeneous and heterogenous domains will be a function of the mineral kinetics, structure of the permeability field, and size of the domain.

10^6 years where weathering rates were compared between discrete fractured networks and gaussian random field permeability distributions (Fig. 6). Where orthoclase was completely depleted in a homogenous simulation after ~6.5×10^5 years, depletion in the correlated random permeability fields required 7.75×10^5 to 10^6 years and depletion in the discrete fracture network fields would take much longer than 10^6 years. As in the studies of Jung and Navarre-Sitchler, depletion of orthoclase occurred first in fast velocity pathways and reaction rates were fast in early time, but as reaction moved from the depleted fractures into low permeability blocks, matrix diffusion controlled the removal of weathering produced solutes and local reaction rates transition from far-from-equilibrium kinetic control to transport-controlled (Fig. 7).

Results from these studies begin to shed light on the role of heterogeneity and variation in fluid travel times on weathering in natural systems. The largest reductions in reaction rates compared to far-from-equilibrium conditions occur in systems with the biggest difference in permeability between fast and slow velocity pathways (Fig. 7). In both studies a single reactive mineral was distributed homogeneously throughout the domain isolating the role of heterogeneity. Where reactive mineral distribution and permeability are correlated, the effects may be dampened or enhanced (e.g., Atchley et al. 2014; Beisman et al. 2015).

CONTROLS ON WEATHERING FRONTS ALONG HILLSLOPES

Hillslopes are an important hydrological and geochemical scale of consideration as this is arguably the scale at which the processes that architect the Critical Zone strongly influence larger scale watershed behavior (McGlynn et al. 2004; McGuire et al. 2005; Maxwell and Kollet 2008a; McGuire and McDonnell 2010; Atchley and Maxwell 2011; Dralle et al. 2018; Rempe and Dietrich 2018). As such, multiple hypotheses about the controlling factors in the formation of regolith have been developed (e.g., Anderson et al. 2013; Rempe and Dietrich 2014; St. Clair et al. 2015; Brantley et al. 2017), the majority of which could be tested by application of reactive transport models. Nevertheless, hillslopes have posed a considerable challenge for RTM applications for several reasons. Although short-term (ca. decadal to millennial) simulations of coupled weathering and transport reactions are feasible (Beisman et al. 2015), to adequately capture the evolution of the regolith requires implementation of geomorphic rules, along with fluid transport and (bio)geochemical processes. In addition, the computational expense of conducting large-scale simulations over time scales at which reaction fronts propagate has been prohibitive, although high-performance computing is ameliorating this problem. A fully integrated model capable of approaching the ca. 15 kyr to 100s of kyr timescales required has not yet been constructed, so reactive transport models at this scale have relied on simplifying assumptions, largely based on the principles developed in the sections above.

As an example of the expansion of 1-D approaches to explicitly test models of hillslope-scale weathering and their potential implications for hydrologic processes, Lebedeva and Brantley (2013) expanded their model for weathering front advance that considers quartz, albite, kaolinite and an aqueous solute component (Na_2Si_2) (see Lebedeva et al. 2010) to model regolith evolution on a convex-upward hillslope assuming only vertical flow of water. To do this they combined a 2-D solute production and transport equation with the solid mass balance for the hillslope that accounts for soil diffusivity. Using this approach, they could then compute the rate of change in material mass and consequently the bedrock surface, or base of the weathering front. Based on the assumption that weathering is isovolumetric, it emerges that the weathering advance rate defines the elevation of unweathered material (Lebedeva and Brantley 2013; Brantley et al. 2017). Examples of steady-state results are shown in Figure 8 for one erosion rate and a range of flow rates assuming a regolith of quartz and albite, where albite weathers to form kaolinite. The volume fraction of mineral (albite) remaining at the top of the hillslope is defined as η_{max}, which is analogous to the extent of reaction, such that $\eta_{max} < 1$ indicates that the mineral is not completely depleted at the ground surface. As expected from 1-D simulations, regolith is thicker when the flow rate is relatively high. It is also evident that while erosion controls the slope of the land surface, the flow rate controls the slope of the bedrock-regolith surface. In simulations where the amount of quartz in the bedrock is increased to reflect the presence of more inert minerals, the regolith thickens, matching field observations. This condition may be typical of shale bedrock comprised of less reactive clay minerals (Jin et al. 2010; Heidari et al. 2017). Recent RTM modeling work also suggests that reactive gases, such as the CO_2 that originates from subsurface respiration, as well as the balance between unsaturated and saturated flow, should strongly influence emergent parameters such as η_{max} in a predictable way (Winnick and Maher 2018). Such connections may be a mechanism for coupling the biosphere to hillslope evolution. Other studies have also pointed out the importance of O_2 supply at depth in controlling the weathering front in rocks where Fe(II)-bearing minerals are abundant (Buss et al. 2008; Bazilevskaya et al. 2015; Winnick et al. 2017). These additional feedbacks, which can also result in strong porosity-permeability feedbacks (Navarre-Sitchler et al. 2009), have not yet been addressed at the scale of hillslopes using a full multi-component model.

Although certain processes may be missing from the model, these simulations nicely demonstrate that it is the combination of climate, reactive mineral content and tectonics that shape the hillslope architecture. Based on their results, Brantley and Lebedeva (2017) argue

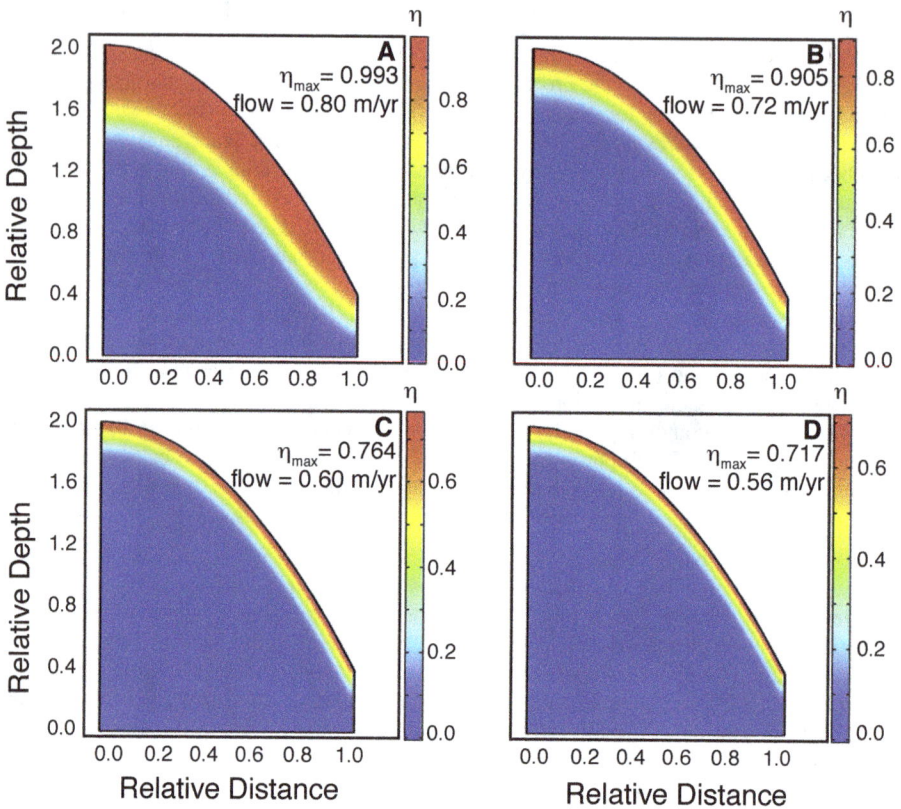

Figure 8. Simulated steady-state hillslopes from Brantley and Lebedeva (2017) for an erosion rate of 4×10^{-5} m/yr and Darcy velocity from 0.8 to 0.56 m/yr (in the vertical direction) for a rock initially containing 40% albite and 50% quartz, with albite weathering to kaolinite (quartz is inert) **(A–D)**. Colors show contours for the extent of albite reaction, η (i.e., the extent of weathering at each depth) with blue (0) representing unweathered bedrock and warmer colors representing greater extent of reaction and thus greater fraction of kaolinite (see key to right of each diagram). The axes are normalized by the specified length of 20 m. Modified from, Fig. 3 of Brantley and Lebedeva (2017).

that there are several emergent properties of the system: (1) the ratio of relief to hillslope length, (2) the distribution of regolith thickness, (3) the extent of weathering at the land surface and (4) the extent of regolith chemical depletion and the related porosity. As a consequence, Brantley and Lebedeva (2017) argue that reaction fronts operate as valves for water based on the permeability contrast that develops between layers. Such valves in turn control the partitioning of precipitation delivered to the hillslope. Given the outstanding questions around the processes that control watershed function and the growing recognition of "catchment co-evolution" (Troch et al. 2015), linking these principles to other watershed characteristics could provide a promising path forward for treating catchment behavior as a product of the co-evolution between regolith development and hydrological process.

Another hypothesis put forward as a control on regolith thickness involves the fractures created in response to topographic stress (St. Clair et al. 2015). Using a different approach and set of assumptions, Anderson et al. (2019) also argue for an important role for climate in shaping the variations in hillslope architecture observed by St. Clair et al. (2015). Whereas Brantley and Lebedeva (2017) in Figure 8 assume dominantly vertical flow, Anderson et al.

(2019) calculate the water table geometry using the Dupuit approximation and use this to partition flow above the water table as vertical drainage, transitioning to lateral flow below the water table. Using a streamline approach for a homogenous porous medium allows them to calculate the amount of dissolution along the stream lines according to the travel time (Atchley et al. 2013). Because their bedrock interface, which acts as a no-flow boundary condition, is set by the base of the stream channel they do not capture the deeper circulation predicted by Tothian models, but nevertheless their results agree well with other solutions at the interface between the unsaturated and saturated zone (e.g., Ameli et al. 2016). Results from end-member simulations for a wet and dry climate are shown in Figure 9, where the water table geometry and flow velocities differ based on the recharge rate but with the base of groundwater flow always coincident with the stream channel bottom (see Fig. 2 of Anderson et al. (2019)). Clearly evident is a transition from surface-parallel weathering to deep weathering that occurs over a relatively narrow threshold of recharge values. As Anderson et al. (2019) further demonstrate, the dry case is actually well represented by a 1-D vertical model, whereas the wet case, where weathering is nearly complete between the ground surface and the bedrock boundary condition, requires a juxtaposition of highly weathered material and fresh parent material.

This example nicely demonstrates that it may be partly to entirely feasible to explain the documented seismic velocity fields in terms of climatic factors. However, the authors assumed a constant hydraulic conductivity, whereas feedbacks could exist between high permeability zones created by tectonically driven fractures and weathering processes, including rapid fluid flow and the introduction of reactive gases, such as CO_2 and O_2. In general, the absence of permeability feedbacks in coupled geomorphic-weathering models is recognized by several authors as the frontier in understanding the connections among the current array of hillslope-scale hypotheses for the controls on subsurface architecture. Collectively, these selected examples demonstrate the utility of expanding the knowledge acquired via reactive transport approaches at the 1-D scale to larger scales.

The contrast between the modeling results in Figure 8 and Figure 9 highlights an important consequence of the quest for larger-scale models of weathering processes: the assumptions implicit within the modeling framework, which are often appropriate simplifications required to address a particular question, will unavoidably bias the results. As integrated geomorphic–hydrologic–geochemical–biological–geomechanical models continue to advance, continued interrogation of the consequences of underlying model assumptions will be critical for linking the growing number of hypothesis regarding the controls on Critical Zone evolution and architecture.

CATCHMENT-SCALE RTM MODELS

Although hillslopes represent a key functional unit for both hydrologists and geochemists to examine processes, river systems average over the many different hillslope units in a catchment. Rivers are often the focus of long-term records of discharge and water quality and these long-term records can provide a window into solute production under varying climatic (e.g., drought compared to inter-drought years) and land-cover (e.g., restoration, massive tree mortality, or deforestation) conditions (e.g., Raymond et al. 2008; Mikkelson et al. 2014). Numerous weathering studies have evaluated river chemistry as a means to quantify catchment-scale weathering (e.g., Gaillardet et al. 1999; Millot et al. 2002; Ibarra et al. 2016). Concentration discharge (C–Q) relationships, where concentrations of solutes are evaluated as a function of total water discharge, provide a window into the geochemical processes that produce solutes (Stallard 1988; Stallard 1992; Moatar et al. 2017; Winnick et al. 2017). C–Q data can be categorized according to the extent of correlation (Godsey et al. 2009), with many catchments displaying chemostatic behavior where solute concentrations remain constant with varying discharge (Clow and Mast 1995; Godsey et al. 2009; Stallard and Murphy 2014;

Dry Case:
Infiltration = 0.02 m/yr
K_{sat} = 0.05 m/day
ϕ = 0.1, θ = 0.08

Wet Case:
Infiltration = 0.12 m/yr
K_{sat} = 0.05 m/day
ϕ = 0.1, θ = 0.084

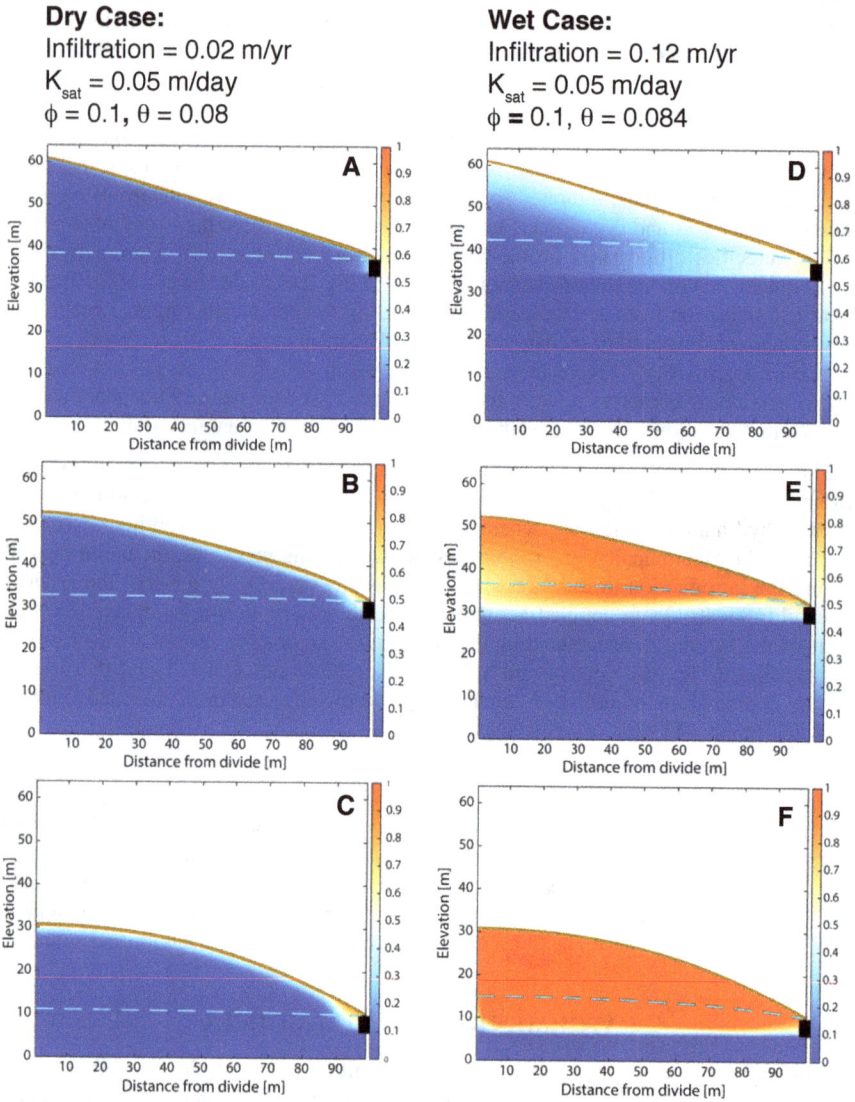

Figure 9. Simulated evolution of hillslopes towards steady state modified from Anderson et al. (2019). Time steps correspond to 50 (**A, D**), 250 (**B, E**), and 1,000 kyr (**C, F**) of weathering for two different recharge (infiltration) scenarios. At 1,000 kyr the weathering fields are at steady state. Colors in simulation correspond to reaction extent. For all simulations, initial topography is linear with slope of 0.25 and becomes parabolic over time. The dashed blue line indicates the water tabler lowering over time as the channel (black bar) also lowers at a rate of 3×10^{-5} m/yr.

Herndon et al. 2015). Chemostatic behavior may result from a number of coupled geochemical and hydrological processes depending on the geochemistry of the element, including buffering of solute concentrations by chemical reactions, transport limitation of reaction rates, and uneven vertical distributions of particular elements (e.g., Evans and Davies 1998; Kirchner et al. 2000; Kirchner et al. 2001; Chanat et al. 2002; Godsey et al. 2009; Clow and Mast 2010; Maher 2011; Stallard and Murphy 2014; Kim et al. 2017).

The integration of geochemical, hydrological, biological, and climatological processes at the catchment scale makes it difficult to isolate and quantify various drivers of weathering, even from well characterized C–Q relationships. Recent work has proposed a conceptual model of catchments that frames them as the product of a co-evolutionary process, resulting in the concept of hydrologic age (Troch et al. 2015). While catchments are often thought of as unique entities, the concept of hydrologic age may allow for organization of catchments according to their starting conditions and several catchment forming factors: climate, geology, tectonics, and time. In effect, the concept of hydrologic age posits that catchments with the same hydrologic age are at the same point along an evolutionary path even if the time it took to achieve this state differs. For example, a catchment in a wet climate compared to a catchment in a dry climate, holding all other factors equivalent, would result in older vs. younger hydrologic age, respectively. This concept is well illustrated in the profile and hillslope examples of the previous section, where climate was a key determinant of the depth and shape of regolith. Expanding weathering concepts developed at profile and hillslope scales, including the temporal evolution of weathering fronts and heterogeneity, to the catchment scale is a critical step in the modeling of weathering processes. Testing concepts of co-evolution using RTMs may help to improve models of catchment scale hydrologic processes, which currently do not integrate the legacy of past conditions into modern representations. As we describe below, new approaches are making this possible.

To scale RTMs of rock weathering from profile or hillslope scale to catchment scale, additional processes and boundary conditions need to be considered. Transitioning from profile to hillslope scale, heterogeneity in rock and regolith structure and mineralogy can be added and 1D vertical flow is expanded to include 2D lateral flow, but other processes controlling the rate and amount of infiltration can be assumed to remain largely constant over a single hillslope. However, when scaling from 2D hillslopes to 3D catchment scale models, spatial distributions of hydrologic processes such as infiltration, runoff, evapotranspiration, and in some cases snow accumulation and melt timing become important controls on water delivery to the subsurface and thus rock weathering (Pelletier et al. 2018). In 3-D catchment models, variations in subsurface structure with larger variations in lithology or orientation of subsurface structure relative to the land surface can be included. For example, in hard rock terranes with oriented joints, fractures, or foliations controlled by regional tectonics, the intersection of these fast fluid flow pathways with the land surface can vary with hillslope aspect, driving differences in water delivery to the subsurface and resulting biogeochemical reactions (Hinckley et al. 2014). Additionally, hydrologic processes can vary with hillslope aspect due to differences in solar radiation that drive differences in soil moisture content (Gomez-Plaza et al. 2001), snow accumulation and melt (Langston et al. 2015), and vegetation types (Istanbulluoglu et al. 2008). Researchers have observed relationships between hillslope aspect and structure of the Critical Zone (St. Clair et al. 2015), however, the driving forces behind these differences are still under investigation. Mechanical models suggest that tectonic unloading and opening of fractures in the subsurface are responsible for variations in depth of regolith (St. Clair et al. 2015) but little is known about the relationships between depth of regolith and position of weathering profiles in varying settings. Reactive transport models at the catchment scale would help elucidate some of the potential controls on catchment scale weathering and generate hypotheses that could be tested with field data. In order to fully simulate weathering at the catchment scale, reactive transport modules are being integrated with distributed hydrologic models that couple subsurface flow, surface flow and vegetation models to account for runoff and evapotranspiration; in some cases land-surface and atmospheric models are coupled to simulate variations in spatial-distribution and timing of precipitation events and temperature changes (e.g., RT-Flux-PIHM (Duffy et al. 2014; Bao et al. 2017) and ParCrunchFlow (Beisman et al. 2015).

One approach to modeling weathering at the catchment scale is to integrate chemical composition across different components of the weathering profile as a function of time using a mass balance or "box" model. One example is the WITCH model developed by Goddéris et al. (2006). The WITCH model accounts for influx into the top of a weathering profile, outflux from the bottom through gravity drainage, solute production in the weathering profile, mineral precipitation, and exchange with the solid phase (Goddéris et al. 2006). To bring an integrated 1D model to a simplified catchment scale, WITCH is coupled to other models that estimate catchment scale fluxes such as ASPECTS, an Atmosphere-Soil-Plant Exchange model of Carbon in Temperate Sylvae. Coupling to ASPECTS allows for tree growth to be integrated into the solute mass balance over several decades. The ASPECTS model calculates photosynthesis, transpiration and stomatal conductance of CO_2 and H_2O. The WITCH-ASPECTS model was applied to the Strengbach catchment in the Vosges Massif, France, demonstrating that the Si mass flux out of the catchment is controlled by mineral dissolution/precipitation in the root zone with poorly crystalline smectite precipitation strongly linked to Mg flux. Model results revealed that Ca concentrations were controlled by trace apatite within the granitic bedrock (Goddéris et al. 2006). This type of modeling approach can provide conceptual constraints on catchment-scale controls on integrated, or flux weighted concentrations, but is not a true 3D model that can parse out processes in space and time within a catchment. Thus, development of reactive transport approaches on a 3D grid with integration of hydrologic, biologic, and atmospheric processes are currently under development.

RT-Flux-PIHM, the Penn State Integrated Hydrologic Model (PIHM, Duffy et al. 2014), with a land-surface interaction module (Flux), and a reactive transport (RT) module (Bao et al. 2017), couples topography and subsurface structure on an unstructured 3D grid that solves for evapotranspiration, infiltration, overland flow, subsurface lateral flow, mineral dissolution and precipitation, and ion exchange with variable precipitation and solar radiation in space and time. RT-Flux-PIHM has been applied to the Susquehanna-Shale Hills catchment (Brantley et al. 2018) to explore the genesis of chemostatic C–Q relationships observed in field data (Li et al. 2017). Model results map temporal and spatial distributions of chlorite dissolution in the watershed that are also a function of discharge. Li et al. (2017) show that chemostatic behavior in Mg C–Q relationships is generated by chlorite dissolution under most conditions, with Mg ion exchange exerting secondary controls during times of intense rainfall or very low discharge.

Recently the Flux-PIHM model was coupled to the WITCH model by Sullivan et al. 2019 to further explore the role of slope aspect on mineral weathering in the Susquehanna-Shale Hills catchment. Slope aspect in catchments controls solar radiation at the land surface, which translates to differences in evapo-transpiration, soil moisture, and vegetation on slopes with different aspects. In order to account for the role of aspect on hydrology and geochemistry Sullivan et al. added a solar radiation module to Flux-PIHM. The use of the WHICH model allows for discretization of geochemical process representation as a function of depth in the soil zone to simulate the role of vegetation and soil processes. A biolifting module was also added to WITCH in order to match depth profiles of elements like potassium, that are cycled by vegetation in the near the surface. Results from the simulations suggest that in this shale bedrock system short-term chemical weathering is more sensitive to clay size distribution than to micro-climate related to slope aspect, despite field observations that suggest deeper weathering on shaded slopes compared to sunny slopes. Sullivan et al. (2019) suggest that slope aspect may play a larger role in long-term weathering and explain the observed differences in clay size distribution between shaded and sunny slopes that drives modern weathering.

Without fully integrated RTM catchment approaches, elucidating geochemical controls on C–Q relationships in space and time or slope aspect on geochemical weathering would be exceedingly difficult. Distributed, fully coupled, integrated reactive transport models like RT-Flux-PIHM are computationally expensive, requiring parallel computing for large-domains or long time-scale simulations, but will increasingly play a central role in identifying the fundamental controls on weathering at the catchment scale in ways that other RTM approaches cannot. Some of the challenges in distributed models that couple modules for different processes include differences in time-steps and spatial-scales for different processes requiring averaging or integration, simplification in subsurface layering and structure, and lack of data needed to fully characterize or parameterize processes (Li et al. 2017).

FRONTIERS IN WEATHERING MODELS

The need for translation across scales is widely recognized but difficult to adopt in both field and modeling studies. In this review, we focus primarily on how knowledge of mineral dissolution, the effect of transport and heterogeneity on reaction rates, and by extension, the propagation of weathering fronts has inspired new understanding of weathering at larger scales, from hillslopes to catchment scales. Traditionally, fewer studies have focused on telescoping down from larger scales. With the proliferation of Critical Zone data, interesting hypotheses surrounding the controls on Critical Zone architecture can be tested with existing modeling approaches. Nevertheless, we see the need for advances in modeling approaches that enable more holistic integration of the various agents that drive weathering. These needs include models that incorporate geomorphic rules in order to allow for the emergence of weathering fronts based on the solid and fluid mass balance and more rigorous descriptions of biota.. Below, we describe several advances that address this integration in order to highlight the frontiers of RTM approaches.

The role of RTM models in the short-term carbon cycle

Many of the examples provided above focus on the long-term evolution (ca. 1–100s of kyrs) of weathering processes that act to balance the carbon cycle over million-year timescales. As such, it is often necessary and appropriate to conceptualize the biosphere as a boundary condition that acts as a constant source of reactive gases (e.g., carbon dioxide an oxygen) and dissolved organic compounds (e.g., organic acids). However, over shorter timescales (1 to 100s of yrs) attention is increasingly focused on the tighter and more dynamic couplings to the biosphere, including the storage of water in deep regolith that supplies trees (Rempe and Dietrich 2018), the role of weathering in the supply of rock-derived nutrients critical for plant growth (Chadwick and Asner 2018; Schuessler et al. 2018), and the potential for secondary minerals, such as clays and oxy(hydr)oxides to serve as vessels for carbon storage (Schmidt et al. 2011; Lehmann and Kleber 2015). Many of these connections are increasingly recognized to play critical roles in the carbon cycle by impacting the ability of vegetation to take up carbon and the capacity of soils to store it. Here, reactive transport models are emerging as critical tools in linking the long-term evolution of weathering described above to shorter-timescale processes driving the biosphere, with models just now beginning to incorporate robust treatment of root growth and nutrient acquisition (Gérard et al. 2017) and the stabilization and turnover of organic carbon in soils (Riley et al. 2014; Dwivedi et al. 2017).

As one example of a coupling, Gérard et al. (2017) coupled a root system architecture model with a reactive transport model to examine P acquisition from the hydroxyapatite and the associated P, Ca, and pH variations introduced into the root zone (Fig. 10). The root model contains provisions for growth, elongation, and branching of roots, as well as decay of existing roots such that each grid cell in the RTM model has a specific root surface density that collectively determines the uptake rate following the Michaelis–Menten equation, but is parameterized in terms of the root surface density and aqueous nutrient concentration. The dissolution of hydroxyapatite, sorption of P, and related aqueous reactions are handled by the reactive transport model. As Gérard et al. (2017) show, sharp gradients develop around the root zone and in response to the geochemical feedbacks. As a result, root-induced acidification increases the efficiency of

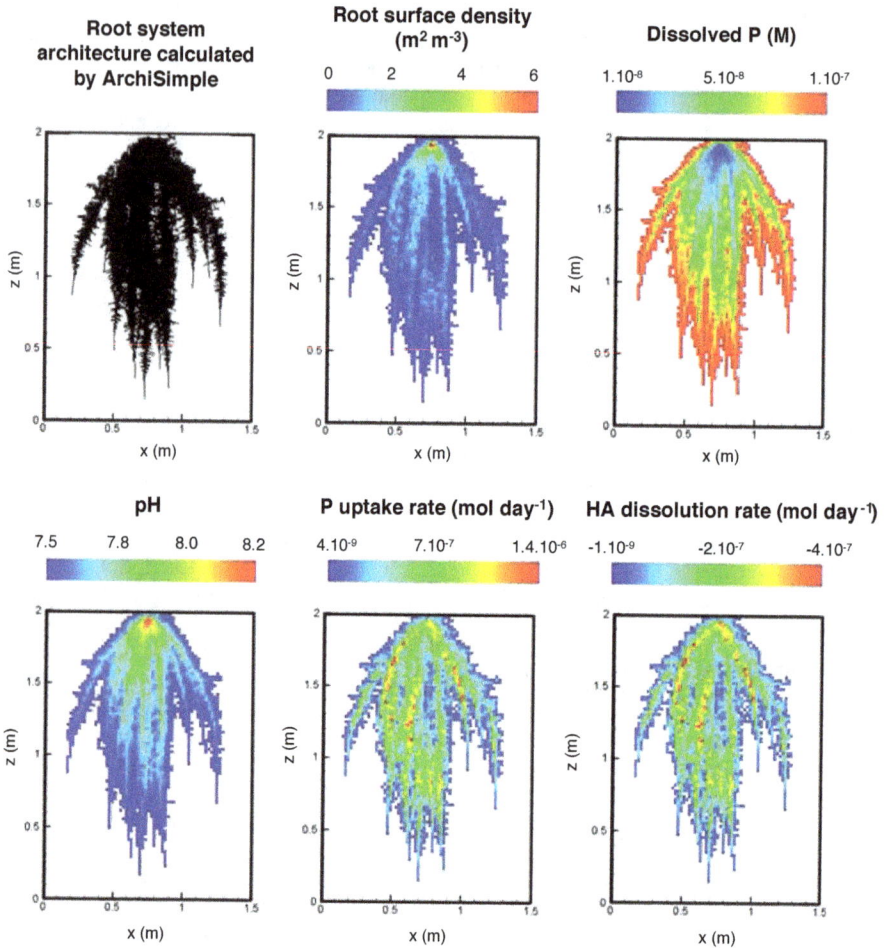

Figure 10. Simulation of root architecture and function using the root model ArchiSimple and reactive transport using the RTM MIN3P. Models results show phosphorus acquisition from hydroxyapatite (HA) via nutrient uptake. Results after 80 days of simulation. [Reproduced with permission from Springer, Gérard et al. (2017) Modelling the interactions between root system architecture, root functions and reactive transport processes in soil. Plant Soil. 413:161–180, Fig. 3.]

P acquisition. Through the application of the coupled modeling approach, the authors are able to uniquely capture the interface between a dynamic root system and the soil matrix in order to provide hypotheses for what types of conditions might be optimal for plant growth.

Another compelling example of the application of RTMs to the short-term carbon cycle is in the area of modeling soil carbon turnover. The fate of soil carbon is controlled by a complex set of interrelated variables, including plant inputs, microbial processing of carbon and nitrogen, soil moisture and texture, temperature, and mineral-associated partitioning (e.g., sorption and occlusion) (Torn et al. 1997; Manzoni and Porporato 2009; Schmidt et al. 2011; Berhe et al. 2012; Manzoni et al. 2012; Moyano et al. 2013; Kleber et al. 2015). In addition, CO_2 produced by autotrophic and heterotrophic respiration is controlled by diffusion and potentially by abiotic reactions, such as carbonate dissolution/precipitation. One of the best metrics of carbon turnover has been radiocarbon, which measures the average turnover time of carbon (Trumbore 2000). Taking a cue from radiocarbon measurements, most models of soil carbon have traditionally treated carbon in terms of operationally defined pools, neglecting some of the underlying biological (plant and microbial) and abiotic (weathering) controls. For example, the age of soil and the accumulation of secondary weathering products has been shown in some environments to be associated with carbon turnover and hence storage (Torn et al. 1997). Recent efforts have demonstrated the utility of a reactive transport approach that allows for treatment of a diverse array of organic molecules and a more rigorous representation of the abiotic controls (Riley et al. 2014; Dwivedi et al. 2017). By developing and applying a reactive transport representation to simulate soil carbon turnover, Riley et al. (2014) argue that observed depth profiles and radiocarbon could be explained by carbon compounds with relatively fast reaction rates such that their net decomposition rate is controlled by vertical transport and protection on mineral surfaces. Given the growing recognition of the importance of mineral surfaces in governing soil organic carbon availability (Torn et al. 1997; Doetterl et al. 2015; Keiluweit et al. 2015) as well as the links to soil transport and deposition (Berhe et al. 2012, 2018; Berhe and Kleber 2013), coupling the evolution of the soil mineralogy and structure over time to soil carbon dynamics is a new frontier that may ultimately contribute to predicting the response of ecosystems to the onslaught of disturbances they are increasingly experiencing. When coupled with models of root architecture, such as the example above, questions surrounding the co-evolution of plants, microbial communities and regolith could potentially be addressed, leading to insights into long-term shifts in weathering that are hypothesized to have been driven by the evolution and response of land plants (Berner 1992; Pagani et al. 2009), but are still poorly constrained.

The role of weathering models in water quality

The majority of the studies we discuss are focused on the factors that control weathering of primary minerals and the implications for Critical Zone architecture. However, accessory minerals in bedrock and sediments can be an important source of groundwater contaminants: well-known examples include arsenic release from pyrite or iron oxides (e.g., Postma et al. 2007; Erban et al. 2013; Neil et al. 2014; Houben et al. 2017) and chromium weathering from ultramafic rocks (e.g., Oze et al. 2007; Izbicki et al. 2015; Manning et al. 2015; McClain and Maher 2016; Vengosh et al. 2016). In some sense, weathering zones behave much like supergene enrichment zones that are well studied from an ore deposits perspective. For redox-sensitive contaminants, steady advancement of the weathering front allows water to penetrate into the bedrock and may facilitate the transport of oxygen, driving oxidation of any redox-sensitive accessory phases (Xu et al. 2000; Buss et al. 2008; Bazilevskaya et al. 2015; Winnick et al. 2017). The accumulation of iron-(hydrox)oxides and other weathering products enhances the sorption of cations, whereas anions such as chromate or selenate may be poorly retained in the weathering zone at or above circumneutral pH and are thus transmitted to surface and groundwater systems (Manning et al. 2013; McClain and Maher 2016). RTMs of this process have been developed to consider the time scales and processes of supergene copper enrichment (Xu et al. 2000), but far less work has been done on other types of weathering reactions and in the development of process representations for trace metal dynamics. In addition, human or climate-driven shifts in the water table may expose unweathered minerals to oxidants, potentially accelerating the release of oxidized weathering products during periods of water table decline (Manning et al. 2013). Increasing water table levels in response to snow melt or other forcings may also drive seasonal redox cycles, as is known to occur in floodplains (Janot et al. 2016). Such seasonal redox cycling, if it promotes the accumulation of reduced products in the weathering zone, could actually limit the penetration of oxygen to the weathering front and thus inhibit weathering in systems such as shales, where pyrite oxidation is thought to play a key role in the initial opening of porosity (Jin et al. 2013). Ultimately, today's groundwater systems contain a legacy of the weathering processes that unfolded over thousands of years.

Integrating this legacy into models of shorter-term processes remains an outstanding challenge. In addition, the redox cycling of trace metals, including their release from accessory phases and partitioning into secondary reservoirs by sorption or co-precipitation require the development of process representations for reactive transport models, a development which would benefit models describing both water quality and redox controls on weathering front evolution.

Models of global weathering

Models of weathering at global scale provide valuable insight into the feedbacks between climate, tectonics, and the biosphere through the coupling of weathering and atmospheric CO_2 levels throughout Earth's history (Kump et al. 2000; Goddéris et al. 2008; Maffre et al. 2018). Analysis of riverine fluxes to the oceans provide key clues, while modern reactive transport models can typically only simulate fluxes over small catchment scales and short periods of time. In contrast, weathering processes that play a role in global climatic and tectonic evolution unfold over spatial scales of $\geq$100s km^2 and time scales of thousands to tens of thousands of years or more. These spatial and temporal scales currently pose a computational barrier for fully coupled three-dimensional reactive transport simulations. Thus, current approaches to scaling riverine fluxes over time do not solve fully coupled geochemical and fluid flow equations but rely on a variety of simplifications and assumptions that allow for calculation of the feedbacks between weathering, tectonics, and climate through empirical approaches. An early example of this approach is the investigation of global change in the late Devonian by Goddéris and Joachimski (2004) using the COMBINE model, where atmospheric CO_2 drawdown by silicate mineral weathering is calculated as a linear function of continental runoff (calibrated to data from Gaillardet et al. 1999) and an exponential function of mean annual temperature and atmospheric CO_2 concentrations. Using a similar modeling approach, Goddéris et al. (2012, 2017) used GEOCLIM to model periods of global climate change. GEOCLIM calculates runoff and surface air temperature that are used in phenomenological relationships to estimate drawdown of atmospheric CO_2 through silicate mineral weathering. In Goddéris et al. (2017), the exposure of weatherable materials at the regolith/rock interface and removal by physical erosion at the surface was added to the model to examine the role of the Hercynian Orogeny in the Late Paleozoic Ice Age. The model results suggested that the uplift associated with the Hercynian Orogeny promoted physical erosion and increased chemical weathering, driving down atmospheric CO_2 levels to those typical for ice age climates. These studies highlight the important links between climate and weathering that continued advances in global weathering models can help elucidate and constrain. Continued development of catchment scale RTMs coupled to erosion, climate, and biological activity along with methods to integrate model results over temporal scales relevant to geologic time will help to evaluate these global-scale models and thus to further our understanding of the connections between the rise of land plants, weathering, tectonics, and climate throughout Earth's history.

CONCLUSIONS

The application of RTMs to chemical weathering has been a critical tool for advancing our understanding of Critical Zone evolution. By enabling constructs developed under laboratory settings to be tested against field-scale data, models have not only supported and expanded on new theories but have also contributed to an inspiring palette of theories about the initiation of weathering, controls on Critical Zone architecture, and links to climate regulation. Reactive transport models will undoubtedly play a role in testing the assumptions and predictions put forth by these theories. This cycle of iteration between theory, observations, and model application has been critical in shaping knowledge of weathering processes, but also in improving how the models are used across other applications. Examples where weathering RTM studies have provided insights into long-standing conundrums include the origins of the lab-field discrepancy, movement and evolution of reaction fronts in natural systems, and potential couplings of weathering to biological processes. As we outline above, model formulations will need to continue to advance to include additional frameworks, such as landscape evolution frameworks to understand geologic time scale weathering problems, climate models to evaluate future response of weathering to a changing climate, and biological models to interrogate shorter time-scale couplings between the biosphere and abiotic processes.

ACKNOWLEDGMENTS

KM would like to acknowledge support from NSF (EAR-1254156) and the DOE Office of Science SBR program (DE-SC0014556 and DE-SC001815). AN-S would like to acknowledge support from NSF-CAREER program (EAR-1554502) and from the DOE Office of Basic Energy Sciences (DE-SC0018647). We are grateful to the editors of this volume, Jenny Druhan and Christophe Tournassat, and to two anonymous reviewers whose comments greatly improved this manuscript.

REFERENCES

Aasgard P Helgeson HC (1982) Thermodynamic and kinetic constraints on reaction-rates among minerals and aqueous-solutions .1. Theoretical considerations. Am J Sci 282:237–285

Ameli AA, Amvrosiadi N, Grabs T, Laudon H, Creed IF, McDonnell JJ, Bishop K (2016) Hillslope permeability architecture controls on subsurface transit time distribution and flow paths. J Hydrol 543:17–30

Amundson R, Richter DD, Humphreys GS, Jobbagy EG, Gaillardet J (2007) Coupling between biota and earth materials in the Critical Zone. Elements 3:327–332

Anderson SP, von Blanckenburg F, White AF (2007) Physical and chemical controls on the Critical Zone. Elements 3:315–319

Anderson RS, Anderson SP, Tucker GE (2013) Rock damage and regolith transport by frost: an example of climate modulation of the geomorphology of the critical zone. Earth Surf Processes Landforms 38:299–316

Anderson RS, Rajaram H, Anderson SP (2019) Climate driven coevolution of weathering profiles and hillslope topography generates dramatic differences in critical zone architecture. Hydrol Processes 33:4–19

Arvidson RS, Ertan IE, Amonette JE, Luttge A (2003) Variation in calcite dissolution rates: A fundamental problem? Geochim Cosmochim Acta 67:1623–1634

Atchley AL, Maxwell RM (2011) Influences of subsurface heterogeneity and vegetation cover on soil moisture, surface temperature and evapotranspiration at hillslope scales. Hydrogeol J 19:289–305

Atchley AL, Maxwell RM, Navarre-Sitchler AK (2013) Using streamlines to simulate stochastic reactive transport in heterogeneous aquifers: Kinetic metal release and transport in CO2 impacted drinking water aquifers. Adv Water Resour 52:93–106

Atchley AL, Navarre-Sitchler AK, Maxwell RM (2014) The effects of physical and geochemical heterogeneities on hydro-geochemical transport and effective reaction rates. J Contam Hydrol 165:53–64

Bandstra J, Buss HL, Campen RK, Liermann LJ, Moore J, Hausrath E, Navarre-Sitchler AK, Jang J-H, Brantley SL (2008) Appendix: Compilation of mineral dissolution rates. In: Kinetics of Water–Rock Interaction. Brantley SL, Kubicki JD, White AF, (eds). Springer, New York, p 737–818

Banfield JF, Ferruzzi GG, Casey WH, Westrich HR (1995) HR-TEM study comparing naturally and experimentally weathered pyroxenoids. Geochim Cosmochim Acta 59:19–31

Bao C, Li L, Shi Y, Duffy C (2017) Understanding watershed hydrogeochemistry: 1. Development of RT-Flux-PIHM Water Resour Res 53:2328

Bazilevskaya E, Rother G, Mildner DFR, Pavich M, Cole D, Bhatt MP, Jin L, Steefel CI, Brantley SL (2015) How oxidation and dissolution in diabase and granite control porosity during weathering. Soil Sci Soc Am J 79:55–73

Beckingham LE, Mitnick EH, Steefel CI, Zhang S, Voltolini M, Swift AM, Yang L, Cole DR, Sheets JM, Ajo-Franklin JB, DePaolo DJ (2016) Evaluation of mineral reactive surface area estimates for prediction of reactivity of a multi-mineral sediment. Geochim Cosmochim Acta 188:310–329

Beckingham LE, Steefel CI, Swift AM, Voltolini M, Yang L, Anovitz LM, Sheets JM, Cole DR, Kneafsey TJ, Mitnick EH, Zhang S (2017) Evaluation of accessible mineral surface areas for improved prediction of mineral reaction rates in porous media. Geochim Cosmochim Acta 205:31–49

Beerling DJ, Berner RA (2005) Feedbacks and the coevolution of plants and atmospheric CO2. PNAS 102:1302–1305

Beisman JJ, Maxwell RM, Navarre-Sitchler AK, Steefel CI, Molins S (2015) ParCrunchFlow: an efficient, parallel reactive transport simulation tool for physically and chemically heterogeneous saturated subsurface environments. Comput Geosci 19:403–422

Berner EK, Berner RA, Moulton KL (2003) Plants and mineral weathering: present and past. *In:* Treatise on Geochemistry: Surface and Ground Water, Weathering, and Soils. Vol 5. Drever JI, (ed) Elsevier, Amsterdam, p 169–188

Berhe AA, Harden JW, Torn MS, Kleber M, Burton SD, Harte J (2012) Persistence of soil organic matter in eroding versus depositional landform positions. J Geophys Res-Biogeosci 117:G02019

Berhe AA, Kleber M (2013) Erosion, deposition, and the persistence of soil organic matter: mechanistic considerations and problems with terminology. Earth Surf Processes Landforms 38:908–912

Berhe AA, Barnes RT, Six J, Marin-Spiotta E (2018) Role of soil erosion in biogeochemical cycling of essential elements: Carbon, nitrogen, and phosphorus. Ann Rev Earth Planet Sci 46:521–548

Berner RA (1992) Weathering, plants, and the long-term carbon cycle. Geochim Cosmochim Acta 56:3225–3231

Berner RA (1995) Chemical weathering and its effect on atmospheric CO_2 and climate. *In:* Chemical Weathering Rates of Silicate Minerals. Vol 31. p 565–583

Berner RA, Kothavala Z (2001) GEOCARB III: A revised model of atmospheric CO_2 over phanerozoic time. Am J Sci 301:182–204

Beven K, Germann P (1982) Macropores and water flow in soils. Water Resour Res 18:1311–1325

Birkeland PW, Burke RM (1988) Soil catena chronosequences on eastern Sierra Nevada moraines, California, USA Arct Alp Res 20:473–484

Blum JD, Erel Y (1995) A silicate weathering mechanism linking increases in marine Sr-87/Sr-86 with global glaciation. Nature 373:415–418

Bouchez J, Gaillardet J, Lupker M, Louvat P, France-Lanord C, Maurice L, Armijos E, Moquet J-S (2012) Floodplains of large rivers: Weathering reactors or simple silos? Chem Geol 332:166–184

Bouchez J, Von Blanckenburg F, Schuessler JA (2013) Modeling novel stable isotope ratios in the weathering zone. Am J Sci 313:267–308

Brantley S, Lebedeva MI (2011) Learning to read the chemistry of regolith to understand the Critical Zone. Ann Rev Earth Planet Sci 39:387–416

Brantley SL, White AF (2009) Approaches to modeling weathered regolith. Rev Mineral Geochem 70:435–484

Brantley SL, Lebedeva MI, Balashov VN, Singha K, Sullivan PL, Stinchcomb G (2017) Toward a conceptual model relating chemical reaction fronts to water flow paths in hills. Geomorphology 277:100–117

Brantley SL, White T, West N, Williams JZ, Forsythe B, Shapich D, Kaye J, Lin H, Shi Y, Kaye M, Herndon E(2018) Susquehanna Shale Hills Critical Zone Observatory: Shale Hills in the context of Shaver's Creek Watershed. Vadose Zone J 17:180092

Bullen T, White A, Blum A, Harden J, Schulz M (1997) Chemical weathering of a soil chronosequence on granitoid alluvium 2. Mineralogic and isotopic constraints on the behavior of strontium. Geochim Cosmochim Acta 61:291–306

Buss HL, Sak PB, Webb SM, Brantley SL (2008) Weathering of the Rio Blanco quartz diorite, Luquillo Mountains, Puerto Rico: Coupling oxidation, dissolution, and fracturing. Geochim Cosmochim Acta 72:4488–4507

Cai Z, Wen H, Komarneni S, Li L (2018) Mineralogy controls on reactive transport of Marcellus Shale waters. Sci Tot Environ 630:1573–1582

Cerling TE (1991) Carbon dioxide in the atmosphere—evidence from Cenozoic and Mesozoic paleosols. Am J Sci 291:377–400

Chadwick KD, Asner GP (2018) Landscape evolution and nutrient rejuvenation reflected in Amazon forest canopy chemistry. Ecol Lett 21:978–988

Chadwick OA, Derry LA, Vitousek PM, Huebert BJ, Hedin LO (1999) Changing sources of nutrients during four million years of ecosystem development. Nature 397:491–497

Chanat JG, Rice KC, Hornberger GM (2002) Consistency of patterns in concentration-discharge plots. Water Resour Res 38:1147

Clow DW, Mast MA (1995) Composition of precipitation, bulk deposition, and runoff at a granitic bedrock catchment in the Loch Vale watershed, Colorado, USA. *In:* Biogeochemistry of Seasonally Snow-Covered Catchments (Proceedings of a Boulder Symposium, July 1995). IAHS Pub no. 228:235–242

Clow DW, Mast MA (2010) Mechanisms for chemostatic behavior in catchments: Implications for CO_2 consumption by mineral weathering. Chem Geol 269:40–51

Damköhler G (1936) Einflüsse der Strömung, Diffusion und des Wärmeüberganges auf die Leitstung von Reationsöfen. Zeitschrift für Electroche 42:846–862

Daval D, Calvaruso C, Guyot F, Turpault MP (2018) Time-dependent feldspar dissolution rates resulting from surface passivation: Experimental evidence and geochemical implications. Earth Planet Sci Lett 498:226–236

Doetterl S, Stevens A, Six J, Merckx R, Van Oost K, Pinto MC, Casanova-Katny A, Muñoz C, Boudin M, Venegas EZ, Boeckx P (2015) Soil carbon storage controlled by interactions between geochemistry and climate. Nat Geosci 8:780–783

Dralle DN, Hahm WJ, Rempe DM, Karst NJ, Thompson SE, Dietrich WE (2018) Quantification of the seasonal hillslope water storage that does not drive streamflow. Hydrol Processes 32:1978–1992

Drever JI (1994) The effect of land plants on weathering rates of silicate minerals. Geochim Cosmochim Acta 58:2325–2332

Drever JI, Stillings LL (1997) The role of organic acids in mineral weathering. Colloids Surf A 120:167–181

Druhan JL, Maher K (2017) The influence of mixing on stable isotope ratios in porous media: A revised Rayleigh model. Water Resour Res 53:1101–1124

Druhan JL, Steefel CI, Williams KH, DePaolo DJ (2013) Calcium isotope fractionation in groundwater: Molecular scale processes influencing field scale behavior. Geochim Cosmochim Acta 119:93–116

Druhan JL, Steefel CI, Conrad ME, DePaolo DJ (2014) A large column analog experiment of stable isotope variations during reactive transport: I A comprehensive model of sulfur cycling and δ^{34}S fractionation. Geochim Cosmochim Acta 124:366–393

Duffy C, Shi YN, Davis K, Slingerland R, Li L, Sullivan PL, Goddéris Y, Brantley SL (2014) Designing a suite of models to explore Critical Zone function. Procedia Earth Planet Sci 10:7–15

Dwivedi D, Riley WJ, Torn MS, Spycher N, Maggi F, Tang JY (2017) Mineral properties, microbes, transport, and plant-input profiles control vertical distribution and age of soil carbon stocks. Soil Biol Biochem 107:244–259

Dzombak DA, Morel FM (1990) Surface Complexation Modeling. Wiley Interscience, New York

Erban LE, Gorelick SM, Zebker HA, Fendorf S (2013) Release of arsenic to deep groundwater in the Mekong Delta, Vietnam, linked to pumping-induced land subsidence. PNAS 110:13751–13756

Evans C, Davies TD (1998) Causes of concentration/discharge hysteresis and its potential as a tool for analysis of episode hydrochemistry. Water Resour Res 34:129–137

Fletcher RC, Buss HL, Brantley SL (2006) A spheroidal weathering model coupling porewater chemistry to soil thicknesses during steady-state denudation. Earth Planet Sci Lett 244:444–457

Gaillardet J, Dupre B, Louvat P, Allegre CJ (1999) Global silicate weathering and CO_2 consumption rates deduced from the chemistry of large rivers. Chem Geol 159:3–30

Gaillardet J, Millot R, Dupre B (2003) Chemical denudation rates of the western Canadian orogenic belt: the Stikine terrane. Chem Geol 201:257–279

Ganor J, Reznik IJ, Rosenberg YO (2009) Organics in water–rock interactions. Rev Mineral Geochem 70:259–369

Gérard F, Blitz-Frayret C, Hinsinger P, Pages L (2017) Modelling the interactions between root system architecture, root functions and reactive transport processes in soil. Plant Soil 413:161–180

Germann PF, Zimmermann M (2005) Directions of preferential flow in a hillslope soil. Quasi-steady flow. Hydrol Processes 19:887–899

Giambalvo ER, Steefel CI, Fisher AT, Rosenberg ND, Wheat CG (2002) Effect of fluid–sediment reaction on hydrothermal fluxes of major elements, eastern flank of the Juan de Fuca Ridge. Geochim Cosmochim Acta 66:1739–1757

Goddéris Y, Joachimski MM (2004) Global change in the Late Devonian: modelling the Frasnian-Famennian short-term carbon isotope excursions. Palaeogeogr Palaeocl 202:309–329

Goddéris Y, Francois LM, Probst A, Schott J, Moncoulon D, Labat D, Viville D (2006) Modelling weathering processes at the catchment scale: The WITCH numerical model. Geochim Cosmochim Acta 70:1128–1147

Goddéris Y, Donnadieu Y, Tombozafy M, Dessert C (2008) Shield effect on continental weathering: Implication for climatic evolution of the Earth at the geological timescale. Geoderma 145:439–448

Goddéris Y, Donnadieu Y, Lefebvre V, Le Hir G, Nardin E (2012) Tectonic control of continental weathering, atmospheric CO_2, and climate over Phanerozoic times. C R Geosci 344:652–662

Goddéris Y, Donnadieu Y, Carretier S, Aretz M, Dera G, Macouin M, Regard V (2017) Onset and ending of the late Palaeozoic ice age triggered by tectonically paced rock weathering. Nat Geosci 10:382

Godsey SE, Kirchner JW, Clow DW (2009) Concentration-discharge relationships reflect chemostatic characteristics of US catchments. Hydrol Processes 23:1844–1864

Gomez-Plaza A, Martinez-Mena M, Albaladejo J, Castillo VM (2001) Factors regulating spatial distribution of soil water content in small semiarid catchments. J Hydrol 253:211–226

Green CT, Stonestrom DA, Bekins BA, Akstin KC, Schulz MS (2005) Percolation and transport in a sandy soil under a natural hydraulic gradient. Water Resour Res 41:W10414

Harman CJ, Cosans CL (2019) A low-dimensional model of bedrock weathering and lateral flow co-evolution in hillslopes: 2. Controls on weathering and permeability profiles, drainage hydraulics, and solute export pathways. Hydrol Processes 33:1168–1190

Heidari P, Li L, Jin LX, Williams JZ, Brantley SL (2017) A reactive transport model for Marcellus shale weathering. Geochim Cosmochim Acta 217:421–440

Hellmann R, Tisserand D (2006) Dissolution kinetics as a function of the Gibbs free energy of reaction: An experimental study based on albite feldspar. Geochim Cosmochim Acta 70:364–383

Hellmann R, Wirth R, Daval D, Barnes JP, Penisson JM, Tisserand D, Epicier T, Florin B L HR (2012) Unifying natural and laboratory chemical weathering with interfacial dissolution and reprecipitation: a study based on the nanometer-scale chemistry of fluid-silicate interfaces. Chem Geol 294–295:203–216

Herndon EM, Dere AL, Sullivan PL, Norris D, Reynolds B, Brantley SL (2015) Landscape heterogeneity drives contrasting concentration-discharge relationships in shale headwater catchments. Hydrol Earth Syst Sci 19:3333–3347

Hilley GE, Porder S (2008) A framework for predicting global silicate weathering and CO_2 drawdown rates over geologic time-scales. PNAS 105:16855–16859

Hinckley ELS, Ebel BA, Barnes RT, Anderson RS, Williams MW, Anderson SP (2014) Aspect control of water movement on hillslopes near the rain-snow transition of the Colorado Front Range. Hydrol Processes 28:74–85

Holbrook WS, Riebe CS, Elwaseif M, Hayes JL, Basler-Reeder K, Harry DL, Malazian A, Dosseto A, Hartsough PC, Hopmans JW (2014) Geophysical constraints on deep weathering and water storage potential in the Southern Sierra Critical Zone Observatory. Earth Surf Processes Landforms 39:366–380

Houben GJ, Sitnikova MA, Post VEA (2017) Terrestrial sedimentary pyrites as a potential source of trace metal release to groundwater—A case study from the Emsland, Germany. Appl Geochem 76:99–111

Huang WH, Keller WD (1970) Dissolution of rock-forming silicate minerals in organic acids - Simulated first-stage weathering of fresh mineral surfaces. Am Mineral 55:2076–2094

Ibarra DE, Caves JK, Moon S, Thomas DL, Hartmann J, Chamberlain CP, Maher K (2016) Differential weathering of basaltic and granitic catchments from concentration–discharge relationships. Geochim Cosmochim Acta 190:265–293

Istanbulluoglu E, Yetemen O, Vivoni ER, Gutierrez-Jurado HA, Bras RL (2008) Eco-geomorphic implications of hillslope aspect: Inferences from analysis of landscape morphology in central New Mexico. Geophys Res Lett 35:6

Izbicki JA, Wright MT, Seymour WA, McCleskey RB, Fram MS, Belitz K, Esser BK (2015) Cr(VI) occurrence and geochemistry in water from public-supply wells in California. Appl Geochem 63:203–217

Jamtveit B, Hammer Ø (2012) Sculpting of rocks by reactive fluids. Geochem Perspect 1:341–481

Janot N, Lezama Pacheco JS, Pham DQ, O'Brien TM, Hausladen D, Noël V, Lallier F, Maher K, Fendorf S, Williams KH, Long PE (2016) Physico-chemical heterogeneity of organic-rich sediments in the Rifle Aquifer, CO: Impact on uranium biogeochemistry. Environ Sci Technol 50:46–53

Jin L, Ravella R, Ketchum B, Bierman PR, Heaney P, White T, Brantley SL (2010) Mineral weathering and elemental transport during hillslope evolution at the Susquehanna/Shale Hills Critical Zone Observatory. Geochim Cosmochim Acta 74:3669–3691

Jin L, Mathur R, Rother G, Cole D, Bazilevskaya E, Williams J, Carone A, Brantley S (2013) Evolution of porosity and geochemistry in Marcellus Formation black shale during weathering. Chem Geol 356:50–63

Jones DL (1998) Organic acids in the rhizosphere—a critical review. Plant Soil 205:25–44

Jung H, Navarre-Sitchler A (2018a) Physical heterogeneity control on effective mineral dissolution rates. Geochim Cosmochim Acta 227:246–263

Jung H, Navarre-Sitchler A (2018b) Scale effect on the time dependence of mineral dissolution rates in physically heterogeneous porous media. Geochim Cosmochim Acta 234:70–83

Keiluweit M, Bougoure JJ, Nico PS, Pett-Ridge J, Weber PK, Kleber M (2015) Mineral protection of soil carbon counteracted by root exudates. Nat Clim Chang 5:588–595

Kim H, Dietrich WE, Thurnhoffer BM, Bishop JKB, Fung IY (2017) Controls on solute concentration-discharge relationships revealed by simultaneous hydrochemistry observations of hillslope runoff and stream flow: The importance of Critical Zone structure. Water Resour Res 53:1424–1443

Kirchner JW, Feng XH, Neal C (2000) Fractal stream chemistry and its implications for contaminant transport in catchments. Nature 403:524–527

Kirchner JW, Feng XH, Neal C (2001) Catchment-scale advection and dispersion as a mechanism for fractal scaling in stream tracer concentrations. J Hydrology 254:82–101

Kleber M, Eusterhues K, Keiluweit M, Mikutta C, Mikutta R, Nico PS (2015) Mineral–organic associations: formation, properties, and relevance in soil environments. Adv Agron 130:1–140

Kump L, Brantley SL, Arthur M (2000) Chemical weathering, atmospheric CO_2, and climate. Ann Rev Earth Planet Sci 28:611–667

Langston AL, Tucker GE, Anderson RS, Anderson SP (2015) Evidence for climatic and hillslope-aspect controls on vadose zone hydrology and implications for saprolite weathering. Earth Surf Processes Landforms 40:1254–1269

Lasaga AC (1984) Chemical kinetics of water-rock interactions. J Geophys Res 89:4009–4025

Lasaga AC (1998) Kinetic Theory in the Earth Sciences. Princeton University Press, Princeton NJ

Lawrence C, Harden J, Maher K (2014) Modeling the influence of organic acids on soil weathering. Geochim Cosmochim Acta 139:487–507

Lebedeva MI, Brantley SL (2013) Exploring geochemical controls on weathering and erosion of convex hillslopes: beyond the empirical regolith production function. Earth Surf Processes Landforms 38:1793–1807

Lebedeva MI, Brantley SL (2017) Weathering and erosion of fractured bedrock systems. Earth Surf Processes Landforms 42:2090–2108

Lebedeva MI, Fletcher RC, Balashov VN, Brantley SL (2007) A reactive diffusion model describing transformation of bedrock to saprolite. Chem Geol 244:624–645

Lebedeva MI, Fletcher RC, Brantley SL (2010) A mathematical model for steady-state regolith production at constant erosion rate. Earth Surf Processes Landforms 35:508–524

Lehmann J, Kleber M (2015) The contentious nature of soil organic matter. Nature 528:60–68

Li L, Peters CA, Celia MA (2004) Upscaling geochemical reaction rates using pore-scale network modeling Adv Water Resour 29:1351–1370

Li L, Steefel CI, Yang L (2008) Scale dependence of mineral dissolution rates within single pores and fractures. Geochim Cosmochim Acta 72:360–377

Li L, Maher K, Navarre-Sitchler A, Druhan J, Meile C, Lawrence C, Moore J, Perdrial J, Sullivan P, Thompson A, Jin L (2017) Expanding the role of reactive transport models in critical zone processes. Earth Sci Rev 165:280–301

Lichtner PC (1988) the quasi-stationary state approximation to coupled mass-transport and fluid-rock interaction in a porous-medium. Geochim Cosmochim Acta 52:143–165

Lichtner PC (1993) Scaling properties of time-space kinetic mass transport equations and the local equilibrium limit. Am J Sci 293:257–296

Luttge A, Arvidson RE, Fischer C (2013) A stochastic treatment of crystal dissolution kinetics. Elements 9:183–188

Ma L, Jin L, Brantley SL (2011) How mineralogy and slope aspect affect REE release and fractionation during shale weathering in the Susquehanna/Shale Hills Critical Zone Observatory. Chem Geol 290:31–49

Maffre P, Ladant J-B, Moquet J-S, Carretier S, Labat D, Goddéris Y (2018) Mountain ranges, climate and weathering. Do orogens strengthen or weaken the silicate weathering carbon sink? Earth Planet Sci Lett 493:174–185

Maher K (2010) The dependence of chemical weathering rates on fluid residence time. Earth Planet Sci Lett 294:101–110

Maher K (2011) The role of fluid residence time and topographic scales in determining chemical fluxes from landscapes. Earth Planet Sci Lett 312:48–58

Maher K, Chamberlain CP (2014) Hydrologic regulation of chemical weathering and the geologic carbon cycle. Science 343:1502–1504

Maher K, Druhan J (2014) Relationships between the transit time of water and the fluxes of weathered elements through the critical zone. Procedia Earth Planet Sci 10:16–22

Maher K, von Blanckenburg F (2016) Surface ages and weathering rates from Be-10 (meteoric) and $^{10}Be/^9Be$: Insights from differential mass balance and reactive transport modeling. Chem Geol 446:70–86

Maher K, Mayer KU (2019) Tracking diverse minerals, hungry organisms, and dangerous contaminants using reactive transport models. Elements 15:101–106

Maher K, Steefel CI, DePaolo DJ, Viani BE (2006) The mineral dissolution rate conundrum: Insights from reactive transport modeling of U isotopes and pore fluid chemistry in marine sediments. Geochim Cosmochim Acta 70:337–363

Maher K, Steefel CI, White AF, Stonestrom DA (2009) The role of reaction affinity and secondary minerals in regulating chemical weathering rates at the Santa Cruz Soil Chronosequence, California. Geochim Cosmochim Acta 73:2804–2831

Maher K, Johnson NC, Jackson A, Lammers LN, Torchinsky AB, Weaver KL, Bird DK, Brown GE, Jr. (2016) A spatially resolved surface kinetic model for forsterite dissolution. Geochim Cosmochim Acta 174:313–334

Maloszewski P, Zuber A (1982) Determining the turnover time of groundwater systems with the aid of environmental tracers. 1. Models and their applicability. J Hydrol 57:207–231

Manning AH, Verplanck PL, Caine JS, Todd AS (2013) Links between climate change, water-table depth, and water chemistry in a mineralized mountain watershed. Appl Geochem 37:64–78

Manning AH, Mills CT, Morrison JM, Ball LB (2015) Insights into controls on hexavalent chromium in groundwater provided by environmental tracers, Sacramento Valley, California, USA Appl Geochem 62:186–199

Manzoni S, Porporato A (2009) Soil carbon and nitrogen mineralization: Theory and models across scales. Soil Biol Biochem 41:1355–1379

Manzoni S, Taylor P, Richter A, Porporato A, Agren GI (2012) Environmental and stoichiometric controls on microbial carbon-use efficiency in soils. New Phytologist 196:79–91

Maxwell RM, Kollet SJ (2008a) Interdependence of groundwater dynamics and land-energy feedbacks under climate change. Nat Geosci 1:665–669

Maxwell RM, Kollet SJ (2008b) Quantifying the effects of three-dimensional subsurface heterogeneity on Hortonian runoff processes using a coupled numerical, stochastic approach. Adv Water Resour 31:807–817

McClain CN, Maher K (2016) Chromium fluxes and speciation in ultramafic catchments and global rivers. Chem Geol 426:135–157

McGlynn BL, McDonnel JJ, Brammer DD (2002) A review of the evolving perceptual model of hillslope flowpaths at the Maimai catchments, New Zealand. J Hydrology 257:1–26

McGlynn BL, McDonnell JJ, Seibert J, Kendall C (2004) Scale effects on headwater catchment runoff timing, flow sources, and groundwater-streamflow relations. Water Resour Res 40:W07504

McGuire KJ, McDonnell JJ (2010) Hydrological connectivity of hillslopes and streams: Characteristic time scales and nonlinearities. Water Resour Res 46:W10543

McGuire KJ, McDonnell JJ, Weiler M, Kendall C, McGlynn BL, Welker JM, Seibert J (2005) The role of topography on catchment-scale water residence time. Water Resour Res 41

Meyerhoff SB, Maxwell RM (2011) Quantifying the effects of subsurface heterogeneity on hillslope runoff using a stochastic approach. Hydrogeol J 19:1515–1530

Mikkelson KM, Bearup LA, Navarre-Sitchler AK, McCray JE, Sharp JO (2014) Changes in metal mobility associated with bark beetle-induced tree mortality. Environ Sci Processes Impacts 16:1318–1327

Millot R, Gaillardet J, Dupre B, Allegre CJ (2002) The global control of silicate weathering rates and the coupling with physical erosion; new insights from rivers of the Canadian Shield. Earth Planet Sci Lett 196:83–98

Moatar F, Abbott BW, Minaudo C, Curie F, Pinay G (2017) Elemental properties, hydrology, and biology interact to shape concentration-discharge curves for carbon, nutrients, sediment, and major ions. Water Resour Res 53:1270–1287

Molins S, Trebotich D, Steefel CI, Shen CP (2012) An investigation of the effect of pore scale flow on average geochemical reaction rates using direct numerical simulation. Water Resour Res 48:W03527

Molins S (2015) Reactive interfaces in direct numerical simulation of pore-scale processes. Rev Mineral Geochem 80:461–481

Molins S, Knabner P (2019) Multiscale approaches in reactive transport modeling. Rev Mineral Geochem 85:27–48

Moore J, Lichtner PC, White AF, Brantley SL (2012) Using a reactive transport model to elucidate differences between laboratory and field dissolution rates in regolith. Geochim Cosmochim Acta 93:235–261

Moyano FE, Manzoni S, Chenu C (2013) Responses of soil heterotrophic respiration to moisture availability: An exploration of processes and models. Soil Biol Biochem 59:72–85

Murphy WM, Oelkers EH, Lichtner PC (1989) Surface-reaction versus diffusion control of mineral dissolution and growth rates in geochemical processes. Chem Geol 78:357–380

Navarre-Sitchler A, Brantley S (2007) Basalt weathering across scales. Earth Planet Sci Lett 261:321–334

Navarre-Sitchler A, Cole D, Rother G, Brantley S (2008) Evolution of micro-porosity during weathering of basalt. Geochim Cosmochim Acta 72:A673-A673

Navarre-Sitchler A, Steefel CI, Yang L, Tomutsa L, Brantley SL (2009) Evolution of porosity and diffusivity associated with chemical weathering of a basalt clast. J Geophys Res Earth Surf 114:14

Navarre-Sitchler A, Steefel CI, Sak PB, Brantley SL (2011) A reactive-transport model for weathering rind formation on basalt. Geochim Cosmochim Acta 75:7644–7667

Navarre-Sitchler A, Brantley SL, Rother G, Steefel C, Emmanuel S, Anovitz L (2015) How porosity increases during incipient weathering of crystalline silicate rocks. Rev Mineral Geochem 80:331–354

Neil CW, Yang YJ, Schupp D, Jun YS (2014) Water chemistry impacts on arsenic mobilization from arsenopyrite dissolution and secondary mineral precipitation: implications for managed aquifer recharge. Environ Sci Technol 48:4395–4405

Noguchi S, Tsuboyama Y, Sidle RC, Hosoda I (1999) Morphological characteristics of macropores and the distribution of preferential flow pathways in a forested slope segment. Soil Sci Soc Am J 63:1413–1423

Oelkers EH, Schott J, Devidal JL (1994) The effect of aluminum, pH, and chemical affinity on the rates of aluminosilicate dissolution reactions. Geochim Cosmochim Acta 58:2011–2024

Ortoleva P, Merino E, Moore C, Chadam J (1987) Geochemical self-organization .1. Reaction–transport feedbacks and modeling approach. Am J Sci 287:979–1007

Oze C, Bird DK, Fendorf S (2007) Genesis of hexavalent chromium from natural sources in soil and groundwater. PNAS 104:6544–6549

Pagani M, Caldeira K, Berner R, Beerling DJ (2009) The role of terrestrial plants in limiting atmospheric CO_2 decline over the past 24 million years. Nature 460:85–U94

Palandri JL, Karakha YK (2004) A compilation of rate parameters of water-mineral interaction kinetics for application to geochemical modeling. US Geological Survey Open File Report 2004–1068

Pandey S, Rajaram H (2016) Modeling the influence of preferential flow on the spatial variability and time-dependence of mineral weathering rates. Water Resour Res 52:9344–9366

Parsekian AD, Singha K, Minsley BJ, Holbrook WS, Slater L (2015) Multiscale geophysical imaging of the Critical Zone. Rev Geophys 53:1–26

Pelletier JD, Barron-Gafford GA, Breshears DD, Brooks PD, Chorover J, Durcik M, Harman CJ, Huxman TE, Lohse KA, Lybrand R, Meixner T (2013) Coevolution of nonlinear trends in vegetation, soils, and topography with elevation and slope aspect: A case study in the sky islands of southern Arizona. J Geophys Res Earth Surf 118:741–758

Pelletier JD, Barron-Gafford GA, Gutiérrez-Jurado H, Hinckley EL, Istanbulluoglu E, McGuire LA, Niu GY, Poulos MJ, Rasmussen C, Richardson P, Swetnam TL (2018) Which way do you lean? Using slope aspect variations to understand Critical Zone processes and feedbacks. Earth Surf Processes Landforms 43:1133–1154

Porta GM, Riva M, Guadagnini A (2012) Upscaling solute transport in porous media in the presence of an irreversible bimolecular reaction. Adv Water Resour 35:151–162

Porta GM, Chaynikov S, Thovert JF, Riva M, Guadagnini A, Adler PM (2013) Numerical investigation of pore and continuum scale formulations of bimolecular reactive transport in porous media. Adv Water Resour 62:243–253

Postma D, Larsen F, Hue NTM, Duc MT, Viet PH, Nhan PQ, Jessen S (2007) Arsenic in groundwater of the Red River floodplain, Vietnam: Controlling geochemical processes and reactive transport modeling. Geochim Cosmochim Acta 71:5054–5071

Raymond PA, Oh NH, Turner RE, Broussard W (2008) Anthropogenically enhanced fluxes of water and carbon from the Mississippi River. Nature 451:449–452

Rempe DM, Dietrich WE (2014) A bottom-up control on fresh-bedrock topography under landscapes. PNAS 111:6576–6581

Rempe DM, Dietrich WE (2018) Direct observations of rock moisture, a hidden component of the hydrologic cycle. Proc Natl Acad Sci USA 115:2664–2669

Riebe CS, Hahm WJ, Brantley SL (2017) Controls on deep critical zone architecture: a historical review and four testable hypotheses. Earth Surf Processes Landforms 42:128–156

Riley WJ, Maggi F, Kleber M, Torn MS, Tang JY, Dwivedi D, Guerry N (2014) Long residence times of rapidly decomposable soil organic matter: application of a multi-phase, multi-component, and vertically resolved model (BAMS1) to soil carbon dynamics. Geosci Model Dev 7:1335–1355

Sak PB, Fisher DM, Gardner TW, Murphy K, Brantley SL (2004) Rates of weathering rind formation on Costa Rican basalt. Geochim Cosmochim Acta 68:1453–1472

Sanchez-Vila X, Dentz M, Donado LD (2007) Transport controlled reaction rates under local non-equilibrium conditions. Geophys Res Lett 34: L10404

Schmidt MW, Torn MS, Abiven S, Dittmar T, Guggenberger G, Janssens IA, Kleber M, Kögel-Knabner I, Lehmann J, Manning DA, Nannipieri P (2011) Persistence of soil organic matter as an ecosystem property. Nature 478:49–56

Schuessler JA, von Blanckenburg F, Bouchez J, Uhlig D, Hewawasam T (2018) Nutrient cycling in a tropical montane rainforest under a supply-limited weathering regime traced by elemental mass balances and Mg stable isotopes. Chem Geol 497:74–87

Siirila ER, Maxwell RM (2012) Evaluating effective reaction rates of kinetically driven solute in large-scale, statistically anisotropic media: Human health risk implications. Water Resour Res 48:W04527

St. Clair J, Moon S, Holbrook WS, Perron JT, Riebe CS, Martel SJ, Carr B, Harman C, Singha K, Richter D (2015) Geophysical imaging reveals topographic stress control of bedrock weathering. Science 350:534–538

Stallard RF (1988) Weathering and erosion in the humid tropics. *In:* Physical and Chemical Weathering in Geochemical Cycles. Lerman A, Meybeck M, (eds). Kluwer Academic Publishers, p 225–246

Stallard R (1992) Tectonic processes, continental freeboard, and the rate-controlling step for continental denudation. *In:* Global Biogeochemical Cycles. Butcher S, (ed) Academic Press, London, 93–121

Stallard RF, Murphy SF (2014) A unified assessment of hydrologic and biogeochemical responses in research watersheds in eastern Puerto Rico using runoff-concentration relations. Aquat Geochem 20:115–139

Steefel CI, Maher K (2009) Fluid–rock interaction: A reactive transport approach. Rev Mineral Geochem 70:485–532

Steefel CI, DePaolo DJ, Lichtner PC (2005) Reactive transport modeling: An essential tool and a new research approach for the Earth sciences. Earth Planet Sci Lett 240:539–558

Steefel CI, Druhan JL, Maher K (2014) Modeling coupled chemical and isotopic equilibration rates. Procedia Earth Planet Sci 10:208–217

Steefel CI, Appelo CA, Arora B, Jacques D, Kalbacher T, Kolditz O, Lagneau V, Lichtner PC, Mayer KU, Meeussen JC, Molins S (2015) Reactive transport codes for subsurface environmental simulation. Comput Geosci 19:445–478

Stillings LL, Drever JI, Brantley SL, Sun YT, Oxburgh R (1996) Rates of feldspar dissolution at pH 3–7 with 0–8 mM oxalic acid. Chem Geol 132:79–89

Sullivan PL, Goddéris Y, Shi Y, Gu X, Schott J, Hasenmueller EA, Kaye J, Duffy C, Jin L, Brantley SL (2019) Exploring the effect of aspect to inform future earthcasts of climate-driven changes in weathering of shale. J Geophys Res Earth Surf:2017JF004556

Torn MS, Trumbore SE, Chadwick OA, Vitousek PM, Hendricks DM (1997) Mineral control of soil organic carbon storage and turnover. Nature 389:170–173

Troch PA, Lahmers T, Meira A, Mukherjee R, Pedersen JW, Roy T, Valdes-Pineda R (2015) Catchment coevolution: A useful framework for improving predictions of hydrological change? Water Resour Res 51:4903–4922

Trumbore S (2000) Age of soil organic matter and soil respiration: Radiocarbon constraints on belowground C dynamics. Ecol Appl 10:399–411

Vengosh A, Coyte R, Karr J, Harkness JS, Kondash AJ, Ruhl LS, Merola RB, Dywer GS (2016) Origin of hexavalent chromium in drinking water wells from the piedmont aquifers of north carolina. Environ Sci Technol Lett 3:409–414

Vidic NJ (1998) Soil-age relationships and correlations: comparison of chronosequences in the Ljubljana Basin, Slovenia and USA. Catena 34:113–129

von Blanckenburg F, Bouchez J (2014) River fluxes to the sea from the ocean's $^{10}Be/^9Be$ ratio. Earth Planet Sci Lett 387:34–43

von Blanckenburg F, Bouchez J, Wittmann H (2012) Earth surface erosion and weathering from the $^{10}Be(meteoric)/^9Be$ ratio. Earth Planet Sci Lett 351:295–305

von Blanckenburg F, Bouchez J, Ibarra DE, Maher K (2015) Stable runoff and weathering fluxes into the oceans over Quaternary climate cycles. Nat Geosci 8:538–U146

Waldbauer JR, Chamberlain CP (2005) Influence of uplift, weathering and base cation supply on past and future CO_2 levels. *In:* Ecological Studies: A History of Atmospheric CO_2 and Its Effects on Plants, Animals and Ecosystems Vol 177. Ehleringer et al. (ed)s Springer Verlag, p 166–184

Wanner C, Sonnenthal EL, Liu XM (2014) Seawater δ^7Li: A direct proxy for global CO_2 consumption by continental silicate weathering? Chem Geol 381:154–167

Werth CJ, Cirpka OA, Grathwohl P (2006) Enhanced mixing and reaction through flow focusing in heterogeneous porous media. Water Resour Res 42:1–10

West AJ, Bickle MJ, Collins R, Brasington J (2002) Small-catchment perspective on Himalayan weathering fluxes. Geology 30:355–358

White AF, Brantley SL (2003) The effect of time on the weathering of silicate minerals: why do weathering rates differ in the laboratory and field? Chem Geol 202:479–506

White AF, Blum AE, Schulz MS, Bullen TD, Harden JW, Peterson ML (1996) Chemical weathering rates of a soil chronosequence on granitic alluvium 1. Quantification of mineralogical and surface area changes and calculation of primary silicate reaction rates. Geochim Cosmochim Acta 60:2533–2550

White AF, Bullen TD, Schulz MS, Blum AE, Huntington TG, Peters NE (2001) Differential rates of feldspar weathering in granitic regoliths. Geochim Cosmochim Acta 65:847–869

White AF, Schulz MS, Vivit DV, Blum AE, Stonestrom DA, Harden JW (2005) Chemical weathering rates of a soil chronosequence on granitic alluvium III Hydrochemical evolution and contemporary solute fluxes and rates. Geochim Cosmochim Acta 69:1975–1996

White AF, Schulz MS, Vivit DV, Blum A, Stonestrom DA, Anderson SP (2008) Chemical weathering of a marine terrace chronosequence, Santa Cruz, California I: interpreting the long-term controls on chemical weathering based on spatial and temporal element and mineral distributions. Geochim Cosmochim Acta 72:36–68

White AF, Schulz MS, Stonestrom DA, Vivit DV, Fitzpatrick J, Bullen T, Maher K, Blum AE (2009) Chemical weathering of a marine terrace chronosequence, Santa Cruz, California II: Controls on solute fluxes and comparisons of long-term and contemporary mineral weathering rates. Geochim Cosmochim Acta 73:2769–2803

Willenbring JK, von Blanckenburg F (2010) Meteoric cosmogenic beryllium-10 adsorbed to river sediment and soil: Applications for Earth-surface dynamics. Earth Sci Rev 98:105–122

Winnick MJ, Maher K (2018) Relationships between CO_2, thermodynamic limits on silicate weathering, and the strength of the silicate weathering feedback. Earth Planet Sci Lett 485:111–120

Winnick MJ, Carroll RWH, Williams KH, Maxwell RM, Dong WM, Maher K (2017) Snowmelt controls on concentration-discharge relationships and the balance of oxidative and acid-base weathering fluxes in an alpine catchment, East River, Colorado. Water Resour Res 53:2507–2523

Xu TF, White SP, Pruess K, Brimhall GH (2000) Modeling of pyrite oxidation in saturated and unsaturated subsurface flow systems. Transp Porous Media 39:25–56

Zahasky C, Thomas D, Matter J, Maher K, Benson SM (2018) Multimodal imaging and stochastic percolation simulation for improved quantification of effective porosity and surface area in vesicular basalt. Adv Water Resour 121:235–244

Zhu C (2005) In situ feldspar dissolution rates in an aquifer. Geochim Cosmochim Acta 69:1435–1453

Reviews in Mineralogy & Geochemistry
Vol. 85 pp. 381-418, 2019
Copyright © Mineralogical Society of America

13

Watershed Reactive Transport

Li Li

Department of Civil and Environmental Engineering
The Pennsylvania State University
University Park, PA 16803
United States

lili@engr.psu.edu

WATERSHEDS AS HYDRO-BIOGEOCHEMICAL REACTORS

A watershed (also drainage basin, river basin, or catchment) is defined as "…the area that topographically appears to contribute all the water that passes through a specified cross section of a stream (the outlet)" (Dingman 2015). In this chapter, I choose to use the term "watershed" as it is a broadly used one; it should be understood more as small watersheds or catchments. Watersheds are the fundamental units that support river networks, the blood vessels at Earth's surface ultimately draining into the ocean.

Watersheds are complex hydro-biogeochemical reactors. They receive water, mass, and energy, transport them to distinct compartments, and transform them into various forms (Fig.1). Hydrological processes partition precipitation to the atmosphere, to the ground surface, and to the subsurface, eventually entering streams. Similarly, plants translate sunlight, water, CO_2, and nutrients into organic matters (leaves, stems, roots) that fall and deposit in soil. As water routes through soils, it interacts with roots, microbes, and reactive gases (i.e., CO_2 and O_2), releases solutes, and ultimately transporting them out of watersheds. The water flow (discharge) and solute concentrations measured at stream outlets therefore reflect convoluted signature of ecohydrological and biogeochemical coupling. The process interactions and feedbacks are dictated not only by hydroclimatic forcing but also the architecture of watersheds, in particular the above-ground characteristics such as land cover and surface topography, as well as the below-ground structure including soil depth, soil type, geology, and root architecture. The external forcing and internal idiosyncrasies dictate the magnitude, timing, and spatial distribution of water flow and chemistry (Chorover et al. 2011; Brooks et al. 2015), giving rise to non-linear emergent behaviors that are unique of watershed reactive transport processes.

Understanding feedbacks and non-linearity requires process-based models that integrate multiple interacting processes. Such integration however does not come easily with traditional boundaries of hydrology versus biogeochemistry relevant disciplines. As will be discussed later in this chapter, relevant model development has been advancing along two separate lines: hydrology models that solve for water storage and fluxes at the watershed scale and beyond (Fatichi et al. 2016), and reactive transport models (RTMs) that center on aqueous and solid concentration changes arising from transport and multi-component biogeochemical reactions typically in "closed" groundwater systems without much interactions with "open" watersheds directly receiving precipitation and sun light (Steefel et al. 2015; Li et al. 2017b). This comes along with a history of hydrologists often trained as physicists studying fluid mechanics, and biogeochemists typically grow up as geologists, chemists, or environmental engineers.

There are however considerable needs to reach beyond disciplinary boundaries and integrate the two lines to develop watershed reactive transport models, not only to gain

1529-6466/19/0085-0013$05.00 (print)
1943-2666/19/0085-0013$05.00 (online)

http://dx.doi.org/10.2138/rmg.2018.85.13

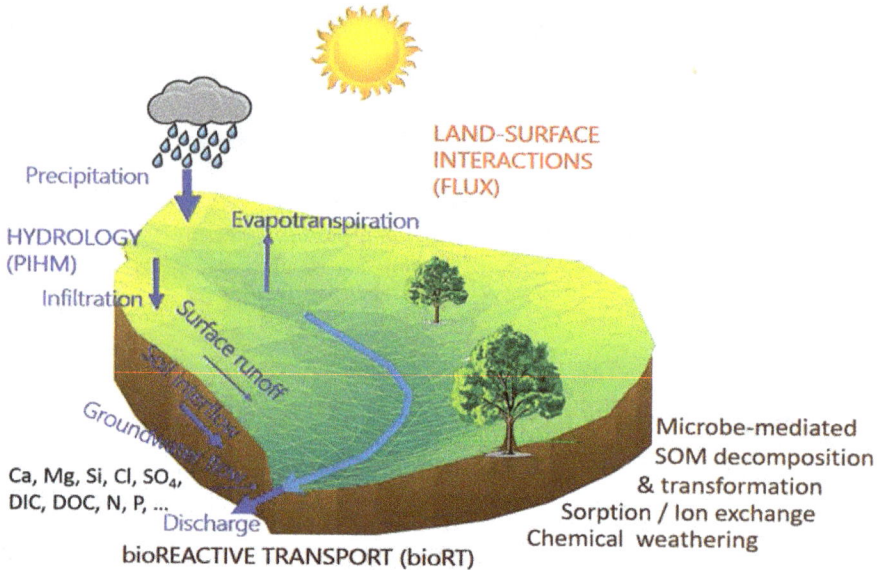

Figure 1. Schematic representation of a watershed as a hydro-biogeochemical reactor and major processes represented in the watershed reactive transport model bioRT-Flux-PIHM (modified based on Li et al. 2017a and Zhi et al. 2019).

fundamental understanding of watersheds as complex systems, but also to solve today's pressing global environmental challenges. Natural systems such as watersheds do not set up artificial disciplinary boundaries; processes relevant to different disciplines all occur simultaneously. Research questions at the interfaces or "ecotones" of different disciplines can stimulate answers that shed light on puzzling observations arising from non-linear emerging behaviors. In addition, as the pace of climate change and human perturbation accelerates, a wide spectrum of water-related hazards (e.g., hurricanes, flooding, droughts, melting glaciers) loom (IPCC 2013; Fan et al. 2014; Zhao et al. 2016; Roque-Malo and Kumar 2017). Large hydrological events bring out excessive water and disproportionally large pulses of "stored" contaminants that deteriorate aquatic and ecosystem health (Raymond and Saiers 2010; Huntington et al. 2016). On the other hand, droughts induce water-borne diseases (Perez-Saez et al. 2017) and impair water quality (Ejarque et al. 2018). Problems related to excessive nutrient export, including eutrophication and hypoxia in rivers, lakes, and coastal areas worldwide, have lingered for decades and will continue to do so, calling for advanced tools for prediction and for management (Royer et al. 2006; Seitzinger et al. 2010; Van Cappellen and Maavara 2016; Van Meter et al. 2017). As such, it is essential to possess the capabilities of forecasting not only for water quantity (flow), which is well underway with models such as the National Water Model (Cosgrove et al. 2016) (https://water.noaa.gov/about/nwm), but also for water quality (chemistry).

Although solute transport and reaction modules have been developed as add-ons to hydrological models at the watershed scale (e.g., Arnold and Soil 1994; Donigian Jr et al. 1995; Santhi et al. 2001), they typically have relatively crude representation without rigorous incorporation of reaction thermodynamics and kinetics theory in geochemistry and biogeochemistry. Reactive transport models have been brought to the watershed scale only recently (Yeh et al. 2006; Bao et al. 2017). The goal of this chapter is to illustrate this development and their promises in facilitating the understanding of watershed processes. To do so it is important to bring together the hydrology and biogeochemistry communities on fundamental concepts and processes. I therefore first lay out concepts in watershed hydrology and biogeochemistry, as surface hydrological processes may sound foreign to geochemists; similarly, biogeochemical processes may appear alien to hydrologists. I will then illustrate examples of hydrological and biogeochemical coupling and the use of watershed reactive transport models in offering mechanistic understanding. The chapter ends at thoughts of existing knowledge gaps and future directions that can potentially be tackled with watershed RTMs.

FUNDAMENTALS OF WATERSHED HYDROLOGY

Watersheds differ from groundwater systems in that they are "open" systems receiving energy, rainfall and snow from the atmosphere. This brings in significant temporal dynamics or non-stationary that depends on hydroclimatic conditions. In addition, watersheds are spatially heterogeneous not only with distinct subsurface zones of soil, weathered rock, and parent rocks, but also with topography, land cover (e.g., barren soil versus forests, grassland, urban), and river networks. Flow complexity therefore originates from temporal dynamics, variations in subsurface properties such as porosity and permeability (e.g., highly permeable soils versus tight rocks), as well as slopes, land cover, and aspect (sunny or shady sides) (Buttle 1998; Bishop and Seibert 2015). This section does not intend to cover the full spectrum of watershed processes. Rather, I will introduce fundamental aspects that are relevant to later discussions on biogeochemical processes.

Water balance

A myriad of processes determine where precipitation (P, rain + snow) goes and how it partitions into different parts of a watershed. It can evaporate through soils or transpire through plants, returning to the atmosphere. It can be intercepted by vegetation leaves or fall on the ground; it can flow directly on the ground surface to the stream or infiltrate into soils. Within soil it can penetrate vertically through the unsaturated zone, form interflow toward the stream or channel into macropores. Some soil water may cross the soil–weathered bedrock interfaces and flow downward into groundwater aquifers, which may eventually re-enter rivers and streams, although not necessarily in the same watershed.

Here I will only coarsely delineate these processes without getting into the intricacies of these processes. Considering a watershed as a control volume or a water processing box, we have precipitation (P) as the input and evaporation (E), and transpiration (TR) and discharge (Q) as the output (Fig. 1). This leads to a simple water balance equation (e.g., Kirchner 2009):

$$\frac{dS}{dt} = P(t) - ET(t) - Q(t) \tag{1}$$

where S, the water storage, is in units of depth (e.g., mm of water, or volume of water per unit area, L), and $P(t)$, $ET(t)$ ($= E(t) + TR(t)$), and $Q(t)$ are the rates of precipitation, evapotranspiration, and discharge, respectively, in units of depth per time (e.g., mm of water per hour, L/t, essentially area-normalized water volume per time). All terms in Equation (1) are a function of time and represent average values over an entire watershed. The equation assumes that water loss into aquifers is negligible. Equation (1) essentially follows the mass conservation principle and states that changes in water storage in a watershed depends on the temporal dynamics of water input (P) and outputs (ET and Q). Solving this equation needs constitutive equations rsuch as storage-discharge relationships. An extensively used example is the power law form $Q = aS^b$ (Horton 1936) (Brutsaert and Nieber 1977; Wittenberg 1999). Whereas precipitation is driven by meteorological forcings, a plethora of watershed features determine the partitioning between ET and Q, as explained below.

Evapotranspiration (ET). Evapotranspiration is the summation of evaporation (from open water, bare soil, and vegetated surfaces), transpiration (from within plant leaves), and sublimation from ice and snow surfaces. *ET* is regulated by land surface interactions that dictate how land surface and atmosphere exchange energy and water. Another important concept is the climatic potential *ET* (*PET*), defined as "the rate at which evapotranspiration would occur from a large area completely and uniformly covered with growing vegetation with access to an unlimited supply of soil water and without advection or heat-storage effects." (Dingman 2015). Although somewhat ambiguous in that different types of vegetation have different transpiration demand, *PET* is considered as a measure of "drying power" of a watershed influenced by both climate and land cover.

A widely used relationship, Budyko equation, relates the long-term ratio of evaporative index *ET/P* to the aridity index *PET/P*. Various forms of Budyko equation exist (Pike 1964; Budyko 1974; Fu 1981; Gentine et al. 2012). An example is as follows:

$$\frac{ET}{P} = \frac{1}{\left[\left(\frac{P}{PET}\right)^{\beta} + 1\right]^{1/\beta}} \tag{2}$$

Here β is dimensionless and modifies the curvature of *ET* as a function *P* and *PET* (Fig. 2). This equation says (empirically) that the long-term average of *ET* is a function of water supply (*P*) and available energy estimated as the evaporative demand for water by the atmosphere (*PET*). Under arid conditions when $P \ll PET$, *ET* converges toward *P*, implying that *ET* is limited by water supply (Fig. 2, water limit line). Alternatively, under humid conditions when $P \gg PET$, *ET* is limited by atmospheric demand and collapses toward *PET* (Fig. 2, energy limit line). Budyko's work and other follow-up studies have shown that under steady state conditions without significant groundwater inputs, losses, or storage changes, hydrological systems often operate close to either energy or water constraints. In addition to climate influence, vegetation modulates *ET* via dynamical relations between soil moisture in the rooting zone, evapotranspiration rates, and rainfall interception, therefore regulating Budyko's original energy-based prediction of the evaporative fraction (Porporato et al. 2004; Gentine et al. 2012). *ET* and *PET* can be estimated based on ground-based meteorological observations and remote sensing data from the MODerate Resolution Imaging Spectroradiometer (MODIS) (Mu et al. 2007). Precipitation data are available from various databases (e.g., WorldClim (Hijmans et al. 2005), NLDAS-2 (the North American Land Data Assimilation Systems phase 2 (https://ldas.gsfc.nasa.gov/nldas/NLDAS2forcing.php)) or measurements from local weather stations.

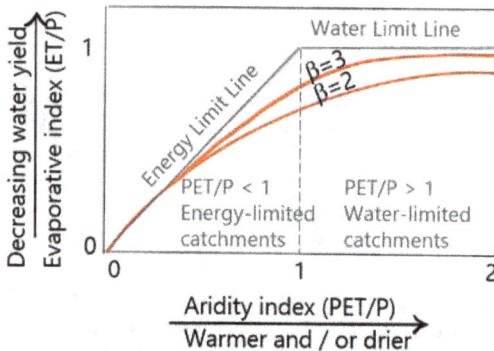

Figure 2. A Budyko diagram (evaporative vs. aridity index). The solid grey lines represent energy and water limits to the evaporative index, and the dashed line represents the original theoretical Budyko curve (after Budyko 1974).

Discharge (Q). Assuming negligible water loss to deep aquifers, the net difference between *P* and *ET* is discharge (stream flow) (Fig. 1). The generation of stream flow is an integrated outcome of surface runoff, soil interflow, and shallow groundwater contributing to streams. *Surface runoff* (also known as overland flow) occurs when excess water flows over the Earth's surface because soil is fully saturated, or rain arrives more quickly than the rate of infiltration, or impervious areas (roofs and pavement) cannot infiltrate most of water. The generation of shallow soil *interflow* depends on multi-phase flow dynamics between water and air in the unsaturated zone (via Richard's equation), and the formation of perched water table and saturated zone where water is sufficient to flow toward a stream (Darcy's law). The formation of such flow depends on soil properties as described by the water retention curve (Van Genuchten 1980):

$$\theta(h) = \theta_r + \frac{\theta_s - \theta_r}{\left[1 + \left(\alpha|h|\right)^n\right]^{1-1/n}} \tag{3}$$

where $\theta(h)$ is the water retention ([$L^3 L^{-3}$]), $|h|$ is the water pressure (or water head, L), θ_s and θ_r are the saturated and residual water content ([$L^3 L^{-3}$]), respectively; by definition is the same as porosity; α and n are parameters defining the shape of the water retention curve and are generally determined by soil properties. Two example water retention curves are shown in Figure 3. Generally sandy soils tend to leak water quickly and have a "flatter" shape, whereas the water holding capacity of clay (θ_r) is higher than sandy soil. The porosity of clayey soil (θ_s) is much lower than that of sandy soil, therefore leading to a smaller water content range ($\theta_s - \theta_r$) that drives flow.

Note that stream water can also come from the groundwater below the soil–weathered bedrock interface, which often forms base flow under dry conditions. Note that this may be different from regional groundwater in deep subsurface that is not connected to the stream. The contribution of *groundwater* to streamflow vary depending on both climate and watershed characteristics such as topography (e.g., slope), land cover, and lithology (Beck et al. 2013; Welch and Allen 2014). Discharge essentially sums up three major components: surface runoff, shallow soil interflow, and shallow groundwater flow that is connected to streams. Their contributions to stream water vary depending on hydrological regimes. Under dry conditions, stream flow is often dominated by groundwater as the base flow. Under wet conditions, stream flow is typically

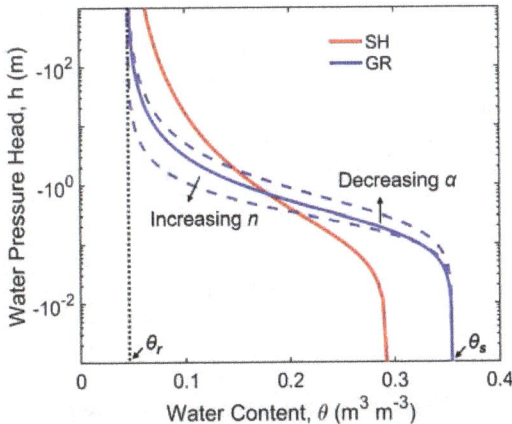

Figure 3. Two representative water retention curves for clayey soils in the Shale Hills catchment (SH) and for sandy soils in the Garner Run catchment (GR) in the Susquehanna Shale Hills Critical Zone Observatory (SSHCZO). Dashed lines demonstrate the effects of increasing *n* from 1.6 (GR value) to 2.0, and decreasing α from 5.0 to 3.0 m^{-1}. The range of water content ($\theta_s - \theta_r$) is a measure of potentially mobile water storage capacity per unit soil depth (paper in review).

composed primarily of soil interflow and / or some surface runoff. These end member water sources have very different water chemistry such that their proportions determine the stream water composition at different times, as will be discussed in the section *Watershed response to hydrological changes: Concentration discharge relationships*

Water storage (S). Equation (1) prescribes that water storage, or water content, is determined by the dynamics of water input (*P*) and output (*ET* and *Q*). Direct measurements of water storage is challenging and often require large field campaigns (Lin et al. 2006; Tromp-van Meerveld and McDonnell 2006). As a result, it is often indirectly estimated based on discharge and / or tracer measurements (Seibert et al. 2011). These observations have led to several major understandings. First, although it is challenging to accurately quantify the total storage, we know that it can be grouped into dynamic storage that is responsive to transient hydrological conditions, and passive storage that is much less responsive (McNamara et al. 2011). Dynamic water store leaves a watershed rapidly as "young" water whereas passive water storage lingers for substantially longer time (Barnes and Bonell 1996; Dunn et al. 2010; Benettin et al. 2015a; Sprenger et al. 2018). Second, it has been shown that dynamic water storage is often dramatically smaller (often > an order of magnitude) than the total storage inferred from average soil porosity (Birkel et al. 2011). Such immense reduction has been attributed to catchment characteristics that lead to low connectivity, a measure of the linkage between disparate regions of the hillslope to the stream via subsurface water flow (Jencso et al. 2009; Bracken et al. 2013). Tracer or isotope signatures often bear signatures of the relative size of the two different pools. A much smaller variation of water isotope signature in streams compared to that of rainfall indicate significant damping and large mixing volumes and water storage. Kirchner (2009) demonstrated that changes in water storage often strongly relate to changes in stream discharge, such that a storage–discharge relationship can be used directly to reconstruct the time series of precipitation and evapotranspiration in some watersheds. This develops from a tradition in catchment hydrology using estimates of storage change in linear or nonlinear reservoirs from water balance calculations and is the foundation for streamflow simulation in many conceptual rainfall–runoff models (Bergstrom 1976).

Modeling watershed hydrological processes

Rainfall-runoff models have been used extensively as conceptual, lumped models to estimate streamflow hydrograph for rainfall events and are well documented in textbooks (Beven 2011; Dingman 2015). The development of process-based watershed models date back fifty years (Crawford and Linsley 1966; Freeze and Harlan 1969). Equation (1) is essentially a simple one box model based on water balance and can be solved for water storage and water fluxes. A slightly more complex model with two boxes (an upper and lower box) has also been formulated based on the conceptualization of the two types of water storage to represent the dynamics of "fast and small" reservoir versus the passive, "slow and large" reservoir (Benettin et al. 2015a; Kobierska et al. 2015; Kirchner 2016b). This two-box model has been shown to better capture the temporal non-stationary of streamflow than the one-box model (Kirchner 2016b).

The hydrology community has developed and utilized models in a spatially distributed manner (distributed models) for more than five decades (Freeze and Harlan 1969; James 1972; Jarboe and Haan 1974; Bergstrom 1976; Abbott et al. 1979, 1986; Beven 1989; Quinn et al. 1991; Singh 1995; VanderKwaak and Loague 2001; Gan et al. 2006; McDonnell et al. 2007; Qu and Duffy 2007; Kumar et al. 2009; Therrien et al. 2010; Fatichi et al. 2016). With the integration of surface energy balance, the recent introduction of land surface processes into hydrological models marks a new advance toward more accurate representation of evapotranspiration (Maxwell and Miller 2005; Shi et al. 2013; Davison et al. 2014). These hydrology models have been brought to continental scales for water forecasting and earth system modeling (Clark et al. 2016; Fan et al. 2019). For example, the prediction of stream flows in the continental U.S. is now accessible through the National Water Model (https://water.noaa.gov/about/nwm).

These spatially distributed models have been criticized for over parameterization, equifinality, or parameter non-uniqueness, i.e., different sets of parameters can produce the same model output (Beven and Freer 2001). Along similar lines, model parameters calibrated for one watershed are often not transferrable to other watersheds, or even to periods other than calibration data in the same watershed (van der Linden and Woo 2003; Heuvelmans et al. 2004; Li et al. 2012; Smith et al. 2016). This issue of parameter transferability prevents the prediction of stream flow in ungauged river basins where data are not available (Sivapalan 2003; Hrachowitz et al. 2013). The hydrology community generally recognizes these challenges (Sivapalan 2017). In many cases however spatially distributed models are our only resort, particularly when spatial and temporal patterns are important drivers of water flow (Fatichi et al. 2016).

Here I use the Penn State Integrated Hydrology Model (PIHM) as an example of a distributed surface hydrology model to illustrate the governing equations (Qu and Duffy 2007). The land surface interaction processes are simulated in the Flux module using the Noah Land Surface Model (LSM) that is coupled to PIHM (Chen and Dudhia 2001; Ek et al. 2003; Shi et al. 2013). A variation of the PIHM code, FIHM, has more complex subsurface structure representation (Kumar et al. 2009). A suite of modules for processes of interests to multiple disciplines (Duffy et al. 2014) have been developed, including LE for landscape evolution (Zhang et al. 2016b), Cycles for agroecosystem functioning (Stöckle 2003; Kemanian 2010), and RT for biogeochemical reactive transport (Bao et al. 2017), all with Flux-PIHM being the backbone of the hydrology code.

PIHM applies the semi-discrete finite volume scheme by integrating the partial differential form of governing equations over a three dimensional control volume, thereby converting them to ordinary differential equations (ODEs) of individual prismatic elements. It solves for five water storages in each prismatic element i: above-ground storage in vegetation canopy, snow, and ground surface (water on land surface that forms surface runoff), and below-ground storage in unsaturated and saturated zones (Qu and Duffy 2007). Here I show the equations of water storage for the last three items. The equation for the surface water storage in element i is as follows:

$$\frac{dS_{\text{sf},i}}{dt} = p_{\text{net},i} - I_i - E_{\text{sf},i} - \sum_{j=1,N_{i,1}}^{N_{i,3}} q_{\text{sf},ij} \qquad (4)$$

where S_{sf} is the water storage above the land surface (L). The rates of net precipitation, i.e., precipitation not intercepted by canopy (p_{net}), infiltration from land surface to unsaturated zone (I), evaporation from surface water (E_{sf}), and lateral surface flow from element i to j ($q_{\text{sf},ij}$) are all normalized by the base area of the finite volume [L^3/(L^2 t)]; $N_{i,1-3}$ is the index of the neighboring elements of i. The $q_{\text{sf},ij}$ is calculated based on a diffusion wave approximation of the 2D St. Venant equation (Gottardi and Venutelli 1993).

The governing equations for the water storages in the unsaturated and saturated zones of element i are as follows:

$$\frac{dS_{\text{u},i}}{dt} = I_i - R_i - E_{\text{u},i} - E_{\text{t},i} \qquad (5)$$

$$\frac{dS_{\text{s},i}}{dt} = R_i - q_{\text{bedrock},i} - \sum_{j=N_{i,1}}^{N_{i,3}} q_{\text{s},ij} \qquad (6)$$

where S_{u} and S_{s} are the water storage in the unsaturated and saturated zones, respectively; R is the recharge rate downward from the unsaturated into the saturated zone (L/t); E_{u} is the evaporation from the unsaturated zone; E_t is the transpiration (L/t); q_{bedrock} is the downward flow rate into bedrock [L/t]; $q_{\text{s},ij}$ is the lateral flow rate [L/t] from element i to j in the saturated zone, which is calculated using Darcy's law.

This system of ODEs for individual elements is assembled to form a global ODE system, which is solved by an ODE implicit solver. Note that if we add Equations (4)–(6) and the unpresented equations for snow and tree canopy, the vertical water flux terms between different vertical zones (surface, unsaturated, saturated) within an individual element (e.g., infiltration, recharge) will be cancelled and we are left with an equation for total water storage with changes depending on the total input and output of the element i. If we add these equations for all elements in the whole watershed, the lateral fluxes between elements (q_{ij}) cancel out and we will have a mass balance equation for the whole watershed that is essentially Equation (1).

An example of hydrological processes in the Shale Hills catchment

Here I use an example calculation using Flux-PIHM in the Shale Hills, a first order catchment in the Susquehanna Shale Hills Critical Zone Observatory (SSHCZO) in central Pennsylvania, U.S.A (Brantley et al. 2018; Li et al. 2018). The mean annual temperature is 10°C with a mean annual precipitation of ~ 1000 mm. Extensive field studies have characterized the topography, hydrological properties and mineral composition (Lin 2006; Jin et al. 2010, 2011b; Ma et al. 2010; Brantley et al. 2013b).

Figure 4 shows the partition of precipitation into several major components. From April 1st to Dec. 31st 2009, the precipitation was 0.9 meter. Assuming no deep groundwater draining bypassing the stream, the model estimates that ~40% of precipitation contributes to stream discharge, and ~ 60% to *ET* (Fig. 4A). SSHCZO is hydrologically responsive

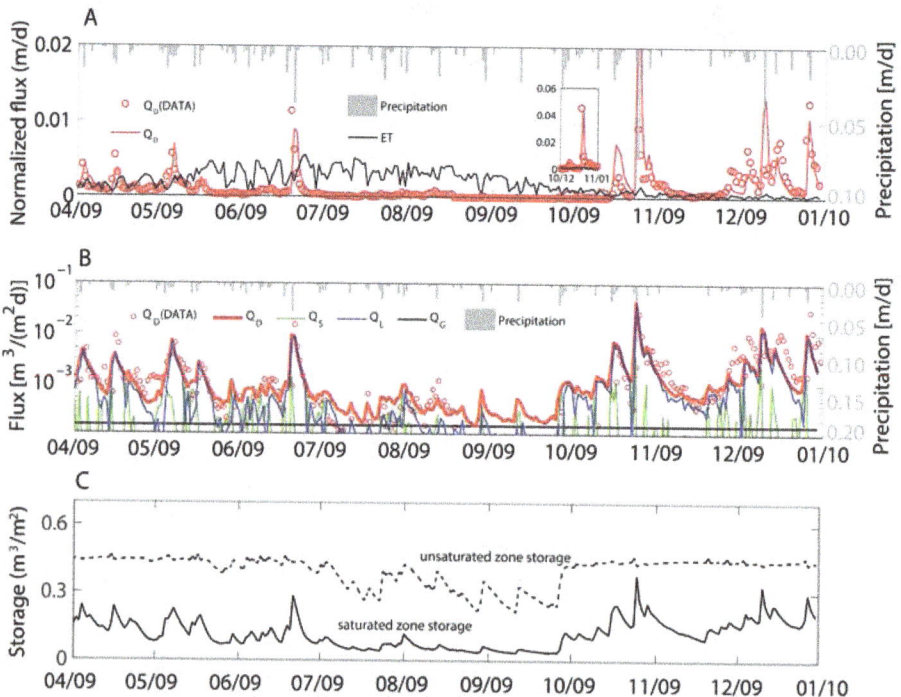

Figure 4. Time series (in 2009, in Shale Hills) of **A:** measured daily precipitation [m/d], simulated and measured discharge QD normalized by watershed area [m/d] and simulated ET [m/d]; **B:** water fluxes (surface runoff QS, lateral flow QL, and groundwater flow QG); **C:** simulated net watershed water storage in the saturated and unsaturated zones [m³/m²]. Precipitation is similar across seasons; water storages are higher in spring and winter and lower in the summer due to higher ET in summer. [Used by permission of Wiley, Li et al. (2017a) Understanding watershed hydrogeochemistry: 2. Synchronized hydrological and geochemical processes drive stream chemostatic behavior. Water Resour Res 53:2346–2367].

with stream discharge closely following intensive precipitation (Fig. 4B). In general, large rainfall events lead to more discharge and less *ET*. A large rainfall event occurred on Oct. 24th resulting in the highest discharge in 2009. Surface runoff showed up as short-lived pulses during rainfalls, followed by soil water lateral flow (interflow). Annually this lateral flow contributes to ~ 70–80% of the discharge, whereas surface runoff contributes ~ 10–20%. Groundwater from below the bedrock is not calculated (in this version of the code) and is assumed constant across the year contributing a total of ~ 10% to the stream, based on field observations (Jin et al. 2014; Sullivan et al. 2016). Although the precipitation is relatively invariable over the year, water storage is lower in the summer due to the higher ET as a result of higher temperature and the presence of tree leaves. This is consistent with the observed water level drop at groundwater monitoring wells (Sullivan et al. 2016).

FUNDAMENTALS OF SOIL BIOGEOCHEMISTRY

Imagine that you walk in a forest on a beautiful autumn afternoon. The sun light and specks of blue sky sneak in between green and orange leaves. You hear the satisfying crunch as you step on the thick layer of dry leaves and branches on the ground. You notice that the shallow soil is the territory of living things and organic matter (OM, roots, fallen leaves, and microbe). The stabilization of OM can occur via sorption on clays or separation from water and microbe (Rasmussen et al. 2018). With exposure to water and other reactive chemicals, soil OM (SOM) can also decompose, followed by the release and reuse of organic carbon, nitrogen, and phosphorous, often called biogenic solutes (Schlesinger and Bernhardt 2013) (Fig. 5), as well as other elements. SOM decomposition is a redox reaction where OM is oxidized and electron acceptors (e.g., O_2, NO_3, Fe(III) oxides, sulfate) are reduced. These reactions have long been recognized as complex and mostly microbe-mediated (von Lützow et al. 2007; Conant et al. 2011; Lehman and Kleber 2015).

Organisms in shallow soils modulate chemical weathering and production of geogenic solutes by transferring water and nutrients among solid, aqueous, and biological reservoirs through root growth and exudation, and through associated fungal and microbial activities. The partial pressure of soil CO_2 (pCO_2), a product of soil respiration, is typically 1–2 orders higher than the atmospheric level in vegetated lands (Romero-Mujalli et al. 2018) and therefore plays a major role setting the acidity of soil waters (Drever 1989, 1997; Heidari et al. 2017). The presence of O_2 drives oxidative weathering of Fe(II)-bearing minerals such as biotite and pyrite, creating fractures and porosity that allow water infiltration and weathering of other minerals (Molins and Mayer 2007; Buss et al. 2008; Bazilevskaya et al. 2013).

Water plays a central role in all these processes. It serves as a medium for the transport of reactants and products, sets the soil–water–microbe contact area, and influences the levels of reactive gases (e.g., CO_2, O_2). The thin soil mantling the Earth's surface is hydrologically more responsive to meteorological events (rainfall, snow, and temperature) compared to weathered and parent bedrocks at depth that harbor groundwater. Much remains to be learned about feedbacks among water flow, transport, weathering, and microbe-mediated processes, and how they couple to the long-term evolution of the critical zone and to global elemental cycles. Interested readers are referred to literature on the global cycle of C (Berner 1999; Falkowski et al. 2000; Wang et al. 2010), N (Galloway et al. 2004; Fowler et al. 2013), and P (Filippelli 2008; Oelkers et al. 2008). Here I briefly introduce important microbe-mediated biogeochemical reactions and chemical weathering.

Biogeochemical reactions

Carbon (C). Carbon exists in both inorganic and organic forms. Inorganic carbon includes CO_2 gas, dissolved or total inorganic C (DIC, TIC, in water), and carbonate rocks (solids). Organic C occurs in ecosystems (plants, animals, other living things), in water (dissolved organic C (DOC)), and in soil (soil organic C (SOC)). SOM represents the largest terrestrial pool of organic carbon (Stockmann et al. 2013). It can decompose partially into smaller organic molecules that dissolve in water, i.e., DOC, or oxidize completely into CO_2 gas or dissolved inorganic carbon (DIC). With coexisting divalent cations (e.g., Ca, Mg), DIC can also precipitate and become carbonate minerals. Hence soil C decomposition can release CO_2 back into the atmosphere and changes CO_2 level (Davidson and Janssens 2006), or releases DOC and DOM to surface water. These processes occurs in soils and also as dissolved carbon transport laterally from land to ocean. They represent a small term in regional carbon budgets compared with gross primary production and respiration and are traditionally thought as negligible in the global carbon cycle. Recent literature however has revealed that these processes are important in assessing net changes in terrestrial C storage and carbon cycle (Battin et al. 2009; Regnier et al. 2013), challenging the previous paradigm and suggesting the need to include them in global carbon models.

Nitrogen (N). The SOM decomposition releases organic nitrogen ($R\text{-}NH_2$), after which complex biotic and abiotic reactions ensue. As shown in Figure 5, organic N can become mineralized and transform into ammonia, which can further become oxidized into forms of high oxidation states through a chain of reactions. Inorganic N is notorious for having 7 oxidation states with ammonium (NH_4^+ and NH_3) at its lowest oxidation state of -3 up to nitrate (NO_3^-, N_2O_5) at +5, with various forms in between (N_2, N_2O, NO, N_2O_3 (NO_2^-), NO_2). Some of the gaseous forms emit back to the atmosphere and are in fact greenhouse gases (Saha et al. 2017; Maavara et al. 2018). Denitrification requires anoxic conditions and does not occur as much in soils owing to the pervasive presence of O_2; it can become prevalent however under extremely wet conditions and in O_2-depleted groundwater systems. The significance of nitrogen resides in its role not only as a limiting element for the growth of organisms but also as a component in fertilizers that feed crops and ultimately human population. Its export from agriculture lands into aquatic systems cause water quality problems such as eutrophication worldwide (Rabalais et al. 2001; Davies and Neal 2007; Conley et al. 2009).

Phosphorous (P). Earth's biological systems have depended on P since the beginning of life (Nealson and Rye 2003). Phosphate is central to the functioning of the adenosine triphosphate (ATP), the keystone for metabolism and the most abundant biomolecule in nature (Schlesinger 1997); the phosphate ester bridges bind the helix strands of DNA; all cell membranes are made up with phospholipids. In soils, P can be in organic form (e.g., leaves), sorbed (on fine soil particles),

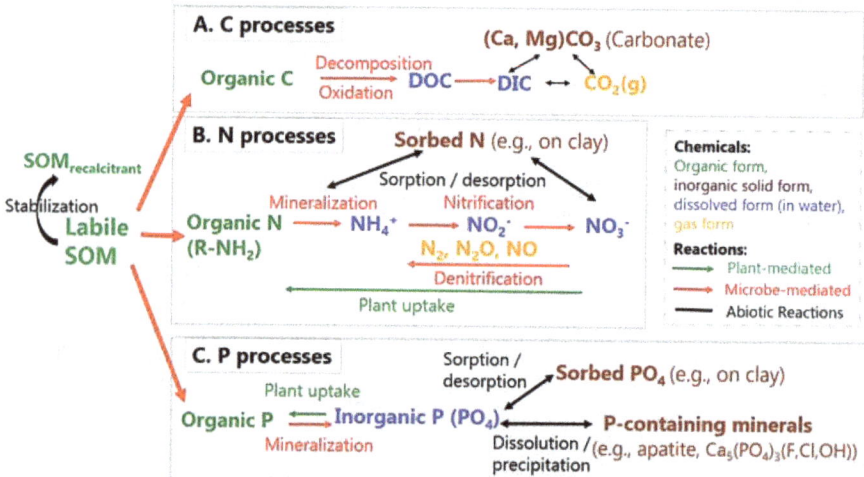

Figure 5. Labile SOM can become stabilized through sorption on clay and separation from reactants. It can also decompose into inorganic forms, transitioning between different phases.

dissolved in water, or in solid forms as P-containing minerals (Fig. 5). The transformation between different forms occurs through various bio-mediated or abiotic reactions. Unlike C and N that are abundant in the atmosphere, all P on Earth comes from the minuscule quantities of P-containing rock on Earth's crust (0.09 wt%) (Filippelli 2002). The most abundant P-containing mineral is apatite $Ca_5(PO_4)_3(F, Cl, OH)$. Once liberated via rock dissolution, P is mostly locked in organic forms (e.g., inositol phosphates, phospholipids, nucleic acids, phosphoproteins). It is barely soluble so it binds on and transports together with soil particles in the form of orthophosphate or pyro-diphosphate. To feed the ever-increasing population in Earth, human has mined rock and manufactured P-containing fertilizer. Excessive leaching from agriculture soils, together with deforestation and soil erosion, have accelerated P delivery from the deep subsurface to the surface water and the ocean, exacerbating the issue of P limitation on Earth.

Writing microbe-mediated reactions in light of microbial energetics. Microbe-mediated redox reactions involve organic carbon as electron donors that become oxidized and electron acceptors that are reduced. This process produces energy that sustains and grows microorganisms. Electron acceptors in natural environments include, for example, O_2, nitrate, iron-containing and manganese-containing minerals, and sulfate (Fierer et al. 2003; Dunn et al. 2006; Philippot et al. 2009). This order represents the descending amount of energy that microorganisms can derive from mediating these reactions, often called *biogeochemical redox ladder* (Van Cappellen and Gaillard 1996; Thullner et al. 2005; Borch et al. 2010). That is, per unit organic carbon oxidized to the same product, microorganisms can glean more energy by reducing O_2 compared to reducing NO_3.

To take into account of the mass and energy balances of these reactions, one will need to consider microbial energetics and thermodynamics and microorganisms as part of reaction products (VanBriesen 2002; Xiao and VanBriesen 2006; Xiao, 2008). Microorganisms cannot grow if the energy derived from these redox reactions are not sufficient (Jin and Bethke 2003; Jin and Bethke 2007). The amount of energy needed for microbial cell maintenance is relatively similar, such that microorganisms using redox reactions that are high in the biogeochemical redox ladder (e.g., oxidation) have more left over energy to channel into cell synthesis and growth. Therefore, values of f_s, the fraction of energy used for microbial cell growth, decrease as we go down the biogeochemical redox ladder from aerobic oxidation to sulfate reduction. That is, less microbial cells are produced as we go down the redox ladder with the same amount of electron donor.

Details of how to write such reaction equations are given in Chapter 2 of Rittmann and McCarthy (2001). Table 1 shows a few example redox reaction equations with microbial cells as a product and acetate as the electron donor (Li et al. 2009, 2010; Cheng et al. 2016) using the chemical formula $C_5H_7O_2N$ for microbe. Note that R is the overall reaction equation incorporating the half reactions of electron donor (R_d), electron acceptor (R_a) and cell synthesis (R_c). The values of f_e and f_s are the fractions of energy that partition into energy consumption reaction (anabolic pathway that consumes energy for cell maintenance) $R_e = R_a - R_d$ and cell synthesis reaction (the catabolic pathway that generate new microbial cells) $R_s = R_c - R_d$, respectively. The overall reaction $R = f_e R_e + f_s R_s = f_e(R_a - R_d) + f_s(R_c - R_d)$, where $f_e + f_s = 1.0$. If your do not care about microbe production and cell growth, the reaction can be written simply as $R = R_a - R_d$. In that case, the aerobic reaction is simply $\frac{1}{4}O_2 + \frac{1}{8}CH_3COO^- = \frac{1}{4}HCO_3^- + \frac{1}{8}H^+$.

Microbe-mediated reaction kinetics. SOM is often conceptualized and modeled as pools with different decomposition rates and turnover times (Ostle et al. 2009; Thornton et al. 2009). An extensively used three-pool model includes a readily degradable (labile) pool with residence times less than 5 years; a slowly degrading pool with residence times of decades; and a relatively stable pool, with residence times between 10^3–10^5 years (Trumbore 1993; Trumbore et al. 1995; Marin-Spiotta et al. 2009). The kinetics of microbe-mediated reactions

Table 1. Examples of biogeochemical reactions (with acetate as a representative dissolved organic carbon and $C_5H_7O_2N$ for microbial formula).

Aerobic oxidation	
R_a	$\frac{1}{4}O_2 + H^+ + e^- = \frac{1}{2}H_2O$
R_d	$\frac{1}{4}HCO_3^- + \frac{9}{8}H^+ + e^- = \frac{1}{8}CH_3COO^- + \frac{1}{2}H_2O$
R_c	$\frac{1}{4}HCO_3^- + \frac{1}{20}NH_4^+ + \frac{6}{5}H^+ + e^- = \frac{1}{20}C_5H_7O_2N + \frac{13}{20}H_2O$
R	$0.100\,O_2 + 0.125\,CH_3COO^- + 0.030\,NH_4^+$ $\rightarrow 0.030\,C_5H_7O_2N + 0.100\,HCO_3^- + 0.090\,H_2O + 0.005\,H^+$ with $f_e=0.4$, $^*Y=0.60$

Denitrification	
R_a	$\frac{1}{5}NO_3^- + \frac{6}{5}H^+ + e^- = \frac{1}{10}N_2 + \frac{3}{5}H_2O$
R_d	$\frac{1}{4}HCO_3^- + \frac{9}{8}H^+ + e^- = \frac{1}{8}CH_3COO^- + \frac{1}{2}H_2O$
R_c	$\frac{1}{4}HCO_3^- + \frac{1}{20}NH_4^+ + \frac{6}{5}H^+ + e^- = \frac{1}{20}C_5H_7O_2N + \frac{13}{20}H_2O$
R	$0.090\,NO_3^- + 0.125\,CH_3COO^- + 0.0275\,NH_4^+ + 0.075\,H^+$ $\rightarrow 0.0275\,C_5H_7O_2N + 0.1125\,HCO_3^- + 0.1275\,H_2O + 0.045\,N_2$ with $f_e=0.45$, $Y=0.55$

Iron reduction	
R_a	$FeOOH(s) + 3H^+ + e^- = Fe^{2+} + 2\,H_2O$
R_d	$\frac{1}{4}HCO_3^- + \frac{9}{8}H^+ + e^- = \frac{1}{8}CH_3COO^- + \frac{1}{2}H_2O$
R_c	$\frac{1}{4}HCO_3^- + \frac{1}{20}NH_4^+ + \frac{6}{5}H^+ + e^- = \frac{1}{20}C_5H_7O_2N + \frac{13}{20}H_2O$
R	$FeOOH(s) + 0.208\,CH_3COO^- + 0.033\,NH_4^+ + 1.925\,H^+$ $\rightarrow 0.033\,C_5H_7O_2N + 0.250\,HCO_3^- + 1.600\,H_2O + Fe^{2+}$ with $f_e=0.60$, $Y=0.40$

Sulfate reduction	
R_a	$\frac{1}{8}SO_4^{2-} + \frac{9}{8}H^+ + e^- = \frac{1}{8}HS^- + \frac{1}{2}H_2O$
R_d	$\frac{1}{4}HCO_3^- + \frac{9}{8}H^+ + e^- = \frac{1}{8}CH_3COO^- + \frac{1}{2}H_2O$
R_c	$\frac{1}{4}HCO_3^- + \frac{1}{20}NH_4^+ + \frac{6}{5}H^+ + e^- = \frac{1}{20}C_5H_7O_2N + \frac{13}{20}H_2O$
R	$0.125\,SO_4^{2-} + 0.13525\,CH_3COO^- + 0.004375\,NH_4^+ + 0.0065\,H^+$ $\rightarrow 0.004375\,C_5H_7O_2N + 0.250\,HCO_3^- + 0.013\,H_2O + 0.125\,HS^-$ with $f_e=0.92$, $Y=0.48$

Note: * In these reactions, the ratio of the biomass carbon number (C) in the biomass ($C_5H_7O_2N$) produced per C of electron donor is the yield coefficient Y that quantifies the produced biomass per mass of organic carbon consumed. For example, for the aerobic oxidation reaction in Table 1, $0.125\,CH_3OO^-$ ($0.125 \times 2C$ as CH_3OO^- has 2 carbon) generates $0.030\,C_5H_7O_2N$ ($0.030 \times 5C$ as $C_5H_7O_2N$ has 5 carbon). Therefore the yield coefficient is $0.030 \times 5C / (0.125 \times 2C) = 0.60$. Generally, large f_s values lead to high yield coefficients. Values of f_s vary in the range of 0.6–0.75 for aerobic oxidation, 0.55–0.7 for denitrification, and 0.08–0.30 for sulfate reduction (Rittmann and McCarthy (2001)).

can be described by the general dual Monod rate law, reflecting the need for both electron donor and acceptor in these reactions (Monod 1949):

$$r = \mu_{max} C_{C_5H_7O_2N} \frac{C_D}{K_{m,D}+C_D} \frac{C_A}{K_{m,A}+C_A} \quad (7)$$

Here μ_{max} is the rate constant (mol/t/microbe cell), $C_{C_5H_7O_2N}$ is the concentration of microorganisms (cells/L^3), C_D and C_A are the concentrations of electron donor and acceptor (mol/L^3), respectively. The $K_{m,D}$ and $K_{m,A}$ are the half-saturation coefficients of the electron donor and acceptor (mol/m^3), respectively; they are the concentrations at which half of the maximum rates are reached for the electron donor and acceptor, respectively. If an electron donor or acceptor is not limiting, it means that $C_D \gg K_{m,D}$ or $C_A \gg K_{m,A}$, so that the term $\frac{C_D}{K_{m,D}+C_D}$ or $\frac{C_A}{K_{m,A}+C_A}$ is essentially 1, lending to a rate that only depends on the abundance of microorganisms or one of the chemicals.

In natural subsurface where multiple electron acceptors coexist, the biogeochemical redox ladder dictates the sequence of redox reactions. That is, aerobic oxidation occurs before denitrification, which in turn occurs before iron reduction, because microorganisms harvest more energy in aerobic oxidation than denitrification. Inhibition terms are introduced to account for the sequence of redox reactions as follows:

$$r = \mu_{max} C_{C_5H_7O_2N} \frac{C_D}{K_{m,D}+C_D} \frac{C_A}{K_{m,A}+C_A} \frac{K_{I,H}}{K_{I,H}+C_H} \quad (8)$$

Here $K_{I,H}$ is the inhibition coefficient for the inhibiting chemical H. The inhibition term is *1* (not inhibiting) only when $C_H \ll K_{I,H}$. In a system where oxygen and nitrate coexist, which is common in agriculture lands, aerobic oxidation occurs first before denitrification. The denitrification rates can be represented by:

$$r_{NO_3} = \mu_{max} C_{C_5H_7O_2N} \frac{C_D}{K_{m,D}+C_D} \frac{C_{NO_3}}{K_{m,NO_3}+C_{NO_3}} \frac{K_{I,O_2}}{K_{I,O_2}+C_{O_2}} \quad (9)$$

Here C_{NO_3} is the concentration of nitrate, K_{I,O_2} is the inhibition coefficient of O$_2$, or the O$_2$ concentration at which it inhibits denitrifcation. This rate law ensures that denitrification kicks in substantially only when O$_2$ is depleted to a level that $C_{O_2} \ll K_{I,O_2}$, such that the term $\frac{K_{I,O_2}}{K_{I,O_2}+C_{O_2}}$ approaches 1.0. If an electron acceptor that is lower in the redox ladder than nitrate also exist, then multiple inhibition terms are needed. For example, for iron oxide, we write the following:

$$r_{Fe(OH)_3} = \mu_{max,Fe(OH)_3} C_{C_5H_7O_2N} \frac{C_D}{K_{m,D}+C_D} \frac{C_{Fe(OH)_3}}{K_{m,Fe(OH)_3}+C_{Fe(OH)_3}} \frac{K_{I,O_2}}{K_{I,O_2}+C_{O_2}} \frac{K_{I,NO_3}}{K_{I,NO_3}+C_{NO_3}} \quad (10)$$

Here K_{I,O_2} is the inhibition coefficient of NO$_3$ or the NO$_3$ concentration at which it inhibits iron reduction. The additional nitrate inhibition term means that iron reduction occurs at significant rates only when both oxygen and nitrate are low compared to their corresponding inhibition coefficients.

Rates in natural soils. The dual-Monod and inhibition term are important under conditions where electron donors and acceptors are limited. In shallow soil, O$_2$ is prevalent except under wet conditions with little pore space for air. Anoxic conditions can develop in local environments such as dead-end pores where water is saturated for a long time and not easily flows out. Under conditions organic carbon and O_2 are abundant, the rate law is simplified to the following form assuming microorganism concentrations are relatively constant:

$$r = \mu_{max}Af(T)f(S_w) \tag{11}$$

where μ_{max} (mol/L^2/t) here depends on the original μ_{max} (mol/L^2/t) in previous equations but also on the concentrations of microbe. That is, here we lump the rate constant with microbial abundance. The A is the SOC surface area (L^2) as an approximation of SOC content, and $f(T)$ and $f(S_w)$ describe the temperature and soil moisture dependence, respectively. For temperature dependence, a Q_{10}-based form (Friedlingstein et al. 2006; Regnier et al. 2013) is commonly used: $f(T) = Q_{10}^{|T-20|/10}$, where Q_{10} is the relative increase in reaction rates when temperature increases by 10 °C (Davidson and Janssens 2006). The $f(S_w)$ accounts for the nonlinear dependence of rates on soil moisture. A simple form of $f(S_w) = (S_w)^\varepsilon$ where ε is the saturation exponent (a typical ε value is 2) is often used. More complex forms of $f(S_w)$ considering both water limitation under dry conditions and O_2 limitation under wet conditions have been proposed (Yan et al. 2018). It has also been suggested that SOC decomposition depends strongly on the depth distribution of SOC (Seibert et al. 2009), which is sometime accounted with a depth function:

$$r = \mu_{max}Af(T)f(S_w)f(Z_w) \tag{12}$$

where Z_w is the water table depth (m). An example is $f(Z_w) = \exp(-Z_w/b_m)$ (Weiler and McDonnell 2006). Here b_m is the declining coefficient describing the gradient of SOC content over depth.

Chemical weathering

Chemical weathering is the process that transforms primary minerals (such as silicates), those formed in the cooling and solidification of molten mass, into secondary minerals such as clays, essentially turning rocks into soils (White and Brantley 1995). In this process, primary minerals dissolve and leach out cations whereas secondary minerals precipitate. At a short time scale (days to years), these reactions change water chemistry. Over long time scales, these reactions alter solid phase mineralogical compositions and physical properties. Soils are the relatively weathered materials so primary minerals are more abundant at depth, the specific distribution of which depend on the position of reaction fronts, or the transition between weathered and unweatherred zones (Brantley et al. 2013b). Still, even the relatively slow-dissolving clay continues to leach out cations when solubilized with infiltrating meteoric water at disequilibrium (Heidari et al. 2017). Chemical weathering has been a rich research topics in geochemistry that has accumulated an impressive literature documenting reaction thermodynamics (Johnson et al. 1992; Wolery 1992) and kinetics, i.e., rate laws that prescribe rate dependence on aqueous chemistry (Blum and Stillings 1995; Oelkers 2001; Brantley 2008). Here I only briefly describe the widely-used Transition State theory (TST) rate law and list a few representative reactions in Table 2. Interested readers are referred to other chapters in this book and earlier RiMG volumes for in depth coverage of the topic.

Silicates and carbonate are among the most common primary rocks on Earth's surface (Moosdorf et al. 2010). Note that in Table 2, carbonate is listed as both primary and secondary minerals, as it can precipitate and form as a secondary mineral. An example of silicate weathering is K-feldspar dissolving out solutes such as Al^{3+} and SiO_2 (Reaction (4) toward the right), which further precipitate and form kaolinite (Reaction (5) toward the left). The net reaction is $2\,KAlSi_3O_8(s) + 2\,H^+ \leftrightarrow Al_2Si_2O_5(OH)_4(s) + 2\,K^+ + 4\,H_2O + 4\,SiO_2(aq)$, essentially K-feldspar transforming into kaolinite while leaching out K and SiO_2. Note that all reactions except pyrite dissolution and ferrihydride precipitation are reversible, meaning the reactions can go either direction depending on aqueous chemistry and their reaction thermodynamics. The widely accepted Transition State Theory (TST) based rate laws prescribe this reversibility as follows (Lasaga 1998):

Table 2. Example Chemical Weathering Reactions

Primary mineral dissolution	
Quartz	$SiO_2(s) \leftrightarrow SiO_2(aq)$
Carbonate	$CaCO_3(s) \leftrightarrow Ca^{2+} + CO_3^{2-}$
Pyrite	$FeS_2(s) + H_2O + 3.5 O_2 \rightarrow 2 H^+ + 2 SO_4^{2-} + Fe^{2+}$
K-feldspar	$KAlSi_3O_8(s) + 4 H^+ \leftrightarrow Al^{3+} + K^+ + 2 H_2O + 3 SiO_2(aq)$
Secondary mineral precipitation	
Kaolinite	$Al_2Si_2O_5(OH)_4(s) + 6 H^+ \leftrightarrow 2 Al^{3+} + 2 SiO_2(aq) + 5 H_2O$
Ferrihydride	$Fe(OH)_3(s) + 2 H^+ \leftarrow 0.25 O_2(aq) + Fe^{2+} + 2.5 H_2O$
Carbonate	$CaCO_3(s) \leftrightarrow Ca^{2+} + CO_3^{2-}$

$$r = A(k_{H^+} a_H^{n_H} + k_{H_2O} + k_{OH} a_{OH}^{n_{OH}}) \left(1 - \frac{IAP}{K_{eq}} \right) \tag{13}$$

Here r is the mineral reaction rate (mol/L^3/t), A is the mineral surface area per unit volume (which depends on mineral abundance (L^2/L^3)), k (k_H, k_{H_2O}, k_{OH}) are rate constants under acidic, neutral, and alkaline conditions (mol /L^2 /t), the activities (a) are for H$^+$, water, and OH$^-$, n (n_H, n_{H_2O}, n_{OH}) are the exponents of activities, IAP is the ion activity product, and K_{eq} is the equilibrium constant. The saturation index IAP/K_{eq} quantifies how far the aqueous phase is from equilibrium. At equilibrium, IAP/K_{eq} equals 1.0. When the saturation index is less than one, the mineral dissolves. Note that in shallow soils with unsaturated water, the surface area A needs to be modified with a dependence on soil moisture, as minerals only dissolve when solubilized in water. An additional term such as f(S_w) = (S_w)^g is necessary to account for that effect, similar to the way we consider soil moisture effects on SOC decomposition. The equilibrium constants determine how much a mineral can dissolve in water (thermodynamics), whereas the rate constants control how fast reactions reach equilibrium.

The exception in Table 2, pyrite dissolution and ferrihydrite precipitation, are redox reactions that involve changes in oxidation states of involved solutes. These reactions cause acid mine drainage and the yellowish water that is enriched with Fe(OH)$_3$ (Druschel et al. 2004). These reactions are irreversible and follow different rate laws (Williamson and Rimstidt 1994; Rimstidt and Vaughan 2003).

WATERSHED REACTIVE TRANSPORT MODELLING

A brief history of reactive transport modeling

Multi-component Reactive Transport Models (RTMs) originated in the 1980s and have been extensively used in the subsurface geochemistry community (Chapman 1982; Chapman et al. 1982). RTMs couple flow and transport calculation within a full biogeochemical thermodynamic and kinetic framework (Steefel et al. 2015), therefore explicitly tracing spatial and temporal patterns of geochemical species in fluid and solid phases. Built upon the theoretical framework of reaction thermodynamics and kinetics (Lichtner 1985, 1988), RTM development advanced rapidly in the 1990s as illustrated by the emergence of various RTM codes that have become extensively used in the past decades (Ortoleva et al. 1987; Yeh and Tripathi 1989; Steefel and Lasaga 1994; Bethke 1996; Lichtner et al. 1996; Van Cappellen and Wang 1996; Xu et al. 1999; White and Oostrom 2000; Mayer et al. 2002; Hammond et al. 2014).

RTMs have since been utilized as an integration and interpretation tool across diverse environments involving both porous and fractured media (as reviewed in MacQuarrie and Mayer 2005; Steefel et al. 2005; Sprocati et al. 2019). They have simulated a plethora of processes, including tracer transport, mineral dissolution and precipitation, ion exchange, surface complexation, as well as biotic processes such as microbe-mediated redox reactions, and biomass growth and decay. RTMs have been applied to understand processes on topics including chemical weathering (Bolton et al. 1996; Maher et al. 2009a; Brantley and Lebedeva 2011; Moore et al. 2012; Heidari et al. 2017), biogeochemical cycling (Regnier et al. 1997; Dale et al. 2008; Krumins et al. 2013; Ng et al. 2017), environmentally bioremediation (Li et al. 2010; Yabusaki et al. 2011; Druhan et al. 2012), natural attenuation (Mayer et al. 2001; Liu et al. 2008), geological carbon sequestration (Xu et al. 2003; Atchley et al. 2013; Brunet et al. 2013; Navarre-Sitchler et al. 2013; Tutolo et al. 2015), nuclear waste storage (Saunders and Toran 1995; Soler and Mader 2005), and energy production (Audigane et al. 2007; Qiao et al. 2015).

RTMs have been primarily applied in groundwater and deep subsurface systems at spatial scales from pores (Kang et al. 2006; Li et al. 2008; Fang et al. 2011; Molins et al. 2014; Scheibe et al. 2015; Sprocati et al. 2019) to field scales at tens to hundreds of meters (Li et al. 2011; Beaulieu et al. 2011, 2015; Navarre-Sitchler et al. 2013). Regional scale RTMs have recently been linked to global vegetation models to understand the role of climate change in controlling weathering over periods of 10^2 to 10^3 years (Goddéris et al. 2006, 2013; Roelandt et al. 2010). Only recently RTM has been introduced to the watershed scale, where hydrological conditions are transient (Yeh et al. 2006; Bao et al. 2017).

Reactive transport equations at the watershed scale

A watershed reactive transport model needs to couple hydrological processes, land-surface interactions with multi-component reactions to capture the dynamics of water and biogeochemical interactions. The governing equations of reactive transport processes are in fact not that different from those for groundwater systems in that it is determined by advective and dispersive / diffusive transport and reactions. The major difference is that now water flow is dictated by watershed hydrology that is open to changes in meteorological conditions and other watershed-relevant characteristics. The watershed reactive transport code BioRT-Flux-PIHM (Bao et al. 2017; Li et al. 2017a; Zhi et al. 2019) follows the semi-discrete finite volume approach discussed earlier for PIHM, for an arbitrary solute m in an arbitrary prismatic element i, an example reactive transport equation is as follows (Bao et al. 2017):

$$V_i \frac{d(S_{w,i} C_{m,i})}{dt} = \sum_{j=1}^{n} \left(A_{ij} D_{ij} \frac{C_{m,j} - C_{m,i}}{L_{ij}} - q_{ij} C_{m,j} \right) + R_{m,i}, \quad m = 1, n_{\text{tot}} \tag{14}$$

where V_i is the total volume of the element i; $S_{w,i}$ is the soil water content (L³ water/L³ porous medium volume) that can be calculated based on water storage from a hydrological model; $C_{m,i}$ is the aqueous concentration of solute m, mol/L³; j is the index of neighbor elements of i sharing grid interfaces with i; A_{ij} is the grid interface area (L²) shared by i and its neighbor grid j; D_{ij} is the combined dispersion/diffusion coefficient (L²/t) normal to the shared surface A_{ij}; L_{ij} is the distance between the center of i and its neighbor element j; q_{ij} is the flow rate across the shared interface A_{ij}, L³/t; R_m is the total rate of kinetically controlled reactions that involve solute m, mol/s; n_{tot} is the total number of solutes that are kinetically controlled.

Note that Equation (14) can be written for any solutes that are of interests with their corresponding rate laws represented in R_m. It describes processes that lead to mass changes of solute m. The first two terms in the right hand side describes the conservative (non-reactive) transport and the last term describes reaction rates. If a solute participates in multiple kinetically-controlled reactions, this rate is the summation of several reaction rates with different rate laws. For example, for a geogenic solute Ca, it can dissolve out of a silicate mineral such as anorthite adding Ca to the water phase but also can precipitate as calcite which is a sink term removing Ca out of the water phase. The R_m then is the net rate of addition and removal following the TST rate law in Equation (13). For DOC or nutrient species coming out of the decomposition of Soil Organic Matter (SOM), the rate follows Equation (12) with parameters specific to reactions in Table 2. Note that DOC is produced by SOC decomposition but can also be consumed in other microbe-mediated reactions as the electron donor in processes such as denitrification. The R_m term in this case would be the net rate of the two microbe-mediated redox reactions (SOC decomposition and denitrification) following their specific rate laws with corresponding rate parameters.

The module Flux-PIHM solves for water storage and fluxes, which are used to drive reactive transport calculation for solute concentrations in the BioRT module. The code is written in a way that users can define solutes, reaction type, and rate laws of interests in the input file, enabling the flexibility based on the need of the users. The code outputs spatial and temporal distributions of solute concentrations and reaction rates of interest. Below I show examples to illustrate how hydrological conditions influence water chemistry at the watershed scale.

Examples of hydological and biogeochemical coupling

How does **ET** *influence water chemistry?* In the Shale Hills example discussed earlier (Fig. 4), although precipitation is relatively similar across the year, summer is dry due to higher ET. These transient hydrological conditions have profound impacts on water chemistry. The non-reactive tracer, chloride (Cl), originates from precipitation, whereas Mg leaches out of clay weathering and participates in ion exchange. In the dry summer, only a small proportion of the catchment, mostly swales and the very vicinity of the stream, is connected to the stream (Fig. 6A). Most Cl is ensnared in isolated pockets of soil water, leading to escalated concentrations that are more than one order of magnitude higher than rainfall concentrations (~2 µmol/L) and those in other locations connected to the stream (Fig. 6B). In wet spring and winter, large hillslope areas are connected to the stream, allowing Cl to export rapidly such that chloride concentration ([Cl]) is relatively low and is spatially homogeneous. In other words, [Cl] varies spatially and temporally because of the water dynamics. When *ET* is low, [Cl] is relatively low and homogeneously distributed across the entire catchment owing to a well-connected, wet watershed that rapidly flushes out Cl. When *ET* is high, flow pathways close up, leading to "islands" of elevated [Cl] and a much more heterogeneous concentration field. Across the catchment, [Cl] is high in "old" water but they are almost irrelevant to the stream water because of the disconnection between stream and hillslope.

The Cl mass in the watershed is determined by the balance between rainfall input and discharge output. Low discharge and connectivity in the summer result in negligible Cl export and therefore Cl accumulation within the watershed. Intense rainfall and snowmelt later mobilize trapped Cl, plummeting to ~50% of the total Cl mass, as shown in Figure 7E in Li et al. (2017a). The increase of Cl concentrations with higher *ET* and lower recharge is well documented and serves as the basis for the Cl mass balance method for estimating groundwater recharge (Rice and Hornberger 1998; Semenov and Zimnik 2015).

Water storage influences Mg concentrations ([Mg]) by modulating the wetted surface area and clay dissolution rates (notably chlorite). Chlorite dissolves faster in swales and valley floor with ample water and connectivity to the stream (Fig. 6C). Averaged dissolution rates at the watershed scale were estimated to drop by about half in the dry summer compared to spring. Notably, [Mg] in soil water does not increase as much as [Cl] in the summer; it also does not vary spatially as much as [Cl], primarily because cation exchange effectively acts as a buffer that dampens concentration fluctuations (Clow and Mast 2010; Herndon et al. 2015). The [Mg] on exchange sites is highest in the valley floor (Fig. 6D) because convergent water flow continuously brings Mg mass fluxes from the upslope.

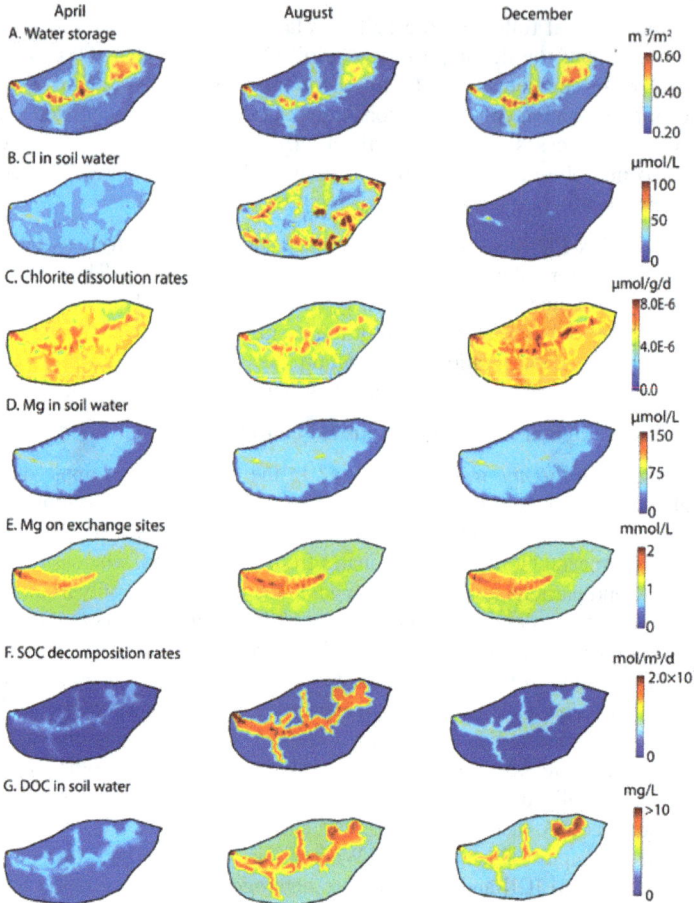

Figure 6. Spatial distribution of water, solutes, and reactions rates on 04/01, 08/01, 12/01 in 2009. **A:** total water storage; **B:** [Cl] in soil water; **C:** chlorite dissolution rates; **D:** [Mg] in soil water; E) [Mg] on exchange sites; **F:** SOC decomposition rates; **G:** [DOC] in soil water. Modified from Li et al. (2017a).

The DOC derives from SOC (or SOM) decomposition, the rate of which depends on water content, temperature, and depth of water table (Eqn. 13). Its spatial distribution mostly mirrors that of the soil moisture in Figure 6A with relatively high rates in swales, which is similar to chlorite dissolution and suggests the dominant control of water content. They however differ from those of chlorite dissolution patterns because of their dependence on water table ($f(Z_w)$). The shallower water table in swales leads to high $f(Z_w)$ values and high SOC rates, amplifying the difference from the planar hillslopes. The DOC concentrations are particularly high in streams in all seasons, because streams receive the produced DOC from the land.

This example illustrates the intimate linkage between water storage, reactions, and solute export. For Cl originating from wet and dry deposition, concentrations in soil water are higher as the watershed dries up at high ET. For the reactive Mg, although clay dissolution is driven by the wetness of mineral surfaces that largely depend on ET, ion exchange reactions buffer its concentration fluctuations, leading to much lower extent of seasonal and spatial variations. For DOC, although it is produced in land that is similar to Mg, its not-as-strong sorption on soils leads to higher spatial and temporal variations in concentrations. In general, however, the variations in solute concentrations are much smaller than the order of magnitude change in discharge.

How do extreme hydrological events influence water chemistry?

Snowmelt and flooding. Seasonal snow covers 30% of the Earth's land surface, 98% of which is in the Northern Hemisphere, specifically in North America and Eurasia (Robinson et al. 1993; Cheng et al. 2018). In western USA, mountain Snow packs are estimated to feed about three fourth of the freshwater, earning them the name of "water tower" (Cayan 1996). The hydrology of these mountains is snow-dominated, with a distinct snow melting season in late spring to early summer contributing to > 70% of the annual water budget. The significance of snowmelt however does not merely lie in its sheer large water volumes. Snow melting also plays an outsize role in flushing out solutes and contaminants. To illustrate this, Figure 7 shows as an example of the Coal Creek watershed, a representative high elevation watershed (2,700 to 3,700 m, 53 km^2) in the central Rocky Mountains of Colorado, with an annual mean temperature around 0.9 °C and average annual rainfall and snowfall of ~ 600 mm and 550 mm, respectively. The watershed has been mined for ~ 100 years in the past with lingering metal-leaching mine tailings at the site, which brings out the eminent water quality issue as Coal Creek supplies drinking water for the skiing town Crested Butte (Zhi et al. 2019).

Figure 7 shows a pronounced discharge peak during snowmelt from early May to late June, 2016. DOC remains at low concentrations under base flow and elevates to a maximum of ~5 mg/L at the discharge peak. In contrast, geogenic cations (Na, Ca, and Mg) exhibit high concentrations under base flow and low concentrations at high flow, demonstrating a dilution trend that is opposite of the DOC pattern. Trace metals (Zn, Mn, and Cd) derived from abandoned mine tailings and natural deposits approach their maxima at early flushes before the discharge peak. Total solute export (mass/d) estimated using the USGS Load Estimator (LOADEST) show that discharge from early May to early July accounts for nearly 80% of the annual discharge (Zhi et al. 2019). In this period, 90% of annual DOC export occurs, compared to 70% and 75% for geogenic solutes and trace metals, respectively. In essence, snow melt is the hot moment of the year not only for water quantity but also for water quality: solute and contaminants have outsized export in this period.

These temporal impacts are in fact similar to flooding, storm water, and melting glaciers (Saberi et al. 2019). Excessive water in large hydrological events has been shown to flush out disproportionally large pulses of "stored" DOC and contaminants that deteriorate water

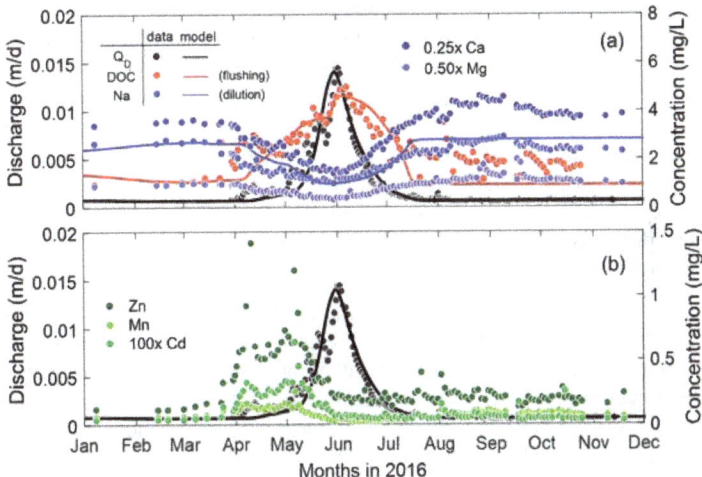

Figure 7. Temporal patterns of stream chemistry: (**a**) biogenic species (doc) and geogenic species (Na, Ca, and Mg); (**b**) trace metals (Zn, Mn, and Cd) (Zhi et al. 2019).

quality and ecosystem health (Raymond and Saiers 2010; Huntington et al. 2016). Rapid changes in discharge during storms and snowmelt can result in shifts in source waters and pronounced variations in DOC concentration, occurring concomitantly with a shift in the optimal properties of DOM that indicates a transition towards higher molecular weight, more aromatic DOM composition (O'Donnell et al. 2010; Voss et al. 2015).

Droughts. Under changing climate conditions, precipitation experiences shifts from snow to rainfall, accompanied by shrinking snow packs and dwindling summer flow in mountainous watersheds (Berghuijs et al. 2014; Rhoades et al. 2018). Such hydrological shift, coupled with increasing temperature, can fundamentally alter biogeochemical reaction rates and water quality (Smith et al. 2005; Laudon et al. 2012; Milner et al. 2017). The effects of droughts on stream water composition however have remained poorly understood (Larned et al. 2010). Harjung et al. (2018) recently examined the impacts of droughts on DOM using a myriad of drought conditions in stream-side flumes. Their work showed production and release of DOM from biofilms and leaf up by 50% during droughts, accompanied by changes in absorbance and fluorescence properties suggesting more labile DOM for microbial degradation. These observations are consistent with measured DOC concentrations and DOM properties in arid areas with prolonged droughts (Jones et al. 1996; Vazquez et al. 2011). Droughts generate disconnected areas with distinct hot spots, amplifying spatial heterogeneities as observed in Figure 6.

RESEARCH FRONTIERS

Drivers of chemical weathering in natural systems

As discussed earlier in the section *Fundamentals of soil biogeochemistry*, chemical weathering and organic matter decomposition follow different rate laws. These rate laws are derived from theories derived considering only chemistry with with parameters measured typically in well-mixed systems without mass transport limitation. Natural systems are almost never well-mixed (Dentz et al. 2011; Loschko et al. 2018). In particular, at the watershed scale, hydrological processes, often shaped by both external forcing and internal structure characteristics, influence reactant concentrations, distance to equilibrium, and redox state. Although large datasets of concentrations and discharges exist and allude to rates at the watershed scale (Bluth and Kump 1994; White and Blum 1995), mechanistic understanding of controls of reaction rates at the watershed scale is generally lacking.

The role of reactive gases and roots. In shallow subsurface, chemical weathering is influenced not only by the factors outlined in the TST rate law (Eqn. 13), but also by the transient nature of soil moisture and characteristics. The gas of O_2 can drive the oxidative weathering of pyrite-containing shale, creating fractures and porosity that allow water infiltration and weathering of other minerals (Moore et al. 2019). Oxidation of pyrite also produces sulfuric acid, which can enhance mineral weathering and create permeable pathways for water and solutes (Brantley et al. 2013a). The presence and abundance of O_2 therefore regulate how fast reaction fronts progress as O_2-filled water infiltrate through pyrite-rich subsurface (Bolton et al. 2006; Heidari et al. 2017).

Soil CO_2 is typically 1–2 orders magnitude higher in vegetated soils compared to the atmospheric CO_2 level because of soil and root respiration, therefore driving soil water acidity to a level of pH around 4.0 and accelerating weathering of silicates and carbonates (Hinsinger et al. 2003; Hasenmueller et al. 2015). Organic acids from root exudates also speed up dissolution by increasing soil water acidity and by forming complexity with cations, effectively elevating mineral solubility (Leake et al. 2008; Lawrence et al. 2014). These "biogeochemistry" aspects of root effects have been extensively discussed (Taylor et al. 2009; Brantley et al. 2011; Thorley et al. 2015). Roots of different plants (e.g., grass-, shrub-, and wood-lands) can additionally leave distinct imprints on the relative distribution of soil carbon with depth. For example, the relatively deep root distributions of shrubs can lead to soil carbon profiles deeper than those in grassland (Jackson et al. 1996), which may help maintain high soil CO_2 (Oh et al. 2005) and low pH level at depth that is in contact with unweathered, more reactive primary minerals.

Plant roots may also affect soil structure, hydrological redistribution, and different patterns of water flow and soil moisture exchange (Le and Kumar 2017). Macropores originated from root channels have been estimated to account for about 70% of the total macropores (Noguchi et al. 1997). Their structure highly depend on plant species (Beven and Germann 1982, 2013). In grasslands, the lateral, dense spread of roots near surface promotes the development of horizontal macropores and granular or "sandy" texture soil aggregates that facilitate near surface lateral flow (Oades 1993; Cheng et al. 2011). In contrast, forests with deep and thick roots tend to have high proportion of macropores and high connectivity in deep soil (Canadell et al. 1996; Nardini et al. 2016), leading to smaller water-holding capacities and deeper water penetration. In other words, deeper roots in forests can lead to branching, or partitioning of water more to the depth, which can pump CO_2-rich water to deeper zones and enhance the contact between CO_2 and primary minerals at depth. As an example, woody encroachment into grassland and associated rooting differences have been attributed to elevated chemical weathering fluxes in carbonate terrains (Sullivan et al. 2018). Much remains to be learned about feedbacks among weathering, transport, and biologically-mediated processes with realistic subsurface structures. Watershed RTMs can be useful to distinguish biogeochemical versus hydrological impacts of root structure on chemical weathering.

Dynamic reactive surface area. The surface area A in Equations (12) and (13) is essential in determining rates and is arguably the most uncertain parameter in the rate laws. For SOC decomposition under aerobic conditions, this surface area quantifies the contact between reactants, i.e., microbe, water, SOC, and O_2. For mineral dissolution, this A quantifies the effective mineral surface that is actually reacting. Five to six orders of magnitude discrepancy between rates of mineral reactions measured in laboratory versus field context (Blum and Stillings 1995; Drever and Clow 1995; Sverdrup et al. 1995; White 1995; White and Brantley 1995, 2003; Richards and Kump 2003). This rate discrepancy has been attributed to two major categories of factors. One category includes processes and conditions that directly affect the "reactivity" of mineral surfaces. Examples include mineral surface roughness or fractalness (Navarre-Sitchler and Brantley 2007), passivation or armoring layers as a result of clay coating or secondary mineral precipitation (Helgeson 1971; Nugent et al. 1998; Maher et al. 2009b), leached layers that act as diffusion barriers (Luce et al. 1972), and water films between grains that vary under different temperature and pressure conditions (Renard et al. 1997). The second category includes factors that regulate fluid compositions. For example, water flow controls reaction rates by regulating the extent of deviation from equilibrium (Maher 2010; Salehikhoo et al. 2013). The amount of wetted surface under unsaturated conditions, as well as concentrations of catalyzing or inhibiting aqueous species, also play a significant role (Oelkers et al. 1994; Lawrence et al. 2014).

In addition, porous medium characteristics—the details of pore structure and distribution of porous medium properties (e.g., permeability and mineral abundance)—can lead to large spatial variations in flow rates and deviations from equilibrium, therefore modifying the overall "effective" reaction rates (Kang et al. 2006; Li et al. 2007; Molins et al. 2012; Al-Khulaifi et al. 2017; Jung and Navarre-Sitchler 2018) or even change the reaction direction (Brunet et al. 2016). For example, a mineral that does not have any coating can be bathed in a fluid at equilibrium in the low permeability zone of a heterogeneous medium and therefore is not effectively dissolving. In fact, any factors that influence reaction kinetics (surface area, concentrations of catalyzing or inhibiting species, and distance from equilibrium) can contribute to the rate discrepancy. In natural systems, all these factors operate simultaneously resulting in much smaller effectively-dissolving surface areas than the measured Brunauer–Emmett–Teller (BET) or imaged surface area, although the dominance of different factors may vary with conditions; e.g., (White and Peterson 1990; Nugent et al. 1998; Zhu 2005; Jin et al. 2011a).

At the watershed scale, the "effective" surface area is influenced not only by the subsurface structure, not also by how much water in different parts of the watershed is connected to the stream. As shown in Figure 6, although minerals are dissolving everywhere, not all areas are connected to the stream and have active flow that drives to disequilibrium. The hydrological regime has a tremendous impact on weathering rates (Torres et al. 2015, 2017), which are controlled via effective surface area at the watershed scale. How are they relate to reaction rates at the watershed scale? Watershed RTM can be used to answer these questions.

Watershed response to hydrological changes: Concentration discharge relationships

Although the integration of reactive transport and watershed hydrology in process-based models is nascent, field measurements have in fact documented decades of stream flow and water chemistry data worldwide. Many rivers in the USA, for example, the Mississippi River and many of its tributaries, have water quality data dating back to 1950s (https://waterdata. usgs.gov/nwis?). A direct measure of biogeochemical responses to hydrological changes is the concentration (C) and discharge (Q) relationships measured at river and stream mouths. CQ relationships have been extensively used to understand watershed structure and response to external forcing (Winterdahl et al. 2016) and to quantify fluxes of chemical weathering (White and Blum 1995) and nutrient export (Zhang et al. 2016a).

Studies of CQ date back more than five decades on event-based CQ hysteresis (Johnson et al. 1969; Hooper et al. 1990; Evans and Davies 1998). The power law relationship $C=aQ^b$ remains the extensively used form (Godsey et al. 2009) to characterize CQ patterns, although recent work has proposed more complex characterization (Thompson et al. 2011; Moatar et al. 2017; Zhang and Ball 2017). Chemostasis is defined as relatively small variations in concentrations compared to discharge with b values within -0.1 and 0.1 (Musolff et al. 2015). Chemodynamic patterns are characterized as having high absolute b values (> 0.1), with positive b values demonstrating flushing or enrichment behavior whereas negative b illustrating dilution.

A big puzzle in this area is the contrasting CQ relationships that have been observed for different solutes: similar CQ patterns have been observed for different solutes; contrasting CQ relationships have been observed for the same solute in different watersheds (Herndon et al. 2015; Musolff et al. 2017; Abbott et al. 2018; Zarnetske et al. 2018). Geogenic species (e.g., Na, Mg, Si) derived from chemical weathering have commonly demonstrated chemostasis (Godsey et al. 2009; Sullivan et al. 2018). Flushing behaviors are most commonly observed for DOC (Boyer et al. 1997; Zarnetske et al. 2018). For nutrients, chemostatic or biogeochemical stationary behaviors have been commonly observed in agricultural lands due to their large legacy store from decades of fertilizer applications (Basu et al. 2010; Van Meter et al. 2017) whereas dilution behavior occurs under conditions where groundwater is the major source of nutrient export (Miller et al. 2016). An example is shown in Figure 8 for the Coal Creek watershed discussed earlier. Here biogenic solutes (DOC, TN, and TON) exhibit flushing behavior, geogenic species manifest dilution behavior, and all metals show hysteresis loops.

Several schools of thoughts in the literature explain contrasting CQ patterns. Chemostasis has been attributed to hydrologically responsive dissolution rates (via wet surface area) as discharge increases (Li et al. 2017a), to buffering capacity of ion exchange reactions (Clow and Mast 2010) (Clow and Mast 2010), and to the approach to equilibrium as residence time increases (Maher 2011; Ameli et al. 2017). Flushing behaviors have been interpreted as due to rising water table tapping shallow soil zones with enriched organic carbon and nutrients, increased hydrological connectivity between uplands and stream under high flow conditions, and biotic control of vegetation and temperature (Seibert et al. 2009; Andrews et al. 2011; Mei et al. 2012). In contrast, dilution patterns have been attributed to high solute concentrations from groundwater diluted by

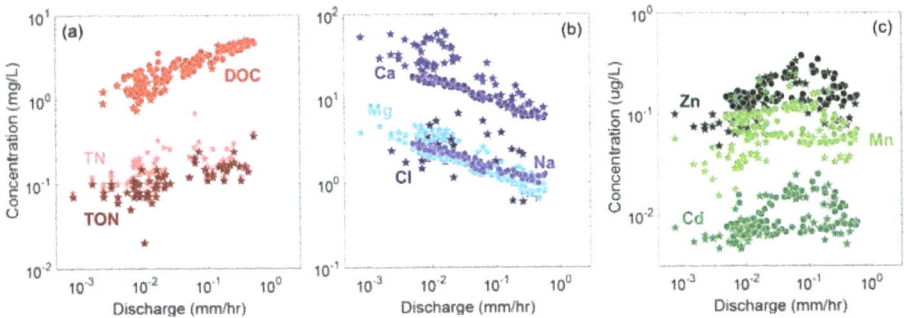

Figure 8. Concentration-discharge relationships in Coal Creek, CO: **(a)** flushing pattern of DOC, TN (Total Nitrogen), and TON (Total Organic Nitrogen); **(b)** dilution pattern of major cations and chloride; **(c)** clockwise loops of trace metals. Filled and star markers represent high frequency measurements in 2016 and long-term low frequency measurements, respectively (Zhi et al. 2019).

rain water at high discharge (Miller et al. 2016; Li et al. 2017a), flow paths bypassing soil pore water with high concentration, and rapid depletion of solute store (Herndon et al. 2015b; Hoagland et al. 2017). A unifying framework has yet to emerge to interpret the diverse observations.

Using bioRT-flux-PIHM and Monte-Carlo simulations using a wide range of conditions, Zhi et al. (2019) illustrates that contrasting CQ patterns can be explained by the switch of source waters under dry and wet conditions, and the source water chemistry contrasts arising from spatial heterogeneity of source materials in the subsurface. In particular, stream water mixes different types of source water, primarily shallow soil water (e.g., perched water table), deep (relatively) groundwater, or water from riparian zones and hillslope areas. The relative contribution of different source waters vary drastically under dry and wet conditions. As shown Figure 6, under wet conditions, stream water is dominated by shallow soil water in largely connected watersheds; under dry conditions, groundwater or water in the very vicinity of streams predominate. The chemistry of these waters differ dramatically. Shallow soils are characterized by rich life and high SOC and DOC content, whereas deeper subsurface groundwater brings high concentrations of geogenic species at depths.

Solutes that are enriched in shallow soils (e.g., DOC) therefore have flushing patterns because their concentrations increase as water table rises under wet conditions and taps shallow soils; in contrast, those abundant at depths show up at high concentrations in base flow during dry times, demonstrating dilution patters. These insights have been encapsulated in a general relationship between CQ slope b and ratios of soil versus ground water concentrations as the two end-member source waters ($C_{ratio} = C_{sw}/C_{gw}$): $b = \dfrac{\delta_b\, C_{ratio}}{C_{ratio,1/2} + C_{ratio}} + b_{min}$ (Zhi et al. 2019), where b_{min} is the minimum b value, δ_b is the difference between maximum and minimum

b values, and $C_{ratio,1/2}$ is the concentration ratio when the b values equal the average of minimum and maximum b values. Disparate CQ patterns of 11 solutes (DOC, DP, NO_3, K, Si, Ca, Mg, Na, Al, Mn, and Fe) from three watersheds (i.e., Coal Creek, Shale Hills, and Plynlimon) with different climate and geology conditions follow this general relationship, indicating its promise of quantifying CQ patterns for a wide range of solutes in watersheds of diverse characteristics. This underscores the essential role of shallow–deep biogeochemistry contrasts in shaping CQ patterns. This comes along similar line to another study that contributes distinct CQ patterns as rising from spatial structure heterogeneity of source waters and travel time that allows transformation reactions to occur within a watershed (Musolff et al. 2017). Similarly, spatial distributions in the horizontal direction or the proximity of solute source zones controls their connectivity to the stream under different flow regimes and therefore also regulates stream chemistry (Tunaley et al. 2017). The general relationship will need to be tested in diverse watersheds with additional data on soil and groundwater chemistry and subsurface structure, and with watershed RTMs that can distinguish the effects of individual processes and properites.

Reaction rates at the watershed scale: linking travel time, age, and reactions

In homogeneous groundwater systems, chemical weathering rates and geogenic concentrations have been shown to hinge on dimensionless numbers such as the Damkohler number (Maher 2011; Salehikhoo et al. 2013; Salehikhoo and Li 2015), defined as the relative time scales of reactions (to reach equilibrium, τ_{eq}) versus advective transport (i.e., residence time τ_r). If the time it takes to reach equilibrium is longer than the time that water stays in a system, i.e., $\tau_{eq} > \tau_r$, solute concentrations leaving the system depend on flow rates; if $\tau_{eq} < \tau_r$, solutes reach and remain at equilibrium concentrations at the system outlet and do not vary with flow rates. In heterogeneous systems where fast-dissolving minerals (e.g., carbonate) reside in low-permeability zones and slow-dissolving minerals (e.g., quartz) in high permeability zones, solute concentrations and weathering rates have been shown to depend similarly on Damkohler number but need to additionally consider the relative magnitude (ratio) of the contact time between water and reacting minerals ($\tau_{ad,r}$) versus the overall residence time in the entire domain (τ_a) (Wen and Li 2018). This is because only a fraction of water flow, not the total flow, channels through the low-permeability reactive zones and dissolves reacting minerals (Wen and Li 2017). Hence the dissolution rates not only depend on dissolution kinetics and surface area, but also on overall flow rates and characteristics of spatial distribution and permeability contrasts between reactive and non-reactive zones. These studies accentuate the key linkage between water contact time and dissolution rates.

Similar framework has yet to be developed at the watershed scale. How do biogeochemical rates at the watershed scale relate to different time scales? How do they differ from groundwater systems with steady-state flow? The catchment hydrology community has been grappling with the question "how old is stream water" using water isotopes and non-reactive tracers (e.g., Cl) for decades and have gleaned ample knowledge on travel time at the watershed scale. Transit (or travel) time, defined as the difference in the time rain water enters a catchment to the time it exits, has been studied extensively as "a fundamental catchment descriptor that reveals information about storage, flow pathways and source of water in a single characteristic." (McGuire et al. 2005; McGuire and McDonnell 2006). Measurements and modeling of stable water isotopes and tracer (e.g., Cl) have estimated mean and distribution of transit times to understand where stream water comes from and how old it is under diverse climate, topography, soil property conditions (McGuire et al. 2005; Soulsby et al. 2006; Tetzlaff et al. 2009; Godsey et al. 2010; Birkel et al. 2011). The field has also evolved from assuming temporal stationary (McGuire and McDonnell 2006), to the recent development of time-variant transit time distribution (TTD) theory with StorAge Selection function (SAS), recognizing the impacts of non-stationary of (or, temporally variable) rainfall events and water storage (Botter et al. 2011; van der Velde et al. 2012; Harman 2015; Rinaldo et al. 2015). There has also been considerable advocate on using the young water fraction to characterize catchment travel time (Jasechko et al. 2016; Kirchner 2016a). In addition to travel time, other concepts have emerged as key measures of watershed hydrology. In particular, hydrological connectivity, defined as the extent of connection from hillslope uplands to streams (McGlynn and McDonnell 2003a, b), has been shown to govern water flow and travel times (Western et al. 2001; Stieglitz et al. 2003; McGuire and McDonnell 2006; Jencso et al. 2009, 2010; Gooseff et al. 2017), and can determine the extent and rates of reactions.

These concepts and quantifications have yet to be linked to biogeochemical measures at the watershed scale. As discussed earlier, concentration discharge measures abound, from which the solute export rates and potentially biogeochemical reaction rates can be inferred.

These rates and concentrations have started to be linked to flow and travel time at the watershed scale (Benettin et al. 2015b; Ameli et al. 2017; Musolff et al. 2017). As an example, the use of general age theory for reactive solutes (N species: ammonia and nitrate) and comparison to a non-reactive solute in agricultural lands have found that the age of N is in fact higher in top soil and in low depression area because of the ion exchange reaction that retards N and the tile drains that mobilize N from immobile zones (Woo and Kumar 2016, 2019). In addition, biogeochemistry reactive interfaces (Krause et al. 2017; Li et al. 2017b), defined as where water flows and converge from contrasting geological, ecohydrological or biogeochemical compartments, are "hot spots" (McClain et al. 2003) where rates are often orders of magnitude higher than the rest of a domain (Krause et al. 2013; Harvey and Gooseff 2015; Salehikhoo and Li 2015). Example include surface water–ground water mixing zones, hyporheic zones at river–land-sediment interfaces, spring and seep emergence zones, subsurface high-low permeability boundaries, and riparian zones at land-river boundaries (Öquist et al. 2009; Stonedahl et al. 2010; Hefting et al. 2013; Gooseff et al. 2016; Dwivedi et al. 2018). With the capability of explicitly calculating these quantities, watershed RTMs can help unravel key linkages between these concepts and control of biogeochemical transformation at the watershed scale.

CONCLUSIONS AND LOOKING FORWARD

This chapter focuses on the fundamentals of watershed hydrology and biogeochemistry and the recent development of watershed RTMs. As has been shown here, biogeochemical reactive transport is water-driven and depends strongly on the dynamics of water fluxes including ET, and different source waters flowing to streams under different hydrological regimes. Unlike groundwater systems where flow is spatially heterogeneous and at steady state, watershed hydrological processes are spatially heterogeneous and temporally dynamic. This adds additional layers of complexity in understanding process coupling and feedbacks at the watershed scale.

Model development in watershed hydrology and biogeochemistry have leaped forward without much interactions until very recently. Observations that record water and water chemistry however have accumulated for decades owing to the long-term water monitoring programs worldwide. In fact, we are now at an exciting time when a deluge of data exist. The past decades have witnessed rapid advances in technology and unprecedented generation of earth surface data from remote sensing from satellites (Hrachowitz et al. 2013; McCabe et al. 2017). Within the U.S. alone, such programs include the Critical Zone Observatories (CZOs), the Long-Term Ecological Research (LTER), the Great Lake Ecological Observatory Network (GLEON), the National Ecological Observatory Network (NEON), and the testbed research sites supported by the Department of Energy Subsurface Biogeochemical Research program (e.g., East River Watershed (Hubbard et al. 2018)). Internationally a myriad of observatories exist (Bogena et al. 2018; Gaillardet et al. 2018). Water and water chemistry data are collected regularly through research and observation networks, presenting unprecedented opportunities for integrated understanding (Baatz et al. 2018). There is also a growing recognition that more integration is needed as we move forward (Grimm et al. 2003; Hrachowitz et al. 2016).

These rich data and process-based modeling tool will enable us to answer questions about linkages between travel time, water age, and biogeochemical rates, and interrogate their functional relationships and underlying mechanisms. Such efforts can guide field campaigns: instead of measuring everything everywhere, we may focus on end-members in distinct zones (surface water, soil water, and groundwater) that dominate different hydrological regime; we may need to focus on hot spots (e.g., riparian zones) and hot moments (e.g., storm events, snow melt, droughts) where biogeochemical transformation rates are disproportionally high. It is also important to go beyond individual watersheds, as each watershed has its own hydroclimatic conditions and idiosyncrasies (e.g., land cover, soil type, lithology, and topography) but general principles often underlie the apparent contrasts and offer common threads. Synthesis studies that explore a large number of watersheds are needed to answers questions including: how the linkages and functional relationships developed in individual watersheds vary under diverse climate, land cover, and geology conditions? What are the first order control and watershed characteristics that can be used for prediction? Answers to these questions will ultimately mount to general patterns, principles, and classification. Watershed RTMs can help guide the field campaigns but also set up digital watersheds and interrogate the role of external forcings and internal watershed idiosyncrasies.

Cross-site understanding can also facilitate understanding cross-scale behavior, i.e., the connection between small-scale physics and large-scale behavior (scaling). The issue of scaling is a major challenge and has been extensively discussed in hydrology and other disciplines (Beven 1989; Levin 1992; Sivapalan et al. 2003; Hrachowitz et al. 2013). Reactive transport models at the pore scale have run at the resolution of microns using synchrotron X-ray tomographic images and direct numerical simulation (Wen et al. 2016; Molins et al. 2019). At the watershed scale, models have been run at a spatial resolution as high as one meter using Light Detection and Ranging (LIDAR) images (Le and Kumar 2017; Woo and Kumar 2017). These models illuminate the important role of small-scale spatial characteristics such as microtopography, depression area, and ponding, in regulating chemical weathering and the age of nitrogen in agriculture soils. The issue of upscaling however almost always arises when prediction is needed at larger spatial scales. The resolution of the National Water Model is approximately ~ 1 km^2; the resolution of Earth system Models is often considerably coarser. An individual watershed may be one homogeneous grid block or even smaller than one grid block in large scale models. To represent watershed reactive transport in these models, we need to go beyond spatial heterogeneities and process complexity that obscure effects of individual processes. Such upscaled understanding can not only reduce computational cost and facilitate explicit incorporation of feedback schemes, but also hold the key to unlock a unifying hydro-biogeochemical theory. Ultimately, going up and down scales are complementary and depend on the needs of answering target research questions.

ACKNOWLEDGMENT

The author acknowledges the support from the NSF EAR – 1331726 (S. Brantley) for the Susquehanna Shale Hills Critical Zone Observatory (SSHCZO) and the Department of Energy Subsurface Biogeochemical Research program DE-SC0016221 (L. Li) for the development of bioRT-flux-PIHM and the initiation of watershed reactive transport modeling. The author acknowledges collaborators and students heavily involved in the code development, including Chris Duffy, Chen Bao, Wei Zhi, and Yuning Shi. The author appreciates the hospitality and stimulation of Andrea Rinaldo, Paolo Benettin, and other ECHO laboratory members at the Ecole Polytechnique Fédérale de Lausanne (EPFL) while developing this chapter during sabbatical. Comments from the Editors Jennifer Druhan and Christopher Tournassat and an anonymous reviewer have improved the chapter.

REFERENCES

Abbott M, Warren I, Jensen KH, Jønch-Clausen T, Havnø K (1979) Coupling of unsaturated and saturated zone models. *In:* 19th Congress of the IAHR

Abbott MB, Bathurst JC, Cunge JA, O'Connell PE, Rasmussen J (1986) An introduction to the European hydrological system—Systeme Hydrologique Europeen,"SHE", 1: History and philosophy of a physically-based, distributed modelling system. J Hydrol 87:45–59

Abbott BW, Gruau G, Zarnetske JP, Moatar F, Barbe L, Thomas Z, Fovet O, Kolbe T, Gu S, Pierson-Wickmann AC, Davy P (2018) Unexpected spatial stability of water chemistry in headwater stream networks. Ecol Lett 21:296–308

Al-Khulaifi Y, Lin Q, Blunt MJ, Bijeljic B (2017) Reaction rates in chemically heterogeneous rock: coupled impact of structure and flow properties studied by X-ray microtomography. Environ Sci Technol 51:4108–4116

Ameli AA, Beven K, Erlandsson M, Creed IF, McDonnell JJ, Bishop K (2017) Primary weathering rates, water transit times, and concentration-discharge relations: A theoretical analysis for the critical zone. Water Resour Res 53:942–960

Andrews DM, Lin H, Zhu Q, Jin L, Brantley SL (2011) Hot spots and hot moments of dissolved organic carbon export and soil organic carbon storage in the Shale Hills catchment. Vadose Zone J 10:943–954

Arnold J, Soil G (1994) SWAT (Soil and Water Assessment Tool). Grassland, Soil and Water Research Laboratory, USDA, Agricultural Research Service

Atchley AL, Maxwell RM, Navarre-Sitchler AK (2013) Using streamlines to simulate stochastic reactive transport in heterogeneous aquifers: Kinetic metal release and transport in CO_2 impacted drinking water aquifers. Adv Water Resour 52:93–106

Audigane P, Gaus I, Czernichowski-Lauriol I, Pruess K, Xu T (2007) Two-dimensional reactive transport modeling of CO2 injection in a saline aquifer at the Sleipner site, North Sea. Am J Sci 307:974–1008

Baatz R, Sullivan PL, Li L, Weintraub SR, Loescher HW, Mirtl M, Groffman PM, Wall DH, Young M, White T, Wen H, Zacharias S, Kühn I, Tang J, Gaillardet J, Braud I, Flores AN, Kumar P, Lin H, Ghezzehei T, Jones J, Gholz HL, Vereecken H and Van Looy, K. (2018) Steering operational synergies in terrestrial observation networks: opportunity for advancing Earth system dynamics modelling. Earth Syst Dynam 9:593–609

Bao C, Li L, Shi Y, Duffy C (2017) Understanding watershed hydrogeochemistry: 1. Development of RT-Flux-PIHM. Water Resour Res 53:2328–2345

Basu NB, Destouni G, Jawitz JW, Thompson SE, Loukinova NV, Darracq A, Zanardo S, Yaeger M, Sivapalan M, Rinaldo A, Rao PS (2010) Nutrient loads exported from managed catchments reveal emergent biogeochemical stationarity. Geophys Res Lett 37: L23404

Battin TJ, Luyssaert S, Kaplan LA, Aufdenkampe AK, Richter A, Tranvik LJ (2009) The boundless carbon cycle. Nat Geosci 2:598

Bazilevskaya E, Lebedeva M, Pavich M, Rother G, Parkinson DY, Cole D, Brantley SL (2013) Where fast weathering creates thin regolith and slow weathering creates thick regolith. Earth Surf Processes Landforms 38:847–858

Beaulieu E, Goddéris Y, Labat D, Roelandt C, Calmels D, Gaillardet J (2011) Modeling of water-rock interaction in the Mackenzie basin: Competition between sulfuric and carbonic acids. Chem Geol 289:114–123

Beck HE, van Dijk AIJM, Miralles DG, de Jeu RAM, Bruijnzeel LA, McVicar TR, Schellekens J (2013) Global patterns in base flow index and recession based on streamflow observations from 3394 catchments. Water Resour Res 49:7843–7863

Beisman J, Maxwell R, Navarre-Sitchler A, Steefel C, Molins S (2015) ParCrunchFlow: an efficient, parallel reactive transport simulation tool for physically and chemically heterogeneous saturated subsurface environments. Comput Geosci:1–20

Benettin P, Kirchner JW, Rinaldo A, Botter G (2015a) Modeling chloride transport using travel time distributions at Plynlimon, Wales. Water Resour Res 51:3259–3276

Benettin P, Bailey SW, Campbell JL, Green MB, Rinaldo A, Likens GE, McGuire KJ, Botter G (2015b) Linking water age and solute dynamics in streamflow at the Hubbard Brook Experimental Forest, NH, USA. Water Resour Res 51:9256–9272

Berghuijs WR, Sivapalan M, Woods RA, Savenije HHG (2014) Patterns of similarity of seasonal water balances: A window into streamflow variability over a range of time scales. Water Resour Res 50:5638–5661

Bergstrom S (1976) Development and application of a conceptual runoff model for Scandinavian catchments. Department of Water Resources Engineering, Lund Institute of Technology, University of Lund

Berner RA (1999) A new look at the long-term carbon cycle. GSA Today 9:6

Bethke C (1996) Geochemical Reaction Modeling : Concepts and Applications. Oxford University Press, New York

Beven K (1989) Changing ideas in hydrology—the case of physically-based models. J Hydrol 105:157–172

Beven KJ (2011) Rainfall-Runoff Modelling: The Primer. John Wiley & Sons

Beven K, Germann P (1982) Macropores and water-flow in soils. Water Resour Res 18:1311–1325

Beven K, Freer J (2001) Equifinality, data assimilation, and uncertainty estimation in mechanistic modelling of complex environmental systems using the GLUE methodology. J Hydrol 249:11–29

Beven K, Germann P (2013) Macropores and water flow in soils revisited. Water Resour Res 49:3071–3092

Birkel C, Soulsby C, Tetzlaff D (2011) Modelling catchment-scale water storage dynamics: reconciling dynamic storage with tracer-inferred passive storage. Hydrol Process 25:3924–3936

Bishop K, Seibert J (2015) A primer for hydrology: the beguiling simplicity of water's journey from rain to stream at 30. Hydrol Process 29:3443–3446

Blum AE, Stillings LL (1995) Feldspar dissolution kinetics. Rev Mineral Geochem 31:291–351

Bluth GJS, Kump LR (1994) Lithologic and climatologic controls of river chemistry. Geochim Cosmochim Acta 58:2341–2359

Bogena HR, White T, Bour O, Li X, Jensen KH (2018) Toward better understanding of terrestrial processes through long-term hydrological observatories. Vadose Zone J 17:180194

Bolton EW, Lasaga AC, Rye DM (1996) A model for the kinetic control of quartz dissolution and precipitation in porous media flow with spatially variable permeability: Formulation and examples of thermal convection. J Geophysi Res Solid Earth 101:22157–22187

Bolton EW, Berner RA, Petsch ST (2006) The weathering of sedimentary organic matter as a control on atmospheric o2: ii. theoretical modeling. Am J Sci 306:575–615

Borch T, Kretzschmar R, Kappler A, Cappellen PV, Ginder-Vogel M, Voegelin A, Campbell K (2010) Biogeochemical redox processes and their impact on contaminant dynamics. Environ Sci Technol 44:15–23

Botter G, Bertuzzo E, Rinaldo A (2011) Catchment residence and travel time distributions: The master equation. Geophys Res Lett 38:L11403

Boyer EW, Hornberger GM, Bencala KE, McKnight DM (1997) Response characteristics of DOC flushing in an alpine catchment. Hydrol Process 11:1635–1647

Bracken LJ, Wainwright J, Ali GA, Tetzlaff D, Smith MW, Reaney SM, Roy AG (2013) Concepts of hydrological connectivity: Research approaches, pathways and future agendas. Earth Sci Rev 119:17–34

Brantley SL (2008) Kinetics of Mineral Dissolution. *In*: Kinetics of Water–Rock Interaction. S.L. Brantley JDK, A.F. White (eds.) Springer, New York, p 151–196

Brantley SL, Lebedeva M (2011) Learning to read the chemistry of regolith to understand the critical zone. Ann Rev Earth Planet Sci 39:387–416

Brantley SL, Megonigal JP, Scatena FN, Balogh-Brunstad Z, Barnes RT, Bruns MA, Van Cappellen P, Dontsova K, Hartnett HE, Hartshorn AS, Heimsath A (2011) Twelve testable hypotheses on the geobiology of weathering. Geobiology 9:140–165

Brantley SL, Lebedeva M, Bazilevskaya E (2013a) Relating weathering fronts for acid neutralization and oxidation to pCO_2 and pO_2. Treatise Geochem 6:327–352

Brantley SL, Holleran ME, Jin L, Bazilevskaya E (2013b) Probing deep weathering in the Shale Hills Critical Zone Observatory, Pennsylvania (USA): the hypothesis of nested chemical reaction fronts in the subsurface. Earth Surf Processes Landforms 38:1280–1298

Brantley SL, White T, West N, Williams JZ, Forsythe B, Shapich D, Kaye J, Lin H, Shi Y, Kaye M, Herndon E, Davis K, He Y, Eissenstat D, Weitzman J, DiBiase RA, Li L, Reed W, Brubaker K, and Gu, X (2018) Susquehanna Shale Hills Critical Zone Observatory: Shale Hills in the Context of Shaver's Creek Watershed. Vadose Zone J 17:180092

Brooks PD, Chorover J, Fan Y, Godsey SE, Maxwell RM, McNamara JP, Tague C (2015) Hydrological partitioning in the critical zone: Recent advances and opportunities for developing transferrable understanding of water cycle dynamics. Water Resour Res 51:6973–6987

Brunet JL, Li L, Karpyn ZT, Kutchko B, Strazisar B, Bromhal G (2013) Dynamic evolution of compositional and transport properties under conditions relevant to geological carbon sequestration. Energy Fuels 27:4208–4220

Brunet J-PL, Li L, Karpyn ZT, Huerta NJ (2016) Fracture opening or self-sealing: Critical residence time as a unifying parameter for cement–CO_2–brine interactions. Int J Greenhouse Gas Control 47:25–37

Brutsaert W, Nieber JL (1977) Regionalized drought flow hydrographs from a mature glaciated plateau. Water Resour Res 13:637–643

Budyko MI (1974) Climate and Life. Academic Press, New York, NY

Buss HL, Sak PB, Webb SM, Brantley SL (2008) Weathering of the Rio Blanco quartz diorite, Luquillo Mountains, Puerto Rico: Coupling oxidation, dissolution, and fracturing. Geochim Cosmochim Acta 72:4488–4507

Buttle JM (1998) Chapter 1—Fundamentals of small catchment hydrology. *In:* Isotope Tracers in Catchment Hydrology. Kendall C, McDonnell JJ, (eds). Elsevier, Amsterdam, p 1–49

Canadell J, Jackson RB, Ehleringer JR, Mooney HA, Sala OE, Schulze ED (1996) Maximum rooting depth of vegetation types at the global scale. Oecologia 108:583–595

Cayan DR (1996) Interannual climate variability and snowpack in the western United States. J Clim 9:928–948

Chapman BM (1982) Numerical simulation of the transport and speciation of nonconservative chemical reactants in rivers. Water Resour Res 18:155–167

Chapman BM, James RO, Jung RF, Washington HG (1982) Modelling the transport of reacting chemical contaminants in natural streams. Marine Freshwater Res 33:617–628

Chen F, Dudhia J (2001) Coupling an advanced land surface-hydrology model with the Penn State-NCAR MM5 modeling system. Part I: Model implementation and sensitivity. Monthly Weather Rev 129:569–585

Cheng JH, Zhang HJ, Wang W, Zhang YY, Chen YZ (2011) Changes in preferential flow path distribution and its affecting factors in southwest China. Soil Sci 176:652–660

Cheng Y, Hubbard CG, Li L, Bouskill N, Molins S, Zheng L, Sonnenthal E, Conrad ME, Engelbrektson A, Coates JD, Ajo-Franklin JB (2016) Reactive transport model of sulfur cycling as impacted by perchlorate and nitrate treatments. Environ Sci Technol 50:7010–7018

Cheng Y, Hubbard CG, Zheng L, Arora B, Li L, Karaoz U, Ajo-Franklin J, Bouskill NJ (2018) Next generation modeling of microbial souring—Parameterization through genomic information. Int Biodeterioration Biodegradation 126:189–203

Chorover J, Troch PA, Rasmussen C, Brooks PD, Pelletier JD, Breshears DD, Huxman TE, Kurc SA, Lohse KA, McIntosh JC, Meixner T (2011) How water, carbon, and energy drive critical zone evolution: The Jemez-Santa Catalina Critical Zone Observatory. Vadose Zone J 10:884–899

Clark MP, Schaefli B, Schymanski SJ, Samaniego L, Luce CH, Jackson BM, Freer JE, Arnold JR, Moore RD, Istanbulluoglu E, Ceola S (2016) Improving the theoretical underpinnings of process-based hydrologic models. Water Resour Res 52:2350–2365

Clow DW, Mast MA (2010) Mechanisms for chemostatic behavior in catchments: Implications for CO_2 consumption by mineral weathering. Chem Geol 269:40–51

Conant RT, Ryan MG, Ågren GI, Birge HE, Davidson EA, Eliasson PE, Evans SE, Frey SD, Giardina CP, Hopkins FM, Hyvönen R (2011) Temperature and soil organic matter decomposition rates—synthesis of current knowledge and a way forward. Global Change Biol 17:3392–3404

Conley DJ, Paerl HW, Howarth RW, Boesch DF, Seitzinger SP, Havens KE, Lancelot C, Likens GE (2009) Controlling eutrophication: Nitrogen and phosphorus. Science 323:1014–1015

Cosgrove BGD, Clark EP, Cui Z, Dugger AL, Feng X, Karsten LR, Khan S, Kitzmiller D, Lee HS, Liu Y, McCreight JL, Newman AJ, Oubeidillah A, Pan L, Pham C, Salas F, Sampson KM, Sood G, Wood A, Yates DN, Yu W (2016) An Overview of the National Weather Service National Water Model, San Francisco, CA, United States, AGU, Fall Meeting 2016, abstract #H42B-05

Crawford NH, Linsley RK (1966) Digital simulation in hydrology: Stanford watershed model IV Stanford University Tech Rept 39, 210 PP, 41 RE

Dale AW, Regnier P, Knab NJ, Jorgensen BB, Van Cappellen P (2008) Anaerobic oxidation of methane (AOM) in marine sediments from the Skagerrak (Denmark): II. Reaction-transport modeling. Geochim Cosmochim Acta 72:2880–2894

Davidson EA, Janssens IA (2006) Temperature sensitivity of soil carbon decomposition and feedbacks to climate change. Nature 440:165–173

Davies H, Neal C (2007) Estimating nutrient concentrations from catchment characteristics across the UK. Hydrol Earth System Sci 11:550–558

Davison J, Hwang H, Sudicky E, Lin J (2014) Development of a fully integrated water cycle model: HydroGeoSphere-Weather Research and Forecasting (HGS-WRF). AGU, Fall Meeting 2014, H33G-0915

Dentz M, Le Borgne T, Englert A, Bijeljic B (2011) Mixing, spreading and reaction in heterogeneous media: A brief review. J Contam Hydrol 120:1–17

Dingman SL (2015) Physical Hydrology. Waveland Press, Inc., Long Grove, Illinois

Donigian Jr A, Bicknell B, Imhoff J, Singh V (1995) Hydrological simulation program-Fortran (HSPF). Computer Models Watershed Hydrol:395–442

Drever JI (1989) The Geochemistry of Natural Waters. Prentice-Hall, Inc., Englewood Cliffs, NJ

Drever JI (1997) Weathering Processes. *In*: Geochemical processes, weathering and groundwater recharge in catchments. Saether OM P. dC, (eds). A.A. Balkema, p 3–19

Drever J, Clow D (1995) Weathering rates in catchments. Rev Mineral Geochem 31:463–483

Druhan JL, Steefel CI, Molins S, Williams KH, Conrad ME, DePaolo DJ (2012) Timing the onset of sulfate reduction over multiple subsurface acetate amendments by measurement and modeling of sulfur isotope fractionation. Environ Sci Technol 46:8895–8902

Druschel GK, Baker BJ, Gihring TM, Banfield JF (2004) Acid mine drainage biogeochemistry at Iron Mountain, California. Geochem Trans 5:13–32

Duffy C, Shi Y, Davis K, Slingerland R, Li L, Sullivan PL, Goddéris Y, Brantley SL (2014) designing a suite of models to explore critical zone function. Procedia Earth Planet Sci 10:7–15

Dunn RM, Mikola J, Bol R, Bardgett RD (2006) Influence of microbial activity on plant-microbial competition for organic and inorganic nitrogen. Plant Soil 289:321–334

Dwivedi D, Arora B, Steefel CI, Dafflon B, Versteeg R (2018) Hot spots and hot moments of nitrogen in a riparian corridor. Water Resour Res 54:205–222

Ejarque E, Khan S, Steniczka G, Schelker J, Kainz MJ, Battin TJ (2018) Climate-induced hydrological variation controls the transformation of dissolved organic matter in a subalpine lake. Limnol Oceanogr 63:1355–1371

Ek M, Mitchell K, Lin Y, Rogers E, Grunmann P, Koren V, Gayno G, Tarpley J (2003) Implementation of Noah land surface model advances in the National Centers for Environmental Prediction operational mesoscale Eta model. J Geophys Res: Atmospheres 108(D22):8851

Evans C, Davies TD (1998) Causes of concentration/discharge hysteresis and its potential as a tool for analysis of episode hydrochemistry. Water Resour Res 34:129–137

Falkowski P, Scholes RJ, Boyle EE, Canadell J, Canfield D, Elser J, Gruber N, Hibbard K, Högberg P, Linder S, Mackenzie FT (2000) The global carbon cycle: a test of our knowledge of earth as a system. Science 290:291–296

Fan F, Bradley RS, Rawlins MA (2014) Climate change in the northeastern US: regional climate model validation and climate change projections. Clim Dyn 43:145–161

Fan Y, Clark M, Lawrence DM, Swenson S, Band LE, Brantley SL, Brooks PD, Dietrich WE, Flores A, Grant G, Kirchner JW (2019) Hillslope Hydrology in Global Change Research and Earth System Modeling. Water Resour Res 55:1737–1772

Fang Y, Scheibe TD, Mahadevan R, Garg S, Long PE, Lovley DR (2011) Direct coupling of a genome-scale microbial in silico model and a groundwater reactive transport model. J Contam Hydrol 122:96–103

Fatichi S, Vivoni ER, Ogden FL, Ivanov VY, Mirus B, Gochis D, Downer CW, Camporese M, Davison JH, Ebel B, Jones N (2016) An overview of current applications, challenges, and future trends in distributed process-based models in hydrology. J Hydrol 537:45–60

Fierer N, Schimel JP, Holden PA (2003) Variations in microbial community composition through two soil depth profiles. Soil Biol Biochem 35:167–176

Filippelli GM (2002) The global phosphorus cycle. Rev Mineral Geochem 48:391–425

Filippelli GM (2008) The global phosphorus cycle: past, present, and future. Elements 4:89–95

Fowler D, Coyle M, Skiba U, Sutton MA, Cape JN, Reis S, Sheppard LJ, Jenkins A, Grizzetti B, Galloway JN, Vitousek P (2013) The global nitrogen cycle in the twenty-first century. Philos Trans R Soc B: BiolSci 368

Freeze RA, Harlan R (1969) Blueprint for a physically-based, digitally-simulated hydrologic response model. J Hydrol 9:237–258

Fu BP (1981) On the calculation of the evaporation from land surface. Scientia Atmospherica Sinica 5:23–31

Gaillardet J, Braud I, Hankard F, Anquetin S, Bour O, Dorfliger N, De Dreuzy JR, Galle S, Galy C, Gogo S, Gourcy L (2018) OZCAR: The French network of critical zone observatories. Vadose Zone J 17

Galloway JN, Dentener FJ, Capone DG, Boyer EW, Howarth RW, Seitzinger SP, Asner GP, Cleveland C, Green P, Holland E (2004) Nitrogen cycles: past, present, and future. Biogeochemistry 70:153–226

Gan TY, Gusev Y, Burges SJ, Nasonova O, Andréassian V, Hall A, Chahinian N, Schaake J (2006) Performance comparison of a complex physics-based land surface model and a conceptual, lumped-parameter hydrological model at the basin-scale. Large sample basin experiments for hydrological model parameterization: results of the model parameter experiment (MOPEX): IAHS Publ. 307:196–208

Gentine P, D'Odorico P, Lintner BR, Sivandran G, Salvucci G (2012) Interdependence of climate, soil, and vegetation as constrained by the Budyko curve. Geophys Res Lett 39: L19404

Goddéris Y, Francois LM, Probst A, Schott J, Moncoulon D, Labat D, Viville D (2006) Modelling weathering processes at the catchment scale: The WITCH numerical model. Geochim Cosmochim Acta 70:1128–1147

Goddéris Y, Brantley SL, Francois LM, Schott J, Pollard D, Deque M, Dury M (2013) Rates of consumption of atmospheric CO2 through the weathering of loess during the next 100 yr of climate change. Biogeosciences 10:135–148

Godsey SE, Kirchner JW, Clow DW (2009) Concentration–discharge relationships reflect chemostatic characteristics of US catchments. Hydrol Process 23:1844–1864

Godsey SE, Aas W, Clair TA, De Wit HA, Fernandez IJ, Kahl JS, Malcolm IA, Neal C, Neal M, Nelson SJ, Norton SA (2010) Generality of fractal 1/f scaling in catchment tracer time series, and its implications for catchment travel time distributions. Hydrol Process 24:1660–1671

Gooseff MN, Van Horn D, Sudman Z, McKnight DM, Welch KA, Lyons WB (2016) Stream biogeochemical and suspended sediment responses to permafrost degradation in stream banks in Taylor Valley, Antarctica. Biogeosciences 13:1723–1732

Gooseff MN, Wlostowski A, McKnight DM, Jaros C (2017) Hydrologic connectivity and implications for ecosystem processes—Lessons from naked watersheds. Geomorphology 277:63–71

Gottardi G, Venutelli M (1993) A control-volume finite-element model for two-dimensional overland flow. Adv Water Resour 16:277–284

Grimm NB, Gergel SE, McDowell WH, Boyer EW, Dent CL, Groffman P, Hart SC, Harvey J, Johnston C, Mayorga E, McClain ME (2003) Merging aquatic and terrestrial perspectives of nutrient biogeochemistry. Oecologia 137:485–501

Hammond G, Lichtner P, Mills R (2014) Evaluating the performance of parallel subsurface simulators: An illustrative example with PFLOTRAN. Water Resour Res 50:208–228

Harjung A, Ejarque E, Battin T, Butturini A, Sabater F, Stadler M, Schelker J (2018) Experimental evidence reveals impact of drought periods on dissolved organic matter quality and ecosystem metabolism in subalpine streams. Limnol Oceanogr 64:46–60

Harman CJ (2015) Time-variable transit time distributions and transport: Theory and application to storage–dependent transport of chloride in a watershed. Water Resour Res 51:1–30

Harvey J, Gooseff M (2015) River corridor science: Hydrologic exchange and ecological consequences from bedforms to basins. Water Resour Res 51:6893–6922

Hasenmueller EA, Jin L, Stinchcomb GE, Lin H, Brantley SL, Kaye JP (2015) Topographic controls on the depth distribution of soil CO_2 in a small temperate watershed. Appl Geochem 63:58–69

Hefting MM, van den Heuvel RN, Verhoeven JTA (2013) Wetlands in agricultural landscapes for nitrogen attenuation and biodiversity enhancement: Opportunities and limitations. Ecol Eng 56:5–13

Heidari P, Li L, Jin L, Williams JZ, Brantley SL (2017) A reactive transport model for Marcellus shale weathering. Geochim Cosmochim Acta 217:421–440

Helgeson HC (1971) Kinetics of mass transfer among silicates and aqueous solutions. Geochim Cosmochim Acta 35:421–469

Herndon EM, Dere AL, Sullivan PL, Norris D, Reynolds B, Brantley SL (2015) Landscape heterogeneity drives contrasting concentration-discharge relationships in shale headwater catchments. Hydrol Earth System Sci 19:3333–3347

Heuvelmans G, Muys B, Feyen J (2004) Evaluation of hydrological model parameter transferability for simulating the impact of land use on catchment hydrology. Phys Chem Earth, Parts A/B/C 29:739–747

Hijmans RJ, Cameron SE, Parra JL, Jones PG, Jarvis A (2005) Very high resolution interpolated climate surfaces for global land areas. Int J Clim 25:1965–1978

Hinsinger P, Plassard C, Tang C, Jaillard B (2003) Origins of root-mediated pH changes in the rhizosphere and their responses to environmental constraints: A review. Plant Soil 248:43–59

Hoagland B, Russo TA, Gu X, Hill L, Kaye J, Forsythe B, Brantley SL (2017) Hyporheic zone influences on concentration–discharge relationships in a headwater sandstone stream. Water Resour Res 53:4643–4667

Hooper RP, Christophersen N, Peters NE (1990) Modelling streamwater chemistry as a mixture of soilwater end-members — An application to the Panola Mountain catchment, Georgia U.S.A. J Hydrol 116:321–343

Horton RE (1936) Natural stream channel-storage. Eos, Trans Am Geophys Union 17:406–415

Hrachowitz M, Savenije HHG, Blöschl G, McDonnell JJ, Sivapalan M, Pomeroy JW, Arheimer B, Blume T, Clark MP, Ehret U, Fenicia F, Freer JE, Gelfan A, Gupta HV, Hughes DA, Hut RW, Montanari A, Pande S, Tetzlaff D, Troch PA, Uhlenbrook S, Wagener T, Winsemius HC, Woods RA, Zehe E and Cudennec C (2013) A decade of Predictions in Ungauged Basins (PUB)—a review. Hydrol Sci J 58:1198–1255

Hrachowitz M, Benettin P, van Breukelen BM, Fovet O, Howden NJK, Ruiz L, van der Velde Y, Wade AJ (2016) Transit times—the link between hydrology and water quality at the catchment scale. WIREs Water 3:629–657

Hubbard SS, Williams KH, Agarwal D, Banfield J, Beller H, Bouskill N, Brodie E, Carroll R, Dafflon B, Dwivedi D, Falco N (2018) The East River, Colorado, watershed: a mountainous community testbed for improving predictive understanding of multiscale hydrological–biogeochemical dynamics. Vadose Zone J 17:180061

Huntington TG, Balch WM, Aiken GR, Sheffield J, Luo L, Roesler CS, Camill P (2016) Climate change and dissolved organic carbon export to the Gulf of Maine. J Geophys Res: Biogeosci:121:2700–2716

IPCC (2013) Climate Change 2013: The Physical Science Basis. Contribution of Working Group I to the Fifth Assessment Report of the Intergovernmental Panel on Climate Change. In Book Climate Change 2013: The Physical Science Basis. Contribution of Working Group I to the Fifth Assessment Report of the Intergovernmental Panel on Climate Change. Cambridge, United Kingdom and New York, NY, USA

Jackson RB, Canadell J, Ehleringer JR, Mooney HA, Sala OE, Schulze ED (1996) A global analysis of root distributions for terrestrial biomes. Oecologia 108:389–411

James LD (1972) Hydrologic modeling, parameter estimation, and watershed characteristics. J Hydrol 17:283–307

Jarboe JE, Haan C (1974) Calibrating a water yield model for small ungaged watersheds. Water Resour Res 10:256–262

Jasechko S, Kirchner JW, Welker JM, McDonnell JJ (2016) Substantial proportion of global streamflow less than three months old. Nat Geosci 9:126–129

Jencso KG, McGlynn BL, Gooseff MN, Wondzell SM, Bencala KE, Marshall LA (2009) Hydrologic connectivity between landscapes and streams: Transferring reach- and plot-scale understanding to the catchment scale. Water Resour Res 45:W04428

Jencso KG, McGlynn BL, Gooseff MN, Bencala KE, Wondzell SM (2010) Hillslope hydrologic connectivity controls riparian groundwater turnover: Implications of catchment structure for riparian buffering and stream water sources. Water Resour Res 46:W10524

Jin QS, Bethke CM (2003) A new rate law describing microbial respiration. Appl Environ Microbiol 69:2340–2348

Jin Q, Bethke CM (2007) The thermodynamics and kinetics of microbial metabolism. Am J Sci 307:643–677

Jin L, Ravella R, Ketchum B, Bierman PR, Heaney P, White T, Brantley SL (2010) Mineral weathering and elemental transport during hillslope evolution at the Susquehanna/Shale Hills Critical Zone Observatory. Geochim Cosmochim Acta 74:3669–3691

Jin L, Andrews DM, Holmes GH, Lin H, Brantley SL (2011a) Opening the "black box": water chemistry reveals hydrological controls on weathering in the Susquehanna Shale Hills Critical Zone Observatory. Vadose Zone J 10:928–942

Jin L, Rother G, Cole DR, Mildner DF, Duffy CJ, Brantley SL (2011b) Characterization of deep weathering and nanoporosity development in shale—A neutron study. Am Mineral 96:498–512

Jin L, Ogrinc N, Yesavage T, Hasenmueller EA, Ma L, Sullivan PL, Kaye J, Duffy C, Brantley SL (2014) The CO_2 consumption potential during gray shale weathering: Insights from the evolution of carbon isotopes in the Susquehanna Shale Hills critical zone observatory. Geochim Cosmochim Acta 142:260–280

Johnson NM, Likens GE, Bormann FH, Fisher DW, Pierce RS (1969) A Working Model for the Variation in Stream Water Chemistry at the Hubbard Brook Experimental Forest, New Hampshire. Water Resour Res 5:1353–1363

Johnson JW, Oelkers EH, Helgeson HC (1992) {SUPCRT92}: a software package for calculating the standard molal thermodynamic properties of minerals, gases, aqueous species, and reactions from 1 to 5000 bar and 0 to 1000 °C. Computers Geosci 18:899–947

Jones JB, Fisher SG, Grimm NB (1996) A long-term perspective of dissolved organic carbon transport in Sycamore Creek, Arizona, USA. Hydrobiologia 317:183–188

Jung H, Navarre-Sitchler A (2018) Physical heterogeneity control on effective mineral dissolution rates. Geochim Cosmochim Acta 227:246–263

Kang QJ, Lichtner PC, Zhang DX (2006) Lattice Boltzmann pore-scale model for multicomponent reactive transport in porous media. J Geophys Res-Solid Earth 111

Kemanian AR, Stöckle CO (2010) C-Farm: a simple model to evaluate the carbon balance of soil profiles. Eur J Agron 32:22–29

Kirchner JW (2009) Catchments as simple dynamical systems: Catchment characterization, rainfall-runoff modeling, and doing hydrology backward. Water Resour Res 45:W02429

Kirchner JW (2016a) Aggregation in environmental systems – Part 1: Seasonal tracer cycles quantify young water fractions, but not mean transit times, in spatially heterogeneous catchments. Hydrol Earth Syst Sci 20:279–297

Kirchner JW (2016b) Aggregation in environmental systems – Part 2: Catchment mean transit times and young water fractions under hydrologic nonstationarity. Hydrol Earth Syst Sci 20:299–328

Kobierska F, Jonas T, Kirchner JW, Bernasconi SM (2015) Linking baseflow separation and groundwater storage dynamics in an alpine basin (Dammagletscher, Switzerland). Hydrol Earth System Sci 19:3681–3693

Krause S, Tecklenburg C, Munz M, Naden E (2013) Streambed nitrogen cycling beyond the hyporheic zone: Flow controls on horizontal patterns and depth distribution of nitrate and dissolved oxygen in the upwelling groundwater of a lowland river. J Geophys Res: Biogeosciences 118:54–67

Krause S, Lewandowski J, Grimm NB, Hannah DM, Pinay G, McDonald K, Martí E, Argerich A, Pfister L, Klaus J, Battin T (2017) Ecohydrological interfaces as hot spots of ecosystem processes. Water Resour Res 53:6359–6376

Krumins V, Gehlen M, Arndt S, Van Cappellen P, Regnier P (2013) Dissolved inorganic carbon and alkalinity fluxes from coastal marine sediments: model estimates for different shelf environments and sensitivity to global change. Biogeosciences 10:371–398

Kumar M, Duffy CJ, Salvage KM (2009) A second-order accurate, Finite volume–based, Integrated Hydrologic Modeling (FIHM) framework for simulation of surface and subsurface flow. Vadose Zone J 8:873–890

Larned ST, Datry T, Arscott DB, Tockner K (2010) Emerging concepts in temporary-river ecology. Freshwater Biol 55:717–738

Lasaga AC (1998) Kinetic Theory in the Earth Sciences. Princeton University Press, Princeton

Laudon H, Buttle J, Carey SK, McDonnell J, McGuire K, Seibert J, Shanley J, Soulsby C, Tetzlaff D (2012) Cross-regional prediction of long-term trajectory of stream water DOC response to climate change. Geophys Res Lett 39: L18404

Lawrence C, Harden J, Maher K (2014) Modeling the influence of organic acids on soil weathering. Geochim Cosmochim Acta 139:487–507

Le PVV, Kumar P (2017) Interaction between ecohydrologic dynamics and microtopographic variability under climate change. Water Resour Res 53:8383–8403

Leake JR, Duran AL, Hardy KE, Johnson I, Beerling DJ, Banwart SA, Smits MM (2008) Biological weathering in soil: the role of symbiotic root-associated fungi biosensing minerals and directing photosynthate-energy into grain-scale mineral weathering. Mineral Mag 72:85–89

Lehmann J, Kleber M (2015) The contentious nature of soil organic matter. Nature 528:60–68

Levin SA (1992) The problem of pattern and scale in ecology. Ecology 73:1943–1967

Li L, Peters CA, Celia MA (2007) Effects of mineral spatial distribution on reaction rates in porous media. Water Resour Res 43:W01419

Li L, Steefel CI, Yang L (2008) Scale dependence of mineral dissolution rates within single pores and fractures. Geochim Cosmochim Acta 72:360–377

Li L, Steefel CI, Williams KH, Wilkins MJ, Hubbard SS (2009) Mineral transformation and biomass accumulation associated with uranium bioremediation at Rifle, Colorado. Environ Sci Technol 43:5429–5435

Li L, Steefel CI, Kowalsky MB, Englert A, Hubbard SS (2010) Effects of physical and geochemical heterogeneities on mineral transformation and biomass accumulation during biostimulation experiments at Rifle, Colorado. J Contam Hydrol 112:45–63

Li L, Gawande N, Kowalsky MB, Steefel CI, Hubbard SS (2011) Physicochemical heterogeneity controls on uranium bioreduction rates at the field scale. Environ Sci Technol 45:9959–9966

Li CZ, Zhang L, Wang H, Zhang YQ, Yu FL, Yan DH (2012) The transferability of hydrological models under nonstationary climatic conditions. Hydrol Earth System Sci 16:1239–1254

Li L, Bao C, Sullivan PL, Brantley S, Shi Y, Duffy C (2017a) Understanding watershed hydrogeochemistry: 2. Synchronized hydrological and geochemical processes drive stream chemostatic behavior. Water Resour Res 53:2346–2367

Li L, DiBiase RA, Del Vecchio J, Marcon V, Hoagland B, Xiao D, Wayman C, Tang Q, He Y, Silverhart P, Szink I (2018) Investigating the effect of lithology and agriculture at the Susquehanna Shale Hills Critical Zone Observatory (SSHCZO): The Garner Run and Cole Farm subcatchments. Vadose Zone J 17:180063

Li L, Maher K, Navarre-Sitchler A, Druhan J, Meile C, Lawrence C, Moore J, Perdrial J, Sullivan P, Thompson A, Jin L (2017b) Expanding the role of reactive transport models in critical zone processes. Earth Sci Rev 165:280–301

Lichtner PC (1985) Continuum model for simultaneous chemical-reactions and mass-transport in hydrothermal systems. Geochim Cosmochim Acta 49:779–800

Lichtner PC (1988) The quasi-stationary state approximation to coupled mass-transport and fluid–rock interaction in a porous-medium. Geochim Cosmochim Acta 52:143–165

Lichtner PC, Steefel CI, Oelkers EH, Sabatier UP, Suarez D, Simdnek J (1996) Reactive Transport in Porous Media. Rev Mineral Vol. 34, Min Soc Am.

Lin H (2006) Temporal stability of soil moisture spatial pattern and subsurface preferential flow pathways in the Shale Hills Catchment. Vadose Zone J 5:317–340

Lin HS, Kogelmann W, Walker C, Bruns MA (2006) Soil moisture patterns in a forested catchment: A hydropedological perspective. Geoderma 131:345–368

Liu C, Zachara JM, Qafoku NP, Wang Z (2008) Scale-dependent desorption of uranium from contaminated subsurface sediments. Water Resour Res 44:W08413

Loschko M, Wöhling T, Rudolph DL, Cirpka OA (2018) Accounting for the decreasing reaction potential of heterogeneous aquifers in a stochastic framework of aquifer-scale reactive transport. Water Resour Res 54:442–463

Luce RW, Bartlett RW, Parks GA (1972) Dissolution kinetics of magnesium silicates. Geochim Cosmochim Acta 36:35–50

Ma L, Chabaux F, Pelt E, Blaes E, Jin L, Brantley S (2010) Regolith production rates calculated with uranium-series isotopes at Susquehanna/Shale Hills Critical Zone Observatory. Earth Planet Sci Lett 297:211–225

Maavara T, Lauerwald R, Laruelle GG, Akbarzadeh Z, Bouskill NJ, Van Cappellen P, Regnier P (2018) Nitrous oxide emissions from inland waters: Are IPCC estimates too high? Global Change Biology 25: 473–448

MacQuarrie KTB, Mayer KU (2005) Reactive transport modeling in fractured rock: A state-of-the-science review. Earth Sci Rev 72:189–227

Maher K (2010) The dependence of chemical weathering rates on fluid residence time. Earth Planet Sci Lett 294:101–110

Maher K (2011) The role of fluid residence time and topographic scales in determining chemical fluxes from landscapes. Earth Planet Sci Lett 312:48–58

Maher K, Steefel CI, White AF, Stonestrom DA (2009a) The role of reaction affinity and secondary minerals in regulating chemical weathering rates at the Santa Cruz soil chronosequence, California. Geochim Cosmochim Acta 73:2804–2831

Maher K, Steefel CI, White AF, Stonestrom DA (2009b) The role of reaction affinity and secondary minerals in regulating chemical weathering rates at the Santa Cruz Soil Chronosequence, California. Geochim Cosmochim Acta 73:2804–2831

Marin-Spiotta E, Silver WL, Swanston CW, Ostertag R (2009) Soil organic matter dynamics during 80 years of reforestation of tropical pastures. Global Change Biology 15:1584–1597

Maxwell RM, Miller NL (2005) Development of a coupled land surface and groundwater model. J Hydrometeorol 6:233–247

Mayer KU, Benner SG, Frind EO, Thornton SF, Lerner DN (2001) Reactive transport modeling of processes controlling the distribution and natural attenuation of phenolic compounds in a deep sandstone aquifer. J Contam Hydrol 53:341–368

Mayer KU, Frind EO, Blowes DW (2002) Multicomponent reactive transport modeling in variably saturated porous media using a generalized formulation for kinetically controlled reactions. Water Resour Res 38:13-11-13-21

McCabe MF, Rodell M, Alsdorf DE, Miralles DG, Uijlenhoet R, Wagner W, Lucieer A, Houborg R, Verhoest NE, Franz TE, Shi J (2017) The future of Earth observation in hydrology. Hydrol Earth Syst Sci 21:3879–3914

McClain ME, Boyer EW, Dent CL, Gergel SE, Grimm NB, Groffman PM, Hart SC, Harvey JW, Johnston CA, Mayorga E, McDowell WH (2003) Biogeochemical hot spots and hot moments at the interface of terrestrial and aquatic ecosystems. Ecosystems 6:301–312

McDonnell J, Sivapalan M, Vaché K, Dunn S, Grant G, Haggerty R, Hinz C, Hooper R, Kirchner J, Roderick M (2007) Moving beyond heterogeneity and process complexity: A new vision for watershed hydrology. Water Resour Res 43:W07301

McGlynn BL, McDonnell JJ (2003a) Quantifying the relative contributions of riparian and hillslope zones to catchment runoff. Water Resour Res 39:1310

McGlynn BL, McDonnell JJ (2003b) Role of discrete landscape units in controlling catchment dissolved organic carbon dynamics. Water Resour Res 39:1090

McGuire KJ, McDonnell JJ, Weiler M, Kendall C, McGlynn BL, Welker JM, Seibert J (2005) The role of topography on catchment-scale water residence time. Water Resour Res 41

McGuire KJ, McDonnell JJ (2006) A review and evaluation of catchment transit time modeling. J Hydrol 330:543–563

McNamara JP, Tetzlaff D, Bishop K, Soulsby C, Seyfried M, Peters NE, Aulenbach BT, Hooper R (2011) Storage as a Metric of Catchment Comparison. Hydrol Process 25:3364–3371

Mei Y, Hornberger GM, Kaplan LA, Newbold JD, Aufdenkampe AK (2012) Estimation of dissolved organic carbon contribution from hillslope soils to a headwater stream. Water Resour Res 48:W09514

Miller MP, Tesoriero AJ, Capel PD, Pellerin BA, Hyer KE, Burns DA (2016) Quantifying watershed-scale groundwater loading and in-stream fate of nitrate using high-frequency water quality data. Water Resour Res 52:330–347

Milner AM, Khamis K, Battin TJ, Brittain JE, Barrand NE, Füreder L, Cauvy-Fraunié S, Gíslason GM, Jacobsen D, Hannah DM, Hodson AJ (2017) Glacier shrinkage driving global changes in downstream systems. PNAS 114:9770–9778

Moatar F, Abbott BW, Minaudo C, Curie F, Pinay G (2017) Elemental properties, hydrology, and biology interact to shape concentration-discharge curves for carbon, nutrients, sediment, and major ions. Water Resour Res 53:1270–1287

Molins S, Mayer KU (2007) Coupling between geochemical reactions and multicomponent gas and solute transport in unsaturated media: A reactive transport modeling study. Water Resour Res 43: W05435

Molins S, Trebotich D, Steefel CI, Shen C (2012) An investigation of the effect of pore scale flow on average geochemical reaction rates using direct numerical simulation. Water Resour Res 48:W03527

Molins S, Trebotich D, Yang L, Ajo-Franklin JB, Ligocki TJ, Shen C, Steefel CI (2014) Pore-scale controls on calcite dissolution rates from flow-through laboratory and numerical experiments. Environ Sci Technol 48:7453–7460

Molins S, Soulaine C, Prasianakis NI, Ladd AJC, Starchenko V, Roman S, Trebotich D, Tchelepi HA, Steefel CI (2019) Simulation of mineral dissolution at the pore scale with evolving fluid-solid interfaces: Review of approaches and benchmark problem set. HAL Id : hal-01998494, version 1

Monod J (1949) The Growth of Bacterial Cultures. Ann Rev Microbiol 3:371–394

Moore J, Lichtner PC, White AF, Brantley SL (2012) Using a reactive transport model to elucidate differences between laboratory and field dissolution rates in regolith. Geochim Cosmochim Acta 93:235–261

Moore OW, Buss HL, Dosseto A (2019) Incipient chemical weathering at bedrock fracture interfaces in a tropical critical zone, Puerto Rico. Geochim Cosmochim Acta 252:61–87

Moosdorf N, Hartmann J, Dürr HH (2010) Lithological composition of the North American continent and implications of lithological map resolution for dissolved silica flux modeling. Geochem Geophys Geosystems 11:Q11003

Mu Q, Heinsch FA, Zhao M, Running SW (2007) Development of a global evapotranspiration algorithm based on MODIS and global meteorology data. Remote Sensing Environ 111:519–536

Musolff A, Schmidt C, Selle B, Fleckenstein JH (2015) Catchment controls on solute export. Adv Water Resour 86:133–146

Musolff A, Fleckenstein JH, Rao PSC, Jawitz JW (2017) Emergent archetype patterns of coupled hydrologic and biogeochemical responses in catchments. Geophys Res Lett 44:4143–4151

Nardini A, Casolo V, Dal Borgo A, Savi T, Stenni B, Bertoncin P, Zini L, McDowell NG (2016) Rooting depth, water relations and non-structural carbohydrate dynamics in three woody angiosperms differentially affected by an extreme summer drought. Plant Cell Environ 39:618–627

Navarre-Sitchler A, Brantley S (2007) Basalt weathering across scales. Earth Planet Sci Lett 261:321–334

Navarre-Sitchler AK, Maxwell RM, Siirila ER, Hammond GE, Lichtner PC (2013) Elucidating geochemical response of shallow heterogeneous aquifers to CO_2 leakage using high-performance computing: Implications for monitoring of CO_2 sequestration. Adv Water Resour 53:45–55

Nealson KH, Rye R (2003) Evolution of Metabolism, *In:* W.H. Schlesinger, H.D. Holland, K.K. Turekian (Eds.), Treatise on Geochemistry, Volume 8, 41–68, Elsevier

Ng G-HC, Yourd AR, Johnson NW, Myrbo AE (2017) Modeling hydrologic controls on sulfur processes in sulfate-impacted wetland and stream sediments. J Geophys Res: Biogeosciences 122:2435–2457

Noguchi S, Tsuboyama Y, Sidle RC, Hosoda I (1997) Spatially distributed morphological characteristics of macropores in forest soils of Hitachi Ohta Experimental Watershed, Japan. J Forest Res 2:207–215

Nugent MA, Brantley SL, Pantano CG, Maurice PA (1998) The influence of natural mineral coatings on feldspar weathering. Nature 395:588–591

O'Donnell JA, Aiken GR, Kane ES, Jones JB (2010) Source water controls on the character and origin of dissolved organic matter in streams of the Yukon River basin, Alaska. J Geophys Res: Biogeosciences 115:G03025

Oades JM (1993) The role of biology in the formation, stabilization and degradation of soil structure. Geoderma 56:377–400

Oelkers EH (2001) General kinetic description of multioxide silicate mineral and glass dissolution. Geochim Cosmochim Acta 65:3703–3719

Oelkers EH, Schott J, Devidal J-L (1994) The effect of aluminum, pH, and chemical affinity on the rates of aluminosilicate dissolution reactions. Geochim Cosmochim Acta 58:2011–2024

Oelkers EH, Valsami-Jones E, Roncal-Herrero T (2008) Phosphate mineral reactivity: from global cycles to sustainable development. Mineral Mag 72:337–340

Oh NH, Kim HS, Richter DD (2005) What regulates soil CO_2 concentrations? A modeling approach to CO_2 diffusion in deep soil profiles. Environ Eng Sci 22:38–45

Öquist MG, Wallin M, Seibert J, Bishop K, Laudon H (2009) Dissolved inorganic carbon export across the soil/stream interface and its fate in a boreal headwater stream. Environ Sci Technol 43:7364–7369

Ortoleva P, Merino E, Moore C, Chadam J (1987) Geochemical self-organization I. Feedbacks, quantitative modeling. Am J Sci 287:979–1007

Ostle NJ, Smith P, Fisher R, Woodward FI, Fisher JB, Smith JU, Galbraith D, Levy P, Meir P, McNamara NP, Bardgett RD (2009) Integrating plant–soil interactions into global carbon cycle models. J Ecol 97:851–863

Perez-Saez J, Mande T, Larsen J, Ceperley N, Rinaldo A (2017) Classification and prediction of river network ephemerality and its relevance for waterborne disease epidemiology. Adv Water Resour 110:263–278

Philippot L, Bru D, Saby NPA, Čuhel J, Arrouays D, Šimek M, Hallin S (2009) Spatial patterns of bacterial taxa in nature reflect ecological traits of deep branches of the 16S rRNA bacterial tree. Environ Microbiol 11:3096–3104

Pike JG (1964) The estimation of annual run-off from meteorological data in a tropical climate. J Hydrol 2:116–123

Porporato A, Daly E, Rodriguez-Iturbe I (2004) Soil water balance and ecosystem response to climate change. Am Nat 164:625–632

Qiao C, Li L, Johns RT, Xu J (2015) Compositional modeling of dissolution-induced injectivity alteration during CO_2 flooding in carbonate reservoirs. SPE journal

Qu Y, Duffy CJ (2007) A semidiscrete finite volume formulation for multiprocess watershed simulation. Water Resour Res 43:W08419

Quinn P, Beven K, Chevallier P, Planchon O (1991) The prediction of hillslope flow paths for distributed hydrological modelling using digital terrain models. Hydrological processes 5:59–79

Rabalais NN, Turner RE, Wiseman WJ (2001) Hypoxia in the Gulf of Mexico. J Environ Qual 30:320–329

Rasmussen C, Heckman K, Wieder WR, Keiluweit M, Lawrence CR, Berhe AA, Blankinship JC, Crow SE, Druhan JL, Pries CE, Marin-Spiotta E (2018) Beyond clay: towards an improved set of variables for predicting soil organic matter content. Biogeochemistry 137:297–306

Raymond PA, Saiers JE (2010) Event controlled DOC export from forested watersheds. Biogeochemistry 100:197–209

Regnier P, Wollast R, Steefel CI (1997) Long-term fluxes of reactive species in macrotidal estuaries: Estimates from a fully transient, multicomponent reaction-transport model. Mar Chem 58:127–145

Regnier P, Friedlingstein P, Ciais P, Mackenzie FT, Gruber N, Janssens IA, Laruelle GG, Lauerwald R, Luyssaert S, Andersson AJ, Arndt S (2013) Anthropogenic perturbation of the carbon fluxes from land to ocean. Nature Geosci 6:597–607

Renard F, Ortoleva P, Gratier JP (1997) Pressure solution in sandstones: influence of clays and dependence on temperature and stress. Tectonophysics 280:257–266

Rhoades AM, Ullrich PA, Zarzycki CM (2018) Projecting 21st century snowpack trends in western USA mountains using variable-resolution CESM. Clim Dyn 50:261–288

Rice KC, Hornberger GM (1998) Comparison of hydrochemical tracers to estimate source contributions to peak flow in a small, forested, headwater catchment. Water Resour Res 34:1755–1766

Richards PL, Kump LR (2003) Soil pore-water distributions and the temperature feedback of weathering in soils. Geochim Cosmochim Acta 67:3803–3815

Rimstidt JD, Vaughan DJ (2003) Pyrite oxidation: a state-of-the-art assessment of the reaction mechanism. Geochim Cosmochim Acta 67:873–880

Rinaldo A, Benettin P, Harman CJ, Hrachowitz M, McGuire KJ, van der Velde Y, Bertuzzo E, Botter G (2015) Storage selection functions: A coherent framework for quantifying how catchments store and release water and solutes. Water Resour Res 51:4840–4847

Rittmann BE, McCarty PL (2001) Environmental Biotechnology: Principles and Applications. McGraw-Hill, New York

Robinson DA, Dewey KF, Richard R, Heim J (1993) Global snow cover monitoring: an update. Bull Amer Meteorol Soc 74:1689–1696

Roelandt C, Goddéris Y, Bonnet MP, Sondag F (2010) Coupled modeling of biospheric and chemical weathering processes at the continental scale. Global Biogeochem Cycles 24:GB2004

Romero-Mujalli G, Hartmann J, Börker J, Gaillardet J, Calmels D (2018) Ecosystem controlled soil-rock pCO2 and carbonate weathering – Constraints by temperature and soil water content. Chem Geol (In Press) doi:10.1016/j.chemgeo.2018.01.030

Roque-Malo S, Kumar P (2017) Patterns of change in high frequency precipitation variability over North America. Sci Rep 7:10853

Royer TV, David MB, Gentry LE (2006) Timing of riverine export of nitrate and phosphorus from agricultural watersheds in Illinois: Implications for reducing nutrient loading to the Mississippi River. Environ Sci Technol 40:4126–4131

Saberi L, McLaughlin RT, Ng GHC, La Frenierre J, Wickert AD, Baraer M, Zhi W, Li L, Mark BG (2019) Multi-scale temporal variability in meltwater contributions in a tropical glacierized watershed. Hydrol Earth Syst Sci 23:405–425

Saha D, Rau BM, Kaye JP, Montes F, Adler PR, Kemanian AR (2017) Landscape control of nitrous oxide emissions during the transition from conservation reserve program to perennial grasses for bioenergy. GCB Bioenergy 9:783–795

Salehikhoo F, Li L (2015) The role of magnesite spatial distribution patterns in determining dissolution rates: When do they matter? Geochim Cosmochim Acta 155:107–121

Salehikhoo F, Li L, Brantley SL (2013) Magnesite dissolution rates at different spatial scales: The role of mineral spatial distribution and flow velocity. Geochim Cosmochim Acta 108:91–106

Santhi C, Arnold JG, Williams JR, Dugas WA, Srinivasan R, Hauck LM (2001) Validation of the swat model on a large river basin with point and nonpoint sources. J Am Water Resour Assoc 37:1169–1188

Saunders JA, Toran LE (1995) Modeling of radionuclide and heavy metal sorption around low- and high-{pH} waste disposal sites at {Oak Ridge, Tennessee}. Appl Geochem 10:673–684

Scheibe TD, Murphy EM, Chen X, Rice AK, Carroll KC, Palmer BJ, Tartakovsky AM, Battiato I, Wood BD (2015) An analysis platform for multiscale hydrogeologic modeling with emphasis on hybrid multiscale methods. Groundwater 53:38–56

Schlesinger WH (1997) Biogeochemistry. An Analysis of Global Change. Academic Press, San Diego, London, Boston, New York, Sydney, Tokyo, Toronto

Schlesinger WH, Bernhardt ES (2013) Biogeochemistry: An Analysis of Global Change. Academic Press

Seibert J, Bishop K, Nyberg L, Rodhe A (2011) Water storage in a till catchment. I: Distributed modelling and relationship to runoff. Hydrol Process 25:3937–3949

Seibert J, Grabs T, Köhler S, Laudon H, Winterdahl M, Bishop K (2009) Linking soil- and stream-water chemistry based on a Riparian flow-concentration integration model. Hydrol Earth Syst Sci 13:2287–2297

Seitzinger S, Mayorga E, Bouwman A, Kroeze C, Beusen A, Billen G, Van Drecht G, Dumont E, Fekete B, Garnier J (2010) Global river nutrient export: A scenario analysis of past and future trends. Global Biogeochem Cycles 24: GB0A08

Semenov MY, Zimnik EA (2015) A three-component hydrograph separation based on relationship between organic and inorganic component concentrations: a case study in Eastern Siberia, Russia. Environ Earth Sci 73:611–620

Shi Y, Davis KJ, Duffy CJ, Yu X (2013) Development of a coupled land surface hydrologic model and evaluation at a critical zone observatory. J Hydrometeorol 14:1401–1420

Singh VP (1995) Computer Models of Watershed Hydrology. Water Resources Publications

Sivapalan M (2003) Prediction in ungauged basins: a grand challenge for theoretical hydrology. Hydrol Process 17:3163–3170

Sivapalan M (2017) From engineering hydrology to earth system science: milestones in the transformation of hydrologic science. Hydrol Earth Syst Sci Discuss 2017:1–48

Sivapalan M, Blöschl G, Zhang L, Vertessy R (2003) Downward approach to hydrological prediction. Hydrol Process 17:2101–2111

Smith JO, Smith P, Wattenbach M, Zaehle S, Hiederer R, Jones RJA, Montanarella L, Rounsevell MDA, Reginster I, Ewert F (2005) Projected changes in mineral soil carbon of European croplands and grasslands, 1990–2080. Global Change Biol 11:2141–2152

Smith T, Hayes K, Marshall L, McGlynn B, Jencso K (2016) Diagnostic calibration and cross-catchment transferability of a simple process-consistent hydrologic model. Hydrol Processes 30:5027–5038

Soler JM, Mader UK (2005) Interaction between hyperalkaline fluids and rocks hosting repositories for radioactive waste: Reactive transport simulations. Nucl Sci Eng 151:128–133

Soulsby C, Tetzlaff D, Rodgers P, Dunn S, Waldron S (2006) Runoff processes, stream water residence times and controlling landscape characteristics in a mesoscale catchment: An initial evaluation. J Hydrol 325:197–221

Sprocati R, Masi M, Muniruzzaman M, Rolle M (2019) Modeling electrokinetic transport and biogeochemical reactions in porous media: a multidimensional Nernst-Planck-Poisson approach with PHREEQC coupling. Adv Water Resour 127:134–147

Sprenger M, Tetzlaff D, Buttle J, Laudon H, Leistert H, Mitchell CPJ, Snelgrove J, Weiler M, Soulsby C (2018) Measuring and modeling stable isotopes of mobile and bulk soil water. Vadose Zone J 17:170149

Steefel CI, Lasaga AC (1994) A coupled model for transport of multiple chemical species and kinetic precipitation/dissolution reactions with application to reactive flow in single phase hydrothermal systems. Am J Sci 294:529–592

Steefel CI, DePaolo DJ, Lichtner PC (2005) Reactive transport modeling: An essential tool and a new research approach for the Earth Sciences. Earth Planet Sci Lett 240:539–558

Steefel CI, Appelo CA, Arora B, Jacques D, Kalbacher T, Kolditz O, Lagneau V, Lichtner PC, Mayer KU, Meeussen JC, Molins S (2015) Reactive transport codes for subsurface environmental simulation. Comput Geosci 19:445–478

Stieglitz M, Shaman J, McNamara J, Engel V, Shanley J, Kling GW (2003) An approach to understanding hydrologic connectivity on the hillslope and the implications for nutrient transport. Global Biogeochem Cycles 17:1105

Stöckle CO, Donatelli M, Nelson R, (2003) CropSyst, a cropping systems simulation model. Eur J Agron 18:289–307

Stockmann U, Adams MA, Crawford JW, Field DJ, Henakaarchchi N, Jenkins M, Minasny B, McBratney AB, De Courcelles VD, Singh K, Wheeler I (2013) The knowns, known unknowns and unknowns of sequestration of soil organic carbon. Agric Ecosystems Environ 164:80–99

Stonedahl SH, Harvey JW, Wörman A, Salehin M, Packman AI (2010) A multiscale model for integrating hyporheic exchange from ripples to meanders. Water Resour Res 46:W12539

Sullivan P, Stops MW, Macpherson GL, Li L, Hirmas DR, Dodds WK (2018) How landscape heterogeneity governs stream water concentration-discharge behavior in carbonate terrains (Konza Prairie, USA). Chem. Geol. doi: 10.1016/j.chemgeo.2018.12.002 (in press)

Sullivan PL, Hynek SA, Gu X, Singha K, White T, West N, Kim H, Clarke B, Kirby E, Duffy C, Brantley SL (2016) Oxidative dissolution under the channel leads geomorphological evolution at the Shale Hills catchment. Am J Sci 316:981–1026

Sverdrup H, Warfvinge P, Blake L, Goulding K (1995) Modelling recent and historic soil data from the Rothamsted Experimental Station, UK using SAFE. Agric Ecosystems Environ 53:161–177

Taylor LL, Leake JR, Quirk J, Hardy K, Banwart SA, Beerling DJ (2009) Biological weathering and the long-term carbon cycle: integrating mycorrhizal evolution and function into the current paradigm. Geobiology 7:171–191

Tetzlaff D, Seibert J, McGuire KJ, Laudon H, Burns DA, Dunn SM, Soulsby C (2009) How does landscape structure influence catchment transit time across different geomorphic provinces? Hydrol Process 23:945–953

Therrien R, McLaren R, Sudicky E, Panday S (2010) HydroGeoSphere: A three-dimensional numerical model describing fully-integrated subsurface and surface flow and solute transport. Groundwater Simulations Group, University of Waterloo, Waterloo, ON

Thompson SE, Basu NB, Lascurain J, Jr., Aubeneau A, Rao PSC (2011) Relative dominance of hydrologic versus biogeochemical factors on solute export across impact gradients. Water Resour Res 47:W00J05

Thorley RM, Taylor LL, Banwart SA, Leake JR, Beerling DJ (2015) The role of forest trees and their mycorrhizal fungi in carbonate rock weathering and its significance for global carbon cycling. Plant Cell Environ 38:1947–1961

Thornton PE, Doney SC, Lindsay K, Moore JK, Mahowald N, Randerson JT, Fung I, Lamarque JF, Feddema JJ, Lee YH (2009) Carbon-nitrogen interactions regulate climate-carbon cycle feedbacks: results from an atmosphere-ocean general circulation model. Biogeosciences 6:2099–2120

Thullner M, Van Cappellen P, Regnier P (2005) Modeling the impact of microbial activity on redox dynamics in porous media. Geochim Cosmochim Acta 69:5005–5019

Torres MA, West AJ, Clark KE (2015) Geomorphic regime modulates hydrologic control of chemical weathering in the Andes-Amazon. Geochim Cosmochim Acta 166:105–128

Torres MA, Moosdorf N, Hartmann J, Adkins JF, West AJ (2017) Glacial weathering, sulfide oxidation, and global carbon cycle feedbacks. PNAS 114:8716–8721

Tromp-van Meerveld HJ, McDonnell JJ (2006) Threshold relations in subsurface stormflow: 2. The fill and spill hypothesis. Water Resour Res 42:W02411

Trumbore SE (1993) Comparison of carbon dynamics in tropical and temperate soils using radiocarbon measurements. Global Biogeochem Cycles 7:275–290

Trumbore SE, Davidson EA, Barbosa de Camargo Pn, Nepstad DC, Martinelli LA (1995) Belowground cycling of carbon in forests and pastures of eastern Amazonia. Global Biogeochem Cycles 9:515–528

Tunaley C, Tetzlaff D, Soulsby C (2017) Scaling effects of riparian peatlands on stable isotopes in runoff and DOC mobilisation. J Hydrol 549:220–235

Tutolo BM, Luhmann AJ, Kong X-Z, Saar MO, Seyfried Jr WE (2015) CO$_2$ sequestration in feldspar-rich sandstone: Coupled evolution of fluid chemistry, mineral reaction rates, and hydrogeochemical properties. Geochim Cosmochim Acta 160:132–154

Van Cappellen P, Gaillard J (1996) Biogeochemical dynamics in aquatic sediments. Rev Mineral Geochem 34:335–376

Van Cappellen P, Maavara T (2016) Rivers in the Anthropocene: Global scale modifications of riverine nutrient fluxes by damming. Ecohydrol Hydrobiol 16:106–111

Van Cappellen P, Wang YF (1996) Cycling of iron and manganese in surface sediments: a general theory for the coupled transport and reaction of carbon, oxygen, nitrogen, sulfur, iron, and manganese. Am J Sci 296:197–243

van der Linden S, Woo M-k (2003) Transferability of hydrological model parameters between basins in data-sparse areas, subarctic Canada. J Hydrol 270:182–194

van der Velde Y, Torfs PJJF, Zee SEATM, Uijlenhoet R (2012) Quantifying catchment-scale mixing and its effect on time-varying travel time distributions. Water Resour Res 48:W06536

Van Genuchten MT (1980) A closed form equation for predicting the hydraulic conductivity of unsaturated soils. Soil Sci Soc Am J 44:892–898

Van Meter KJ, Basu NB, Van Cappellen P (2017) Two centuries of nitrogen dynamics: Legacy sources and sinks in the Mississippi and Susquehanna River Basins. Global Biogeochem Cycles 31:2–23

VanBriesen JM (2002) Evaluation of methods to predict bacterial yield using thermodynamics. Biodegradation 13:171–190

VanderKwaak JE, Loague K (2001) Hydrologic-response simulations for the R-5 catchment with a comprehensive physics-based model. Water Resour Res 37:999–1013

Vazquez E, Amalfitano S, Fazi S, Butturini A (2011) Dissolved organic matter composition in a fragmented Mediterranean fluvial system under severe drought conditions. Biogeochemistry 102:59–72

von Lützow M, Kögel-Knabner I, Ekschmitt K, Flessa H, Guggenberger G, Matzner E, Marschner B (2007) SOM fractionation methods: Relevance to functional pools and to stabilization mechanisms. Soil Biol Biochem 39:2183–2207

Voss BM, Peucker-Ehrenbrink B, Eglinton TI, Spencer RG, Bulygina E, Galy V, Lamborg CH, Ganguli PM, Montlucon DB, Marsh S, Gillies SL (2015) Seasonal hydrology drives rapid shifts in the flux and composition of dissolved and particulate organic carbon and major and trace ions in the Fraser River, Canada. Biogeosciences 12:5597–5618

Wang YP, Law RM, Pak B (2010) A global model of carbon, nitrogen and phosphorus cycles for the terrestrial biosphere. Biogeosciences 7:2261–2282

Weiler M, McDonnell JRJ (2006) Testing nutrient flushing hypotheses at the hillslope scale: A virtual experiment approach. J Hydrol 319:339–356

Welch LA, Allen DM (2014) Hydraulic conductivity characteristics in mountains and implications for conceptualizing bedrock groundwater flow. Hydrogeol J 22:1003–1026

Wen H, Li L (2017) An upscaled rate law for magnesite dissolution in heterogeneous porous media. Geochim Cosmochim Acta 210:289–305

Wen H, Li L (2018) An upscaled rate law for mineral dissolution in heterogeneous media: The role of time and length scales. Geochim Cosmochim Acta 235:1–20

Wen H, Li L, Crandall D, Hakala A (2016) Where lower calcite abundance creates more alteration: enhanced rock matrix diffusivity induced by preferential dissolution. Energy Fuels 30:4197–4208

Western AW, Blöschl G, Grayson RB (2001) Toward capturing hydrologically significant connectivity in spatial patterns. Water Resour Res 37:83–97

White AF (1995) Chemical weathering rates of silicate minerals in soils. Rev Mineral Geochem 31:407–461

White AF, Blum AE (1995) Effects of climate on chemical weathering in watersheds. Geochim Cosmochim Acta 59:1729–1747

White AF, Brantley SL (1995) Chemical weathering rates of silicate minerals: an overview. Rev Mineral Geochem 31:1–583

White AF, Brantley SL (2003) The effect of time on the weathering of silicate minerals: why do weathering rates differ in the laboratory and field? Chem Geol 202:479–506

White MD, Oostrom M (2000) STOMP subsurface transport over multiple phases version 2.0 theory guide (No. PNNL-12030). Pacific Northwest National Lab.(PNNL), Richland, WA (United States)

White AF, Peterson ML (1990) Role of reactive-surface-area characterization in geochemical kinetic-models. Acs Symposium Series 416:461–475

Williamson MA, Rimstidt JD (1994) The kinetics and electrochemical rate-determining step of aqueous pyrite oxidation. Geochim Cosmochim Acta 58:5443–5454

Winterdahl M, Laudon H, Lyon SW, Pers C, Bishop K (2016) Sensitivity of stream dissolved organic carbon to temperature and discharge: Implications of future climates. J Geophys Res: Biogeosci 121:126–144

Wittenberg H (1999) Baseflow recession and recharge as nonlinear storage processes. Hydrol Process 13:715–726

Wolery TJ (1992) EQ3/6: A software package for geochemical modeling of aqueous systems: Package overview and installation guide (version 7.0). Lawrence Livermore National Laboratory Livermore, CA

Woo DK, Kumar P (2016) Mean age distribution of inorganic soil-nitrogen. Water Resour Res 52:5516–5536

Woo DK, Kumar P (2017) Role of micro-topographic variability on the distribution of inorganic soil-nitrogen age in intensively managed landscape. Water Resour Res 53:8404–8422

Woo DK, Kumar P (2019) Impacts of subsurface tile drainage on age–concentration dynamics of inorganic nitrogen in soil. Water Resour Res 55:1470–1489

Xiao JH, VanBriesen JM (2006) Expanded thermodynamic model for microbial true yield prediction. Biotechnol Bioeng 93:110–121

Xiao J, VanBriesen JM (2008) Expanded thermodynamic true yield prediction model: adjustments and limitations. Biodegradation 19:99–127

Xu T, Samper J, Ayora C, Manzano M, Custodio E (1999) Modeling of non-isothermal multi-component reactive transport in field scale porous media flow systems. J Hydrol 214:144–164

Xu TF, Apps JA, Pruess K (2003) Reactive geochemical transport simulation to study mineral trapping for CO_2 disposal in deep arenaceous formations. J Geophys Res-Solid Earth 108(B2):2071

Yabusaki SB, Fang Y, Williams KH, Murray CJ, Ward AL, Dayvault RD, Waichler SR, Newcomer DR, Spane FA, Long PE (2011) Variably saturated flow and multicomponent biogeochemical reactive transport modeling of a uranium bioremediation field experiment. J Contam Hydrol 126:271–290

Yan Z, Bond-Lamberty B, Todd-Brown KE, Bailey VL, Li S, Liu C, Liu C (2018) A moisture function of soil heterotrophic respiration that incorporates microscale processes. Nat Commun 9:2562

Yeh G-T, Huang G, Cheng H-p, Zhang F, Lin H-c, Edris E, Richards D (2006) A first principle, physics based watershed model: WASH123D. *In*: Watershed Models. Singh VP, Frevert DK, (eds). CRC, Boca Raton

Yeh GT, Tripathi VS (1989) A critical evaluation of recent developments in hydrogeochemical transport models of reactive multichemical components. Water Resour Res 25:93–108

Zarnetske JP, Bouda M, Abbott BW, Saiers J, Raymond PA (2018) Generality of hydrologic transport limitation of watershed organic carbon flux across ecoregions of the United States. Geophys Res Lett 45:702–711

Zhang Q, Ball WP (2017) Improving riverine constituent concentration and flux estimation by accounting for antecedent discharge conditions. J Hydrol 547:387–402

Zhang Q, Harman CJ, Ball WP (2016a) An improved method for interpretation of riverine concentration-discharge relationships indicates long-term shifts in reservoir sediment trapping. Geophys Res Lett 43:10215–10224

Zhang Y, Slingerland R, Duffy C (2016b) Fully-coupled hydrologic processes for modeling landscape evolution. Environ Modell Software 82:89–107

Zhao G, Gao H, Cuo L (2016) Effects of urbanization and climate change on peak flows over the San Antonio River Basin, Texas. J Hydrometeorol 17:2371–2389

Zhi W, Li L, Brown W, Dong W, Kaye J, Steefel C, Williams KH (2019) Distinct Water Chemistry Shapes Contrasting Concentration - Discharge Patterns. Water Resour Res (in press). doi: 10.1029/2018WR024257

Zhu C (2005) In situ feldspar dissolution rates in an aquifer. Geochim Cosmochim Acta 69:1435–1453

Reviews in Mineralogy & Geochemistry
Vol. 85 pp. 419-457, 2019
Copyright © Mineralogical Society of America

RTM for Waste Repositories

Olivier Bildstein

Commissariat à l'Énergie Atomique et aux Énergies Alternatives (CEA)
Direction de l'Energie Nucléaire (DEN)
Cadarache
DTN, SMTA, LMTE
F-13108 Saint-Paul-Lez-Durance
France

olivier.bildstein@cea.fr

Francis Claret

Bureau de Recherches Géologiques et Minières
3 Avenue Guillemin
F-45060, Orléans Cedex 2
France

francis.claret@brgm.fr

Pierre Frugier

Commissariat à l'Énergie Atomique et aux Énergies Alternatives (CEA)
Direction de l'Energie Nucléaire (DEN)
Marcoule
DTCD, SECM, LCLT
F-30207 Bagnols-sur-Cèze Cedex
France

pierre.frugier@cea.fr

INTRODUCTION

A need for numerical simulations

Power generation plays an important role in global warming (Audoly et al. 2018) and although the use of nuclear power in the energetic mix can be seen as sustainable (Brook et al. 2014; Knapp and Pevec 2018), nuclear energy is debated in many countries (Meserve 2004; Cici et al. 2012). Over 50 years of nuclear energy and the use of radioactive material in nuclear research and in industrial, medical and other applications have left a legacy of different kinds of nuclear waste awaiting final disposal worldwide. To ensure very long-term isolation to protect the environment and ensure the safety of the future generations (Linsley and Fattah 1994; Hummel and Schneider 2005), geological disposal facilities ('repositories') are considered to be the most suitable solution. When the activity in waste is relatively low and half-lives are less than about 30 a, i.e. for intermediate- and low-level waste, near-surface disposal is often considered to be adequate. For waste with a higher activity and/or longer life, i.e. for high-level long-lived waste (HLLW), the concept of geological disposal of radioactive waste emerged in the late 1950s (Hess 1957; De Marsily et al. 1977; Apted and Ahn 2010; Chapman and Hooper 2012). Since then, many international research and development programs have been launched to study deep argillaceous formations, granitic rocks and salt formations as potential host rocks for radioactive waste disposal (Landais and Aranyossy 2011)(Fig. 1).

1529-6466/19/0085-0014$05.00 (print)
1943-2666/19/0085-0014$05.00 (online)

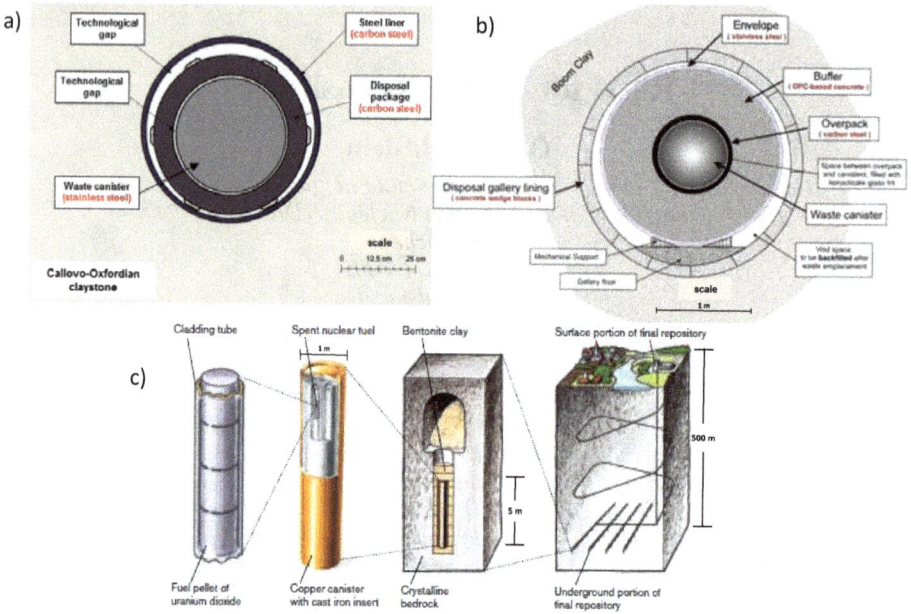

Figure 1. Schematic view of repository concepts/disposal cells in clay host rock **a)** without concrete (modified from Andra 2005), **b)** with concrete Supercontainer (modified from Ondraf/Niras 2009), and **c)** in granitic host-rock (modified from SKB 2006).

Based on their outcomes, there is a broad technical consensus that geologic disposal will meet the safety requirement to minimize safety and environmental impacts, now and far into the future (Grambow and Bretesché 2014), meaning a hundred thousand to million years into the future. Demonstrating safety over time comparable to geological timescales relies on a rigorous, complex and iterative scientific approach referred to as "long-term behavior science" that is based upon three pillars: experiments, modeling, and natural/archaeological analogs (Poinssot and Gin 2012; Dillmann et al. 2014; Alexander et al. 2015; Martin et al. 2016). Today, numerical models are ubiquitous tools that can be used either for long-term predictive evaluation of solute transport or design optimization of nuclear waste geologic repositories (Tsang et al. 1994; Johnson et al. 2017). They are also used to make predictive multi-physical assessments within a timeframe and space scale larger than experiments can cover (Bredehoeft 2003; Bildstein and Claret 2015). These numerical simulations require integrating, in a consistent framework, an increasing amount of scientific knowledge (Geckeis and Rabung 2008) acquired for each of the individual components of such repositories. This implies considering couplings of different non-linear processes, applied to a wide range of materials with contrasting properties as a function of time and space in ever-larger systems. Reactive transport modeling (Yeh and Tripathi 1989) has established itself as a powerful, versatile and essential tool for tackling the complexity of the multi-space and temporal-scale issues engineers and scientists face. This includes describing the evolution of the components constituting the so-called "multiple-barrier system" between the waste matrix and the biosphere (Apted and Ahn 2010; Chapman and Hooper 2012). Predicting how (i) the waste matrices (e.g., glass, bitumen, cement), (ii) waste overpacks (e.g., metal canisters, concrete), (iii) engineered barriers such as bentonite (Sellin and Leupin 2013), (iv) and natural geological barriers will evolve with time in response to physical and chemical perturbations is of prime importance for performance and safety evaluations of the repository concepts. This chapter summarizes recent improvements and discusses future challenges in the application of reactive transport to deep geological nuclear waste disposal focusing on disposal in clay formations.

A need to consider coupled processes

The long-term safety of geological disposal is based on the multi-barrier concept consisting of a combination of natural (host rocks) and engineered (waste form, package, backfill and seal materials) barriers that will have mutual interactions during the lifetime of the repository. Predicting the effects of these interactions entails understanding, evaluating and prioritizing the pertinent thermal, hydraulic, mechanical, chemical and radiological (THMCR) processes (Fig. 2).

Thermal effects arise principally from heat generated by waste that is both dissipated in the over-packs and the geological medium, and evacuated by air ventilation (Benet et al. 2014a).The hydraulic stage and gas-related effects are a combination of repository resaturation and of gas generation. (e.g., H_2 production by anaerobic corrosion of metals or through water radiolysis). The hydraulic transient is also impacted upon by the ventilation systems used during the operation period to ensure adequate gallery aeration for the workers and for the infrastructure. It will create heat and vapor exchanges with walls and storage packages, pressure variations and evaporation, and sometimes vapor condensation (Benet et al. 2014b). Mechanical effects can be induced during construction and excavation of underground drifts that will cause both tunnel convergence (Lisjak et al. 2015) and damage to the rock in the vicinity of the opening, with the formation of an associated excavated disturbed zone (EDZ) (Armand et al. 2014). There are also numerous chemical effects as foreign materials like borosilicate glasses (Poinssot et al. 2010; Gin et al. 2015), metallic canisters (King and Shoesmith 2010), and concrete (Alonso et al. 2010) that are introduced into the repository will induce chemical gradients across the repository components (Nagra 2002; Andra 2005, 2009). Because of these chemical gradients, perturbations such as pH and redox changes may alter the performance of the barriers over time

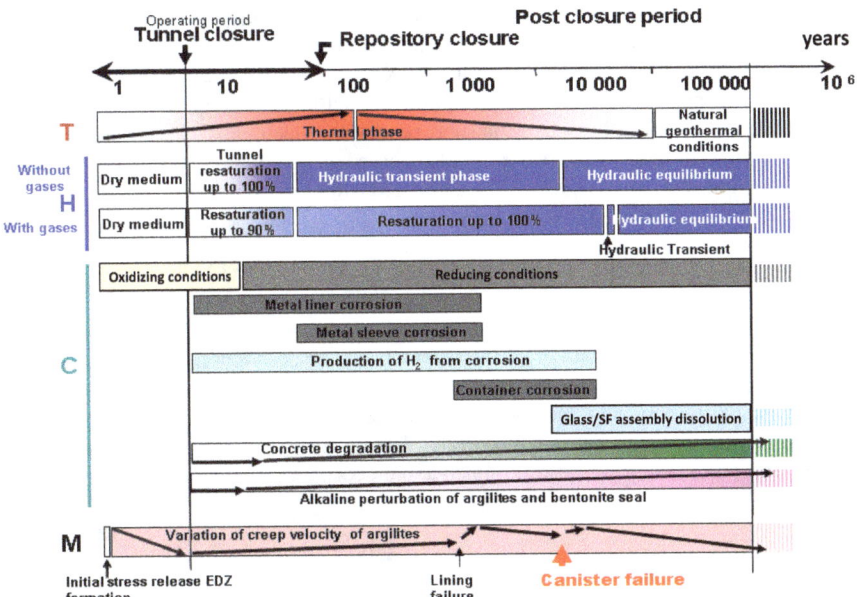

Figure 2. Phenomenological evolution of the repository in high-level long-lived waste (HLLW) and intermediate-level long-lived waste (ILLW) quarters showing timescale of THMC processes (modified from Andra 2005).

(Bildstein and Claret 2015). Last but not least, radiological effects also exist in association with the radiological inventory of waste independently from the technological solution, retreatment or direct disposal of irradiated fuels (Odorowski et al. 2017) or after reprocessing (Gong et al. 1999; Berner et al. 2013). Moreover, an additional challenge arises because all of the THMC phenomena described above are coupled. The coupling of physical and chemical phenomena can be either weak or strong and may vary in time and space scales. As an illustration, excavation induces a mechanical rock effect producing an EDZ. The change in rock microstructure in the EDZ compared to pristine rock might impact radionuclide (RN) transport parameters if this damage zone acts as a preferential pathway for fluid flow and RN transport. However, issues are more complicated as the EDZ can heal and seal itself depending on hydraulic conditions (Thatcher et al. 2016). Another example of a coupled process is the evolution of redox conditions, a key parameter for RN solubility (Duro et al. 2014) which can be modified in connection with hydraulic and mechanical process. The oxidation event that will also occur in the repository near field (Matray et al. 2007; Craen et al. 2008; Vinsot et al. 2013; De Windt et al. 2014; Vinsot et al. 2014) as result of excavation and ventilation will change redox conditions. These conditions could also been modified if radiological effects are considered. The radiolysis effects of water under alpha irradiation simultaneously producing oxidizing species (e.g., hydrogen peroxide), and reducing species like hydrogen might play a role when competing with redox active species from the ambient environment (Odorowski et al. 2017). Many other examples of features and processes and their relevance in performance assessment and safety cases are given by (Bernier et al. 2017), including microbiological effects. These combined processes may be detrimental or favorable to the overall performance of a repository system over time. Since the combinations will occur on time-scales that are not accessible to experimentation, we must develop modeling approaches that can help to predict how barriers will evolve in time and space (Claret et al. 2018b). Reactive transport modeling is probably one of the most efficient techniques used to account for and quantify the complexity of coupling processes over a long period of time. Many codes exist (see Steefel et al. 2015b for a recent review) and to our knowledge, none of them are able to integrate all coupled THMCR effects. However, since the 34th volume of *Reviews in Mineralogy* (dedicated to reactive transport in porous media) was published about 20 years ago, numerical codes have continuously improved and are capable of representing more and more complex situations. Active topics of research dealing with RTM include the development of pore scale and hybrid, or multiple continua, models to capture the scale dependence of coupled reactive transport processes (Steefel et al. 2005). In this chapter, most of the reviewed studies that are described are based on continuum models (Lichtner 1996), although other kinds of approaches will be addressed at the end.

REACTIVE TRANSPORT MODELING

Governing equations

The reactive transport constitutive equations rely on the description of the porous medium of interest (e.g., one of the component of the multi-barrier system) at the continuum scale, with respect to its macroscopically measurable properties such as permeability, dispersivity and diffusability (Steefel et al. 2014, 2015b). In this framework, a generic equation for advective/ dispersive transport in the liquid phase coupled to bio-geochemical reactions can be written:

$$\frac{\partial(\phi S_L \Psi_i)}{\partial t} = \nabla \left(\phi S_L D_i^* \nabla \Psi_i \right) - \nabla \left(q \Psi_i \right) - \sum_{r=1}^{Nr} v_{ir} R_r - \sum_{m=1}^{Nm} v_{im} R_m - \sum_{s=1}^{Ns} v_{is} R_s - \sum_{l=1}^{Nl} v_{il} R_l \quad (1)$$

$$\Psi_i = C_i + \sum_{j=1}^{N_x} v_{ji} \cdot C_j \quad (2)$$

where ϕ is the porosity ($m^3_{void} \cdot m^{-3}_{medium}$), S_L is the liquid saturation (unitless), $\emptyset_i$ ($mol \cdot m^{-3}_{water}$) is the total concentration term that integrates the distribution between primary species (with concentration C_i) and secondary species ($\sum_{j=1}^{N_x} v_{ji} \cdot C_j$), D_i^* is the diffusion coefficient for species i in the porous media ($m^2 \cdot s^{-1}$), q is the volumetric (or Darcy flux) of water ($m^3_{water} \cdot m^{-2}_{medium} \cdot s^{-1}$), and R_r, R_m, R_s, and R_l are the aqueous phase, mineral, surface, and gas reactions ($mol \cdot m^{-3}_{medium} \cdot s^{-1}$) respectively. v_{ji} is the number of moles of component i in one mole of secondary species j. v_{ik} with $k = r$, m or l is the number of moles of component i in one mole of phase k.

These equations can be expanded to consider (i) transport in phases other than the liquid (e.g., gases), (ii) a dispersion term, or (iii) the Nernst–Planck Equation, instead of Fick's Equation, to model diffusional transport. A comprehensive description of these equations can be found in Steefel et al. (2015b). Recently, the significance of the parameters in Equations (1) and (2), their relationship in case of mutual dependence (e.g., diffusion and saturation permeability and saturation) and the way they can be approximated using different empirical or based on theory laws have been reviewed (Claret et al. 2018b). Therefore, it will not be detailed again here. Equation (1) also highlights the fact that a porosity has to be defined for the considered materials. However, some of them (e.g., steel and other metals, glass) are not porous materials, and, unfortunately, reactive transport models are, by definition and according to Equation (1), not suited to include non-porous media (Bildstein et al. 2007; Claret et al. 2018b). To unravel this problem, the volume clearance that exists at the interface between the canister or the glass and the surrounding material, also known as the "technological gap", is often included in the numerical cells that represent these materials.

The critical need for thermodynamic databases

Inherently, to model reactivity and couple it with transport, geochemical databases are needed. To illustrate this point, let us consider the transport of contaminants in a repository and its relationship with geochemistry. Speciation of the considered element will be a key parameter controlling the species distribution among the different phases and their solubility limits. While the solute species can diffuse with the porewater through the barriers, their precipitation represents an appreciable retardation mechanism (Wanner 2007). Speciation also controls sorption processes that can induce retardation mechanism for transport (Grenthe 1991; Grambow 2008). Nonetheless, it is not only a question of speciation. Transport is also governed by porosity change, and this is why thermodynamic databases should also account for minerals that constitute the barriers and secondary minerals that are able to precipitate when different barriers are in contact. In the field of nuclear energy, database comprehensiveness and consistency are key issues that have been initiated by Nuclear Energy Agencies and later prioritized at the national level (see Ragoussi and Costa 2019 for a review). In addition, the pivotal importance of databases for RTM calculation has recently been the focus of a special issue of *Applied Geochemistry* (Kulik et al. 2015). Within the framework of national radioactive waste disposal projects, database development has focused on consistent data for actinides, lanthanides, and chemotoxics, among other potential contaminants. When modeling disposal conditions, there is also a need to describe the evolution of the materials constituting the barriers. This research effort combines new thermodynamic data acquisition, especially for clay minerals (Gailhanou et al. 2007, 2009, 2012, 2013, 2017; Blanc et al. 2015a) and cement phases (Roosz et al. 2018), critical data selection for repository materials such as cement phases (Blanc et al. 2010; Walker et al. 2016; Lothenbach et al. 2019) and how to evaluate their capacity to describe interactions between materials (Blanc et al. 2015a). In addition, kinetic databases with kinetic rate parameters are being built (Marty et al. 2015a), as well as a sorption database (Brendler et al. 2003). An important constraint for kinetic rate parameters is consistency with the associated thermodynamic database for stoichiometry and solubility values, so that we can calculate how rates depend on saturation states.

Modeling materials and reactive interfaces

Studies on deep geological storage show that most of the physical and chemical reactivity will be concentrated close to the interface between the different materials. A first challenge therefore consists of predicting the extent of the alteration and the distance at which the perturbations will progress into the barriers and, in case of failure, into the waste matrices. Another challenge is to predict the evolution of the material properties in these barriers resulting from their alteration, and to assess the impact on the overall performance and safety of the system.

A non-exhaustive list of interfaces is given in Table 1 with a description of their occurrence in different disposal concepts including clay formations (e.g., in Belgium, France, Spain, and Switzerland) and crystalline host rocks (note that bentonite is used as a barrier between the containers and granite in this case, e.g., in Canada, Spain, and Switzerland). Another degree of complexity is added in HLLW repository concepts from Sweden and Finland where the use of copper containers is envisaged, as well as in France where a bentonitic cement may be emplaced at the extrados of the steel liner to avoid early corrosion in acidic conditions.

Concerning waste matrices, only the specific behavior of vitrified waste (glass) will be detailed in this chapter. For the alteration of spent fuel (SF), the authors refer the reader to the recent review of these aspects by (Ewing 2015). Waste matrices for ILLW *per se* are not detailed in this chapter, but most of the conclusions will be applicable to cementitious and metallic waste types. For bituminous and organic waste, the authors refer the reader to the modeling work of Sercombe et al. (2006), von Schenck and Källström (2014), and De Windt et al. (2015).

The direct interaction of materials with groundwater is not treated specifically in this chapter, even though the authors acknowledge that many interesting studies related to waste disposal in fractured hard rocks, where this issue is particularly relevant, have been conducted. In the disposal concepts that rely upon a chemical buffer of high pH (mainly ILLW), the issue of the interaction of groundwater with cement/concrete has been addressed (see (Wilson et al. 2018) for a recent review). RTM studies of cement-groundwater interactions have been reported with results concerning pH evolution, and chemical and mechanical degradation of concrete barriers (Höglund 2001, 2013; Bamforth et al. 2012; Cronstrand 2014; Wilson et al. 2018). In HLW disposal concepts, the bentonite buffer surrounding waste packages will evolve with time due to progressive reaction with saturating ambient groundwater, including ion exchange, dissolution-precipitation of accessory (gypsum, halite quartz, calcite) and clay minerals potentially affecting properties such as the hydraulic conductivity and the swelling pressure. RTM studies regarding these aspects have evolved from simple exchange models to more complex 1- and 2-D problem setup, including the effect of advection in a fracture intersecting the disposal cell (e.g., Arcos et al. 2003, 2008; Sena et al. 2010; Benbow et al. 2019).

Perturbations from atmospheric O_2 and CO_2 during the construction and ventilation stages of the repository life constitute a particular type of interface, a physical contact with air, where the aggressive agent is a gaseous component instead of interstitial water and minerals.

The modeling of the behavior of each type of material and the relevant interfaces (Table 1) will be detailed in this chapter with a particular emphasis on the evolution of the RTM approaches and achievements over the past two decades.

MATERIALS IN PHYSICAL CONTACT WITH AIR

This "interface" is characterized by the direct physical contact between materials and air containing reactive gases (essentially O_2 and CO_2). In this situation, porous media are generally partially saturated with water and therefore fast diffusional transport of gas (four orders of magnitude faster than in water) can possibly occur into the pore network. The water saturation

Table 1. Types of material interfaces in different repository concepts (SF, bituminous, and cementitious waste are not detailed here).

Material interface	HLLW in clay formation	HLLW in crystalline formation	ILLW in clay formation
Clay–atmospheric O_2	Construction phase (disposal cell)	Bentonite handling and emplacement	Construction phase (disposal cell)
Concrete–atmospheric CO_2	Construction and ventilation phase (wells, tunnel)	Construction and ventilation phase (tunnel)	Construction and ventilation phase (wells, tunnel, disposal cell, container)
Iron–clay	Corrosion phase (metallic structure components, container)	Corrosion in bentonite after failure of copper container	—
SF–clay	SF alteration after failure of container	SF alteration in bentonite after failure of copper container	—
Glass–iron	Matrix alteration after water intrusion	—	—
Glass–(iron)–clay	Matrix alteration after container corrosion	Matrix alteration after container corrosion	—
Iron–concrete	Metallic structure corrosion (tunnel) and overpack corrosion in supercontainer (Belgian concept)	—	Container and metallic structure corrosion
Concrete–clay	Repository lifetime (wells, tunnel, disposal cell) and supercontainer in Belgian concept	—	Repository lifetime (wells, tunnel, disposal cell, container)
Glass–concrete–(clay)	Matrix alteration after container corrosion in Belgian supercontainer concept	—	—
Bituminous, cementitious, metallic waste–concrete–(clay)	—	—	Matrix alteration after container corrosion

conditions will change with time resulting from the resaturation process with water coming from the surrounding host-rock and invading the tunnels, shafts, and wells. Resaturation will be initially impeded by ventilation during the operation phase and then by the generation of gas that may accumulate in the disposal cells (mainly H_2) after closure. These two stages will produce very different geochemical environments, with contrasting reactivity, during which oxic vs. anoxic redox conditions will prevail. Long timespans with oxic conditions are to be avoided to limit the corrosion rate and to make sure reduced conditions are settled when radionuclides are released (because their oxidized forms tend to be more mobile). Calculating the duration of this stage is therefore important, but it is not thought to be very long once the disposal cell is closed (tens to thousands of years) due to the numerous oxygen consuming reactions that will occur in the vicinity of the disposal cell such as iron corrosion, dissolution of reducing mineral, and microbial respiration (see following sections).

The duration of the resaturation stage is difficult to predict because it strongly depends on the relative importance of the two counteracting processes (Marschall et al. 2005). It has been studied by conducting numerical simulations at disposal cell scale where the focus was to predict the transport and fate of hydrogen. The resaturation time calculated in the near field for different repository concepts varies in a very broad range from hundreds to hundreds of thousand years. It was found to be strongly linked to the diffusivity of dissolved hydrogen and the H_2 production rate in the HLLW cells (Xu et al. 2008; Poller et al. 2011; Senger et al. 2011; Enssle et al. 2014; Brommundt et al. 2014; Sedighi et al. 2015) and was even longer in the ILLW cells where the degradation of waste matrices is also susceptible to produce H_2 (Talandier et al. 2006; Poller et al. 2011; Avis et al. 2013). Note that flow and reactive transport in unsaturated conditions have also been extensively studied and modeled, though in a very different geological context and repository concept, in the framework of the Yucca Mountain project (USA); in this concept, the HLLW repository was envisaged in a fractured volcanic tuff formation at a depth approximately 300 m above the water table (e.g., Glassley et al. 2003; Spycher et al. 2003).

Corrosion of carbon steel components in oxic conditions

In repository conditions, metallic iron is not thermodynamically stable in the presence of water, and the corrosion of metallic iron occurs both in oxic (operation phase) and reduced conditions (post-closure period). In the first case, corrosion leads to the release of ferric iron and hydroxide in solution (Reaction 3), whereas in the second case corrosion produces aqueous ferrous iron, hydrogen and hydroxide (Reaction 4).

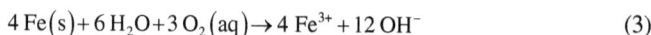

$$4\,Fe(s) + 6\,H_2O + 3\,O_2(aq) \rightarrow 4\,Fe^{3+} + 12\,OH^- \qquad (3)$$

This reaction leads either to a pH increase with an increase in redox potential or to a pH increase with a decrease in redox potential. While anoxic corrosion is often taken into account in RTM studies (see dedicated section below), oxic corrosion is seldom tackled, despite a much higher corrosion rate (Féron et al. 2008; Johnson and King 2008) and the development of more aggressive mechanisms such as pitting or crevice corrosion (Barnichon et al. 2018). This phenomenon is also observed in experiments with sequential aerobic–anaerobic conditions (Sherar et al. 2011; El Hajj et al. 2013).

There are few modeling studies dealing with corrosion during the oxic phase in the context of a disposal cell. The first attempts were dedicated to the assessment of the evolution of O_2 and H_2 content during the ventilation period using a 2D domain containing the container-EDZ-argillites system (Hoch and Wendling 2011). They used a TH(C) model where the only reaction is iron corrosion, and the model uses a water saturation-dependent formulation for the corrosion rate in oxic (consuming O_2) and anoxic (producing H_2) conditions. The results showed that oxygen diffusing from the tunnel is rapidly consumed by corrosion (with a rate ~100 μm/y) and the propagation front was limited to ~10 m (in a 40 m disposal cell). With the production of H_2 on the other side (corrosion rate of 10 μm/y), the mole fraction increased to 0.3 at the end of the cell, but the fraction of both gases reached only ~0.1 where they mixed.

In their study, De Windt et al. (2014) added the kinetics of pyrite oxidative dissolution as an oxygen consuming reaction. They first checked the robustness of their RTM approach to reproduce an in situ experiment conducted in a borehole in the Tournemire URL (France). The predicted mineralogical evolution (goethite/magnetite) qualitatively matched the *in situ* experiment. Applying the model to the repository scale, they show that the gas diffusion coefficient in the partially saturated zones plays a major role in the location of the oxidizing/reducing front inside the engineered barriers with oxic corrosion of steel on one side and anoxic steel corrosion accompanied by hydrogen production on the other side. A similar study was performed by (Bond et al. 2013), matching the sulfate concentration from pyrite dissolution in an *in situ* experiment in the Mont-Terri underground research laboratory (Switzerland).

Recent studies tend to integrate more phenomenological models to determine the corrosion rate in oxic and anoxic conditions. The oxic corrosion model takes into account the thickness of the water film on the iron surface, and the diffusion of O_2 in a goethite/lepidocrocite layer as the rate-limiting step (Hoerlé et al. 2004). The onset of corrosion happens when a critical amount of water condenses on the iron surface. The results of calculations at the disposal cell scale show in particular that corrosion concentrates in locations where the relative humidity is highest, especially where water "leaks" from fractures, with corrosion rates up to 200 µm/y locally.

Oxic transient: impact on claystone

Argillaceous rocks react with meteoric carbon dioxide (CO_2) and oxygen (O_2) from the atmosphere and nested chemical reaction fronts form in the subsurface in response to acid–base and redox reactions (Brantley et al. 2013). In a geological repository, similar phenomena will occur in the anaerobic host rock during construction (excavation, drilling operations) and operations (gallery ventilation). Under these conditions, the prevailing reducing conditions will be perturbed and redox-sensitive minerals (e.g., Fe-bearing minerals) will react. The expected chemical changes have been recently reviewed (Bildstein and Claret 2015) and the main weathering effects to be expected are recalled here. In claystone, oxidation mainly affects pyrite, which is ubiquitously found in the mineral assemblage. As it oxidizes with exposure to oxygen and water, pyrite releases sulfate and protons. The overall reaction under oxic conditions is expressed as:

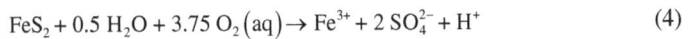

$$FeS_2 + 0.5\ H_2O + 3.75\ O_2(aq) \rightarrow Fe^{3+} + 2\ SO_4^{2-} + H^+ \tag{4}$$

While the release of ferric ions causes the precipitation of iron (oxy)hydroxides, carbonates will play a major role in the buffering of proton release concomitantly affecting the population of the clay exchanger. At the laboratory scale, flow-through oxidation experiments conducted on COx tailings can be reproduced using RTM studies (Claret et al. 2018a). Recently an *in situ* experiment (OXITRAN) has been set up in the Tournemire underground research laboratory in which the time evolution of oxygen partial pressure in a measurement chamber isolated from the atmosphere has been recorded (Barnichon et al. 2018). Although pyrite is present and can buffer an oxygen plume while CO_2 is produced (Vinsot et al. 2017), oxygen was never completely depleted in the test chambers. The key controlling parameters are the thickness of the Excavated Disturbed Zone and the ratio between the effective diffusion coefficient and the oxygen consumption first-order effective rate. In contrast, the 'Full-Scale Emplacement' (FE) experiment that was initiated at the Mont Terri rock laboratory aims to simulate, as realistically as possible, the construction, waste emplacement, backfilling and early post-closure evolution of a spent fuel/vitrified high-level waste disposal tunnel according to the Swiss repository concept. First results indicate rapid oxygen consumption at locations not affected by O_2 inflow from the access tunnel (Müller et al. 2018). In addition to oxygen diffusion, drilling is also associated with desaturation and water evaporation, leading to increased salt concentration in the pore water (Zheng et al. 2008; Lerouge et al. 2013). A multiphase flow and reactive transport model of a ventilation experiment performed on Opalinus Clay indicates that changes in the clay mineral porosity caused by oxidation and the associated mineral dissolution/precipitation may seem weak (Zheng et al. 2008). The pore water seeping into the drifts and the gallery (e.g., after closure of the repository) will however interact with the oxidation products and the salts inherited by water evaporation. This more saline water will interact first with the repository materials.

Atmospheric concrete carbonation

This interface is relevant for concrete waste packages and cell structures (in most ILLW and the Belgian HLLW concepts) and for concrete components found in parts of other HLLW repositories (tunnels). Studies concerning this interface have benefited from the work that has been conducted for the atmospheric carbonation of buildings and civil engineering structures (see Ashraf 2016, for a review). In the carbonation process, the main chemical reaction is the acid attack caused by CO_2 dissolving in water and triggering the dissolution of portlandite and the initial calcium-silicate-hydrate (C-S-H) phases to form calcium carbonate minerals (calcite, aragonite, vaterite) (Reaction 5), and new C-S-H phases with progressively decreasing Ca/Si ratio (amorphous silica being the ultimate alteration product, Reaction 6). Sulfate-bearing minerals are also affected by carbonation with the initial ettringite dissolving to form basanite ($CaSO_4 \cdot 0.5H_2O$) or gypsum ($CaSO_4 \cdot H_2O$), depending on the relative humidity.

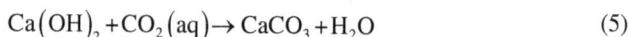

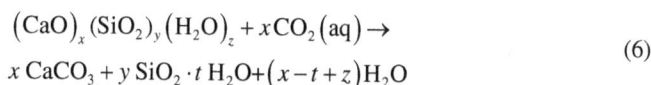

$$Ca(OH)_2 + CO_2(aq) \rightarrow CaCO_3 + H_2O \tag{5}$$

$$(CaO)_x(SiO_2)_y(H_2O)_z + xCO_2(aq) \rightarrow \\ x\,CaCO_3 + y\,SiO_2 \cdot t\,H_2O + (x - t + z)H_2O \tag{6}$$

The models taking into account the coupling of CO_2 diffusion in the gas phase with the drying process are of particular interest for atmospheric carbonation at the disposal cell scale. During the ventilation period, air will be forced into the tunnels at a constant, low relative humidity (~40%) at a temperature close to 25 °C (or may be slightly heated, up to 40 °C, in preliminary concepts from Andra 2005). The drying process is mainly controlled by water flow out of the concrete, the driving force being water evaporation at the surface. It can be described via full multiphase RT models or alternatively with Richards' Equation (depending on the concrete properties).

Most of the experiments conducted up to 2005 focused on carbonation "kinetics," looking for the location of the carbonation front (identifying the pH front at ~9 with the phenolphthalein test) without providing quantitative mineralogical information. These experiments were modeled with simplified chemical models (e.g., Bary and Sellier 2004; Burkan Isgor and Razaqpur 2004). Only recently have experimental studies looked more closely at mineralogical changes, starting with portlandite and calcite, then adding aragonite and vaterite (Drouet 2010; Drouet et al. 2018), and ettringite, gypsum, and basanite (Auroy et al. 2018). The first modeling work actually using RTM was proposed by (Bary and Mügler 2006) with the objective of matching the portlandite and calcite profiles observed in experiments. They used a full multiphase flow code and developed a shrinking core model for the dissolution of portlandite and precipitation of calcite with a diffusion-limited rate. They also took into account the changes in porosity and the effect on permeability (through Kozeny–Carman relationship). The model only solves for the mass conservation of water, calcium and carbonate without full chemical treatment. Modeling results reproduce the observed carbonation front and a decrease in porosity for a series of experiments for three cement types at three different liquid saturations (0.65, 0.80, and 0.95). This model was then extended to account for the complete chemistry by Leterrier and Bary (2011), but was only calibrated on the carbonation front for the same experiment. A similar modeling approach was used by (Park 2008), although without the shrinking core model.

RTM studies of concrete carbonation at the disposal cell scale are scarce, especially those covering typical ventilation times (~100 years) and coupling carbonation with concrete drying (Trotignon et al. 2011; Thouvenot et al. 2013). In these two studies, the geometries of the waste package and the tunnel structure were simplified and handled as a 1D problem. The evolution of the complete mineralogy of concrete is investigated (see dedicated section below) and the problem is treated with full multiphase flow and transport using TOUGHREACT. It includes the feedback of porosity and phase saturation on diffusion using the Millington-Quirk relationship (with specific parameters for concrete) and a slipping factor to account for the larger gas-intrinsic permeability compared to water (Thiery et al. 2007; Zhang et al. 2015). Results show that the carbonation of concrete develops over 1 to 10 cm/100 years depending on the concrete properties and that, in the carbonated zone, the primary minerals dissolve all the way to the "ultimate" secondary minerals (calcite, amorphous silica, gypsum, gibbsite, ferric hydroxide), with no intermediate minerals (e.g., C-S-H). In contrast with previous results, porosity changes in these simulations are not significant (a few percent at most). It is noteworthy that these results are sensitive to the way the atmospheric cell in contact with concrete is implemented in codes as it has a direct impact on the flux of gaseous CO_2 entering into concrete: size of the cell, diffusive properties in the gas and the aqueous phase.

The modeling results do not integrate some of the features observed in experiments. For instance, C-S-H carbonation may overrun CH carbonation (Groves et al. 1990), which is not predicted in numerical simulations. Some portlandite also remains, even after a long time in some experiments (Drouet 2010; Auroy et al. 2018; Drouet et al. 2018). Further tests with the shrinking core model may succeed in reproducing these observations. In combination with this phenomenon, the decrease of reactivity when liquid saturation is below 0.3–0.4 has been observed (Thiery 2005; Thiery et al. 2013) and implemented by Thouvenot et al. (2013) but to calibrate such a function in codes requires more data.

Another noteworthy feature of this type of simulation is that the computing time can be quite long. For instance, due to the lack of an implicit scheme for the coupling of transport in the gas phase and RT in the aqueous phase in TOUGHREACT, the CPU time can be one to several months for 100 years of physical time (Trotignon et al. 2011; Thouvenot et al. 2013). Note that a recent development proposes an interesting "look-up table approach" for the two-phase reactive transport, which could partly help solve the simulation CPU time (Huang et al. 2018).

Finally, some important effects of carbonation on the properties of concrete that have been largely observed are not actually taken into account in modeling (Czarnecki and Woyciechowski 2015; Ashraf 2016; Savija and Lukovic 2016): changes in porosity, transport properties, micro- and macro-mechanical properties, shrinkage and cracking. Experimental data has been recently acquired on changes in water retention properties and permeability (Auroy et al. 2013, 2015, 2018). The impact of carbonation is not the same for the different types of concrete (Auroy et al. 2015), which would also constitute an interesting challenge for RTM.

THE IRON–CLAY INTERFACE

The iron–clay interface is studied in almost all countries where deep geological disposal of HLLW is envisaged. Significant amounts of metallic components (drift liner, structure, and containers) are used to ensure mechanical resistance of the waste package and to retard contact of waste matrices with water during the initial thermal stage, whereas clay barriers are used to retard RN migration. Although a variety of metallic alloys are envisaged in repositories (King and Shoesmith 2010; King 2014), the corrosion of these components in anoxic conditions is often simplified in the modeling studies into iron corrosion and takes place according to Reaction (7).

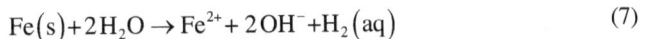

$$Fe(s) + 2H_2O \rightarrow Fe^{2+} + 2OH^- + H_2(aq) \tag{7}$$

First modeling studies at the scale of the disposal cell

The first modeling studies concerning this interface attempted to implement the knowledge acquired in experiments on iron–clay interactions in deep geological conditions initially conducted in batch systems with powdered materials starting in the late 1990's (see Bildstein and Claret 2015 for a recent review). These experiments focused on understanding the processes, identifying the corrosion products (magnetite, siderite) and the secondary minerals resulting from the alteration of clay minerals (Fe-serpentine, and zeolites and Fe-chlorite at higher temperature), and quantifying the corrosion rate (in the order of 1 μm/year). The calculations presented in the study of Montes-H et al. (2005) were performed at the scale of the bentonite barrier perturbed by a source of iron on one side and *in situ* groundwater on the other side; mineral dissolution used the kinetics law, precipitation occurred at equilibrium, looking in particular at the montmorillonite-to-chlorite conversion.

Modeling studies then turned to a geometry closer to that of a disposal cell by including the iron canister into the calculation domain, adding a full kinetics approach, and focusing on porosity change and feedback on diffusional properties (Bildstein et al. 2006). The authors obtained the first result on porosity clogging after 5 000 years at the interface with bentonite and 15,000 years in the iron grid cell at the interface with the Callovo–Oxfordian claystone.

Both studies showed converging results in terms of pH increase, Eh decrease, and mineralogical evolution, although the presence of iron in the calculation domain produced stronger changes in the second study: pH from an initial circum-neutral value to ~10.5 and Eh from an initial −200 mV down to −800 mV during corrosion.

First modeling attempts to fit batch experiment results

Because significant uncertainties in the parameters used in the previous studies were highlighted (surface area, kinetics, evolution of the corrosion rate), the first experimental study to include modeling results came early with the objective of calibrating these parameters, at least partly, by matching the mineral paragenesis and the chemical indicators (pH, Eh, iron aqueous concentration) observed in the experiments. De Combarieu et al. (2007) used a mean corrosion rate (1.4 µm/year) measured in his batch experiments at 90 °C using powdered materials and iron foils. Modeling results matched the experimental data (pH = 10.5, Eh = −700 mV, precipitation of magnetite, Fe-serpentine, and Fe-silicate) using reactive surface areas calculated based on the particle size for primary minerals and the local equilibrium assumption for secondary minerals.

In a study looking at the diffusion of iron in bentonite, Hunter et al. (2007) were the first to introduce ion exchange and surface complexation in the modeling approach, using the corrosion rate measured in their experiments at 30 °C. Simple 1D calculations were performed to model the profile of iron measured into the bentonite; matching the amount of iron required making assumptions on the diffusion coefficient, surface complexation and magnetite precipitation.

Finally, Pena et al. (2008) introduced the first attempt to model a variable corrosion rate using a semi-analytical approach, hypothesizing that corrosion is controlled by diffusion through a growing magnetite film. The authors were able to match the data reported for corrosion in bentonite in batch experiments by (Smart et al. 2006) on a time scale of 0.5 year using a solid phase diffusion coefficient of 10^{-20} m^2/s for the Fe^{2+} and 10^{-19} m^2/s for H$_2$, H$_2$O, and OH$^-$ species.

A decade of RTM evolution of simulations at the scale of the disposal cell

Later, a long series of modeling studies came back to the scale of the disposal cell; see review in Claret et al. (2018a) for a detailed description of the modeling set-up of the different studies.

While the previous studies focused on corrosion rate and mineral paragenesis, the subsequent studies at the disposal cell put the focus on full ion exchange and surface complexation with proton and Fe^{2+} (Samper et al. 2008; Wersin et al. 2008). While a 1D grid is used by Wersin et al. (2008) for the iron-bentonite system, the first 2D-axisymetric calculations are proposed by Samper et al. (2008) with an application to the iron–bentonite–granite system. Note that in both studies, the mineralogical system is simplified (no reactivity for clay minerals) and a constant corrosion rate is used in both cases, the emphasis being put on the role of sorption. The results show the importance of the surface protonation of bentonite, which reduces the pH increase during corrosion (from 11 to 9 in the bentonite), and a decrease of porosity in bentonite. The mass balance of iron shows that the precipitation of magnetite, and to a lesser extent siderite, accounts for most of the iron immobilization (iron sorbed on complexation surface is anecdotal). Note that

Samper et al. (2008) introduced an interesting "progressive", cell-by-cell, corrosion process. This is also the first study to perform a thorough sensitivity analysis of uncertain parameters (corrosion rate, protonation sites, and surface complexation vs. exchange of Fe^{2+}).

In a series of simulations with a constant corrosion rate, Savage et al. (2010) tried to tackle the tricky issue of the long-term evolution of the Fe-minerals in the mineralogical assemblage in the iron-bentonite system by using evidence for mineral parageneses from analogous natural systems. They allowed for faster minerals to precipitate initially and other more stable minerals to take over through nucleation and Oswald ripening, based on the early work of (Steefel and Vancappellen 1990). This is the only work where the authors were able to describe a magnetite → cronstedtite → berthierine → chlorite mineralogical sequence at the iron–clay interface over a period of 1 My (according to experimental results reviewed by Mosser-Ruck et al. 2010).

It was only a couple of years after the first attempt to develop a model corrosion rate depending on the chemical environment by Peña et al. (2008), that non-constant iron corrosion rates were introduced in large-scale calculations, and to date only Wilson et al. (2015) have implemented a corrosion rate controlled by diffusion. In most other cases, a simpler model was adopted, in the form of a "standard" rate law, i.e. with a term considering the departure from equilibrium considering reaction (3). This assumption results in a progressive decrease of the corrosion rate, usually by an order of magnitude over a period of 100 000 years (Marty et al. 2010; Lu et al. 2011; Ngo et al. 2014). Note that a decrease of the corrosion rate is also achieved, to some extent, by considering that the reactive surface area depends on the amount of iron and on porosity (Bildstein et al. 2006; Savage et al. 2010; Wersin and Birgersson 2014). Interestingly, in most of these studies (except Lu et al. 2011 and Wilson et al. 2015), ion exchange and surface complexation were not included in the calculations. This choice was motivated either by the fact that the focus was put on other processes (mineralogical changes, effect of transport), or because explicit surface reactions were considered to create mass by double-counting the counter ions when the mineral-bearing surface sites dissolves away in significant amounts (see sensitivity analysis in (Bildstein et al. 2012). This problem was recently treated by creating specific thermodynamic data for surface-bearing minerals to reconcile ion exchange and dissolution/precipitation processes (Benbow et al. 2019). However, note that very few data exist for surface reactions at temperature higher than 25 °C, which undermines the robustness of the interpretation.

Non-isothermal calculations were introduced into simulations quite recently by imposing a variable temperature field calculated by a TH code (Bildstein et al. 2012) or using a full THC modeling tool (Samper et al. 2016; Mon et al. 2017). This additional feature usually tends to stiffen the calculations, because RT processes and temperature are not implicitly coupled and temperature changes affect both the reactive processes (creating a thermodynamic disequilibrium and modifying kinetics) and transport.

Table 2 provides a synoptic view of the results of these studies and shows that the most abundant corrosion product predicted by the models in the long term is magnetite, sometimes with Fe-carbonates (siderite), and Fe-silicates (greenalite), sometimes incorporating Al (berthierine, cronstedtite). Primary minerals in clay are often destabilized in favor of Fe-phyllosilicates or zeolites if they are allowed to precipitate. Numerical studies often differ on the precise nature of the secondary minerals. The transformation of clay minerals into Fe-chlorite, and the timing, very much depend on whether (i) it is included as a secondary mineral, in which case it is the most stable phase and precipitates from the beginning of the simulation (e.g., Marty et al. 2010, or (ii) it results from a ripening process, taking the place of a precursor mineral in Savage et al. 2010). One of the most sensitive parameter remains the corrosion rate. The extent of the perturbation is always predicted to be limited to a few centimeters, up to 20 centimeters into the clay barrier.

Table 2. Recent RTM studies at the disposal scale with salient results.

Interfaces	Maximal perturbation extent	Main corrosion products	Main secondary minerals	Ref.
Iron–bentonite (100 °C)	(not explicit)	(not explicit)	Chlorite	[1]
Iron–claystone	5 cm	Magnetite, cronstedtite clogging after 16,000 y	Chamosite, Fe-smectite, scolecite, Ca-zeolite	[2]
Iron–bentonite (50 °C)	5 cm	clogging after 5,000 y		
Iron–bentonite (100 °C)	7 cm	Magnetite	Fe^{2+} exchange only	[3]
Iron–bentonite (100 °C)	few cm	Magnetite	Cronstedtite, berthierine	[4]
Iron–bentonite (100 °C)	15 cm	Magnetite clogging after 100,000 y	Fe-chlorite, Fe-saponite	[5]
Iron–bentonite (^{T} not explicit)	(not explicit)	Magnetite → croenstedtite → berthierine → chlorite	Berthierine, cronstedtite	[6]
Iron–bentonite (25 °C)	(not explicit)	Magnetite	Siderite	[7]
Iron–claystone (variable T)	20 cm	Magnetite (Ca-siderite, greenalite)	Vermiculite, saponite, pyrrhotite	[8]
Iron–claystone Iron–bentonite (100 °C)	15 cm 10 cm	Magnetite	Greenalite, Fe-saponite, Fe-chlorite, bethierine	[9]
Iron–bentonite (25 °C)	2 cm	Test cases with only $Fe_3(OH)_7$, $Fe(OH)_2$, goethite or magnetite	Na-phillipsite, chabazite, berthierine	[10]
Iron–bentonite (70 °C)	2 cm	(not explicit)	Berthierine, greenalite, Fe-saponite	[11]
Iron–bentonite (variable T)	up to 14 cm	Magnetite	Fe-phyllosilicates, zeolites	[12]
Iron–bentonite (variable T)	1 cm	Magnetite	Brucite, gypsum	[13]

References

[1] Montes-H et al. 2005 [2] Bildstein et al. 2006 [3] Samper et al. 2008 [4] Wersin et al. 2008 [5] Marty et al. 2010 [6] Savage et al. 2010 [7] Lu et al. 2011 [8] Bildstein et al. 2012 [9] Ngo et al. 2014 [10] Wersin and Birgersson 2014 [11] Wilson et al. 2015 [12] Samper et al. 2016 [13] Mon et al. 2017.

Concerning porosity clogging, a major assumption was made in some simulations: because the phenomenology is not known when porosity vanishes, porosity update was disabled in order to reach the end of the corrosion stage (~50,000 years) (Bildstein et al. 2012). A complete inhibition of the corrosion process has never been observed in experiments, even if a dense corrosion product layer is often identified. In addition, the ubiquity of magnetite in simulation results as the dominant corrosion product in the long term is questioned by many experimental results and archeological analogs. These issues, along with recent experimental work (Martin et al. 2008; Bourdelle et al. 2014, 2017), motivated the modelers to come back to modeling experimental results.

A recent return to modeling experimental results

Two of the most recent modeling studies revisited interactions at the experimental scale to refine the understanding and modeling of this interface. In the first work, Ngo et al. (2015) modeled iron-COx claystone interaction results obtained from the batch experiment with powdered material at 90 °C for 90 days (Bourdelle et al. 2014). Using a mean corrosion rate (determined from the experiment), standard and diffusion-controlled kinetics for mineral dissolution, and the local equilibrium assumption for precipitation, they were able to match the low pH (value ~7) and the evolution of aqueous species concentrations. They also matched the set of secondary minerals observed at the end the experiment: Ca-saponite, and greenalite. Interestingly, no magnetite was observed or simulated in these conditions. This particular feature, also observed by Bourdelle et al. (2017), was attributed to the high reactive surface area in the powdered system, producing a high precipitation rate for greenalite in the experiment (and matched by the modeling with equilibrium precipitation). Note that the simulation also predicted the transient precipitation of chukanovite, a Fe-hydroxyl-carbonate recently observed in experiments (Schlegel et al. 2010) and archeological analogs (Saheb et al. 2012), which disappears after the corrosion stage to form Ca-saponite.

In the second study, the modeling of iron corrosion in a Callovo–Oxfordian claystone block (machined from a core plug) at 90 °C for 2 years (Martin et al. 2008) was conducted by Bildstein et al. (2016). The objective was to reproduce the mineralogical paragenesis identified by (Schlegel et al. 2014) using the evolution of the corrosion rate measured in the experiment using electrochemical impedance spectroscopy (from 100 to 0.1 μm/year). The sequence of minerals observed from the contact with iron towards the claystone was as follows: magnetite, Fe-silicate, and Ca-siderite. This sequence could not be matched with the standard simulation setup (typical parameters for mineral dissolution and precipitation kinetics data, evolution of diffusion coefficient using Archie's law). The sensitivity analysis performed on the kinetic parameters (quartz dissolution, precipitation) was unsuccessful in generating the correct sequence. It was particularly difficult to maintain magnetite at the iron surface as the corrosion rate decreased with time. The observed mineral paragenesis could only be reproduced by using a cell-by-cell corrosion process and by attributing very slow diffusional transport properties to pre-corroded cells (i.e. by using a large value for the cementation factor). This layer "isolated" the iron surface from the claystone, allowing specific chemical conditions to develop at this location favoring magnetite precipitation.

The RTM simulations of iron corrosion will evolve in the short future to integrate more phenomenological corrosion models that calculate the evolution of the corrosion rate as a function of the geochemical conditions. Such electrochemical corrosion models have already been developed (Bataillon et al. 2010; King et al. 2014) and have to be coupled to RTM codes. This approach will be useful to simulate and interpret the existing laboratory experiments and also *in situ* results obtained recently (Necib et al. 2016; Schlegel et al. 2016, 2018).

THE CLAY CONCRETE INTERFACE

A brief materials mineralogy overview and associated chemical gradient at the materials interface

Before dealing with the interaction between clay and concrete, it might be interesting to recall briefly the mineralogy of these two materials. In a deep geological disposal, clay materials can be the clay-rock itself where clay minerals are the main constituents of these rocks, but also the bentonites used for the construction of engineered barriers (Sellin and Leupin 2013). The mineralogy of clay-rocks is complex, meaning that its clay fraction not only contains pure clay mineral end-members, such as kaolinite, smectite and illite, but also mixed layer minerals (Claret et al. 2004). In addition, carbonate minerals with a range of chemical compositions and structures are present (see for example (Lerouge et al. 2013), and pyrite, along with organic matter and quartz (Gaucher et al. 2004a; Jenni et al. 2014; Zeelmaekers et al. 2015). Proportions of total phyllosilicates, carbonates, quartz, pyrite and organic matter can be found in the literature for various clay formations (e.g., Opalinus Clay, Boom Clay, Callovian-Oxfordian formation, and Boda Clay, in Altmann et al. (2012)) and can be very different from bentonites (e.g., Ufer et al. 2008; Savage and Cloet 2018).

The mineralogical composition of cementitious materials is also complex. Concrete is a composite material made of a porous matrix (the hydrated binder) filled with water, into which are embedded filler materials such as quartz and calcite, which act as a granular skeleton. As the hydration reaction proceeds due to the cement–water interaction, the anhydrous phases, calcium silicates (C3S or C2S) and calcium aluminates (C3A and C4AF) are converted into hydrates such as portlandite, C-S-H (calcium silicate hydrate), ettringite, monosulfate or monocarbonate. This decreases the bulk porosity, since the molar volume of the hydrates is much larger than that of the anhydrous phases (Van Damme et al. 2013; Gaboreau et al. 2017; Claret et al. 2018b). Concrete has a long history that began in the pre-Roman age, and its formulation, which was greatly improved at the beginning of nineteenth century with the invention of Portland cement, has recently become more and more sophisticated. Within the framework of nuclear waste disposal, such sophistication has been introduced with the development of low alkaline concrete (Codina et al. 2008). The idea behind this development was to improve concrete compatibility with the repository environment, yet it remains a high-strength concrete. On one hand, low alkaline concrete (often called "low pH" cement) has a lower alkali content than that of ordinary Portland-based cement material, which may reduce the pH gradient at the interface and therefore the changes in clay in contact with the concrete. On the other hand, it has a low-heat hydration temperature, which minimizes the microcracking that can have negative consequences on cement's long-term durability. One must also keep in mind that whatever the formulation, the hydration of cement-based material is kinetically driven, and thus the composition of its pore water evolves with its curing time.

While the measured pH in concrete is in the range 10.4 to 13.6 depending on time and formulation (Lothenbach and Wieland 2006; Luke and Lachowski 2008; García Calvo et al. 2010; Lothenbach 2010; Bach et al. 2012; Lothenbach et al. 2012, 2014), the pH of pore water in clay materials is generally in the range of 7 to 9 (Bradbury and Baeyens 2003, 2009; Wersin 2003; Bildstein and Claret 2015; Lerouge et al. 2018). Therefore, even for low-pH concrete material, the pH difference at the interface may be as much as two pH units. When the pH is above 13, the alkali (Na^+, K^+) concentrations in pore water are higher in cement material than in clay-rocks. The reverse is true for K^+ when the cement material is made of low-pH cement. Still, in relation to the pH values, partial CO_2 pressures are far lower in cement materials than in the pore water of clay-rocks, where partial pressures of CO_2 that are higher than the atmospheric pressure ($\sim 10^{-3}$ to $\sim 10^{-2}$ bars) are reported (Gailhanou et al. 2009; Lassin et al. 2016). Redox conditions are reducing in both claystone (Brendler et al. 2003) and in reinforced steel cement materials (Aréna et al. 2018). Nevertheless, in the latter, a strong Eh gradient exists from the steel surface outward to the bulk concrete. As described just above, a steep geochemical gradient exists at the clay–concrete interface. In the 90's, early calculations based on mass balance assumptions only, and reported in Gaucher and Blanc (2006) and Savage et al. (2007), led to the conclusion that 0.2–1 m^3 of bentonite are needed to buffer the chemical perturbation created by 1 m^3 of concrete. If true, this conclusion would have been problematic for the storage concepts that rely on the properties of unaltered clay materials, and this explains why so much effort has been put in reactive transport modeling studies of both experiments that mimic clay–concrete interfaces and the long-term evolution of clay–concrete interfaces.

Reactive transport modeling of laboratory and in situ scale experiments

Before their use for large-scale simulation and long-term evolution, reactive transport models and computer codes have to be tested against experimental data in order to test and improve their robustness. Experimental data can be obtained in the laboratory with experimental apparatus that mimic storage situations (e.g., column experiment), *in situ* to better reproduce field conditions (this is possible thanks to many underground research laboratories that are in operation) or gathering data from natural analogs for which hydrogeology and geochemical conditions reproduced phenomena that are expected in the repository. In Table 2, reactive transport studies that are compared against experimental data dealing with clay–concrete interaction are reported. The focus of our chapter is claystone, which is why experimental studies that deal with high-pH plume but in granite (e.g., (Soler 2003; Pfingsten et al. 2006; Soler and Mader 2007)) have not been reported here. In addition, when some geochemical modeling is made without coupling chemistry and transport (e.g., Lalan et al. 2016) these references are also not mentioned as the focus is on reactive transport. Among the thirteen references discussed in Table 2, three deal with column laboratory experiments (two focusing on the same experiment), four are *in situ* experiments and four focus on the same natural cement analog. The latter is the Maqarin natural analog in the north of Jordan (Khoury et al. 1985; Alexander et al. 1992; Khoury et al. 1992; Chaou et al. 2017). In this area, hyperalkaline groundwater compositions are the product of low-temperature leaching of an assemblage of natural cement minerals produced as a result of high-temperature/low-pressure metamorphism of marls (i.e. clay biomicrites) and limestones. On the hydraulic downstream of the cement zone, alkaline groundwaters circulated through fractures within the biomicrite clay. On the edge of the fractures, calcite, kaolinite, silica, low amounts of illite, albite and organic matter dissolved. Within the fractures, different opening/clogging stages may have occurred, leading to a complex mineralogical pathway. Two of the modeled cement claystone interfaces were sampled during the Cement–Opalinus Clay Interaction (CI) Experiment at the Mont Terri rock laboratory (Jenni et al. 2017). This is a long-term passive diffusion-reaction experiment between contrasting materials of relevance to engineered barrier systems/near-field for deep disposal of radioactive waste in claystone (Opalinus Clay). The sampled interfaces have been (and are still) extensively characterized from the mineralogical and petrophysical point of view (Jenni et al. 2014, 2017; Dauzeres et al. 2016; Lerouge et al. 2017). The two other *in situ* interfaces were sampled in the Tournemire underground research laboratory. Again, these interfaces have been extensively investigated (Tinseau et al. 2006; Gaboreau et al. 2011; Bartier et al. 2013). In one of the column laboratory experiments, bentonite was exposed on one face to a solution that mimic cementitious pore water, while the opposite face was a homo-ionization solution. In the second, concrete and bentonite plugs were juxtaposed. For the column laboratory experiments, heat was applied either in isothermal conditions (60 °C or 90 °C) or by applying a thermal gradient. These are the only studies in where temperature plays a role. Except for (i) the hydration concrete-bentonite column test, where non-saturated conditions and thermal gradients imply the use of a THMC coupled model, (ii) the natural cement analog and one *in situ* experiment where fractures play a role, a diffusive regime was the main dominant transport process in all other experiments. The column experiments and *in situ* experiments were sized in the 10 cm range, whereas for the Maqarin natural analog site a ~100 m fracture was considered in simulation, but matrix diffusion perpendicular to the fracture was in the 10 cm range. In most of the publications described in Table 2, the thermodynamic data used come from existing databases that are either merged together (i.e. a clay-oriented database with a cement-oriented database) or completed with some missing data (e.g. zeolite, M-S-H). Depending on the reactive transport code used, the C-S-H representation is made either by considering a discrete calcium to silica ratio (e.g. 1.6, 1.2 and 0.8), or by considering a jennite-tobermorite solid solution. The latter approach was used for the studies that used GEM (Wagner et al. 2012; Kulik et al. 2013) for the reactive part. The main findings

Table 2. Review of the experiments that mimic clay/concrete interaction and that have been modeled using reactive transport modeling.

Experiment	Main conclusion	Ref.
	In qualitative agreement with observations at the Maqarin site, the simulations predicted that ettringite with lesser amounts of hillebrandite and tobermorite are the dominant alteration products formed at the expense of the primary silicates in the rock matrix and fracture. Depending on the rate constant for secondary mineral precipitation reactions (either the same in both the rock matrix and fracture or one order of magnitude higher in the fracture compared to the rock matrix), the simulations suggested two possible scenarios for porosity reduction (either the matrix first or the fracture first).	[1]
Mineral fluid interactions occurring at the Maqarin site in Jordan (natural cement analog)	Simulations indicated that the pore clogging caused by precipitation of ettringite and C-S-H minerals occurred after several hundred years at a distance of 5–10 mm from contact with the hyper-alkaline solution. Sensitivity analysis shows that clay minerals controlled the availability of Al, which is needed for ettringite and C-S-H phase precipitation. Therefore, temporal evolution of porosity changes was controled by clay dissolution.	[2]
	Major secondary minerals (e.g., ettringite, C-S-H) were controlled by the dissolution of primary silicates. Extension of porosity reduction along the fracture and in the fracture-wall rock interface depended on assumptions regarding flow velocity and composition of the high-pH solution.	[3]
	RTM predicted that ettringite, thaumasite, jennite and tobermorite dominate the fracture filling materials. Alteration of the marl led to scolecite, ettringite, C-S-H and a small amount of sepiolite. If an armored layer was considered in the fracture, only calcite and jennite precipitate. Fracture sealing is complex.	[4]
For about 2 years, a cylindrical compacted bentonite column was in contact with a 0.6 $MgCl_2$ solution and either a young cement water (YCW, pH = 12.4 at 60 °C) or an evolved cement water (ECW, pH = 11.5 at 60 °C)	Major mass transfer and mineral transformation pathway (including the diffusion of alkaline cations through ion exchange reaction) experimentally observed were reproduced by RTM. For the most reactive system (YCW), partial dissolution of montmorillonite and precipitation of Mg-Silicate, hydrotalcite and brucite were predicted	[5]
	In this case modeling was blind because it was done prior to the experiment. The model reproduced well the porosity decrease induced by secondary mineral precipitation as well as ion exchange. The kinetic dissolution model played an important role in the outcome of the simulations.	[6]
Two cylinders of bentonite and concrete were put into contact for 54 months in hermetic cells that allowed both heating up to 100 °C on the bottom (bentonite side) and water circulation on the top.	On-line measured temperature, relative humidity data, water content and porosity data were well reproduced by the THMC model. Except for ettringite and C-S-H phase, for which the predicted precipitation was smaller than experimental data, RTM captured the main trend of the mineralogical pathway. Increasing the specific surface area of ettringite improved the model.	[7]

Table 2 (cont'd). Review of the experiments that mimic clay/concrete interaction and that have been modeled using reactive transport modeling.

Experiment	Main conclusion	Ref.
A 15-year in situ claystone /Ordinary Portland Concrete (OPC) interface was sampled in the Tournemire Underground Research Laboratory	While profiles calculated at local equilibrium or using kinetics are very similar in concrete, kinetics really improved the representation of experimental data in the claystone. Mineralogical transformation simulated in the claystone matrix and in the fracture filling were quite comparable as observed on the field. C-S-H, ettringite and carbonates (predominance of metastable vaterite over calcite) were modeled at the OPC/claystone interface, followed by a zone of clay-likes phases and calcite and a last zone with precipitation of calcite (going deeper in the fracture).	[8]
An eighteen-year in situ experiment putting CEM II and Toarcian claystone in contact at Tournemire Underground Research Laboratory was dismantled, analyzed and modeled.	RTM was used to test one hypothesis, namely the introduction of a sedimentary fluid into the macroporosity of the cement paste to explain the formation of part of these secondary phases. Simulations indicate that such transport could have occurred near the argillite/cement paste contact at a very early stage. After this stage, the transport was reversed and "cementitious" fluids flowed from the cement paste to the argillite. Experimental observations of C-S-H neoformation in the claystone, carbonate formations at the interface were well reproduced by simulation.	[9]
Five and 2.5-year in situ interaction of two different low-pH cements (ESDRED and LAC) with Opalinus Clay (OPA)	RTM corroborated calcite precipitation, C-S-H decalcification and M-S-H formation at the interface.	[10]
Five-year in situ interaction of Ordinary Portland Cement (OPC) with Opalinus Clay (OPA)	RTM reproduced well the decalcification of the cement at the interface, the Mg enrichment in OPA detached from the interface and the sulfur enrichment in the OPC detached from the interface.	[11]
Six-year interaction between a hardened Portland cement and water-conducting shear zone. In situ experiment conducted at the Grimsel Test Site in Switzerland	Two RTM models were used, one in 1D and one in 2D. They did not properly represent all the concentration evolutions as a function of time. However, the main parameters like pH and Ca were captured. Both models evidenced dissolution of the fault gouge minerals. The 2D model indicated an associated secondary mineral precipitation and a porosity reduction. Ettringite, C-A-S-H, and hydrotalcite also precipitated.	[12]
	In comparison to Chaparro et al. (2017), a 3D model was used here. Although some elements' concentration as a function of time (like Mg) were poorly reproduced, RTM captured the essential features of the cement leaching (e.g., portlandite dissolution) and carbonate mineral precipitation. In addition, the model results highlighted uncertainties surrounding cement solid dissolution rates and rates of secondary mineral formation.	[13]

References
[1] Steefel and Lichtner 1998 [2] Shao et al. 2013 [3] Soler 2016 [4] Watson et al. 2016 [5] Fernandez et al. 2010 [6] Watson et al. 2009 [7] Samper et al. 2018 [8] De Windt et al. 2008 [9] Bartier et al. 2013 [10] Dauzeres et al. 2016 [11] Jenni et al. 2017 [12] Chaparro et al. 2017 [13] Watson et al. 2018

and conclusions of the various studies are given in Table 2; therefore, they will not be detailed in the text. However, general conclusions can be drawn. Although some discrepancies between models and experiments could appear because new or completed datasets become available after modeling (Chaou et al. 2017), reactive transport modeling shows a great capability for reproducing the experiments (e.g., mineralogical transformation pathway and clogging processes). It is also a very useful tool for performing sensitivity analyses of input parameters and testing hypotheses. One of the most critical parameters that has been described in reported studies deals with kinetics and reactive surface areas that can play a large role in sequential minerals' appearance or disappearance, as well as the localization of porosity reduction.

Reactive transport modeling of the long-term evolution of clay–concrete interfaces

As previously stated, reactive transport modeling is a powerful tool for describing and reproducing phenomena that occur at both the time and space experimental scale. However, these scales are at least four or five orders of magnitude lower than the one we have to deal with for the long-term evolution of clay–concrete interfaces with geometries that are relevant for the repository gallery scale. In that case, reactive transport models are used to bridge the scale and make predictions about long-term evolution. Our literature review (which might not be fully exhaustive) found about twenty reactive transport studies addressing these issues during the past two decades (Table 3). In these studies, anionic exclusion in clay and multi-components diffusion were not considered. The complexity of the phenomena that were modeled lay more in considering an exhaustive mineralogy and steep chemical gradients that are not always easy to handle from a numerical point of view, even for modern reactive transport codes. The number of reactive transport codes used was also considerable: PRECIP (Savage et al. 2002), GIMRT (Lichtner 1996), HYTEC (van der Lee et al. 2003), PHREEQC (Parkhurst and Appelo 1999),TOUGHREACT (Xu et al. 2004 ; Burnol et al. 2006), ALLIANCES (PHREEDC/MT3D, (Montarnal et al. 2007), CORE^{2D} V4 (Samper et al. 2003), CRUNCH (Bildstein and Claret 2015), OpenGeoSys-GEM (Kolditz et al. 2012), ORCHESTRA (Meeussen 2003), MIN3P-THCm (Mayer et al. 2002). A detailed description of most of the above-mentioned codes and their history can be found in Steefel et al. (2015b). Many databases have also been used: EQ3/6 (Wolery 1983), CEMDATA2007 (Lothenbach and Wieland 2006; Lothenbach and Winnefeld 2006), Nagra/PSI database (Hummel et al. 2002), THERMOCHIMIE (Blanc et al. 2015a,b), THERMODDEM (Bartier et al. 2013). The geometries (1D or 2D Cartesian, 1D radial, 1D or 2D axisymmetric and 2D with cylindrical coordinates), the mesh size (usually in the cm range), the transport parameters, the way mineral dissolution or precipitation was accounted for (either at local equilibrium and/or in kinetics) have been recently reviewed (Claret et al. 2018b). Otherwise, in the case of the mass balance calculations reported earlier in this chapter, all reported studies are consistent with very limited spatial extension, in terms of mineral precipitation and dissolution, of cement perturbation in clay materials and vice versa. Not all the reported studies consider exactly the same starting mineralogical assemblages, the same secondary potential mineral phases, the same transport properties, or the same geometries (see (Table 3). This is inherent to the starting conceptualization, but also to the quality and completeness of the databases that have been improved with time, as well as the code capability. For example, in earlier calculations, cement material was not represented as a porous medium but rather as an alkaline plume imposed as a boundary condition on one side of the clay domain. Therefore, in order to evaluate the impact of different modeling assumptions among available studies, Marty et al. (2014) carried out calculations with a consistent set of data and input parameters arranged with increasing order of complexity (e.g., the considered geometry, the representation of porous media, the choice of secondary minerals phases). This standardized approach allowed for a proper comparison of numerical results and showed that modeled reaction pathways were mostly independent of the modeling assumptions for simulations carried out in the presence of water-saturated conditions. To compare the results of the various simulations after 100,000 years, concrete and clay degradation indicators were selected. In the concrete zone, the simulations were

Table 3. Review of the experiments that mimic clay/concrete interaction and that have been modeled using react

Interfaces	Maximal perturbation extent	Main mineralogical changes
Hyperalkaline fluids-bentonite (25 and 70 °C)	60 cm after 1000 yr	C-S-H neoformation close to the interface, while zeolites and sheet s▮ Some growth of primary bentonite minerals (analcime, chalcedony, c observed under certain conditions. Porosity clogging.
Hyperalkaline fluids fractured marl (25 °C)	6 m along the fracture after 5000 yr for the higher pH	Replacement of dolomite by calcite and precipitation of secondary m▮ analcime, natrolite and tobermorite led to decreased porosity in the fr
Ordinary Portland cement-Toarcian claystone (25 °C)	Depends on the selected minerals but alkaline plume is a buffer close to the interface	Close to the interface portlandite dissolution, C-S-H, brucite (hydrota on, the interface dolomite dissolve and calcite precipitate. Depending secondary phases such as illite and zeolite might occur. Clogging occurs after 2500 years in the claystone, whereas porosity i▮ interface.
Waste pckage-MX80 bentonite-Portland cement liner/wall-OPA claystone	1 m in bentonite and 0.5 m in cement after 100,000 yr	Portlandite dissolution with sequential precipitation of C-S-H with d▮ brucite, calcite, hydrotalcite and illite in relatively low concentration▮ whereas kaolinite and quartz are partly dissolved. Change in the exchanger population.
Hyperalkaline fluids-COx claystone (25 °C)	70 cm after 100,000 yr	First a change in the exchanger population (Na is replaced by K and ▮ of the montmorillonite occurs. Between the illitized zone and the cor▮ precipitate. Finally, cement phases (e.g. tobermorite, ettringite) repla▮ leading to porosity reduction.
Portland concrete-COx claystone Non isothermal and non-saturated condition		Tobermorite precipitation induced porosity clogging after 900 yr. In ▮ dissolve and monosulfoaluminate is transformed into ettringite. In cl▮ while illite and a calcic saponite are precipitating to the detriment of
CEM I/ CEM V cement paste-either interstitial COx pore water or claystone (25 °C)	1.3 m after 400,000 yr for CEM I 0.2 m after 400,000 yr for CEM V	For CEM I, the porosity reduction is driven by zeolites, illite, quartz For CEM V, secondary smectite and calcite precipitation are describ▮

Table 3. (cont'd). Review of the experiments that mimic clay/concrete interaction and that have been mo

Interfaces	Maximal perturbation extent	Main mineralogical
Bentonite–concrete-Callovo Oxfordian claystone (25 °C)	Not explicit	Portlandite dissolution and C-S-H precipitation. In clay, an observed. Montmorillonite dissolved but remains despite th
Portland cement-either interstitial COx pore water or claystone 25 °C	Less than 2 m after 100,000 yr	Cement–interstitial CO_x porewater: Portlandite dissolution, brucite, sepiolite and hydrotalcite representing the final stag Cement–CO_x: Portlandite dissolution, reduction of Ca/Si ra hydrogarnet and brucite precipitate. Dissolution of illite anc precipitation. Both porosity opening and reduction depending on the posi and illite and quartz kinetic parameters are key parameters.
Bentonite–concrete–claystone 25 °C	~30 cm in both concrete and clay side	Dissolution of primary silicate minerals (e.g. montmorilloni phillipsite, illite and calcite. In concrete, portlandite dissolu precipitation. Porosity clogging localization and appearance (local equilibrium, kinetic rates).
Portland cement–FEBEX bentonite 80 °C	few cm after 100 000 yr	Precipitation of hydroxides, zeolites secondary clay mineral bentonite.
Portland cement–COx claystone 25 °C	~30 cm in both concrete and clay side	Dissolution of primary silicate minerals (e.g. montmorilloni phillipsite, illite and calcite. In concrete, portlandite dissolu precipitation. Porosity clogging localization and appearance (local equilibrium, kinetic rates).
CEM I cement–COx claystone	Non isothermal, non saturated conditions	Portlandite dissolution and calcite precipitation. C-S-H with precipitation. Brine formation and salt (e.g. syngenite, burke where drying occurs.
Bentonite-ESRED "low pH" concrete–OPA claystone 25 °C	Few tens cm both in ESDRED and OPA or less 10 cm at bentonite-ESDRED interface	In bentonite, the main mineralogical transformations are the pr replacement of the [Na,Mg]-montmorillonite by a [Ca,K]-mon In OPA claystone, two alteration regions are observed (inclusio kaolinite and illite are transformed into a substitute for smectite space and time, quartz, calcite and pyrite may dissolve, but part possible. In the concrete liner zone close to the interface, the complete de predicted, accompanied by the precipitation of hydrotalcite, zec

e transport modeling.

Table 3. (cont'd). Review of the experiments that mimic clay/concrete interaction and that have been modeled u

Interfaces	Ref	Maximal perturbation extent	Main mineralogical changes
OPA or Effingen Member or Palfris formation-CEM I, concrete (to tobermorite), ettringite, and tobermorite and zeolite	[7] [8]	10 cm in claystone 20 cm in concrete (depends slightly on the considered claystone)	Montmorillonite and portlandite dissolution. In claystone, phillipsite (Na, K or Ca), illite, calcite, hydromagnetite, In concrete, C-S-H, ettringite, calcite precipitation. Clogging occurs in all cases.
Portland cement-interstitial Bonne pore water 25 °C		Few mm	Portlandite dissolution, Ca/Si ratio decreases in C-S-H. Dissolution o calcite precipitation leads to the clogging of porosity.
precipitation of saponite, Hyperalkaline fluide Portland cement-COx claystone	[9]	Depends on the selected indicators and the scenario (from m to cm range)	In the concrete, zone alteration front is followed based on portlandite claystone dolomite, montmorillonite are used.
ion phases (e.g. C-S-H) in CEM I concrete-COx claystone	[10]	Not explicit	Interest of the paper lies in the benchmark. All codes described the s interface as well as clogging.
precipitation of saponite, Cement backfill representative of crystalline rock	[11]		For the higher ionic strength (0.48 mol/kg), cement alteration (e.g. h chloride-rich solid (e.g. Friedel's salt) precipitation, along with some after 78 years. In contrast, the lower ionic strength (0.005 mol/kg) le of calcite, along with saponite, and minor amounts of ettringite, with
and katoite	[12]		Thaumasite could also be a byproduct of sulfate attack on the cemen

References

[1] Savage et al. 2002 [2] Soler 2003 [3] De Windt et al. 2004 [4] Gaucher et al. 2004b [5] Burnol et al. 2006 [6] Trotignon et al. 2006 [7] Montar and 2008 [10] Fernandez et al. 2010 [11] Marty et al. 2009 [12] Trotignon et al. 2011 [13] Berner et al. 2013 [14] Ferrand et al. 2014 [15] Blanc et al.

close to the interface, ling on the coordinate in calcite and illite is also

Ca/Si-ratio C-S-H phases is gypsum and calcite.

compared by examining both the total portlandite-dissolution (accounting for both hydrolysis and carbonation) and (ii) the ettringite-precipitation (volumetric variations above twice the initial volume) front. The dolomite-dissolution front, the smectite-dissolution front (volumetric variations below half of the initial volume) and the pH plume in the clay barrier (pH>9) were chosen to analyze clay-rock alterations. Dolomite is one of the most destabilized minerals in the claystone (Callovo–Oxfordian formation in this study) due to secondary phase formation (e.g., saponite) incorporating magnesium in their structural formulas, while montmorillonite (bearing hydroxyl groups) contributes to the buffer capacity. The criterion chosen for pH corresponds to a limit above which the dissolution rates of many aluminum silicate phases increase significantly. Regardless of the complexity of the simulation, after 100,000 years of interaction, not one of the above criteria indicates a perturbation higher than one meter. For the cases that account for full complexity, the perturbation extension is even 5 times lower. Still, in saturated conditions and besides the simulation conceptualization, the accuracy and numerical stability of the reactive transport codes have been successfully benchmarked (Blanc et al. 2015b). While simulations conducted in saturated conditions are numerous, simulations accounting for non-saturated and non-isothermal conditions are rare, making their results more difficult to evaluate. In these simulations, the main uncertainty lies in the period necessary to reach complete saturation inside the concrete. Considering an initial saturation of 30 %, this period is estimated to be approximately 2,000 years (Burnol et al. 2006), but it may depend on the grid size resolution and on the power exponent of the Millington relationship that describes dependency of pore aqueous and gas diffusion coefficients as a function of saturation and porosity (Trotignon et al. 2011). From the mineralogical point of view, the transformation pathways in concrete were found to be similar to those simulated in saturated conditions (e.g., portlandite degradation, carbonation, Ca to Si ratio decrease in C-S-H), except for sulfate/carbonate salts being deposited where drying takes place.

OTHER INTERFACES WITHOUT EXTENSIVE RTM STUDIES

The glass–(iron)–clay interface

This interface is studied in countries where the spent fuel is retreated, in total or in part, by removing and recycling uranium and plutonium, e.g., in Belgium, France, Germany, Japan, Russia, UK, and the USA (Gin et al. 2013). In this case, borosilicate glasses are developed as the waste matrix for HLLW. Note that preliminary studies have also been conducted for the use of glass for ILLW (e.g., in France, UK and the USA). This interface is complex, since iron may be present if steel corrosion is still going on when water enters the canister. Once corrosion is finished, corrosion products will be present and have an influence on glass alteration. RN can be considered as traces and are not thought to have a great impact on the glass alteration process.

Numerical studies of glass alteration at the space scale of the disposal cell (50 m) and the corresponding time scale (100,000 years) are scarce (Bildstein et al. 2007, 2012). In these simulations, glass alteration follows a very simple "operational" model in which the alteration proceeds in two stages, after a time lag corresponding to the time before canister failure (700 years): a first initial phase where the alteration rate (r_0) is high and a second stage where a residual, much lower, rate has been established ($\sim r_0/10,000$). The feedback between glass alteration and the chemical environment is indirectly taken into account by the duration of the initial stage, which stops when silica derived from the glass has saturated the "sorption" capacity of the system close to glass (essentially the corrosion products). In these simulations, iron corrosion starts at the beginning of the simulations so that from 700 to 45,000 years, when corrosion is completed, it proceeds simultaneously with glass alteration favoring the precipitation of Fe-silicates.

The most noticeable evolution in the RTM approach to glass alteration in these two studies concerns the complexity of the geochemical and mineral system. Bildstein et al. (2007) took into account only four simple mineral phases as glass alteration products, including one

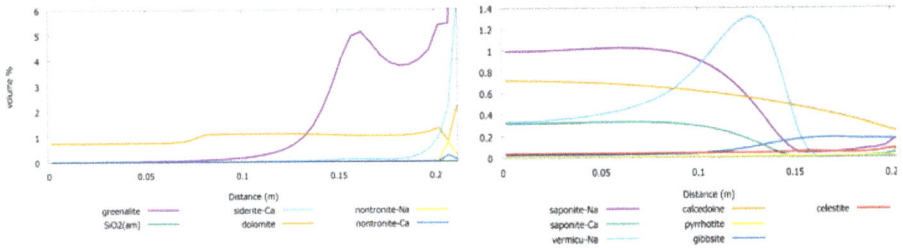

Figure 3. Profile of glass alteration products at 100,000 years in a 21 cm radius glass cylinder (Bildstein et al. 2012). Results obtained after alteration 5% of the initial volume of glass. Chalcedony, saponite, and vermiculite dominated in the center of the glass **(left plot)**. Greenalite and nontronite dominate towards the interface with iron, and Ca-siderite **(right plot)**.

zeolite. The second paper (Bildstein et al. 2012) includes 26 secondary minerals, with 14 glass alteration products, some of them determined from experimental studies (including Fe-silicates and Fe-aluminosilicates)(Figure 3). An explicit limitation of the H_2 partial pressure (hydrogen is considered to form a gas phase when pH_2 reaches 60 bar) was also taken into account as well as the presence of an excavation damaged zone with degraded transport properties in contact with the canister. Finally, to approach realistic conditions, the thermal gradient evolutions determined from THM calculations were included in the calculations, with temperature-dependent thermodynamic, kinetic and transport properties.

The relative simplicity of the simulations, in terms of glass behavior, can be explained by the fact that the alteration of glass in the presence of complex mineralogical environments has been only studied extensively in the last decade, allowing for specific models for glass alteration to be developed and calibrated.

Along with transport by diffusion, the dissolution of corrosion products and clay minerals and the precipitation of secondary crystallized minerals should be the main processes governing the long-term fate of glass. Dissolved cations such as Fe^{2+}, Mg^{2+}, Ni^{2+} have been shown to precipitate with silicon (if the pH is sufficiently high), which is the major element of the glass protective amorphous layer, and sustain glass dissolution (Aréna et al. 2016, 2017, 2018). The influence of Mg-rich Callovo–Oxfordian groundwater on glass dissolution has been modeled and fitted to the experimental results in Jollivet et al. (2012). The mechanism hypothesized by Rebiscoul et al. (2015) for modeling the long term alteration rate of glass in the presence of magnetite is slow magnetite dissolution followed by iron silicate precipitation. The ability of magnesium-rich minerals like hydromagnesite, dolomite and the argillite clay fraction to provide magnesium and sustain silicon consumption and glass alteration has been experimentally studied and modeled by (Debure et al. 2012, 2013, 2018). The effect of clay on glass alteration has often been described with a partition coefficient (Kd) of silicon in the clay, reflecting the complexity of the underlying dissolution/precipitation mechanisms (Godon et al. 1989; Gin et al. 2001; Pozo et al. 2007).

Current research in glass alteration modeling focuses on the description of the amorphous layer (formation, structure, composition, solubility) and on the capacity of the amorphous layer to lower the glass dissolution rate (Gin et al. 2016, 2018). Following an approach initiated by Bourcier et al. (1994), the hypothesis of a backward reaction of formation of the amorphous layer from the bulk fluid composition is made in order to build a thermodynamic description of the amorphous layer (McGrail and Chick 1984; Daux et al. 1997; Frugier et al. 2009; Rajmohan et al. 2010; Steefel et al. 2015a; Frugier and Godon 2018). The amorphous layer composition and solubility are currently described by concatenating database minerals or hypothetical phases that account for both the amorphous layer composition and the fluid composition at steady state.The GRAAL model (Frugier 2008; Frugier and Godon 2018) computes the local amount of protective layer to calculate the local thickness of the protective layer and deduce the glass alteration rate.

The remaining challenge is to couple the specific model for glass alteration (e.g., GRAAL) with large-scale RTM simulations including the materials present in the near field of the disposal cell, i.e. to ensure compatibility between the two modeling systems.

The glass–concrete–(clay) interface

This interface plays a role in deep geological repository, especially in the Belgian concept of concrete supercontainer. Once the overpack (carbon steel) and primary container (stainless steel) have failed, glass will be in direct physical contact with concrete. There are currently no RTM studies of this interface at the scale of the disposal cell, to best of the authors' knowledge. This is partly due to the challenge of understanding the complex behavior of glass in water and its interactions with materials such as concrete, which has only been studied in recent years.

The glass–concrete interface has been a growing field of interest since Belgium proposed the supercontainer concept for the disposal of high-level wastes (Kursten and Druyts 2008). A thick concrete buffer is meant to maintain high pH, to preserve the integrity of the carbon steel overpack at least during the waste's thermal phase. France is studying the option of filling the gap between the rock wall and the casing with bentonite/cement grout. The grout thickness is dimensioned to compensate for acidic fluids coming from the oxidation of iron and sulfur in the clay occurring during the facility's operational period. With high pH for long periods of time, the Belgian concept relies more on the overpack's lifetime, whereas the French concept takes more advantage of the confinement properties of the waste glass. However, both require understanding, deterministic models, and long-term quantification relative to glass alteration close to concrete.

Alteration of glass at a pH higher than 10 results in the precipitation of other secondary crystallized minerals, especially zeolites, as well as C-S-H phases. Zeolite precipitation has been proven to sustain glass corrosion and to change the composition of the protective amorphous layer at the glass surface (Ribet and Gin 2004; Ferrand et al. 2013, 2014; Mercado-Depierre et al. 2013, 2017).

The higher the pH, the faster the zeolite nucleation-growth process and the slower the induction time required for zeolite surface to be high enough to impact the glass dissolution rate (Fournier et al. 2017). Zeolite precipitation consumes Si and Al, two key constituents in the amorphous layer, and therefore sustains glass dissolution. However, zeolites also consume alkali, which in return lowers both the pH and the zeolite precipitation rate. Zeolite-seeded experiments have been a major tool for investigating those experimental conditions where a mineral's precipitation kinetics drive the glass dissolution rate (Fournier et al. 2017). The first time-dependent geochemical modeling of alteration resumption has been achieved with the GRAAL model (Frugier et al. 2017). Only a limited number of modeling studies of glass–cement interactions exist and none of them integrated the transport component. Liu et al. (2015) report simulations of glass alteration in hyperalkaline solution from their 300-day experiments at the laboratory scale, showing a diffusion-controlled glass alteration rate at 30 °C and a shift towards a surface reaction-controlled rate after 100 days at 70 °C. Baston et al. (2017) performed batch type scoping thermodynamic modeling on the interactions between two illustrative vitrified ILLW products and two cementitious backfill, predicting limited changes in mineralogy and properties of the two materials. Improving knowledge about i) amorphous layer composition and solubility at high pH and ii) zeolite precipitation kinetics is still required before applying the model to the complex chemistry and the long time scales of geological disposal.

The iron–concrete interface in anoxic conditions

This interface will be encountered in the Belgian concept with the concrete supercontainers (steel envelope and overpack) and also in all type of repository concepts where steel reinforced concrete is envisaged. Although experimental studies on iron corrosion in concrete have been performed in the context of waste disposal (e.g., L'Hostis et al. 2011; Kursten et al. 2017), no RTM studies have been found in the literature concerning this interface in anoxic conditions.

In the highly alkaline interstitial solution of Portland concrete, the corrosion of carbon steel is characterized by a low corrosion rate (passive mode) in the order of 0.1 µm/yr to 1 µm/yr (Smart et al. 2013; Chomat et al. 2017; Kursten et al. 2017). This corrosion mode is due to the formation of a thin protective layer, with corrosion products similar to those observed at the iron–clay interface: magnetite dominates, accompanied by small amounts of hematite. The stability of this layer depends highly on the solution chemistry: pH, buffering effect, carbonate and sulfate content (Chomat et al. 2017). Corrosion can therefore resume when the conditions are no longer favorable, i.e. when concrete does not buffer the pH at high values (typically upon degradation by carbonation).

Since corrosion models are at the heart of this problematic, significant advances will be achieved in the near future by coupling RTM codes with electrochemical corrosion models (Bataillon et al. 2010; Macdonald et al. 2011).

GENERAL CONCLUSION AND PERSPECTIVES

Over the last two decades, the use of RTM in simulations of the long-term behavior of materials in waste repositories has evolved to consider geochemical systems of increasing complexity, driven by the improvement of numerical codes (more functionalities), and their efficiency (solver, parallelization), as well as the accumulation of data acquired to feed thermodynamic and kinetics databases. Overall, these simulations provide a better understanding of physical and chemical change and how materials interact; they give more confidence in the prediction of the durability of materials by converging on the alteration extent that can be expected at the interfaces between different materials over long periods of time. Benchmarking RTM codes on problems related to waste repositories (e.g., Marty et al. 2015b) and continuous database development (Blanc et al. 2012, 2015b; Giffaut et al. 2014; Lothenbach et al. 2019) also improve the robustness of the predictions. This is not to say that we fully understand all the physical and chemical phenomena or that modeling is fully representative of all the processes occurring in such complex systems. These are two different aspects to be considered which call for different treatment: either acquiring more scientific knowledge through experiments, or new developments or approaches in numerical codes. Often, both issues have to be tackled together.

An example of the "lack of knowledge" issue is the question of how redox conditions are controlled in deep geological environments (e.g., see the results of redox measurements in European project RECOSY (Duro et al. 2014). Oxygen is a very reactive chemical compound, readily producing oxic conditions along its diffusive pathway during the operational stage (see dedicated section in this chapter). These oxic conditions will not remain long after the closure of the disposal cells and the onset of the corrosion phase. The re-establishment of reduced conditions in the host rock is expected within a few hundreds to thousands of years. Moreover, reduced conditions do not mean reducing conditions, i.e. the formation of reactive reductants: the reactivity of hydrogen is low, sulfate reduction is thought to be very slow and only catalyzed by microorganisms (Truche et al. 2009) and argillaceous rocks only contain small amounts of ferric compounds (e.g., in the Callovo–Oxfordian claystone). The question therefore remains about the buffering capacity of the host rock with respect to additional oxidative perturbations once the hydrogen has diffused away from the cell disposal. Such perturbations could come from the alteration of glass that will release significant amounts of ferric species. This is of course a crucial issue for the concomitant migration of RNs. It is not clear whether the host rock will contain enough reactive reductants (ferrous and/or sulfide compounds, such as pyrite) to maintain reduced conditions at this stage. Most RTM codes are ready to simulate this type of situation, since redox couples can usually be deactivated to define only the reactive ones (even H^+/H_2 in post-corrosion conditions).

Another recurrent issue in RTM is the value of reactive surface area that should be used for minerals in the simulation of these systems (see discussion in Li 2019, this volume). An example of assessment of the reactive surface of glass dissolving in water is given by Fournier et al. (2016), who demonstrated that the use of geometrical surface area provides a consistent approach whenever it can be easily calculated. When trying to apply this approach to porous media, and even more so to evolving porous media, the question is inevitably linked to the concepts of porosity network, accessibility to water, and physical and chemical heterogeneities (e.g., Landrot et al. 2012, Noiriel et al. 2012). Variable porosity is the result of the volume balance between mineral dissolution and precipitation (Seigneur et al. 2019, this volume); clogging phenomena are predicted at the interfaces between reactive materials in the repository because secondary minerals take up more volume than primary minerals. This is particularly true for metallic components, especially if technological gaps are rapidly filled by swelling clay upon rehydration (Wilson et al. 2015). This behavior is also predicted at the interface between concrete and clay. No consensus has been reached today about how the RT processes are affected by porosity clogging, i.e. how the macroscopic mineral reaction rate changes when porosity vanishes. This is a domain where RT simulations at the pore scale could help decipher the mechanisms controlling the dissolution and precipitation rate (Molins and Knabner 2019, this volume). A corollary aspect of this question brings us to the way we consider the evolution of transport properties in (strongly) altered zones. For instance, the diffusion properties in the corrosion layer, which controls the mineral sequence of corrosion products, had to be set to very low values to reproduce the observed paragenesis (Bildstein et al. 2016). In these situations, Kozeny–Carman and Archie's law fail to reproduce experimental observations and thermodynamic properties such as solubility are also modified if pores reach micrometric size (e.g. Emmanuel and Berkowitz 2007, Mürmann et al. 2013, Bergonzi et al. 2016, Rajyaguru et al. 2019). Although the precipitation of minerals, in general and in particular in this type of pores, should theoretically be described in RTM as a sequence of steps including nucleation, scavenging (or ripening), and growth, this approach has only been used by Savage et al. (2010) in the context of waste repositories. This type of model has been developed in the last decade and could be used more widely, providing that the necessary data are acquired (Fritz and Noguera 2009, Noguera et al. 2006, 2016). Combined with the thermodynamic approach in small pores, they may also shed new light on the processes controlling mineral reactivity during porosity clogging.

Recognizing that the combination of heat release in HLLW disposal cells and the production of hydrogen gas (also in the ILLW cells) will significantly delay the resaturation of the repository near field, more THC processes have been implemented to perform simulations with realistic temperature and water saturation changes during the lifetime of the repository. Multiphase RTM has been conducted, first looking at TH processes with a source of hydrogen (e.g., Xu et al. 2008, Zheng et al. 2008, Senger et al. 2011, Treille et al. 2012). It has progressively integrated more complex chemistry (e.g., pyrite oxidation in De Windt et al. 2014; bentonite illitization in Zheng et al. 2015), but full chemical treatment has been limited so far mainly to concrete carbonation (Trotignon et al. 2011, Thouvenot et al. 2013). The coupling between the fast diffusive transport of very reactive gas has proven to be a challenge for numerical codes. In this regard, the RTM of waste repositories could benefit from the experience return of modeling in other systems such as CO_2 sequestration (Sin and Corvisier 2019, this volume). Another aspect of unsaturated porous media that is not yet widely treated in RTM relates to the dependence of chemical reactivity with regard to the water content, in particular looking at how the mineral and gas solubility and the aqueous speciation relate to capillary pressure at low water saturation (Lassin et al. 2005, 2011, 2016). This process may be enhanced by the fact that some chemical reactions, such as steel corrosion and glass alteration, consume water; this coupled effect is only starting to be fully integrated in RTM codes (Seigneur et al. 2018).

Finally, even though large-scale 3D RTM calculations started to be tractable very recently thanks to improvements in computer calculation capacity, studies looking at waste repositories in this way are scarce (Trinchero et al. 2017). The main reason remains the large scale of the problems considered: metric to kilometric spatial domain, time period of 100,000 to millions of years. This explains, at least partly, why numerical simulations in this context have not yet fully benefited from recent advancements in RTM such as the diffusion in charged porous media (Tournassat and Steefel 2019, this volume). This feature would be particularly appropriate for clay materials encountered in deep geological storage to account for different "porosities" accessible to anions and cations. The same statement is applicable for modeling at the pore scale, which could be combined with the macroscopic approach, for instance, for problems involving porosity clogging.

REFERENCES

Alexander WR, Dayal R, Eagleson K, Eikenberg J, Hamilton E, Linklater CM, McKinley IG, Tweed CJ (1992) A natural analogue of high pH cement pore waters from the Maqarin area of northern Jordan. II: Results of predictive geochemical calculations. J Geochem Explor 46:133–146

Alexander WR, Reijonen HM, McKinley IG (2015) Natural analogues: studies of geological processes relevant to radioactive waste disposal in deep geological repositories. Swiss J Geosci 108:75–100

Alonso MC, García Calvo JL, Hidalgo A, Fernández Luco L (2010) Development and application of low-pH concretes for structural purposes in geological repository systems. *In:* Geological Repository Systems for Safe Disposal of Spent Nuclear Fuels and Radioactive Waste. Ahn J, Apted MJ, (eds). Woodhead Publishing, p 286–322

Altmann S, Tournassat C, Goutelard F, Parneix JC, Gimmi T, Maes N (2012) Diffusion-driven transport in clayrock formations. Appl Geochem 27:463–478

Andra (2005) Dossier 2005: Argile Synthesis. Evaluation of the Feasibility of a Geological Repository in an Argillaceous Formation Meuse/Haute-Marne Site. National Agency for Radioactive Waste Management, Paris.

Andra (2009) JALON 2009 HA-MAVL - Options de conception du stockage en formation géologique profonde. Report C.NSY.ASTE.08.0429.A, National Agency for Radioactive Waste Management, Paris

Apted M, Ahn J (2010) Multiple-barrier geological repository design and operation strategies for safe disposal of radioactive materials. *In:* Geological Repository Systems for Safe Disposal of Spent Nuclear Fuels and Radioactive Waste. Ahn J, Apted MJ, (eds). Woodhead Publishing, p 3–28

Arcos D, Bruno J, Karnland O (2003) Geochemical model of the granite–bentonite–groundwater interaction at Äspö HRL (LOT experiment). Appl Clay Sci 23:219–228

Arcos D, Grandia F, Domènech C, Fernandez AM, Villar MV, Muurinen A, Carlsson T, Sellin P, Hernan P (2008) Long-term geochemical evolution of the near field repository: Insights from reactive transport modelling and experimental evidences. J Contam Hydrol 102:196–209

Aréna H, Godon N, Rebiscoul D, Podor R, Garces E, Cabie M, Mestre JP (2016) Impact of Zn, Mg, Ni and Co elements on glass alteration: Additive effects. J Nucl Mater 470:55–67

Aréna H, Godon N, Rébiscoul D, Frugier P, Podor R, Garcès E, Cabie M, Mestre JP (2017) Impact of iron and magnesium on glass alteration: Characterization of the secondary phases and determination of their solubility constants. Appl Geochem 82:119–133

Aréna H, Rébiscoul D, Podor R, Garcès E, Cabie M, Mestre JP, Godon N (2018) Impact of Fe, Mg and Ca elements on glass alteration: Interconnected processes. Geochim Cosmochim Acta 239:420–445

Armand G, Leveau F, Nussbaum C, de La Vaissiere R, Noiret A, Jaeggi D, Landrein P, Righini C (2014) Geometry and Properties of the Excavation-Induced Fractures at the Meuse/Haute-Marne URL Drifts. Rock Mech Rock Eng 47:21–41

Ashraf W (2016) Carbonation of cement-based materials: Challenges and opportunities. Constr Build Mater 120:558–570

Audoly R, Vogt-Schilb A, Guivarch C, Pfeiffer A (2018) Pathways toward zero-carbon electricity required for climate stabilization. Appl Energy 225:884–901

Auroy M, Poyet S, Le Bescop P, Torrenti JM (2013) Impact of carbonation on the durability of cementitious materials: water transport properties characterization. EPJ Web of Conferences 56 01008

Auroy M, Poyet S, Le Bescop P, Torrenti J-M, Charpentier T, Moskura M, Bourbon X (2015) Impact of carbonation on unsaturated water transport properties of cement-based materials. Cem Concr Res74:44–58

Auroy M, Poyet Sp, Le Bescop P, Torrenti J-M, Charpentier T, Moskura Ml, Bourbon X (2018) Comparison between natural and accelerated carbonation (3% CO_2): Impact on mineralogy, microstructure, water retention and cracking. Cem Concr Res 109:64–80

Avis J, Suckling P, Calder N, Walsh R, Humphreys P, King F (2013) T2GGM: A coupled gas generation model for deep geologic disposal of radioactive waste. Nucl Technol 187:175–187

Bach TTH, Coumes CCD, Pochard I, Mercier C, Revel B, Nonat A (2012) Influence of temperature on the hydration products of low pH cements. Cem Concr Res 42:805–817

Bamforth P, Baston G, Berry J, Glasser F, Heath T, Jackson C, Savage D, Sawanton S (2012) Cement materials for use as backfill, sealing and structural materials in geological disposal concepts. Report SERCO/005125/001, Serco, Harwell, Oxfordshire

Barnichon JD, De Windt L (2018) Understanding oxidizing transient conditions in clayey rocks. Appl Geochem 98:435–447

Bartier D, Techer I, Dauzères A, Boulvais P, Blanc-Valleron M-M, Cabrera J (2013) In situ investigations and reactive transport modelling of cement paste/argillite interactions in a saturated context and outside an excavated disturbed zone. Appl Geochem 31:94–108

Bary B, Mügler C (2006) Simplified modelling and numerical simulations of concrete carbonation in unsaturated conditions. Revue Européenne Génie Civil 10:1049–1072

Bary B, Sellier A (2004) Coupled moisture-carbon dioxide-calcium transfer model for carbonation of concrete. Cem Concr Res 34:1859–1872

Baston G, Heath T, Hunter F, Swanton S (2017) Modelling of cementitious backfill interactions with vitrified intermediate-level waste. Phys Chem Earth, Parts A/B/C 99:121–130

Bataillon C, Bouchon F, Chainais-Hillairet C, Desgranges C, Hoarau E, Martin F, Perrin S, Tupin M, Talandier J (2010) Corrosion modelling of iron based alloy in nuclear waste repository. Electrochim Acta 55:4451–4467

Benbow S, Wilson J, Metcalfe R, Lehikoinen J (2019) Avoiding unrealistic behaviour in coupled reactive-transport simulations of cation exchange and mineral kinetics in clays. Clay Minerals:1–11

Benet L-V, Bouillet C, Wendling J (2014a) Analysis of the ambient conditions in an IL-LLW storage cell in a deep clay repository during the waiting closure period. *In:* Clays in Natural and Engineered Barriers for Radioactive Waste Confinement. Vol 400. Norris S, Bruno J, Cathelineau M, Delage P, Fairhurst C, Gaucher EC, Hohn EH, Kalinichev A, Lalieux P, Sellin P, (eds). p 145–161

Benet L-V, Tulita C, Calsyn L, Wendling J (2014b) Evolution of temperature and humidity in an underground repository over the operation period. *In:* Clays in Natural and Engineered Barriers for Radioactive Waste Confinement. Vol 400. Norris S, Bruno J, Cathelineau M, Delage P, Fairhurst C, Gaucher EC, Hohn EH, Kalinichev A, Lalieux P, Sellin P, (eds). Geol Soc Spec Publ, p 413–426

Bergonzi I, Mercury L, Simon P, Jamme F, Shmulovich K (2016) Oversolubility in the microvicinity of solid–solution interfaces. Phys Chem Chem Phys 18:14874–14885

Berner U, Kulik DA, Kosakowski G (2013) Geochemical impact of a low-pH cement liner on the near field of a repository for spent fuel and high-level radioactive waste. Phys Chem Earth, Parts A/B/C 64:46–56

Bernier F, Lemy F, De Cannière P, Detilleux V (2017) Implications of safety requirements for the treatment of THMC processes in geological disposal systems for radioactive waste. J Rock Mech Geotech Eng 9:428–434

Bildstein O, Claret F (2015) Chapter 5—Stability of Clay Barriers Under Chemical Perturbations. *In:* Developments in Clay Science. Vol Volume 6. Tournassat C, Steefel CI, Bourg IC, Faiza B, (eds). Elsevier, p 155–188

Bildstein O, Trotignon L, Perronnet M, Jullien M (2006) Modelling iron–clay interactions in deep geological disposal conditions. Phys Chem Earth, Parts A/B/C 31:618–625

Bildstein O, Trotignon L, Pozo C, Jullien M (2007) Modelling glass alteration in an altered argillaceous environment. J Nucl Mater 362:493–501

Bildstein O, Lartigue J, Pointeau I, Cochepin B, Munier I, Michau N (2012) Chemical evolution in the near field of HLW cells: interactions between glass, steel and clay-stone in deep geological conditions. 5th ANDRA International Meeting, 22–25 Oct 2012, Montpellier, France

Bildstein O, Lartigue J-E, Schlegel ML, Bataillon C, Cochepin Bt, Munier I, Michau N (2016) Gaining insight into corrosion processes from numerical simulations of an integrated iron-claystone experiment. Geol Soc, London, Spec Publ 443:253–267

Blanc P, Bourbon X, Lassin A, Gaucher EC (2010) Chemical model for cement-based materials: Temperature dependence of thermodynamic functions for nanocrystalline and crystalline C-S-H phases. Cem Concr Res 40:851–866

Blanc P, Lassin A, Piantone P, Azaroual M, Jacquemet N, Fabbri A, Gaucher EC (2012) Thermoddem: A geochemical database focused on low temperature water/rock interactions and waste materials. Appl Geochem 27:2107–2116

Blanc P, Vieillard P, Gailhanou H, Gaboreau S, Gaucher E, Fialips CI, Made B, Giffaut E (2015a) A generalized model for predicting the thermodynamic properties of clay minerals. American Journal of Science 315:734–780

Blanc P, Vieillard P, Gailhanou H, Gaboreau S, Marty N, Claret F, Madé B, Giffaut E (2015b) ThermoChimie database developments in the framework of cement/clay interactions. Appl Geochem 55:95–107

Bond A, Benbow S, Wilson J, Millard A, Nakama S, English M, McDermott C, Garitte B (2013) Reactive and non-reactive transport modelling in partially water saturated argillaceous porous media around the ventilation experiment, Mont Terri. J Rock Mech Geotech Eng 5:44–57

Bourcier WL, Carroll SA, Phillips BL (1994) Constraints on the affinity term for modeling long-term glass dissolution rates. *In:* Scientific Basis for Nuclear Waste Management XVII. Vol 333. Barkatt A, Van Konynenburg RA, (eds). Mater. Res. Soc., Pittsburgh, PA, p 507–512

Bourdelle F, Truche L, Pignatelli I, Mosser-Ruck R, Lorgeoux C, Roszypal C, Michau N (2014) Iron–clay interactions under hydrothermal conditions: Impact of specific surface area of metallic iron on reaction pathway. Chem Geol 381:194–205

Bourdelle F, Mosser-Ruck R, Truche L, Lorgeoux C, Pignatelli I, Michau N (2017) A new view on iron-claystone interactions under hydrothermal conditions (90 °C) by monitoring in situ pH evolution and H_2 generation. Chem Geol 466:600–607

Bradbury MH, Baeyens B (2003) Porewater chemistry in compacted re-saturated MX-80 bentonite. J Contam Hydrol 61:329–338

Bradbury MH, Baeyens B (2009) Experimental and modelling studies on the pH buffering of MX-80 bentonite porewater. Appl Geochem 24:419–425

Brantley SL, Holleran ME, Jin L, Bazilevskaya E (2013) Probing deep weathering in the Shale Hills Critical Zone Observatory, Pennsylvania (USA): the hypothesis of nested chemical reaction fronts in the subsurface. Earth Surface Processes and Landforms 38:1280–1298

Bredehoeft JD (2003) From models to performance assessment: the conceptualization problem. Groundwater 41:571–577

Brendler V, Vahle A, Arnold T, Bernhard G, Fanghänel T (2003) RES3T-Rossendorf expert system for surface and sorption thermodynamics. J Contam Hydrol 61:281–291

Brommundt J, Kaempfer TU, Enssle CP, Mayer G, Wendling J (2014) Full-scale 3D modelling of a nuclear waste repository in the Callovo–Oxfordian clay. Part 1: thermo-hydraulic two-phase transport of water and hydrogen. Geological Society, London, Special Publications 400:SP400.434

Brook BW, Alonso A, Meneley DA, Misak J, Blees T, van Erp JB (2014) Why nuclear energy is sustainable and has to be part of the energy mix. Sustainable Mater Technol 1–2:8–16

Burkan Isgor O, Razaqpur AG (2004) Finite element modeling of coupled heat transfer, moisture transport and carbonation processes in concrete structures. Cement and Concrete Composites 26:57–73

Burnol A, Blanc P, Xu T, Spycher N, Gaucher EC (2006) Uncertainty in the reactive transport model response to an alkaline perturbation in a clay formation. TOUGH Symposium 2006, Berkeley, California

Chaou A, Abdelouas A, Mendilia YE, Martin C (2017) The role of pH in the vapor hydration at 175 °C of the French SON68 glass. Appl Geochem 76:22–35

Chaparro MC, Saaltink MW, Soler JM (2017) Reactive transport modelling of cement-groundwater–rock interaction at the Grimsel Test Site. Phys Chem Earth 99:64–76

Chapman N, Hooper A (2012) The disposal of radioactive wastes underground. Proc Geol Assoc 123:46–63

Chomat L, Amblard E, Varlet J, Blanc C, Bourbon X (2017) Passive corrosion of steel reinforcement in blended cement-based material in the context of nuclear waste disposal. Corrosion Eng Sci Technol 52:148–154

Cici G, Cembalo L, Del Giudice T, Palladino A (2012) Fossil energy versus nuclear, wind, solar and agricultural biomass: Insights from an Italian national survey. Energy Policy 42:59–66

Claret F, Sakharov BA, Drits VA, Velde B, Meunier A, Griffault L, Lanson B (2004) Clay minerals in the Meuse-Haute marne underground laboratory (France): Possible influence of organic matter on clay mineral evolution. Clays Clay Minerals 52:515–532

Claret F, Marty N, Tournassat C (2018a) Modeling the long-term stability of multi-barrier systems for nuclear waste disposal in geological clay formations. *In:* Reactive Transport Modeling: Applications in Subsurface Energy and Environmental Problems. Yitian X, Fiona W, Tianfu X, Carl S, (eds). John Wiley & Sons Ltd., 395–451

Claret F, Grangeon S, Loschetter A, Tournassat C, De Nolf W, Harker N, Boulahya F, Gaboreau S, Linard Y, Bourbon X, Fernandez-Martinez A (2018b) Deciphering mineralogical changes and carbonation development during hydration and ageing of a consolidated ternary blended cement paste. IUCrJ 5:150–157

Codina K, Cau-dit-Coumes C, Le Bescop P, Verdier J, Ollivier JP (2008) Design and characterization of low-heat and low-alkalinity cements. Cem Concr Res38:437–448

Craen MD, Geet MV, Honty M, Weetjens E, Sillen X (2008) Extent of oxidation in Boom Clay as a result of excavation and ventilation of the HADES URF: Experimental and modelling assessments. Phys Chem Earth, Parts A/B/C 33, Supplement 1:S350-S362

Cronstrand P (2014) Evolution of pH in SFR 1. SKB report TR-14–01, Swedish Nuclear Fuel and Waste Management Company, Stockholm, Sweden

Czarnecki L, Woyciechowski P (2015) Modelling of concrete carbonation; is it a process unlimited in time and restricted in space? Bull Polish Acad Sci Tech Sci 63:43–54

Daux V, Guy C, Advocat T, Crovisier JL, Stille P (1997) Kinetic aspects of basaltic glass dissolution at 90 °C : role of aqueous silicon and aluminium. Chem Geol 142:109–126

Dauzeres A, Achiedo G, Nied D, Bernard E, Alahrache S, Lothenbach B (2016) Magnesium perturbation in low-pH concretes placed in clayey environment-solid characterizations and modeling. Cem Concr Res 79:137–150

de Combarieu G, Barboux P, Minet Y (2007) Iron corrosion in Callovo–Oxfordian argilite: From experiments to thermodynamic/kinetic modelling. Phys Chem Earth, Parts A/B/C 32:346–358

De Marsily G, Ledoux E, Barbreau A, Margat J (1977) Nuclear Waste Disposal: Can the Geologist Guarantee Isolation? Science 197:519–527

De Windt L, Pellegrini D, van der Lee J (2004) Coupled modeling of cement/claystone interactions and radionuclide migration. J Contam Hydrol 68:165–182

De Windt L, Marsal F, Tinseau E, Pellegrini D (2008) Reactive transport modeling of geochemical interactions at a concrete/argillite interface, Tournemire site (France). Phys Chem Earth 33:S295-S305

De Windt L, Marsal F, Corvisier J, Pellegrini D (2014) Modeling of oxygen gas diffusion and consumption during the oxic transient in a disposal cell of radioactive waste. Appl Geochem 41:115–127

De Windt L, Bertron A, Larreur-Cayol S, Escadeillas G (2015) Interactions between hydrated cement paste and organic acids: Thermodynamic data and speciation modeling. Cem Concr Res 69:25–36

Debure M, Frugier P, De Windt L, Gin S (2012) Borosilicate glass alteration driven by magnesium carbonates. J Nucl Mater 420:347–361

Debure M, Frugier P, De Windt L, Gin S (2013) Dolomite effect on borosilicate glass alteration. Appl Geochem 33:237–251

Debure M, De Windt L, Frugier P, Gin S (2018) Mechanisms involved in the increase of borosilicate glass alteration by interaction with the Callovian-Oxfordian clayey fraction. Appl Geochem 98:206–220

Dillmann P, Neff D, Féron D (2014) Archaeological analogues and corrosion prediction: from past to future. A review. Corros Eng Sci Technol 49:567–576

Drouet E (2010) Impact de la température sur la carbonatation des matériaux cimentaires—prise en compte des transferts hydriques, Thèse de Doctorat. ENS Cachan.

Drouet E, Poyet Sp, Le Bescop P, Torrenti J-M, Bourbon X (2018) Carbonation of hardened cement pastes: Influence of temperature. Cem Concr Res 115:445–459

Duro L, Bruno J, Grivé M, Montoya V, Kienzler B, Altmaier M, Buckau G (2014) Redox processes in the safety case of deep geological repositories of radioactive wastes. Contribution of the European RECOSY Collaborative Project. Appl Geochem 49:206–217

El Hajj H, Abdelouas A, El Mendili Y, Karakurt G, Grambow B, Martin C (2013) Corrosion of carbon steel under sequential aerobic-anaerobic environmental conditions. Corros Sci 76:432–440

Emmanuel S, Berkowitz B (2007) Effects of pore-size controlled solubility on reactive transport in heterogeneous rock. Geophys Res Lett 34: L06404

Enssle CP, Brommundt J, Kaempfer TU, Mayer G, Wendling J (2014) Full-scale 3D modelling of a nuclear waste repository in the Callovo–Oxfordian clay. Part 2: thermo-hydraulic two-phase transport of water, hydrogen, ^{14}C and [129]I. Geol Soc, London, Spec Publ 400:469–481

Ewing RC (2015) Long-term storage of spent nuclear fuel. Nat Mater 14:252

Fernandez R, Cuevas J, Mader UK (2010) Modeling experimental results of diffusion of alkaline solutions through a compacted bentonite barrier. Cem Concr Res 40:1255–1264

Féron D, Crusset D, Gras J-M (2008) Corrosion issues in nuclear waste disposal. J Nucl Mater 379:16–23

Ferrand K, Lui S, Lemmens K (2013) The interaction between nuclear waste glass dans ordinary Portland cement. Int J Appl Glass Sci 4:328–340

Ferrand K, Liu S, Lemmens K (2014) The effect of ordinary portland cement on nuclear waste glass dissolution. Procedia Mater Sci 7:223–229

Fournier M, Ull A, Nicoleau E, Inagaki Y, Odorico M, Frugier P, Gin S (2016) Glass dissolution rate measurement and calculation revisited. J Nucl Mater 476:140–154

Fournier M, Gin S, Frugier P, Mercado-Depierre S (2017) Contribution of zeolite-seeded experiments to the understanding of resumption of glass alteration. Mater Degradation 1:17

Fritz B, Noguera C (2009) Mineral precipitation kinetics. Rev Mineral Geochem 70:371–410

Frugier P, Godon N (2018) Effet de la stœchiométrie Si-Mg des produits d'altération sur la modélisation GRAAL. CEA report DTCD/SEVT/2018–20, 31 p.

Frugier P, Gin S, Minet Y, Chave T, Bonin B, Godon N, Lartigue JE, Jollivet P, Ayral A, De Windt L, Santarini G (2008) SON68 Nuclear glass dissolution kinetics: Current state of knowledge and basis of the new GRAAL model. J Nucl Mater 380:8–21

Frugier P, Chave T, Gin S, Lartigue JE (2009) Application of the GRAAL model to leaching experiments with SON68 nuclear glass in initially pure water. J Nucl Mater 392:552–567

Frugier P, Fournier M, Gin S (2017) Modeling resumption of glass alteration due to zeolites precipitation. Procedia Earth Planet Sci 17:340–343

Gaboreau S, Prêt D, Tinseau E, Claret F, Pellegrini D, Stammose D (2011) 15 years of in situ cement–argillite interaction from Tournemire URL: Characterisation of the multi-scale spatial heterogeneities of pore space evolution. Appl Geochem 26:2159–2171

Gaboreau S, Prêt D, Montouillout V, Henocq P, Robinet J-C, Tournassat C (2017) Quantitative mineralogical mapping of hydrated low pH concrete. Cem Concr Compos 83:360–373

Gailhanou H, van Miltenburg JC, Rogez J, Olives J, Amouric M, Gaucher EC, Blanc P (2007) Thermodynamic properties of anhydrous smectite MX-80, illite IMt-2 and mixed-layer illite-smectite ISCz-1 as determined by calorimetric methods. Part I: Heat capacities, heat contents and entropies. Geochim Cosmochim Acta 71:5463–5473

Gailhanou H, Rogez J, van Miltenburg JC, van Genderen ACG, Greneche JM, Gilles C, Jalabert D, Michau N, Gaucher EC, Blanc P (2009) Thermodynamic properties of chlorite CCa-2. Heat capacities, heat contents and entropies. Geochim Cosmochim Acta 73:4738–4749

Gailhanou H, Blanc P, Rogez J, Mikaelian G, Kawaji H, Olives J, Amouric M, Denoyel R, Bourrelly S, Montouillout V, Vieillard P (2012) Thermodynamic properties of illite, smectite and beidellite by calorimetric methods: Enthalpies of formation, heat capacities, entropies and Gibbs free energies of formation. Geochim Cosmochim Acta 89:279–301

Gailhanou H, Blanc P, Rogez J, Mikaelian G, Horiuchi K, Yamamura Y, Saito K, Kawaji H, Warmont F, Grenèche JM, Vieillard P (2013) Thermodynamic properties of saponite, nontronite, and vermiculite derived from calorimetric measurements. Am Mineral 98:1834–1847

Gailhanou H, Vieillard P, Blanc P, Lassin A, Denoyel R, Bloch E, De Weireld G, Gaboreau S, Fialips CI, Madé B, Giffaut E (2017) Methodology for determining the thermodynamic properties of smectite hydration. Appl Geochem 82:146–163

García Calvo JL, Hidalgo A, Alonso C, Fernández Luco L (2010) Development of low-pH cementitious materials for HLRW repositories: Resistance against ground waters aggression. Cem Concr Res 40:1290–1297

Gaucher EC, Blanc P (2006) Cement/clay interactions—A review: Experiments, natural analogues, and modeling. Waste Manage 26:776–788

Gaucher E, Robelin C, Matray JM, Negral G, Gros Y, Heitz JF, Vinsot A, Rebours H, Cassagnabere A, Bouchet A (2004a) ANDRA underground research laboratory: interpretation of the mineralogical and geochemical data acquired in the Callovian-Oxfordian formation by investigative drilling. Phys Chem Earth 29:55–77

Gaucher EC, Blanc P, Matray JM, Michau N (2004b) Modeling diffusion of an alkaline plume in a clay barrier. Appl Geochem 19:1505–1515

Geckeis H, Rabung T (2008) Actinide geochemistry: From the molecular level to the real system. J Contam Hydrol 102:187–195

Giffaut E, Grivé M, Blanc P, Vieillard P, Colàs E, Gailhanou H, Gaboreau S, Marty N, Madé B, Duro L (2014) Andra thermodynamic database for performance assessment: ThermoChimie. Appl Geochem 49:225–236

Gin S, Jollivet P, Mestre J, Jullien M, Pozo C (2001) French SON 68 nuclear glass alteration mechanisms on contact with clay media. Appl Geochem 16:861–881

Gin S, Abdelouas A, Criscenti LJ, Ebert WL, Ferrand K, Geisler T, Harrison MT, Inagaki Y, Mitsui S, Mueller KT, Marra JC (2013) An international initiative on long-term behavior of high-level nuclear waste glass. Mater Today 16:243–248

Gin S, Jollivet P, Fournier M, Angeli F, Frugier P, Charpentier T (2015) Origin and consequences of silicate glass passivation by surface layers. Nat Commun 6:6360

Gin S, Neill L, Fournier M, Frugier P, Ducasse T, Tribet M, Abdelouas A, Parruzot B, Neeway J, Wall N (2016) The controversial role of inter-diffusion in glass alteration. Chem Geol 440:115–123

Gin S, Collin M, Jollivet P, Fournier M, Minet Y, Dupuy L, Mahadevan T, Kerisit S, Du J (2018) Dynamics of self-reorganization explains passivation of silicate glasses. Nat Commun 9:2169

Glassley WE, Nitao JJ, Grant CW (2003) Three-dimensional spatial variability of chemical properties around a monitored waste emplacement tunnel. J Contam Hydrol 62–63:495–507

Godon N, Vernaz E, Thomassin JH, Touray JC (1989) Effect of environmental materials on aqueous corrosion of R7T7 glass. *In:* Scientific Basis for Nuclear Waste Management XII. Vol 127. Werner L, (ed) Mater Res Soc, Pittsburgh, PA, p 97–104

Gong WL, Lutze W, Abdelouas A, Ewing RC (1999) Vitrification of radioactive waste by reaction sintering under pressure. J Nucl Mater 265:12–21

Grambow B (2008) Mobile fission and activation products in nuclear waste disposal. J Contam Hydrol 102:180–186

Grambow B, Bretesché S (2014) Geological disposal of nuclear waste: II. From laboratory data to the safety analysis—Addressing societal concerns. Appl Geochem 49:247–258

Grenthe I (1991) Thermodynamics in migration chemistry. Radiochimica Acta 52–53:425–432

Groves GW, Rodway DI, Richardson IG (1990) The carbonation of hardened cement pastes. Adv Cem Res 3:117–125

Hess HH (1957) The Disposal of Radioactive Waste on Land: Report of the Committee on Waste Disposal of the Division of Earthsciences [chairman Harry H. Hess]. National Academy of Sciences, National Research Council

Hoch A, Wendling J (2011) Migration of gases around a cell containing high-activity vitrified wastes during the operational phase. Phys Chem Earth, Parts A/B/C 36:1743–1753

Hoerlé S, Mazaudier F, Dillmann P, Santarini G (2004) Advances in understanding atmospheric corrosion of iron. II. Mechanistic modelling of wet–dry cycles. Corros Sci 46:1431–1465

Höglund L (2001) Modelling of long-term concrete degradation processes in the Swedish SFR repository. SKB report TR-01-08, Swedish Nuclear Fuel and Waste Management Company, Stockholm, Sweden

Höglund L (2013) The impact of concrete degradation on the BMA barrier functions. SKB report TR-13-40, Swedish Nuclear Fuel and Waste Management Company, Stockholm, Sweden

Huang Y, Shao H, Wieland E, Kolditz O, Kosakowski G (2018) A new approach to coupled two-phase reactive transport simulation for long-term degradation of concrete. Constr Build Mater 190:805–829

Hummel W, Schneider JW (2005) Safety of nuclear waste repositories. Chimia 59:909–915

Hummel W, Berner U, Curti E, Pearson F, Thoenen T (2002) Nagra/PSI chemical thermodynamic data base 01/01. Radiochim Acta 90:805–813

Hunter F, Bate F, Heath T, Hoch A (2007) Geochemical investigation of iron transport into bentonite as steel corrodes. SKB report TR-07-09

Jenni A, Mäder U, Lerouge C, Gaboreau S, Schwyn B (2014) In situ interaction between different concretes and Opalinus Clay. Phys Chem Earth, Parts A/B/C

Jenni A, Gimmi T, Alt-Epping P, Mader U, Cloet V (2017) Interaction of ordinary Portland cement and Opalinus Clay: Dual porosity modelling compared to experimental data. Phys Chem Earth 99:22–37

Johnson L, King F (2008) The effect of the evolution of environmental conditions on the corrosion evolutionary path in a repository for spent fuel and high-level waste in Opalinus Clay. J Nucl Mater 379:9–15

Johnson B, Newman A, King J (2017) Optimizing high-level nuclear waste disposal within a deep geologic repository. Ann Oper Res 253:733–755

Jollivet P, Frugier P, Parisot G, Mestre JP, Brackx E, Gin S (2012) Effect of clayey groundwater on the dissolution rate of the simulated nuclear waste glass SON68. J Nucl Mater 420:508–518

Khoury HN, Salameh E, Abdul-Jaber Q (1985) Characteristics of an unusual highly alkaline water from the Maqarin area, northern Jordan. J Hydrol 81:79–91

Khoury HN, Salameh E, Clark ID, Fritz P, Bajjali W, Milodowski AE, Cave MR, Alexander WR (1992) A natural analogue of high pH cement pore waters from the Maqarin area of northern Jordan. I: introduction to the site. J Geochem Explor 46:117–132

King F (2014) Predicting the lifetimes of nuclear waste containers. JOM 66:526–537

King F, Shoesmith DW (2010) 13 - Nuclear waste canister materials, corrosion behaviour and long-term performance in geological repository systems. In: Geological Repository Systems for Safe Disposal of Spent Nuclear Fuels and Radioactive Waste. Ahn J, Apted MJ, (eds). Woodhead Publishing, p 379–420

King F, Kolar M, Keech PG (2014) Simulations of long-term anaerobic corrosion of carbon steel containers in Canadian deep geological repository. Corros Eng Sci Technol 49:455–459

Knapp V, Pevec D (2018) Promises and limitations of nuclear fission energy in combating climate change. Energy Policy 120:94–99

Kolditz O, Bauer S, Bilke L, Böttcher N, Delfs J, Fischer T, Görke U, Kalbacher T, Kosakowski G, McDermott C (2012) OpenGeoSys: an open-source initiative for numerical simulation of thermo-hydro-mechanical/chemical (THM/C) processes in porous media. Environ Earth Sci 67:589–599

Kulik DA, Wagner T, Dmytrieva SV, Kosakowski G, Hingerl FF, Chudnenko KV, Berner UR (2013) GEM-Selektor geochemical modeling package: revised algorithm and GEMS3K numerical kernel for coupled simulation codes. Comput Geosci 17:1–24

Kulik DA, Hummel W, Lützenkirchen J, Lefèvre G (2015) Preface: SI: Geochemical speciation codes and databases. Appl Geochem 55:1–2

Kursten B, Druyts F (2008) Methodology to make a robust estimation of the carbon steel overpack lifetime with respect to the Belgian Supercontainer design. J Nucl Mater 379:91–96

Kursten B, Macdonald DD, Smart NR, Gaggiano R (2017) Corrosion issues of carbon steel radioactive waste packages exposed to cementitious materials with respect to the Belgian supercontainer concept. Corros Eng Sci Technol 52:11–16

L'Hostis V, Amblard E, Blanc C, Miserque F, Paris C, Bellot-Gurlet L (2011) Passive corrosion of steel in concrete in context of nuclear waste disposal. Corros Eng Sci Technol 46:177–181

Lalan P, Dauzeres A, De Windt L, Bartier D, Sammaljarvi J, Barnichon JD, Techer I, Detilleux V (2016) Impact of a 70 °C temperature on an ordinary Portland cement paste/claystone interface: An in situ experiment. Cem Concr Res 83:164–178

Landais P, Aranyossy J-F (2011) Clays in natural and engineered barriers for radioactive waste confinement. Phys Chem Earth, Parts A/B/C 36:1437

Landrot G, Ajo-Franklin JB, Yang L, Cabrini S, Steefel CI (2012) Measurement of accessible reactive surface area in a sandstone, with application to CO_2 mineralization. Chem Geol 318–319:113–125

Lassin A, Azaroual M, Mercury L (2005) Geochemistry of unsaturated soil systems: Aqueous speciation and solubility of minerals and gases in capillary solutions. Geochim Cosmochim Acta 69:5187–5201

Lassin A, Dymitrowska M, Azaroual M (2011) Hydrogen solubility in pore water of partially saturated argillites: Application to Callovo–Oxfordian clayrock in the context of a nuclear waste geological disposal. Phys Chem Earth, Parts A/B/C 36:1721–1728

Lassin A, Marty NCM, Gailhanou H, Henry B, Trémosa J, Lerouge C, Madé B, Altmann S, Gaucher EC (2016) Equilibrium partial pressure of CO_2 in Callovian–Oxfordian argillite as a function of relative humidity: Experiments and modelling. Geochim Cosmochim Acta 186:91–104

Lerouge C, Vinsot A, Grangeon S, Wille G, Flehoc C, Gailhanou H, Gaucher EC, Madé B, Altmann S, Tournassat C (2013) Controls of Ca/Mg/Fe activity ratios in pore water chemistry models of the Callovian–Oxfordian Clay formation. Procedia Earth Planet Sci 7:475–478

Lerouge C, Gaboreau S, Grangeon S, Claret F, Warmont F, Jenni A, Cloet V, Mäder U (2017) In situ interactions between opalinus clay and low alkali concrete. Phys Chem Earth, Parts A/B/C 99:3–21

Lerouge C, Robinet JC, Debure M, Tournassat C, Bouchet A, Fernández AM, Flehoc C, Guerrot C, Kars M, Lagroix F, Landrein P (2018) A deep alteration and oxidation profile in a shallow clay aquitard: Example of the Tégulines Clay, East Paris Basin, France. Geofluids 2018:20

Leterrier N, Bary B (2011) Fully Coupled Unsaturated Hydraulics and Reactive Transport Model for the Simulation of Concrete Carbonation. J Nucl Res Dev 2:11–18

Li L (2019) Watershed reactive transport. Rev Mineral Geochem 85:381–418

Lichtner PC (1996) Continuum formulation of multicomponent-multiphase reactive transport. *In:* Reactive Transp Porous Media. Vol 34. Lichtner PC, Steefel CI, Oelkers EH, (eds). p 1–81

Linsley G, Fattah A (1994) The interface between nuclear safeguards and radioactive waste disposal: Emerging issues. IAEA Bulletin 36:22–26

Lisjak A, Garitte B, Grasselli G, Müller HR, Vietor T (2015) The excavation of a circular tunnel in a bedded argillaceous rock (Opalinus Clay): Short-term rock mass response and FDEM numerical analysis. Tunnelling and Underground Space Technol 45:227–248

Liu S, Ferrand K, Lemmens K (2015) Transport and surface reaction-controlled SON68 glass dissolution at 30 °C and 70 °C and pH=13.7. Appl Geochem 61:302–311

Lothenbach B (2010) Thermodynamic equilibrium calculations in cementitious systems. Mater Struct 43:1413–1433

Lothenbach B, Wieland E (2006) A thermodynamic approach to the hydration of sulphate-resisting Portland cement. Waste Manage 26:706–719

Lothenbach B, Winnefeld F (2006) Thermodynamic modelling of the hydration of Portland cement. Cem Concr Res 36:209–226

Lothenbach B, Rentsch D, Wieland E (2014) Hydration of a silica fume blended low-alkali shotcrete cement. Phys Chem Earth 70–71:3–16

Lothenbach B, Le Saout G, Ben Haha M, Figi R, Wieland E (2012) Hydration of a low-alkali CEM III/B–SiO$_2$ cement (LAC). Cem Concr Res 42:410–423

Lothenbach B, Kulik DA, Matschei T, Balonis M, Baquerizo L, Dilnesa B, Miron GD, Myers RJ (2019) Cemdata 18: A chemical thermodynamic database for hydrated Portland cements and alkali-activated materials. Cem Concr Res 115:472–506

Lu C, Samper J, Fritz B, Clement A, Montenegro L (2011) Interactions of corrosion products and bentonite: An extended multicomponent reactive transport model. Phys Chem Earth, Parts A/B/C 36:1661–1668

Luke K, Lachowski E (2008) Internal composition of 20-year-old fly ash and slag-blended ordinary Portland cement Pastes. J Am Ceram Soc 91:4084–4092

Macdonald DD, Urquidi-Macdonald M, Engelhardt GR, Azizi O, Saleh A, Almazooqi A, Rosas-Camacho O (2011) Some important issues in electrochemistry of carbon steel in simulated concrete pore water Part 1: Theoretical issues. Corros Eng Sci Technol 46:98–103

Marschall P, Horseman S, Gimmi T (2005) Characterisation of gas transport properties of the Opalinus Clay, a potential host rock formation for radioactive waste disposal. Oil Gas Sci Technol – Rev IFP 60:121–139

Martin FA, Bataillon C, Schlegel ML (2008) Corrosion of iron and low alloyed steel within a water saturated brick of clay under anaerobic deep geological disposal conditions: An integrated experiment. J Nucl Mater 379:80–90

Martin LHJ, Leemann A, Milodowski AE, Mader UK, Munch B, Giroud N (2016) A natural cement analogue study to understand the long-term behaviour of cements in nuclear waste repositories: Maqarin (Jordan). Appl Geochem 71:20–34

Marty NCM, Tournassat C, Burnol A, Giffaut E, Gaucher EC (2009) Influence of reaction kinetics and mesh refinement on the numerical modelling of concrete/clay interactions. J Hydrol 364:58–72

Marty NCM, Fritz B, Clément A, Michau N (2010) Modelling the long term alteration of the engineered bentonite barrier in an underground radioactive waste repository. Appl Clay Sci 47:82–90

Marty NM, Munier I, Gaucher E, Tournassat C, Gaboreau S, Vong C, Giffaut E, Cochepin B, Claret F (2014) Simulation of Cement/Clay Interactions: Feedback on the Increasing Complexity of Modelling Strategies. Transp Porous Media 104:385–405

Marty NCM, Claret F, Lassin A, Tremosa J, Blanc P, Madé B, Giffaut E, Cochepin B, Tournassat C (2015a) A database of dissolution and precipitation rates for clay-rocks minerals. Appl Geochem 55:108–118

Marty NC, Bildstein O, Blanc P, Claret F, Cochepin B, Gaucher EC, Jacques D, Lartigue JE, Liu S, Mayer KU, Meeussen JC (2015b) Benchmarks for multicomponent reactive transport across a cement/clay interface. Comput Geosci 19:635–653

Matray JM, Savoye S, Cabrera J (2007) Desaturation and structure relationships around drifts excavated in the well-compacted Tournemire's argillite (Aveyron, France). Eng Geol 90:1–16

Mayer KU, Frind EO, Blowes DW (2002) Multicomponent reactive transport modeling in variably saturated porous media using a generalized formulation for kinetically controlled reactions. Water Resour Res 38:13–11-13–21

McGrail BP, Chick LA (1984) Initial resultats for the experimental evaluation of a nuclear waste reposity source term model. Nucl Technol 69:114–118

Meeussen JCL (2003) ORCHESTRA: An object-oriented framework for implementing chemical equilibrium models. Environ Sci Technol 37:1175–1182

Mercado-Depierre S, Angeli F, Frizon F, Gin S (2013) Antagonist effects of calcium on borosilicate glass alteration. J Nucl Mater 441:402–410

Mercado-Depierre S, Fournier M, Gin S, Angeli F (2017) Influence of zeolite precipitation on borosilicate glass alteration under hyperalkaline conditions. J Nucl Mater 491:67–82

Meserve RA (2004) Global warming and nuclear power. Science 303:433

Molins S, Knabner P (2019) Multiscale approaches in reactive transport modeling. Rev Mineral Geochem 85:27–48

Mon A, Samper J, Montenegro L, Naves A, Fernandez J (2017) Long-term non-isothermal reactive transport model of compacted bentonite, concrete and corrosion products in a HLW repository in clay. J Contam Hydrol 197:1–16

Montarnal P, Mugler C, Colin J, Descostes M, Dimier A, Jacquot E (2007) Presentation and use of a reactive transport code in porous media. Phys Chem Earth 32:507–517

Montes-H G, Fritz B, Clement A, Michau N (2005) Modeling of transport and reaction in an engineered barrier for radioactive waste confinement. Appl Clay Sci29:155–171

Mosser-Ruck R, Cathelineau M, Guillaume D, Charpentier D, Rousset D, Barres O, Michau N (2010) Effects of temperature, ph, and iron/clay and liquid/clay ratios on experimental conversion of dioctahedral smectite to berthierine, chlorite, vermiculite, or saponite. Clays Clay Miner 58:280–291

Müller HR, Garitte B, Vogt T, Köhler S, Sakaki T, Weber H, Spillmann T, Hertrich M, Becker JK, Giroud N, Cloet V (2018) Implementation of the full-scale emplacement (FE) experiment at the Mont Terri rock laboratory. Swiss J Geosci 110:287–306

Mürmann M, Kühn M, Pape H, Clauser C (2013) Numerical simulation of pore size dependent anhydrite precipitation in geothermal reservoirs. Energy Proced 40:107–116

Nagra (2002) Project Opalinus Clay. Models, Codes and Data for Safety Assessment. Demonstration of disposal feasibility for spent fuel, vitrified high-level waste and long-lived intermediate-level waste. NAGRA Technical Report 02–06

Necib S, Linard Y, Crusset D, Michau N, Daumas S, Burger E, Romaine A, Schlegel ML (2016) Corrosion at the carbon steel-clay borehole water and gas interfaces at 85 °C under anoxic and transient acidic conditions. Corrosion Science 111:242–258

Ngo VV, Delalande M, Clément A, Michau N, Fritz B (2014) Coupled transport-reaction modeling of the long-term interaction between iron, bentonite and Callovo–Oxfordian claystone in radioactive waste confinement systems. Appl Clay Sci101:430–443

Ngo VV, Clément A, Michau N, Fritz B (2015) Kinetic modeling of interactions between iron, clay and water: Comparison with data from batch experiments. Appl Geochem 53:13–26

Noguera C, Fritz B, Clément A, Baronnet A (2006) Nucleation, growth and ageing scenarios in closed systems I: A unified mathematical framework for precipitation, condensation and crystallization. J Cryst Growth 297:180–186

Noguera C, Fritz B, Clément A (2016) Kinetics of precipitation of non-ideal solid-solutions in a liquid environment. Chem Geol 431:20–35

Noiriel C, Steefel CI, Yang L, Ajo-Franklin J (2012) Upscaling calcium carbonate precipitation rates from pore to continuum scale. Chem Geol 318–319:60–74

Odorowski M, Jegou C, De Windt L, Broudic V, Jouan G, Peuget S, Martin C (2017) Effect of metallic iron on the oxidative dissolution of UO_2 doped with a radioactive alpha emitter in synthetic Callovian-Oxfordian groundwater. Geochim Cosmochim Acta 219:1–21

Ondraf/Niras (2009) The long-term safety strategy for the geological disposal of radioactive waste, SFC1 level 4 report: second full draft', Report no. NIROND-TR 2009–12E, Ondraf/Niras, Brussels, Belgium.

Park DC (2008) Carbonation of concrete in relation to CO_2 permeability and degradation of coatings. Construction and Building Materials 22:2260–2268

Parkhurst DL, Appelo C (1999) User's guide to PHREEQC (Version 2): A computer program for speciation, batch-reaction, one-dimensional transport, and inverse geochemical calculations.

Peña J, Torres E, Turrero MJ, Escribano A, Martin PL (2008) Kinetic modelling of the attenuation of carbon steel canister corrosion due to diffusive transport through corrosion product layers. Corrosion Science 50:2197–2204

Pfingsten W, Paris B, Soler J, Mäder U (2006) Tracer and reactive transport modelling of the interaction between high-pH fluid and fractured rock: Field and laboratory experiments. Journal of Geochemical Exploration 90:95–113

Poinssot C, Gin S (2012) Long-term Behavior Science: The cornerstone approach for reliably assessing the long-term performance of nuclear waste. J Nucl Mater420:182–192

Poinssot C, Fillet C, Gras JM (2010) 14 - Post-containment performance of geological repository systems: source-term release and radionuclide migration in the near- and far-field environments. *In:* Geological Repository Systems for Safe Disposal of Spent Nuclear Fuels and Radioactive Waste. Ahn J, Apted MJ, (eds). Woodhead Publishing, p 421–493

Poller A, Enssle CP, Mayer G, Croisé J, Wendling J (2011) Repository-scale modeling of the long-term hydraulic perturbation induced by gas and heat generation in a geological repository for high-and intermediate-level radioactive waste: methodology and example of application. Transp Porous Media 90:77–94

Pozo C, Bildstein O, Raynal J, Jullien M, Valcke E (2007) Behaviour of silicon released during alteration of nuclear waste glass in compacted clay. Appl Clay Sci 35:258–267

Ragoussi M-E, Costa D (2019) Fundamentals of the NEA Thermochemical database and its influence over national nuclear programs on the performance assessment of deep geological repositories. J Environ Radioact 196:225–231

Rajmohan N, Frugier P, Gin S (2010) Composition effects on synthetic glass alteration mechanisms: Part 1. Experiments. Chem Geol 279:106–119

Rajyaguru A, L'Hôpital E, Savoye S, Wittebroodt C, Bildstein O, Arnoux P, Detilleux V, Fatnassi I, Gouze P, Lagneau V (2019) Experimental characterization of coupled diffusion reaction mechanisms in low permeability chalk. Chem Geol 503:29–39

Rebiscoul D, Tormos V, Godon N, Mestre JP, Cabie M, Amiard G, Foy E, Frugier P, Gin S (2015) Reactive transport processes occurring during nuclear glass alteration in presence of magnetite. Appl Geochem 58:26–37

Ribet S, Gin S (2004) Role of neoformed phases on the mechanisms controlling the resumption of SON68 glass alteration in alkaline media. J Nucl Mater324:152–164

Roosz C, Vieillard P, Blanc P, Gaboreau S, Gailhanou H, Braithwaite D, Montouillout V, Denoyel R, Henocq P, Made B (2018) Thermodynamic properties of C-S-H, C-A-S-H and M-S-H phases: Results from direct measurements and predictive modelling. Appl Geochem 92:140–156

Saheb M, Berger P, Raimbault L, Neff D, Dillmann P (2012) Investigation of iron long-term corrosion mechanisms in anoxic media using deuterium tracing. J Nucl Mater 423:61–66

Samper J, Yang C, Montenegro L (2003) User's manual of CORE2D Version 4: A code for ground-water flow and reactive solute transport: La Coruña. Universidad de A Coruña, Spain

Samper J, Lu C, Montenegro L (2008) Reactive transport model of interactions of corrosion products and bentonite. Phys Chem Earth, Parts A/B/C 33:S306-S316

Samper J, Naves A, Montenegro L, Mon A (2016) Reactive transport modelling of the long-term interactions of corrosion products and compacted bentonite in a HLW repository in granite: Uncertainties and relevance for performance assessment. Appl Geochem 67:42–51

Samper J, Mon A, Montenegro L, Cuevas J, Turrero MJ, Naves A, Fernandez R, Torres E (2018) Coupled THCM model of a heating and hydration concrete-bentonite column test. Appl Geochem 94:67–81

Savage D, Cloet V (2018) A review of cement–clay modelling, Nagra Working Report NAB 18-24, Nagra, Wettingen, Switzerland.

Savage D, Noy D, Mihara M (2002) Modelling the interaction of bentonite with hyperalkaline fluids. Appl Geochem 17:207–223

Savage D, Walker C, Arthur R, Rochelle C, Oda C, Takase H (2007) Alteration of bentonite by hyperalkaline fluids: A review of the role of secondary minerals. Phys Chem Earth 32:287–297

Savage D, Watson C, Benbow S, Wilson J (2010) Modelling iron-bentonite interactions. Appl Clay Sci 47:91–98

Savija B, Lukovic M (2016) Carbonation of cement paste: Understanding, challenges, and opportunities. Construction and Building Materials 117:285–301

Schlegel ML, Bataillon C, Blanc C, Prêt D, Foy E (2010) Anodic activation of iron corrosion in clay media under water-saturated conditions at 90°C: Characterization of the corrosion interface. Environ Sci Technol 44:1503–1508

Schlegel ML, Bataillon C, Brucker F, Blanc C, Prêt D, Foy E, Chorro M (2014) Corrosion of metal iron in contact with anoxic clay at 90°C: Characterization of the corrosion products after two years of interaction. Appl Geochem 51:1–14

Schlegel ML, Necib S, Daumas S, Blanc C, Foy E, Trcera N, Romaine A (2016) Microstructural characterization of carbon steel corrosion in clay borehole water under anoxic and transient acidic conditions. Corros Sci 109:126–144

Schlegel ML, Necib S, Daumas S, Labat M, Blanc C, Foy E, Linard Y (2018) Corrosion at the carbon steel-clay borehole water interface under anoxic alkaline and fluctuating temperature conditions. Corros Sci 136:70–90

Sedighi M, Thomas HR, Al Masum S, Vardon PJ, Nicholson D, Chen Q (2015) Geochemical modelling of hydrogen gas migration in an unsaturated bentonite buffer. Geol Soc , London, Spec Publ 415:189

Seigneur N, Lagneau V, Corvisier J, Dauzères A (2018) Recoupling flow and chemistry in variably saturated reactive transport modelling - An algorithm to accurately couple the feedback of chemistry on water consumption, variable porosity and flow. Adv Water Resour 122:355–366

Seigneur N, Mayer KU, Steefel CI (2019) Reactive transport in evolving porous media. Rev Mineral Geochem 85:197–238

Sellin P, Leupin OX (2013) The use of clay as an engineered barrier in radioactive-waste management - a review. Clays Clay Miner 61:477–498

Sena C, Salas J, Arcos D (2010) Aspects of geochemical evolution of the SKB near field in the frame of SR-Site. SKB report TR-10–59, Swedish Nuclear Fuel and Waste Management Company, Stockholm, Sweden

Senger R, Ewing J, Zhang K, Avis J, Marschall P, Gaus I (2011) Modeling approaches for investigating gas migration from a deep low/intermediate level waste repository (Switzerland). Transp Porous Media 90:113–133

Sercombe J, Gwinner B, Tiffreau C, Simondi-Teisseire B, Adenot F (2006) Modelling of bituminized radioactive waste leaching. Part I: Constitutive equations. J Nucl Mater 349:96–106

Shao HB, Kosakowski G, Berner U, Kulik DA, Mader U, Kolditz O (2013) Reactive transport modeling of the clogging process at Maqarin natural analogue site. Phys Chem Earth 64:21–31

Sherar BWA, Keech PG, Shoesmith DW (2011) Carbon steel corrosion under anaerobic-aerobic cycling conditions in near-neutral pH saline solutions. Part 2: Corrosion mechanism. Corros Sci 53:3643–3650

Sin I, Corvisier J (2019) Multiphase multicomponent reactive transport and flow modeling. Rev Mineral Geochem 85:143–195

SKB (2006) Long-term safety for KBS-3 repositories at Forsmark and Laxemar – a first evaluation Main Report of the SR-Can project, Technical Report TR-06–09, Svensk Kärnbränslehantering AB, Sweden.

Smart NR, Rance AP, Carlson L, Werme LOC (2006) Further studies of the anaerobic corrosion of steel in bentonite. MRS Proceedings 932:32.31

Smart NR, Rance AP, Fennell PAH, Kursten B (2013) The anaerobic corrosion of carbon steel in alkaline media: Phase 2 results. EPJ Web of Conferences 56

Soler JM (2003) Reactive transport modeling of the interaction between a high-pH plume and a fractured marl: the case of Wellenberg. Appl Geochem 18:1555–1571

Soler JM (2016) Two-dimensional reactive transport modeling of the alteration of a fractured limestone by hyperalkaline solutions at Maqarin (Jordan). Appl Geochem 66:162–173

Soler JM, Mader UK (2007) Mineralogical alteration and associated permeability changes induced by a high-pH plume: Modeling of a granite core infiltration experiment. Appl Geochem 22:17–29

Spycher NF, Sonnenthal EL, Apps JA (2003) Fluid flow and reactive transport around potential nuclear waste emplacement tunnels at Yucca Mountain, Nevada. J Contam Hydrol 62–63:653–673

Steefel CI, Vancappellen P (1990) A new kinetic approach to modeling water–rock interaction—the role of nucleation, precursors, and Ostwald ripening. Geochim Cosmochim Acta 54:2657–2677

Steefel CI, Lichtner PC (1998) Multicomponent reactive transport in discrete fractures: II: Infiltration of hyperalkaline groundwater at Maqarin, Jordan, a natural analogue site. J Hydrol 209:200–224

Steefel CI, DePaolo DJ, Lichtner PC (2005) Reactive transport modeling: An essential tool and a new research approach for the Earth sciences. Earth Planet Sci Lett 240:539–558

Steefel CI, Appelo CA, Arora B, Jacques D, Kalbacher T, Kolditz O, Lagneau V, Lichtner PC, Mayer KU, Meeussen JC, Molins S (2014) Reactive transport codes for subsurface environmental simulation. Comput Geosci:1–34

Steefel C, Beckingham L, Landrot G (2015a) Micro-continuum approaches for modeling pore–scale geochemical processes. Rev Mineral Geochem 80:217–246

Steefel CI, Appelo CA, Arora B, Jacques D, Kalbacher T, Kolditz O, Lagneau V, Lichtner PC, Mayer KU, Meeussen JC, Molins S (2015b) Reactive transport codes for subsurface environmental simulation. Comput Geosci 19:445–478

Talandier J, Mayer G, Croisé J (2006) Simulations of the hydrogen migration out of intermediate-level radioactive waste disposal drifts using tough2. TOUGH Symposium, May 15–17, 2006, Lawrence Berkeley National Laboratory, Berkeley, California

Thatcher KE, Bond AE, Norris S (2016) Engineered damage zone sealing during a water injection test at the Tournemire URL. Environ Earth Sci 75:933

Thiery M (2005) Modelling of atmospheric carbonation of cement based materials considering the kinetic effects and modifications of the microstructure and the hydric state. PhD thesis, Ecole des Ponts ParisTech, Paris

Thiery M, Villain G, Dangla P, Platret G (2007) Investigation of the carbonation front shape on cementitious materials: Effects of the chemical kinetics. Cem Concr Res 37:1047–1058

Thiery M, Dangla P, Belin P, Habert G, Roussel N (2013) Carbonation kinetics of a bed of recycled concrete aggregates: A laboratory study on model materials. Cem Concr Res 46:50–65

Thouvenot P, Bildstein O, Munier I, Cochepin B, Poyet S, Bourbon X, Treille E (2013) Modeling of concrete carbonation in deep geological disposal of intermediate level waste. EPJ Web of Conferences 56:05004

Tinseau E, Bartier D, Hassouta L, Devol-Brown I, Stammose D (2006) Mineralogical characterization of the Tournemire argillite after in situ interaction with concretes. Waste Manage 26:789–800

Tournassat C, Steefel CI (2019) Reactive transport modeling of coupled processes in nanoporous media. Rev Mineral Geochem 85:75–109

Treille E, Wendling J, Trenty L, Loth L, Pépin G, Plas F (2012) Probabilistic analysis based on simulations of the long-term gas migration at repository-scale in a geological repository for high and intermediate level radioactive waste disposal in a deep clay formation. Proceedings TOUGH Symposium, Lawrence Berkeley National Laboratory, Berkeley, California, September 17–19, 2012

Trotignon L, Peycelon H, Bourbon X (2006) Comparison of performance of concrete barriers in a clayey geological medium. Phys Chem Earth 31:610–617

Trotignon L, Devallois V, Peycelon H, Tiffreau C, Bourbon X (2007) Predicting the long term durability of concrete engineered barriers in a geological repository for radioactive waste. Phys Chem Earth 32:259–274

Trotignon L, Thouvenot P, Munier I, Cochepin B, Piault E, Treille E, Bourbon X, Mimid S (2011) Numerical simulation of atmpspheric carbonation of concrete components in a deep geological radwaste disposal during operating period. Nucl Technol 174:424–437

Trinchero P, Puigdomenech I, Molinero J, Ebrahimi H, Gylling B, Svensson U, Bosbach D, Deissmann G (2017) Continuum-based DFN-consistent numerical framework for the simulation of oxygen infiltration into fractured crystalline rocks. J Contam Hydrol 200:60–69

Truche L, Berger G, Destrigneville C, Pages A, Guillaume D, Giffaut E, Jacquot E (2009) Experimental reduction of aqueous sulphate by hydrogen under hydrothermal conditions: Implication for the nuclear waste storage. Geochim Cosmochim Acta 73:4824–4835

Tsang C-F, Gelhar L, de Marsily G, Andersson J (1994) Solute transport in heterogeneous media: A discussion of technical issues coupling site characterization and predictive assessment. Adv Water Resour 17:259–264

Ufer K, Stanjek H, Roth G, Dohrmann R, Kleeberg R, Kaufhold S (2008) Quantitative phase analysis of bentonites by the Rietveld method. Clays Clay Miner 56:272–282

Van Damme H, Pellenq RJM, Ulm FJ (2013) Chapter 14.3 - Cement Hydrates. *In:* Developments in Clay Science. Vol Volume 5. Faïza B, Gerhard L, (eds). Elsevier, p 801–817

van der Lee J, De Windt L, Lagneau V, Goblet P (2003) Module-oriented modeling of reactive transport with HYTEC. Comput Geosci 29:265–275

Vinsot A, Linard Y, Lundy M, Necib S, Wechner S (2013) Insights on desaturation processes based on the chemistry of seepage water from boreholes in the Callovo–Oxfordian argillaceous rock. Procedia Earth Planet Sci 7:871–874

Vinsot A, Leveau F, Bouchet A, Arnould A (2014) Oxidation front and oxygen transfer in the fractured zone surrounding the Meuse/Haute-Marne URL drifts in the Callovian–Oxfordian argillaceous rock. Geol Soc London, Spec Publ 400:207–220

Vinsot A, Lundy M, Linard Y (2017) O_2 consumption and CO_2 production at Callovian–Oxfordian rock surfaces. Procedia Earth Planet Sci 17:562–565

von Schenck H, Källström K (2014) Reactive transport modelling of organic complexing agents in cement stabilized low and intermediate level waste. Phys Chem Earth, Parts A/B/C 70–71:114–126

Wagner T, Kulik DA, Hingerl FF, Dmytrieva SV (2012) GEM-SELEKTOR geochemical modeling package: TSolMod library and data interface for multicomponent phase models. Can Mineral 50:1173–1195

Walker CS, Sutou S, Oda C, Mihara M, Honda A (2016) Calcium silicate hydrate (C-S-H) gel solubility data and a discrete solid phase model at 25 °C based on two binary non-ideal solid solutions. Cem Concr Res79:1–30

Wanner H (2007) Solubility data in radioactive waste disposal. Pure Appl Chem 79:875–882

Watson C, Hane K, Savage D, Benbow S, Cuevas J, Fernandez R (2009) Reaction and diffusion of cementitious water in bentonite: Results of 'blind' modelling. Appl Clay Sci 45:54–69

Watson C, Wilson J, Savage D, Benbow S, Norris S (2016) Modelling reactions between alkaline fluids and fractured rock: The Maqarin natural analogue. Appl Clay Sci 121:46–56

Watson C, Wilson J, Savage D, Norris S (2018) Coupled reactive transport modelling of the international Long-Term Cement Studies project experiment and implications for radioactive waste disposal. Appl Geochem 97:134–146

Wersin P (2003) Geochemical modelling of bentonite porewater in high-level waste repositories. J Contam Hydrol 61:405–422

Wersin P, Birgersson M (2014) Reactive transport modelling of iron-bentonite interaction within the KBS-3H disposal concept: the Olkiluoto site as a case study. Geol Soc London, Spec Publ 400:SP400.424

Wersin P, Birgersson M, Karnland O, Snellman M (2008) Impact of corrosion-derived iron on the bentonite buffer within the KBS-3H disposal concept. SKB report TR-08–34

Wilson JC, Benbow S, Sasamoto H, Savage D, Watson C (2015) Thermodynamic and fully coupled reactive transport models of a steel-bentonite interface. Appl Geochem 61:10–28

Wilson JC, Benbow S, Metcalfe R (2018) Reactive transport modelling of a cement backfill for radioactive waste disposal. Cem Concr Res 111:81–93

Wolery TJ (1983) EQ3NR a computer program for geochemical aqueous speciation-solubility calculations: user's guide and documentation. Lawrence Livermore Nat. Lab. UCRL-53414-report, Livermore, CA, USA.

Xu T, Sonnenthal E, Spycher N, Pruess K (2004) TOUGHREACT user's guide: a simulation program for non-isothermal multiphase reactive geochemical transport in variable saturated geologic media. Lawrence Berkeley National Laboratory Report LBNL-55460, Berkeley, USA

Xu T, Senger R, Finsterle S (2008) Corrosion-induced gas generation in a nuclear waste repository: Reactive geochemistry and multiphase flow effects. Appl Geochem 23:3423–3433

Yang C, Samper J, Montenegro L (2008) A coupled non-isothermal reactive transport model for long-term geochemical evolution of a HLW repository in clay. Environ Geol 53:1627–1638

Yeh GT, Tripathi VS (1989) A critical evaluation of recent developments in hydrogeochemical transport models of reactive multichemical components. Water Resour Res 25:93–108

Zeelmaekers E, Honty M, Derkowski A, Srodon J, De Craen M, Vandenberghe N, Adriaens R, Ufer K, Wouters L (2015) Qualitative and quantitative mineralogical composition of the Rupelian Boom Clay in Belgium. Clay Miner 50:249–272

Zhang Z, Thiery M, Baroghel-Bouny V (2015) Numerical modelling of moisture transfers with hysteresis within cementitious materials: Verification and investigation of the effects of repeated wetting–drying boundary conditions. Cem Concr Res 68:10–23

Zheng L, Samper J, Montenegro L, Mayor JC (2008) Multiphase flow and multicomponent reactive transport model of the ventilation experiment in Opalinus clay. Phys Chem Earth, Parts A/B/C 33, Suppl 1:S186-S195

Zheng L, Rutqvist J, Birkholzer JT, Liu H-H (2015) On the impact of temperatures up to 200 °C in clay repositories with bentonite engineer barrier systems: A study with coupled thermal, hydrological, chemical, and mechanical modeling. Eng Geol 197:278–295

Reviews in Mineralogy & Geochemistry
Vol. 85 pp. 459-498, 2019
Copyright © Mineralogical Society of America

15

Acid Water–Rock–Cement Interaction and Multicomponent Reactive Transport Modeling

Jordi Cama, Josep M. Soler, Carles Ayora

Department of Geosciences
Institute of Environmental Assessment and Water Research (IDAEA-CSIC)
Barcelona 08034
Catalonia
Spain

jordi.cama@idaea.csic.es; josep.soler@idaea.csic.es; caigeo@idaea.csic.es

INTRODUCTION

This chapter addresses the use of multicomponent reactive transport modeling (MCRTM) in an attempt to understand and quantify the interaction between acid water and rocks or Portland cement (mortar, concrete) during and after the injection of CO_2 in deep aquifers (geological CO_2 storage) and in the treatment of acid mine drainage (AMD). Anthropogenic acidification of water occurs in the two cases (Gunter et al. 1993; Nordstrom and Alpers 1999).

In the first case, CO_2-rich water is acidic and out of equilibrium with the reservoir and seal rocks of a geological CO_2 storage system, leading to mineral reactions (e.g. calcite dissolution and gypsum precipitation) and potential changes in porosity and permeability. These adverse effects could even be more profound in the contact between acid waters and Portland cement, because of the high pH conditions in cement (pH > 12). Water acidity is due to the CO_2 enrichment of the formation waters, i.e. production of H_2CO_3

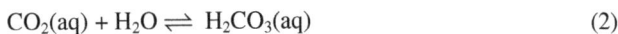

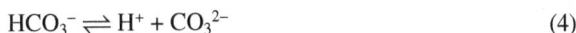

$$CO_2(g) \rightleftharpoons CO_2(aq) \tag{1}$$

$$CO_2(aq) + H_2O \rightleftharpoons H_2CO_3(aq) \tag{2}$$

$$H_2CO_3(aq) \rightleftharpoons H^+ + HCO_3^- \tag{3}$$

$$HCO_3^- \rightleftharpoons H^+ + CO_3^{2-} \tag{4}$$

In the second case, acid mine drainage (AMD) is a result of proton release during the oxidative dissolution of pyrite (FeS_2) and the precipitation of other Fe-bearing sulfides, sulfosalts and oxyhydroxides. The dissolution of pyrite by oxygen can be expressed as

$$FeS_2 + 3.5\,O_2(aq) + H_2O \rightleftharpoons Fe^{2+} + 2\,SO_4^{2-} + 2\,H^+ \tag{5}$$

SO_4^{2-}, Fe(II) and protons are released to solution, and Fe(II) is oxidized to Fe(III)

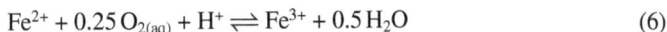

$$Fe^{2+} + 0.25\,O_{2(aq)} + H^+ \rightleftharpoons Fe^{3+} + 0.5\,H_2O \tag{6}$$

Fe(III) may precipitate as schwertmannite (Bigham et al. 1996)

1529-6466/19/0085-0015$05.00 (print)
1943-2666/19/0085-0015$05.00 (online)   http://dx.doi.org/10.2138/rmg.2018.85.15

$$8Fe^{3+} + SO_4^{2-} + 14H_2O \rightleftharpoons Fe_8O_8(OH)_6(SO_4) + 22H^+ \qquad (7)$$

releasing more protons, or as a Fe(III)-hydroxide (ferrihydrite), also releasing protons

$$Fe^{3+} + 3H_2O \rightleftharpoons Fe(OH)_3 + 3H^+ \qquad (8)$$

But at pH lower than 3, Fe(III) remains mainly in solution and acts as another oxidizing agent for pyrite according to

$$FeS_2 + 14Fe^{3+} + 8H_2O \rightleftharpoons 15Fe^{2+} + SO_4^{2-} + 16H^+ \qquad (9)$$

In CO_2-rich waters, acidic conditions can be maintained during long periods owing to the buffering effect of a weak acid (H_2CO_3) during the mineral reactions, whereas acidity in a strong acid (H_2SO_4 in AMD) is rapidly neutralized when protons are consumed by reaction with the solid phases.

Geological storage of CO_2

In the last decade, laboratory flow-through experiments under the conditions of the injection sites have been performed to study the interaction of the CO_2-rich waters with reservoir and seal rocks (Noiriel et al. 2007; Luquot and Gouze 2009; Noiriel et al. 2013, 2015; Luquot et al. 2014; Noiriel 2015; Smith et al. 2013a, 2017; Beckingham et al. 2016; Menke et al., 2016; Rohmer et al. 2016; Loring et al. 2019 and references therein). These authors highlight the effect of mineral dissolution and precipitation reactions on porous/fractured media, flow, permeability and porosity. In several studies, the geochemical variation in the experimental runs has been interpreted using multicomponent reactive transport modeling (MCRTM) (Carroll et al. 2011, 2013; Canal et al. 2012; Hao et al. 2013; Luhmann et al., 2014, 2017; Tutolo et al. 2015a; Luquot et al. 2016; Beckingham et al. 2017; Smith et al. 2017; Dai et al. 2019). The resulting discussions on the effectiveness of the derived kinetic parameters (dissolution rate constant, mineral reactive surface area) and porosity–permeability relationships and their application to the reservoir scale reinforce the need to use MCRTM.

Our research on water–rock interaction in geological carbon sequestration has yielded some experimental and MCRTM results concerning the demonstration test site for CO_2 storage at Hontomín (N Spain) (García-Rios et al. 2014, 2015, 2017; Dávila et al. 2016a, 2016b, 2017; Thaysen et al. 2017). At this site, reservoir and seal formations are mainly constituted by sedimentary rocks (limestone, sandstone and argillaceous limestones (referred to as marls)) with a high content of calcite (67–90 wt%) and some dolomite ($\approx$ 10 wt%). We addressed the quantitative interpretation of the effect of mineral dissolution and precipitation reactions on rock porosity and fracture permeability. Given that reservoir and cap (seal) rock permeability ($k = 10^{-19}$–10^{-18} m^2; Alcalde et al. 2013; Mbia et al. 2014) can be several orders of magnitude smaller than fracture permeability, fluids will flow preferentially through fractures or microcracks, where chemical alteration induced by the CO_2-rich fluids can bring about changes in physical and chemical properties. Since carbonate rocks are common in reservoir and seal formations of potential CO_2-storage sites worldwide (e.g. Midale carbonate unit in Weyburn, Emberley et al. 2004; Kharaka et al. 2013; sedimentary basins, Benson and Cole 2008; Dover 33 reef in the Michigan Basin, Welch et al. 2019), research can help us to better understand the basic reaction mechanisms. Portland cement is placed between the borehole steel casing and the surrounding rock at CO_2 injection sites to prevent gas or fluid leakage. During and after CO_2 injection, the resulting CO_2-rich acid water may deteriorate the cement and favor undesired leaking (Walsh et al. 2014; Iyer et al. 2017). Quantification of geochemical processes at the laboratory scale through MCRTM is therefore necessary to understand their occurrence under the reservoir/seal and wellbore conditions.

Acid mine drainage treatment

Our research aimed at improving passive treatments to remediate the AMD-contaminated streams and rivers in the Iberian Pyrite Belt (IPB) in south-western Spain. MCRTM has been used for more than 20 years to investigate sulfide oxidation rates and the subsequent metal release from mine wastes causing pollution in aquifers downstream (Wunderly et al. 1996; Mayer et al. 2002; Salmon and Malstrom 2004; Brookfield et al. 2006; Malstrom et al. 2008; Pedretti et al. 2017). These models coupled unsaturated flow and chemical reactions involving a number of poorly constrained parameters. As a result, MCRTM has led to qualitative predictions of water compositions and of the relative importance of the multiple processes involved (Xu et al. 2000; Mayer et al. 2002; Acero et al. 2007, 2009). More common is the use of MCRTM in simulating the behavior of remediation actions, such as passive treatment systems, permeable reactive barriers and covers to inhibit sulfide oxidation (Amos et al. 2004; Mayer et al. 2006; Perez-Lopez et al. 2007, 2009; Pabst et al. 2017, 2018). One of the most common passive treatment systems is the anoxic limestone drain (ALD), or systems derived from it (e.g. disperse alkaline substrate (DAS) tanks). These systems are characterized by a mixture of a fine-grained alkaline source such as limestone and an inert matrix with high porosity such as wood shavings (Rötting et al. 2008b). The efficiency of these treatments and the quality of the outflowing water clearly depend on the competition between the rates of water flow and chemical reactions. Therefore, accurate parameterization of the models is crucial to the design of an efficient remediation system.

Of the different parameters describing MCRTM, the mineral reactive surface area is the most difficult to determine. A number of reasons such as the heterogeneous flow paths, rugosity of the solid surfaces, precipitation of neo-formed minerals and the formation of diffusion boundary layers have been proposed to account for the difficulty of upscaling laboratory to field reaction rates (White and Brantley 2003; Steefel et al. 2015a). A considerable effort has recently been devoted to understanding the reactive surface area from the microscale point of view (Noiriel et al. 2009; Gouze and Louquot 2011; Li et al. 2015; Beckingham et al. 2016, 2017). Below we describe a practical methodology to obtain the parameters for reactive transport modeling, especially the experimental determination of the reactive surface area, and its application to the design of a field scale remediation system (Ayora et al. 2013).

Outline

We show how MCRTM is used to simulate the interaction between the acid water and the solid phases controlling the variations in the aqueous chemistry and mineralogical changes in different sets of our laboratory experiments related to geological storage of CO_2 and AMD treatment systems. We discuss the reactivities of the different minerals and the model fitting procedure together with the corresponding adjusted parameters (bulk surface areas, diffusion coefficients in stagnant domains).

In the first section, we present the main characteristics of the laboratory experimental setups and the AMD- treatment plant employed in our studies. The second section presents a brief description of the equations and parameters in the reactive transport codes used (CrunchFlow, Steefel et al. (2015a) and RETRASO, Saaltink et al. (2004)). The following section addresses the quantitative interpretation of the experimental results from rock-fragment-filled columns and fractured-core percolation experiments subjected to injection of CO_2-rich waters. The fourth section deals with the MCRTM that is used to reproduce the results in AMD-Portland cement columns and cores, and in AMD–DAS columns in order to improve field-scale AMD treatment plants. Finally, conclusions and future perspectives concerning the use of MCRTM for the implementation of CO_2 geological storage and AMD passive treatment are given.

LABORATORY EXPERIMENTS AND FULL SCALE SYSTEM

Laboratory-scale experiments

Several sets of column and percolation experiments were conducted to study the anthropogenic acid water–rock/cement interaction under atmospheric P_{CO_2} or at elevated P_{CO_2} pressures (AMD systems or CO_2 storage, respectively).

Experimental setup. In the different setups an injected solution with a desired chemical composition interacted with the solid sample contained in a column (Fig. 1). This approach allowed us to monitor the variation of the aqueous chemistry caused by the ongoing geochemical reactions. The input solution was continuously injected from bottom to top of the samples at a constant flow rate by a peristaltic pump in experiments under atmospheric P_{CO_2} conditions and by a dual-piston pump under elevated P_{CO_2} conditions (Fig. 1a). In the experiments performed under atmospheric pressure, P_{Total} was 1 bar and pCO_2 was $10^{-3.4}$ bar; under elevated total pressures ($P_{Total} = 10$ or 150 bar), P_{CO_2} was equal to P_{Total} when P_{Total} was 10 bar (subcritical P_{CO_2} conditions) and P_{CO_2} ranged between 34 and 130 bar when P_{Total} was 150 bar (supercritical P_{CO_2} conditions).

In a first set of experiments, columns were filled with micrometer- or millimeter-size fragments of the same rock or with alternating layers of rock and Portland cement (Fig. 1b). The approximate porosities (volume of voids between grains) ranged from 45% to 65%. In a second set of experiments, solid samples consisted of cylindrical cores of 9 mm in diameter and 18 mm in length containing an artificially created fracture (Fig. 1c). In a third

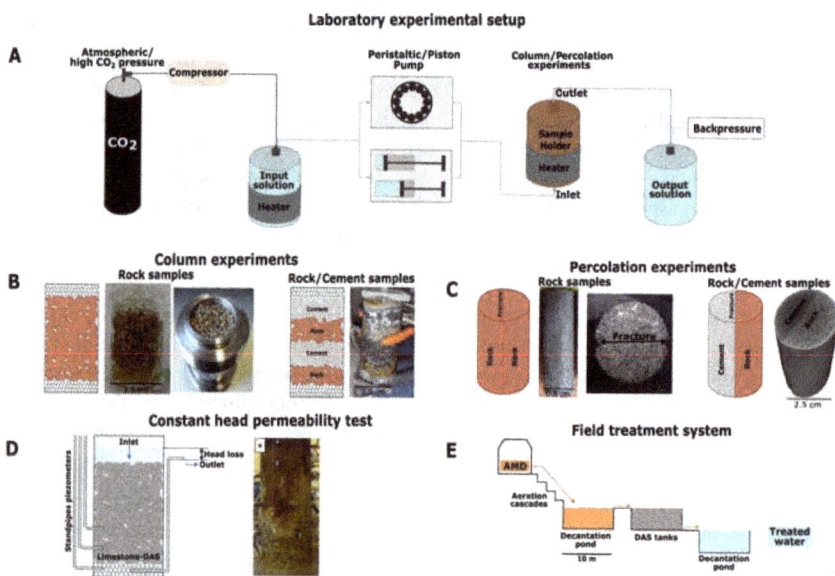

Figure 1. Schemes that show the laboratory experimental setups, different types of experiments and field treatment system: **a)** parts of the experimental setups for column and percolation experiments: CO_2 bottle and compressor, peristaltic or dual piston pump, reacting sample holder, solution collector and backpressure (elevated pressure); **b)** column experiments filled with rock fragments and rock/cement fragments in alternating layers (5 cm long and 2.5 cm in diameter; at the bottom and top, 5-mm layers of silica beads were placed to homogeneously distribute the solutions at the inlet and outlet; **c)** percolation experiments with fractured cores of 2 cm long and 0.8 cm in diameter (rock–rock halves) and 5 cm long and 2.5 cm in diameter (cement–rock halves); **d)** Mixtures of calcite sand and wood shavings columns (15 cm diameter × 50 cm length) with connecting piezometric standpipes to measuring ports; **e)** passive AMD-treatment system showing mine adit, aeration cascades, decantation ponds and DAS tank (see text).

set of experiments, artificially fractured cores of 2.5 cm in diameter and 5 cm in length were composed of half a Portland cement cylinder and half a rock cylinder (Fig. 1c). Experiments that belong to the second and third sets are referred to as percolation experiments. Another type of columns with larger dimensions and with sampling points at different depths was built to measure the hydraulic conductivity of a porous substrate (dispersed alkaline substrate, DAS) by means of constant head tests (Fig. 1d). The head loss can be accurately measured by connecting piezometric standpipes along the column.

Solid samples and aqueous solutions. The reservoir rocks used in our experiments are mainly composed of calcite and dolomite (limestone), and calcite, quartz and microcline (clast-poor sandstone) with a 2–7% matrix porosity. The minerals of the cap rocks (marly limestone, black shale and marl) are calcite, quartz, illite, albite, clinochlore, gypsum, anhydrite and pyrite with a matrix porosity of ca. 7% (Table 1). The mineralogical composition of the hydrated Portland cement (sulfate resistant, CEM I 42.5R/SR) is summarized in Table 1, in which C-S-H (calcium silicate hydrate) and portlandite are the main phases.

For the CO_2 experiments, CO_2-rich solutions were prepared to be compositionally similar to the native saline solution from the Hontomín site (García-Rios et al. 2014; Dávila et al. 2017). Once the solutions were equilibrated with the corresponding P_{CO_2} the input pH was ≈ 3.5 (Table 2). Moreover, to investigate the effect of SO_2 impurities in the CO_2 injection stream on the interaction between CO_2-rich solutions and carbonate rocks, sulfuric acid was added in amounts corresponding to 0.4% SO_2 in the flue gas (Thaysen et al. 2017), anticipating full oxidation of SO_2 to H_2SO_4 (Table 2), with a resulting input pH of ≈ 2. For the AMD columns with only rock, solutions were prepared at pH ≈ 2–3 (H_2SO_4) with Fe(II) and Al (Table 2). For the ones with alternating layers of rock and cement fragments, solutions were prepared at pH ≈ 2 (H_2SO_4) containing Al, Ca, Mg and Fe as major ions and Cd, As, Cu, Zn and Ni as minor components (Table 2). The DAS columns were fed with natural AMD (pH of 2.3 to 3.5) and inflow net acidity of 1350 to 2300 mg L^{-1} as $CaCO_3$.

Table 1. Mineral content (wt%).

Phase	Chemical formula	limestone	dolostone	sandstone	argillaceous marl	hydrated Portland cement
calcite	$CaCO_3$	99		68	71.2–86	4
dolomite	$CaMg(CO_3)_2$		100			
ankerite	$Ca(Fe^{++},Mg)(CO_3)_2$	1		1		
quartz	SiO_2			24	4–9.7	
microcline	$KAlSi_3O_8$			4		
illite	$K_{0.6}Mg_{0.25}Al_{2.3}Si_{3.5}O_{10}(OH)_2$			2	7.1–15.9	
albite	$NaAlSi_3O_8$				0.9–6.5	
clinochlore	$Mg_3Fe_2Si_3Al_2O_{10}(OH)_8$				0.0–3.1	
pyrite	FeS_2				0.2–2	
gypsum	$CaSO_4 \cdot 2H_2O$				1.0–2.0	
anhydrite	$CaSO_4$				0.0–0.5	
C-S-H	$1.67\,Ca(OH)_2 \cdot SiO_2 \cdot H_2O$– $0.83\,Ca(OH)_2 \cdot SiO_2 \cdot 0.83\,H_2O$– SiO_2					47
portlandite	$Ca(OH)_2$					25
ettringite	$Ca_6Al_2(SO_4)_3(OH)_{12} \cdot 26H_2O$					13
Si-hydrogarnet	$Ca_3AlFe(SiO_4)_{0.84}(OH)_{8.64}$					8
hydrotalcite	$Mg_4Al_2(OH)_{14} \cdot 3H_2O$					3

Table 2. Average concentrations and pH of injected solutions.

	CO$_2$-experiments		AMD-experiments	
	only CO$_2$	CO$_2$+H$_2$SO$_4$	AMD-limestone	AMD-layers
injected solution	(mol/kgw)			
Ca^{2+}	4.91×10^{-2}	5.40×10^{-2}		4.00×10^{-3}
SO$_4$$^{2-}$	2.68×10^{-2}	3.00×10^{-2}	1.00×10^{-2}–6.00×10^{-2}	3.20×10^{-2}
Mg^{2+}	3.24×10^{-2}	3.30×10^{-2}		5.30×10^{-3}
K$^+$	1.13×10^{-2}	1.20×10^{-2}		
Na$^+$	3.89×10^{-1}	3.93×10^{-1}		
Cl$^-$	4.98×10^{-1}	5.10×10^{-1}		
Br$^-$	1.14×10^{-2}	1.00×10^{-2}		
CO$_2$	6.16×10^{-1}	6.16×10^{-1}		
Fe^{2+}			3.50×10^{-3}–2.60×10^{-2}	9.06×10^{-3}
Al^{3+}			3.60×10^{-3}–3.60×10^{-2}	2.90×10^{-3}
Cd^{2+}				4.40×10^{-6}
AsO$_4$$^{3-}$				1.20×10^{-5}
Cu^{2+}				2.70×10^{-4}
Zn^{2+}				1.00×10^{-3}
Ni^{2+}				1.25×10^{-5}
pH	3.3	2.3	2.0 and 3.0	2.0

Full scale system

In situ passive remediation technologies are suitable for remediating the drainages of abandoned mines (Fig. 1e). A pilot plant of DAS-calcite technology was operated at Mina Monte Romero (Macías et al. 2012), and a full scale AMD treatment was constructed at Mina Esperanza (Caraballo et al. 2011; Fig. 2). The reactive tank (120 m^2 of surface and 4-m deep) was filled with a 2.5-m layer of limestone–DAS (50% porosity). Water flow through the reactive material was gravity fed from top to bottom (Fig. 1e), emerging finally from the top of a water collecting well.

MULTICOMPONENT REACTIVE TRANSPORT MODELING (MCRTM)

MCRTM at the continuum scale is at an advanced position in terms of its treatment of chemical processes (Steefel et al. 2005). Continuum formulations for reactive transport models and numerical simulations of reactive processes in porous media incorporate rate expressions, which once formulated with reliable kinetic parameters, are capable of accurately describing the mineral dissolution and precipitation reactions (Black et al. 2015; Marty et al. 2015; Molins 2015; Beckingham et al. 2017).

Brief description of MCRTM

The observed mineralogical alterations and changes in solution compositions in the experiments presented in this chapter were modeled using the CrunchFlow or RETRASO reactive transport codes. These codes solve numerically the mass balance of solutes expressed as

10 m

Figure 2. Photographs of the limestone–DAS remediation system to treat AMD from abandoned Mina Esperanza (Iberian Pyrite Belt (IPB), south-western Spain). Numbers show sampling points in Figure 18.

$$\frac{\partial(\phi C_i)}{\partial t} = \nabla \cdot \left(D \, \nabla C_i \right) - \nabla \left(q C_i \right) + R_i \ (i = 1, 2, 3, \ldots, n) \tag{10}$$

where ϕ is porosity, C_i is the concentration of component i (mol m^{-3}), q is the Darcy velocity (m^{3}m^{-2}s^{-1}), R_i is the total reaction rate affecting component i (mol m^{-3} rock s^{-1}) and D is the combined dispersion-diffusion coefficient (m^{2}s^{-1}). The total reaction rate R_i is given by

$$R_i = -\sum_m \nu_{im} R_m \tag{11}$$

where R_m is the rate of precipitation ($R_m > 0$) or dissolution ($R_m < 0$) of mineral m in mol m^{-3} rock s^{-1}, and n_{im} is the number of the moles of i in mineral m.

The reaction rate laws used in the calculations are expressed as

$$R_m = A_m \sum_{\text{terms}} k_{m,T} \, a_{H^+}^{n_{H^+}} f\left(\Delta G_r \right) \tag{12}$$

where A_m is the mineral surface area in m^2 m^{-3} rock, $k_{m,T}$ is the reaction rate constant at the temperature of interest in mol m^{-2} s^{-1}, $a_{H^+}^{n_{H^+}}$ the term describing the effect of pH on the rate, ΔG_r is the Gibbs energy of the reaction, and $f(\Delta G_r)$ is the functional dependence of the dissolution rate on the deviation from equilibrium

$$f\left(\Delta G_r \right) = \left(\left(\frac{IAP}{K_{eq}} \right)^{m_2} - 1 \right)^{m_1} \tag{13}$$

where IAP is the ionic activity product of the solution with respect to the mineral, K_{eq} is the equilibrium constant for the dissolution reaction (ionic activity product at equilibrium) and m_1 and m_2 allow for nonlinear dependencies on the affinity term. The rate constant at temperature T ($k_{m,T}$) is calculated from

$$k_{m,T} = k_{m,25} \exp\left(\frac{E_a}{R} \left(\frac{1}{T_{25}} - \frac{1}{T} \right) \right) \tag{14}$$

where $k_{m,25}$ is the rate constant at 25 °C, E_a is the apparent activation energy of the overall reaction (J mol^{-1}) and R is the gas constant (J mol^{-1} K^{-1}). Change in mineral surface area (A_m in m^2$_{mineral}$ m^{-3}$_{bulk\ rock}$) due to dissolution is given by

$$A_m = A^{\text{initial}} \left(\frac{\phi_m}{\phi_{m(i)}} \right)^{\frac{2}{3}} \left(\frac{\phi}{\phi_{(i)}} \right)^{\frac{2}{3}} \tag{15}$$

Change due to precipitation is given by

$$A_m = A^{\text{initial}} \left(\frac{\phi}{\phi_{(i)}} \right)^{\frac{2}{3}} \tag{16}$$

where $\phi_{m(i)}$ is the initial volume fraction of the mineral m and $\phi_{(i)}$ is the initial porosity of the medium. The term $(\phi/\phi_{(i)})^{2/3}$ results from the expression of the dependence of mineral surface area (m^2 m^{-3}$_{rock}$) on volume fraction for any given grain geometry (e.g. spheres or cubes). This formulation ensures that as the volume fraction of a mineral goes to 0, so does its surface area. Moreover, for both dissolving and precipitating minerals, the term $(\phi/\phi_{(i)})^{2/3}$ demands that the surface area of a mineral in contact with fluid goes to 0 when the porosity of the medium reaches 0. This formulation is employed for primary minerals (i.e., minerals with initial volume fractions > 0). For secondary minerals which precipitate, the value of the initial bulk surface area specified is used as long as precipitation occurs. If this phase later dissolves, the above formulation is employed but with an arbitrary initial volume fraction of 0.01.

If specific surface areas are used instead of bulk surface areas, the evolution of the mineral volume fraction causes the bulk surface area (A_m) to evolve with time according to the following equation

$$A_m = \frac{\phi_m A_{\text{specific}} MW_m}{V_m} \tag{17}$$

where MW_m refers to the molecular weight of the phase, V_m is the molar volume of the solid phase and A_{specific} is the specific surface area.

Numerical discretization

1D simulations were performed in column experiments only filled with rock fragments because the small porosity of the rock matrix ($\approx 2\%$) would not contribute to the whole rock reactivity. The very small porosity translates into negligible diffusion fluxes into or out of the rock fragments compared with the reactivity of the outer surfaces. Conversely, in the column experiments where hydrated Portland cement was present, 2-D models were used to determine the effect of fluid diffusion through cement and its contribution to system reactivity. 2D models were also used for the experiments involving fractured cores.

One-dimensional numerical domains. In the rock-fragment columns, the size of the numerical elements may be the same along the whole 1D domain, but in some cases a finer resolution is needed close to the inlet (Fig. 3a), where the reactivity is highest. The solution is injected in element 1 and collected from the last element.

Two-dimensional numerical domains. To simulate the column experiments with both cement and rock fragments, a number of considerations had to be taken into account. The system was divided into two parts (Trapote et al. 2016): an immobile zone (cement or rock fragments where solute transport takes place only by diffusion) and a mobile zone (pore space between the grains where water circulates; Fig. 3b). The numerical domain (2D model with symmetry around the axis of the cylinder) was designed with two concentric cylindrical zones: an internal zone with mortar or rock (only diffusion), and an external zone filled only with circulating water (Fig. 3b). The radius of the internal zone was equal to the radii of the rock/cement fragments. The thickness of the mobile zone was calculated according to the open pore volume between grains in the experiment. And a model column length was calculated to conserve the same total volume (fragments + water) as in the experiment. A model flow velocity (advection along the external mobile zone) was finally calculated according to the experimental mean residence time of the circulating water in the experiment.

In fractured cores with exclusively rock, only half of the core was considered given the symmetry in the experimental setup (Fig. 3c). In addition, rectangular coordinates were used. The dimensions of the 2D domain were R_y, the length of the core (L) in the y direction, and R_x in the x direction. R_x was computed by considering that half of the cylindrical core section was equivalent to a rectangle with the same area with sides d (core diameter) and R_x (Fig. 3c). The domain was composed of two parts: (1) high permeability zone (fracture, large porosity) and (2) rock matrix (small porosity). The fracture zone (zone 1) was on the left of the rectangular domain, parallel to the flow direction (y axis), and had a thickness equal to half of the experimental fracture aperture (i.e., the first element along the x direction). An appropriate number of elements in the x and y directions was selected in each simulation (see detailed spatial discretization in García-Rios et al. 2017). The model considered advection and dispersion along the fracture. Solute transport in the rock matrix was only by diffusion.

Parameterization

Mineral dissolution and precipitation reactions: equilibrium constants. Modeling used equilibrium constants, stoichiometric coefficients and parameters for activity coefficient calculation from the EQ3/6 database included in the codes (Wolery et al. 1990), with the exception of the equilibrium constant for gypsum at 60 °C which was as measured by

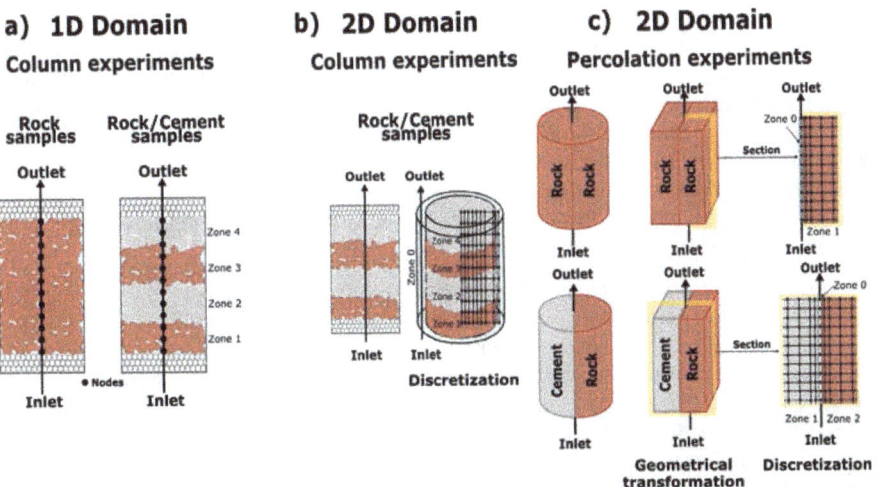

Figure 3. Illustrations showing the conceptual models ((**a**) 1D for columns, (**b**) 2D for columns and (**c**) 2D for percolation experiments), implemented grids, and geometry and boundary conditions of the flow domains. Left and right boundaries are no-flow boundaries. Zone 0 represents mobile zone (saturated void space or fracture) and zones 1–4 represent non-mobile zone (fragments/rock).

García-Rios et al. (2014). The log K_{eq} values for the C-S-H and C-A-S-H gels were obtained from the solid solution model in Kulik and Kersten (2001) and Myers et al. (2015), respectively. The log K_{eq} value and the molar volume for Si-hydrogarnet ($C_3(A,F)S_{0.84}H_{4.32}$) were calculated assuming an ideal solid solution between $C_3AS_{0.84}H_{4.32}$ and $C_3FS_{0.84}H_{4.32}$ using the data from Dilnesa et al. (2011, 2012, 2014). The log K_{eq} values for portlandite, hydrotalcite-OH, ettringite were obtained from Hummel et al. (2002) and cemdata07 database (Matschei et al. 2007; Lothenbach et al. 2008; Schmidt et al. 2008), respectively. The dissolution and precipitation reactions considered in the simulations under all the conditions studied are given in Table 3.

Mineral dissolution and precipitation rates: forms of the rate laws. The values of the rate constants, a_{H^+} terms and m_1 and m_2 exponents in the affinity term of the reaction rate law are normally taken from experimental studies that were compiled in Palandri and Kharaka (2004) and from Domènech et al. (2002), Hamer et al. 2003; Bansdtra et al. (2008), Bibi et al. (2011) and Hellmann et al. (2010) (Table 4). Nevertheless, experiments performed recently to study the kinetics of mineral dissolution/precipitation reactions that are significant for the acid–water–rock interaction have provided new rate laws (calcite: Xu et al. 2012; chlorite: Smith et al. 2013b; Black and Haese 2014; Zhang et al. 2015; smectite: Cappelli et al. 2018). Xu et al. (2012) proposed a calcite dissolution rate that improves the rate–ΔG_r dependence under close-to-equilibrium conditions ($-12 \leq \Delta G_r \leq 1.7$ kJ/mol) with respect to the simplest TST-based rate law ($m_1 = m_2 = 1$ in Equation (13)). Cappelli et al. (2018) showed the importance of calculating specific values for the m and n parameters of the $f(\Delta G_r)$ term (Eqn. 12) in the rate laws in order to better account for the dissolution kinetics of clay minerals. The rate constant for portlandite was $10^{-5.4}$ mol m^{-2} s^{-1} as reported in Bullard et al. (2010). The C-S-H solid solution was discretized into different stoichiometries, ranging from Ca/Si = 1.67 to Ca/Si = 0.0 (Table 3). For the C-S-H gel and the rest of the phases present in the mortar (ettringite, monocarboaluminate, Si-hydrogarnet and hydrotalcite-OH) the rate constants were sufficiently high to allow fast kinetics (local equilibrium).

Mineral reactive surface area. In the continuum models, which make use of the REV (Representative Elementary Volume) approach, the mineral reactive surface area is a bulk parameter. When the specific (BET) surface area of the initial solid can be measured (e.g. micrometer size grains), a resulting total surface area can be calculated. However, when determination of the BET surface area is not possible (e.g. millimeter-, centimeter-size fragments or rock cores), a geometric reactive surface area, calculated by assuming for instance spherical fragments, can provide an estimated value of the reactive surface area. However, model fits to experimental data (i.e. concentrations of elements at the outlet and pH) use the reactive surface areas as an adjustable parameter given a specific mineral reaction rate law. As explained below, relatively small reactive surface areas are required for fast-reacting minerals (e.g. calcite). For secondary phases, fast rates (implemented through the use of large surface areas) result in the assumption of local equilibrium (Table 5).

Flow and transport parameters. The Darcy velocity, q, was computed according to the volumetric flow rates and the appropriate cross-section areas of the columns ($q = 7 \times 10^{-7}$ to 4.5×10^{-6} m^3 m^{-2} s^{-1}) and the fracture apertures in the percolation experiments ($q = 7 \times 10^{-4}$ to 3×10^{-1} m^3 m^{-2} s^{-1}). Mineral reaction rates in stagnant domains (e.g. impermeable rock matrix next to a fracture, unfractured Portland cement) are strongly limited by diffusional solute transport. Effective diffusion coefficients (D_e) can be calculated from empirical relations between D_e and porosity such as Archie's Law, but these D_e values usually have to be adjusted according to measured results. The models also consider a dependency of D_e on the changing porosity such as (Archie's law)

$$D_e = \phi^n D_0 \tag{18}$$

where D_0 is a reference diffusion coefficient and n is the cementation exponent (2–2.5) used

Phases	Reaction	Environment
Calcite	$CaCO_3 + H^+ \leftrightarrow Ca^{2+} + HCO_3^-$	GCS, AMD
Quartz	$SiO_2 \leftrightarrow SiO_2(aq)$	GCS
Illite	$K_{0.6}Mg_{0.25}Al_{2.3}Si_{3.5}O_{10}(OH)_2 + 8H^+ \leftrightarrow 0.6K^+ + 0.25Mg^{2+} + 2.3Al^{3+} + 3.5SiO_2(aq) + 5H_2O$	GCS
Albite	$NaAlSi_3O_8 + 4H^+ \leftrightarrow Na^+ + Al^{3+} + 3SiO_2(aq) + 2H_2O$	GCS
Gypsum	$CaSO_4 \cdot 2H_2O \leftrightarrow Ca^{2+} + SO_4^{2-} + 2H_2O$	GCS, AMD
Clinochlore	$Mg_{2.9}Fe_{2.1}Si_3Al_2O_{10}(OH)_8 + 16H^+ \leftrightarrow 2.9Mg^{2+} + 2.1Fe^{2+} + 3SiO_2(aq) + 2Al^{3+} + 12H_2O$	GCS
Anhydrite	$CaSO_4 \leftrightarrow Ca^{2+} + SO_4^{2-}$	GCS
Pyrite	$FeS_2 + H_2O \leftrightarrow Fe^{2+} + 0.25SO_4^{2-} + 1.75HS^-$	GCS, AMD
Microcline	$KAlSi_3O_8 + 4H^+ \leftrightarrow K^+ + Al^{3+} + 3SiO_2(aq) + 2H_2O$	GCS
Kaolinite	$Al_2Si_2O_5(OH)_4 + 6H^+ \leftrightarrow 2Al^{3+} + 2SiO_2(aq) + 5H_2O$	GCS
$SiO_2(am)$	$SiO_2(am) \leftrightarrow SiO_2(aq)$	GCS
Dolomite	$(CaMg)(CO_3)_2 + 2H^+ \leftrightarrow Ca^{2+} + Mg^{2+} + 2HCO_3^{-}$	GCS
Mesolite	$Ca_{0.657}Na_{0.676}Al_{1.99}Si_{3.01}O_{10} \cdot 2.647H_2O + 7.96H+ \leftrightarrow 0.657Ca^{2+} + 0.676Na^+ + 1.99Al^{3+} + 3.01SiO_2(aq) + 6.627H_2O$	GCS
Stilbite	$Ca_{1.019}Na_{0.136}K_{0.006}Al_{2.18}Si_{6.82}O_{18-7.33}H_2O + 8.72H^+ \leftrightarrow 1.019Ca^{2+} + 0.136Na^+ + 0.006K^+ + 2.18Al^{3+} + 6.82SiO_2(aq) + 11.69H_2O$	GCS
Scolecite	$CaAl_2Si_3O_{10} \cdot 3H_2O + 8H^+ \leftrightarrow Ca^{2+} + 2Al^{3+} + 3SiO_2(aq) + 7H_2O$	GCS
Gismondine	$Ca_2Al_4Si_4O_{16} \cdot 9H_2O + 16H^+ \leftrightarrow 2Ca^{2+} + 4Al^{3+} + 4SiO_2(aq) + 17H_2O$	GCS
Smectite	$Na_{0.33}Mg_{0.33}Al_{1.67}Si_4O_{10}(OH)_2 + 6H^+ \leftrightarrow 0.33Na^+ + 0.33Mg^{2+} + 1.67Al^{3+} + 4SiO_2(aq) + 4H_2O$	GCS
Boehmite	$\gamma\text{-}AlO(OH) + 3H^+ \leftrightarrow Al^{3+} + 2H_2O$	GCS, AMD
Gibbsite	$Al(OH)_3 + 3H^+ \leftrightarrow Al^{3+} + 3H_2O$	GCS, AMD
Diaspore	$\alpha\text{-}AlO(OH) + 3H^+ \leftrightarrow Al^{3+} + 2H_2O$	GCS, AMD
Alunite	$KAl_3(SO_4)_2(OH)_6 + 6H^+ \leftrightarrow K^+ + 2SO_4^{2-} + 3Al^{3+} + 6H_2O$	GCS
Muscovite	$KAl_2(AlSi_3O_{10})(OH)_2 + 10H^+ \leftrightarrow K^+ + 3Al^{3+} + 3SiO_2(aq) + 6H_2O$	GCS
Goethite	$\alpha\text{-}FeO(OH) + 3H^+ \leftrightarrow Fe^{3+} + 2H_2O$	GCS, AMD
Brucite	$Mg(OH)_2 + 2H^+ \leftrightarrow Mg^{2+} + 2H_2O$	GCS, AMD
Aragonite	$CaCO_3 + H^+ \leftrightarrow Ca^{2+} + HCO_3^-$	GCS

Table 3b. Equilibrium reactions and $\log K$ at 25 and 60 °

Phases	Reaction
CSH-1667	$1.67Ca(OH)_2 \cdot SiO_2 \cdot H_2O + 3.34H^+ \leftrightarrow 1.67Ca^{2+} + SiO_2(aq) + 4.34H_2O$
CSH-00	$SiO_2 \leftrightarrow SiO_2(aq)$
CSH-02	$0.23Ca(OH)_2 \cdot 1.16SiO_2 \cdot 0.23H_2O + 0.46H^+ \leftrightarrow 0.23Ca^{2+} + 1.16SiO_2(aq) + 0.69H_2O$
CSH-04	$0.56Ca(OH)_2 \cdot 1.39SiO_2 \cdot 0.56H_2O + 1.12H^+ \leftrightarrow 0.56Ca^{2+} + 1.39SiO_2(aq) + 1.68H_2O$
CSH-06	$1.03Ca(OH)_2 \cdot 1.72SiO_2 \cdot 1.03H_2O + 2.06H^+ \leftrightarrow 1.03Ca^{2+} + 1.72SiO_2(aq) + 3.09H_2O$
CSH-08	$1.82Ca(OH)_2 \cdot 2.27SiO_2 \cdot 1.82H_2O + 3.64H^+ \leftrightarrow 1.82Ca^{2+} + 2.27SiO_2(aq) + 5.46H_2O$
CSH-10	$Ca(OH)_2 \cdot SiO_2 \cdot 0.86H_2O + 2H^+ \leftrightarrow Ca^{2+} + SiO_2(aq) + 2.86H_2O$
CSH-12	$1.20Ca(OH)_2 \cdot SiO_2 \cdot 0.91H_2O + 2.40H^+ \leftrightarrow 1.20Ca^{2+} + SiO_2(aq) + 3.31H_2O$
CSH-14	$1.40Ca(OH)_2 \cdot SiO_2 \cdot 0.95H_2O + 2.80H^+ \leftrightarrow 1.40Ca^{2+} + SiO_2(aq) + 3.75H_2O$
CASH-005	$1.05CaO \cdot 0.025Al_2O_3 \cdot SiO_2 \cdot 1.2H_2O \leftrightarrow 1.05Ca^{2+} + H_2SiO_4^{2-} + 0.05AlO_2^- + 0.05OH^- + 0.175H_2O$
CASH-010	$1.10CaO \cdot 0.05Al2O_3 \cdot SiO_2 \cdot 1.2H_2O \leftrightarrow 1.10Ca^{2+} + H2SiO_4^{2-} + 0.1AlO_2^- + 0.1OH- + 0.15H_2O$
CASH-015	$1.15CaO \cdot 0.075Al_2O_3 \cdot SiO_2 \cdot 1.2H_2O \leftrightarrow 1.15Ca^{2+} + H_2SiO_4^{2-} + 0.15AlO_2^- + 0.15OH^- + 0.125H_2O$
Portlandite	$Ca(OH)_2 + 2H^+ \leftrightarrow Ca^{2+} + H_2O$
Ettringite	$Ca_6Al_2(SO_4)_3(OH)_{12} \cdot 26H_2O \leftrightarrow 6Ca^{2+} + 2Al(OH)_4^- + 3SO_4^{2-} + 4OH^- + 26H_2O$
Hydrogarnet	$Ca_3(Al_{0.5}Fe_{0.5})_2(SiO_4)_{0.84}(OH)_{8.64} + 2.52H_2O \leftrightarrow 3Ca^{2+} + Al(OH)_4^- + Fe(OH)_4^- + 0.84SiO(OH)_3^- + 3.16$
Hydrotalcite	$Mg_4Al_2(OH)_{14} \cdot 3H_2O \leftrightarrow 4Mg^{2+} + 2AlO_2^- + 6OH^- + 7H_2O$

References

[1] Wolery et al. (1990); García-Rios et al. (2014); [3] Kulik and Kersten (2001); [4] Myers et al. (2015); [5] Hummel et al. (2008), Schmidt et al. (2008); [7] Dilnesa et al. (2011, 2012, 2014)

Phases	log K_{eq} (25 °C)	log K_{eq} (60 °C)	Refs.	Acid			Neutral		Basic	
				$k_{m,25}$-rate-H [mol m^{-2} s^{-1}]	E_a-H [kcal mol^{-1}]	$a_{H^+}^{n_{H^+}}$ -H	$k_{m,25}$-rate-n [mol m^{-2} s^{-1}]	E_a-n [kcal mol^{-1}]	$k_{m,25}$-rate-OH [mol m^{-2} s^{-1}]	E_a-OH [kcal mol^{-1}]
Calcite	29.13	26.03	[3]	5.01×10^{-1}	3.44	1.000	6.46×10^{5}	5.62	—	—
Quartz	−1.20	−1.07	[3]	—	—	—	1.02×10^{-14}	20.95	—	—
Illite	1.96	1.77	[3]	2.20×10^{-4}	11.00	0.600	2.50×10^{-13}	13.00	1.26×10^{-57}	15.30
Albite	6.48	5.83	[3]	1.35×10^{-10}	15.52	0.457	9.12×10^{-13}	16.67	1.05×10^{-17}	16.98
Gypsum	13.27	11.96	[3]	—	—	—	1.62×10^{-3}	15.00	—	—
Clinochlore	24.63	22.22	[3]	3.21×10^{-10}	16.00	−0.450	—	—	—	—
Anhydrite	14.58	13.09	[3]	—	—	—	6.45×10^{-4}	3.42	—	—
Pyrite*	18.80	16.82	[3]	3.02×10^{-8}	13.61	−0.500	2.82×10^{-5}	13.61	—	—
Microcline	23.12	20.66	[3]	8.70×10^{-11}	12.40	0.500	3.89×10^{-13}	9.08	—	—
Kaolinite	−8.83	−8.82	[4]	4.90×10^{-12}	15.76	0.777	6.61×10^{-14}	5.31	8.91×10^{-18}	4.28
SiO$_2$(am)	−8.68	−8.64	[4]	—	—	—	1.00×10^{-9}	0.00	—	—
Dolomite	−8.52	−8.73	[4]	6.45×10^{-4}	8.63	0.500	2.95×10^{-8}	12.48	—	—
Mesolite	22.56	20.20	[5]	—	—	—	1.00×10^{-9}	0	—	—
Stilbite	−44.84	−44.64	[6]	—	—	—	1.00×10^{-9}	0	—	—
Scolecite	−29.90	−30.53	[7]	—	—	—	1.00×10^{-9}	0	—	—
Gismondine	−51.14	−50.33	[6]	—	—	—	1.00×10^{-9}	0	—	—
Smectite				—	—	—	1.00×10^{-12}	0	—	—
Boehmite				2.24×10^{-8}	11.35	0.992	4.68×10^{-14}	11.35	2.51×10^{-24}	11.35
Lothenbach				2.24×10^{-8}	11.35	0.992	4.68×10^{-14}	11.35	2.51×10^{-24}	11.35
Diaspore				2.24×10^{-8}	11.35	0.992	4.68×10^{-14}	11.35	2.51×10^{-24}	11.35
Alunite				—	—	—	3.98×10^{-5}	7.65	—	—
Muscovite				1.41×10^{-12}	5.26	0.370	2.82×10^{-14}	5.26	2.82×10^{-15}	5.26
Goethite				—	—	—	1.00×10^{-10}	0.00	—	—
Brucite				1.86×10^{-5}	14.10	0.500	5.75×10^{-9}	10.04	—	—
Aragonite				—	—	—	7.94×10^{-9}	0.00	—	—

...tschei et al. (2007), Lothenbach

Table 4 b. Parameters for the mineral reaction rate la

Phase	Acid			Neutral		
	$k_{m,25}$-rate-H [mol m^{-2} s^{-1}]	E_a-H [kcal mol^{-1}]	$a_{H^+}^{n_{H^+}}$ -H	$k_{m,25}$-rate-n [mol m^{-2} s^{-1}]	E_a-n [kcal mol^{-1}]	$k_{m,25}$-rate-OH [mol m^{-2} s^{-1}]
CSH-1667	—	—	—	1.00×10^{-8}	0	—
CSH-00	—	—	—	1.00×10^{-8}	0	—
CSH-02	—	—	—	1.00×10^{-8}	0	—
CSH-04	—	—	—	1.00×10^{-8}	0	—
CSH-06	—	—	—	1.00×10^{-8}	0	—
CSH-08	—	—	—	1.00×10^{-8}	0	—
CSH-10	—	—	—	1.00×10^{-8}	0	—
CSH-12	—	—	—	1.00×10^{-8}	0	—
CSH-14	—	—	—	1.00×10^{-8}	0	—
CASH-005	—	—	—	1.00×10^{-5}	0	—
CASH-010	—	—	—	1.00×10^{-5}	0	—
CASH-015	—	—	—	1.00×10^{-5}	0	—
Portlandite	—	—	—	3.98×10^{-6}	0	—
Ettringite	—	—	—	1.00×10^{-6}	0	—
Hydrogarnet	—	—	—	1.00×10^{-6}	0	—
Hydrotalcite	—	—	—	1.00×10^{-6}	0	—

Notes:
* Pyrite rate law depends on oxygen concetration ($n_{O_2} = 0.5$)
** Large value of $A \cdot km$ to allow local equilibrium conditions;
† k_{acid} same as gibbsite
‡ $k_{neutral}$ and k_{basic} same as diaspore

References: [1] Xu et al. (2012), Arvidson et al. (2013); [2] Palandri and Kharaka (2004); [3] Bandstra et al. (200 et al. (2015); [6] Domènech et al. (2002); [7] Cama et al. (2000); [8] Acero et al. (2015); [9] Cubillas et al. (200 (2010).

Table 5. Values of mineral reactive surface areas used in the simulations.

Mineral	$A_{mineral}$ $m^2_{mineral} m^{-3}_{bulk\ rock}$	Rock	*P* conditions	Injected solution	Refs.
		dissolution			
calcite	130 ± 50	L	atm, SubC & SupC	CO_2	[1, 3]
	250 → 32	L	SubC	CO_2+H_2SO_4	[3]
	3–30	M	atm	CO_2	[2]
	130 ± 50	M	SubC & SupC	CO_2	[2]
	200–500	M	SubC	CO_2	[3]
	90–500	M	SubC	CO_2+H_2SO_4	[3]
	250–5000	S	SubC	CO_2	[3]
	15 → 7	S	SubC	CO_2+H_2SO_4	[3]
	600	L	SupC	CO_2	[4]
	300000	S	SupC	CO_2	[4]
	20000–250000	M	SupC	CO_2	[5]
	20 → 0.1	L	atm	AMD	[6]
	180	L	atm	AMD	[7]
dolomite	5–10	D	atm, SubC & SupC	CO_2	[1, 3]
quartz	1–36	M, S	atm, SubC & SupC	CO_2, CO_2+H_2SO_4	[2, 3]
	3605	S	SupC	CO_2	[4]
	1000000	M	SupC	CO_2	[5]
aluminosilicates	40–2×10⁸	M	atm, SubC & SupC	CO_2	[2]
illite	8000	M	SubC	CO_2, CO_2+H_2SO_4	[3]
clinochlore	500–9000	M	SubC	CO_2, CO_2+H_2SO_4	[3]
	3.9×10⁹	M	SupC	CO_2	[5]
albite (99.6%)	41	M	SubC	CO_2+H_2SO_4	[3]
albite (0.4%)	5–55×10⁶	M	SubC	CO_2+H_2SO_4	[3]
	3.5×10⁸	M	SupC	CO_2	[5]
microcline (99.85%)	40000	S	SubC	CO_2+H_2SO_4	[3]
microcline (≫ 0.1%)	2–7×10⁶	S	SubC	CO_2+H_2SO_4	[3]
	2180	S	SupC	CO_2	[4]
pyrite	5	M	atm, SubC & SupC	CO_2	[2]
		precipitation			
gypsum	0.01 - 10	L	atm, SubC & SupC	CO_2	[1]
	0.3	D	atm, SubC & SupC	CO_2	[1]
	90	M	atm	CO_2	[2]
	0.02 - 0.09	M	SubC & SupC	CO_2	[2]
	0–0.01 → 0.3–500	M	SubC	CO_2+H_2SO_4	[3]
	0–0.005 → 0.04–5000	S	SubC	CO_2+H_2SO_4	[3]
	100	L	SupC	CO_2	[4]
	10	M	SupC	CO_2	[5]
	10	L	atm	AMD	[6]
secondary phases-M	10000	M	atm, SubC & SupC	CO_2	[2, 3, 5]
secondary phases-S	100000	S	SubC	CO_2, CO_2+H_2SO_4	[3]
goethite	100000	M, S	SubC	CO_2+H_2SO_4	[3]
	0.1	L	atm	AMD	

Notes:

L = limestone; M = marl; S = sandstone; atm = atmospheric pressure; SubC = PTOT = pCO_2 = 10 bar; SupC = PTOT = 150 bar & pCO_2 = 34–37 bar

Ageometric = 900 to 8000 $m^2_{min}\ m^{-3}_{bulk}$; fragment diameter: 0.25–5 mm; Ref. [1–3]

Ageometric = 1000000 $m^2_{min}\ m^{-3}_{bulk}$ in limestone; Ageometric = 750000 $m^2_{min}\ m^{-3}_{bulk}$ in sandstone; fragment diameter: 2.5 mm; Ref. [4]

Ageometric = 200000 $m^2_{min}\ m^{-3}_{bulk}$; fragment diameter: 10 mm; Ref. [5]

Ageometric = 2000 $m^2_{min}\ m^{-3}_{bulk}$; ABET = 600000 $m^2_{min}\ m^{-3}_{bulk}$; fragment diameter: 1–2 mm; Ref. [6]

Ageometric = 4000 $m^2_{min}\ m^{-3}_{bulk}$; fragment diameter: 1–2 mm; Ref. [7]

[1] García-Rios et al. (2014); [2] Dávila et al. (2017); [3] Thaysen et al. (2017); [4] García-Rios et al. (2017); [5] Dávila et al. (2016); [6] Offeddu et al. (2016); [7] Ayora et al. (2013)

aluminosilicates (marl): illite, albite, clinochlore; secondary phases-M (marl): kaolinite, mesolite, stilbite, scolecite, boehmite, gibbsite, diaspore and alunite; secondary phases-S (sandstone): boehmite, gibbsite, diaspore and alunite

CO_2+H_2SO_4 = CO_2 with H_2SO_4 addition

in the simulations (Domenico and Schwartz 1990; Revil and Cathles 1999). The longitudinal dispersivity (α_L) used in the simulations was of the order of 10% of the length of the columns.

In the fractured cores, the initial porosity of the fracture zone was defined to be 100% whereas initial porosity of the rock matrix and cement zones was 5–6% for limestone, sandstone and marl and 11% for cement. Flow field was updated according to porosity, ϕ, and permeability, k, changes. The code solved Darcy's law (neglecting the buoyancy term)

$$q = -\frac{k}{\mu}\nabla P \tag{19}$$

where μ is the dynamic viscosity and P is pressure. Permeability was updated at each time step according to

$$k = k_0 \left(\frac{\phi}{\phi_0}\right)^3 \tag{20}$$

MCRTM IN CO$_2$-RICH WATERS

Column experiments: reservoir and cap rocks

Columns filled with rock fragments provided a porous medium to quantify the geochemical processes (García-Rios et al. 2014; Dávila et al. 2017; Thaysen et al. 2017). Variation of total pressure, P_{CO_2}, temperature and rock mineralogy improved our understanding of the effects of these parameters on the coupled dissolution/precipitation reactions and on porosity, pore structure and permeability of reservoir and cap rocks.

Experimental results showed an increase in pH, an excess of Ca ($\Delta C_{Ca} > 0$) and a deficit in S ($\Delta C_S < 0$) (Fig. 4a,b,c). The experiments with sandstone and marl rocks showed an additional increase in Si and Fe (ΔC_{Si} and $\Delta C_{Fe} > 0$) (Fig. 4d,e). It was observed that the output pH decreased with increasing P_{CO_2} from atmospheric to supercritical CO$_2$ conditions (pH $\approx$ 7–8 and 4–6, respectively).

1D simulations. For the simulations, the rate laws including the values of the rate constants were taken from literature (Table 4). 1D simulations showed that dissolution of calcite in limestone, sandstone and marl and dolomite in dolostone led to the increase in pH and to the excess of Ca. To match the measured [Ca] concentrations the values of calcite and dolomite reactive surface area were reduced one to two orders of magnitude with respect to the calculated geometric ones (Table 5). The resulting range of values was attributable to both rock heterogeneity and rock structure (silicate content, grain size etc.; Noiriel et al. 2009) and provided a good fit to all experimental results, corroborating this modeling approach.

The increase in Δ_{Si} was associated with dissolution of albite and microcline. To reproduce the measured [Si] concentrations, the reactive surface areas of the Si-bearing minerals were larger than the geometric values (Table 5). An explanation for the diminished values of the reactive surface areas of calcite and dolomite (i.e. small reactivities) can be given by the solute transport control of the rates due to diffusion, at pore scale, from or to the carbonate mineral surfaces at low pH (e.g., Sjöberg and Rickard 1984). However, the increase in the reactive surface area values of the aluminosilicates is attributed to different size, shape and surface roughness of these minerals (Deng et al. 2018). The value of quartz reactive surface area was varied in the calculations although the quartz dissolution in this pH range (4–6) is negligible.

To fit the S deficit, which was mainly controlled by gypsum precipitation (Fig. 5), the reactive surface area of gypsum was smaller at elevated pressure than at atmospheric pressure

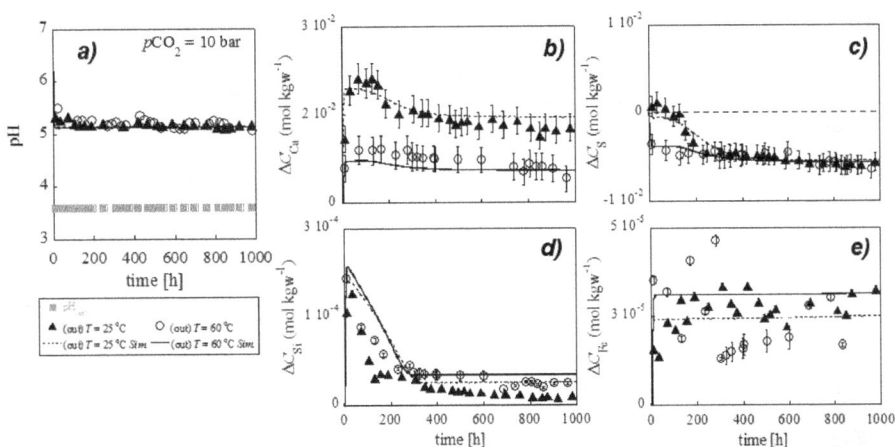

Figure 4. Variations in the measured (symbols) and simulated (*Sim.*; lines) pH and concentrations in the output solutions: **a)** pH, **b)** ΔCa, **c)** ΔS, **d)** ΔSi and **e)** ΔFe with time in a representative marl experiment at 10 bar of pCO_2 and 25 °C and 60 °C and $v_D = 1.2 \times 10^{-6}$ m s^{-1}. [Reproduced from Dávila et al. (2017) Experimental and modeling study of the interaction between a crushed marl caprock and CO$_2$-rich solutions under different pressures and temperatures Chemical Geology 448:26–42, with permission of Elsevier.]

(Table 5). As for the other secondary minerals (e.g. kaolinite, mesolite, stilbite, scolecite, boehmite, gibbsite, diaspore and alunite), their initial reactive surface areas were assumed to be sufficiently high to allow fast precipitation, i.e. local equilibrium conditions (Table 5).

It is worth highlighting that to test the non-linear effect of P on CO$_2$ solubility and mineral equilibria (through the molar volume of solutes) reported by Appelo et al. (2014), simulations using the CrunchFlow and PhreeqC (v.3; Parkhurst and Appelo 2013) codes were compared. Under supercritical conditions, using the same reactive surface area value for calcite dissolution, gypsum did not precipitate in the PhreeqC (v.3) simulation since these calculations considered the rise in the solubility of gypsum to be due to the increased pressure. Therefore, at $pCO_2 > 20$ bar, model calculations should consider the P effect on equilibrium constants to yield more reliable predictions.

Geochemical processes and porosity evolution. Simulations showed that calcite dissolution was greater than that of dolomite, albite, illite, and clinochlore, and the precipitation of gypsum was also much larger than that of any other secondary mineral (kaolinite, mesolite, stilbite, boehmite, gibbsite, diaspore, alunite, muscovite and goethite). The volume of precipitated gypsum was always smaller than the larger volume of dissolved calcite, yielding in all cases a porosity increase. The volume of dissolved limestone was larger than that of dolostone owing to the faster calcite dissolution kinetics. Likewise, a rise in P_{CO_2} results in a drop in pH, which substantially increases the calcite dissolution rate with respect to that of dolomite (Pokrovsky et al. 2005, 2009).

As for the temperature effect, under all pCO_2 conditions, low temperature favored calcite dissolution rate although the calcite dissolution rate constants increase with temperature (up to 100 °C; Pokrovsky et al. 2009). This inverse tendency is attributed to the fact that calcite undersaturation was increased by lowering the temperature. In addition, a rise in CO$_2$ solubility with decreasing temperature (Duan and Sun 2003) also contributes to a faster dissolution rate with decreasing temperature. The volume of precipitated gypsum was barely influenced by temperature variations. Simulations showed that an increase in temperature did not affect the trend of porosity variation along the columns (Fig. 6a) but reduced porosity creation (Fig. 6b).

When raising P_{CO_2}, the calcite dissolution rate increased along the columns because of the direct pH effect on the calcite dissolution rate (buffering effect of dissolved CO$_2$ on pH).

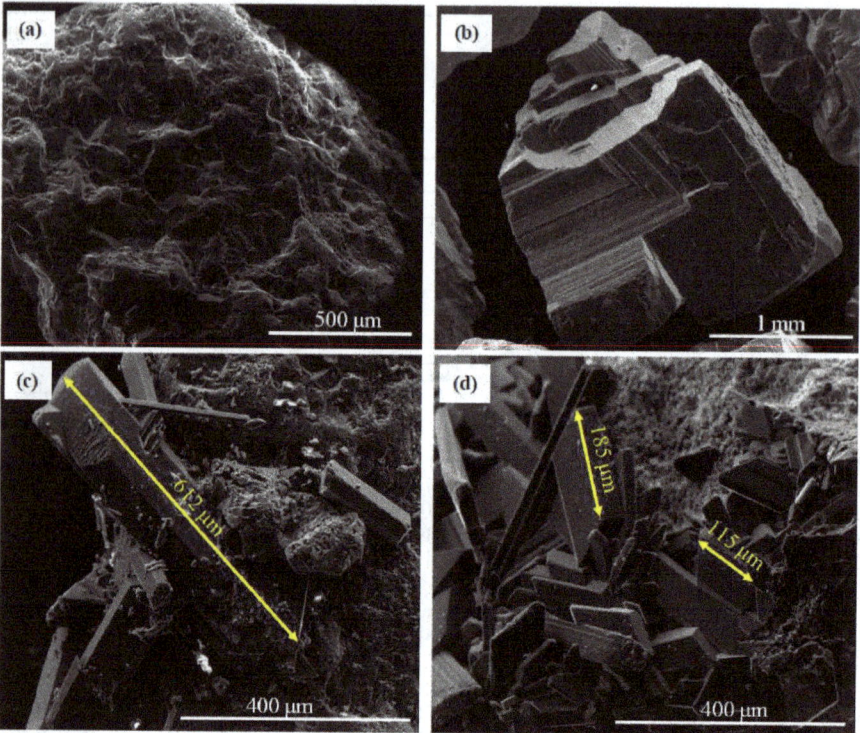

Figure 5. SEM images of unreacted and reacted limestone (**a** and **c**, respectively) and dolostone samples (**b** and **d**, respectively). The unreacted limestone fragments show a rough surface in contrast to a flat and terraced surface of the unreacted dolostone fragments. Gypsum precipitated in the form of needles (see dimensions). [Reproduced from García-Rios et al. (2014), Interaction between CO_2-rich sulfate solutions and carbonate reservoir rocks from atmospheric to supercritical CO_2 conditions: experiments and modeling. Chemical Geology 383:107–122, with permission of Elsevier].

Model results show that under high pCO_2 conditions pH remains acidic (ca. 5 due to the carbonic acid buffer capacity) and the brine is permanently undersaturated with respect to calcite, dolomite, clinochlore, albite and illite, yielding a higher increase in porosity all over the rock-brine contact. As a result, a rise in P_{CO_2} changed the pattern of porosity variation along the column, increasing the distance affected by dissolution (Fig. 6c).

Co-injection of SO$_2$. Oxidation near the well bore promotes formation of sulfuric acid, leading to additional brine acidification (Knauss et al. 2005; Xu et al. 2007). Thaysen et al. (2017) showed that relative to pure CO_2 experiments, the co-injection of 0.4% SO_2 as H_2SO_4 lowered the pH of the injected brine by ca. 1.5 pH units with respect to the pH of ca. 3.6 of H_2SO_4-free brine. The lower brine pH elevated the dissolution of calcite relative to pure CO_2 conditions, which in turn triggered more gypsum precipitation.

In contrast to the CO_2-experiments, in which one value for the reactive surface area of calcite was enough to match the experimental data, for the 1-D modeling of the $CO_2+H_2SO_4$ experiments, stepwise reductions of the calcite reactive surface area over time (e.g. from 250 to 32 $m^2_{mineral}$ $m^{-3}_{bulk\ rock}$ (limestone) or from 15 to 7 $m^2_{mineral}$ $m^{-3}_{bulk\ rock}$ (sandstone); Table 5) were used to simulate an effect of gypsum coating of the calcite surfaces (partial passivation). Daval et al. (2009) and Harrison et al. (2015) showed the effect that surface coating may exert on dissolving phases (Si-rich and Ca-rich layers on wollastonite surface and carbonation on brucite, respectively).

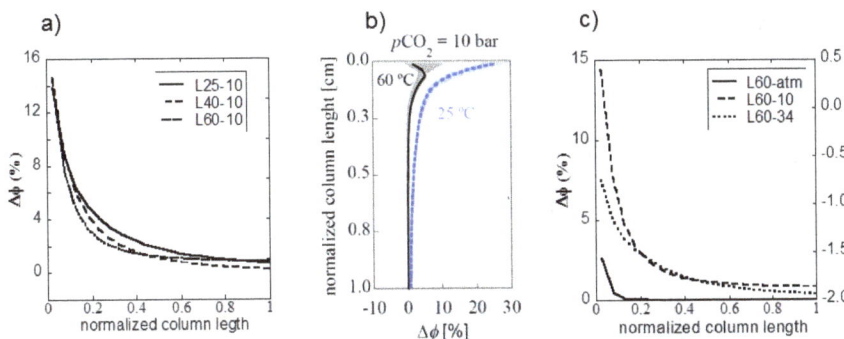

Figure 6. Variation of increase in porosity, $\Delta\Phi\%$, along the normalized column length: **a)** limestone ($pCO_2 = 10$ bar; $T = 25$, 40 and 60°C, and $v_D = 7 \times 10^{-7}$ m s^{-1}.); **b)** marl ($pCO_2 = 10$ bar; $T = 25$ and 60 °C, and $v_D = 1 \times 10^{-6}$ m s^{-1}); **c)** limestone (atmospheric pressure and pCO_2 of 10 and 34 bar of pCO_2; $T = 60$ °C, and $v_D = 4 \times 10^{-6}$ m s^{-1}). [Reproduced from García-Rios et al. (2014), Interaction between CO_2-rich sulfate solutions and carbonate reservoir rocks from atmospheric to supercritical CO_2 conditions: experiments and modeling. Chemical Geology 383:107–122, with permission of Elsevier].

To account for the passivation effect the former authors used a shrinking particle model and the latter ones proposed an empirical function that related the reactive surface area to the extent of brucite conversion. In the current study, the stepwise reduction of calcite area was used to implement the corresponding reduction in the reactivity of calcite. The three studies highlight the relevance of including a surface passivation effect on mineral dissolution in MCRTM.

Simulations showed that the marl rock showed higher reactivity over limestone and sandstone and had the highest gypsum precipitation rate relative to the rate of dissolving calcite. Nevertheless, there were no indications of gypsum coating on calcite grains in the marl experiments. This suggested a different growth mechanism of gypsum in marl versus sandstone and limestone rocks, e.g. a predominance of needle growth versus surface coating, which may be attributed to the high silicate content in the marl rock.

Column experiments: reservoir/cap rocks and Portland cement

Brine circulated through columns filled with alternating layers of fragments of rocks (limestone, sandstone and marl) and hydrated Portland cement at P_{CO_2} of 10 bar (experimental setup (Fig. 1b) and model implementation (Fig. 3b)). These experiments provided a porous medium in which advective transport (inter-grain porosity of ca. 50%) and diffusive transport (cement porosity of ca. 15%) occurred. These elevated P_{CO_2} conditions are suitable for determining the extent of the reactions involved in the interaction between CO_2-acidic brines and carbonate rock/cement, which is crucial for the assessment of wellbore integrity (Carroll et al. 2011).

2D simulations. To match the experimental concentrations of the output solutions the reactive surface area value of calcite (rock) was reduced. Those of C-S-H-1667, ettringite, hydrogarnet and hydrotalcite (cement) were slightly increased. Large values for possible secondary phases (e.g. gypsum, dolomite, muscovite, kaolinite, stilbite, C-S-H with variable Ca/Si ratio, aragonite, alunite and mesolite) allowed local equilibrium. Diffusion through cement fragments was calculated using an initial $D_e = 1 \times 10^{-11}$ m^2 s^{-1}.

Calculations showed that dissolution of portlandite caused an initially elevated pH of ca. 12. After portlandite exhaustion, calcite dissolution buffered the solution pH. The release of Ca and the injected sulfate content were sufficient to allow gypsum precipitation in the rock and cement layers (Fig. 7a). In the cement layers, ettringite and hydrogarnet dissolved partially whereas hydrotalcite and portlandite dissolved totally. Calculated precipitation of

a) rock

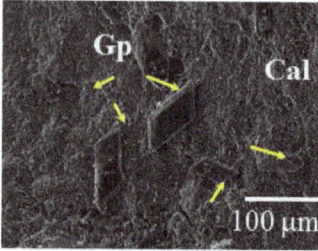

b) cement

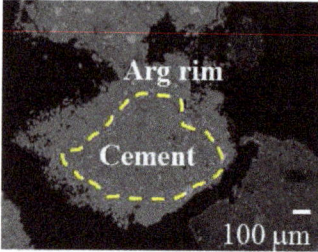

Figure 7. SEM images of (**a**) the rock layer 1 and (**b**) cement layer 2 displaying gypsum needles formed on calcite (limestone) surfaces and an aragonite rim coating a cement-fragment surface; rim thickness is about 100 μm.

aragonite in the two cement layers agreed with the observed formation of aragonite rims on calcite grains (Fig. 7b). In the three cement-rock columns, the volume of dissolved phases exceeded the volume of precipitated ones, yielding increases in porosity. Simulated variation of the resulting porosity displayed different patterns among the experiments (Fig. 8a). Porosity was higher near the inlet in the limestone than in sandstone and marl due to more dissolution of calcite in the former experiment (Fig. 8b). Larger calcite dissolution in the second rock layer in the sandstone and marl experiments enhanced porosity in this region. A marked porosity increase in the cement layers was due to dissolution of portlandite and ettringite, where the volume of precipitated aragonite was relatively smaller.

Therefore, it is deduced that in the carbonate rock–cement contact under elevated P_{CO_2}, calcite dissolution in the rock and portlandite and ettringite dissolution in the Portland cement were the reactions that controlled the changes in mineralogy. Gypsum and aragonite precipitates were the main secondary minerals involved in the porosity evolution.

Percolation experiments: reservoir and cap rocks

García-Rios et al. (2015, 2017) and Dávila et al. (2016a, 2016b) focused on the effects that solution composition and flow rate exert on the hydrodynamic

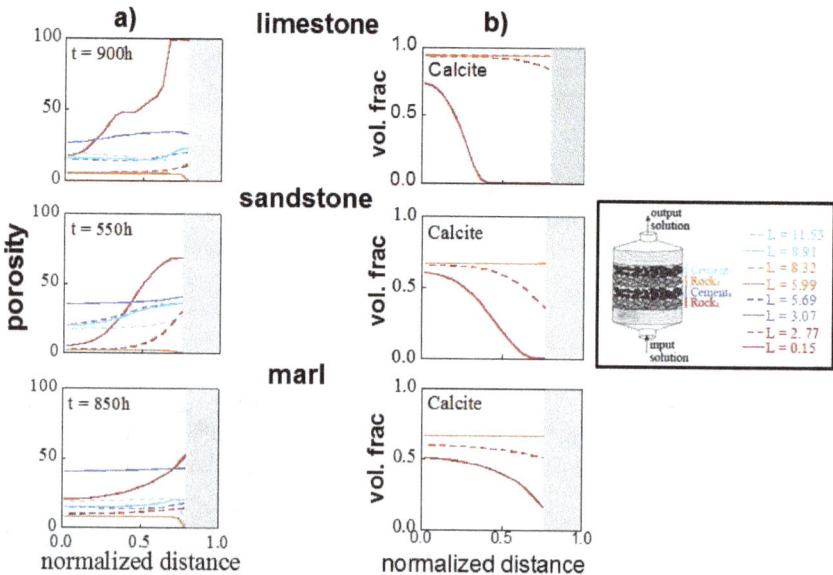

Figure 8. Variation in (**a**) porosity and (**b**) calcite volumetric fraction with respect to normalized distance in the three rocks at the end of experiments (v_D=2.5–5×10^{-6} m s^{-1}). Flux flow in mobile zone (acidic brine in grey); diffusion in non-mobile zone; see Fig. 3b).

and geochemical evolution of artificially fractured cores. Limestone, sandstone and marl samples (Table 1) reacted under supercritical CO_2 conditions. 2D reactive transport modeling was performed to quantify the dissolution and precipitation processes which induced a number of dissolution patterns: face dissolution (only dissolution at the inlet and no further alteration along the fracture), wormhole (preferential flow path) formation or uniform dissolution along the fracture.

Experimental results showed that the output [Ca] and [Si] concentrations were always higher than the input ones, indicating dissolution of calcite, microcline (sandstone) and clinochlore and albite (marl). A S deficit was caused by gypsum precipitation. The volume of dissolved calcite was always larger than that of precipitated gypsum. As regards fracture changes, it was observed that in the limestone experiments (Fig. 9a), a preferential path (wormhole) started in regions with high initial permeability (i.e., fracture surface heterogeneities), and the formed single wormholes could divert and develop branching near the outlet. Fracture surface became rough at slow flow rate. Gypsum precipitated preferentially where calcite surface roughness increased (Fig. 9b). In the sandstone experiments, calcite dissolution left non-dissolved grains of quartz and microcline along the fractures (Fig. 9c), leading to a non-uniform increase in aperture. In the marl experiments (Fig. 9d,e), dissolution of calcite contributed to the formation of a highly porous reacted zone with a width up to 300 mm, made up of non-dissolving quartz, illite and pyrite, partially dissolved silicate grains and precipitated gypsum.

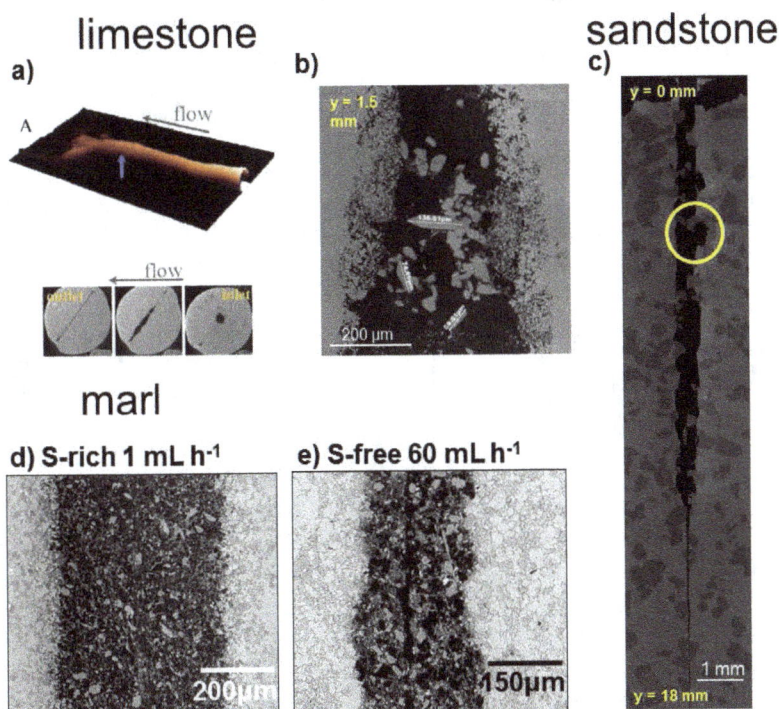

Figure 9. (a) mCT images showing ramified wormhole through fracture and inlet to outlet cross-sections (limestone); (b–e) SEM images of altered fracture regions: (b) gypsum precipitates on dissolving calcite surface (limestone); (c) yellow circle shows a site of grain detachment along the fracture (sandstone); (d–e) altered porous zone along the fracture at 8 mm from the inlet (marl). In (b and c) *y* values indicate distance from the fracture inlet along the flow direction (*y*). [Adapted from García-Rios M, Luquot L, Soler JM, Cama J (2017) The role of mineral heterogeneity on the hydrogeochemical response of two fractured reservoir rocks in contact with dissolved CO_2. Applied Geochemistry 84:202–217, Fig. 4, with permission of Elsevier].

2D simulations. The match between the experimental aqueous chemistry and the model variation was obtained by adjusting the values of the mineral reactive surface areas and the initial effective diffusion coefficients. In the three fractured cores, advection was the dominant mass transport process in the fracture, and the solute transport in the rock matrix was attributed only to diffusion.

In the limestone fractured core, where face dissolution formed, the output [Ca] and [S] concentrations and the resulting porosity with distance normal to fracture (Fig. 10a) were reproduced by reducing the calcite reactive surface area (Table 5) and considering $D_e = 5.6 \times 10^{-13}\,\mathrm{m^2\,s^{-1}}$ and $n = 2.5$ (Eqn. 18) for diffusion through the rock matrix.

Figure 10b compares the calculated porosity variation with distance normal to fracture and the one measured by SEM in the sandstone fracture, where uniform dissolution developed. Although the distance normal to fracture affected by dissolution next to the inlet was slightly shorter in the model, the calculated porosity variation was consistent with the experimental dissolution pattern. To reproduce the experimental data, a larger D_e value ($1.9 \times 10^{-11}\,\mathrm{m^2\,s^{-1}}$) and $n = 2$ (Eqn. 18) a reduced calcite reactive surface area value (Table 5) were considered. The calculated value of geometric surface area was reduced owing to the transport-controlled dissolution rate of calcite at pH < 5.

In the marl simulations, the calcite reactive surface area value was also diminished relative to the geometric value (Table 5). To match the measured [Fe] concentration, the reactive surface area of clinochlore was six orders of magnitude larger than the geometric surface area, which was not unreasonable for a clay mineral (e.g., $4.4 \times 10^8\,\mathrm{m^2_{mineral}\,m^{-3}_{rock}}$ for the clayey caprock at Sleipner; Gaus et al. 2005). A good match for [Si] was achieved by using reactive surface area values of albite and clinochlore that were two to five orders of magnitude larger than the geometric values (Table 5). Dissolution of quartz and illite was negligible. Precipitation of gypsum was reproduced using a value of reactive surface area smaller than the values used for precipitation of clay minerals (smectite and kaolinite), zeolites (mesolite, stilbite, scolecite and gismondine) and aluminum oxyhydroxide (boehmite), which were sufficiently high to allow fast precipitation (local equilibrium).

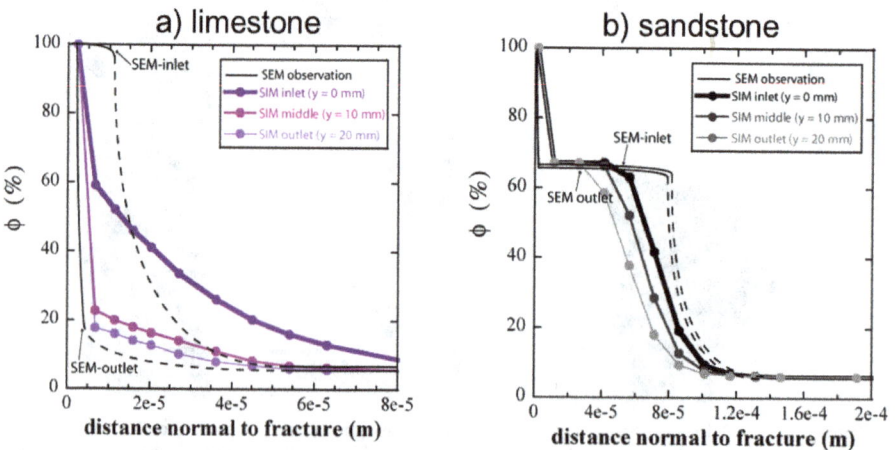

Figure 10. Variation of porosity along the distance to the fracture under 10 bar of pCO_2 in **(a)** limestone and **(b)** sandstone. y indicates distance from inlet (0 mm) to outlet (20 mm). [Adapted from García-Rios M, Luquot L, Soler JM, Cama J (2017) The role of mineral heterogeneity on the hydrogeochemical response of two fractured reservoir rocks in contact with dissolved CO_2. Applied Geochemistry 84:202–217, Fig. 5, with permission of Elsevier].

In the experiment with fast Darcy velocity, $D_e = 6\times10^{-12}\,m^2\,s^{-1}$, which was greater by a factor of 20 than the value used in slow velocity ($D_e = 1–3\times10^{-13}\,m^2\,s^{-1}$). This increase was probably due to a fast flow rate that would result in less transport control, which was reproduced in the model using a large initial D_e. However, in the experiments with high flow rate, a poor match was obtained where a wormhole dissolution pattern formed. The rectangular geometry of the model domain could not reasonably account for the evolution of the cylindrical geometry related to wormholing. Discrepancies between the calculated and measured concentrations were therefore to be expected.

MCRTM: limitations and alternative approaches

Using MCRTM to successfully simulate the experimental results presented above, we encountered a number of problems that should be resolved to improve future modeling.

The transport control of the rates due to diffusion from or to carbonate mineral surfaces at low pH, which is not included explicitly in this type of model, obliged us to diminish the reactive surface areas (i.e., reduction in calcite reactivity) with respect to the estimated geometric value. The range of diminished values in the calcite surface area varied several orders of magnitude to simulate column experiments with a homogeneously distributed pore space and cores with finely characterized short fractures. The variation in calcite reactive surface area was dependent on the fluid residence time. In contrast, reactive surface areas of the Si-bearing minerals were enlarged because of surface roughness. Beckingham et al. (2016) produced Image Perimeter specific surface areas for quartz and Mg- and Al-silicate minerals to quantify the physical surface area in contact with the reacting fluid (pH = 3.2). Therefore, MCRTM of the experimental results from pristine or fractured cores with a heterogeneous pore space or fractures displaying local heterogeneities (surface roughness, high permeable zones) demands a suitable fit of the effective (reactive) mineral surface area.

A stepwise reduction of the calcite reactive surface area was implemented to simulate calcite passivation by gypsum coatings. Since these codes assign average property values (e.g. porosities, surface areas, diffusion coefficients, etc.) to each numerical element (REV approach) and do not consider an explicit representation of mineral grains and their surfaces (pore scale representation), an added limitation due to diffusion to and from the mineral surfaces to the bulk solution was not included. Thus, calcite passivation was only reproduced by reducing calcite reactivity through a decrease in the reactive surface area of calcite.

Modeling the coupled mechanisms of transport and reaction in fractures when localized dissolution (wormhole) occurred is complex. In our case, the rectangular geometry of the model domain could not reasonably account for the evolution of the cylindrical geometry due to wormholing. A new model accounting for the wormhole formation is warranted to interpret the experimental results under fast flow. It is therefore necessary to use numerical approaches which consider local heterogeneities (e.g. Szymczak and Ladd 2011; Soler-Sagarra et al. 2016).

We used 2D reactive transport modeling to simulate fractured cores that underwent face and uniform dissolution. In the marl fracture, when porous layers formed by the non-dissolved minerals remaining behind the dissolution reaction front, non-reactive minerals formed a porous layer which prevented the solution from easily accessing the reactive minerals of the matrix (calcite). This diffusion-controlled phenomenon resulted in a decrease in calcite dissolution. Deng et al. (2016) performed 2.5 D reactive transport modeling to study fracture alteration in a dolostone core (87 wt% dolomite and 10 wt% calcite), taking into account the permeability heterogeneity in the fracture plane and thickness variation of an altered layer. Effective mineral reaction rates as a function of the effective diffusion coefficient were considered as an alternative solution to better understand fracture alteration under different flow regimes (Deng and Spycher 2019, this volume). In their calculations, the D_e value ($10^{-10}\,m^2\,s^{-1}$) was about two orders of magnitude greater than the ones fitted in our simulations for calcite-rich carbonate rocks. Additionally, a cubic law developed for fractures to calculate fracture permeability was used instead of the Kozeny–Carman relationship (Eqn. 20).

It should be pointed out that as shown by Elkhoury et al. (2013), dissolution patterns in fractures resemble those in porous media where dissolution features evolve from face dissolution to wormhole formation when increasing the flow rate (Golfier et al. 2002; Kang et al. 2014). The main difference lies in the fact that the presence of fractures focuses the wormholes along the fractures. However, under high flow rates, preferential flow paths grow uniformly, compared with the formation of ramified wormholes in porous media. In our research with carbonate rocks, the fragments-filled columns and the fractured cores represent highly porous and fractured subsurface media, respectively. Carroll et al. (2013), Smith et al. (2013a), Hao et al. (2013) and Smith et al. (2017) used mostly intact cores with pore-scale heterogeneities to quantify the reactivity of carbonate rocks with a variable mineralogy content (vuggy limestone, marly dolostone and dolostone) under supercritical P_{CO_2} conditions. The outcomes achieved in these studies are very informative for comparison and should be regarded in further modeling. First, it was shown that in limestones wormholes initiated and developed in areas of greater porosity and permeability contrast, following preexisting preferential paths. Taking into account that a fracture is a preferential path, our fractured limestone cores developed wormholes in a similar way subject to high flow rates. Second, two median values for the exponent n in the permeability–porosity relationship (Eqn. 20) were derived: $n = 7$ for vuggy limestone with heterogeneously distributed pore space and $n = 3$ (as the one used in our simulations) for marly dolostone regardless of pore space distribution. 2D simulations of our fractured cores using $n = 3$ together with solute diffusion through the porous rock matrix captured reasonably well the fracture evolution observed when face dissolution and uniform dissolution prevailed. It was perhaps in the wormholing simulations when the difficulties encountered could have been overcome by using a full 3D model together with a higher n value. In their 3D simulations the rate constants for calcite and dolomite were variably diminished up to two orders of magnitude with respect to those reported in the literature (Palandri and Kharaka 2004). This finding is in line with our reduction of the estimated geometric surface area by about two orders of magnitude to account for the diffusion-controlled rate of calcite.

MCRTM IN AMD-WATERS

Interaction between AMD and Portland cement/sedimentary rock

Mitigation and regional control of AMD require the construction of concrete-based structures, such as aeration cascades, tanks to hold the materials of the passive treatment systems (Figs. 1e and 2), and dams to control the level of the rivers. The durability of these concrete-based structures depends very much on the processes arising from the interaction between the concrete and the AMD water. Given that these highly polluted waters have very low pH ($0 < pH < 4$) and high concentrations of sulfate, iron, aluminum and metal(loid)s, the dissolution of the cement phases (e.g. calcium silicate hydrate (C-S-H) and portlandite), precipitation of secondary minerals (gypsum, goethite, schwertmannite, ettringite, etc.) and adsorption of metal(loid)s will be the dominant reactions that determine the fate of the concrete-based structures.

Column experiments. We are studying the effect of these reactions on the mineralogy of cement and rock and to evaluate the consequences for concrete durability. To this end, column experiments with two alternating layers of milimetric-size fragments of sandstone and Portland cement were carried out (Table 1 and Fig. 1b). The columns were filled with two alternating layers of rock and cement with an approximate intergrain porosity of 65% (Fig. 1b). An input solution of pH = 2 (H_2SO_4) containing Al, Ca, Mg and Fe(II) as major ions and arsenate and divalent cations (Zn, Cd and Ni) as minor components was injected at a constant flow rate from the bottom upwards (Table 2).

During the experiments, pH rapidly increased to ≈ 12.5, after which it decreased, remaining at ≈ 6 until the conclusion of the experiments (Fig. 11a). The output Ca concentration exceeded the input one in contrast to that of S, which always showed a deficit. The evolution of the concentrations of the divalent cations (Fe(II), Zn, Cd and Ni) showed almost total depletion of metals at the high pH and partial elimination as pH decreased (Fig. 11b–e). This behavior was attributed to the formation of metal hydroxides at high pH. In contrast, arsenate depletion occurred throughout the experiments (not shown), suggesting adsorption onto Fe-hydroxides (e.g. goethite).

1D simulations of the column experiments corroborated that precipitation of gypsum and aragonite occurred at the expense of total and partial dissolution of portlandite and calcite, respectively. Formation of basaluminite was also calculated (Fig. 11f). These mineralogical changes yielded a marked increase in porosity at the inlet of the columns and a significant decrease in porosity in the first sandstone layer (Fig. 11g). SEM-EDS observations and XRD analyses corroborated the absence of portlandite in the cement matrix and the presence of newly formed gypsum and aragonite. Note that basaluminite identification by XRD is highly difficult given the poor crystallinity of this Al oxyhydroxisulfate (Lozano et al. 2018).

Concrete cores. The AMD-treatment plant at the Esperanza mine in the Iberian Pyrite Belt (Huelva, south Spain) is used as a case study of interaction between AMD and concrete. To this end, concrete cores (2 - 6 cm long and 2.5 cm in diameter) were sampled at the 3-year-old Esperanza II Mine AMD treatment plant. SEM-EDS inspection of thin sections along the concrete core samples showed that alteration mainly occurred at the concrete-AMD interface (top 2–3 mm) (Fig. 12), revealing precipitation of Fe-rich phases on top of the concrete and little concrete alteration, which was not observed further into the cores.

Figure 13 shows the results of a preliminary 1D simulation in which an AMD water of pH 2.7 and rich in sulfate, Al and Fe (II) is in contact with concrete, which is the situation at the ME2 point at Mina Esperanza (Fig. 2). AMD flows continuously along the channel, with a rather constant composition with time (pH = 2.7, [Ca] = 3.0 mM, [Mg] = 5.7 mM, [Si] = 1.5 mM, [Al] = 3.9 mM, [SO$_4$] = 27 mM, [Fe$_{TOT}$] = 9.7 mM), and this water interacts with the concrete at the floor of the channel. In the simulations, the AMD water composition is fixed in the first 2 nodes of the domain (channel), but schwertmannite is allowed to precipitate, as observed in the channel at Mina Esperanza. Solute transport along the concrete is only by diffusion using an initial D_e value of $1.59\,m^2\,s^{-1}$.

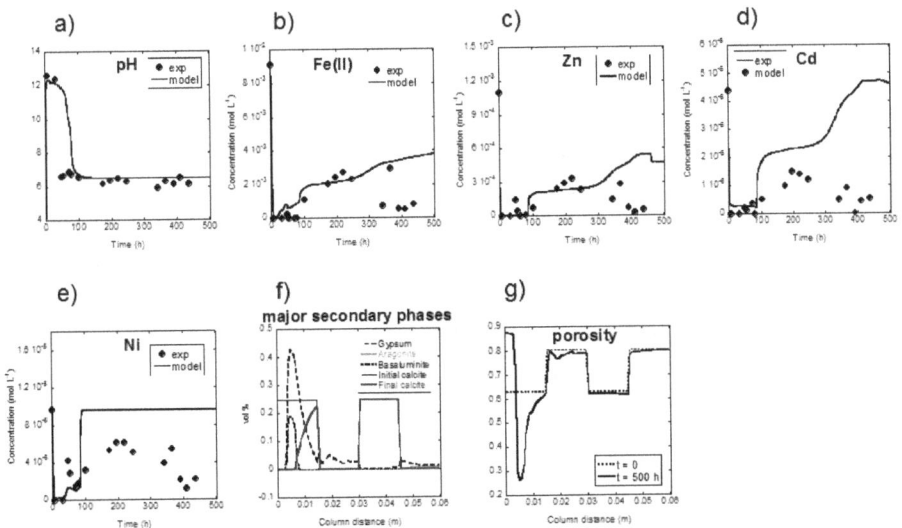

Figure 11. Experimental and modeling results of the sandstone-cement column under atmospheric pressure: (**a–e**) pH and output concentrations vs. time; (**f and g**) variation in volumetric fraction of secondary minerals and porosity along the columns distance at the conclusion of the experiments.

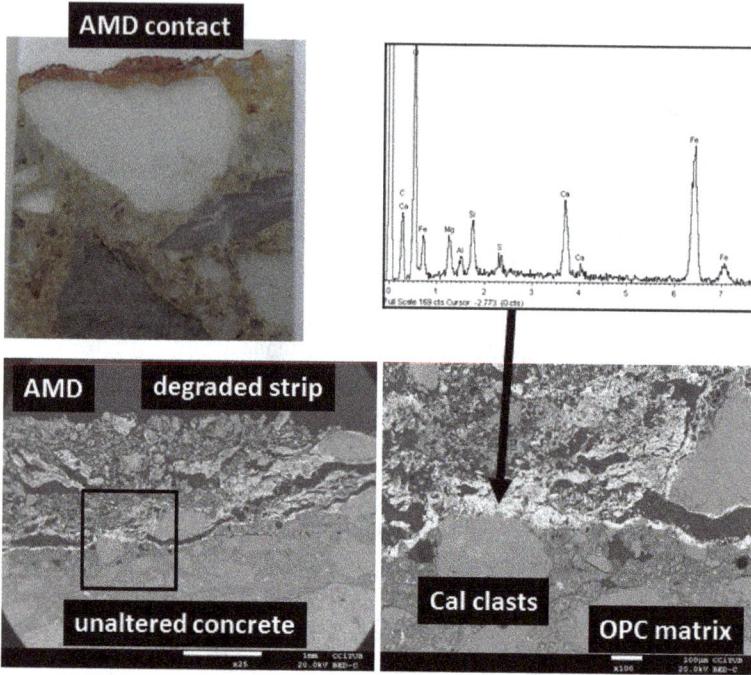

Figure 12. **(Top left)** Photograph of a thin section along the Z axis of core sample (2.5 cm in diameter). AMD contact indicates AMD-upper-core contact. The brownish strip at the top formed during concrete degradation. Concrete clasts are calcite, dolomite and quartz. **(Bottom)** SEM images of thin section along the Z axis of core sample: left: degraded strip at the core-AMD contact suggests alteration of concrete; right: close-up and EDS analysis of the strip-concrete interface related to the presence of Fe-rich phases (schwertmannite), carbonate minerals (C, Ca, Mg), quartz (Si) and ettringite or gypsum (Al, Ca, S).

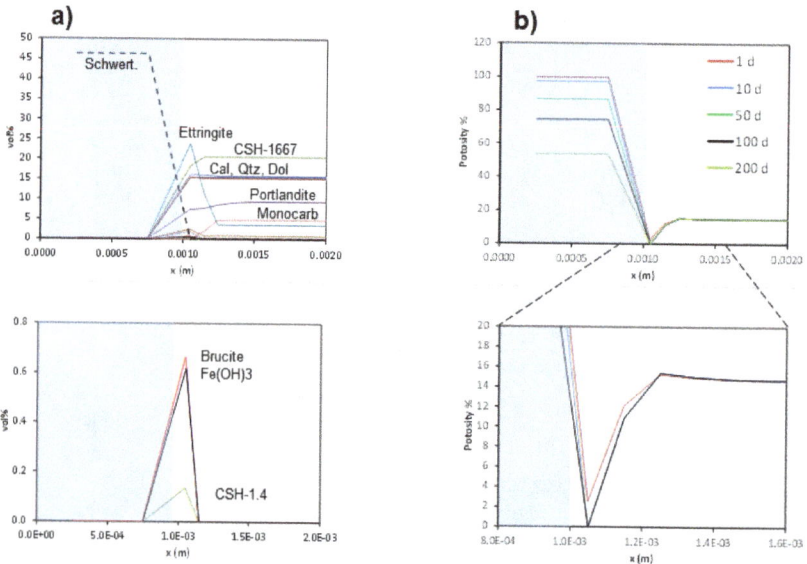

Figure 13. Calculated variation in **(a)** mineral volume fraction and **(b)** porosity along depth in a concrete core exposed to AMD. Alteration is calculated to take place at the top across the first 1.5 mm. AMD is represented in blue.

Schwertmannite precipitation kinetics in the channel was adjusted so that precipitation occurred progressively with time, as observed. In the concrete, the acidic water causes the dissolution of portlandite, but the high sulfate and Al content of AMD promotes the intense precipitation of ettringite, which quickly clogs porosity right next to the interface. The clogging of porosity severely limits concrete alteration (Fig. 13a). Other reactions are some decalcification of C-S-H (dissolution of primary C-S-H and precipitation of C-S-H with lower Ca/Si), precipitation of small amounts of hydrotalcite, calcite, brucite and ferrihydrite, and dissolution of primary monocarboaluminate. The model results showed a very thin alteration zone right at the concrete–AMD interface (Fig. 13b), suggesting that alteration under the aggressive AMD conditions is very limited.

Interaction between AMD and limestone

AMD that flows through the limestone sand used in the treatment systems (e.g. anoxic limestone drain (ALD) or disperse alkaline substrate (DAS) tanks), dissolves calcite, raising pH and alkalinity, yielding trivalent metal retention as Me-oxyhydroxide precipitates (Me = Al and Fe(III)). The efficiency of these passive systems is however limited because secondary mineral precipitation (e.g. gypsum) causes the passivation (armoring) of the limestone grains and clogging of the pores, reducing limestone reactivity and acid neutralization (Caraballo et al. 2009a, 2011; Soler et al. 2008).

To gain further insight into the loss of calcite reactivity due to grain coating or clogging of porosity, Offedu et al. (2015) performed column experiments using limestone (fragment size = 1–2 mm and inter-grain porosity = 49%; Fig. 1b) and synthetic acid solutions (pH of 2–3, H_2SO_4; Table 2) containing Fe(III) and Al with concentrations that fell in the range found in AMD (Nordstrom et al. 2000). X-ray microtomography (mCT) measurements showed how the gypsum coating–calcite passivation (Booth et al. 1997; Offedu et al. 2014) and precipitation of metal oxyhydroxides influenced the porosity changes in the columns (Fig. 14). The dissolution of calcite (limestone) released Ca, which combined with the SO_4 in solution, caused precipitation of gypsum on the calcite surfaces. As a result of passivation, output pH dropped to values close to the input value, metal retention stopped and SO_4 concentrations came close to the initial value (or even higher if gypsum dissolved). Calcite dissolution also caused an increase in pH from 2 to $\approx$ 6–7 (proton consumption), resulting in supersaturation of the solutions with respect to Fe- or Al-oxyhydroxides. Precipitation of metal-oxyhydroxides between the grains (Fig. 14) caused clogging. The overall process is represented by the following reactions

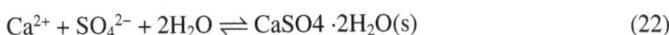

$$CaCO_3 + 2H^+ \rightleftharpoons Ca^{2+} + H_2CO_3 \qquad (21)$$

$$Ca^{2+} + SO_4^{2-} + 2H_2O \rightleftharpoons CaSO4 \cdot 2H_2O(s) \qquad (22)$$

$$Fe^{3+} + 2H_2O \rightleftharpoons FeOOH(s) + 3H^+ \qquad (23)$$

Microtomography examinations were performed at different times (Fig. 14): the four mCT images show the same section close to the column inlet. Initially (d0) only calcite grains were present (light gray), separated from each other by pore space (dark areas). After 4 days (d4), some of the calcite grains were coated by a thin gypsum layer (dark gray) and a whitish phase (goethite as identified by mXRD) filled the pore space. After 8 and 12 days (d8 and d12), the gypsum coatings grew and goethite content increased, yielding column passivation. The contribution of gypsum and goethite precipitates to porosity decrease calculated by mCT image segmentation was ca. 15% (goethite) and ca. 5% (gypsum) from the inlet to the middle of the column.

1D simulations. 1D modeling was performed to simulate the processes occurring in the columns. Since passivation was controlled by gypsum coating on calcite surfaces, calcite reactivity diminished as a result of the loss of calcite reactive surface area. Given that the calcite passivation

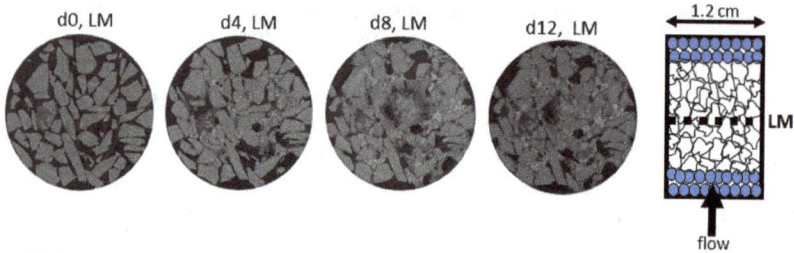

Figure 14. Four mCT images of the same section of a column during the experiment: d0, before reaction; d4, d8 and d12, after 4, 8 and 12 days (passivation at ≈ 300 h). Since the grains were not cemented, relative positions changed slightly in the four images. [Reprinted from Offeddu FG, Cama J, Soler JM, Dávila G, McDowell A, Craciunescu T, Tiseanu I (2015) Processes affecting the efficiency of limestone in passive treatments for AMD: Column experiments. Journal of Environmental Chemical Engineering 3:304–316, Figs. 6, 7, with permission of Elsevier].

was not implemented in the CrunchFlow code, as in the CO_2–H_2SO_4 modeling described above, the calcite reactive surface area was decreased stepwise (Table 5; Fig. 15). The reactive surface areas of goethite and gypsum were assumed to be constant during the experiment (Table 5). This was a model simplification since absolute areas should increase with precipitation. But even with this underestimation of gypsum and goethite reactivity it was necessary to reduce the calcite surface area to match the experimental results. This simple model (stepwise reduction

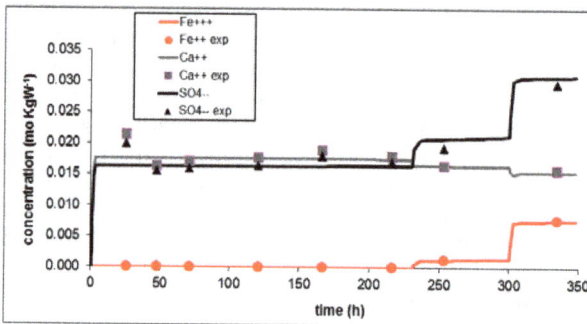

Figure 15. Experimental and modeling results (output concentrations vs. time) from representative column. [Reprinted from Offeddu FG, Cama J, Soler JM, Dávila G, McDowell A, Craciunescu T, Tiseanu I (2015) Processes affecting the efficiency of limestone in passive treatments for AMD: Column experiments. Journal of Environmental Chemical Engineering 3:304–316, Fig. 13, with permission of Elsevier].

of calcite surface area) showed that the experimental results are consistent with a reduction of calcite reactivity induced by the precipitation of gypsum (passivation mechanism).

Interaction between AMD and DAS: column experiments

The performance of an AMD remediation treatment depends on the relationship between the AMD flux and the dissolution rate of the reagent. Prior to building a field-scale system, this relationship must be quantitatively investigated through column experiments and MCRTM. In the columns, the flux is imposed by scaling both the field discharge and the expected surface of the treatment plant. The adequate dissolution rate of the reagent (e.g. calcite), however, can be obtained by selecting the grain size and the reagent-wood shavings ratio. Owing to the non-linear relationship between flux and chemical reactions, MCRTM is necessary to obtain an optimum criterion for the reagent dissolution rate (Eqn. 12).

The dissolution rate constant for calcite was obtained from earlier studies (Arvidson et al. 2003; Table 4). However, given the discrepancies of the orders of magnitude between experimentally and field derived mineral rates (White and Brantley 2003), the adjustment of the reagent reactive surface area value is critical for the modeling, as shown above. This is because gas sorption BET methods or geometric approximations are not useful to estimate the amount of solid surface area in contact with flowing water. Column experiments were built to derive a realistic estimation of the reactive surface area. The columns were filled with a mixture of 1–3 mm calcite fragments and wood shavings with a 1:1 weight ratio. Hydraulic conductivity profiles of the substrates were determined by connecting piezometric standpipes to every pressure measuring port (Fig. 1d). Head-loss between the different ports was then measured at high flow rates by reading the piezometric head difference between adjacent pipes (Rötting et al. 2008a). The water flow was imposed throughout the head difference between the inflow and outflow water, and the main parameters controlling solute transport, porosity and dispersivity were determined by injecting water with a conservative solute and analyzing the breakthrough curve. Once the flux and transport properties are known, the sampling points at different depths of the columns provide solution chemistry data to be reproduced with MCRTM using A_m (Eqn. 12) as the only fitting parameter (e.g. column in Fig. 16). 1D simulations matched the experimental results by diminishing the calcite reactive surface area to 180 $m^2_{mineral}$ $m^{-3}_{bulk\ rock}$ (Table 5).

Calcite dissolution is the main mineral controlling the evolution of the AMD hydrochemistry along the treatment. However, there are some other mineral phases governing the specific processes responsible for Al and Fe removal within the limestone–DAS reactive material. As evidenced in Figure 16a–e, the pH increases in two steps that occur when all the Fe and Al have been removed

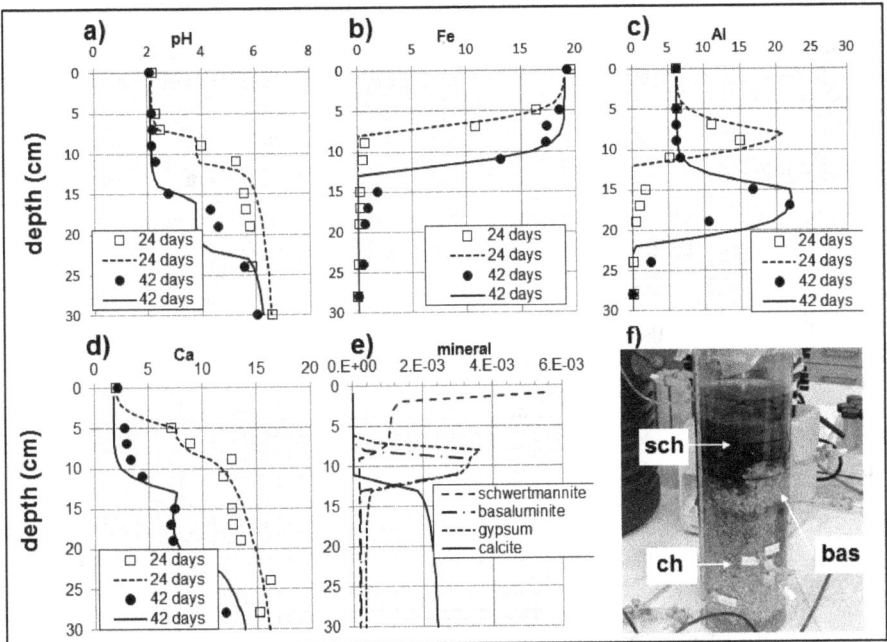

Figure 16. (a–d) Pore water composition of a limestone–DAS column treating an AMD sample (concentration in mM; symbols and lines represent experimental and modeling data, respectively; (e) calculated volumes of the main precipitates after 42 d (m^3_{min} m^{-3}_{column}); (f) photograph of the column after 24 d (sch = schwertmannite; bas = basaluminite and gypsum; ch = limestone).

from the water. Mineral precipitation takes place in two different fronts (Fig. 16f). In the column, from bottom to top the first front corresponds to the precipitation of hydrobasaluminite (bas) and is formed at the expense of the alkalinity produced by calcite dissolution.

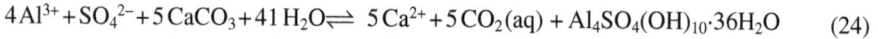

$$4\,Al^{3+}+SO_4^{2-}+5\,CaCO_3+41\,H_2O \rightleftharpoons 5\,Ca^{2+}+5\,CO_2(aq)+Al_4SO_4(OH)_{10}\cdot36H_2O \quad (24)$$

Upstream from the Al front, schwertmannite forms at the expense of the alkalinity produced by dissolution of calcite

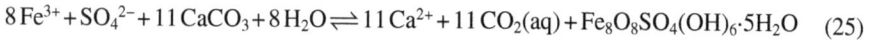

$$8\,Fe^{3+}+SO_4^{2-}+11\,CaCO_3+8\,H_2O \rightleftharpoons 11\,Ca^{2+}+11\,CO_2(aq)+Fe_8O_8SO_4(OH)_6\cdot5H_2O \quad (25)$$

However, when calcite reagent is exhausted, schwertmannite also forms at the expense of the alkalinity released by basaluminite dissolution

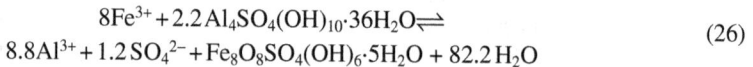

$$8Fe^{3+}+2.2\,Al_4SO_4(OH)_{10}\cdot36H_2O \rightleftharpoons$$
$$8.8Al^{3+}+1.2\,SO_4^{2-}+Fe_8O_8SO_4(OH)_6\cdot5H_2O+82.2\,H_2O \quad (26)$$

The consumption of hydrobasaluminite at the schwertmannite front is also evidenced by the increase in Al concentration far above the inflow concentration (Fig. 16c). This increase in Al concentration is also clearly observed in field-scale treatments (Rötting et al. 2008b; Caraballo et al. 2009a, 2011) and is an additional contribution to downstream hydrobasaluminite formation. This localized precipitation of hydrobasaluminite is a major threat to hydraulic conductivity, as discussed below. In fact, hydrobasaluminite can contain up to 36 water molecules, and it shows a characteristic jelly aspect. Its molar volume is very difficult to measure. An estimate of 482 cm^3 mol^{-1} is obtained from the specific gravity of basaluminite, (2.12 g cm^{-3}), adding the weight of up to 31 water molecules. Accordingly, the reaction of basaluminite formation leads to an increase in the volume of the solid phases and almost reduces porosity and hydraulic conductivity to zero.

Moreover, calcite dissolution will increase Ca concentration in the water (Fig. 16d), and AMD from the Iberian pyrite belt (IPB) often contains thousands of milligram per liter of sulfate. Therefore, gypsum may also precipitate within the reactive substrate and contributes to passivation of calcite surfaces and/or clogging of the pore space.

Interaction between AMD and DAS: field-scale passive remediation treatment

Design of field-scale treatment: role of reactive surface area. Once the reactive transport model is calibrated using the column results, it can be used to simulate different scenarios of a field-scale treatment. The main parameters to vary are the flux of AMD (m^3 m^{-2} s^{-1}), the AMD chemical composition, the proportion of reactive and inert material, and the thickness of the reactive filling. The performance of a hypothetical treatment of a limestone–DAS system (2 m thick × 250 m^2 of surface and initial porosity of 0.45) to treat AMD from Mina Esperanza (pH = 2.8, [Fe(III)] = 749 mg L^{-1}, [Al] = 167 mg L^{-1} and [SO$_4^{2-}$] = 2720 mg L^{-1}) flowing at 1 L s^{-1} was calculated for two cases with different limestone–wood shavings ratios (Fig. 17). The results showed that in both cases (limestone–wood shavings weight ratios of 2:1 and 1:1) the output water with a pH of 6 retained all Fe and Al (not represented) after 2 years of functioning. Decrease in porosity was located at two depths. The first one was caused by precipitation of high amounts of schwertmannite near the surface. It is not difficult to mechanically remove the surface material during the operation of the treatment system. However, the higher reduction in porosity was due to precipitation of hydrobasaluminite (pH ≈ 4). The a priori more efficient 2:1 ratio case led to a reduction in porosity, causing clogging. A sensitivity study with the model showed that a lower limestone proportion (limestone–wood shavings ratio of 1:1) slowed the calcite dissolution rate, expanded the reaction front, and caused less concentrated hydrobasaluminite precipitation and less porosity reduction. Therefore, the 1:1 ratio option is more suited to this case. Lower limestone–wood shavings ratios (e.g. 1:2) that mostly prevent porosity reduction are unable to neutralize the water acidity (not represented).

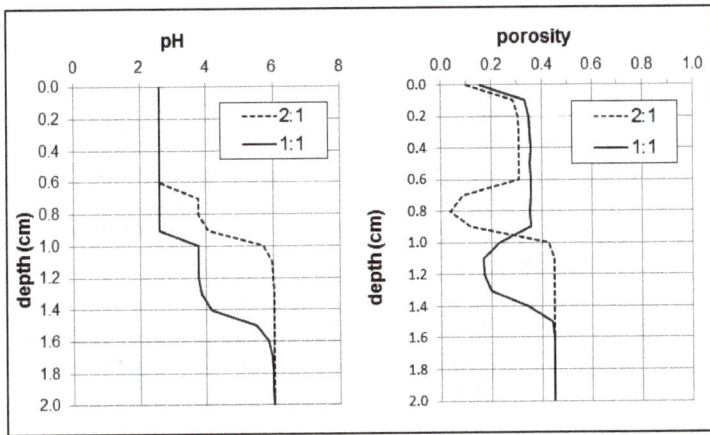

Figure 17. Distribution of pH and porosity within a DAS-calcite system after 2 yr of treating 1 L s^{-1} of AMD (see text for composition). Results from two limestone–wood shavings ratios (2:1 and 1:1) in reactive filling.

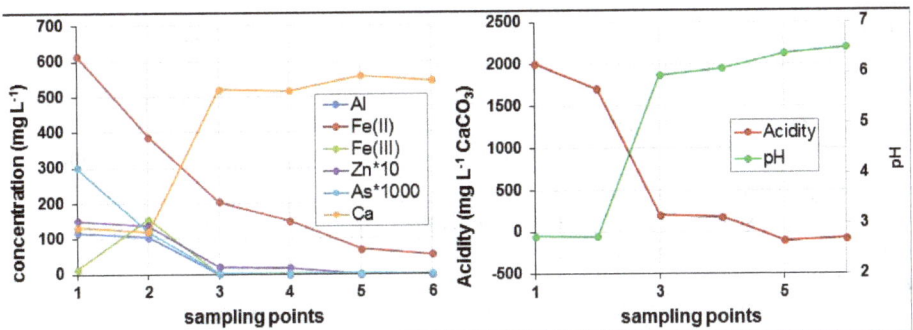

Figure18. Evolution of the main chemical parameters of AMD through the remediation system of Mina Esperanza II after 32 months of treatment (Orden et al. 2018). Location of sampling points is shown in Figure 2.

Field-scale treatment: blind MCRTM prediction. The average chemical composition of water at different stages of the treatment after 32 months of functioning is represented in Figure 18. Water oxidation along the aeration cascades and at the first decantation lagoon (Fig. 2) is evidenced by the decrease in Fe(II) and the increase in Fe(III), which will be subsequently precipitated as schwertmannite. pH shows a very small decrease due to water hydrolysis by Fe(III). Arsenic is retained by schwertmannite with a decrease from 300 to 70 μg L^{-1}. The concentrations of the remaining elements do not undergo systematic variations.

The most important changes in the chemistry of water occur inside the first DAS–calcite tank (between sampling points 2 and 3; Fig. 18). pH increases up to 6 and acidity drops to 200 mg L^{-1} CaCO$_3$. Total iron decreases from 600 to 200 mg L^{-1} (due to faster oxidation at higher pH and schwertmannite precipitation), and the remaining aqueous Fe is all Fe(II). The remaining elements, Al, As and Zn (and the other elements that are not represented) are depleted close to or below detection levels. Arsenic is adsorbed by schwertmannite (Fukushi et al. 2003), Cu is precipitated as cuprite, and REEs are adsorbed on hydrobasaluminite or co-precipitated in fluorite (Ayora et al. 2016). The other divalent metals are probably precipitated as hydrated complex carbonates (Pérez-Lopez et al. 2011). The role of the second DAS-calcite pond is minor. pH rises to 6.5 and the remaining divalent metals are totally removed from water, only Fe remains in solution in concentrations below 100 mg L^{-1}. Fe(II) was in part oxidized to Fe(III) and precipitated as ferrihydrite (see the different color of the two

DAS-calcite tanks in Figure 2). Along the treatment system, sulfate decreases from 3000 to 2000 mg L^{-1} due to precipitation of schwertmannite, hydrobasaluminite and gypsum, but the calcite–DAS is unable to deplete SO$_4$ below gypsum solubility.

From the MCRTM point of view, the most interesting issue is to check the validity of the predicted functioning of the AMD-treatment plant from the results calculated in the column experiments. Since the Mina Esperanza II system is currently in operation, no evolution on the final outflow composition and no signals of treatment exhaustion are observed. Therefore, the dismantled DAS-calcite tank of Mina Esperanza I (Fig. 2) is suitable to make comparisons between the design predictions and observations.

In the first year, the hydrochemical behavior was fairly steady. Only a slow decrease in water pH from 6 to 5.1 occurred during the last months (Fig. 19a). Systematically analyzed total Fe at the output of the system showed ca. 20–30% of Fe removal (Fig. 19b), which is similar to the Fe(III) concentration in the inflow (Caraballo et al. 2011a). Aluminum was completely removed during the first year, and only when the pH of the water outflow was close to 5 was a small decrease in the aluminum removal observed (Fig. 19c). 1D simulations reproduced the chemical composition of the outflow water using the reactive surface area value fitted in the column modeling (Table 5). Two cases with the extreme solution composition during the functioning of the calcite–DAS tank were used in the calculations (pH = 2.35 and 2.96; [FeTOT] = 755 and 1100 mg L^{-1} (40% Fe(III)) and [Al] = 128 and 167 mg L^{-1}; Caraballo et al. 2011). Figure 19 shows that the pH and the measured output Fe and Al concentrations were within the predicted values. Hence, the model successfully predicted the outflow water chemistry for 18 months of the Mina Esperanza I treatment. Discrepancies were observed between the predicted and observed Fe values because Fe(II) was not efficiently removed with only one DAS-reactive system.

As regards the solid phases precipitated within the reactive material, XRD patterns revealed the presence of schwertmannite and goethite as the only detected Fe mineral phase at 0–20 cm depth. No mineral phase was detected in samples at 20–35 and 35–50 cm depth, whereas gypsum and calcite were the only phases detected from 50 to 250 cm depth (Caraballo et al. 2011). Sequential extractions of these samples specifically designed to identify this type of Fe-Al-SO$_4$–solids (Caraballo et al. 2009b) reproduced the XRD results. Thus, the presence of schwertmannite was detected by the amount of Fe recovered in the third step of the sequential extraction, while the Fe analyzed in the fourth step was attributed to goethite dissolution (Fig. 20a). Although not identified by XRD due to its amorphous nature, the presence of hydrobasaluminite was evidenced by Al recovered in the third step of samples between 50 and 210 cm. The retention of an important sulfur concentration in steps 2 and 3 was attributed to

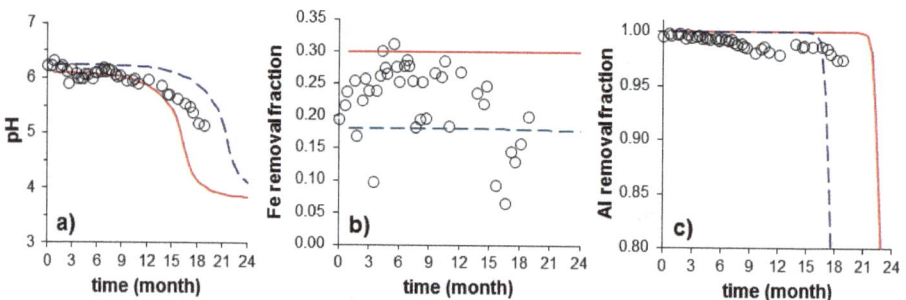

Figure 19. Water chemistry evolution at the outlet of Mina Esperanza I treatment. Symbols represent experimental values, and solid and dotted lines are model predictions under less and more acidic inflow waters. [Reprinted from Ayora C, Caraballo MA, Macías F, Rötting T, Carrera J, Nieto JM (2013) Acid mine drainage in the Iberian Pyrite Belt: 2. Lessons learned from recent passive remediation experiences. Environmental Science Pollution Research 20:7837–7853, Fig. 8, with permission of Elsevier].

adsorbed and structural sulfate in schwertmannite and basaluminite. The absence of recovered Ca in the second step confirmed the complete calcite dissolution along the Fe-precipitation zone. From 50 to 250 cm, the presence of Ca and S in the first step of sequential extraction and Ca in the second step confirmed the existence of gypsum and calcite, respectively. In spite of the inherent difficulties in comparing real and predicted amounts of precipitates on a quantitative

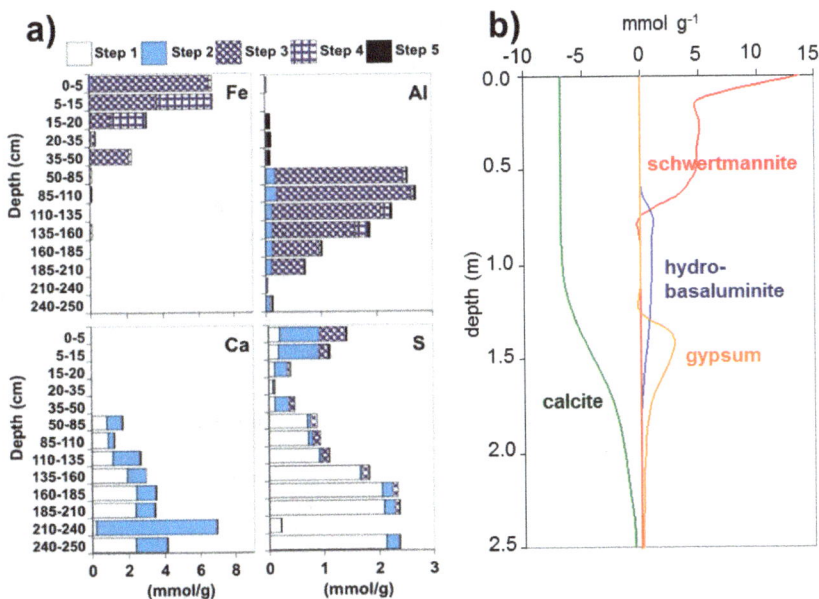

Figure 20. Comparison between the experimental (**a**) and calculated (**b**) distribution of solid phases after 18 month of functioning of the Mina Esperanza treatment. Experimental values (Caraballo et al. 2011) consist of cumulative graphs for the concentration of the main constituents (Fe, Al, Ca, and S) obtained after each step of a sequential extraction. Step 1 = water soluble fraction, Step 2 = exchangeable fraction, Step 3 = poorly ordered oxy-hydroxides, Step 4 = ordered hydroxides and oxides. [Reprinted from Ayora C, Caraballo MA, Macías F, Rötting T, Carrera J, Nieto JM (2013) Acid mine drainage in the Iberian Pyrite Belt: 2. Lessons learned from recent passive remediation experiences. Environmental Science Pollution Research 20:7837–7853, Fig. 9, with permission of Elsevier].

basis, Figure 20b shows a good agreement between the model and the mineralogy and phase distribution observed for 18 months of operation. Therefore, the reactive transport model successfully predicted the treated water chemistry and mineral distribution.

CONCLUSIONS AND REMARKS ON FUTURE WORK

Conclusions

MCRTM is a useful tool to elucidate the reactivity of minerals intervening in geochemical processes. Advances in pore-scale modeling towards precise interpretation of the geochemical processes are promising (Molins and Knabner 2019, this volume). Still, MCRTM is able to quantitatively describe the processes dominating the anthropogenic acid water–rock/cement interaction.

Column experiments using crushed carbonate rocks (porous medium) and 1D simulations that reproduced the experimental aqueous chemistry proved useful in gaining fresh insight into the effects exerted by mineralogy, temperature, P_{CO_2} and sulfate content on the reactivity of reservoir and cap carbonate rocks. The feasibility for predicting changes in porosity was also confirmed. Percolation experiments were used to measure the outflow aqueous chemistry in fractured cores. 2D simulations of experiments with carbonate rocks that developed different types of dissolution patterns were performed to estimate transport and reaction kinetic parameters. It was inferred that by reducing the calcite reactive surface area and by increasing that of the aluminosilicates with respect to the geometric estimates, MCRTM reproduced the geochemical processes in these laboratory media (centimeter scale).

Thus, for the assessment of CO_2 injection and long-term storage in fractured reservoir and cap carbonate rocks at field scale (meter–kilometer scale), predictions can be reliable, (1) when using diminished values of the calcite reactive surface area to take into account the transport-controlled dissolution of calcite together with elevated values of aluminosilicate reactive surface areas and (2) when considering the effect of the residence time on the kinetic mineral rate laws (Wen and Li 2017, 2018) given the much shorter residence times in the experiments (from seconds to hours) compared with those (years) in rock formations during CO_2 injection (Bachu et al. 1994). Moreover, MCRTM yields reliable predictions of long-term behavior in reservoir and cap rock fractures under the P_{CO_2} and temperature of injection and storage sites by providing a field parameterization (dimensions (aperture and length) and morphology of fractures, reactive (accessible) mineral surface areas and fluid flow regime). Without this information, a sensitivity analysis of the variability of these parameters should be performed.

1D and 2D simulations of the columns with rock/cement filling shed light on the effect of acidity in CO_2-rich water and AMD on carbonate rocks and Portland cement. The dominant processes ((i) calcite passivation by gypsum coating and Fe-rich-phase-clogging, (ii) extensive Portland cement alteration by dissolution of portlandite and precipitation of gypsum and aragonite at elevated P_{CO_2}, (iii) precipitation of Fe oxyhidroxides and Al sulfates and (iv) depletion of AMD contaminant metals) indicate the complexity of the acid water–rock/cement interaction. Moreover, MCRTM suggests that secondary mineral precipitation at the AMD–concrete interface prevents the alteration of cement-based materials used in treatment-plant infrastructures.

Limestone disperse-alkaline-substrate (DAS) column experiments are suitable for building a 1D MCRTM with fully determined parameters. The flux is imposed and porosity and dispersivity (transport parameters) are determined by controlled conservative tracer experiments. The chemical composition of the porewater sampled through the column allows us to calibrate the reactive surface area of the porous medium. Thus, a well parameterized dissolution rate law of calcite, which controls the treatment system, guarantees a fully determined MCRTM. If the material in the column is the same as that for the real case, a realistic prediction of the system performance is possible. This is crucial for designing a field-scale treatment to ensure the optimum consumption of the reagent (e.g. calcite) and to avoid or retard clogging by precipitates. However, in AMD remediation systems, the porous medium is tailored according to a precise characterization of its properties in the laboratory. This optimum situation can hardly be extrapolated to complex subsurface porous media.

Remarks on future work

At field scale, hydrogeologists have considerable experience in determining flow properties of aquifers by means of pumping/injection hydraulic tests. Transport properties are also conventionally determined by means of tracer tests using conservative solutes but with more difficulty. Given our knowledge of the kinetics of water-mineral reactions obtained from laboratory experiments, a successful MCRTM would require a prior characterization of the mineral reactive (accessible) surface area under field conditions. As shown by De Gaspari et al. (2014) and Snaebjörnsdóttir et al. (2017), controlled tracer tests using reactive solutes after conventional hydraulic tests constitute a promising way to make applications of MCRTM that are more realistic for natural systems.

At laboratory scale, performance of new experiments and simulations that closely mimic field conditions (rock/cement mineralogy, fluid composition, heterogeneous pore-space distribution in rocks and fractured media, pressure, temperature and flow variability) will lead to MCRTM improvements. Successful predictive simulations of site-specific cases (Balashov et al. 2013; Zheng et al. 2013; Tutolo et al. 2015b; Wolf et al. 2016; Kampman et al. 2017; Gíslason et al. 2019; Xiao et al. 2019) would require (1) constraining kinetic and flow parameters that control the geochemical processes (2) a review of equilibrium and kinetic parameters (Black et al. 2015; Luttge et al. 2019), (3) the incorporation of experimentally tested power-law-based porosity–permeability expressions in porous and fractured media (Gouze and Luquot 2011; Luhmann et al. 2014; Deng et al. 2016; Smith et al. 2017), (4) estimates of accessible mineral surface area (Beckingham et al. 2016, 2017; Deng et al. 2018) and (5) implementation of stress and chemo-mechanical and thermo-mechanical effects (Carroll et al. 2016; Vilarrasa et al. 2019).

Merging these complementary approaches will improve multiscale MCRTM.

ACKNOWLEDGMENTS

This work has been financed by the CGL2016–78783-C2-R and CGL2017–82331-R projects (Spanish Ministry of Economy and Competitiveness), with contribution of FEDER funds, and the 2017SGR 1733 Catalan Government project. We would like to thank Jordi Bellés, Natàlia Moreno, Rafael Bartolí and Mercè Cabanas (IDAEA), Eva Pelegrí and Maite Romero (SCT-Barcelona University) and José Antonio Jiménez (CENIM) for analytical assistance. We are indebted to Francisco Macías (Huelva University) for field sampling assistance. We also wish to thank Georg Kosakowski (PSI, Switzerland) and an anonymous reviewer whose constructive comments have greatly improved the manuscript.

REFERENCES

Acero P, Ayora C, Carrera J (2007) Coupled thermal, hydraulic and geochemical evolution of pyritic tailings in unsaturated column experiments. Geochim Cosmochim Acta71: 5325–5338

Acero P, Ayora C, Carrera J, Saaltink MW, Olivella S (2009) Multiphase flow and reactive transport model in vadose tailings. Appl Geochem 24:1238–1250

Alcalde J, Martí D, Calahorrano A, Marzán I, Ayarza P, Carbonell R, Juhlin C, Pérez-Estaún A (2013) Active seismic characterization experiments of the Hontomín research facility for geological storage of CO$_2$, Spain. Int J Greenh Gas Control 19:785–795

Amos RT, Mayer KU, Blowes DW, Ptacek CJ (2004) Reactive transport modeling of column experiments for the remediation of acid mine drainage. Env Sci Tech 38:3131–3138

Appelo CAJ, Parkhurst, DL, Post, VEA (2014) Equations for calculating hydrogeochemical reactions of minerals and gases such as CO$_2$ at high pressures and temperatures. Geochim Cosmochim Acta 125:49–67

Arvidson RS, Ertan IE, Amonette JE, Luttge A (2003) Variation in calcite dissolution rates: A fundamental problem? Geochim Cosmochim Acta 67:1623–1634

Ayora C, Caraballo MA, Macias F, Rötting T, Carrera J, Nieto JM (2013) Acid mine drainage in the Iberian Pyrite Belt: 2. Lessons learned from recent passive remediation experiences. Environ Sci Pollut Res 20:7837–7853

Ayora C, Macías F, Torres E, Lozano A, Carrero S, Nieto JM, Perez-Lopez R, Fernandez-Martínez A, Castillo-Michel H (2016) Recovery of rare earth elements and yttrium from passive-remediation systems of acid mine drainage. Environ Sci Technol 50:8255–8262

Bachu S, Gunter WD, Perkins EH (1994) Aquifer disposal of CO$_2$: hydrodynamic and mineral trapping. Energy Convers Mgmt 35:269–279

Balashov VN, Guthrie GD, Hakala JA, Lopano CL, Rinstidt JD, Brantley SL (2013) Predictive modeling of CO$_2$ sequestration in deep saline sandstone reservoirs: Impacts of geochemical kinetics. Appl Geochem 30:41–56

Bandstra JZ, Buss HL, Campen RK, Liermann LJ, Moore J, Hausrath EM, Navarre-Sitchler, AK, Jang J-H, Brantley SL(2008) Appendix: Compilation of Mineral Dissolution Rates. Kinetics of Water Rock Interactions. Brantley SL, Kubicki JD, White AF (eds) Springer, p 737–823

Beckingham LE, Steefel CI, Swift AM, Voltolini M, Yang L, Anovitz LM, Sheets JM, Cole DR, Kneafsey TJ, Mitnick EH, Zhang S, Landrot G, Ajo-Franklin JB, DePaolo DJ, Mito S, and Xue Z (2017) Evaluation of accessible mineral surface areas for improved prediction of mineral reaction rates in porous media. Geochim Cosmochim Acta 205:31–49

Beckingham LE, Mitnick EH, Steefel CI, Zhang S, Voltolini M, Swift AM, Yang L, Cole DR, Sheets JM, Ajo-Franklin JB, DePaolo DJ, Mito S, Xue Z (2016) Evaluation of mineral reactive surface area estimates for prediction of reactivity of a multi-mineral sediment. Geochim Cosmochim Acta 188:310–329

Benson SM, Cole DR (2008) CO_2 sequestration in deep sedimentary formations. Elements 4:325–331

Bibi I, Singh B, Silvester E (2011) Dissolution of illite in saline–acidic solutions at 25°C. Geochim Cosmochim Acta 75:3237–3249

Bigham JM, Schwertmann U, Traina SJ, Winland RL, Wolf M (1996) Schwermannite and the chemical modelling of iron in acid sulphate waters. Geochim Cosmochim Acta 60:2111–2121

Black JR, Haese, RR (2014) Chlorite dissolution rates under CO_2 saturated conditions at 50 to 120 °C and 120 to 200 bar CO_2. Geochim Cosmochim Acta 125:225–240

Black JR, Carroll SA, Haese RR (2015) Rates of mineral dissolution under CO_2 storage conditions. Chem Geol 399:134–144

Booth J, Compton RG, Prout K, Payne RM (1997) Gypsum overgrowths passivate calcite to acid attack. J Colloid Interf Sci 192:207–214

Brookfield AE, Blowes DW, Mayer KU (2006) Integration of field measurements and reactive transport modelling to evaluate. J Cont Hydrol 88:1–22

Bullard JW, Enjolras E, George WL, Satterfield SG, Terrill JE (2010) A parallel reaction-transport model applied to cement hydration and microstructure development. Model Simul Mater Sci Eng 18:025007

Canal J, Delgado J, Falcón I, Yang Q, Juncosa R, Barrientos V (2013) Injection of CO_2saturated water through a siliceous sandstone plug from the Hontomin test site (Spain): experiment and modeling. Environ Sci Technol 47:159–167

Cappelli Ch, Yokoyama S, Cama J, Huertas FJ (2018) Montmorillonite dissolution kinetics: Experimental and reactive transport modelling interpretation. Geochim Chosmochim Acta 227:96–122

Caraballo MA, Rötting TS, Macías F, Nieto JM, Ayora C (2009a) Field multi-step calcite and MgO passive system to treat acid mine drainage with high metal concentration. Appl Geochem 24:2301–11

Caraballo MA, Rötting TS, Nieto JM, Ayora C (2009b) Sequential extraction and DXRD applicability to poorly crystalline Fe- and Al-phase characterization from an acid mine water passive remediation system. Am Min 94:1029–1038

Caraballo MA, Macías F, Castillo J, Quispe D, Nieto JM, Ayora C (2011) Hydrochemical performance and mineralogical evolution of a dispersed alkaline substrate (DAS) remediating the highly polluted acid mine drainage in the full scale passive treatment of Mina Esperanza (SW, Spain). Am Mineral 96:1270–1277

Carroll SA, McNab WW, Torres SC (2011) Experimental study of cement–sandstone/shale–brine–CO_2 interactions. Geochem Trans 12:1–19

Carroll SA, Hao Y, Smith MM, Sholokhova Y (2013) Development of scaling parameters to describe CO_2-rock interactions within Weyburn-Midale flow units. Int J Greenh Gas Cont 16:S185-S195

Carroll SA, Carey JW, Dzombak S, Huerta NJ, Li L, Richard T, Um W, Walsh SDC, Zhang L (2016) Review: Role of chemistry, mechanics and transport on well integrity in CO_2 storage environments. Int J Greenh Gas Cont 49:149–160

Chen L, Kang Q, Viswanathan HS, Tao W, (2014) Pore-scale study of dissolution-induced changes in hydrologic properties of rocks with binary minerals. Water Resour Res 50:9343–9365

Dai Z, Viswanathan H, Xiao T, Hakala A, Lopano C, Guthrie G, McPherson B (2019) Reactive transport modeling of geological carbon storage associated with CO_2 and brine leakage. *In*: Science of carbon storage in deep saline formations. Neweell P, Ilgen AG (eds) Elsevier, p 67–86

Daval D, Martinez I, Corvisier J, Findling N, Goffé B, Guyot F (2009) Carbonation of Ca-bearing silicates, the case of wollastonite: Experimental investigations and kinetic modeling. Chem Geol 265:63–78.

Dávila G, Luquot L, Soler JM, Cama J (2016a) Interaction between a fractured marl caprock and CO_2-rich sulfate solution under supercritical CO_2 conditions. Int J Greenh Gas Control 48:105–119

Dávila G, Luquot L, Cama J Soler JM (2016b) 2D reactive transport modeling of the interaction between a marl and a CO_2-rich sulfate solution under supercritical CO_2 conditions. Int J Greenh Gas Control 54:145–149

Dávila G, Cama J, Luquot L, Soler JM, and Ayora C (2017) Experimental and modeling study of the interaction between a crushed marl caprock and CO_2-rich solutions under different pressures and temperatures. Chem Geol 448:26–42

De Gaspari F, Cabeza Y, Luquot L, Rötting T, Saaltink MW, Carrera J (2014) Reactivity of Hontomín carbonate rocks to acidic solution injection: reactive "push-pull" tracer tests results. EGU General Assembly, id.16788

Deng H, Spycher N (2019) Modeling reactive transport processes in fractures. Rev Mineral Geochem 85:49–74

Deng H, Molins S, Steefel C, DePaolo D, Voltolini M, Yang L, Ajo-Franklin J (2016) A 2.5D reactive transport model for fracture alteration simulation. Environ Sci Technol 50:7564–7571

Deng H, Molins S, Trebotich D, Steefel CI, DePaolo D (2018) Pore-scale numerical investigation of the impacts of surface roughness: Upscaling of reaction rates in rough fractures. Geochim Cosmochim Acta 239:374–389

Dilnesa BZ, Lothenbach B, Le Saoût G, Renaudin G, Mesbah A, Filinchuk Y, Wichser A, Wieland E. (2011) Iron in carbonate containing AFm phases. Cement Concr Res 41:311–323

Dilnesa BZ, Lothenbach B, Renaudin G, Wichser A, Wieland E (2012) Stability of monosulfate in the presence of iron. J Am Cer Soc 95:3305–3316

Dilnesa BZ, Lothenbach B, Renaudin G, Wichser A, Kulik D (2014) Synthesis and characterization of hydrogarnet $Ca_3(Al_xFe_{1-x})_2(SiO_4)_y(OH)_{4(3-y)}$. Cement Concr Res 59:96–211

Domènech C, Ayora C, De Pablo J (2002) Oxidative dissolution of pyritic sludge from the Aznalcóllar mine (SW Spain). Chem Geol 190:339–353

Domenico PA, Schwartz FW (1990) Physical and Chemical Hydrogeology. John Wiley and Sons, New York

Duan Z, Sun R (2003) An improved model calculating CO_2 solubility in pure water and aqueous NaCl solutions from 273 to 533 K and from 0 to 2000 bar. Chem Geol 193:257–271

Elkhoury JE, Ameli P, Detwiler RL (2013) Dissolution and deformation in fractured carbonates caused by flow of CO_2-rich brine under reservoir conditions. Int J Greenh Gas Control 16: S203-S215

Fukushi K, Sato T, Yanase N (2003) Solid-solution reactions in As(V) sorption by schwertmannite Environ Sci Technol 37:3581–3586

García-Rios M, Cama J, Luquot L, Soler JM (2014) Interaction between CO_2-rich sulfate solutions and carbonate reservoir rocks from atmospheric to supercritical CO_2 conditions: experiments and modeling. Chem Geol 383:107–122

García-Rios M, Luquot L, Soler JM, Cama J (2015) Influence of the flow rate on dissolution and precipitation features during percolation of CO2-rich sulfate solutions through fractured limestone samples. Chem Geol 414:95–108

García-Rios M, Luquot L, Soler JM, Cama J (2017) The role of mineral heterogeneity on the hydrogeochemical response of two fractured reservoir rocks in contact with dissolved CO_2. Appl Geochem 84:202–217

Gaus I, Azaroual M, Czernichowski-Lauriol I (2005) Reactive transport modelling of the impact of CO_2 injection on the clayey caprock at Sleipner (North Sea). Chem Geol 217:319–337

Gíslason SR, Sigurdardóttir H, Aradóttir ES, Oelkers EH (2018) A brief history of CarbFix: Challenges and victories of the project's pilot phase. Energy Procedia 146:103–114

Golfier F, Zarcone C, Bazin B, Lenormand R, Lasseux D, Quintard M (2002) On the ability of a Darcy-scale model to capture wormhole formation during the dissolution of a porous medium. J Fluid Mech 457:213–254

Gouze P, Luquot L (2011) X-ray microtomography characterization of porosity, permeability and reactive surface changes during dissolution. J Contam Hydrol 120–121:45–55

Gunter WD, Perkins E (1993) Aquifer disposal of CO_2-rich gases: reaction design for added capacity. Energy Convers Mgmt 34:941–948

Hamer M, Graham RC, Amrhein C, Bozhilov KN (2003) Dissolution of ripidolite (Mg, Fe-Chlorite) in organic and inorganic acid solutions. Soil Sci Soc Am J 67:654–661

Hao Y, Smith MM, Sholokhova Y, Carroll SA (2013) CO_2-induced dissolution of low permeability carbonates. Part II: Numerical modeling of experiments. Adv Water Resour 62:388–408

Harrison AL, Dipple GM, Power IM, Mayer U (2015) Influence of surface passivation and water content on mineral reactions in unsaturated porous media: Implications for brucite carbonation and CO_2 sequestration. Geochim Cosmochim Acta 148:477–495

Hellmann R, Daval D, Tisserand D (2010) The dependence of albite feldspar dissolution kinetics on fluid saturation state at acid and basic pH: Progress towards a universal relation. C R Geosci 342:676–684

Hummel W, Berner U, Curti E, Pearson FJ, Thoenen T (2002) Nagra/PSI chemical thermodynamic data base 01/01. Radioch Acta 90:805–813

Iyer J, Walsh SDC, Hao Y, Carroll SA (2017) Incorporating reaction-rate dependence in reaction-front models of wellbore-cement/carbonated-brine systems. Int J Greenh Gas Con 59:160–171

Kampman N, Bertier P, Busch A, Snippe J, Harrington J, Pipich V, Maskell A, Bickle M (2017) Validating reactive transport models of CO_2–brine–rock reactions in caprocks using observations from a natural CO_2 reservoir. Energy Procedia 114:4902–4916

Kang Q, Chen L, Valocchi AJ, Viswanathan HS (2014) Pore-scale study of dissolution induced changes in permeability and porosity of porous media. J Hydrol 517:1049–1055

Kharaka YK, Cole DR, Thordsen JJ, Gans KD, Thomas BR (2013) Geochemical monitoring for potential environmental impacts of geologic sequestration of CO_2. Rev Min Geochem 77:399–430

Knauss KG, Johnson JW, Steefel CI (2005) Evaluation of the impact of CO_2, co-contaminant gas, aqueous fluid and reservoir rock interactions on the geologic sequestration of CO_2. Chem Geol 217:339–350

Kulik DA, Kersten M (2001) Aqueous solubility diagrams for cementitious waste stabilization systems: II, end-member stoichiometries of ideal calcium silicate hydrates solid solutions. J Am Cer Soc 84:3017–3026

Lai P, Moulton K, Krevor S (2015) Pore-scale heterogeneity in the mineral distribution and reactive surface area of porous rocks. Chem Geol 411:260–273

Loring JS, Miller QRS, Thompson CJ, Schaef HT (2019) Experimental studies of reactivity and transformations of rocks and minerals in water-bearing supercritical CO_2. *In*: Science of Carbon Storage in Deep Saline Formations. Neweell P, Ilgen AG (eds) Elsevier p 47–65

Lothenbach B, Matschei T, Möschner G, Glasser FP (2008) Thermodynamic modelling of the effect of temperature on the hydration and porosity of Portland cement. Cement Concr Res 38:1–18

Lozano A, Fernández-Martínez A, Ayora C, Poulain A (2018) Local structure and ageing of basaluminite at different pH values and sulphate concentrations Chem Geol 496:25–33

Luhmann AJ, Kong, XZ, Tutolo, BM, Garapati N, Bagley BC, Saar MO, Seyfried Jr, WE (2014) Experimental dissolution of dolomite by CO_2-charged brine at 100 °C and 150 bar: evolution of porosity, permeability, and reactive surface area. Chem Geol 380:145–160

Luhmann AJ, Tutolo, BM, Chunyang T, Moskowitz BM, Saar MO, Seyfried Jr, WE (2017) Whole rock basalt alteration from CO_2-rich brine during flow-through experiments at 150 °C and 150 bar. Chem. Geol. 453:92–110

Luquot L, Gouze P (2009) Experimental determination of porosity and permeability changes induced by injection of CO_2 into carbonate rocks. Chem Geol 265:148–159

Luquot L, Andreani M, Gouze P, Camps P (2012) CO_2 percolation experiment through chlorite/zeolite-rich sandstone (Pretty Hill Formation – Otway Basin–Australia). Chem Geol 294–295:75–88

Luquot L, Rodriguez O, Gouze P (2014) Experimental characterization of porosity structure undergoing different dissolution regimes. Transp Porous Media 101:507–532

Luquot L, Gouze P, Niemi A, Bensabat J, Carrera J (2016) CO_2-rich brine percolation experiments through Heletz reservoir rock samples (Israel): Role of the flow rate and brine composition. Int J Greenh Gas Control 48:44–58

Luttge A, Arvidson RS, Fischer C, Kurganskaya I (2019) Kinetic concepts for quantitative prediction of fluid-solid interactions. Chem Geol 504:216–235

Macías F, Caraballo MA, Rötting TS, Perez-Lopez R, Nieto JM, Ayora C (2012) From highly polluted Zn-rich acid mine drainage to non-metallic waters: Implementation of a multi-step alkaline passive treatment system to remediate metal pollution. Sci Total Environ 433:323–330

Malmstrom ME, Berglund S, Jarsjo J (2008) Combined effects of spatially variable flow and mineralogy on the attenuation of acid mine drainage in groundwater. Appl Geochem 23:1419–1436

Marty NCM, Cama J, Sato T, Chino D, Villiéras F, Razafitianamaharavo A, Brendlé J, Giffaut E, Soler JM, Gaucher EC, Tournassat C (2011) Dissolution kinetics of synthetic Na-smectite. An integrated experimental approach. Geochim Cosmochim Acta 75:5849–5864

Matschei T, Lothenbach B, Glasser FP (2007) Thermodynamic properties of Portland cement hydrates in the system $CaO-Al_2O_3-SiO_2-CaSO_4-CaCO_3-H_2O$. Cement Concr Res 37:1379–1410

Mayer KU, Frind EO, Blowes DW (2002) Multicomponent reactive transport modeling in variably saturated porous media using a generalized formulation for kinetically controlled reactions. Water Resour Res 38:1–21

Mayer KU, Benner SG, Blowes DW (2006) Process-based reactive transport modeling of a permeable reactive barrier for the treatment of mine drainage. J Cont Hydro 85:195–211

Mbia EN, Frykman P, Nielsen C, Fabricius I, Pickup GE, Bernstone C (2014) Caprock compressibility and permeability and the consequences for pressure development in CO_2 storage sites. Int J Greenh Gas Con 22:139–153.

Menke HP, Andrew MG, Blunt MJ, Bijeljic B (2016) Reservoir condition imaging of reactive transport in heterogeneous carbonates using fast synchrotron tomography – Effect of initial pore structure and flow conditions. Chem Geol 428:15–26

Molins S (2015) Reactive interfaces in direct numerical simulation of pore-scale processes. Rev Min Geochem 80:461–481

Molins S, Knabner P (2019) Multiscale approaches in reactive transport modeling. Rev Mineral Geochem 85:27–48

Myers RJ, L'Hopital E, Provis JL, Lothenbach B (2015) Effect of temperature and aluminium on calcium (alumino) silicate hydrate chemistry under equilibrium conditions. Cem Con Res 68:83–93

Noiriel C (2015) Resolving time-dependent evolution of pore-scale structure, permeability and reactivity using X-ray microtomography. Rev Mineral Geochem 80:247–285

Noiriel C, Madé B, Gouze P (2007) Impact of coating development on the hydraulic and transport properties in argillaceous limestone fracture. Water Resour Res 43:W09406

Noiriel C, Luquot L, Madé B, Raimbault L, Gouze P, van der Lee J (2009) Changes in reactive surface area during limestone dissolution: An experimental and modeling study. Chem Geol 265:160–170

Noiriel C, Gouze P, Madé B (2013) 3D analysis of geometry and flow changes in a limestone fracture during dissolution. J Hydrol 486:211–223

Noiriel C, Steefel CI, Yang L, Bernard D (2015) Effects of pore-scale precipitation on permeability and flow. Adv Water Resour 95:125–137

Nordstrom DK, Alpers CN (1999) Geochemistry of acid mine waters. *In:* Plumlee, G.S., Logsdon, M.J. (Eds.), Reviews in Economic Geology, vol. 6A. The Environmental Geochemistry of Mineral Deposits. Part A. Processes, Methods and Health Issues. Soc. Econ. Geol., Littleton, CO, p. 133–160

Nordstrom DK, Alpers CN, Ptacek CJ, Blowes DW (2000) Negative pH and Extremely Acidic Mine Waters from Iron Mountain, California. Environ Sci Technol 34:254–258

Offeddu FG, Cama J, Soler JM, Dávila G, McDowell A, Craciunescu T, Tiseanu I (2015) Processes affecting the efficiency of limestone in passive treatments for AMD: Column experiments. J Environ Chem Eng 3:304–316

Offedu FG, Cama J, Soler JM, Putnis CV (2014) Direct nanoscale observations of the coupled dissolution of calcite and dolomite and the precipitation of gypsum. Beilstein J. Nanotechnol. 5:1245–1253

Oliva J, De Pablo J, Cortina JL, Cama J, Ayora C (2010) The use of Apatite II (TM) to remove divalent metal ions zinc(II), lead(II), manganese(II) and iron(II) from water in passive treatment systems: Column experiments. J Hazar Mat 184:364–37

Orden S, Macías F, Nieto JM, Ayora C (2018) Tratamiento sostenible de drenaje ácido de mina: Tecnología DAS-calizo en Mina Esperanza (Faja Pirítica Ibérica). Macla 23, http://www.ehu.eus/sem/revista/macla.htm

Pabst T, Molson J, Aubertin M, Bussiere B (2017) Reactive transport modelling of the hydro-geochemical behaviour of partially oxidized acid-generating mine tailings with a monolayer cover. Appl Geochem 78:219–233

Pabst T, Bussiere B, Aubertin M, Molson J (2018) Comparative performance of cover systems to prevent acid mine drainage from pre-oxidized tailings: A numerical hydro-geochemical assessment. J Cont Hydrol 214:39–53

Palandri JL, Kharaka YK (2004) A compilation of rate parameters of water-mineral interaction kinetics for application to geochemical modeling. U.S. Geological Survey

Parkhurst DL, Appelo CAJ, (2013) Description of input and examples for PHREEQC (Version 3)-a computer program for speciation, batch-reaction, one-dimensional transport, and inverse geochemical calculations. U.S. Geological Survey Techniques and methods report. Book 6

Pedretti D, Mayer KU, Beckie RD (2017) Stochastic multicomponent reactive transport analysis of low quality drainage release from waste rock piles: Controls of the spatial distribution of acid generating and neutralizing minerals. J Cont Hydrol 201:30–38

Pérez-Lopez R, Cama J, Nieto JM, Ayora C (2007) The iron-coating role on the oxidation kinetics of a pyritic sludge doped with fly ash. Geochim Cosmochim Acta 71:1921–1934

Pérez-Lopez R, Cama J, Nieto JM, Ayora C, Saaltink MW (2009) Attenuation of pyrite oxidation with a fly ash pre-barrier: Reactive transport modelling of column experiments. Appl Geochem 24:1712–1723

Pérez-López R, Macías F, Caraballo MA, Nieto JM, Román-Ross G, Tucoulou R, Ayora C (2011) Mineralogy and geochemistry of Zn-rich mine-drainage precipitates from an MgO passive treatment system by synchrotron-based X-ray analysis. Environ Sci Technol 45:7826–7833

Pokrovsky OS, Golubev SV, Schott J (2005) Dissolution kinetics of calcite, dolomite and magnesite at 25 °C and 0 to 50 atm pCO$_2$. Chem Geol 217:239–255

Pokrovsky OS, Golubev SV, Schott J, Castillo A (2009) Calcite, dolomite and magnesite dissolution kinetics in aqueous solutions at acid to circumneutral pH, 25 to 150 °C and 1 to 55 atm pCO$_2$: New constraints on CO$_2$ sequestration in sedimentary basins. Chem Geol 265:20–32

Revil A, Cathles III LM (1999) Permeability of shaly sands. Water Resour Res 35:651–662

Rohmer J, Pluymakers A, Renard F (2016) Mechano-chemical interactions in sedimentary rocks in the context of CO2 storage: Weak acid, weak effects? Earth Sci Rev 157:86–110

Rötting TS, Thomas RC, Ayora C, Carrera J (2008a): Passive treatment of acid mine drainage with high metal concentrations using dispersed alkaline substrate. J Environ Quality 37:1741–1751

Rötting TS, Caraballo MA, Serrano JA, Ayora C, Carrera J (2008b) Field application of calcite Dispersed Alkaline Substrate (calcite–DAS) for passive treatment of acid mine drainage with high Al and metal concentrations. Appl Geochem 23:1660–1674

Rötting TS, Ayora C, Carrera J (2008c) Improved passive treatment of high Zn and Mn concentrations using Caustic Magnesia (MgO): Particle size effects. Environ Sci Technol 24:9370–9377

Saaltink MW, Batlle F, Ayora C, Carrera J, Olivella S (2004) RETRASO, a code for modeling reactive transport in saturated and unsaturated porous media. Geol Acta 2:235–251

Salmon SU, Malmstrom ME (2004) Geochemical processes in mill tailings deposits: modelling of groundwater composition. Appl Geochem 19:1–17

Saaltink MW, Batlle F, Ayora C, Carrera J, Olivella S (2004) RETRASO, a code for modeling reactive transport in saturated and unsaturated porous media. Geol Acta 2:235–251

Samson E, Marchand J (2003) Calculation of ionic diffusion coefficients on the basis of migration test results. Mater Struct 36:156–165

Schmidt T, Lothenbach B, Romer M, Scrivener K, Rentsch D, Figi R (2008) A thermodynamic and experimental study of the conditions of thaumasite formation. Cement Concr Res 38:337–349

Sjöberg EL, Rickard DT (1984). Temperature-dependence of calcite dissolution kinetics between 1° C and 62 °C at pH 2.7 to 8.4 in aqueous-solutions. Geochim Cosmochim Acta 48:485–493

Smith MM, Sholokhova Y, Hao Y, Carroll SA (2013a) CO$_2$-induced dissolution of low permeability carbonates. Part I: Characterization and experiments. Adv Water Resour 62:370–387

Smith MM, Wolery TJ, Carroll, SA (2013b) Kinetics of chlorite dissolution at elevated temperatures and CO$_2$ conditions. Chem Geol 347:1–8

Smith MM, Hao Y, Carroll SA (2017) Development and calibration of a reactive transport model for carbonate reservoir porosity and permeability changes based on CO$_2$ core-flood experiments. Int. J. Greenh. Gas Cont. 57:73–88

Snæbjörnsdóttir SO, Oelkers EH, Mesfin K, Aradóttir ES, Dideriksen K, Gunnarsson I, Gunnlaugsson E, Matter JM, Stute M, Gislason SR (2017) The chemistry and saturation states of subsurface fluids during the in situ mineralisation of CO$_2$ and H$_2$S at the CarbFix site in SW-Iceland. Int J Greenh Gas Con 58:87–102.

Soler JM, Mäder KU (2010) Cement-rock interaction: Infiltration of a high-pH solution into a fractured granite core. Geo Acta 3:221–233

Soler JM, Boi M, Mogollon J, Cama J, Ayora C, Nico P (2008) The passivation of calcite by acid mine water. Column experiments with ferric sulfate and ferric chloride solutions at pH 2. Appl Geochem 23:3579–3588

Soler JM, Vuorio M. Hautojärvi A (2011) Reactive transport modeling of the interaction between water and a cementitious grout in a fractured rock. Application to ONKALO (Finland). Appl Geochem 26:1115–1129

Soler-Sagarra J, Luquot L, Martínez-Pérez L, Saaltink MW, De Gaspari F, Carrera J (2016) Simulation of chemical reaction localization using a multiporosity reactive transport approach. Int J Greenh Gas Control 48:59–68

Steefel CI, Appelo CAJ, Arora B, Jacques D, Kalbacher T, Kolditz O, Lagneau V, Lichtner P C, Mayer KU, Meeussen JCL, Molins S, Moulton D, Shao H, Simunek J, Spycher N, Yabusaki SB and Yeh GT (2015a) Reactive transport codes for subsurface environmental simulation. Comput Geosci 19:445–495

Steefel CI, Beckingham LE, Landrot G (2015b) Micro-continuum approaches for modeling pore-scale geochemical processes. Rev Mineral Geochem 80:217–246

Steefel CI, DePaolo DJ, Lichtner PC (2005) Reactive transport modeling: an essential tool and a new research approach for the Earth sciences. Earth Planet Sci Lett 240:539–558

Szymczak P, Ladd AJC (2009) Wormhole formation in dissolving fractures. J Geophys Res 114:B06203

Szymczak P, Ladd AJC, (2011) The initial stages of cave formation: beyond the one dimensional paradigm. Earth Planet Sci Lett 301:424–432

Thaysen EM, Soler JM, Boone M, Cnudde V, Cama J (2017) Effect of dissolved H_2SO_4 on the interaction between CO_2-rich brine solutions and limestone, sandstone and marl. Chem Geol 450:31–43

Torres E, Lozano A, Macias F, Gomez-Arias A, Castillo J, Ayora C (2018) Passive elimination of sulfate and metals from acid mine drainage using combined limestone and barium carbonate systems. J Cleaner Produc 182:114–123

Trapote-Barreira A, Cama J, Soler JM, Lothenbach B (2016) Degradation of mortar under advective flow: column experiments and reactive transport modeling. Cem Conc Res 81:81–93

Tutolo BM, Luhmann AJ, Kong XZ, Saar MO, Seyfried Jr WE (2015a) CO_2 sequestration in feldspar-rich sandstone: Coupled evolution of fluid chemistry, mineral reaction rates, and hydrogeochemical properties. Geochim Cosmochim Acta 160:132–154

Tutolo BM, Kong XZ, Seyfried Jr WE, Saar MO (2015b) High performance reactive transport simulations examining the effects of thermal, hydraulic, and chemical (THC) gradients on fluid injectivity at carbonate CCUS reservoir scales. Int J Greenh Gas Con 39:285–301

Vilarrasa V, Makhnenko RY Rutquist J (2019) Field and laboratory studies of geomechanical response to the injection of CO_2. *In*: Science of carbon storage in deep saline formations. Neweell P, Ilgen AG (eds) Elsevier, p 159–178

Vuorinen TL, Laurila T (2005) Interfacial reactions between lead-free solders and common base materials. Mat Sci Eng 49:1–60

Walsh SDC, Mason EH, Du France WL, Carroll SA (2014) Experimental calibration of a numerical model describing the alteration of cement/caprock interfaces by carbonated brine. Int J Greenh Gas Con 22:176–188

Welch SA, Sheets JM, Place MC, Saltzman MR, Edwards CT, Gupta N, Cole DR (2019) Assessing geochemical reactions during CO_2 injection into an oil-bearing reef in the Northern Michigan basin. Appl. Geochem. 100:380–392

Wen H, Li L (2017) An upscaled rate law for magnesite dissolution in heterogeneous porous media. Geochim Cosmochim Acta 210:289–305

Wen H, Li L (2018) An upscaled rate law for mineral dissolution in heterogeneous media: The role of time and length scales. Geochim Cosmochim Acta 235:1–20

White AF, Brantley SL (2003) The effect of time on the weathering of silicate minerals: why do weathering rates differ in the laboratory and field? Chem Geol 202:479–506

Wolery TJ, Jackson KJ, Bourcier WL, Bruton CJ, Viani BE, Knauss KG, Delany JM (1990) Current status of the EQ3/6 software package for geochemical modeling. *In:*Chemical Modeling of Aqueous Systems II. Melchior C, Bassett RL (ed), Am Chem Soc Sympos Ser, p 104–116

Wolf JL, Niemi A, Bensabat J, Rebscher D (2016) Benefits and restrictions of 2D reactive transport simulations of CO_2 and SO_2 co-injection into a saline aquifer using TOUGHREACTV3.0-OMP. Int J Greenh Gas Con 54:610–626

Wunderly MD, Blowes DW, Frind EO, Ptacek CJ (1996) Sulfide mineral oxidation and subsequent reactive transport of oxidation products in mine tailings impoundments: A numerical model. Water Resour Res 32:3173–3187

Xiao T, McPherson B, Esser R, Jia W, Moodie N, Chu S, Lee S-Y. (2019) Forecasting commercial-scale CO_2 storage capacity in deep saline reservoirs: Case study of Buzzard's bench, Central Utah. Com Geosci

Xu TF, White SP, Pruess K, Brimhall GH (2000) Modeling of pyrite oxidation in saturated and unsaturated subsurface flow systems. Trans Porous Med 39:25–56

Xu T, Apps JA, Pruess K, Yamamoto H (2007) Numerical modeling of injection and mineral trapping of CO_2 with H_2S and SO_2 in sandstone formation. Chem Geol 242:319–346

Xu J, Fan C, Teng, HH (2012) Calcite dissolution kinetics in view of Gibbs free energy, dislocation density, and pCO_2. Chem Geol 322–323:11–18

Zhang S, Yang L, DePaolo DJ, Steefel CI (2015) Chemical affinity and pH effects on chlorite dissolution kinetics under geological CO_2 sequestration related conditions. Chem Geol 396:208–217

Zheng L, Spycher N, Birkholzer J, Xu T, Apps J, Kharaka Y (2013) On modeling the potential impacts of CO_2 sequestration on shallow groundwater: Transport of organics and co-injected H_2S by supercritical CO_2 to shallow aquifers. Int J Greenh Gas Con 14:113–127

Reviews in Mineralogy & Geochemistry
Vol. 85 pp. 499-528, 2019
Copyright © Mineralogical Society of America

Industrial Deployment of Reactive Transport Simulation: An Application to Uranium *In situ* Recovery

Vincent Lagneau

MINES ParisTech, PSL Research University
Centre de Géosciences
Fontainebleau
France

Vincent.lagneau@mines-paristech.fr

Olivier Regnault and Michaël Descostes

Orano
Paris La Défense
France

olivier.regnault@orano.group; michael.descostes@orano.group

INTRODUCTION

The development of reactive transport soared during the 1990's, driven by the necessity to demonstrate the long-term efficiency of radioactive waste repositories (Bildstein et al. 2019; Claret, 2019; Cama et al. 2019, both this volume). The approach, based on a rigorous description of processes and their coupling, provides a basis of confidence to bridge the gap in time and space between knowledge gained in laboratory experiments and the dimension and lifetime of a high-level waste repository. Reactive transport codes progressively increased in complexity, as more processes were included to account e.g., for variably saturated flow, heat transport and more complex chemical processes (Steefel et al. 2015). In the meantime, improved algorithms and computer power opened the way for larger simulations.

Reactive transport simulation now embraces a large array of applications in the Earth and environmental sciences. Thus, in the field of contaminant hydrology, such studies address problem from the interface scale, to quantify migration and retention mechanisms, up to the scale of a watershed integrating multilayer reservoirs and their inherent spatial variability. Over the last decades, extensive monitoring and characterization studies allowed for better calibration of processes, parameters determination and more generally better quantitative understanding of sites (e.g., Hammond and Lichtner 2010; Li et al. 2010; Hammond et al. 2011; Arora et al. 2016) or on less emblematic sites (e.g., Mangeret et al. 2012; Estublier et al. 2013). As a result, better understanding of mechanisms, quantification of competing retention and migration processes were achieved. They allow for the estimation of plume migration, or the evaluation of remediation strategies.

In these applications, a unique application site is characterized, then a model is built in order to draw conclusions and propose recommendations. Apart from academic or agencies sponsored studies, the industry is also interested by this approach: estimation of reservoir properties in the oil and gas industry (Xiao and Jones 2006), evolution of gas storage sites (e.g., CO_2 in Audigane et al. 2007, or H_2 in Saínz García et al. 2017), evaluation of impacts of the mining industry (Metschies and Jenk 2011; Molson et al. 2012; Neuner and Fawcett 2015).

1529-6466/19/0085-0016$05.00 (print)
1943-2666/19/0085-0016$05.00 (online)

http://dx.doi.org/10.2138/rmg.2019.85.16

Another type of use is also relevant: industrial deployment of reactive transport within operation with a view to support operators in decision-making. Operating conditions are then very different: models should be built on a regular basis on numerous application cases, sometimes relying on incomplete sets of data. Additionally, this type of application only makes sense if it can be integrated in the regular workflow of operations. Hence, timing (to keep pace with the operations), and interfacing (with other aspects of production) are critical. Finally, such deployment can only be performed if the value of the simulations for the operation can be demonstrated.

An illustration of industrial use of reactive transport in operations is given: exploitation of uranium by *in situ* recovery (ISR). The technique is best suited for large, low grade, medium depth deposits. It consists in dissolving the metal in place by circulating solutions, using a large number of injection and production wells to cover the surface of the field. Relying on the circulation of fluids and their interaction with the deposit, the technique is particularly suitable for reactive transport analysis (Nguyen et al. 1983; Regnault et al. 2012). Although models can be built to understand the general behavior of such systems, when quantitative assessment of a unit production is wanted, the size of the exploitation and the strong spatial variability inherent to the formations require to build models specific to the local conditions (geology and constraints of exploitation) in each portion of the deposit.

After a brief description of the ISR technique, the use of reactive transport simulation is exemplified on the KATCO mine. The modeling approach is presented, including the data necessary to calibrate the model. The conditions to transform a mostly academic application into an operational tool are commented; some examples of use in operation are then proposed. Finally, in addition to exploitation, reactive transport modeling is also relevant for controlling environmental impacts. Although this application is more traditional, it shows that reactive transport modeling deployed in this context can create the conditions for an integrated approach to better identify post-operational constraints, or even to test alternative exploitation strategies limiting the need for active remediation *a posteriori*.

IN SITU RECOVERY

In situ recovery (ISR) mining is defined as the extraction of a metal from its host formation by the circulation of chemical solutions (lixiviants) using injection and extraction wells from the surface (Fig. 1). The solution is treated in surface plants to recover the metal. Reagents are then adjusted in the solution before recirculation in the ore formation (IAEA 2001). ISR was developed simultaneously by the USA and the USSR in the early 1960's for the recovery of uranium in secondary ore bodies in permeable aquifers. It has become one of the standard production methods, with a share exceeding 50% of world uranium production in 2014 (IAEA 2016). The technique is particularly suited for large ore bodies, at intermediate depth and low concentrations, like roll-front deposits (Dahlkamp 1993; Kyser 2014; Saunders et al. 2016). High permeability and presence of confining formations for the deposit are also desirable.

Uranium solubility increases markedly under oxidizing conditions, at low or high pH. Complexation with several anions further enhances solubility: particularly relevant for *in situ* context are SO_4^2 and CO_3^{2-}. Two technological choices are then available (Bhargava et al. 2015): acid (usually sulfuric acid) or alkaline (higher pH carbonate or bicarbonate solutions) leaching. In both cases, oxidizing agents can be added: air or oxygen, hydrogen peroxide, ferric iron and others.

The selection of technique is guided by regulatory aspects (mostly driven by remediation process opportunities), geological considerations (notably presence of mineral buffers, e.g., calcite) and economical optimization. However, acid leaching (Kazakhstan, Australia) is *de facto* the preferred technique, representing up to 96% of total uranium produced by ISR.

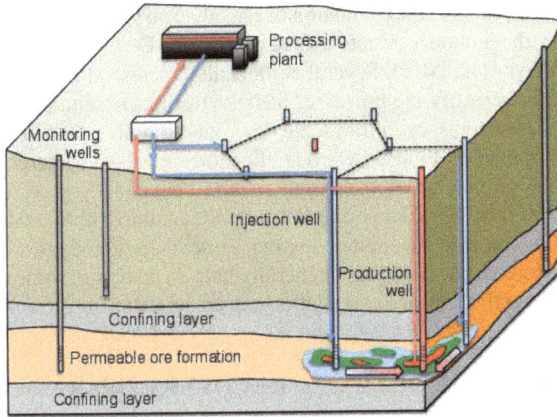

Figure 1. Schematic view of an ISR operation.

The fundamental advantage of ISR over other mining techniques is reduced costs: it is by far the most cost effective extraction technique (Kidd 2009). By bringing the reagents directly in the ore body, bulk material handling is reduced to the minimum. The absence of a mining fleet, milling capacity and heavy preparatory work (removal of the overburden for open-pit mines, development of access shafts for underground mining) drastically reduces upfront capital cost (Heili 2018). ISR mining plans are also highly flexible: new fields can be quickly developed or alternatively production rates can be reduced without substantial impact on production costs. Not only this progressive development allows for reduced return on investment time, it also provides for dynamic adjustment to the demand.

Reduced environmental footprint is another key advantage of ISR (IAEA 2005). Since liquid is moved rather than rock, disturbance at the surface is minimized (IAEA 2001). Hence, a major benefit is the absence of tailings; the quantity of solids generated by the exploitation is mostly limited to cuttings from the (extensive) well drilling. Other interesting features are limited water consumption, limited impact on landscape, and low impact on air quality (no milling operations). The major concern during operations is then the risk of contamination of groundwater by the leaching solution (Mudd 2001). Surface risks can be limited operations by the application of best practices; migration from the ore formation is limited when ISR is applied only in confined aquifers.

Closure operations are also facilitated since no legacy sites are created. The key point in post-mining is the restoration of water quality in the ore formation, and its *a priori* demonstration. Provided that good practices are followed and depending on the local geology and presence or absence of buffer minerals, monitored passive remediation or active restoration (e.g., pump and treat) can be chosen (IAEA 2016).

With the technical challenges and economic potential associated with ISR, this domain has triggered a lot of research interest (Märten et al. 2013; IAEA 2016; Le Beux et al. 2018): e.g., better assessment of the ore geometry, development of monitoring tools, increased valorization of production data, improvement of well engineering, economic and hydraulic optimization of mining plan. Simulation also plays an important role: hydrological software tools are used for well-field planning and operational control. Simulations, coupled with careful characterization of the formation, can help maximize the contact between the leaching fluid and the ore, reduce the risk of fluid migration out of the well-field, and micromanage injection rates including role reversal between injectors and producers.

Along these developments, reactive transport modeling has been identified as a significant tool to quantify impacts, support the evaluation of remediation techniques, and help demonstrate their compatibility with regulatory targets (Kalka et al. 2006; Descostes et al. 2014; Johnson and Tutu 2016; Dangelmayr et al. 2017). Several authors also mention the use of reactive transport to help understand and quantify the processes at stake during exploitation (Kalka et al. 2006), although full 3D reactive transport at the scale of production units is still scarce (Regnault et al. 2014, 2017). The model developed in two applications is presented hereby. The quantitative results of the simulation can be used to improve the understanding of processes and their coupling: this offers in itself some opportunities to guide tentative optimizations. Furthermore, the simulation can also be used by engineering departments to directly test optimization ideas, improve production forecasting, and more generally help in planning: some examples illustrate how intensive use of reactive transport simulation can create value during and after exploitation.

SIMULATION OF *IN SITU* RECOVERY OPERATIONS

Reactive transport model for the KATCO Mine

The KATCO mining company is a joint venture between Orano Mining (51%) and Kazakhstan national mining company Kazatomprom. It specializes in the *in situ* recovery of uranium, in roll-front type formations, in Southern Kazakhstan. With a 4,000 t uranium yearly capacity, it became in 2009 the world largest ISR mine. Its total production since 2006 amounts to 36,000 t uranium. KATCO operates on two distinct sites in the Chu-Saryssu basin: Tortkuduk and Muyunkum, 40 km apart, each with their own well-field (4,000 wells in activity at any given time) and separation plant.

The construction of reactive transport simulations at the block scale relies on a combination of hydrogeochemical description of the reservoir, and identification of main processes. Spatial variability is inherently high in this type of deposits; its huge impact on the shape of the recovery curve imposes a careful evaluation and integration in the simulation workflow. All simulations were performed using the HYTEC code developed at MINES ParisTech (van der Lee et al. 2003; Lagneau and van der Lee 2010).

Hydrogeochemical description. The concession around the Muyunkum deposit is composed of several aquifers containing roll-front accumulations, permeable sandy formations from the Paleogene era (Petrov 1998). Particularly, the Uyuk aquifer is an Eocene sandy aquifer, mildly cemented, with a low content in calcite (<0.1 wt%), approximately 400 m deep and 30 m thick. The aquifer presents a high spatial variability, both vertically and laterally, as a result of its sedimentation history: coarse to fine sands channels deposited under tidal influence, presence of clay intercalations stemming from continental flooding plains.

Hydrogeological characterizations show porosity and permeability in the sand ~20–25% and 10^{-4} m/s respectively (ben Simon 2011). With head gradients around 1‰, the natural regional flow velocity is around 5 m/y, from the oxidized to the reduced part of the aquifer through the uranium bearing roll-front.

Hydrogeochemical analyses show a clear discrimination between oxidized (pH ~7.6, Eh~50 mV/SHE), and mineralized or reduced zones in the aquifer (pH ~7.8, Eh~−150 mV/SHE). The solid is a mildly cemented arkosic sandstone (Munara 2012; Robin et al. 2015a), with composition around 70% quartz, 10% feldspar, 5% micas, 5–10% clay minerals (mostly smectite, and kaolinite). Authigenic smectite is mostly composed of Beidellite, with a significant content in Fe(III) (Robin et al. 2015b). Other minerals include goethite or pyrite (<1% in the oxidized or reduced part of the aquifer respectively), and low concentration of calcite (<0.1%). Finally, uranium-bearing minerals are found in the roll-front, mostly uraninite and coffinite, at relatively low concentrations (~0.1%).

Initial porosity is around 20% in the sandstone. Porosity evolution remains small in the reservoir, limited to 1 to 2%, with the weak reactivity of most minerals. Actively reactive minerals (calcite, uraninite) represent a very small volume fraction. Likewise, precipitation of secondary minerals (mostly gypsum and possibly aluminum hydroxysulfates) has a limited impact on porosity if precipitation if spread over the reservoir, although chemical clogging can occur when precipitations are concentrated close to the production well.

Initial aquifer water composition was prescribed using base line monitoring: geochemistry is controlled by equilibrium with the host rock minerals, with total dissolved solids around 1 g/L at temperature around 20 °C. The injection solution composition evolves through the simulation. An initial acidification phase uses aquifer water acidified with sulfuric acid at typically 20 g/L. When pH drops below 2 at the producers, the block is connected to the plant. Fluid composition then stems from the multiple circulations of the fluid over the whole well field: the result is an accumulation of dissolved elements in the solution (only uranium I stripped at the plant). Typical solutions reach up to e.g., 0.5 g/L in Ca, or 1 g/L in Al. The model uses compositions in line with analyses at the plant, and acid content is adjusted daily according to the operator prescription: acid content is adjusted so that pH at the production well remains between 1 and 2, to ensure Fe(III) mobility and limit precipitation of secondary mineral; usually, the operator starts production with high acid content (typically 20 g/L H_2SO_4), then acid content is reduced for older, less producing, blocks (down to 5 g/L).

Chemical processes. Uranium dissolution during acidic leaching is dominated by the oxidation of U(IV) minerals (e.g., uraninite UO_2, coffinite $USiO_4$) by Fe^{3+} in the bearing solution. The origin of ferric iron is multiple: local dissolution of gangue minerals (goethite, beidellite), recirculation of Fe^{3+} from the well field, or surface active regeneration of Fe^{3+} using peroxide (e.g., Heathgate operations in Southern Australia, Märten 2006) or other oxidants. Low pH (typically <2) is required to allow for ferric iron mobility.

The governing reaction for uranium dissolution is then:

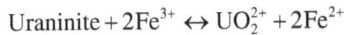

$$\text{Uraninite} + 2Fe^{3+} \leftrightarrow UO_2^{2+} + 2Fe^{2+}$$

$$\text{Coffinite} + 2Fe^{3+} \leftrightarrow UO_2^{2+} + SiO_2 + 2Fe^{2+}$$

The reaction is controlled by the availability of ferric iron, solubility of uranium (in which complexation, notably with, SO_4^{2-} can play an important role) and kinetics. In this case, it was verified that oxidant availability controls uranium dissolution, so that the specific (reduced uranium) mineral is not a determining factor. In other cases, several uranium-bearing mineral should be carefully considered, including possibly oxidized uranium minerals (e.g., schoepite in other geologic contexts).

Reactions with gangue minerals are also key, since they can control pH or local sources of ferric iron (Robin et al. 2016):

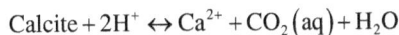

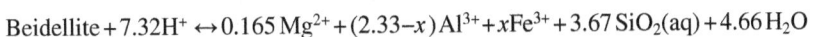

$$\text{Calcite} + 2H^+ \leftrightarrow Ca^{2+} + CO_2\left(aq\right) + H_2O$$

$$\text{Beidellite} + 7.32H^+ \leftrightarrow 0.165\,Mg^{2+} + (2.33-x)\,Al^{3+} + xFe^{3+} + 3.67\,SiO_2(aq) + 4.66\,H_2O$$

Sorption can also play an important role: clay surface sites can constitute an important sink term for protons. However, this effect is limited mostly to the initial acidification phase of the exploitation, when pH is driven from initial near neutral conditions to less than 2. Surface sites then remain mostly saturated in protons, independently of pH evolution during the whole exploitation phase. Their influence is again crucial post exploitation, when the remaining buffer capacity of the aquifer drives pH above 4.

Column tests are ubiquitous in ISR operations, a quick, easy means to determine the leachability of a sample ore. As already mentioned, direct use of the results to infer production curves should be exerted with care, since they are only representative of the conditions in which they were performed. Moreover, representativeness of a (small) sample compared to large heterogeneous ore bodies is limited. Finally, preservation and preparation of samples can significantly alter the behavior of the system; e.g., ben Simon et al. (2014) used a mixture of different samples from different parts of ore formation; moreover, oxidation between sample collection and test column led to a significant evolution: very low pH due to pyrite oxidation with air O_2, oxidation of nearly half the uranium in the system. Nevertheless, these column tests offer perfect opportunities to test and calibrate the reactive paths and kinetics in controlled conditions. The chemical system can be further tuned using production data.

Hydrodynamic model. The ISR process relies on the recycling of the leaching solutions: after recovery at the producers, the solution is treated in the surface plant to strip uranium, then the reagents concentration (acidity, oxidizers) is adjusted before re-injection in the well field. Over time, the dissolved content of the solution increases. At such low pH, Ca, Mg or Al content can reach values as high as 0.5 to 1 g/L, and over 20 g/L for sulfate. Therefore, despite high flow rates imposed at the wells (several m^3/h over 10 m screens), density differences between the leaching solution and the initial aquifer water are sufficient to create downwards migration: this was verified with observation wells in test production cells and confirmed by sensitivity analysis flow simulations (Bonnaud et al. 2014).

A gravity driven (saturated) flow solver is then used for the simulations. Density can be deduced from solution composition. The 3D model is built using the geometry of the well field (position of the wells, length and depth of the screens), on the whole thickness of the aquifer. Source terms are prescribed at each well, both in terms of flow-rate and solution composition: ideal conditions can be used, or they can be set to reproduce actual production data for history matching exercises. Boundary conditions can be imposed to allow for regional flow, although it has been demonstrated that the effect is limited.

Spatial variability. The ore genesis process is responsible for a high spatial variability of the system. Sedimentation creates a first stage of heterogeneity, with sand channels of varying grain size and discontinuous clay interlayers. Flow and oxidation in the aquifer during the creation of the roll-front leads to a second stage of heterogeneity. Indeed, the roll-front is an interface between an oxidized upstream and a reduced downstream aquifer. The migration of the interface is therefore directly linked to the local reducing buffering capacity of the aquifer (organic matter, pyrite) and the flux of oxidizers i.e., the flow field (and *in fine* spatially variable permeability and porosity field). The sequential processes account for a very complex 3D shape for the roll-front. Far from a typical roll-front shape, the mineralized zone is a complex 3D structure, immerged in oxidized and reduced zones and intersected by low permeability clay barriers. Finally, uranium grade within the mineralized envelop is also variable, again a result of the complex history of the front migration.

The correlation length for uranium grade is typically a few 10's meters, similar to the size of the production cells (radius 42 m). Therefore, facies and uranium grade distribution play a huge role in the shape of the response function of the production cell. Particularly, smoother distributions tend to yield sharper production curves. On the contrary, high heterogeneity can present high concentration clusters that take longer to dissolve (mainly due to hydrodynamic control), resulting in very elongated residual production (Lagneau et al. 2018). Another aspect of the variability within the ore body is the diversity of response over the permit. Indeed, local differences in geometry of the mineralized envelop, grade distribution or clay content can lead to very different evolution of pH or uranium concentrations in the producers.

A consequence of the variability is that homogeneous models, or even 1D and 2D descriptions come short to being fully predictive. Reactive transport simulations must therefore tackle a 3D description and incorporate an accurate description of the geology, with a block model both in terms of geochemical facies, permeability, and ore grade.

Model description. The workflow was applied on a particular block on the KATCO operation: Block A1. The block is composed of 8 adjacent production cells. Due to operation choices, mostly to enhance production in rich areas, several production cells contain 2 or more producers. In areas with two levels of mineralization separated by clay interlayers, multiple screens were also positioned. The model geometry then takes into account 18 producers and 43 injectors. Production data were used to individually constrain the flow rate and injection solution in each well. Both prescriptions evolve through the operations. Solution composition (particularly acid content) is a key lever for the operator. Injection and production flow-rates are regularly adjusted by the operator; they are also prone to modifications due to well evolution (clogging), maintenance (work-over), and flow allocation constrained at the mine scale by the plant specification.

One realization of the block model was chosen to constrain the variability. The implications of this simplification are discussed further.

The geochemical model was adjusted against production history, mostly pH, ferric iron balance and uranium concentration in the producers. During this process, an evaluation of the hierarchy of processes was performed. The aim was to determine the main controlling reactions; second order reactions were then removed in order to reduce the complexity of the system. The minimal mineralogical assemblage and associated kinetic parameters (from Palandri and Kharaka 2004; Robin et al. 2016) for each chemical unit are given Tables 1 and 2.

Table 1. Mineralogical composition of the model at the block scale.

Mineral	Oxidized	Mineralized	Reduced
Uraninite	–	Variable	–
Goethite	0.44 wg‰	–	–
Calcite	–	0.23 wg‰	–
Cristobalite	660 wg‰	830 wg‰	830 wg‰
Beidellite	23 wg‰	31 wg‰	31 wg‰
Kaolinite	95 wg‰	20 wg‰	20 wg‰

Table 2. Kinetic parameters associated to the model minerals (from Palandri and Kharaka 2004; Robin et al. 2016).

Mineral	Surface (cm^2/g)	$Log_{10}k$ $(mol/m^2/s)$	Catalyst
Calcite	Thermodynamic equilibrium		
Uraninite	500	$\pm 3.2\times10^{-8}$	$\left[H^+\right]^{0.37}\cdot\left[O_2(aq)\right]^{0.31}$
Goethite	10^4	$\pm 1\times10^{-7}$	
Cristobalite	2000	$\pm 3.2\times10^{-14}$	
Beidellite	6×10^6	$\pm 0.93\times10^{-14}$	$\left[H^+\right]^{0.22}$
Kaolinite	500	$\pm 4.9\times10^{-12}$	$\left[H^+\right]^{0.777}$

The simulation gives access to a wide array of results: e.g., Figure 2 shows the uraninite concentration in the system after 600 days of exploitation. The progression of uranium recovery is noticeable by the depletion within the limit of the block. The yellow cloud shows the extension of the acidification (pH< 1.8). This 3D representation is useful to identify and localize possible defects in the exploitation scheme: poor leaching, residual high pH areas, slow dissolving high grade clusters, …

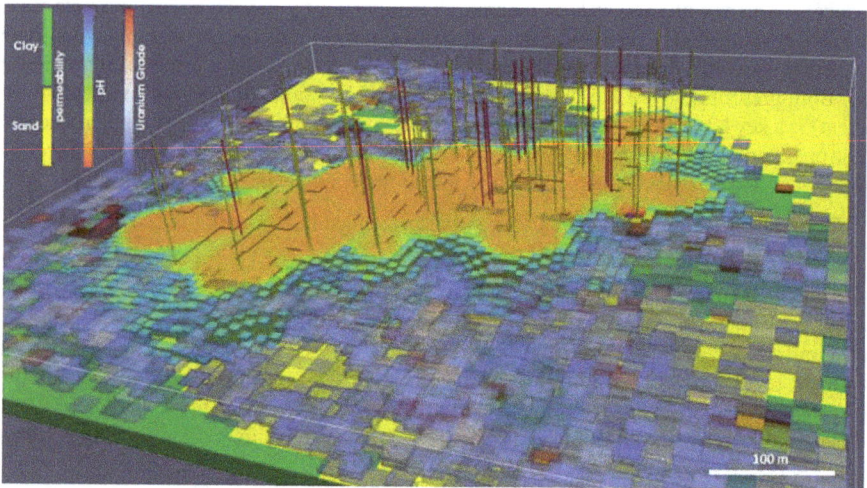

Figure 2. Simulation result for Block A1 after 600 days: 3D map of uraninite concentrations and extension of the low pH plume (pH< 8 in the yellow cloud). Block size ~200 m, reservoir thickness ~30 m, the vertical shafts represent the wells position.

Another option is to plot the concentrations at the production wells, for each producer or in the collector (i.e., a mixing of all producer wells in the block). This visualization is particularly relevant for the operator as it corresponds to the only information available on site to monitor the exploitation (daily analysis of the solution). Such representation is proposed in Figure 3: pH and dissolved uranium concentration in the collector of all the producers from Block A1. This was used to adjust the parameters of the model against production data.

As can be seen, a very good fit was obtained, even after various events during the exploitation. The adjustment is facilitated since most parameters are constrained by measures (permeability, mineral concentrations), databases (solubility constants, rate constants), geometry of the system (well field, block model), and operations (flow-rates, composition of the leaching solution). In the end, the adjustment was performed on poorly determined parameters: calcite concentration (too low in the system to allow for an accurate determination), beidellite reactive surface area, and ferric iron content of the beidellite. The adjustment is fairly easy to perform, since the three parameters have mostly independent impacts on the system. Indeed, calcite concentration is the first determining parameter (with porosity) on the time to obtain the pH drop. Beidellite (kinetically controlled) dissolution is the major long-term acid consumer, so that beidellite surface can be adjusted on the pH values or acid balance after one year. Finally, once beidellite kinetics is adjusted, its content in ferric iron controls the release of oxidizer within the block and the increase in uranium dissolution.

Interestingly, three parameters only are then sufficient to reproduce the complexity of the reaction of the system to multiple events: evolution due to the acidic and oxidizing solution, modification of the acid content in the injected solution, modification of flow-rates, and redrilling.

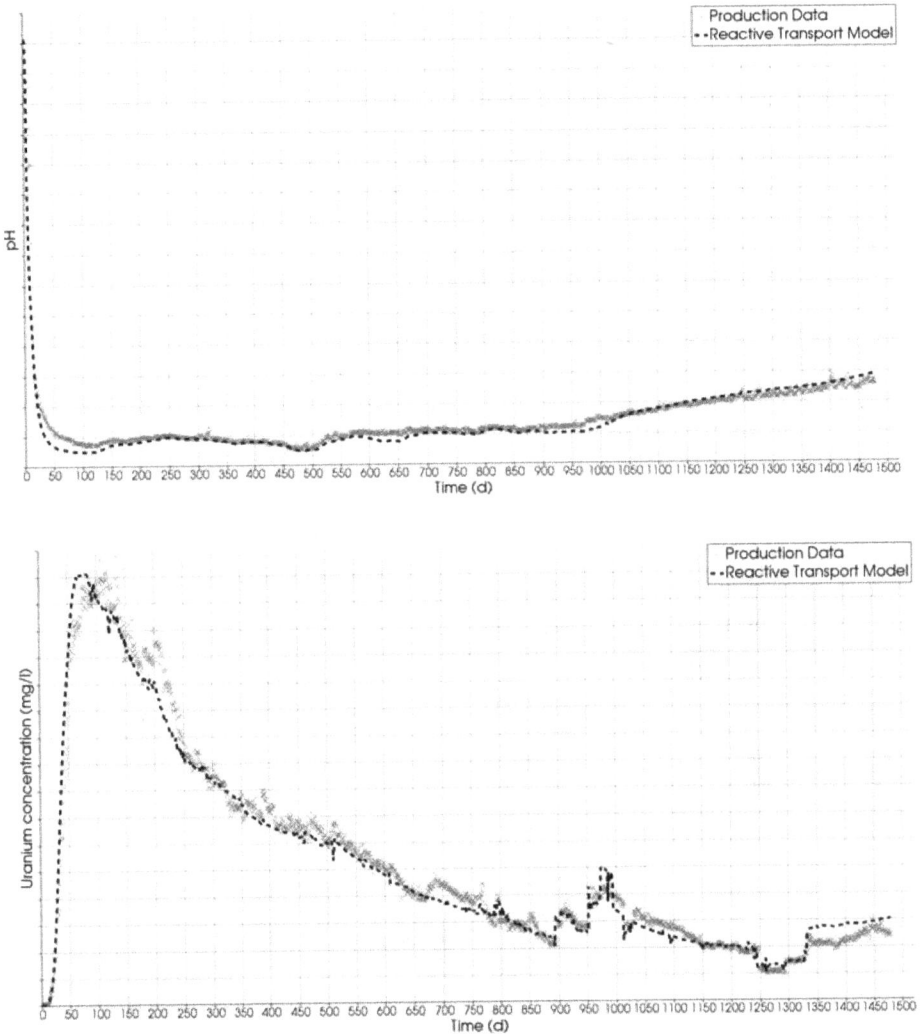

Figure 3. History matching on the production Block A1. Solutions from all the producers are mixed to create an equivalent collector of the block: pH (pH drops from initial 8 to around 1 after acidification) and uranium concentration (linear scale, a few 10s to 100s mg/L) of the resulting solution.

Validation. The model was verified by application to another block in the vicinity: Block A2. For this application, the model was obviously adapted: one realization of the block model over Block A2 area, position of the wells, exploitation scenario (flow-rates, composition of the leaching solution). However, the geochemical model calibrated from Block A1 was used without any adjustment. Figure 4 shows the simulation results for Block A2: the model is very close to the production data, providing a good validation of the model.

Interestingly, the two blocks behave very differently. Block A2 displays a sharper uranium peak, with high peak concentrations: these differences are mostly due to the local distribution of uranium in both blocks. Also, the two blocks were exploited sequentially, so that part of the acidic plume from Block A1 invaded block Block A2, resulting in lower initial pH.

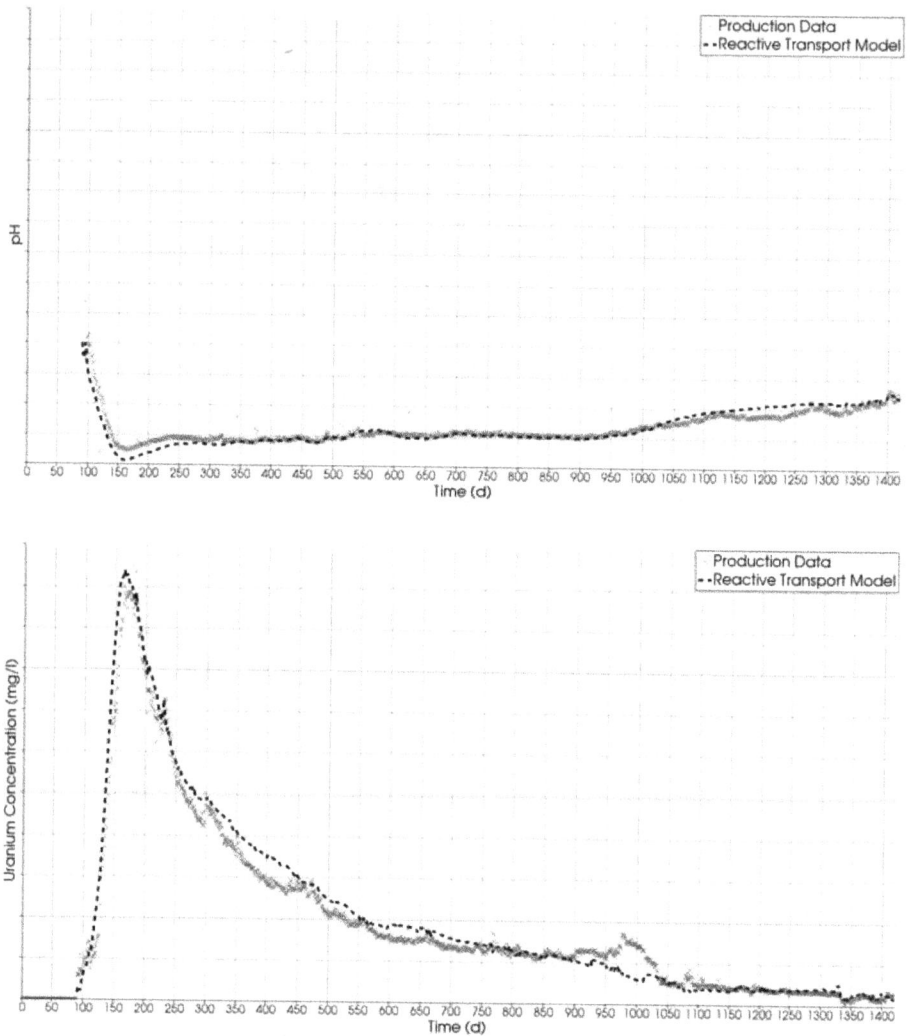

Figure 4. Direct application of the model calibrated on block Block A1 to block Block A2. No calibrations were performed – the differences in pH and uranium behavior stem from a different geometry (wells position, geology) and exploitation scenario (flow-rate, composition of the leaching solution).

Limits. The strength of the model relies on its roots in processes and geometric description of the system. As a result, very few adjustment parameters are needed. The operator currently strives to acquire these missing parameters in order to increase the predictive capacity of the model. Also, although chemistry evolves over the width of the mining concession (~20 km), the wavelength is sufficiently large that adjustment on a block can be used without modification in adjoining blocks. However, the adjustment needs to be performed again in farther areas, where chemical specification may be significantly different.

It was mentioned that spatial variability was key to control the shape of production curve. And yet, correct adjustments were possible using a single realization of the block model. Use of another realization would have given a similarly good adjustment, perhaps with marginally altered parameters, provided the total amount of uranium in the block was similar. There seems to be a contradiction between the two remarks: spatial variability is key but simulation results seem independent of the realization. This conflict disappears when looking as the relative dimension of the range of the geostatistical model for uranium grade (~ 50 m) and the block (250 –400 m). The block size is well above the range, so that stochastic variability is averaged over the size of the block. In the end, at the block scale, the structure of the distribution is more important than the actual distribution. This is not true at the size of a unit cell (radius 42 m): the simulation result for a specific producer (as opposed to the collector of the block) changes radically for different realizations. Langanay et al. (2018) provide some insights into the impact of geological uncertainty on the simulated uranium recovery.

Industrialization. The demonstration provides solid foundations for the reactive transport simulation of ISR operations. A direct benefit is a better understanding and quantification of the processes at stake over the lifetime of a block. Indeed, the information available to the operator is very limited: after the initial geologic reconstruction, the main flux of information during operations comes from the monitoring of flow-rates, composition of injected fluids, and chemical analysis of produced solutions for each well on a daily basis. This information is very useful for operation management: at the mine scale, the operator needs to arbitrate resource allocation (acid consumption and overall flow manageable by the plant). However, deconvolution of fluid composition is hard to perform, since it stems both from the fluid evolution along a flow tube and the mixing of all the flow tubes converging to a producer.

Once the model is validated, it can be used to evaluate the controlling processes in the system. For instance, kinetically controlled beidellite dissolution is the first acid consumer on the long term, over 80% of acid balance over the lifetime of a block. The weak catalyzing effect of protons, and conditions always far equilibrium at pH< 2, mean that dissolution rate is only weakly dependent on pH. Acid consumption (and operational costs) could therefore be reduced using faster injection rates. However, this should be compared to the impact of lower local production of ferric iron associated with beidellite dissolution. Another investigating concerns the dissolution regime of uraninite: depending on the position in the production cell, prescribed flow-rates and availability of reactants (Fe^{3+}), uraninite dissolution can be controlled by kinetics or hydrodynamics. Here again, quantitative analysis can provide useful insights on a possible lever to optimize uranium production.

This kind of quantitative analysis using reactive transport codes, based on carefully devised model of geologic systems, is already performed in a large array of applications. However, a more intensive use of simulations can benefit the operations. Indeed, it is clear that a block behavior is heavily constrained by the local geology. A simulation calibrated on a specific block cannot therefore be used to illustrate a *typical* production curve. Going beyond generic recommendations, systematic simulations for each block are therefore needed. In this case, academic codes reach their limits.

The strength of a code developed in an academic environment is its versatility: continuous development offers a wide array of options to choose from, which allowed performing the demonstration. The downside of the versatility is a complex combination of options (and keywords in text-based input files), most of them probably not needed for the specific application. Also, although importing input parameters from the operator into the code is manageable for a single application, it can become cumbersome for routine use. Finally, systematic simulations are typically performed by engineering teams. Rapid turnover and dispersion between multiple tasks imply that training should be minimal.

Early in the development of the project, the need for an engineer tool (as opposed to a research code) was identified. A graphical user interface is obviously required, but not sufficient. Several specifications were identified: clear interface to minimize training, easy integration in the operator workflow, presentation of outputs to provide ready-to-use data. A GUI for HYTEC, specifically devised for ISR needs was then created: HYSR. The interface only proposes options relevant to the simulation of ISR exploitation. Moreover, the controls are organized to match the practice of engineers; they do not necessarily reflect numerical simulation thought-patterns. For instance, the interface does not ask for initial or boundary conditions. Rather, it imports a block model or well specifications, which are the standard data an ISR engineer is used to manipulate. Imported data, directly from the operator workflow, are converted by the preprocessor to launch the reactive transport simulation.

Output data management is also a key element to facilitate the use in operational conditions. 3D visualization of the simulation results is still possible, using ParaView (Ahrens et al. 2005). This allows for fast evaluation of leaching efficiency, particularly with application of preselected filters. HYSR also transforms the outputs into production curves: the curves show information at the producer wells or globally for the block, which directly relate to production data. Also, mass balances are performed at the scale of the block. Finally, the net present value (NPV) of the block is proposed: the calculation is based on a cost model integrating capital costs (construction of the block, connection to the plant), operation costs (power use, fraction load at the plant), reagent costs (acid consumed), and revenue (uranium produced). The combination of output formats offers different opportunities for the operator: identification of poorly productive areas to help propose corrective strategies, direct comparison to production data for validation or use in mining plan elaboration, or a robust means to compare alternative exploitation strategies.

Examples of application. Mine planning for ISR exploitation consists in a predictive simulation of the sequence of well-field operations and construction works needed to reach the annual target of production. Planning also needs to respect regulatory constraints and financial limits imposed by the shareholders. The operator differentiates short- and long-term planning. Short-term planning (12–18 months) is the foundation for construction and revision of the company's budget for the current or following year. Long-term planning is a projection for the whole life expectancy of the mine. It provides a guide for long-term exploitation strategy and data to dimension infrastructures and associated CAPEX. In both cases, the quality of the mining plan depends on the reliability of estimated behavior of the system.

Short-term planning. In the Katco operations, lifetime of exploitation blocks is between three and five years, with around 60 blocks running at any given time. Past experience of one-year projections shows that mature running blocks (i.e., in quasi stationary state after ramp-up) amount to roughly 80% of extracted uranium and 70% of acid consumption. The prediction of the end-of-life of mature blocks is therefore a major input data for short-term planning: it drives the gap between expected and target production, and therefore the sequence of new blocks needed to reach the target. Predictions of block end-of-life behavior also allows anticipating block closure, on criteria like expected uranium concentration or recovery rate. This information is needed to allocate resources (total flow capacity of the plant or main pipes) over the well-field and assess potential availability for new blocks. A correct assessment of the future behavior of currently operated blocks is therefore key for the robustness of the provisional budget allocated for new blocks development: earthworks on future exploited zones, wells digging, and development of hydraulic network.

The well field behavior is monitored and simulated at each technological block, for three main parameters: flow in the block, dissolved uranium concentration at the collector, and acid consumption. Traditionally in ISR operations, evolution is evaluated independently for the three parameters. Acid consumption is based on empirical coefficients (e.g., mass of acid consumed by mass of uranium produced), based on supposedly similar areas of the exploitation. Evolution of uranium concentration in the leach solution is based on analytical functions using parameters adjusted on production data: the extrapolation of these functions is used as predictor.

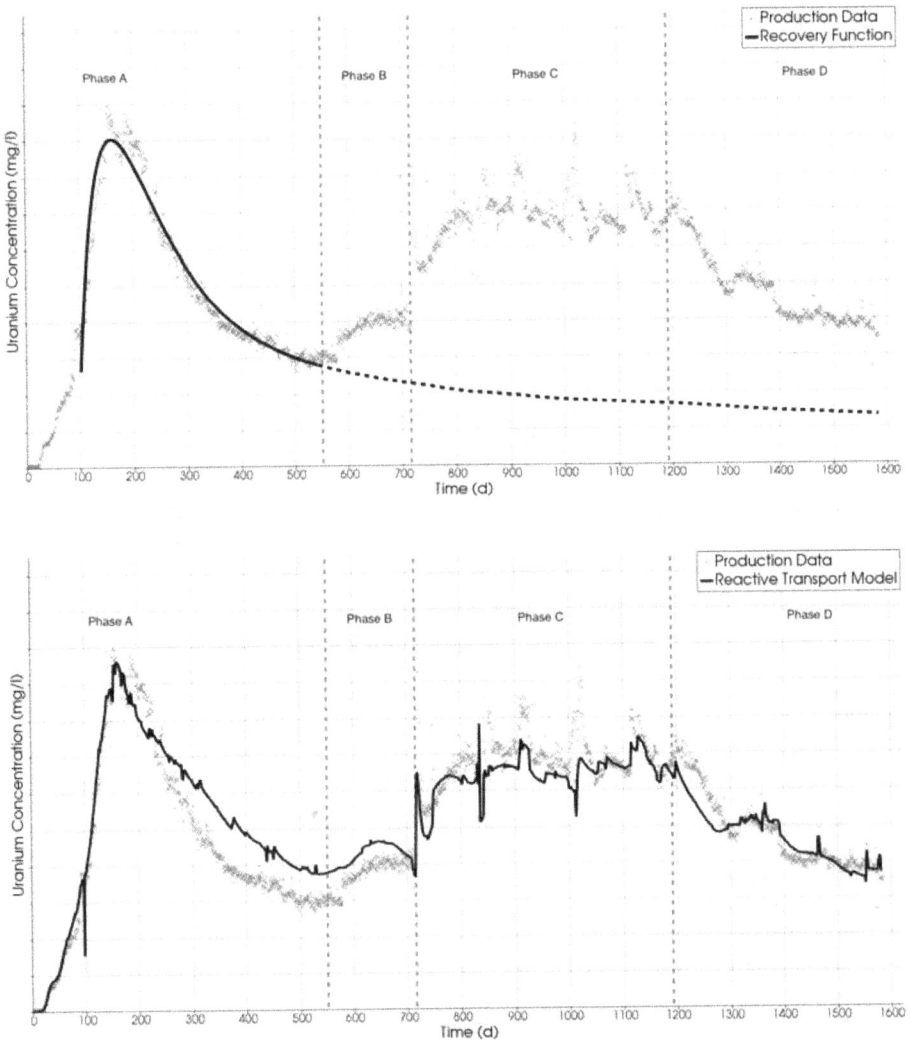

Figure 5. Uranium concentration for Block B (linear scale). The operator tested several optimization strategies, resulting in frequent modifications of operating conditions. Classical analytical functions to predict the block behavior works for the first period only (top). Integration of the evolving operating conditions in the reactive transport simulations allows for a correct prediction on the whole life of the block (bottom).

Such approach is acceptable as long as operating conditions are maintained: i.e., small evolution of the flow-rate and small variations of the acid prescription curve in the injected solution. Following the standard conditions, technological blocks display the characteristic bell curve (e.g., Figures 3 and 4 for blocks Block A1 and A2). However, this approach looses robustness when changing or complex operating conditions are applied. And yet, such modifications, planned (e.g., acid availability due to overall allocation decision) or provoked (e.g., additional wells, optimization decision) by the operator should be taken into account to allow for a precise short-term prediction.

The 3D reactive transport simulation is here particularly adapted. It offers the possibility to precisely adapt evolving conditions and evaluate their impact on uranium concentration and acid consumption. Better anticipation of alternative strategies (optimization, reaction to unforeseen constraints) can therefore be integrated in the short-term mining plan. Most common actions to increase uranium recovery are increasing acid content in the injected solution, and improving flow rates. Improved flow rates in a production cell can be obtained by work-over on ageing production wells or by additional wells drilling.

An example of reactive transport simulation used in short-term planning is illustrated on block Block B (Fig. 5). The block life started with a classical production curve for the first 18 months, with a sharp uranium peak followed by a rapid decrease (phase A). Since the block had important estimated residual reserve, the operator tested several strategies to increase uranium yield. First, the operator decided to increase acid content in the injected solution (phase B), which stabilized uranium concentration at the collector. The operator then decided to improve production flow rate by repeated work-over on several production wells (to counter well clogging), and drilling of additional production wells (phase C). The impact was an important increase in uranium concentration at the collector, which stayed on a plateau for two years. Finally, when estimated optimal recovery rate was reached, the operator decided to reduce the amount of acid injected in the block (phase D), which resulted in a rapid decrease of uranium concentration.

The analytical function, with parameters adjusted on production data, yields correct results during the first phase of exploitation (Fig. 5, top), but fails when operating conditions deviate from the standard procedure (phase B, C).

On the other hand, the reactive transport simulation produces a correct estimation of the whole production life (with a unique set of parameters): Fig. 5, bottom. The link between injected acid and uranium concentration is correctly anticipated thanks to the geochemical model used: the increased acidity in phase B induces lower pH, faster kinetics for beidellite dissolution, and finally increased Fe^{3+} availability. Uranium oxidative dissolution in the ore body is then increased resulting in high uranium concentration in the producers. The opposite is observed when the operator decreases the acidity in phase D.

The model provides correct estimation of the impact of evolving velocity field within the ore body (phase C). Improved injection and production rates increase the acidification of the ore, including mineralized areas far from the producers. Also, production of dissolved uranium is improved by minimizing uranium-rich solution loss, particularly in the lower parts of the reservoir. Finally, additional production wells in high-grade areas of the block stabilize uranium concentration in the collector by adding high yield lines to the overall leached solution. The simulation correctly predicts the impact of this combination of factors. The 3D approach is here a necessity: the importance of uranium grade spatial variability is demonstrated here by the impact of local modifications (additional wells) on the overall behavior of the block.

Long-term planning. The strength of reactive transport in short-term planning is the capacity to rapidly anticipate the impact of fluctuating operational conditions at the block scale. For long-term planning, the key is to produce simulations, on a large scale, to help propose a sequence of production all the way to the end of exploitation.

By construction, long-term planning is mostly constrained by the predictive behavior of future blocks. Geological and grade data on these areas is usually coarse: well-spacing on the exploration grid (typically 100–400 m) is larger than at the technological stage (around 40 m). Moreover, the planning team cannot rely on the operational feedback on undeveloped areas: at best, a pilot operation can provide some indication on acid consumption or uranium yield on a small scale (a few cells) and on a limited operation time (several months). In these conditions, semi-analytical recovery functions are unsatisfactory: indeed, their parameterization relies on (unavailable) production data or analogy on similar blocks produced in the same area.

On the other hand, reactive transport simulation can be used even if the model is not completely constrained: a 3D geological model can be derived from (large grid) exploration data, uranium and gangue mineral behavior can be inferred from the geochemical model based on (limited number) mineralogical analyses. The simulation provides an interesting tool to illustrate and quantify the impact on production of certain features of the ore body. Thus, geometry has a huge impact, which can be calibrated by simulations: thickness of the reservoir, density of clay interlayers, morphology of the mineralization and localization in the reservoir. For instance, recovery is facilitated for compact ore bodies in confined low thickness reservoir. Conversely, dispersed mineralization in a thick reservoir leads to reactant dispersion, increased acid consumption, and lower recovery due to uranium loss.

Figure 6 illustrates the behavior of technological blocks with different positions of the ore in relation with the reservoir boundaries: basal mineralization in contact with the underlying clay formation, perched mineralization close to the top of the reservoir, and finally stacked mineralization with a complex vertical superimposition of roll-fronts. Uranium recovery is fast and complete for the basal mineralization (Fig. 7). The leaching solution, with higher density than reservoir water, is guided to the production wells by the bottom impermeable layer, through the mineralization. On the contrary, perched mineralization leads to slow, incomplete leaching. In this configuration, correct acidification of the ore is complicated by downwards flow due to density gradients between the leaching solution and the reservoir, resulting in incomplete dissolution of the uranium bearing minerals. Also, reservoir volume below the mineralization and the bottom impermeable barrier increases dispersion, so that a fraction of the dissolved uranium remains in the reservoir, and decreases uranium recovery rate. Stacked mineralization display intermediate behavior both in term of dynamic and recovery rate.

SIMULATION TO MINIMIZE ENVIRONMENTAL IMPACT

Reactive transport modeling can also be used as a powerful tool to assess the environmental footprint of an ISR mining site. As previously stated, the potential issues lie mostly within the aquifer targeted by ISR. Low environmental impact for all surface operations can be achieved by careful use of industrial best practices: clean well-field operation, well and plant decommissioning, removal of networks. Therefore, the remaining leaching solutions within the aquifer can be considered as the main source of chemicals of interest to be treated. Stakeholders are rightfully cautious about operators' capacity to guarantee aquifer quality after the end of exploitation (Mudd 2001; Saunders et al. 2016). Modeling can play a role to help demonstrate that post-exploitation environmental targets can be reached.

The first role of this demonstration is to comply with local regulations and the companies own targets. It can help plan a monitoring and restoration strategy, including an estimation of costs. For instance, Clay (2015) insists on the low amount of published data and the importance to include evolution and migration processes in the definition of rehabilitation targets. Finally, the demonstration is fundamental to obtain the social license, a prerequisite to any new project development. With time scales from 10 years to a century after exploitation and spatial range over 10 km (extension of the deposit), the use of reactive transport simulation is fundamental to quantify the demonstration. The demonstration is based on initial conditions data (baseline) including the inherent variability of roll-front systems, a good understanding and quantification of the physicochemical processes of evolution in the wake of the exploitation, and the identification of the key possible contaminants.

The objective of the remediation strategy revolves around the exploited aquifer, in relation with down-gradient outlets, and the demonstration that a potential water resource is not compromised. Pre-exploitation water conditions are usually poor quality (high contents in metal, uranium and its daughter elements, sometimes high salinity), resulting from the ore genesis mechanisms (Sodov et al. 2016). The physicochemical quality target to be reached (and demonstrated) depends on the local regulation: usually same class of water quality, or concentration range as close to the baseline as possible.

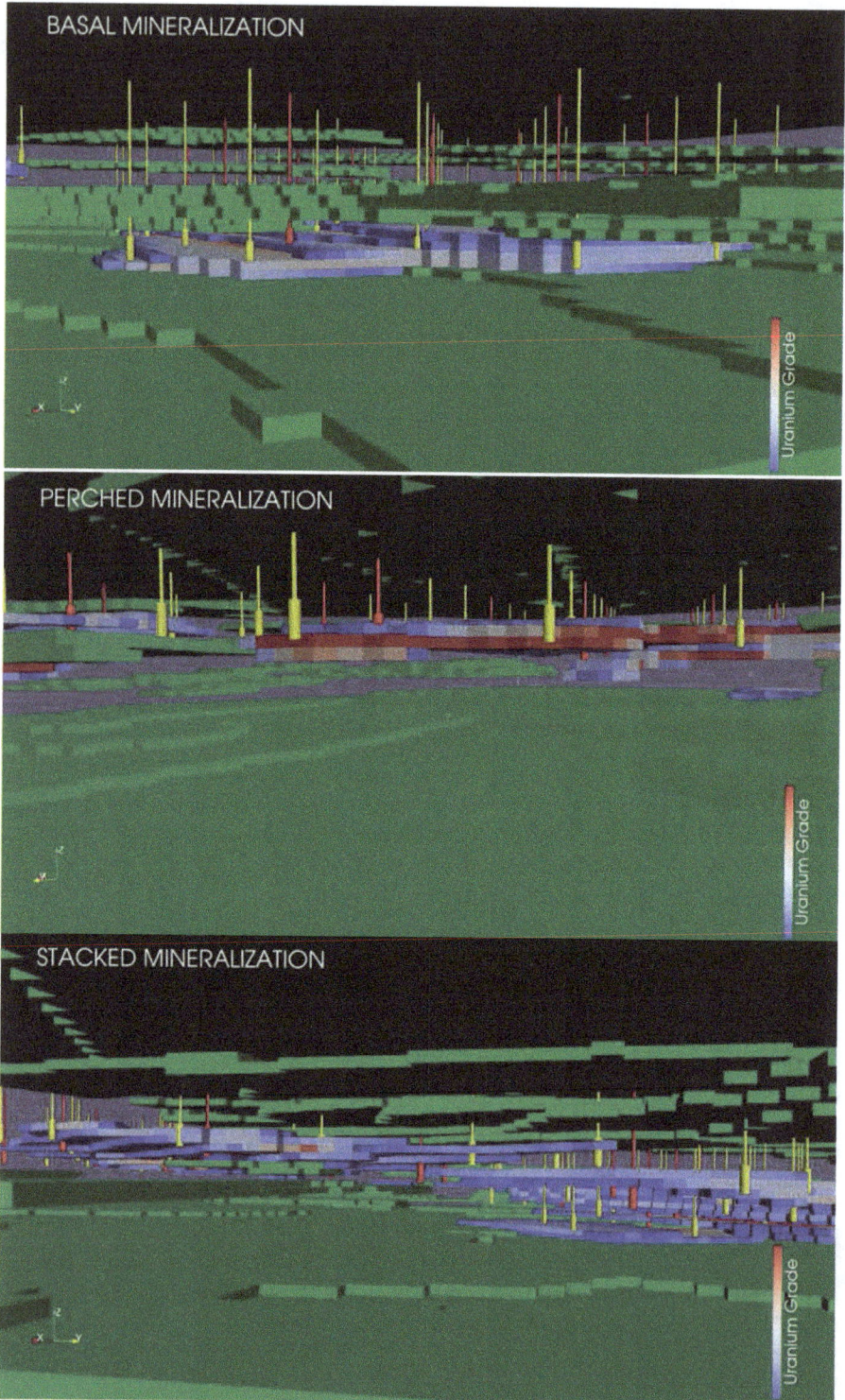

Figure 6. Three types of ore geometry considered for simulations in a long-term planning view: basal, perched and stacked mineralization.

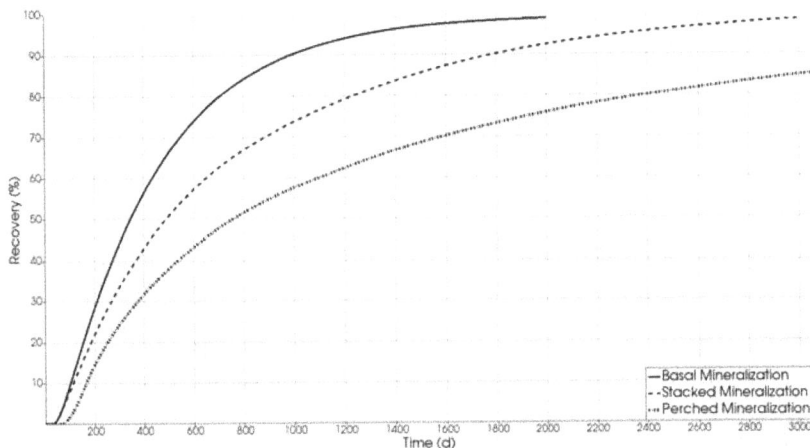

Figure 7. Simulation results for three characteristic types of ore geometry: evolution of uranium recovery rate. Basal mineralization is much more favorable with a fast and complete recovery.

In ISR operations specifically, the buffering capacity of the aquifer is largely reduced by the circulation of the leaching solution. Also, recycling of the solution in the reservoir (with uranium only stripping at the plant) tends to increase the concentration in all the dissolved elements including metals, creating a saline plume. Radioactive daughter products of uranium decay (mostly ^{226}Ra) can be mobilized by the leaching solution and accumulated in the reservoir. The actual hierarchy of chemicals of potential concern depends on the geology and the choice of ISR technology. For instance, alkaline ISR does not create pH singularities, however the high complexation capacity of carbonate uranyl increases the effective solubility of residual uranium. In acidic ISR, the main chemicals of potential concern are the pH plume, sulfate and residual uranium.

The use of reactive transport modeling for such purpose is closer to the type of simulations mentioned in the introduction. However, the industrial context brings several specificities. The operator can start testing remediation strategies in the early stage of development of the project, with a view to optimizing the process while maintaining the targets. Here, pilots can play an interesting role: central in the technical and economic demonstration of the viability of the project, pilots developed prior to the operations can be monitored so that experience on the natural evolution of the system is obtained very early in the mining development. Another opportunity is to use the model in a global optimization perspective. Traditionally, mining operators decouple exploitation technique and remediation. However, particularly in ISR, some industrial choices during exploitation can have important impacts on the remediation costs: notably management of the hydraulic balance and confinement of the saline plume during exploitation. A global model, integrating exploitation and post-exploitation phase allows for a global evaluation, taking into account the specific source term resulting from production choices and their evolution under several remediation strategies.

Geochemical mechanisms involved in post acidic ISR mining

Chemicals of Concern. According to best environmental assessment practices, the chemicals of potential concerns (COPCs) for an ISR operation are defined according to the pre-mining groundwater conditions, the type of ISR process and the post-mining groundwater quality targets agreed upon by the mining company and the regulators. However, it is possible to assess the changes in chemical composition of the aquifer after mining operations, which will help in the remediation solutions to be deployed. Depending on the buffering properties of the aquifer and the water composition target, one may count on to attenuate the residual solution. Indeed, concentrations of the main chemicals of concern will tend to decrease due to dilution on the one hand, and geochemical reactions on the other hand. These contaminants will undergo, natural attenuation while migrating down gradient with regional groundwater.

Three main chemicals of potential concern are addressed specially in acidic ISR context: sulfate, acidity (pH) and uranium. At the end of the exploitation, residual SO_4-enriched solutions will have to be managed. In the mined part of the aquifer, the acidic plume remains in a mostly pH-buffers depleted aquifer. Finally, depending on the industrial optimization targets, residual uranium concentration can still be above the initial baseline. Moreover, acidification of the ore body also induces the increase in other elements concentrations. Concurrently to the dissolution of the uranium minerals, the oxidizing solutions lead to the dissolution of redox sensitive metals (transition metals) and radioactive decay products of uranium (mainly ^{226}Ra). Lastly, dissolution and exchange reactions at the surface of clay minerals participate to the increase of the salinity with Al, Ca, Mg, Na, K concentrations reaching locally the g/L range. The pregnant solutions should therefore be considered as a saline and oxidizing plume.

It is important to pinpoint the distribution of the different chemicals of potential concern according to the aqueous and the solid compartments. Historically, the solid fraction, e.g., neoformed minerals or elements sorbed on the surface of minerals exhibiting high sorption properties, were most of the time disregarded. This retroactively explains why most of remediation operations based on pump & treat strategies failed or were costly as only the aqueous fraction of the chemicals of concern is targeted. Injection of fresh waters tends to dissolve newly-formed minerals such as SO_4 bearing minerals (gypsum, alunite, jarosite…) known to be less soluble in the physico-chemical conditions prevailing during the production stage (high SO_4 concentrations). The desorption of metals from the surface of clay minerals also depends on the concentrations of other competitive cations which differ drastically between the fresh waters injected during the pump and treat process and the pregnant solutions remaining after the acidification stage. This is illustrated in Figure 8, where the distribution of uranium between solid, aqueous and sorbed species is simulated for a pilot at the end of production. Aqueous uranium represents only less than 20% of the total uranium stock while exchangeable uranium accounts for more than 40%. This stock is therefore available and may be desorbed when fresh waters are renewed during the pump and treat operations. This result clearly highlights the importance of the reactive transport modeling as a tool to help the mining operator in building a remediation strategy.

The geochemical modeling strategy should therefore be seen as assessing the buffering capacity of the aquifer submitted to a saline, oxidizing, and acidic plume. It is important to carefully analyze the main geochemical mechanisms involved in the fate of the different chemicals of potential concern. These data will help in the definition of the remediation strategy, from passive remediation, e.g., natural attenuation, to active remediation operations such as pump and treat or bioremediation (Descostes et al. 2014).

A special attention is put on the development of the geochemical model. Indeed, even if the area of interest is the same as for the production optimization, the key-geochemical reactions involved in the environmental impact can significantly differ. The model previously presented satisfies the constraints of the operator and focuses mainly on the oxidation of uranium and the acid consumption. In other words, geochemical reactions that may be considered as secondary for the production of uranium may become relevant when trying to assess the migration of chemicals of concern at larger time and space scale.

During the exploitation stage, an emphasis is made on the competition between kinetics of major reactions and flow rates in the order of several cubic meters per hour. Meanwhile,

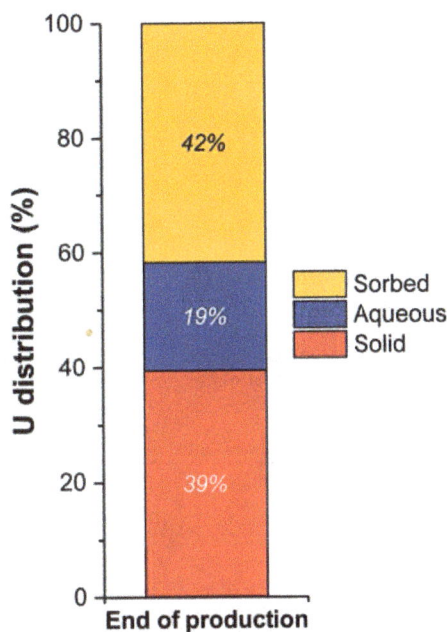

Figure 8. Distribution of U simulated at the end of production in a technological block. The U solid compartment stands for the remaining mineralization not extracted at the end of the operations (2.5% of the initial U).

when considering longer duration for the environmental post-mining remediation, migration of chemicals of concern is mainly driven by the very low natural flow of the aquifer (~ m/y). Less extreme ranges of pH, compared to production phase, also reduce the impact of kinetics.

One typical example is the consideration of sorption on clay minerals. Mineralogical characterization demonstrated the presence of clay minerals (smectite and kaolinite) within the host formation (Robin et al. 2015a). Their micromorphology and spatial distribution, localized mainly in the porosity of sandstone, allow them to be also involved in the regulation of the water chemistry. Indeed, they exhibit significant cationic sorption properties (Robin et al. 2015b, 2017) and low solubility in acidic conditions (Robin et al. 2016). During the acidification stage, smectite minerals occurring in the porosity sorb protons (Fig. 9); the amount of sorbed protons remains negligible when comparing the total amount of acid used during the lifetime of a technological block (several years). However, once the mining operations cease, proton exchange from sorption sites contributes to maintain moderately acidic conditions, between pH 4-5 depending on the slow renewal of cations from circulating fresh waters. This is well illustrated in Figure 10, where the pH measured in a production well was measured during and after an ISR test. In such case, the acidification was performed for approximately 200 days. The environmental survey performed after the ISR test indicates an increase of pH after several years, but values remain buffered at pH 4. Such trend can only be simulated if sorption is considered on clay minerals (de Boissezon et al. 2017). Finally, sorption, in these physico-chemical conditions, is also a key mechanism governing the mobility of U, ^{226}Ra and other metals (Reinoso-Maset and Ly 2016; Robin et al. 2017).

Another difference between the geochemical models developed for production and remediation, lies in the number of chemical elements to be considered and their respective range of concentration. In the production geochemical model, only major elements are taken into account with of course the reactivity of uranium. Concurrently, in the geochemical model devoted to the assessment of the environmental footprint, even if SO_4 and in a lesser extent acidity can be seen

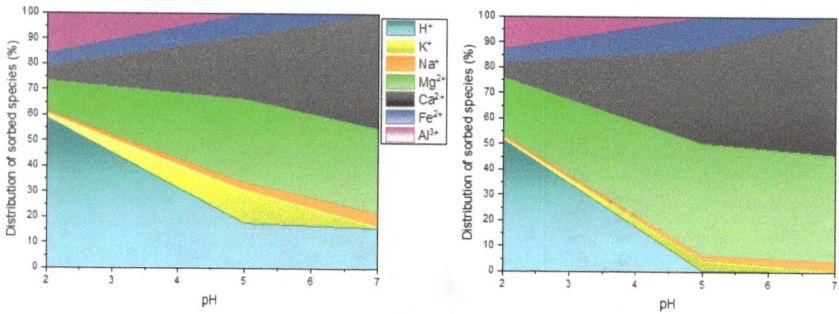

Figure 9. Illustration of the sorption of major cations (H^+, Na^+, K^+, Ca^{2+}, Mg^{2+}, Fe^{2+} and Al^{3+}) on smectite-type surface minerals such as beidellite (left) and montmorillonite (right) in ISR context using a multisite model. Protons can be significantly trapped for pH < 5.

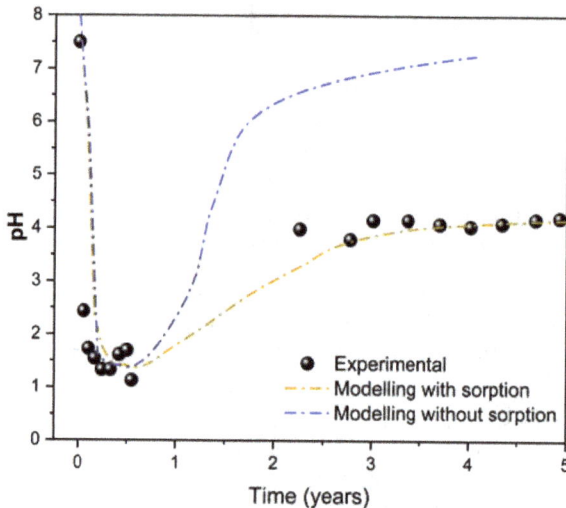

Figure 10. pH monitoring during and after an acidic ISR test. Environmental survey indicates an increase of pH although values remain buffered at pH 4. Reactive transport simulation with and without sorption on clay minerals highlights the importance of sorption on pH evolution.

obviously as major elements, the fate of residual uranium and other redox sensitive metals, has to be modeled as trace elements, or even ultra-trace elements in the special case of ^{238}U radioactive decay products (mainly ^{226}Ra). Hence, the long term evolution modeling of major elements constitutes a necessary preliminary step before attempting to model the fate of the trace elements.

Lastly, the major difference between the two modeling approaches is the time and space scales of interest. During the production stage, modeling focuses on short durations and on very restricted areas delimited by the localization of the injectors and producing wells (block of production scale). In such approach, local heterogeneities in the permeability distribution or in uraninite distribution may induce significant discrepancies in the predicted production curve. This is why a 3D description of the permeability and the mineralogy is unavoidable. This requirement is less clear for the assessment of the environmental footprint: it depends on the objectives of the simulation, more precisely on the timescale of interest.

The evaluation of the extension of the acidic plume in the vicinity of a technological block during the production phase (e.g., due to hydraulic imbalance) is controlled by flow, which derives from spatial heterogeneity. The environmental piezometer network localized in the vicinity of each technological block allows detecting very shortly such intrusions. For this application, the modeling strategy needs to rely on a 3D approach, with careful analysis of the block model; i.e., production modeling can be used directly in the optimization of the environmental footprint (see Fig. 2).

On the other hand, longer durations and distances are investigated for post-mining simulation, up to the kilometer scale or more. In these conditions, the vertical extension of the aquifer is negligible. Also, the knowledge of geometry outside the exploited area is reduced so that permeability and facies distribution are less constrained. Both reasons lower the need for 3D modeling. Furthermore, although this is not a justification in itself, the geochemical complexity of the system leads to CPU intensive models, so that 2D is easier to handle: e.g., a faster 2D approach can help test different remediation scenarios and assess their efficiency. This 2D/3D modeling strategy is consistent with a far field/near field modeling devoted to different aims and is essential when discussing environmental impact of mining activities with stakeholders (Fig. 11).

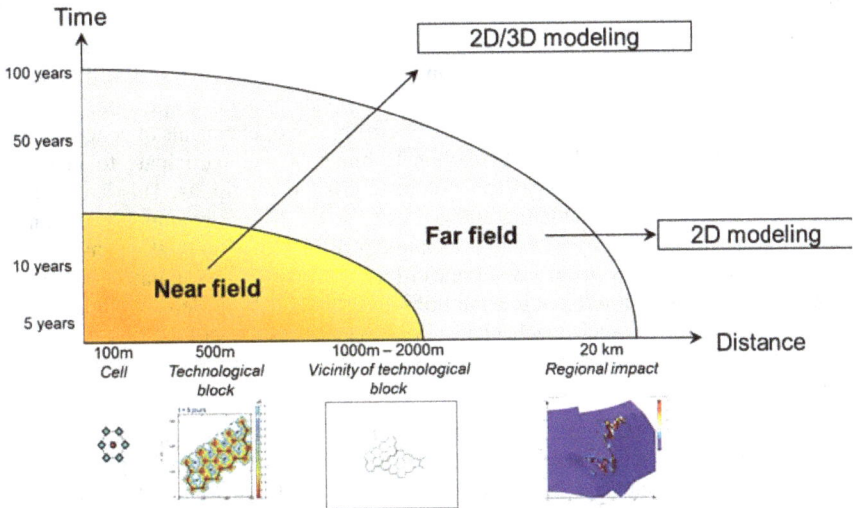

Figure 11. Modeling strategy devoted to the environmental footprint of ISR mining technique.

Main geochemical mechanisms involved in environmental impact assessment

Sulfate may be considered as the main chemical of concern regarding its chemistry. Remaining aqueous sulfate within the technological block porosity is expected in a first stage to be diluted in the aquifer but also to precipitate in SO_4-bearing minerals (gypsum, alunite, jarosite...) in the vicinity of the mined technological blocks.

Precipitation is favored by the general increase of concentrations in cationic species (Ca^{2+}, Al^{3+} and other cations) inherent in the acidification of the ore body. Indeed, as previously indicated, calcium is released in solution consecutively to the dissolution of calcite but also to exchange reactions on clay minerals according to the overall reactions:

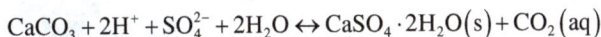

$$CaCO_3 + 2H^+ + SO_4^{2-} + 2H_2O \leftrightarrow CaSO_4 \cdot 2H_2O(s) + CO_2(aq)$$

$$X \equiv Ca^{2+} + 2H^+ + 2SO_4^{2-} + 2H_2O \leftrightarrow 2X \equiv H^+ + CaSO_4 \cdot 2H_2O(s)$$

where $X \equiv Ca^{2+}$ and $X \equiv H^+$ arbitrarily stand for the Ca^{2+} and H^+ sorbed at the surface of smectite type minerals (see Robin et al. 2015 and Reinoso-Maset and Ly 2014 for more details). Partial dissolution of alumino-silicate minerals provides sufficient aqueous aluminium to favor the precipitation of alunite type minerals. These reactions are expected to occur shortly after the closure of a technological block. Natural attenuation may be effective mainly thanks to aquifer dispersion, while SO_4-bearing minerals precipitation lowers aqueous SO_4 concentration during and right after the end of production. On the other hand, high solubility SO_4-bearing minerals are slowly dissolved with the renewal of the pore volume by fresh water, therefore maintaining sulfate at relatively high concentration.

Additionally to the dilution of sulfate downstream the geological block, its immobilization is expected according to the prevailing reducing conditions typical of the classical geochemical architecture of a roll front. Taking into account the regional hydraulic gradient, mitigation of the residual solutions will occur mainly downstream, where reducing conditions are found. These conditions are confirmed and explained by the presence of native sulfate-reducing and metal-reducing bacteria involved in the biogeochemical reactions regulating the water composition (Coral et al. 2018). (Bio)reduction of sulfate into sulfide is known to be favorable to the formation of insoluble metallic sulfide minerals such as pyrite ($FeS_2(s)$), therefore participating to the decrease of SO_4 concentration.

Acidic conditions inherited from the mining operations will slightly evolve to mid-acidic conditions (around pH 4-5) in a first stage. Such evolution is linked to the slow dissolution of clays minerals and exchange reactions at the surface of clay minerals as previously illustrated (Fig. 10). For longer duration (> several years), as the plume of chemicals of concern migrates downstream, additional dissolution of carbonate minerals will participate to increase the pH and to favor a slow return to pH value close to pristine conditions. This is illustrated in Figure 12, where predictive modeling allowed us to assess the pH potential evolution at the kilometer scale in the vicinity of several technological blocks. Note that the modeling was performed before mining operations started in this area; here, the simulation evaluates the natural evolution towards mid-acidic conditions and locally a hypothetical return to pristine pH values. Validation of such mechanisms was obtained from laboratory experiments and field observations as shown in Figure 10 and in the literature (Yazikov and Zabaznov 2002; Kayukov 2005; Jeuken et al. 2009; Schmitt et al. 2013; Dong et al. 2016).

Residual aqueous uranium is expected to be immobilized by precipitation into newly-formed minerals, stable in geochemical conditions imposed by the ISR process, and by sorption on clay minerals (see dedicated references for sorption above). These reactions are expected to occur shortly after the closure of a technological block as shown in Figure 13. For longer duration (e.g., in the range of years), a reestablishment of reducing conditions is

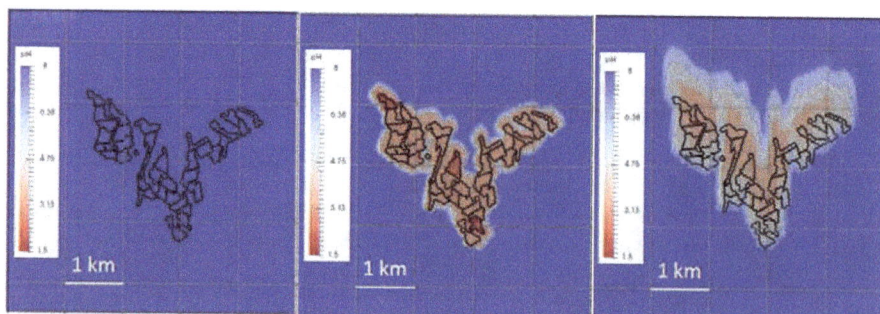

Figure 12. Predictive modeling of the pH at the kilometer scale in the vicinity of technological blocks: pH before exploitation (left), at the end of the mining operations (center) and several years after the end of exploitation in absence of active remediation operations.

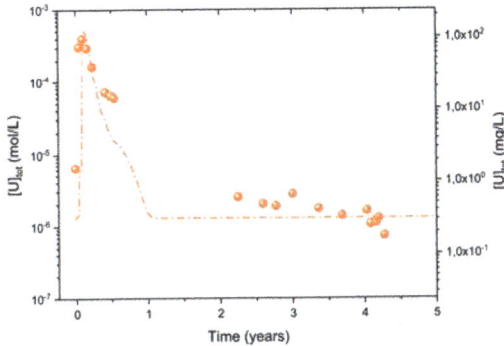

Figure 13. Modeling of the aqueous U concentration in a producer well during and after an acidic ISR test. Acidification lasted 200 days. Environmental survey indicates a rapid decrease of concentration some years in the range of the natural background.

reasonably expected in agreement with the natural redox conditions occurring downstream the ore body as mentioned for the discussion related to the fate of sulfate. These conditions will favor the reduction of uranium and its stabilization under the form of highly insoluble uranium (IV) minerals like uraninite ($UO_2(s)$). The same geochemical mechanisms are proposed to operate for other redox sensitive metals released during the ISR process. Reducing conditions are favorable to immobilize uranium and most metallic elements as their respective solubility is usually 10^3 to 10^4 times lower than in oxidizing conditions (Schmitt et al. 2013).

An emphasis is proposed here on the mobility of ^{226}Ra as one of the main radioactive decay products of ^{238}U. According to the amount of dissolved uranium, using a ^{238}U/^{226}Ra ratio of 1, one may expect identical aqueous concentrations, expressed in Bq/L. Field measurements show lower ^{226}Ra activities than expected (see Fig.14), indicating that more than 90% of the ^{226}Ra is trapped, even in acidic conditions. Through its specific activity and its relatively long half-life (1620 y), ^{226}Ra must be considered as ultra-trace element with typical concentrations in the range of ppb to ppt. Therefore, its mobility in natural context is principally governed by sorption at the

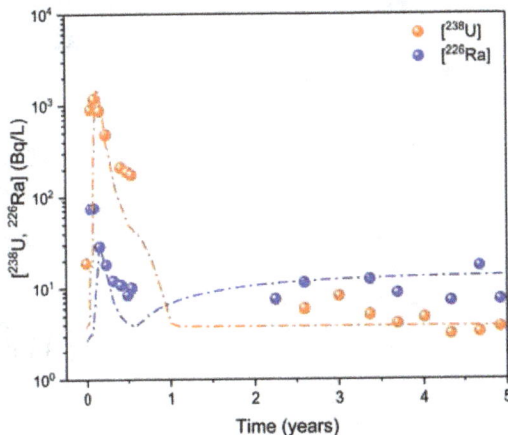

Figure 14. Simulation of aqueous ^{226}Ra activity in a producer well during and after an acidic ISR test. Acidification lasted 200 days. Environmental survey indicates ^{226}Ra retention according to the values expected from measured ^{238}U activities. Reactive transport simulation with coprecipitation of ^{226}Ra in barite and sorption on clay minerals highlights the importance of ancillary minerals on the fate of ^{226}Ra.

surface of ancillary minerals, natural organic matter and coprecipitation in SO_4-bearing minerals (see among others references dedicated to the mobility of ^{226}Ra in mining context: Lestini et al. 2013; Sajih et al. 2014; Reinoso-Maset and Ly 2016; Robin et al. 2017; Bordelet et al. 2018, Lestini et al. 2019). No proper Ra mineral phase has ever been found in natural environments.

The mobility of ^{226}Ra was successfully simulated in acidic ISR context taking into account three main ^{226}Ra hosting minerals: uraninite, which according to its age may be in secular equilibrium with ^{226}Ra (typically if the mineralization is older than 2 My), barite ($BaSO_4$), and montmorillonite. Such ancillary minerals were observed during and after the acidification stage illustrating their stability (Fig. 15). Also, their dispersion within the pore space allows maximum contact and reaction with circulating fluids. These minerals are known to be efficient traps for ^{226}Ra, especially barite (Curti et al. 2010; Brandt et al. 2015). Even at low concentration, with typical barite grade around 1 to 3 mg/kg, this mineral may contain up to 10^{11} Bq/L of ^{226}Ra, i.e., several order of magnitude higher than concentrations observed during acidic ISR.

Reactive transport simulations of ^{226}Ra were able to reproduce environmental survey data, only when considering both ^{226}Ra coprecipitation in barite and sorption on clay minerals:

$$(1-x)Ba^{2+} + x^{226}Ra + SO_4^{2-} \leftrightarrow Ba_{1-x}{}^{226}Ra_xSO_4(s)$$

$$(1-x)Ba^{2+} + x^{226}Ra + SO_4^{2-} \leftrightarrow Ba_{1-x}{}^{226}Ra_xSO_4(s)$$

where $X \equiv Ra^{2+}$ and $X \equiv H^+$ arbitrarily stand for $^{226}Ra^{2+}$ and H^+ sorbed at the surface of smectite type minerals (see Robin et al. 2017 for more details).

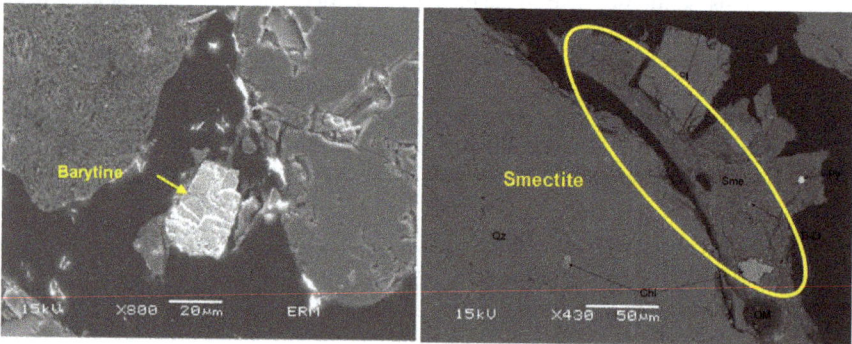

Figure 15. Occurrence of barite ($BaSO_4$) and montmorillonite (right) in ISR context observed before and after acidification suspected to play important role in the fate of ^{226}Ra.

Implications for an industrial remediation solution strategy. Depending on remediation targets, a dedicated strategy has to be developed by the mining operator. To assess its efficiency and have preliminary associated costs, reactive transport modeling appears also as an interesting complementary tool. Among the remediation strategies, one may distinguish on the one hand passive solutions (usually referred to as natural attenuation) and on the other hand active solutions (pump and treat, bioremediation...).

Natural attenuation relies mainly on the buffering properties of the aquifer to attenuate the oxidative, saline and mid-acidic plume inherited by the acidic ISR mining operations. This remediation strategy was deployed beyond the framework of acidic ISR context, mostly for contaminated aquifers (Christensen et al. 2004, Jørgensen et al. 2010). Among its advantages, one may cite the cost as it relies mainly on an environmental monitoring associated to predictive reactive transport modeling. Natural attenuation in acidic context is however scarcely described even if some examples are available in the scientific literature (Yazikov and Zabaznov 2002; Fyodorov 2002; Bakarzhiyev et al. 2004; IAEA 2005; Dong et al. 2016). Uncertainties on duration and spatial extension appear as the main disadvantages of such remediation strategy. Duration can be shortened using additional pumping operations, which led to the concept of enhanced natural attenuation (Park et al. 2007). In mildly cemented arkosic sandstone where smectite type minerals are found, pH mitigation may drive the duration of natural attenuation as previously discussed. Sulfate appears also as the other chemicals of concern, which governs the overall duration of mitigation, since precipitation of gypsum like minerals is limited in time and space (see further Fig.16): indeed, fairly soluble SO_4 minerals do not constitute a stable sink for sulfate. Concurrently to dispersion, reducing conditions usually observed downstream the natural hydraulic gradient of a roll front constitute the major natural mechanisms involved in the attenuation of SO_4.

Pump and treat like solutions encompass several technical processes, which are mostly mining site specific. It may be summarized according to several steps. First, residual solutions are pumped out and then treated on dedicated water treatment stations on surface (usually by reverse osmosis). Once the targeted chemical quality is reached, waters are reinjected in the leached part of the aquifer. Additional steps or sequences can be adapted with the injection of a reducing agent to prevent the oxidation of sensitive redox elements, and/or a mixing step with fresh waters (IAEA 2001, 2005). Return of experience and associated reactive transport simulations both highlight the limited efficiency of pump and treat based solutions (Anastasi and Williams 1984; Hall 2009). Indeed, these solutions focus only on the aqueous fraction of the chemicals of concern, which may be in specific case a minor part of the targeted elements (see Fig. 8). To avoid the dissolution of the residual trapped target elements, the remediation strategies should therefore also consider the overall geochemical equilibriums, in the reservoir including the history of production.

All these aspects should benefit from reactive transport modeling, with a special emphasis on the link between production and remediation stages. This is illustrated in Figure 16 where simulation of the environmental footprint was investigated, integrating the initial phase of production of several technological blocks. Simulation was achieved on a $20 \times 20 km^2$ area according to a sequential acidification of the technological blocks. Precipitation of gypsum is predicted on the edges of each technological block, which contributes to limit the spreading of SO_4.

Recently, alternative remediation strategies have been proposed. They were developed initially and generally out of the scope of ISR. They rely on the catalysis of natural geochemical reactions to reduce the uncertainty on the duration needed for natural attenuation (Jove Colon et al. 2000). Such approach requires a full understanding of the geochemical equilibriums prevailing before the exploitation of an ISR mine. Amongst the main geochemical processes naturally regulating the aqueous concentrations of major and trace elements in the aquifer before ISR mining operations, microbial activity plays an essential role in establishing the contrasting redox conditions through the roll front. This is especially the case for the reducing conditions observed within the mineralized zone and downstream, where sulfato-reducing bacteria are present (Coral et al. 2018). Taking into account the low solubility of uranium and other redox sensitive metals in reducing conditions, a return to reducing conditions after mining operations can be targeted using relevant bacteria. Hence, depending on the use of native bacteria or exogenous bacteria, reducing conditions can potentially be reached in shorter durations than the ones related to natural attenuation. Concurrently, acidity and SO_4 are also targeted. Such alternative remediation solutions fall into a bioremediation strategy with on the one hand, biostimulation and on the other hand, bioaugmentation. Frequently used for organic compounds, several industrial tests have been launched on uranium-polluted aquifers and (conventional) uranium mining sites (Wu et al. 2006a,b; Groudev et al. 2008, 2010; Yabusaki et al. 2014; Long et al. 2015). These bioremediation techniques are still under study, mostly in laboratory or small-scale pilots, with no industrial scale demonstration on ISR sites. Once more, reactive transport modeling can be

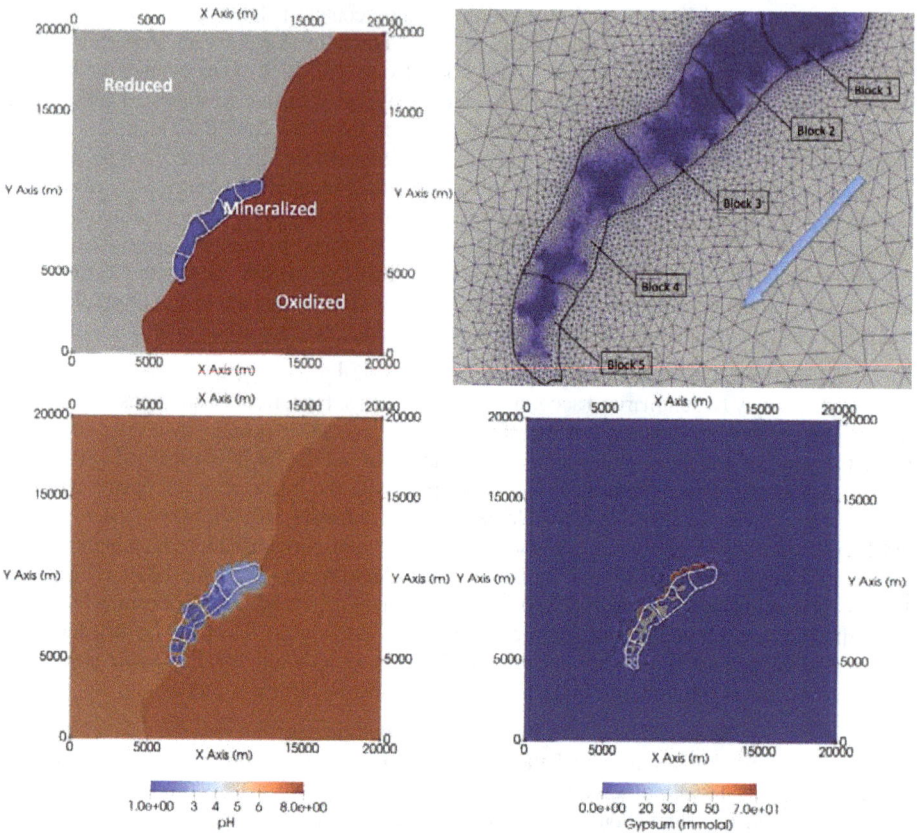

Figure 16. Simulation coupling production stage of 6 technological blocks (sequence of production follows the blue arrow) and associated environmental footprint 6 years after acidification. Acidic to mid-acidic conditions are still observed within the technological blocks while gypsum precipitation is predicted in the vicinity of the leached area downstream the natural hydraulic gradient. Area of interest and associated mesh are also given.

used in the understanding and interpretation of bioremediation tests from the laboratory scale, in column experiments for instance, up to the field test (see for instance Yabuzaki et al. 2007).

Industrial use of reactive transport modeling for post-mining. All these examples can be summarized in a general remediation strategy adapted to acidic ISR where reactive transport modeling is a key tool to assess the efficiency of the retained remediation solution. The specific remediation strategy (among others natural attenuation, pump and treat and bioremediation) depends on the local geology, mining technology, and environmental targets defined by local regulation.

1. Prior to the closure of any geological block, a predictive modeling of the remediation solution is provided. Such model should merge, both the production and remediation stages in order to take into account the production history. Economical aspects are considered according to the design of the environmental survey network and the duration of monitoring.

2. Deployment of a dedicated environmental survey, and monitoring of the efficiency of the remediation solution over a reasonable period of time to evaluate if predictive modeling agrees with observations.

3. If there is a good consistency between the modeling and the environmental monitoring then maintain the environmental survey with an adapted chronic (decreasing frequency).

4. If no agreement between modeling and observations is found, then the model should be updated (including using new data from the monitoring phase), another monitoring phase should be launched, and eventually the remediation solution should be adapted.

LESSONS LEARNT FOR INDUSTRIALIZATION

Several applications above illustrated how an academic, versatile, reactive transport code like HYTEC can be used in an industrial context. The fast pace imposed by operations, need for quick results, imposed the use of a specific graphical interface, specially devised for this application: HYSR. The code is presently tested, in operation conditions, on the Katco ISR mine.

An important lesson from this project is the benefit of co-construction. After the initial identification of an industrial problem and the possibility to address it with reactive transport tools, a close cooperation was built between an academic team, with the knowledge of the tool and expertise to optimize it, and an industrial team, with the knowledge of industrial context and issues and the driver to improve their simulation capacity. Also, access to an extensive mass of data was key in the early stage of the project to build, calibrate, and optimize the model. As the project moved from R&D department to technical department and finally operations, it became clear the code had to evolve from a versatile, keyword-based input files to a more intuitive interface. GUI is of course essential, but a key was to calibrate the interface to the needs of the engineers: the GUI is basically devised so that ISR engineers interact with the code using the frame of mind of their trade: inputs organize various data of the project and proposed exploitations scenarios, outputs are displayed following the need of production engineers including their financial conversion for comparison purpose.

ACKNOWLEDGMENT

This work was funded by the industrial chair "*in situ* recovery of uranium", a program supported by the French National Research Agency and Orano. The authors also wish to thank Katco and the Mongolian project team, for fruitful discussions and access to years of operation data. Finally, the authors wish to thank many people involved in the simulation project, particularly Gwenaële Petit, Hélène de Boissezon, Tatiana Okhulkova, and Marie Mazurier, and more generally in the ISR project, Valérie Langlais, Nicolas Fiet and Anthony Le Beux.

REFERENCES

Ahrens J, Geveci B, Law C (2005) ParaView: an end-user tool for large-data visualization. *In:* Visualization Handbook. Hansen CD, Johnson CR (eds), Butterworth-Heinemann, p 717–731

Anastasi FS, Williams RE (1984) Aquifer restoration at uranium in situ leach sites. Int J Mine Water 34:29–37

Arora B, Dwivedi D, Hubbard SS, Steefel CI, Williams KH (2016) Identifying geochemical hot moments and their controls on a contaminated river floodplain system using wavelet and entropy approaches. Environ Modell Software 85:27–41

Audigane P, Gaus I, Czernichowski-Lauriol I, Pruess K, Xu T (2007) Two-dimensional reactive transport modeling of CO_2 injection in a saline aquifer at the Sleipner site, North Sea. Am J Sci 307:974–1008

Bakarzhiyev AC, Bakarzhiyev Y, Babak MI, Makarenko MM (2004) Perspective of exploitation of new sandstone type deposits by ISL method and environmental impact from uranium deposits mined out by in situ leaching in Ukraine. *In:* Recent Developments in Uranium Resources and Production with Emphasis on In Situ Leach Mining. Technical report IAEA-TECDOC-1396

Ben Simon R (2011) Tests de lessivage acide de minerais d'uranium et modélisations géochimiques des réactions; application à la recuperation minière *in situ* (ISR). PhD Dissertation, MINES ParisTech, Fontainebleau, France

Ben Simon R, Thiry M, Schmitt JM, Lagneau V, Langlais V, Bélières M (2014) Kinetic reactive transport modelling of column tests for uranium In Situ Recovery (ISR) mining. Appl Geochem 51:116–129

Bhargava SK, Ram R, Pownceby M, Grocott S, Ring B, Tardio J, Jones L (2015) A review of acid leaching of uraninite. Hydrometallurgy 151:10–24

Bildstein O, Claret F, Frugier P (2019) RTM for waste repositories. Rev Mineral Geochem 85:419–457

Bonnaud E, Lagneau V, Regnault O, Fiet N (2014) Reactive transport simulation applied on uranium ISR: effect of the density driven flow. *In:* UMH VII – Uranium Mining and Hydrology, p 699–704, Sept 21–25, Freiberg, Germany

Bordelet G, Beaucaire C, Descostes M, Phrommavanh V (2018) Chemical reactivity of natural peat towards U and Ra. Chemosphere 202:651–660

Brandt F, Curti E, Klinkenberg M, Rozov K, Bosbach D (2015) Replacement of barite by a $(Ba,Ra)SO_4$ solid solution at close-to-equilibrium conditions: A combined experimental and theoretical study. Geochim Cosmochim Acta 155:1–15

Cama J, Soler JM, Ayora C (2019) Acid water–rock–cement interaction and multicomponent reactive transport modeling. Rev Mineral Geochem 85:459–498

Christensen O, Cassiani G, Diggle P, Ribeiro Jr P, Andreotti G (2004) Statistical estimation of the relative efficiency of natural attenuation mechanisms in contaminated aquifers. Stochastic Environ Res Risk Assess 18:339–350

Clay J (2015) Groundwater restoration research at an In Situ Recovery (ISR) uranium mine. National Mining Association (NMA) Uranium Recovery Workshop 2015, Denver, Colorado, USA.

Coral T, Descostes M, de Boissezon H, Bernier-Latmani R, de Alencastro F, Rossi P (2018) Microbial communities associated with uranium in-situ recovery mining process are related to acid mine drainage assemblages. Sci Total Environ 628–629:26–35

Curti E, Fujiwara K, Iijima K, Tits J, Cuesta C, Kitamura A, Glaus MA, Müller W (2010) Radium uptake during barite recrystallization at 23 ± 2°C as a function of solution composition: An experimental Ba-133 and Ra-226 tracer study. Geochim Cosmochim Acta 74:3553–3570

Dahlkamp FJ (1993) Uranium Ore Deposits. Springer, Berlin

Dangelmayr MA, Reimus PW, Wasserman NL, Punsal JJ, Johnson RH, Clay JT, Stone JJ (2017) Laboratory column experiments and transport modeling to evaluate retardation of uranium in an aquifer downgradient of a uranium in-situ recovery site. Appl Geochem 80:1–13

de Boissezon H, Levy L, Jakymiw C, Descostes M (2017) Remediation of a uranium acidic In Situ Recovery mine: from laboratory to field experiments using reactive transport modelling. Migration 16th International Conference on the Chemistry and Migration Behaviour of Actinides and Fission Products in the Geosphere, Barcelona, Spain

Descostes M, de Boissezon H, Fiet N (2014) R&D studies devoted to enhance the natural attenuation as solution for the post-mining remediation of ISR operation. Uranium and Mining Hydrogeology VII, Freiberg, Germany.

Dong Y, Xie Y, Li G, Zhang J (2016) Efficient natural attenuation of acidic contaminants in a confined aquifer. Environ Earth Sci 75:595–602

Estublier A, Fornel A, Parra T, Deflandre JP (2013) Sensitivity study of the reactive transport model for co_2 injection into the Utsira saline formation using 3d fluid flow model history matched with 4D seismic. Energy Procedia 37:3574–3582

Fyodorov GV (2002) Uranium production and the environment in Kazakhstan. *In:* The Uranium Production Cycle and the Environment. IAEA C&S Papers Series 10/P, 1p 91–198

Hall S (2009) Groundwater Restoration at Uranium In-situ Recovery Mines, South Texas coastal plain. Technical report USGS 2009–1143

Hammond GE, Lichtner PC (2010) Field-scale model for the natural attenuation of uranium at the Hanford 300 Area using high-performance computing. Water Resour Res 46:W09527

Hammond GE, Lichtner PC, Rockhold ML (2011) Stochastic simulation of uranium migration at the Hanford 300 Area. J Contam Hydrol 120–121:115–128

Heili W (2018) Key lessons learnt from the application of ISR to uranium. ALTA 2018 ISR Symposium Proceedings, Perth, Australia

IAEA (2001) Manual of acid *in situ* leach uranium mining technology. IAEA-TECDOC-1239, IAEA, Vienna, Austria

IAEA (2005) Guidebook on environmental impact assessment for *in situ* leach mining projects. IAEA-TECDOC-1428, IAEA, Vienna, Austria

IAEA (2016) *In situ* leach uranium mining: an overview of operations. IAEA Technical report NF-T-1.4, IAEA, Vienna, Austria

Jeuken B, Kalka H, Maerten H, Nicolai J, Woods P (2009) Uranium ISR Mine Closure – General Concepts and Model-based Simulation of Natural Attenuation for South Australian Mine Sites. URAM-2009, International Symposium on Uranium Raw Material for the Nuclear Fuel Cycle, IAEA, Vienna, Austria

Johnson RH, Tutu H (2016) Predictive reactive transport modeling at a proposed uranium *in situ* recovery site with a general data collection guide. Mine Water Environ 35:369–380

Jørgensen KS, Salminen JM, Björklöf K (2010) Monitored natural attenuation. *In:* Cummings S (eds) Bioremediation. Methods in Molecular Biology (Methods and Protocols) 599:217–233. Humana Press

Kalka H, Märten H, Kahnt R (2006) Dynamical models for uranium leaching – production and remediation cases. *In:* Uranium in the Environment. Merkel BJ, Hasche-Berger A (eds). Springer, Berlin, Heidelberg, p 235–245

Kayukov P (2005) APPENDIX VI. Kanzhugan environmental rehabilitation after closure. *In:* Guidebook on environmental impact assessment for in situ leach mining projects. IAEA Technical report IAEA-TECDOC-1428:101–117, IAEA, Vienna, Austria.

Kidd S (2009) Uranium mining – what method works best. Nucl Eng Int 54:12–13

Kyser K (2014) Uranium ore deposits. *In:* Treatise of Geochemistry, Geochemistry of Mineral Deposits, second ed, Vol 13. Scott, S.D. (ed). Elsevier, Amsterdam, p 489–513

Lagneau V, van der Lee J (2010) Operator-splitting-based reactive transport models in strong feedback of porosity change: The contribution of analytical solutions for accuracy validation and estimator improvement. J Contamin Hydrol 12:118–129

Lagneau V, Regnault O, Okhulkova T, Le Beux A (2018) Predictive simulation and optimization of uranium in situ recovery using 3D reactive transport simulation at the block scale. ALTA 2018 ISR Symposium Proceedings, Perth, Australia

Langanay J, Romary T, Lagneau V, Petit G (2018) Scenario Reduction and dimension reduction in uranium ore deposit mining simulations by In Situ Recovery. 12th conference on geostatistics for environmental applications, July 2018, Belfast, IR

Le Beux A, Regnault O, Joubert G, Garcia Vasquez C, Fiet N (2018) Optimization of low-grade uranium *in situ* leaching in a roll-front deposit, Kazakhstan. ALTA 2018 ISR Symposium Proceedings, Perth, Australia

Lestini L, Beaucaire C, Vercouter T, Descostes M (2013) Radium uptake by recrystallized gypsum: An incorporation study. Procedia Earth Planet Sci 7:479–482

Lestini L, Beaucaire C, Vercouter T, Ballini M, Descostes M (2019) Role of trace elements in the 226-Radium incorporation in sulfate minerals (gypsum and celestite). ACS Earth Space Chem 3:295–304

Li L, Steefel CI, Kowalsky MB, Englert A, Hubbard SS (2010) Effects of physical and geochemical heterogeneities on mineral transformation and biomass accumulation during biostimulation experiments at Rifle, Colorado. J Contamin Hydrol 112:45–63

Mangeret A, De Windt L, Crançon P (2012) Reactive transport modelling of groundwater chemistry in a chalk aquifer at the watershed scale. J Contamin Hydrol 138–139:60–74

Märten H, Marsland-Smith A, Ross J, Haschke M, Kalka H, Schubert J (2013) From advanced geophysical surveying and borehole logging to optimized uranium ISR technology – fiction or reality? The AusIMM international mining conference, 11–12 June 2013, Darwin, Australia

Märten H (2006) Environmental Management and Optimization of In-situ-Leaching at Beverley. *In:* Uranium in the Environment. Merkel BJ, Hasche-Berger A (ed) Springer, Berlin, p 537–546

Metschies T, Jenk U (2011) Implementation of a modeling concept to predict hydraulic and geochemical Conditions during flooding of a deep mine. *In:* The New Uranium Mining Boom. Merkel B, Schipek M (ed) Springer, Berlin, Heidelberg

Molson J, Aubertin M, Bussière B (2012) Reactive transport modelling of acid mine drainage within discretely fractured porous media: Plume evolution from a surface source zone. Environ Modell Software 38:259–270

Mudd GM (2001) Critical review of acid in-situ leach uranium mining: 1-USA and Australia. Environ. Geol. 41:390–403

Munara A (2012) Formation des gisements d'uranium de type roll: approche minéralogique et géochimique du gisement uranifère de Muyunkum (Basin de Chu-Sarysu, Kazakhstan). PhD Dissertation, Université de Lorraine, Nancy

Neuner M and Fawcett S (2015) Reactive Transport Model of the Carbonate-Evaporite Elk Point Group Underlying the Athabasca Oil Sands. 10th international conference on Acid Rock Drainage & IMWA annual conference, Santiago, Chile

Nguyen VV, Pinder GF, Gray WG, Botha JF (1983) Numerical simulation of uranium in-situ mining. Chem Eng Sci 38:1855–1862

Palandri JL, Kharaka YK (2004) A compilation of rate parameters of water-mineral interaction kinetics for application to geochemical modeling. U.S. Geol Surv Open File Report 2004–1068, 64p

Park DK, Ko NY, Lee KK (2007) Optimal groundwater remediation design considering effects of natural attenuation processes: pumping strategy with enhanced-natural-attenuation. Geosci J 11:377–385

Petrov N (1998) Epigenetic stratified-infiltration uranium deposits of Kazakhstan (in Russian). Geol Kazakhstan, 2:22–39

Regnault O, Lagneau V, Fiet N, Langlais V (2012) Reactive transport simulation of Uranium ISL at the block scale: a tool for testing designs and operation scenarios. IAEA Technical Meeting on Optimization of In Situ Leach (ISL) Uranium Mining Technology, 15–18 April 2013, Vienna, Austria.

Regnault O, Lagneau V, Fiet N (2014) 3D reactive transport simulations of uranium *in situ* leaching: forecast and process optimization. Uranium Mining and Hydrogeology, 21–25 Sept 2014, Freiberg, Germany

Regnault O, Petit G, Lagneau V, Le Beux A, Fiet N (2017) 3D reactive transport simulation of uranium in situ leaching. The AusIMM International Uranium Conference, 6–7 June 2017, Adelaide, Australia

Reinoso-Maset E, Ly J (2014) Study of major ions sorption equilibria to characterise the ion exchange properties of kaolinite. J Chem Eng Data 59:4000–4009

Reinoso-Maset E, Ly J (2016). Study of uranium(VI) and radium(II) sorption at trace level on kaolinite using a multi-site ion exchange model. J Environ Radioact 157:136–148

Robin V, Hebert B, Beaufort D, Sardini P, Tertre E, Regnault O, Descostes M (2015a) Occurrence of authigenic beidellite in the Eocene transitional sandy sediments of the Chu-Saryssu basin (South-Central Kazakhstan). Sediment Geol 321:39–48

Robin V, Tertre E, Beaufort D, Regnault O, Sardini P, Descostes M (2015b) Ion exchange reactions of major inorganic cations (H^+, Na^+, Ca^{2+}, Mg^{2+} and K^+) on beidellite: experimental results and new thermodynamic database. Towards a better prediction of contaminant mobility in natural environments. Appl Geochem 59:74–84

Robin V, Tertre E, Regnault O, Descostes M (2016) Dissolution of beidellite in acidic solutions: ion exchange reactions and effect of crystal chemistry on smectite reactivity. Geochim Cosmochim Acta 180:97–108

Robin V, Tertre E, Beaucaire C, Descostes M, Regnault O (2017) Experimental data and assessment of predictive modeling for radium ion exchange on swelling clay minerals with a tetrahedral charge. Appl Geochem 85:1–9

Saínz-Garcia A, Abarca E, Rubi V, Grandia F (2017) Assessment of feasible strategies for seasonal underground hydrogen storage in a saline aquifer. Int J Hydrogen Energy 42:1665–16666

Sajih M, Bryan ND, Vaughan DJ, Descostes M, Phrommavanh V, Nos J, Morris K (2014) Adsorption of Radium and Barium on Goethite and Ferrihydrite: A kinetic and surface complexation modelling study. Geochim Cosmochim Acta 146:150–163

Saunders JA, Pivetz BE, Voorhies N, Wilkin RT (2016) Potential aquifer vulnerability in regions down-gradient from uranium *in situ* recovery (ISR) sites. J Environ Manage 183:67–83

Schmitt JM, Descostes M, Polak C (2013) L'exploitation par ISR des gisements d'uranium de type Roll front : des interactions multiples avec les eaux souterraines. Géologues 179:54–58

Sodov A, Gaskova O, Vladimirov A, Battushig A, Moroz E (2016) Spatial distribution of uranium and metalloids in groundwater near sandstone-type uranium deposits, Southern Mongolia. Geochem J 50:393–401.

Steefel CI, Appelo CAJ, Arora B, Jacques D, Kalbacher T, Kolditz O, Lagneau V, Lichtner PC, Mayer KU, Meeussen JCL, Molins S, Moulton D, Shao H, Šimůnek J, Spycher N, Yabusaki SB, Yeh GT (2015) Reactive transport codes for subsurface environmental simulation. Comput Geosci 19:445–478

van der Lee J, de Windt L, Lagneau V, Goblet P (2003) Module oriented modeling of reactive transport with HYTEC. Comput Geosci 29:265–275

Xiao Y, Jones GD (2006) Reactive transport modeling of carbonate and siliciclastic diagenesis and reservoir quality prediction. Soc Petrol Eng 101669

Yazikov VG, Zabaznov VU (2002) Experience with restoration of ore-bearing aquifers after in situ leach uranium mining. *In:* Proceedings of The Uranium Production Cycle and the Environment Conference, IAEA, 2nd-6th October, Vienna, Austria. IAEA C&S Papers Series 10/P

RiMG Series

HISTORY OF RiMG

Volumes 1–38 were published as *"Reviews in Mineralogy"* (ISSN 0275-0279). Volumes 1-6 originally appared as *"Short Course Notes"* (no ISSN). The name was changed to *"Reviews in Mineralogy & Geochemistry"* (RiMG) (ISSN 1529-6466) starting with Volume 39. Paul Ribbe was sole editor for volumes 1–41. He was joined by Jodi Rosso as series editor for volumes 42–53 in the RiMG series submitted through the Geochemistry Society. With his retirement, Jodi Rosso became sole editor for volumes 54–79 in the RiMG series. With Jodi Rosso's move to Executive Editor of Elements magazine, Ian Swainson became Series Editor starting with volume 80.

HOW TO PUBLISH IN RiMG

RiMG volumes are based on topics that have been proposed and appoved by the MSA Council or Geochemical Society Board of Directors. If you have an idea for a future RiMG volume, or a short course accompanied by a RiMG volume, you should read the Short Course Guide which describes how to develop and propose a topic for consideration for either case. Proposals should be submitted to the Short Course Committee (http://www.minsocam.org/ msa/SC/SCCommittee.html). Contributions to an appoved volume are by invitation only.

A listing of the previous volume numbers and their volume editors is below. Selecting the titles at http://www.minsocam.org/msa/RIM/index2.html gives you to a detailed description of each volume, the table of contents, and any errata or supplementary material

Previous volumes can be ordered from https://msa.minsocam.org/orders.html.

Volume 57: *Micro- and Mesoporous Mineral Phases*

2005   ISBN 0-939950-69-3;
ISBN13 978-0-939950-69-0   G Ferraris, S Merlino   i-xiii + 448 pp

Volume 56: *Epidotes*

2004   ISBN 0-939950-68-5;
ISBN13 978-0-939950-68-3   A Liebscher, G Franz   i-xviii + 628 pp

Volume 55: *Geochemistry of Non-Traditional Stable Isotopes*

2004   ISBN 0-939950-67-7;
ISBN13 978-0-939950-67-6   CM Johnson, BL Beard, F Albarede   i-xvi + 454 pp

Volume 54: *Biomineralization*

2003   ISBN 0-939950-66-9;
ISBN13 978-0-939950-66-9   PM Dove, JJ De Yoreo, S Weiner   i-xiv + 381 pp

Volume 53: *Zircon*

2003   ISBN 0-939950-65-0;
ISBN13 978-0-939950-65-2   JM Hanchar, PWO Hoskin   i-xviii + 500 pp

Volume 52: *Uranium-Series Geochemistry*

2003   ISBN 0-939950-64-2;
ISBN13 978-0-939950-64-5   B Bourdon, GM Henderson, CC Lundstrom, SP Turner,   i-xx + 656 pp

Volume 51: *Plastic Deformation of Minerals and Rocks*

2002   ISBN 0-939950-63-4;
ISBN13 978-0-939950-63-8   S Karato, H-R Wenk   i-xiv + 420 pp

Volume 50: *Beryllium Mineralogy, Petrology, and Geochemistry*

2002   ISBN 0-939950-62-6;
ISBN13 978-0-939950-62-1   E Grew   i-xii + 691 pp

Volume 49: *Appications of Synchrotron Radiation in Low-Temperature Geochemistry and Environmental Science*

2002   ISBN 0-939950-61-8;
ISBN13 978-0-939950-61-4   PA Fenter, ML Rivers, NC Sturchio, SR Sutton,   i-xxii + 579 pp

Volume 48: *Phosphates Geochemical, Geobiological, and Materials Importance*

2002   ISBN 0-939950-60-X;
ISBN13 978-0-939950-60-7   ML Kohn, J Rakovan, JM Hughes   i-xvi + 742 pp

Volume 47: *Nobel Gases in Geochemistry and Cosmochemistry*

2002   ISBN 0-939950-59-6;
ISBN13 978-0-939950-59-1   DP Porcelli, CJ Ballentine, R Wieler   i-xviii + 844 pp

Volume 46: *Micas: Crystal Chemistry & Metamorphic Petrology*

2002   ISBN 0-939950-58-8;
ISBN13 978-0-939950-58-4   A Mottana, FP Sassi, JB Thompson, Jr., S Guggenheim   i-xiv + 499 pp

Volume 45: *Naturnal Zeolites: Occurrence, Properties, Appications*

2001   ISBN 0-939950-57-X;
ISBN13 978-0-939950-57-7   DL Bish, DW Ming   i-xiv + 654 pp

Volume 44: *Nanoparticles and the Environment*

2001   ISBN 0-939950-56-1;
ISBN13 978-0-939950-56-0   JF Banfield ,A Navrotsky   i-xiv + 349 pp

Volume 43: *Stable Isotope Geochemistry*

2001   ISBN 0-939950-55-3;
ISBN13 978-0-939950-55-3   JW Valley, D Cole   i-x11 + 531 pp

Volume 42: *Molecular Modeling Theory: Applications in the Geosciences*

2001   ISBN 0-939950-54-5;
ISBN13 978-0-939950-54-6   RT Cygan, JB Kubicki   i-v +662 pp

Volume 41: *High-Temperature and High-Pressure Crystal Chemistry*

2001   ISBN 0-939950-53-7;   RM Hazen, RT Downs   i-viii + 596 pp
        ISBN13 978-0-939950-53-9

Volume 40: *Sulfate Minerals - Crystallography, Geochemistry, and Environmental Significance*

2000   ISBN 0-939950-52-9;   CN Alpers, JL Jambor, DK Nordstrom   i-xii + 608 pp
        ISBN13 978-0-939950-52-2

Volume 39: *Transformation Processes in Minerals*

2000   ISBN 0-939950-51-0;   SAT Redfern, MA Carpenter,   i-x + 361 pp
        ISBN13 978-0-939950-51-5

Volume 38: *Uranium: Mineralogy, Geochemistry and the Environment*

1999   ISBN 0-939950-50-2;   PC Burns, R Finch   i-xvi + 679 pp
        ISBN13 78-0-939950-50-8

Volume 37: *Ultrahigh-Pressure Mineralogy: Physics and Chemistry of the Earth's Deep Interior*

1998   ISBN 0-939950-48-0;   R Hemley   i-xx + 671 pp
        ISBN13 978-0-939950-48-5

Volume 36: *Planetary Materials*

1998   ISBN 0-939950-46-4;   JJ Papike   i-xx + 864 pp
        ISBN13 978-0-939950-46-1

Volume 35: *Geomicrobiology: Interaction Between Microbes and Minerals*

1997   ISBN 0-939950-45-6;   JF Banfield ,KH Nealson   i-xvi + 448 pp
        ISBN13 978-0-939950-45-4

Volume 34: *Reactive Transport in Porous Media*

1997   ISBN 0-939950-45-6;   PC Lichtner, CI Steefel, EH Oelkers   i-xiv + 438 pp
        ISBN13 978-0-939950-45-4

Volume 33: *Boron Mineralogy, Petrology and Geochemistry*

1996   ISBN 0-939950-41-3;   LM Anovitz, ES Grew   i-xx + 864 pp
        ISBN13 978-0-939950-41-6

Volume 32: *Structure, Dynamics and Properties of Silicate Melts*

1995   ISBN 0-939950-39-1;   JF Stebbins, PF McMillan, DB Dingwell i-xvi + 616 pp
        ISBN13 978-0-939950-39-3

Volume 31: *Chemical Weathering Rates of Silicate Minerals*

1995   SBN 0-939950-38-3;   AF White, SL Brantley   i-xvi + 583 pp
        ISBN13 978-0-939950-38-6

Volume 30: *Volatiles in Magmas*

1994   ISBN 0-939950-36-7;   MR Carroll, JR Holloway   i-xviii + 517 pp
        ISBN13 978-0-939950-36-2

Volume 29: *Silica: Physical Behavior, Geochemistry and Materials Appcations*

1994   ISBN 0-939950-35-9;   PJ Heaney, CT Prewitt, GV Gibbs   i-xviii + 606 pp
        ISBN13 978-0-939950-35-5

Volume 28: *Health Effects of Mineral Dusts*

1993   ISBN 0-939950-33-2;   GD Guthrie, Jr., BT Mossman   i-xvi + 584 pp
        ISBN13 978-0-939950-33-1

Volume 27: *Minerals and Reactions at the Atomic Scale: Transmission Electron Microscopy*

1992   ISBN 0-939950-32-4;   PR Buseck   i-xvi + 516 pp
        ISBN13 978-0-939950-32-4

Volume 26: *Contact Metamorphism*

1991   ISBN 0-939950-31-6;   DM Kerrick   i-xvi + 672 pp
        ISBN13 978-0-939950-31-7

Volume 25: *Oxide Minerals:Petrologic and Magnetic Significance*

1991 ISBN 0-939950-30-8; DH Lindsley i-xiv + 509 pp
ISBN13 978-0-939950-30-0

Volume 24: *Modern Methods of Igneous Petrology: Understanding Magmatic Processes*

1990 ISBN 0-939950-29-4; J Nicholls, JK Russell i-viii + 314 pp
ISBN13 978-0-939950-29-4

Volume 23: *Mineral–Water Interface Geochemistry*

1990 ISBN 0-939950-28-6; MF Hochella, Jr., AF White i-xvi + 603 pp
ISBN13 978-0-939950-28-7

Volume 22: *The Al₂SiO₅ Polymorphs*

1990 ISBN 0-939950-27-8; DM Kerrick i-xii + 406 pp
ISBN13 978-0-939950-27-0

Volume 21: *Geochemistry and Mineralogy of Rare Earth Elements*

1989 ISBN 0-939950-25-1; BR Lipin, GA McKay i-x + 348 pp
ISBN13 978-0-939950-25-6

Volume 20: *Modern Powder Diffraction*

1989 ISBN 0-939950-24-3; DL Bish, JE Post i-xii + 369 pp
ISBN13 978-0-939950-24-9

Volume 19: *Hydrous Phyllosilicates (exclusive of micas)*

1988 ISBN 0-939950-23-5; SW Bailey, i-xiii + 725 pp
ISBN13 978-0-939950-23-2

Volume 18: *Spectroscopic Methods in Mineralogy and Geology*

1988 ISBN 0-939950-22-7; FC Hawthorne i-xvi + 512 pp
ISBN13 978-0-939950-22-5

Volume 17: *Thermodynamic Modeling of Geological Materials: Minerals, Fluids and Melts*

1987 ISBN 0-939950-21-9; ISE Carmichael, HP Eugster i-xiv + 499 pp
ISBN13 978-0-939950-21-8

Volume 16: *Stable Isotopes in High Temperature Geological Processes*

1986 ISBN 0-939950-20-0; JW Valley, HP Taylor, Jr., JR O'Neil i-xvi + 570 pp
ISBN13 978-0-939950-20-1

Volume 15: *Mathematical Crystallography*

1985 ISBN 0-939950-19-7; MB Boisen, Jr., GV Gibbs i-xii + 460 pp
ISBN13 978-0-939950-19-5

Volume 14: *Microscopic to Macroscopic*

1985 ISBN 0-939950-18-9; SW Kieffer, A Navrotsky i-x + 428 pp
ISBN13 978-0-939950-18-8

Volume 13: *Micas*

1984 ISBN 0-939950-17-0; SW Bailey i-xii + 584 pp
ISBN13 978-0-939950-17-1

Volume 12: *Fluid Inclusions*

1984 ISBN 0-939950-16-2; Edwin Roedder i-vi + 646 pp
ISBN13 978-0-939950-16-4

Volume 11: *Carbonates: Mineralogy and Chemistry*

1983, ISBN 0-939950-15-4; RJ Reeder i-xii + 399 pp
1990 ISBN13 978-0-939950-15-7

Volume 10: *Characterization of Metamorphism through Mineral Equilibria*

1982 ISBN 0-939950-12-X; JM Ferry i-xiv + 397 pp
ISBN13 978-0-939950-12-6

Volume 9B: *Amphiboles and Other Hydrous Pyriboles—Mineralogy*

| 1982 | ISBN 0-939950-11-1;
ISBN13 978-0-939950-11-9 | DR Veblen, PH Ribbe | i-x + 390 pp |

Volume 9A: *Amphiboles: Petrology and Experimental Phase Relations*

| 1981 | ISBN 0-939950-10-3;
ISBN13 978-0-939950-10-2 | DR Veblen, PH Ribbe | i-xii + 372 pp |

Volume 8: *Kinetics of Geochemical Processes*

| 1981 | ISBN 0-939950-08-1;
ISBN13 978-0-939950-08-9 | AC Lasaga, RJ Kirkpatrick | i-x + 398 pp |

Volume 7: *Pyroxenes*

| 1980 | ISBN 0-939950-07-3;
ISBN13 978-0-939950-07-2 | CT Prewitt | i-x + 525 pp |

Volume 6: *Marine Minerals*

| 1979 | ISBN 0-939950-06-5;
ISBN13 978-0-939950-06-5 | RG Burns | i-x + 380 pp |

Volume 5: *Orthosilicates*

| 1980 | ISBN 0-939950-13-8;
ISBN13 978-0-939950-13-3 | RG Burns | i-xii + 450 pp |

Volume 4: *Mineralogy and Geology of Natural Zeolites*

| 1977 | ISBN 0-939950-04-9;
ISBN13 978-0-939950-04-1 | FA Mumpton | i-xii + 233 pp |

Volume 3: *Oxide Minerals*

| 1976 | ISBN 0-939950-03-0;
ISBN13 978-0-939950-03-4 | D Rumble, III | i-3 + 706 pp |

Volume 2: *Feldspar Mineralogy*

| 1975
1983 | ISBN 0-939950-14-6;
ISBN13 978-0-939950-14-0 | PH Ribbe | i-vii + 362 pp |

Volume 1: *Sufide Mineralogy*

| 1974 | ISBN 0-939950-01-4;
ISBN13 978-0-939950-01-0 | PH Ribbe | i-v + 301 pp |

www.ingramcontent.com/pod-product-compliance
Lightning Source LLC
Chambersburg PA
CBHW052010230326
41598CB00078B/2243